Handbook of
Public Water
Systems

Handbook of Public Water Systems

Culp/Wesner/Culp

Edited by

Robert B. Williams
Gordon L. Culp

VNR VAN NOSTRAND REINHOLD COMPANY
New York

Published by Van Nostrand Reinhold Company Inc.
115 Fifth Avenue
New York, New York 10003

Van Nostrand Reinhold Company Limited
Molly Millars Lane
Wokingham, Berkshire RG11 2PY, England

Van Nostrand Reinhold
480 Latrobe Street
Melbourne, Victoria 3000, Australia

Macmillan of Canada
Division of Gage Publishing Limited
164 Commander Boulevard
Agincourt, Ontario M1S 3C7, Canada

15 14 13 12 11 10 9 8 7 6 5 4 3 2 1

Library of Congress Cataloging-in-Publication Data
Main entry under title:

Handbook of public water systems.

(Van Nostrand Reinhold environmental engineering series)
Bibliography: p.
Includes index.
1. Municipal water supply—United States.   2. Water
quality—United States.   3. Water—Purification.
I. Culp/Wesner/Culp.   II. Series.
TD223.H27   1986   628.1   85–26458
ISBN 0–442–21597–5

Handbook Contributors

Manuscript Editors and Coordinators
Gordon L. Culp
Robert B. Williams

Authors
Henry H. Benjes, Jr.
Bruce E. Burris
Gordon L. Culp
Russell L. Culp
William F. Ettlich
Robert C. Gumerman, Ph.D.
Sigurd P. Hansen
Mark S. Montgomery, Ph.D.
David J. Reardon
Michael R. Rothberg
Joseph U. Tamburini
George M. Wesner, Ph.D.
Bruce R. Willey
Robert B. Williams

Manuscript Preparation
May L. Bray
Candice E. Cornell
Marie A. Filippello
Judy A. Hinrichs
Judy E. Ross
I. Jean Wagy
Dawn M. Wisniewski

Editing
Candice E. Cornell
Shaula P. Smith

Graphics
Teri D. Boon
Daryl G. Cornell
Robert R. Livingston
Bruce M. Person
G. Kathy Sady

Preface

Public water systems today face challenges that could not have been envisioned just a few years ago. Trace quantities of pollutants that only recently could even be measured are now considered potential threats to the public health. Many of America's water supply systems were installed in the early 1900s; not only must they now face these new challenges, but the condition of many systems has deteriorated to the point that soon they must be rebuilt. It is likely that the next decade or two will bring major changes to many of our existing public water systems. Development and protection of our water sources will receive a great deal of attention as we discover the insidious results of past waste disposal techniques. Many of our storage and distribution systems will require rehabilitation or expansion. Concerns over trace quantities of organic compounds will require application of processes not widely used in the past in conventional water treatment. Escalating costs and scarcity of energy will require concerted efforts to minimize the energy consumed in water treatment. The competing demands for upgrading of our nation's highways, bridges, airports, and other public systems will require the use of innovative treatment techniques that will improve treatment capacity and performance at a fraction of the cost of building all-new facilities.

We can take pride in the fact that the development and use of water treatment techniques has eliminated once rampant waterborne diseases in the United States; however, the challenges and opportunities of the next few years to further improve our public water supplies will require that we marshal all of our knowledge and resources. This book provides a comprehensive source of information on all aspects of public water systems—from the point of source development to the delivery of the final product at the consumer's tap. We hope that practitioners will find a single source of information on all of these aspects to be useful. Integrating our knowledge of past practice with the tremendous resource of information from modern, scientific studies on water problems will enable us to solve virtually any water-quality problem. It was with this goal in mind that we prepared this book, which brings together information on historic practice and information from current research, evaluation, and application of new techniques for design and operation of public water systems. It is the hope of the authors that the material presented in this book will be of interest and benefit to engineers, chemists, bacteriologists, biologists, treatment plant operators, utility managers, administrators, and others faced with the daily tasks of furnishing or assuring supplies of high-quality water to the public. Also, we anticipate that the book will prove to be valuable as a reference for teachers and students who are interested in studying methods for enhancement of water quality.

The authors of this book are fortunate to have been involved in the development of several new and improved water treatment processes. They also have had the opportunity to incorporate these new methods into water treatment plant designs and to observe the results of full-scale plant operation. These experiences are heavily drawn upon in this book. The work of many other investigators using other new methods is also cited, in order to present the most accurate possible updating of currently available treatment methods.

This book is the result of the efforts of many individuals who were members of the consulting engineering firm of Culp/Wesner/Culp at the time of their contribution to this text. A list of the contributors to this handbook appear on page v. Their work is acknowledged with appreciation.

Special thanks are given to equipment manufacturers who so generously supplied illustrations for the text. The encouragement of our families and colleagues throughout the preparation of the book is greatly appreciated.

GORDON L. CULP
ROBERT B. WILLIAMS

Contents

Chapter 1
Criteria and Standards for Improved Potable Water Qualities

HISTORICAL NOTE

Baker reports that the search for pure water is not a recent idea, but began in ancient times.[1] Writings in Sanskrit (circa 2000 B.C.) medical lore described early water treatment practices as follows:

Impure water should be purified by being boiled over a fire, or being heated in the sun, or by dipping a heated iron into it, or it may be purified by filtration through sand and coarse gravel and then allowed to cool.[1]

Drawings of equipment to clarify liquids have been discovered on Egyptian walls that date back to 1500 B.C. These drawings and the medical lore described above exemplify modern water treatment practices of sand filtration and disinfection. Hippocrates (460 to 354 B.C.), the father of medicine, focused attention on the importance of water in maintaining public health, and suggested that rainwater be boiled and strained before being drunk.

Dr. John Snow's epidemiological studies linked contaminated water supplies at the Broad Street pump to the 1854 cholera outbreak in London. Dr. Snow also conducted epidemiological studies on the water supplied by the Southwark and Vauxhall Company and the Lambeth Company. The former company obtained water from the Thames River in the middle of London, in an area polluted with sewage, whereas the latter obtained Thames River water upstream of London. Studies in an area served by both companies showed that the people receiving water from the Lambeth Company had a low incidence of cholera, whereas those served by the Southwark and Vauxhall Company had a high incidence. Because all other environmental factors were the same for both groups, Snow concluded that cholera was being spread by the water supply.

Other early epidemics include the typhoid outbreaks in the United States in Butler, Plymouth, New Haven, Nanticoke, and Reading, which involved over 39,000 cases. Of these individuals, more than 360 died.[2] These examples, coupled with the germ theory of disease, which became firmly established as a result of work by Louis Pasteur and others, made it clear that some drinking waters contain impurities that can cause diseases in consumers.

Water treatment as we know it can be traced to the filtration of municipal water supplies in Scotland in 1804, and later in England in 1829.[3] Experiments on water filtration were conducted in the United States in the late 1880s and early 1890s.[4] An experiment station was established in Lawrence, Massachusetts, by the Massachusetts State Board of Health in 1887, after an outbreak of typhoid, and a sand filter was installed there. Other filtration experiments were conducted at Louisville, Kentucky, in 1895 to 1897. By 1900 there were about 10 sand filter plants, and by 1907 there wree 33.[4] Hazen noted that filtration resulted in a 99 percent reduction of the bacteria present in the water supplies.[4]

The next major step in assuring water quality was the introduction of chlorine to the water. This was first done in Jersey City, New Jersey, in late 1908. The dramatic effect of chlorination was a marked decrease of typhoid and other waterborne diseases. This subject is discussed in more detail in Chapter 11.

An article in the *Sacramento Bee* (June 24, 1984) identified the results of a survey by *Prevention Magazine* on the crucial components of longevity of human life. More than 100 experts in the health field were questioned as to the actions people should take to safeguard their well-being. The first six recommendations were:

- Do not smoke.
- Do not smoke in bed.
- Wear seat belts.
- Do not drive while drinking.
- Have a reliable smoke detector.
- Live where drinking water is of acceptable quality.

The importance of safe drinking water is demonstrated by the results of this survey, in which water quality was placed sixth out of 24 items by a panel of experts.

WATER QUALITY

The high quality of public water supplies in the United States has long been a source of local and national pride. Travelers drink water from the tap wherever they may be, with no question of its safety. Conformance to the mandatory requirements of the U.S. Public Health Drinking Water Standards has proved to be an acceptable measure of water safety in the past. Restrictions on the quantity of water used and the occurrence of waterborne diseases in this country are not the serious problems that they are elsewhere in the world.

Despite this generally good record, water utilties are currently subjected to a great deal of criticism in the press, by the public, and in the Congress. There are many reasons for this. In many cases, water service is not nearly as good as it could and should be. Many community water systems are antiquated and do not meet present-day standards of quality and service. There is a lack of planning for future needs, as well as a need to integrate an existing proliferation of small separate water systems. Local, state, and federal efforts at surveillance and monitoring, operator and manager training, and water research fall far short of today's needs. There is public demand for improved water quality.

EXISTING DEFICIENCIES IN WATER SUPPLIES

The findings of the U.S. Public Health Service Community Water Supply Study in 1969 were a severe shock to laypersons and professionals alike. Approximately 17 percent of all public water supplies failed to meet one or more of the mandatory quality standards; 25 percent did not meet one or more of the recommended quality standards; more than 50 percent had major deficiencies in supply, storage, or distribution facilities; 90 percent failed to meet the bacteriological standards; and 90 percent had no cross-connection control programs. Conditions were particularly bad in the very small systems. Many were substandard in almost every respect.

The State of Washington, Division of Health, has accumulated some interesting statewide statistics in this regard. The Washington findings are similar to conditions prevailing in other states. It was learned that some 20 to 50 Washington communities, including some of the largest cities, are in need of additional filtration treatment. The exact needs are not known because of uncertainty that has arisen recently about the quality of supplies derived from so-called protected watersheds. The easy public access to any and all areas, including protected watersheds, that has been provided by the advent of all-terrain vehicles, trail bikes, and snowmobiles has made it very difficult or impossible to assure the safety of subsurface water supplies, which have no facilities for filtration treatment. Pending federal regulations that would require filtration of all surface water supplies appear to be a step in the right direction.

The study reported that 55 supplies, including all of the largest cities in the state, have open distribution reservoirs for storage of finished water. In the judgment of the Washington Division of Health, open distribution reservoirs are incompatible with present good public health practice, and they must now be covered under state regulations.

The Washington Division of Health is now requiring all cities to submit a comprehensive water system plan to identify present and ten-year future needs. It is anticipated that one-third of the cities already have most of the necessary data for such a plan; one-third will have a considerable amount of updating to do; and the other one-third have virtually none of the data required for advance planning, and will have to start from scratch. Two of the major unmet planning needs are coordinated planning among neighboring water utilities and minimization of the number of new systems being constructed.

Statistics compiled by the State of Washington also reveal how little effort is being expended in the surveillance of water supplies by local, state, and federal regulatory agencies, as compared to similar efforts in air and stream pollution control. Expenditures for air pollution control were 10 times that for water supply monitoring and 200 times greater than for stream pollution control. The Washington study concluded that the incidence of disease is no longer an acceptable measure of the adequacy of public water supply systems, and that there is a need to make this clear to the public, the press, and even the waterworks industry.

In addition to the deficiencies already mentioned with respect to public health, there are aesthetic considerations that deserve mention because of their importance to the consumer. Many water supplies contain sufficient iron or manganese to cause staining of laundry and plumbing fixtures, or even of building facings within the range of lawn sprinklers. Some waters are too hard to be satisfactory for many uses, or have corrosive or scaling tendencies that adversely affect plumbing in homes or factories. One of the most common complaints about water quality concerns taste and odor. Bad-tasting water drives many consumers to other sources of drinking water, such as private wells, cisterns, or bottled water. More often than not, such alternate sources of drinking water are of questionable bacteriological quality.

TRENDS IN RAW WATER QUALITY

Increased Water Withdrawals

Surface water sources are subject to ever increasing withdrawals to supply a growing population, industry, and agriculture. Heavy withdrawals of water for consumptive use, such as irrigation, decrease stream flows available for downstream dilution of wastewaters. In the case of underground water supply sources, excessive withdrawals may have an adverse effect on the chemical quality of the supply, such as increasing the content of iron or manganese or of total dissolved solids.

Waste Discharges

Ever increasing quantities of domestic, industrial, and agricultural wastewaters are being discharged directly or indirectly to water supply sources. Domestic sewage continues to add increasing quantities of bacteria, viruses, algal nutrients, substances with a high oxygen demand, suspended and dissolved solids, and taste- and odor-producing substances to water supply sources. Return irrigation waters in some locations are producing substantial increases in the total dissolved solids content of receiving streams and aquifers, and many of these substances, such as sodium, sulfate, chloride, and others, are not removed by conventional treatment methods or other economically feasible means.

Serious changes are also produced by industries that are discharging a wide variety of new complex chemical pollutants to the waters of the nation. A recent listing of refractory organic wastes—some of which have been found in drinking water—is presented in Table 1-1, as an illustration of the complex waste materials that now may be entering our water supplies.

With adequate treatment of these increased wastewater flows by modern techniques prior to discharge, raw water quality will not suffer, and, in fact, may be substantially improved in many cases. Unfortunately, the construction of pollution control facilities is not keeping pace with increased flows. The rate of construction is so far behind today's needs that many receiving waters are being progressively degraded with the passing of each year. This lack of essential facilities is a matter of grave national concern.

Wastewater Treatment Deficiencies

Many existing sewage and industrial waste treatment plants that discharge their treated waters to water supply sources do not achieve their planned treatment efficiency. In the past, less attention has been given to plant reliabil-

Table 1-1. Refractory Industrial Wastes.*

Acetone	Dichloroethyl ether
Benzene**	Dinitrotoluene**
2-Benzothiozole	Ethylbenzene**
Borneol (bornyl alcohol)	Ethylene dichloride**
Bromobenzene	2-Ethylhexanol
Bromochlorobenzene	Guaiacol (methoxy phenol)
Bromophenylphenyl ether	Isoborneol
Butylbenzene	Isocyanic acid**
Camphor	Isopropylbenzene
Chlorobenzene	Methylbiphenyl
Chloroethyl ether	Methylchloride**
Chloroform	Nitrobenzene**
Chloroethyl ethyl ether	Styrene**
Chloronitrobenzene	Tetrachloroethylene
Chloropyridine	Trichloroethane
Dibromobenzene	Toluene**
Dichlorobenzene	Veratrole (1,2-dimethoxybenzene)

* Reprinted with permission from *Environmental Science and Technology,* January, 1973, American Chemical Society.
** In trace amounts, these compounds also have been found to impart taste and odor to drinking water supplies.

ity, backup, and standby facilities in the design of wastewater treatment plants than is commonly devoted to the design of power generation stations or water systems, for example. In many of the plants, monitoring of effluent quality is inadequate or nonexistent, and control of plant operations and the efficiency of treatment are very poor.

In view of this situation, it is obvious that there is a great opportunity to improve raw water quality and the safety of public water supplies through better operation of existing wastewater treatment plants, the expansion of plants to handle current sewage flows and provide more reliable operation, the addition of facilities for advanced waste treatment, adequate financing of operations, the training of plant operators, and better surveillance of plant efficiencies. The Federal Construction Grants Program under the Clean Water Act has gone a long way to construct additional wastewater treatment capacity and clean up the nation's waters. However, recent reports still show that many of these plants are not meeting their design standards, and, in some instances, are discharging poorly treated wastewater. The sad fact is that very little improvement in raw water quality is occurring, and the prudent course for advance planning is to anticipate even further deterioration of raw water quality, and to plan, construct, and operate water purification facilities accordingly.

Some Special Problems

The interaction of many water quality control problems and proposed solutions can create even more complex or new problems for the water treatment industry. An example is offered by the seemingly separate problems of control of heavy metal discharges and the reduction of phosphorus discharges to watercourses.

Because of the high toxicity of mercury, disclosures of the discharge of substantial quantities of mercury in wastes from paper mills to the water environment, and its fate there, aroused a great deal of interest on the part of both water purveyors and the general public. It also raised the question of the possible undetected presence of other heavy metals in water supplies.

The possible presence of trace quantities of heavy metals in water supplies would be affected by the use of NTA (nitrilotriacetic acid) as a substitute or supplement for phosphorus compounds in the manufacture of synthetic detergent washing products. NTA is a chelating agent that has the property of increasing the solubility of heavy metal ions in water. It also can prevent the formation of insoluble salts of metal ions. Thus, NTA could cause serious interference with wastewater treatment and water purification. If sequestered, the heavy metal ions—some of them quite toxic—could escape removal by coagulation, sedimentation, filtration, or other processes that otherwise would be effective in their elimination. As a result, potentially toxic ions could pass into receiving streams. If present in sufficient concentration in surface water supply sources, NTA could similarly reduce the efficiency of water purification plants.

As a result of the discharge of increased quantities and new types of pollutants in wastewaters, water purveyors are faced with a myriad of new problems in plant design and operation. Many persons are concerned that the factor of safety that public water supplies have enjoyed in the past is being reduced to a dangerous extent by current trends in the degradation of raw water quality. Fortunately, many professionals in the waterworks field are seriously concerned with these problems and are actively working within the limits of their resources to find practical solutions to them. There also is considerable public support for improving present conditions.

IMPROVING WATER QUALITY

In the face of the probable continued degradation of raw water quality, the prospects for improvement of finished water quality may not appear bright. Fortunately, this is not the case. Despite the difficulties posed by increased levels of pollution, it is still possible and feasible to improve water quality substantially. Much of this potential improvement can be realized by judicious application of conventional purification processes, but even further gains can be had through processes such as:

- Mixed-media filtration combined with use of alum or polymer as a filter aid.
- Proper instantaneous and thorough mixing of chlorine and the water being treated to provide better disinfection.
- The use of granular activated carbon for removal of trace organic compounds and for complete color removal and taste and odor control.
- Practical applications of shallow depth sedimentation (tube settling).
- The use of polymers to improve coagulation, flocculation, and settling.
- The provision of methods for recycling or satisfactory disposal of sludges.
- The use of methods for continuously monitoring and controlling water treatment processes. These systems and others are explored in detail in the following sections.

Objectives of Water Treatment

The primary objective of water treatment for public supply is to take water from the best available source and to subject it to processing that will assure that it is always safe for human consumption and is aesthetically acceptable to the consumer. For water to be safe for human consumption, it must be free of pathogenic organisms or other biological forms that may be harmful to health, and it should not contain concentrations of chemicals that may be physiologically harmful. To provide safe water, the treatment plant must be properly designed and skillfully operated. As already pointed out, the task of furnishing safe water in the future in many instances may be made more difficult because of the necessity of treating a raw water of poorer quality.

The general requirements of an aesthetically acceptable water are that it be cool, clear, colorless, odorless, and pleasant to the taste; also it should not stain, form scale, or be corrosive. Treatment plants must be designed to produce water of uniformly good quality despite variations in raw water quality and plant throughput. Because the consumer is interested in the quality of the water at his tap rather than that at the treatment plant, precautions must be taken to preserve water quality in the distribution system and to control water quality from tests of tap water samples as well as plant samples. Many of the advantages of a high-quality water supply are difficult to express in terms of exact economic return, but they are, without question, quite substantial. An ample supply of high-quality water assures good public relations and favors industrial and community growth.

1962 Standards

Until 1977, the minimum standards for water quality were the U.S. Public Health Service Drinking Water Standards of 1962. These standards are in-

Table 1–2. Summary of 1962 U.S. Public Health Service Drinking Water Standards.

	MANDATORY REQUIREMENTS	
CONSTITUENT	LIMIT (mg/l EXCEPT AS SHOWN)	RANGE OF CONCENTRATIONS FOUND IN NATIONAL SURVEY** (mg/l EXCEPT AS SHOWN)
Arsenic (As)	0.05	<0.03–0.10
Barium (Ba)	1.0	0–1.55
Boron (B)	5.0	0–3.28
Cadmium (Cd)	0.01	<0.2–3.94
Chromium (hexavalent)	0.05	0–0.079
Coliform organisms	1/100 ml	2,000/100 ml
Cyanide (CN^{-2})	0.10	0–0.008
Fluoride (F^-)	Varies,* 0.8 to 1.7	<0.2–4.40
Gross beta activity	1,000 $\mu\mu c/l$	154 $\mu\mu c/l$
Lead (Pb)	0.05	0–0.64
Selenium (Se)	0.01	0–0.07
Silver (Ag)	0.05	0–0.03

* Dependent on annual average of maximum daily air temperatures.
** 1969 Community Water Supply Study by USPHS, 2,595 distribution samples.

cluded here for historical perspective. The 1962 standards were promulgated for use in regulating the quality of water supplied to common carriers engaged in interstate commerce. The standards set mandatory limits for certain chemical constituents and recommended concentrations for others, including some radioactive elements. The bacteriological standards set limits for coliform organisms and prescribed methods for the collection and laboratory analysis of water samples, including the frequency thereof.

The minimum number of water samples per month to be collected for bacteriological examination varied according to population served by the system from 2 per month for 2,000 or fewer people, to 100 per month for 100,000 people, to 300 per month for 1,000,000 persons. The fact that the total number of bacteriological samples collected and analyzed was often less than the minimum required for significant results was one of the most common reasons for failure to meet the Drinking Water Standards. Tables 1–2 and 1–3 summarize the 1962 standards.

Table 1–3. Summary of 1962 U.S. Public Health Service Drinking Water Standards.

CONSTITUENT	RECOMMENDED REQUIREMENTS RECOMMENDED LIMIT (mg/l EXCEPT AS SHOWN)	RANGE OR MAXIMUM CONCENTRATION FOUND IN NATIONAL SURVEY* (mg/l EXCEPT AS SHOWN)
Alkyl benzene sulfonate (ABS)	0.5	0–0.41
Arsenic (As)	0.01	0–3.28
Chloride (Cl⁻)	250	<1.0–1,950
Color	15 units	49 units
Copper (Cu)	1.0	0–8.35
Carbon chloroform extract (CCE)	0.200	0.0008–0.56
Cyanide (CN^{-2})	0.01	0–0.008
Fluoride (F⁻)	Varies, 0.8–1.7	<0.2–4.40
Iron (Fe)	0.3	26.0
Manganese (Mn)	0.05	1.32
Nitrate (NO_3)	45	0.1–127.0
Phenols	0.001	
Radium-226	3 pc/l	0–135.9 pc/l
Strontium-90	10 pc/l	0–1.0 pc/l
Sulfate (SO_4)	250	<1–770
Total dissolved solids (TDS)	500	2,760
Zinc (Zn)	5	0–13.0
Turbidity		
Chlorination treatment only	5 NTU	53 NTU
Water treatment plants	1 NTU	
Odor, threshold number	3	

* 1969 Community Water Supply Study by USPHS, 2,595 distribution system samples.

SAFE DRINKING WATER ACT

On December 16, 1974, President Ford signed into law the Safe Drinking Water Act, which was recorded as Public Law 93–523 (PL 93–523). This act gave the administrator of the Environmental Protection Agency (EPA) the authority to control the quality of the drinking water in public water systems through the development of regulations, or by other methods. The act required a three-stage mechanism to establish comprehensive regulations (and standards) for drinking water quality:

- National Interim Primary Drinking Water Regulations were to be promulgated to protect the public health. This would be accomplished using generally available technology and treatment techniques.
- The National Academy of Sciences (NAS) was to conduct a study on the human health effects of exposure to contaminants in drinking waters. This study would "consider only what is required for protection of the public health, not what is technologically or economically feasible or reasonable."
- Revised National Primary Drinking Water Regulations were to be promulgated, based on the NAS report, establishing maximum contaminant levels, to be set at levels sufficient to prevent the occurrence of any known or anticipated adverse health effects with an adequate margin of safety.

The EPA promulgated interim regulations based on the 1962 Public Health Service Standards included in Tables 1–2 and 1–3, as revised by the EPA Advisory Committee on the Revision and Application of the Drinking Water Standards. The interim regulations became effective on June 24, 1977, and contained maximum contaminant levels (MCL's) and monitoring requirements for microbiological contaminants, ten inorganic chemicals, six organic compounds, radionuclides, and turbidity. These are shown in Table 1–4. The NAS completed its series of reports on human health effects in 1980.[5] Prior to discussing the water quality standards, it is useful to include a few of the definitions listed by the NAS.

Definitions

Some of the more relevant definitions are presented below, as abstracted from the Safe Drinking Water Act.

- "Contaminant" means any physical, chemical, biological, or radiological substance or matter in water.
- "Maximum contaminant level" means the maximum permissible level

Table 1–4. Contaminants and MCL
in the National Interim Primary Drinking Water Regulations.*

CONTAMINANT	MCL
Inorganic	(mg/l)
Arsenic	0.05
Barium	1.0
Cadmium	0.01
Chromium	0.05
Lead	0.05
Mercury	0.002
Nitrate (as N)	10.0
Selenium	0.01
Silver	0.05
Fluoride	
At ≤53.7°F (≤12.0°C)**	2.4
At 53.8–58.3°F (12.1–14.6°C)	2.2
At 58.4–63.8°F (14.7–17.6°C)	2.0
At 63.9–70.6°F (17.7–21.4°C)	1.8
At 70.7–79.2°F (21.5–26.2°C)	1.6
At 79.3–90.5°F (26.3–32.5°C)	1.4
Organic	(mg/l)
Endrin	0.0002
Lindane	0.004
Toxaphene	0.005
Methoxychlor	0.1
2,4-D	0.1
2,4,5-TP (Silvex)	0.01
TTHM	0.1
Radiological	(pCi/l)
Alpha Emitters	
Radium-226	5
Radium-228	5
Gross alpha activity	15
(excluding radon and uranium)	
Beta and Photon Emitters†	
Tritium	20,000
Strontium-90	8
Turbidity (NTU)	1††

* Reference 5.
** Average annual maximum daily air temperature.
† Based on a water intake of 2 l/d. If gross beta particle
activity exceeds 50 pCi/l, other nuclides should be identified
and quantified on the basis of a 2-l/d intake.
†† One turbidity unit based on a monthly average. Up to
5 NTU may be allowed for the monthly average if it can
be demonstrated that no interference occurs with disinfection
or microbiological determination.

of a contaminant in water that is delivered to the free-flowing outlet of the ultimate user of a public water system, except in the case of turbidity where the maximum permissible level is measured at the point of entry to the distribution system. Contaminants added to the water under circumstances controlled by the user, except those resulting from corrosion of piping and plumbing caused by water quality, are excluded from this definition.

- "Municipality" means a city, town, or other public body created by or pursuant to state law, or an Indian tribal organization authorized by law.
- "Person" means an individual, corporation, company, association, partnership, state, municipality, or federal agency.
- "Public water system" means a system for the provision to the public of piped water for human consumption, if such system has at least 15 service connections or regularly serves an average of at least 25 individuals daily at least 60 days out of the year. Such term includes (1) any collection, treatment, storage, and distribution facilities under control of the operator of such system and used primarily in connection with such system; and (2) any collection or pretreatment storage facilities not under such control that are used primarily in connection with such system. A public water system is either a community water system or a noncommunity water system.
- "Community water system" means a public water system that serves at least 15 service connections used by year-round residents or regularly serves at least 25 year-round residents.
- "Noncommunity water sytem" means a public water system that is not a community water system.

Suggested No-Adverse-Response Level (SNARL)

The Safe Drinking Water Committee developed a methodology to establish the appropriate maximum concentration of a constituent that is not toxic for human consumption. The basic methodology was to apply an "uncertainty factor" to known or measured data from either human or animal test specimens. For corroborating data from human and animal data, the uncertainty factor is smaller. As the available data become less precise, or there is little or no information, the uncertainty factor is increased significantly. The resulting value of the uncertainty factor multiplied by the measured constituent concentration causing toxic results is termed the SNARL.

The following assumptions were used to assign the uncertainty factor when computing either the acute or chronic SNARL's:[5]

- An uncertainty (safety) factor of 10 was used when good chronic or acute human exposure data were available and supported by chronic or acute data in other species.
- A factor of 100 was used when good chronic or acute toxicity data were available for one or more species.
- A factor of 1,000 was used when the acute or chronic toxicity data were limited or incomplete.

Acute Exposure. Acute toxicity has been defined in terms of suggested no-adverse-response levels (SNARL's) for acute exposures of 24 hours or 7 days. The SNARL's are computed by assuming that the entire chemical ingestion was supplied by drinking water during either the 24-hour or the 7-day period. SNARL's were developed only for the instances where human exposure data or sublethal animal data were available. The SNARL's should not be used to estimate the potential hazards for periods exceeding 7 days, and they are not a guarantee of complete safety. Also, the SNARL's are based on exposures to a single chemical, and do not account for possible synergistic effects of other contaminants. The safety factor, or degree of uncertainty, used in computing the SNARL's reflects the degree of confidence with regard to the available data.

Chronic Exposures. A SNARL was computed for chronic exposures for only those chemicals that are *not* known or suspected carcinogens. The majority of SNARL's were estimated using data from studies covering most of the lifetime of the experimental animal. The chronic SNARL's were developed by assuming that 20 percent of the chemical of concern was ingested through the drinking water. These values must not be used as maximum contaminant intakes. For certain chemicals there is adequate evidence of carcinogenicity, and in these cases a risk estimate is developed rather than a SNARL.

Safety and Risk Assessment

The recommendations of the Safe Drinking Water Committee were limited to the specific risk of cancer. However, the same considerations apply to both mutagenesis and teratogenesis. Furthermore, the committee only considered carcinogens whose mechanisms involve somatic mutations. Four principles are enumerated below that were considered in assessing the irreversible effects of continuous exposures to carcinogenic substances at low dose rates:[5]

- *Principle 1:* Effects in animals, properly qualified, are applicable to man.
- *Principle 2:* Methods do not now exist to establish a threshold for long-term effects of toxic agents.

- *Principle 3:* The exposure of experimental animals to toxic agents in high doses is a necessary and valid method of discovering possible carcinogenic hazards in man.
- *Principle 4:* Material should be assessed in terms of human risk, rather than as "safe" or "unsafe."

The SNARL's for certain chemicals are shown in Table 1–5, for both acute and chronic effects. Also included are carcinogenic risk estimates for some of the chemicals.

Table 1–5. Summation of Acute and Chronic Exposure Levels and Carcinogenic Risk Estimates for Chemicals Reviewed.

CHEMICAL	SUGGESTED NO-ADVERSE-RESPONSE LEVEL (SNARL), mg/l, BY EXPOSURE PERIOD*			UPPER 95% CONFIDENCE ESTIMATE OF LIFETIME CANCER RISK PER $\mu g/l$**
	24-HOUR	7-DAY	CHRONIC	
Acrylonitrile				1.3×10^{-6}
Benzene		12.6		
Benzene hexachloride	3.5	0.5		
Cadmium		0.08	0.005	
Carbon tetrachloride	14	2.0		
Dichlorodifluoromethane	350	5.6		
1,2-Dichloroethane				7.0×10^{-7}
Epichlorohydrin	0.84	0.53		
Ethylene dibromide				9.1×10^{-6}
Methylene chloride	35	5.0		
Polychlorinated biphenyl	0.35	0.05		
Tetrachloroethylene	172	24.5		1.4×10^{-7}
1,1,1-Trichloroethane	490	70	3.8	
Trichloroethylene	105	15		
Trichlorofluoromethane	88	8		
Toluene	420	35	0.34	
Uranium	3.5	0.21		
Xylenes	21	11.2		
Bromide	1,400	224	2.3	
Catechol	2.2			
Chlorine dioxide			0.38	
Chlorite		0.21		
Chloroform	22	3.2		
Dibromochloromethane	18			
2,4-Dichlorophenol			0.7	
Hexachlorobenzene		0.03		2.9×10^{-5}
Iodide	115.5	16.5	1.19	
Resorcinol	11.7	0.5		

* See text for details on individual compounds.
** See Reference 4 for details.

Water Quality Standards

The proposed water quality standards have been established in terms of Primary and Secondary Standards, and are presented in terms of Primary and Secondary Maximum Contaminant Levels (MCL's). The Primary MCL's are shown in Table 1–4, and the Secondary MCL's are included in Table 1–6. A few introductory comments are made below concerning the two sets of MCL's, after which the rationale for the selection of each set is summarized.

Primary Maximum Contaminant Levels. As already outlined, the Primary MCL's (PMCL's) are subdivided into:

- Inorganic chemicals
- Organic chemicals
- Radiological agents
- Turbidity
- Bacteriological contaminants

The five categories are described briefly below.

Inorganic Chemicals. The standards for all but the chemical fluoride are straightforward, having only a single value. Fluoride has several values, which are related to atmospheric temperatures. Fluoride is effective against dental caries over a relatively narrow range of concentrations. Because people drink more water in warm climates, the concentration of fluoride must be reduced there so that the total intake of fluoride remains about the same.

Table 1–6. Secondary Maximum Contaminant Levels for Drinking Water.

CONTAMINANT	LEVEL
Chloride	250 mg/l
Color	15 color units
Copper	1 mg/l
Corrosivity	Noncorrosive
Foaming agents	0.5 mg/l
Iron	0.3 mg/l
Manganese	0.05 mg/l
Odor	3 TON
pH	6.5 to 8.5
Sulfate	250 mg/l
Total dissolved solids (TDS)	500 mg/l
Zinc	5 mg/l

Organic Chemicals. Seven organic chemicals are included in the PMCL's. The first four organic chemicals listed in Table 1–4 are insecticides and the next two are herbicides, which typically are used to control aquatic growth. These substances enter the water supplies through industrial discharges during manufacture and/or runoff following precipitation.

Recently, trihalomethane (THM) was added to the PMCL's. THM's are by-products of water disinfection using chlorine, bromine, or iodine, in the presence of humic and fulvic compounds. This subject is discussed in detail in Chapter 14, where alternative methods are presented for controlling or reducing THM concentrations.

Radiological. The radiological PMCL's are subdivided into man-made and naturally occurring contaminants. Limitations on man-made radiological contaminants are established for water systems serving more than 100,000 customers. These contaminants enter the water supply principally through nuclear fallout, and are higher at higher ground elevations.

Naturally occurring radiological contaminants enter water systems via receiving water that has passed over radioactive elements in the environment. Water that contains more than 5 pCi/l of radium-226 and radium-228 cannot be used as a water source.

Turbidity. Turbidity in water can be an indirect health hazard. Turbidity is known to:

- Interfere with disinfection because it can form a shield around disease-causing organisms.
- Make it difficult to maintain a chlorine residual in the water system.
- Interfere with bacteriological testing of the water.
- Be aesthetically displeasing.

The PMCL applies only to surface waters. The turbidity limit of 1 NTU can be increased to 5 NTU at the discretion of the respective state health boards for a specific location.

The 2-day average turbidity may not exceed 5 NTU. The 2-day limit is included to protect against high, short-term turbidity levels that could threaten the bacteriological quality of the product water.

Bacteriological (Coliform Organisms). Coliform organisms are used as indicators of the bacteriological quality of the water. The presence of coliform bacteria in a water sample indicates the probability of disease-causing bacteria also being present in the sample. The PMCL for coliform organisms, shown in Table 1–7, depends on whether the membrane filter technique or the fermentation tube technique is utilized.

Table 1-7. Maximum Contaminant Levels for Coliform Organisms.

DETENTION TECHNIQUE USED	NUMBER OF SAMPLES EXAMINED PER MONTH*	MAXIMUM NUMBER OF COLIFORM BACTERIA
Membrane filter		1/100 ml shall be the arithmetic mean of all samples examined each month.
	Fewer than 20	4/100 ml shall be present in no more than one sample.
	20 or more	4/100 ml shall be present in no more than 5% of all samples examined each month.
Fermentation tube 10-ml standard portions		Coliforms shall not be present in more than 10% of the portions in any month.
	Fewer than 20	Coliforms shall not be present in three or more portions in more than one sample.
	20 or more	Coliforms shall not be present in three or more portions in more than 5% of the samples.
100-ml standard portions		Coliforms shall not be present in more than 60% of the portions in any month.
	Fewer than 5	Coliforms shall not be present in five portions in more than one sample.
	5 or more	Coliforms shall not be present in five portions in more than 20% of the samples.

* Based on population.

Secondary Maximum Contaminant Levels. The Secondary MCL's (SMCL's) are for substances that may be a nuisance to the consumer at high concentrations. These substances adversely affect the aesthetic quality of the drinking water, but health implications may result if their concentrations are substantially in excess of the recommended values.

The SMCL's are guidelines, and are not enforced by the EPA, although they represent reasonable and prudent goals for drinking water quality. The individual states are encouraged to set standards based on these values. The actual standards established by a state may differ from the goals, depending on specific, local conditions. The aesthetic quality of drinking waters is important to the consumer, and should be considered in developing a water supply or setting standards.

The EPA does not have the authority to require that SMCL's be met.

However, it does have the responsibility to inform the state whenever a potable water sample does not conform to the SMCL's. The intent of the regulations is to ensure that the importance of the aesthetic quality of public water supplies is understood.

RATIONALE FOR PMCL SELECTION

Inorganic Chemicals

Arsenic. Arsenic is widely distributed across the United States in low concentrations, as shown in Table 1–8. Except for local exceptions, where arsenic concentrations could be traced to specific causes, there are only minor regional differences in the average values, or in the percentage of contaminated samples. Natural sources, such as the erosion of rocks, are thought to account for most of the arsenic in surface water and groundwaters. Arsenic is present in the earth's crust in concentrations averaging 2 mg/l. It is concentrated in shales, clays, phosphorites, coals, sedimentary iron ore, and manganese ores.[5]

Water-soluble arsenicals are readily absorbed through the gastrointestinal tract, lungs, and skin. Pentavalent arsenic, both organic and inorganic, is absorbed more readily than the trivalent form. Arsenic primarily enters the liver, kidneys, intestinal wall, spleen, and lungs.

Toxic Effects. Human exposure to arsenic of a sufficient concentration to cause toxicosis (poisoning) is usually a result of ingestion of contaminated food or drink. Adverse health effects have not been reported from ingestion of arsenic at concentrations of 0.1 mg/l or less. There is a wide variation in toxicity of various arsenical formulations. The arsenic compounds are listed below in descending order of toxicity:[5]

- Arsines (trivalent, inorganic or organic)
- Arsenite (inorganic)
- Arsenoxides (trivalent with two bonds to oxygen)
- Arsenate (inorganic)
- Pentavalent arsenicals, such as arsenic acids
- Arsenium compounds (four organic groups with a positive charge on arsenic)
- Metallic arsenic

Arsine gas results in the destruction of the red-cell membranes. Inorganic arsenite produces rapid collapse, shock, and death. The major characteristics of acute poisoning are substantial gastrointestinal damage and cardiac abnormalities. Symptoms may be evident within 8 minutes if arsenic is ingested

Table 1-8. Regional Summary of Arsenic in U.S. Surface Waters.*

REGION	MAXIMUM μg/l	MINIMUM μg/l	MEDIAN μg/l	PROPORTION, <10 μg/l %	PROPORTION, >10 μg/l %
New England and Northeast	60	<10	<10	80	20
Southeast	1,110	<10	<10	70	30
Central	140	<10	<10	75	25
Southwest	10	<10	<10	87	13
Northwest	30	<10	<10	86	14

* Reference 9.

in a drink, but at up to 10 hours if it is solid and taken with food. The acute fatal dose of arsenic trioxide for a human is in the range of 70 to 180 mg, or 0.76 to 1.95 mg/kg of body weight for a 70-kg person.

Studies on the carcinogenicity of arsenic have shown mixed results. It is generally accepted that contaminants other than arsenic were responsible for the cancers. However, some studies have linked skin cancer to arsenic-contaminated drinking water.

There are some apparent beneficial effects. Recent studies show that rats require small quantities of arsenic for improved hair coat and growth. Earlier studies also postulated that humans require some minimal amount of arsenic.

The recommended limit of 0.05 mg/l was based on acute toxicity. However, the available epidemiological evidence that arsenic is associated with skin cancer may require that the limit be reduced in the future.

Barium. Barium is one of the alkaline earth metals, and occurs naturally in virtually all surface waters examined.[5] Typical concentrations are in the range of 2 to 340 μg/l, with an average of 43 μg/l. The 100 largest cities in the United States were found to have a median barium concentration of 43 μg/l, with a maximum value of 340 μg/l.

A common form of barium is barium sulfate (barite). The normally low solubility of barium sulfate increases in the presence of chloride and other anions. Barium is stable in dry air, but readily oxidized by humid air or water.[6]

Toxic Effects. Barium is very toxic when its soluble salts are ingested. The human fatal dose of barium chloride is about 0.8 to 0.9 g, or about 550 to 600 mg. of barium. The human digestive system is permeable to barium, which is transferred to the blood plasma. Acute barium poisoning exerts a "strong, prolonged stimulant action on all muscles, including cardiac and smooth muscle of the gastrointestinal tract and bladder."[5] In small doses, barium causes an increase in blood pressure.

Chronic effects of barium ingestion have not been demonstrated, in either food or drinking water. The drinking water MCL was derived from the 8-hour weighted maximum allowable concentration in industrial air of 0.5 mg/m³. Based on this value, and some other assumptions, a value of 2 mg/l was computed. To provide for additional safety for more susceptible humans, the MCL value of 1 mg/l was selected.

Cadmium. Cadmium is a silvery-white, soft metal that is in the same periodic group as zinc and mercury. Cadmium is found wherever zinc is located in nature, and is widely distributed in the earth's crust, although not in large quantities. Cadmium in the environment results principally from industrial sources such as electroplating facilities, textile and chemical industries, and others. Cadmium and zinc also are found in the soils and tailings around mines and smelters, which may result in localized, high concentrations in adjacent waters. In unpolluted waters, cadmium concentrations are about 1 μg/l. This is consistent with the chemistry of systems with a normal alkaline pH, because of the low solubility of carbonate and hydroxide.[5]

Cadmium intake each day from air, water, food, and cigarettes is estimated to be in the range of 40 to 190 μg/d, depending on diet, environment, and whether or not the person is a smoker.[5] Cadmium tends to accumulate in the body, particularly in the renal cortex.

Toxic Effects. Overexposure to cadmium (in excess of 50 μg/m³) has resulted in bronchitis, emphysema, anemia, and renal stones.[5] It is a very toxic element, and is not an essential nutrient for humans. Toxicity results from either acute or chronic exposures. The general population adsorbs cadmium principally through the gastrointestinal tract. Its major effects probably will be on the kidneys. An important determinant in cadmium toxicity is the zinc:cadmium ratio, which is about 1:100 in most foodstuffs, and is higher in meats. Some evidence indicates that sodium also affects the toxicity of cadmium. Certain experimental studies show that zinc is an important protective factor against cadmium toxicity.

There is limited evidence that cadmium can be carcinogenic in rats and mice. Epidemiological surveys on older cadmium workers indicated an increased incidence of prostate cancer in comparison to the general population.[5] However, these studies were limited in scope, and firm conclusions regarding cancer in humans cannot be drawn. There are no data linking cadmium to mutagenicity, but it is known to be teratogenic in rats.

The MCL of 0.01 mg/l for cadmium was based on avoiding health problems. However, questions remain about the effect of cadmium on vitamin D metabolism and the effect of soft water on the accumulation of cadmium in drinking water at the tap.

Chromium.   Chromium can exist as Cr(II) through Cr(VI) in the environment, but Cr(III), the trivalent form, predominates in natural waters. Chromium is present in U.S. rivers in varying amounts, ranging from zero to a high of 112 μg/l. However, in many surveys more than 50 percent of the sampled rivers did not contain chromium, a finding consistent with the fact that little chromium goes into solution unless the pH is quite low. Chromium also is found in foods, in the air, and in most biological systems. Because Cr(III) can be oxidized and Cr(VI) can be reduced, it has been recommended by some that water quality standards should be based on total chromium, rather than on hexavalent chromium.

The average daily intake of chromium varies widely, based on diet and location. One estimate ranges from 5 to 115 μg/l/d with an average of 60 to 65 μg/l/d. A second study showed a range of 5 to 500 μg/l/d with an average of 280 μg/l/d.[5] Chromium has been found to be beneficial in humans, based on a study of older people, diabetics, pregnant women, and malnourished children.[5]

Toxic Effects.   The toxicity of chromium depends upon its valence. Trivalent chromium is low in toxicity, and is the nutritional form of chromium. Cr(III) is so low in toxicity that a wide margin of safety exists between the amount normally ingested and the amount needed to induce undesirable effects. The National Academy of Sciences Committee on Chromium stated: "Compounds of chromium in the trivalent state have no established toxicity. When taken by mouth, they do not give rise to local or systematic effects and are poorly absorbed. No specific effects are known to result from inhalation. In contact with the skin, they combine with proteins in the superficial layers, but do not cause ulceration."[5]

Hexavalent chromium, Cr(VI), is significantly more toxic than Cr(III), and does not have any nutritional value. Normally, Cr(VI) compounds cause irritation and corrosion. Cr(VI) may be absorbed through ingestion, contact with the skin, and inhalation. Cr(VI) causes hemorrhage of the gastrointestinal tract, ulceration of the nasal septum, and cancer of the respiratory tract from inhalation.

Chromium has been shown to be carcinogenic in laboratory animals, through the development of sarcomas. An IARC* work group concluded, "There is no evidence that nonoccupational exposure to chromium constitutes a cancer hazard."[5] The proposed chromium limit of 0.05 mg/l is less than the level at which no adverse health effects were observed. Because chromium is a strong skin sensitizer, and because of the possibility of dermal effects from immersion in waters containing chromium, potable water supplies should be limited to the stated MCL.

---

* IARC = International Agency for Research on Cancer.

Fluoride. The most electronegative element is fluoride, which exists naturally in the fluoride form. It occurs principally in the earth's crust in the form of fluorite (CaF_2) and fluorapatite ($Ca_{10}(PO_4)_6F_2$), and is the 17th most abundant element.[5] Small amounts of fluoride are present in most soils, except those that have been leached.

A 1969 study found that fluoride concentration ranged from 0.2 to 4.4 mg/l in water supplies.[5] Fluoride concentrations greater than the recommended limits were found in 52 systems. Fluoride has been added to drinking water for more than 30 years to reduce the potential for dental caries. (The methods and chemicals used for this addition are described in Chapter 20.) The EPA has endorsed controlled additions of fluoride to domestic waters because small amounts, about 1 mg/l, have a beneficial effect in preventing tooth decay, particularly among children. The amount ingested is important; so it is significant that fluoride consumed in drinking water changes with atmospheric temperature—consumers tend to drink more water as the outside temperature increases.

Fluoride is present to some degree in most foods, but in widely varying concentrations. Foods that are high in fluoride include fish (particularly those fish that are eaten with the bones, such as sardines), tea, and some wines. One study on natural water sources found the following fluoride concentrations.[5]

- Rivers: 0.0 to 6.5 mg/l
- Lakes: up to 1,625 mg/l
- Groundwaters: 0.0 to 35.1 mg/l
- Seawater: average value of 1.2 mg/l

Recent studies have shown significant differences in fluoride intakes from foods prepared in communities having fluoridated water supplies compared to those with nonfluoridated water. The fluoride intake from food in the fluoridated communities was in the range of 1.6 to 3.4 mg/l, with an average of 2.6 mg/l. Food in nonfluoridated communities had fluoride intakes in the range of 0.8 to 1.9 mg/l, with an average of 0.9 mg/l. The higher values are explained in part by the inclusion of tea and coffee in the diet.

Fluoride is present in the air in insignificant amounts, where it does not contribute significantly to human intake, amounting to only a few hundredths of a milligram per day.[5] However, industrial exposure to fluoride-containing dusts has been a problem in several parts of the world. This is caused by operation-process fluoride-containing minerals. Ventilation and emission controls have reduced the impact of these sources.

Fluoride has not been shown, unequivocally, to be an essential element for human nutrition, except to the extent that it reduces the incidence of dental caries. Adequate and safe intakes of fluoride have been estimated as follows:[5]

- Infants less than 6 months: 0.1 to 0.5 mg/d
- Infants from 6 to 12 months: 0.2 to 1.0 mg/d
- Children from 1 to 3 years: 0.1 to 1.0 mg/d
- Children from 4 to 6 years: 1.0 to 2.5 mg/d
- Children from 7 years to adulthood: 1.5 to 2.5 mg/d
- Adults: 1.5 to 4.0 mg/d

These levels are thought to be sufficient to protect against dental caries and osteoporosis.

Toxic Effects.   Acute toxicity due to fluoride rarely occurs in humans. An acute dose acts swiftly on the gastrointestinal mucosa causing vomiting, abdominal pain, convulsions, and other effects. A lethal dose is unlikely to occur.

Chronic toxic effects generally are recognized as having two forms:

- Mottling of teeth enamel (dental fluorosis)
- Skeletal fluorosis

The most prevalent of the two is mottling of the teeth, particularly among children. Mottling has been observed at intakes as low as 0.8 to 1.6 mg/l.[5] Crippling skeletal fluorosis has been observed at fluoride intake levels of 3 mg/l. Based on available studies, mottling of teeth occasionally can be expected in hotter regions of the United States if the water contains 1 mg/l or more of fluoride, and in other areas if the fluoride concentration is 2 mg/l or more.

Lead.   Lead is present in natural waters (rivers and lakes) throughout the world, ranging in concentration from 1 to 10 μg/l. However, there is some evidence that much of this lead ends up in sedimentary deposits. The use of lead pipes in service lines in the presence of corrosive waters has resulted in higher lead concentrations at the tap than in the finish water at the plant. Great care must be exercised to avoid corrosive waters.

Lead and its compounds also contaminate the environment as a result of mining and processing activities. These include the manufacture of batteries, the increased use of lead as a gasoline additive, and, to a lesser extent, the use of leaded paints.

Lead primarily occurs as sulfide (galena) in rocks, as well as in the forms of oxides and potassium feldspar. Lead carbonate is commonly seen in the oxidized zone of lead ores. The aqueous solubility of lead ranges from 0.5 mg/l in soft water to 0.003 mg/l in hard water.[7] Lead is likely to be dissolved from distribution piping or plumbing by water that is low in hardness, bicarbonate, and pH, and high in dissolved oxygen and nitrate. Lead contamination in water supplies is believed to come principally from water pipes.

Toxic Effects.   Lead has no known beneficial or nutritional value for humans. It is a toxic constituent that accumulates in humans and animals, but acute lead poisoning of consumers is extremely rare. Lead does not cause gastrointestinal symptoms within hours of ingestion, unlike such other metals as cadmium and iron.

Mutagenic and teratogenic effects have been produced in experimental animals, but have not been documented for man. The main chronic adverse effects of lead are those produced in the nervous system, the hematopoietic system, and the kidneys.

The most common forms of lead poisoning are anemia, intestinal cramps, loss of appetite, and fatigue. Symptoms of these forms of lead poisoning take time to develop and become evident. Children have been known to suffer irreversible brain damage as a result of lead intoxication.

Because of the other sources of lead exposure, it has been suggested that the present MCL for lead of 0.05 mg/l may not provide a satisfactory or acceptable margin of safety. Less than 4 percent of analyzed samples exceed the MCL; and most of those that do, do so because of corrosive water action. It has been suggested that the MCL be lowered to protect children (infants) and fetuses.

Mercury.   Mercury, which is one of the least abundant elements in the earth's crust, has three oxidation states, as follows:[7]

- Elemental mercury
- Mercurous compounds
- Mercuric compounds

Mercury is found in more than trace amounts in at least 30 ores and throughout the environment, but is a nonessential element in human nutrition. Industrial and agricultural applications may cause increased concentrations above natural levels in water, soils, and air in local areas. This is particularly apt to be the case around chlor-alkali manufacturing plants and industrial processing with mercurial catalysts. The concentration of mercury in the air is ill-defined because insufficient data are available for analysis.

The toxicity of mercury is well known and documented. There are specific dramatic cases of mercury poisoning in humans, including those at Minamata

Bay, Japan. Other incidents have been reported in Iraq, Pakistan, and Guatemala.

Toxic Effects.   The comparative toxicity of inorganic mercury salts, such as mercurous chloride (calomel), is related to their absorption. The fact that certain microorganisms can convert inorganic and organic forms of mercury to the very toxic methyl- or dimethylmercury, has made any form of mercury potentially dangerous in the environment.

Mercury intoxication can be acute or chronic. Acute intoxication usually is the result of self-inflicted or accidental exposure. Acute poisoning results in pharyngitis, gastroenteritis, vomiting, and bloody diarrhea initially, followed by systemic effects such as anuria with uremia, stomatitis, ulcerative-hemorrhage colitis, nephritis, hepatitis, and circulatory collapse. The inhalation of mercury vapor or dusts leads to the typical symptoms of mercury poisoning with lesions of the mucous membranes.

Chronic mercury poisoning results from exposure to small amounts of mercury over extended periods of time. Typically, chronic poisoning by inorganic mercurials is associated with industrial exposures, while poisoning from organic mercurials is the result of accidents or environmental contamination. Workers who are continually exposed to inorganic mercury are susceptible to chronic mercurialism. Chronic alkyl mercury poisoning, known as Minamata disease, is insidious because the onset of mercurialism can take weeks or even years. This type of poisoning is characterized by major neurological symptoms that lead to permanent damage or death.[8]

Epidemiological studies can be used to determine the safe quantity of mercury ingestion. The lowest concentration associated with methylmercury toxic symptoms is 0.2 μg/g. This corresponds to prolonged continuous exposure of 0.3 mg/70 kg body weight/day. Using a margin of safety of 10, the maximum intake from all sources (air, water, and food) is 0.03 mg/person/day.

The MCL for mercury is 0.002 mg/l, which is seldom exceeded by mercury in drinking water. Using this concentration and an intake of 2 l/d, mercury intake is a total of 4 μg/d. For the toxic methylmercury form, the daily intake would be about 0.1 percent of this value. At this level, the potential hazard to consumers is insignificant.

Nitrates.   In aerobic surface waters all inorganic and organic forms of nitrogen eventually will be converted biologically to the nitrate form. The principal sources of nitrogen in surface waters included runoff from fertilized agricultural lands, feedlot runoff, municipal and industrial wastewater discharges, leachate from sanitary landfills, atmospheric fallout, decaying vegetation, and others. Nitrogen in the groundwater is derived from leachate from fertilized, irrigated agriculture and septic tank discharges. Nitrate concentrations in the groundwater have been measured at concentrations significantly above the recommended limit of 45 mg/l as nitrate.

The majority of human intake of nitrogen is from food, rather than from water supplies. The mean food intake is estimated at 100 mg/d, most in the form of vegetables.[5] Also, nitrate is secreted in the saliva, with the mean value being about 40 mg/d. Of this, about 25 percent is reduced to nitrite. These quantities must be included in the analysis of nitrate/nitrite toxicity.

Toxic Effects. Serious and occasionally fatal poisonings in infants have occurred following ingestion of waters containing nitrates in excess of 10 mg/l nitrate nitrogen as N. Two health hazards are identified with nitrate-contaminated waters:

• Induction of methemoglobinemia (oxygen deprivation), especially in infants.
• Possible formation of carcinogenic nitrosamines.

Acute toxicity of nitrate is caused by its rapid reduction to nitrite in the stomach. The nitrite then converts hemoglobin (blood pigment that carries oxygen) to methemoglobin. Methemoglobin does not act as an oxygen carrier, and consequently anoxia and death may ensue. This phenomenon seems only to affect infants up to about 3 months. It is reasoned that older children and adults are not so susceptible because the ratio of fluid intake to body weight is significantly lower in them than it is in infants. The MCL of 10 mg/l NO_3-N was based on epidemiological studies that showed this value to be the limit above which methemoglobinemia could occur. At this level, there appears to be little margin of safety for some infants.

The carcinogenic effects of nitrate are unclear. Some epidemiological studies have correlated gastric cancer to waters containing nitrate. However, more study is needed to develop a scientific basis to support conclusions on the carcinogenic effect of nitrates. The nitrate may be converted to N-nitroso compounds. The steps are speculated to be as follows:

• Reduction of nitrate to nitrite.
• Reaction of nitrite with secondary amines or amides in food or water to form N-nitroso compounds.
• Carcinogenic reaction of N-nitroso compounds.

Selenium. Selenium is present in the environment in soils, some salt deposits, and rocks. The concentration of selenium in groundwaters and surface waters is directly related to the presence of selenium in the rocks and soils. Selenium concentrations vary from 1 to 2,000 $\mu g/l$. The higher concentration of selenium results from runoff from seleniferous soils. Typical surface water values are likely to be about 10 $\mu g/l$, unless they are affected by the presence of seleniferous soils.

Selenium occurs naturally in four oxidation states: selenide ($-II$), elemental selenium (0), selenite ($+IV$), and selenate ($+VI$). Environmental contamination probably is minimized because a majority of organic selenium is selenide, which decomposes to elemental selenium, which is not absorbed. The forms that occur most frequently in water are selenite and selenate. Selenate is taken up by plants and may reach toxic concentrations. Selenite salts are less soluble than the selenates, and are reduced to elemental selenium under acidic conditions.

Selenium is present in consumable foodstuffs, ranging from 6 to over 500 $\mu g/l$, depending on the specific food. Selenium is an essential nutrient for domestic animals, but toxic to them when ingested in amounts over 1 mg/ kg food intake.

Toxic Effects. Some selenium compounds are toxic to humans, with hydrogen selenide being one of the most irritating and toxic compounds. The poisoning symptoms are similar to those of arsenic poisoning. In industrial situations, human exposure is through the skin and lungs as a result of exposure to dust or fumes.

Chronic exposures to selenium, either by ingestion or through inhalation of dust and fumes, have resulted in depression, nervousness, occasional dermatitis, gastrointestinal disturbances, giddiness, and a garlic odor.[5] Epidemiological studies have shown an increase in the incidence of dental caries in children when small amounts of selenium were ingested as part of their diets.

Although selenium is toxic to humans and animals, it is usually the result of accidental exposure. The MCL for selenium is 0.01 mg/l, which was established by the Public Health Service (PHS) in 1962 in response to the carcinogenic effect on animals. The dietary requirement of selenium is in the range of 0.04 to 0.10 mg/kg of food eaten. Also, signs of toxicity are evidenced at 0.7 to 7.0 mg/d.[8] Allowing for the dietary intake of selenium, and adding the contribution from water ingestion, a margin of safety of 3 results.

Silver. Silver is a rare element, and has a low solubility of 0.1 to 10 mg/l, depending on pH and chloride concentration. Trace amounts of silver are found in raw and finished water, from both natural sources of silver and industrial wastes.

Silver has a bactericidal action, and has been used as a water disinfectant, an application that is discussed in more detail in Chapter 11. This use represents one possible source of silver in public water supplies. Examination of water supplies in large cities revealed trace quantities of silver, with a median concentration of 2.3 $\mu g/l$.

Toxic Effects.   Silver is a nonessential element, providing no beneficial effects from its ingestion in trace amounts. Acute toxicity can result from large single doses, and can be fatal. Poisoning victims experience pulmonary edema after exhibiting anorexia and anemia.[5]

Chronic toxicity takes the form of an unsightly blue-gray discoloration of the skin, mucous membranes, and eyes, which is called argyrosis or argyria. Apparently, besides the cosmetic changes, there are no physiologic effects. Ingestion of trace amounts of silver or silver salts results in its accumulation in the body, particularly the skin and eyes. There is some evidence that changes to the kidneys, liver, and spleen can occur.

The exact quantity of silver that accumulates in the body is not known, but probably is about 50 percent. On the basis of the MCL of 50 $\mu g/l$, and a 2-l/d water intake, 50 $\mu g/d$ of silver would be accumulated. After 55 years, 1 g would be accumulated, which would indicate a borderline argyria. Also, there is a high adsorbability of silver bound to sulfur components of foodstuffs cooked in silver-containing waters.

Organic Chemicals

Cyclodiene Insecticides.   The cyclodiene insecticides are endrin, dieldrin, chlordane, heptachlor, and heptachlor epoxide, and are derived from hexachlorocyclopentadiene, which is produced using the Diels-Alder, or diene, reaction.[5] It has been estimated that about 600 million pounds of these highly chlorinated, cyclic organic compounds have been dispersed in the environment since their development in 1944. Not much is known about their fate in the environment, but traces of them and their stable epoxide oxidation products are ubiquitous. A study of U.S. rivers showed that the cyclodienes are found virtually everywhere. The following average concentrations were measured:

Aldrin: <0.001 to 0.006 ppb
Dieldrin: 0.08 to 0.122 ppb
Endrin: 0.008 to 0.214 ppb
Heptachlor: 0.0 to 0.0031 ppb
Heptachlor epoxide: <0.001 to 0.008 ppb
DDT: 0.008 to 0.144 ppb

The highest concentrations generally were found in the lower Mississippi Basin.

Degradation products of the insecticides are neurotoxic poisons, whose metabolic pathways are largely unknown and complex. The cyclodienes have been used principally as preemergence soil insecticides for the control of rootworms, wireworms, cutworms, and others, and for the control of termites, ants, and the boll weevil and bollworm.

The MCL's for chlorinated hydrocarbons (cyclodienes) are based on extrapolation to humans using concentrations that caused minimal toxic effects in animals. A margin of safety (MS) factor is then applied to these doses to further reduce the risk. The margin of safety factors used to establish MCL's are based on the following:

- Existing safe levels found in humans: MS = 0.1.
- Animal data corroborated by some human data: MS = 0.01.
- Animal data alone available: MS = 0.002.

The computation of acceptable levels of exposure for drinking waters must consider exposures from all sources.

Endrin. Endrin (1,2,3,4,10-hexachloro-6,7-epoxy-1,4,4a,5,6,7,8,8a-octahydro-endo-1,4-endo-5,8 dimethanonoaphthalene) exists in the *endo-endo* configuration. Endrin isomerizes in light, and is degraded to 9-ketoendrin, 9-hydroxyendrin, and 5-hydroxyendrin in animals.

Toxic Effects. Illness and death of humans have been observed after poisoning during the manufacture, spraying, or accidental ingestion of cyclodienes. Typical symptoms include stimulation of the central nervous system, blurred vision, insomnia, dizziness, involuntary muscular movements, and generally bad health. Severe poisoning is characterized by epileptiform convulsions and personality changes in humans, which can recur for 2 to 4 months after cessation of exposure.

Workers in plants manufacturing endrin had encephalograms suggesting brain stem injuries, which usually returned to normal after 6 months. Epidemics of poisoning have occurred after consumption of bread contaminated with endrin.

The cyclodiene compounds are extraordinarily toxic. Endrin is so toxic that it is registered as a rodenticide, and its dermal toxicity is equivalent to its oral toxicity.

There are no available data proving endrin is mutagenic, although other cyclodienes are carcinogenic. Endrin did not exhibit carcinogenic effects in rats that were fed it for 2 years, although it appears to have ill effects on reproduction, as it was found to be teratogenic in rats.

Lindane. Lindane is a commercial insecticide containing at least 99 percent of the lowest-melting-point and most reactive isomer of benzene hexachloride (BHC) (the γ isomer). BHC is the common name for the mixed isomer

1,2,3,4,5,6-hexachlorocyclohexane. The γ isomer for lindane is obtained by selective crystallization. The relatively high water solubility and vapor pressure (10 mg/l and 0.14 torr, respectively) cause it to have a relatively low persistence in the environment.

Toxic Effects.   Many cases of exposure to lindane have resulted in the development of aplastic anemia. However, a satisfactory animal model of this condition has not been found, and a specific relationship between lindane and aplastic anemia cannot be stated with certainty. Also, two cases of leukemia have been reported, although the causal relationship is inconclusive.

The most toxic of the BHC isomers, lindane excites the central nervous system, causing hyperirritability, convulsions, and death due to the collapse of the respiratory system. The single dose LD_{50} in rats varies between 88 and 300 mg/kg, while the value for dogs is in the range of 15 to 25 mg/kg.

Administration of lindane under chronic doses showed that the γ isomer is considerably less toxic than other isomers. The liver and kidneys of the test animals were damaged. The lowest concentration of lindane causing significant liver injuries was 100 ppm, while no effects were noted at concentrations below 50 ppm.

Generally, lindane is not considered to be mutagenic. Dosages of lindane administered to rats showed no evidence of carcinogenicity, although the average life-span was significantly reduced. Invertebrates have proved to be more sensitive to lindane than vertebrates, based on chronic bioassay tests. The midge is the most sensitive at 2.2 μg/l, and the *Daphnia magua* the least sensitive at 11 μg/l. Trout are sensitive to lindane, having a 96-hour LD_{50} of 2 μg/l. The fact that lindane can accumulate in fish and mollusk tissues led the FDA to limit the quantity of lindane in fish tissue to 0.3 mg/kg.

Because adequate data for humans are not available, the drinking water standard would be based on 0.002 (or 1/500) of the no-effect level of the most sensitive test animal. Using the dietary value for dogs of 15 mg/kg, or 0.3 mg/kg of body weight, the equivalent value for humans is 0.004 mg/kg, or 4 μg/l, in the water.

Toxaphene.   Toxaphene is a complex mixture of largely uncharacterized chlorinated camphene derivatives, which is the most often used but least understood organochlorine insecticide. Its makeup includes at least 175 possible compounds, fewer than 10 of which are known.

This product was used extensively as foliage insecticide for various food, feed, and fiber crops. Residues of toxaphene are found in the tissues of animals fed on contaminated food, and in grains and seeds. Toxaphene is known to be persistent in the environment, particularly in soils, as is the case with many organochlorine insecticides. Toxaphene is $C_{10}H_{10}Cl_8$- technical chlorinated camphene, 67 to 69 percent chlorine.

Toxic Effects. Toxaphene poisoning in man has been reported, but acute poisoning is rare. The poisoning has occurred as a result of accidental exposures during agricultural use, and from voluntary human studies. Inhalation of toxaphene by human volunteers showed no evidence of toxicity in the range of 0.4 to 250 ppb.

Chronic studies on a variety of animals showed significant changes in the liver. Kidney damage in dogs and rats has also been reported. A reproductive study showed no evidence of mutagenicity, carcinogenicity, or teratogenicity. Lakes treated with toxaphene remained toxic to fish for periods of up to 5 years.

2,4-Dichlorophenol. 2,4-Dichlorophenol (2,4-D) is a herbicide that has been used widely for the control of aquatic weeds in many lakes and streams. Because many of these streams and lakes are used as a water source or public water supply, it is important to establish a standard to protect the consumer. 2,4-D is slightly soluble in water, and the highest concentration measured in water supplies was 36 μg/l. 2,4-D is hydrolyzed to the corresponding acid in the body, yielding 2,4-dichlorophenoxyacetic acid.

Toxic Effects. Extensive studies have not been completed to determine the effects of 2,4-D on humans. Several studies have been completed on test animals, particularly mice and rats. The acute oral LD_{50} dose is 1.63 g/kg in mice and 4.5 g/kg in male rats, while for the acute subcutaneous dose, the LD_{50} is 1.73 g/kg. One study showed slight abnormalities in liver histopathology.

There is little available information on the chronic toxicity effects of 2,4-D. No studies have been completed on its mutagenicity and teratogenicity, and only limited work has been performed to determine carcinogenicity. A dose of 312 mg/kg promoted papillomas and carcinomas in mice. The maximum no-effect chronic dose in test animals has been estimated to be 100 mg/kg/d. The SNARL can be computed as 0.7 mg/l for a 70-kg consumer who drinks 2/l/d, and if 20 percent of the compound is ingested in the water.

2,4,5-Trichlorophenol. This compound, 2,4,5-trichlorophenol (2,4,5-TP), is similar to 2,4-D, and was a widely used herbicide for the control of aquatic weeds in rivers and lakes. 2,4,5-TP has been banned for all aquatic uses.

Toxic Effects. Tests on rats using 2,4,5-TP showed abnormalities in a majority of their litters and fetuses. Effects included cleft palate and cystic kidneys.

The no-effect level in mice was found to be 21.5 mg/kg, while a 4.5 mg/kg dose in rats caused some effects in the rat. Tests also showed that the 2,4,5-TP contained the contaminant dioxin, which is highly toxic to experimental animals; however, purified 2,4,5-TP has produced teratogenic effects in hamsters and rats at high doses. Because 2,4-D and 2,4,5-TP are very similar, their toxicities are of the same order and magnitude. Using the data for 2,4-D, an acute oral dose of 2,4,5-TP would be in the range of 3,000 to 4,000 mg. Based on a feeding study of mice, the no-effect level was established as 5 mg/kg/d.

Methoxychlor. Methoxychlor, or 2,2-*bis*-(*p*-methoxyphenyl)-1,1, 1-trichloroethane, is similar to DDT, and was first introduced in 1945. It is used as an insecticide of very low mammalian toxicity for home and garden, on domestic animals, elm bark beetle, and others. The half-life of methoxychlor in water is about 46 days, and no residues were detected in samples from finished drinking water from the Mississippi and Missouri rivers.

Toxic Effects. Methoxychlor is not bioconcentrated and stored in animal tissues, unlike DDT. Methoxychlor differs significantly from DDT, being much more biodegradable than DDT.

The highest concentration of methoxychlor to have minimal or no long-term effects on humans is 20 mg/kg/d. Methoxychlor is one of the safest of the insecticides, requiring over 6,000 mg/kg for an acute oral LD_{50}. For mice the value is 2,900 mg/kg, and for monkeys it is 2,500 mg/kg.

Methoxychlor results in growth retardation in test animals, but no gross pathologic changes were found when they were fed at high dosages, above 2,500 ppm. In dogs, high doses yielded convulsions, but no effects were observed at 1 g/kg and below. There are no available data for teratogenic effects, but mutagenicity tests gave negative results, and extremely high doses in rats resulted in testicular tumors.

Because the limited data for humans can be corroborated using test animals, the margin of safety factor is 1/100. The safe level for drinking water supplies is based on a 70-kg consumer who drinks 2 l/d, and ingests 20 percent of the total intake from water. The safe value is 0.14 mg/l when a 2.0 mg/kg dose is used.

Total Trihalomethanes. The formation and the removal of trihalomethanes are discussed in Chapter 14. Trihalomethanes are formed by the reaction of free chlorine with certain naturally occurring organic compounds. Total trihalomethane (TTHM) is defined as the sum of the concentrations of trichloromethane (chloroform, $CHCl_3$), tribromomethane (bromoform, $CHBr_3$), bromodichloromethane ($CHBrCl_2$), and dibromochloromethane ($CHBrCl$). The MCL of 0.1 mg/l for TTHM's became final on November 29, 1979.

Toxic Effects. Chloroform is absorbed rapidly throughout body tissues, and is metabolized to CO_2, $Cl-$, phosgene, and other compounds. The test animals, mice, rats, and monkeys, have similar metabolic functions to humans with regard to chloroform.

Acute toxic effects of chloroform in humans and test animals are hepatic and renal lesions and damage, including necrosis and cirrhosis.[5] However, there are no data to determine mutagenicity, and chloroform is not highly teratogenic, although it has been shown to be embryotoxic.

Acute toxic responses to THM's have not been documented, although the potential effects should be of concern. Human epidemiological studies yield inconclusive evidence regarding carcinogenic effects, although positive correlations have been shown. The studies evaluated cancer mortality data rather than incidence data. However, because of limitations in the epidemiological studies, the evidence cannot lead to form conclusions about THM's and cancer mortality. The studies did show statistically significant correlations between THM's and cancer of the bladder, and chloroform and cancer of the rectum, bladder, and large intestine.

Risk assessments made by various federal science groups have estimated incremental risks associated with chloroform in drinking water. The MCL of 0.1 mg/l was based on a 2-l/d water consumption for 70 years.

Radioactivity

Radioactivity is the ability of a substance to emit positively or negatively charged particles, and sometimes also electromagnetic radiation, by the disintegration of atomic nuclei. Alpha (α) particles have a mass of 4, and are doubly charged ions of helium. Beta (β) particles move at about the speed of light, and are negatively charged (electrons). Gamma (γ) rays are electromagnetic radiation (photons), and travel at the speed of light. The radioactive standards are related to the amounts of spontaneous radiation emitted by both natural and man-made contaminants. The natural ionizing radiation includes cosmic rays and products of the decay of radioactive materials in the earth's crust and atmosphere. Part of the radiation is from external sources, and part is due to the inhalation and/or ingestion of contaminated air, food, and water.

Radiation forms are measured in three ways, which are:

- The *roentgen,* which is the amount of radiation absorbed in air at a certain location.
- The *rad,* which is the amount of radioactivity of a gram of material (note that doses in the range of 600 to 700 rads are dangerous to humans).
- The *rem,* which indicates the degree of health hazard and is equal to the rad multiplied by a factor for potential hazard.

Sources of Radioactivity. Traces of radioactivity are usually found in all drinking water supplies. The radioactivity varies, depending on the radio-chemical composition of the soil and rock strata over or through which the raw water passes. Several natural and artificial radionuclides have been found in water, but the majority of the radioactivity is from a small number of nuclides and their decay products, as follows:

- Low linear energy transfer (LET) emitters: potassium-40 (^{40}K), tritium (^3H), carbon-14 (^{14}C), and rubidium-87 (^{87}Rb).
- High LET, alpha-emitting radionuclides: radium-226 (^{226}Ra), the daughters of radium-228 (^{228}Ra), polonium-210 (^{210}Po), uranium (U), thorium (Th), radon-220 (^{220}Rn), and radon-222 (^{222}Rn).

Low LET radiation results from cosmic ray interactions with atmospheric oxygen and nitrogen, yielding tritium. Carbon-14 is also due to cosmic ray interactions with atmospheric nitrogen. Potassium-40 is a naturally occurring radionuclide, and is found as a constant percentage (0.0118 percent) of total potassium.

High LET radiation includes that produced from the decay of uranium-238 and thorium-232, which are both widely distributed throughout the earth's crust. The majority of the high LET emitters are alpha-emitters. The natural alpha-emitters that are found in drinking water are the bone seekers, of which radium-226 and the daughters of radium-228 have the greatest potential for harming consumers.

Artificial radiation results from the atmospheric testing of nuclear weapons, and the subsequent fallout of contaminated material. As a result, all surface water sources today contain some degree of radiation. However, since the Nuclear Test Ban Treaty of 1963, there has been a substantial decrease in radioactive fallout. Other sources of artificial radiation include wastes from radiopharmaceuticals and the use and processing of nuclear fuels for power generation.

Toxic Effects. Numerous studies have shown that developing mammals are more radiosensitive than adults. The harmful effects of radiation include increased tumor incidence, death, and development of abnormalities. Developmental and teratogenic effects of radionuclides would not be measurable from the amount of radiation received from drinking waters. The doses in drinking water represent one five-thousandth of the lowest dose at which developmental effects are seen in animals.

The impact on genetic diseases is thought to be small, with an estimated increase of 0.0098 case per million live births per year. It is estimated that there are between 4.5 and 45 cases of cancer per million people, depending on the risk model used. The Safe Drinking Water Committee concludes that the radiation associated with most water supplies is a small proportion of the normal background levels to which humans are exposed, and that it is virtually impossible to measure adverse health effects. However, mathematical models have been developed to project radioactive doses, which have been confirmed through statistical correlations. Ingestion of radium-226 at 5 pCi/l by 1 million people results in 1.5 fatalities per year, while 4 mrem annual total body exposure causes 2 to 4 cancer fatalities per 1 million people.

Turbidity

Suspended constituents in drinking waters include organic solids, viruses, bacteria, algae, and other substances, which are present in finely divided particles. Sizes can range from 1 nm (10^{-9} m) to as much as 100 μm (10^{-4} m). The majority of these particles are clay, which does not have a direct effect on health, whereas particles such as asbestos minerals may be hazardous to human health when ingested. The fine particles present in water are measured by the degree of light scattering and the absorbing properties of the particles, and are termed the turbidity of the water. Turbidity also causes cloudy water that is aesthetically displeasing.

Most of the particles, although not harmful in themselves, may affect the quality of the water indirectly, as they can act as a vehicle for the concentration, transportation, and release of other contaminants, such as bacteria and viruses.

Asbestos.   The removal of asbestos is the topic of Chapter 16, and is discussed only briefly here. Epidemiological studies of workers have shown an increase in death rates from gastrointestinal cancer through the inhalation of asbestos fibers. Similar studies in areas that have asbestos in the water supplies do not indicate any increase of cancer over other areas not having asbestos-contaminated water. Animal studies have shown that fiber length and diameter affect the carcinogenic response, but the same effects have not been evident in human studies.

Indirect Health Effects.   Clay and organic particles have large surface areas, and adsorb ions, molecules, and biological matter. The presence of clay particles in water is the natural result of surface runoff after precipitation. The clay particles can adsorb organic and inorganic toxic materials, bacteria, and viruses and release them later, unless the clay particles are removed.

MCL.   Turbidity in water can interfere with:

- Disinfection, by creating a potential shield for disease-causing organisms.
- Maintaining an effective chlorine residual.
- Bacteriological testing of the water.

The MCL was based on minimizing the potential for hazard to human health. The MCL is basically 1 NTU, with a maximum of 5 NTU for a 2-day average. In some cases of high asbestos concentrations in raw water, it may be necessary to reduce turbidity to 0.02 NTU or less for good asbestos removal. This is discussed in detail in Chapter 16.

Bacteria

Pathogenic organisms appear in small numbers in water supplies. Their isolation is unlikely and difficult. They include disease-causing bacteria, viruses, protozoans, fungi, and worms. An indirect approach to measuring microbial hazards is needed, and this requires that an indicator organism be used to indicate the presence of disease-causing constituents. Such an indicator organism should behave as follows:

- Be applicable to all types of water.
- Be present in sewage and polluted waters when pathogens are present.
- Correlate quantitatively with the amount of pollution.
- Be present in greater numbers than pathogens.
- Have no aftergrowth in water.
- Have a greater survival time than pathogens.
- Be absent from unpolluted waters.
- Be easily detected by simple laboratory tests in the shortest time consistent with accurate results.
- Have constant characteristics.
- Be harmless to humans and animals.

The coliform group of organisms nearly fulfills these criteria.

Coliform Bacteria.   Coliform bacteria have been used for many years as indicators of the sanitary quality of water bodies. The organism usually used is *Escherichia coli* because it is found in the human intestine. The use of total coliforms is not necessarily indicative of the presence of disease-causing organisms because some strains of the coliform group have a wide distribution in the environment. One advantage of using the coliform group as an indicator of fecal pollution is that the coliform generally survive longer than pathogenic bacteria in the water environment. Also, pathogenic bacteria are generally more sensitive to disinfection than coliforms. The same is not true for viruses, which survive longer than bacterial pathogens.

There are two acceptable methods of measuring the coliform group of organisms: (1) the membrane filter test and (2) the fermentation tube test. MCL's have been established for each of these tests based on the number of monthly samples taken, which in turn depends on the community population. The MCL's established are strict, but there is still no absolute assurance that there are no pathogens present. The MCL's provide confidence that pathogens are not present, and hence the probability of waterborne disease is significantly decreased. The reduction in this century of waterborne diseases such as typhoid fever, cholera, and others is indicative of the validity of this confidence. The proposed standards for coliform counts are included in Table 1–7.

RATIONALE FOR SMCL SELECTION

Chloride

Chloride is not harmful to people in reasonable concentrations, but it is corrosive when its concentration is more than 250 mg/l. At concentrations above 400 mg/l, chloride imparts a salty taste to water, which can be objectionable to consumers. The taste threshold varies with individual sensitivity, the presence of other substances, and individual acclimatization to the use of a particular water. Chloride can be removed using one of the demineralization processes, such as reverse osmosis or electrodialysis, that are described in Chapter 13. However, proper aquifer or water source selection can help minimize the concentrations of chloride.

Color

Color is related to consumer acceptance of water, rather than to its safety. At 15 color units (CU's) the appearance of water becomes objectionable to many persons, who often turn to alternative, possibly less safe, supplies of drinking water.

Color may be the result of organic materials in the water, which could cause the generation of trihalomethanes and other organohalogen compounds. Color can also be caused by manganese, copper, and iron.

Copper

Copper is frequently found in surface waters and some groundwaters. In one study, over 74 percent of 1,500 rivers surveyed contained it. Copper salts are highly soluble in low-pH waters, but may hydrolyze and precipitate in water of normal alkalinity.

Copper is recognized as an essential element in human metabolism. Copper deficiency is characterized by anemia, loss of hair pigment, reduced growth, and loss of arterial elasticity. However, copper is a gastrointestinal tract irritant and can impart an undesirable taste to drinking water. Chronic toxicity due to low intakes (<1 mg/l) of copper has not been identified in humans.

A small number of individuals suffer from an inherited disease, called Wilson's disease, that causes liver damage due to an inability to metabolize copper. Wilson's disease can be arrested by the use of chelating agents.

The interim standard of 1 mg/l for drinking water is based on avoiding undesirable tastes. The taste threshold varies from 1 to 5 mg/l. The U.S. standard is in contrast to the European standard of 0.05 mg/l, which was established on the basis of fixture discoloration.

At low alkalinity, copper in water can be toxic to aquatic organisms. At high alkalinity, copper is complexed by anions, which reduce its toxicity. The critical effect of copper is its heightened toxicity to young or juvenile fish. Concentrations of copper below 25 μg/l are not toxic to fish.

Corrosivity

Water corrosivity is a complex phenomenon related to pH, dissolved oxygen, alkalinity, total dissolved solids, and some other factors. Corrosive water is undesirable because it causes metals to dissolve and can stain plumbing fixtures. It should be avoided if other sources are available. Alternatively, it must be treated to limit its corrosive characteristics.

Corrosive drinking waters have the following impacts:

- Health effects
- Economic impacts
- Aesthetic effects

The health impacts are related to metals such as lead, cadmium, and copper dissolving and going into solution. The limits for lead and cadmium are covered under the primary regulations, and copper is included under the secondary regulations. The economic impacts are related to damage caused to metal piping and fixtures. The aesthetic impacts were mentioned above. These factors are sufficient cause to include corrosivity in the secondary regulations. Many corrosive indices have been proposed, but there is no general agreement on a single index that would unfailingly indicate corrosive water. Therefore, the standard simply requires that the water be noncorrosive.

Foaming Agents

Foaming is caused by the presence of detergents and similar substances. Foaming water is unsightly, and is not suitable for ingestion. The foaming potential of water is measured by the quantity of methylene blue-active substances (MBAS) present. The MBAS tests indicate the presence of alkyl sulfonates, which can also cause an oily taste in water. Foaming substances can be removed by activated carbon treatment.

Iron

Iron in water is not likely to have any toxicologic significance because the quantity consumed in the daily diet far exceeds the quantity that causes objectional tastes and staining. Iron imparts a yellowish color and a bitter or astringent taste to water, and it appreciably affects the taste of beverages made from water. Iron also causes a brownish stain on laundered goods, plumbing fixtures, and buildings in lawn sprinkler range. It is easily removed from water by conventional treatment schemes, which are described in Chapter 19. If the iron is the result of corrosive waters in iron or steel pipes, then by adjusting the chemical balance of the water, this problem can be eliminated.

Manganese

Manganese has similar characteristics to iron in natural waters, but is less abundant than iron in the environment. Thus, the manganese concentration in water is lower than that of iron. Neurologic effects of manganese, which have been attributed to manganese dust inhalation, have not been reported after its oral ingestion in humans. Manganese is an essential trace element for humans, and plays an important role in enzyme systems. Limiting the manganese concentration in water for aesthetic reasons will preclude any physiologic effects from excessive intake. Manganese produces a brownish or purplish discoloration in water and laundered goods (when oxidized) and also impairs the taste of drinking water and beverages such as coffee or tea.

At concentrations above the secondary limit, manganese occasionally can cause a buildup of coatings in distribution piping. It is desirable for domestic use and imperative for some industrial purposes to remove all manganese from water. However, there are practical difficulties in reducing manganese to a residual concentration of less than 0.05 mg/l by treatment. (Methods of manganese removal are discussed in Chapter 19.)

Odor

Odor is an aesthetic quality of water that is important to consumers and industries in such consumable products as food, beverages, and pharmaceuticals. Odor is measured in terms of the dilution required to remove it, as expressed by the threshold odor number (TON). The lower the number is, the more odor-free the water. The determination of the TON value is made by a panel of consumers, as is explained in detail in Chapter 18. Reaction products of disinfection, such as chlorophenols, and phenols are often the cause of odors. Algae and fungi in water may cause a musty odor. The removal of odors, which is discussed in Chapter 18, can be achieved using activated carbon and aeration.

pH

The pH of domestic waters is limited to the neutral range of 6.5 to 8.5 units. At values outside this range, water can be corrosive or cause precipitation of metal salts. Corrosive characteristics are evident at pH values below 6.5, and low pH's also cause the release of toxic metals. At pH values above 8.5, precipitation and scaling can result, which in turn are associated with taste and odor problems. Higher pH's reduce the efficiency of chlorine, and the rate of trihalomethane formation increases, as discussed in Chapters 11 and 14, respectively. To minimize the potential for these problems, the pH is restricted to a neutral range.

Sulfate

Sulfates do not adversely affect the health of consumers at concentrations below 500 mg/l. However, the taste threshold for sulfate in water is about 300 to 400 mg/l for most persons. At concentrations above 600 mg/l, the only physiological consequence of sulfate is a laxative effect (diarrhea). The laxative effects of sulfate typically do not affect regular users of water, but travelers are particularly susceptible to them. High concentrations of sulfate can cause scaling in boilers and heat exchangers. Sulfate can be removed using demineralization processes, such as reverse osmosis and electrodialysis, which are described in Chapter 13.

Total Dissolved Solids (TDS)

The total dissolved solids (TDS) value is indicative of the aesthetic quality of drinking waters, and suggests the presence of specific objectional ions that may cause taste and odor problems, or have physiologic effects, as already described. Many of the ions included in the TDS measurement have been discussed separately in this chapter. Total hardness or calcium hardness, corrosive characteristics, and metal precipitates are also measures of highly mineralized water. The dissolved solids impart a distinct taste at values above 750 mg/l, although the objectional tastes are only noticed by occasional users. A TDS value of 500 mg/l should minimize the effects of taste, the laxative effects, and other problems associated with specific ions. Dissolved solids can be removed by the demineralization processes described in Chapter 13.

Zinc

Zinc is ubiquitous in foods, water, and the general environment, and is an essential and beneficial element in human metabolism. However, zinc has a low bioavailability in many foods, particularly plant seeds, so that zinc deficiencies have resulted in some humans. The National Academy of Sciences Food and Nutrition Board has determined that an adult zinc dose of 15 mg/l is needed, and for children over a year old, the requirement is 10 mg/l.

Concentrations of zinc of 30 mg/l or more impart a strong astringent taste and a milky appearance to water. There have been reports of acute toxicity at intakes of 40 to 50 μg/l, but there are no known chronic effects. Zinc can be removed from water using conventional treatment processes and ion exchange. However, the zinc content in water is frequently the result of corrosive water passing through galvanized pipes, and can be controlled by chemically adjusting the water to remove the corrosive characteristics. Control of corrosion will also limit the introduction of cadmium and lead into water.

INDUSTRIAL WATER QUALITY REQUIREMENTS

General

The quality requirements for industrial water are generally consistent with public demands for clear, attractive water of moderate mineral content, free from iron and manganese; but certain requirements may exceed those of public water supplies. Usually this means that the industry concerned must provide additional treatment at its plant site. However, if the industrial needs are known and considered in the municipal water treatment plant, they sometimes can be met by additional processing there and without the need for supplemental treatment by the industry.

Air Conditioning

Water for air conditioning purposes should be as cold as possible, and low in hydrogen sulfide, iron, and manganese.

Baking

Additional water quality requirements for baking beyond those of the standards are for total iron and manganese less than 0.2 mg/l, hydrogen sulfide less than 0.2 mg/l, and no taste or odor.

Boiler Feed Water

General boiler feed water requirements, which exceed those for public water supplies, are given in Table 1–9.

Brewing

For brewing light beer, special water quality requirements include: iron less than 0.1 mg/l, manganese less than 0.1 mg/l, TDS below 500 mg/l, alkalinity under 75 mg/l as $CaCO_3$, hydrogen sulfide less than 0.2 mg/l, no taste or odor, no chlorine, sodium chloride less than 275 mg/l, and pH 6.5 to 7.0.

Canning

Water quality requirements for canning are hardness 25 to 75 mg/l as $CaCO_3$ (for legumes), iron under 2.0 mg/l, manganese less than 2.0 mg/l, hydrogen sulfide less than 1.0 mg/l, and no taste or odor.

Carbonated Beverages

Recommended water quality includes turbidity less than 2.0 NTU, organic color plus oxygen consumed less than 10 mg/l, hardness under 200 mg/l as $CaCO_3$, iron less than 1.0 mg/l, manganese less than 0.2 mg/l, TDS less than 850 mg/l, alkalinity less than 50 mg/l as $CaCO_3$, no taste or odor, and hydrogen sulfide less than 0.2 mg/l.

Cooling Water

The total hardness of cooling water should be 50 mg/l as $CaCO_3$ or less, hydrogen sulfide less than 5 mg/l, and iron and manganese each less than 0.5 mg/l. The water should not be corrosive or slime-forming.

Food Processing

Water from public water supplies meeting the standards is generally satisfactory for food processing, provided iron and manganese are each less than 0.2 mg/l and the water is free of objectionable taste and odor.

Ice Making

For ice making, water should have the following special characteristics: turbidity less than 5.0 NTU, color under 5, alkalinity as $CaCO_3$ 30 to 50 mg/l, TDS less than 300 mg/l, total iron and manganese less than 0.2 mg/l, and

Table 1-9. General Quality Tolerances for Boiler Feed Water.

BOILER PRESSURE PSI	COLOR	DISSOLVED OXYGEN ml/l	HARDNESS mg/l AS CaCO$_3$	pH	TDS mg/l	Al$_2$O$_3$ mg/l	SiO$_2$ mg/l	CO$_3$ mg/l	HCO$_3$ mg/l	OH mg/l
0–150	80	2.0	75	8.0+	3,000–1,000	5.0	40	200	50	50
150–250	40	0.2	40	8.5+	2,500–500	0.5	20	100	30	40
250 and up	5	0.0	8	9.0+	1,500–100	0.05	5	40	5	30

silica less than 10 mg/l. Calcium bicarbonate is especially troublesome, and magnesium bicarbonate tends to produce a greenish color in the ice. Free carbon dioxide is beneficial in preventing cracking of the ice. Calcium, magnesium, and sodium sulfates and chlorides should each be less than 300 mg/l in order to prevent white butts in the ice cakes.

Laundering

Public water supplies conforming to the standards are satisfactory for laundry use, provided iron and manganese are each less than 0.2 mg/l. Hardness less than 50 mg/l as $CaCO_3$ is preferred.

Plastic Manufacture

To make clear, uncolored plastics, water is required to have turbidity and color each less than 2.0, iron and manganese each less than 0.02 mg/l, and TDS less than 200 mg/l.

Paper Making

To make the highest grades of light paper, the following water quality requirements should be met: turbidity and color each under 5, total hardness as $CaCO_3$ less than 50 mg/l, TDS under 200 mg/l, and total iron and manganese under 0.1 mg/l; also, uniformity of composition and temperature are desirable.

Rayon (Viscose)

For pulp production, turbidity and color each should be less than 5, hardness less than 8 mg/l as $CaCO_3$ (hard water coats fibers), iron less than 0.05 mg/l, manganese not over 0.03 mg/l, TDS less than 100 mg/l, total alkalinity under 50 mg/l, and hydroxide alkalinity under 8 mg/l as $CaCO_3$, copper less than 5 mg/l, SiO_2 under 25 mg/l, and Al_2O_3 not over 8 mg/l. For rayon manufacturing, no iron or manganese can be tolerated, turbidity must be less than 0.3 NTU, and the pH must be in the range of 7.8 to 8.3.

Tanning

For use in tanneries, water should have a total hardness less than 135 mg/l, total alkalinity under 135 mg/l, and hydroxide alkalinity less than 8 mg/l as $CaCO_3$. Iron and manganese each should be under 0.2 mg/l, and the pH 8.0.

Textiles

General textile manufacture requires water with a total hardness less than 20 mg/l as $CaCO_3$ and iron and manganese each less than 0.2 mg/l. Residual alumina must be less than 0.5 mg/l in order to prevent uneven dying, as alum is a mordant. The chemical composition of the water should not vary. In production of cotton bandages, the color must be under 5 and calcium, magnesium, soluble organic matter, and suspended matter may be objectionable; also, the iron and manganese content should be zero.

REFERENCES

1. Baker, M. N., *The Quest for Pure Water,* American Water Works Association, New York, 1948.
2. Rosenau, M. J., *Preventative Medicine and Hygiene,* 5th ed., Appleton-Century, New York, 1935.
3. Klein, L., *Aspects of Water Pollution,* Academic Press, Inc., New York, 1957.
4. *Drinking Water and Health,* Safe Drinking Water Committee, National Academy of Sciences, Washington, D.C., 1977.
5. *Drinking Water and Health,* Vols. 1–5, Safe Drinking Water Committee, National Academy Press, Washington, D.C., 1980, 1982, 1983.
6. Benefield, Larry D., Judkins, Joseph F., and Weand, Barron L., *Process Chemistry for Water and Wastewater Treatment,* Prentice-Hall, Inc., Englewood Cliffs, N.J., 1982.
7. Faust, Samuel D., and Aly, Ossman M., *Chemistry of Water Treatment,* Ann Arbor Science, Woburn, Mass., 1983.
8. EPA, *National Interim Drinking Water Regulations,* Office of Water Supply, EPA-570/9–76–003.
9. Durum, W. H., Hem, J. D., and Heidel, S. G., "Reconnaissance of Selected Minor Elements in Surface Waters of the United States," October, 1970, U.S. Geological Survey Circular 643, Washington, D.C., 1971.

Chapter 2
Specific Contaminant Removal Methodologies

INTRODUCTION

This chapter presents information on the removal of the following inorganic contaminants from water: arsenic, barium, cadmium, chromium, copper, fluoride, lead, mercury, nitrate, selenium, silver, and zinc. All of these contaminants except copper and zinc are included in the National Interim Primary Drinking Water Regulations, and the Maximum Contaminant Levels (MCL's) are shown in Table 1–4 (p. 11). Copper and zinc are included in the Secondary Drinking Water Regulations, and their MCL's are 1.0 and 5.0 mg/l, respectively.

Additional information on ion exchange treatment for barium, fluoride, and nitrate is presented in Chapter 17. Iron and manganese removal processes are described in Chapter 19.

Several conventional water treatment techniques are effective for the removal of inorganic contaminants. A summary of the best treatment methods for removal of nine of the ten inorganic chemicals (excluding fluoride) is abstracted from an EPA report on water treatment methods and is shown in Table 2–1.[1] These data indicate that chemical clarification, with iron salts, alum, or lime, is the most generally applicable method for removal of inorganic chemicals. Figures 2–1, 2–2, and 2–3 show effective removal of inorganic contaminants using iron, alum, and lime coagulation, respectively.

Results of a study on raw and treated water at 12 plants in California and Colorado were reported by Zemansky,[2] and data on four metals—chromium, copper, lead, and zinc—are summarized in Table 2–2. Concentrations of metals in the raw waters are higher than might be expected because supplies known to be contaminated with industrial discharges were purposely selected for the study. However, cadmium and silver were below detectable concentrations at the 12 plants studied.

There have been several studies on the use of activated carbon to remove inorganic chemicals from water. Smith[3] and Sigworth and Smith[4] report that arsenic, chromium, copper, mercury, and silver may be significantly removed by adsorption on activated carbon. Lead and iron in the oxidized state also may be adsorbed to some extent by activated carbon. The removal of mercury by powdered activated carbon has been investigated in several

Table 2–1. Most Effective Methods for Removal of Inorganic Chemicals in Primary Drinking Water Regulations.*

CONTAMINANT	MOST EFFECTIVE METHODS
Arsenic^{+3}	Chemical clarification (oxidation prior to treatment required)
	Ferric sulfate, pH 6 to 8
	Alum, pH 6 to 7
	Lime
Arsenic^{+5}	Chemical clarification
	Ferric sulfate coagulation, pH 6 to 8
	Alum coagulation, pH 6 to 7
	Lime
Barium	Lime clarification, pH 10 to 11
	Ion exchange
Cadmium	Ferric sulfate clarification, above pH 8
	Lime clarification
Chromium^{+3}	Chemical clarification
	Ferric sulfate, pH 6 to 9
	Alum, pH 7 to 9
	Lime
Chromium^{+6}	Ferrous sulfate clarification, pH 7 to 9.5
Lead	Chemical clarification
	Ferric sulfate, pH 6 to 9
	Alum, pH 6 to 9
	Lime
Mercury, inorganic	Ferric sulfate clarification, pH 7 to 8
Mercury, organic	Granular activated carbon
Nitrate	Ion exchange
Selenium^{+4}	Ferrous sulfate clarification, pH 6 to 7
	Ion exchange
	Reverse osmosis
Selenium^{+6}	Ion exchange
	Reverse osmosis
Silver	Chemical clarification
	Ferric sulfate, pH 6 to 9
	Alum, pH 6 to 8
	Lime

* Reference 1.

studies; the work by Thiem and others is summarized in an EPA report.[1,5] Many contaminants can be removed by ion exchange or reverse osmosis. Table 2–3 and Table 2–4 are presented to illustrate the upper limiting raw water concentrations that can be treated by ion exchange and reverse osmosis and still yield a product water that meets the MCL.

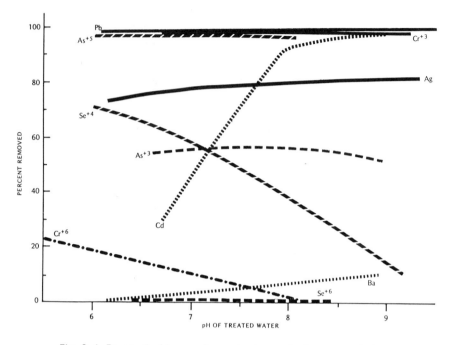

Fig. 2-1. Removal of inorganic contaminants by iron coagulation.[1]

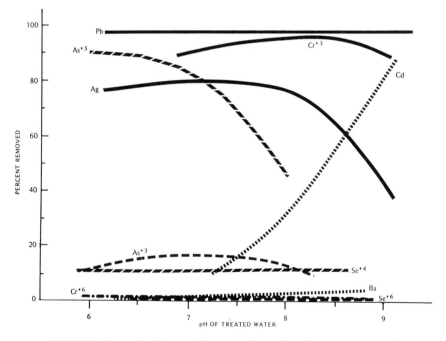

Fig. 2-2. Removal of inorganic contaminants by alum coagulation.[1]

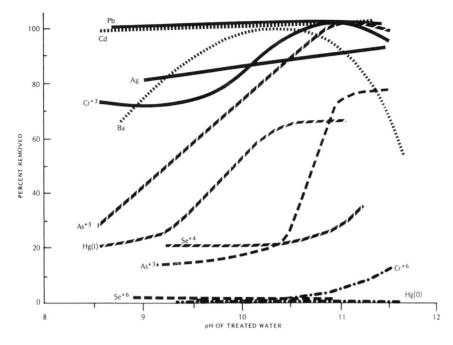

Fig. 2–3. Removal of inorganic contaminants by lime softening.[1]

CONTAMINANT DESCRIPTIONS

The 12 inorganic contaminants are described in the following paragraphs. The predominant species of each are identified, and the effect on human consumption is outlined. Finally, the various water treatment techniques are discussed for each one. The effects of some chemicals are described as either acute or chronic. Acute toxicity refers to effects occurring in a short time, such as severe poisoning. Chronic toxicity refers to effects occuring over an extended time period. These would include carcinogenic (tumor-forming), mutagenic (affecting heredity), and teratogenic (producing deformed off-spring) diseases.

Arsenic

The principal forms of arsenic in water are the anions arsenite (AsO_2^{-1}, arsenic valence +3) and arsenate (AsO_4^{-3}, arsenic valence +5). The ingestion of as little as 100 mg of arsenic can result in severe poisoning.

In general, inorganic arsenic is more toxic to humans and animals than are organic arsenic compounds, and the trivalent form (arsenite) is more toxic than the pentavalent form (arsenate). Chronic arsenic exposure results in various kidney, liver, and bone disorders, and there is some evidence that

Table 2-2. Trace Metal Occurrence and Removal at 12 Water Treatment Plants in Colorado and California.*,**

METAL	RAW WATER CONCENTRATION, mg/l		AVERAGE PERCENT REMOVALS				AVERAGE PRODUCT WATER CONCENTRATION, mg/l
	AVERAGE	HIGH	MICROSTRAINER†	CLARIFIER	FILTER	ENTIRE PLANT	
Chromium	0.022	0.084	25	35	15	31	0.015
Copper	0.030	0.160	14	26	37	49	0.015
Lead	0.012	0.042	3	27	29	32	0.008
Zinc	0.078	0.538	30	36	37	48	0.041

* Reference 2.
** Conversion factors may be found in Appendix A.
† Only 4 of the 12 plants were equipped with microstrainers.

Table 2–3. Upper Limiting Raw Water Concentrations of Various Contaminants That Can Be Treated by Ion Exchange without Exceeding the MCL.*

CONTAMINANT TO BE REMOVED	UPPER LIMITING RAW WATER CONCENTRATION	MCL	REMARKS
Arsenic, trivalent	Unknown	0.05 mg/l	Activated alumina
Barium	45 mg/l. Generally by blending of raw and finished water for corrosion and hardness control	1.0 mg/l	Softening resins
Fluoride	pH-dependent (best @ pH = 5.5 to 6)	1.4–2.4 mg/l	Activated alumina
Inorganic mercury	0.1 mg/l	0.002 mg/l	Cation and anion resins
Organic mercury	0.1 mg/l	0.002 mg/l	Cation and anion resins
Nitrate, as N	50 mg/l	10.0 mg/l	NO_3-selective resin
Selenium, quadrivalent	0.33 mg/l	0.01 mg/l	—
Selenium, hexavalent	0.33 mg/l	0.01 mg/l	—

* Reference 1.

Table 2–4. Upper Limiting Raw Water Concentrations of Various Contaminants That Can Be Treated by Reverse Osmosis without Exceeding the MCL.*

CONTAMINANT TO BE REMOVED	UPPER LIMITING RAW WATER CONCENTRATION	MCL
Arsenic, trivalent	0.33 mg/l	0.05 mg/l
Barium	45.0 mg/l	1.0 mg/l
Chromium, hexavalent	0.4 mg/l	0.05 mg/l
Lead	0.4 mg/l	0.05 mg/l
Nitrate, as N	67 mg/l	10 mg/l
Selenium, quadrivalent or hexavalent	0.33 mg/l	0.05 mg/l
Silver	0.83 mg/l	0.05 mg/l

* Reference 1.

arsenic is carcinogenic. In many cases where arsenic has been detected in natural waters, it has been found in concentrations exceeding the MCL of 0.05 mg/l. Arsenic concentrations in streams have exceeded the MCL far more than any other potentially toxic constituent has.

Arsenic is found in the environment in several forms and from several sources. It occurs in rocks in many forms including arsenides, and is present in some mineral veins as native arsenic and as oxides of arsenic. Other sources of arsenic in the environment include: (1) combustion of fossil fuels; (2) pesticides, herbicides, and insecticides; (3) mine tailings; (4) by-products of copper, gold, and lead refining; (5) other industrial processes; and (6) some phosphate detergent builders and presoaks.

The solubility of arsenic depends upon its chemical form and pH. Elemental arsenic is essentially insoluble in water, but many of the arsenates are highly soluble. It has been suggested that arsenic may be oxidized from 0 to +3 to +5, that ligand exchange reactions with OH^-, H_2O, and S^{-2} may occur, and that removal may occur through precipitation, sorption, and biological uptake.[6] In a particular water, arsenic may be present in various chemical forms that will react differently from each other in treatment processes.

It is believed that chemical oxidation of arsenite is not an important process in natural waters, but that rapid oxidation is catalyzed by bacteria. In fresh surface waters there is likely to be a partitioning of arsenic between the oxidation states. Depending on the arsenic sources, the relative rates of chemical and bacterial oxidation of As^{+3}, the rate of bacterial reduction of As^{+5}, and the presence or absence of an anaerobic zone, the As^{+3} content could be large or small. Although the concentrations of the species are important, there is little information on the forms present in water and wastewater. No measurements in freshwaters are reported, and information about the rates of the reactions is insufficient to make predictions.[6]

Water Treatment.   Pentavalent arsenic can be removed from water by conventional chemical clarification with lime, alum, or iron salts as coagulants. Trivalent arsenic is not effectively removed by chemical clarification; therefore, it must be oxidized to the pentavalent form with an oxidant such as chlorine, ozone, or potassium permanganate prior to clarification. Based on laboratory and pilot tests, a small 40,000 gpd (150 m³/d) water treatment plant was constructed in Taiwan to remove arsenic from groundwater. During a 4-month period in 1969, a range in raw water arsenic concentrations of 0.36 to 0.56 mg/l was reduced to trace levels by a treatment system consisting of chlorination, chemical clarification with ferric chloride, and filtration.[7]

It is possible to obtain nearly complete removal of anionic forms of arsenic (arsenite and arsenate) by anion exchange resins. Laboratory experiments indicate that arsenic also can be removed by activated alumina. It can be removed by reverse osmosis and adsorption on activated carbon as well, although quantitative data are lacking. Activated carbon is reported to have a high adsorption potential for pentavalent arsenic.[4]

Barium

Barium has toxic effects on the heart, blood vessels, and nerves. Soluble barium salts are very toxic, whereas the insoluble compounds are generally nontoxic. There is no information available on the chronic effects of low levels of barium ingested over a prolonged period of time. Barium is included in the National Interim Primary Drinking Water Regulations because of the seriousness of its toxic effect. In most surface water supplies and groundwaters barium generally is not a problem, and is present only in trace amounts. However, up to 10 mg/l have been found in some water supplies, compared to the MCL of 1 mg/l.

In mineral form, barium often occurs as barium sulfate (barite), which has a low solubility. Because most natural waters contain sulfate, barium will dissolve only in trace amounts. However, barium becomes increasingly soluble in the presence of chloride and other anions and cations in dilute solutions.[8]

Barium as barite is used principally in drilling muds for oil and gas well-drilling operations. Barite and other barium compounds also are used in the production of glass, paints, rubber, ceramics, and many chemicals. However, very little is known with respect to actual concentrations of barium in various industrial effluents.[9]

Water Treatment. Barium can be removed effectively from water either by lime softening or by cation exchange. Lime softening pilot plant tests on water containing 10 to 12 mg/l of barium at pH's of 9.2, 10.5, and 11.6, resulted in removals of 84, 93, and 82 percent, respectively.[1] Samples from two full-scale lime softening plants showed removals of 88 and 95 percent, where the raw water contained barium concentrations of 7.5 and 17.4, respectively. A full-scale ion exchange softening plant achieved a 98 percent barium reduction, from 11.7 to 0.18 mg/l.[10]

Cadmium

Cadmium in the body is believed to be a nonbeneficial, nonessential element of high acute and chronic toxic potential. The concentration, not the absolute amount, is believed to determine the acute toxicity, and equivalent cadmium concentrations in water are considered more toxic than in food. The USGS data for various streams in the United States indicate that cadmium concentrations may range from less than 0.002 mg/l to a high of 0.120 mg/l. A typical value for cadmium, however, is expected to be near 0.002 mg/l, as compared to the drinking water MCL of 0.010 mg/l. Consequently, cadmium generally should not present a major problem in most water supplies.

Although only trace quantities of cadmium are likely to be found in natural waters, significant quantities may be introduced into a water supply from a number of sources. The electroplating industry is the largest user of cadmium metal, and currently is the major source of cadmium waste.[9] Other principal sources of cadmium wastewaters include waste sludges from paint manufacturing, paint residue left in used paint containers, and wash water used in battery manufacturing. Other industrial processes that constitute potential sources of cadmium wastewaters include metallurgical alloying, ceramic manufacturing, textile printing, chemical industries, and mine drainage.

Water Treatment. Waters containing up to 0.50 mg/l cadmium can be treated by lime softening to achieve the MCL of 0.010 mg/l. Laboratory experiments and pilot plant studies on lime softening showed that greater than 98 percent removal in the 8.5 to 11.3 pH range could be achieved on groundwater containing 0.3 mg/l of cadmium.[1]

Cadmium removal by ferric sulfate and alum clarification is effective, although somewhat less effective than removal by lime softening. For ferric sulfate clarification, 90 percent removal is possible at pH values above 8. Data on alum clarification indicate that cadmium removal is dependent on both the turbidity and the pH of the raw water.

At initial cadmium concentrations ranging from 0.010 to 0.10 mg/l, cadmium may be removed coincidentally during treatment of high-coliform waters, and waters with moderate to high turbidity. However, proper pH conditions must be maintained and sufficient coagulant must be used to achieve satisfactory results.

Chromium

Chromium has not been proved to be an essential or beneficial element in the body, but some studies indicate it may be essential in minute quantities. Chromium can exist in water as both the trivalent (Cr^{+3}) and hexavalent (Cr^{+6}) ions, the hexavalent state being the more toxic to humans.

Naturally occurring chromium usually is found in the form of chromite (Cr_2O_3), or as chrome iron ore ($FeO \cdot Cr_2O_3$). In natural waters, only traces of chromium in the trivalent state normally occur unless the pH is very low. However, under strong oxidizing conditions, chromium can be converted to the hexavalent state and can occur as chromate (CrO_4^{-2}) or dichromate ($Cr_2O_7^{-2}$), both of which forms normally indicate pollution by industrial waste.

Because of its broad industrial applications, chromium has been introduced into natural waters from a number of sources. The major source of waste chromium is the metal plating industry, with the automobile parts manufacturers being among the largest producers. Other major applications of chromium include alloy preparation, tanning, corrosion inhibition, wood preservation, electroplating, and pigments for inks, dyes, and paints.

The hexavalent form of chromium is more toxic and also is more difficult to remove from water and wastewater than the trivalent form. Because hexavalent chromium is more difficult to treat than the trivalent form, hexavalent chromium treatment often involves reduction to the trivalent state prior to removal.

Water Treatment. Research by Zemansky indicates that chromium is slightly removed by microstraining and clarification, but generally is not removed by filtration.[2] The study showed that an average 31 percent chromium removal occurred in the 12 water treatment plants studied.

Methods available specifically to treat water for trivalent chromium include chemical clarification with alum, iron salts, or lime. Laboratory studies show that chromium removal by lime clarification is highly dependent on pH.[1] Optimum removal (greater than 98 percent) occurs in a pH range of 10.6 to 11.3. The same removal efficiency is possible over a larger pH range (6.5 to 9.3) using ferric sulfate. Alum coagulation in the same study achieved 90 percent removal in the 6.7 to 8.5 pH range. Greater than 99 percent removal of hexavalent chromium is possible using a special ferrous sulfate coagulation process in which pH adjustment of the 6.5 to 9.3 range is made several minutes after coagulation. This pH adjustment procedure following coagulation provides time to reduce Cr^{+6} to Cr^{+3} prior to floc formation.

Other methods of chromium removal suggested in the literature include reverse osmosis and adsorption by activated carbon, although no actual pilot plant or full-scale plant operation data are available. Activated carbon is reported to have the ability to reduce hexavalent chromium to the trivalent form.[4]

Copper

Copper is a fairly common trace element in natural water. Copper salts such as sulfate and chloride are very soluble in low-pH waters; but in water of normal alkalinity these salts hydrolyze, and copper may be precipitated. In rocks, copper occurs most commonly as a sulfide, which may be oxidized in the process of weathering. Some of the copper may go into solution as sulfate, although if there is sufficient carbon dioxide in the water, a significant amount would be precipitated as carbonate. Small quantities may enter the water supply by solution of copper and brass water pipes in contact with water. Copper salts added to open reservoirs for algae control may provide another source of copper in water supplies.

The principal source of copper in industrial wastewater is metal cleaning and plating baths. Copper mining wastes and acid mine drainage also contribute significant quantities of dissolved copper to waste streams. Other sources of copper wastewater include pulp and paper mills, wood preserving, fertilizer manufacturing, petroleum refining, paints and pigments, steel works and foundries, and nonferrous metal works and foundries.

Water Treatment. Research by Zemansky indicates that copper is slightly removed by microstraining, and that increased removal occurs with clarification and filtration.[2] In plants where significant amounts of copper were removed in the clarification process, no appreciable increase in removal was evident when clarification was followed by filtration. The study showed that an average of 49 percent copper removal occurred in 12 plants using these treatment techniques.

Laboratory tests on water with low concentrations of copper show some adsorption on activated carbon at a pH greater than 6.[4] Removal efficiency, however, is reduced as the pH is lowered, to a point where no copper removal is achieved at pH 3.

Ion exchange and reverse osmosis are effective methods for removal of copper. Research reported by Furukawa indicated that 99 percent copper removal could be achieved by reverse osmosis.[11]

Fluoride

Fluoride is a normal constituent of all diets and is an essential nutrient. The only harmful effect of excessive fluoride in drinking water observed in the United States is spotting or mottling of the enamel of teeth. An optimum level of fluoride in the diet will help prevent tooth decay. The optimum fluoride level in drinking water depends on the climate because the amount of water consumed is affected by the air temperature.

Most fluorides are low in solubility; consequently, naturally occuring fluorides in a water supply usually are quite limited. As a salt, fluoride is moderately insoluble. There are a considerable number of fluoride-bearing minerals that constitute a source for fluoride in some waters of the United States, but, in general, fluoride is not often found in excessive concentrations.

Soluble fluorides from industrial wastewaters are much more common than naturally occurring fluorides. Industries that may discharge significant quantities of fluoride include glass and plating, fertilizer manufacturing, aluminum processing, electronics manufacturing, ceramics manufacturing, aircraft industries, and mineral and mining operations.[12]

Water Treatment. There are a variety of treatment methods available for removing fluorides from water, including ion exchange, chemical precipitation with alum and/or lime, electrodialysis, and reverse osmosis. Waters containing excessive fluoride ion can be treated for reduction to the MCL by ion exchange using synthetic resins or activated alumina. Efficiency of fluoride removal using these materials, however, is pH-dependent. The optimum pH for fluoride removal is usually near 5.5. Harmon and Kalichman report several successful full-scale treatment plants in California using these media.[13]

Fluoride may be removed coincidentally in lime softening of high magnesium water. The actual fluoride reduction accomplished by this technique is dependent upon both the initial fluoride concentration and the amount of magnesium removed in the softening process.

A study of chemical clarification for fluoride removal concluded that: (1) alum at pH levels of 6.2 to 6.4 was one of the more effective methods tested; (2) fluoride is adsorbed by magnesium hydroxide; and (3) iron salts are not effective for fluoride removal.[14]

Electrodialysis and reverse osmosis were among the treatment techniques evaluated for the removal of fluoride from groundwater in Arizona. Based on small-scale tests, it was estimated that a four-stage electrodialysis system could reduce the TDS from 1,100 mg/l to about 50 mg/l and reduce the fluoride concentration from 6 mg/l to 1 mg/l. For the same raw water, reverse osmosis would produce water with higher TDS (100 mg/l) but lower fluoride concentration (0.6 mg/l) than water treated by electrodialysis.

Lead

Lead is toxic in both acute and chronic exposures. The major risk of lead in water is to small children, in whom it may cause brain and kidney damage and other disorders. Because of its widespread presence in the environment, the MCL in drinking water was established as low as practicable.

Lead occurs in rocks primarily in sulfide or oxide forms, and lead concentrations of 0.4 to 0.8 mg/l can occur in natural waters as a result of the solution of lead-bearing minerals.[8] The extensive use of lead compounds as gasoline additives has greatly increased the availability of lead for solution in natural waters through automobile exhaust emissions. Studies by Hem and Durum suggest that the geographic pattern for lead in rainwater has a significant effect on the lead content of surface waters.[15]

Major industrial sources of lead include storage battery manufacture, printing pigments, fuels, photographic materials, and matches and explosives manufacturing. The storage battery industry is the single largest consumer of lead, followed by the petroleum industry in providing lead compounds for gasoline additives. Lead mines, mining, and smelting also are sources of lead wastewater.

Lead hydroxide and carbonate solubilities tend to control the concentration of lead that may be present in natural waters; the solubility limit is low for more alkaline and moderately mineralized waters.[8,15] Lead concentrations in drinking water (having a pH near 8 and bicarbonate concentration in excess of 100 mg/l) are likely to meet the regulations.

Water Treatment. Because of its insolubility in the carbonate and hydroxide forms, lead is relatively easy to remove from water by conventional treatment methods.[15] Detectable concentrations of lead were found in 95 percent of 12 conventional water treatment plants.[2] The efficiency of lead removal achieved by these plants averaged 32 percent. Ferric sulfate and alum clarification are capable of achieving 80 to 95 percent removal from raw water containing up to 10 mg/l of lead in a pH range of 6 to 10. Lead can also be removed effectively by lime clarification. Solutions of lead nitrate and lead acetate showed poor adsorption on activated carbon at pH 2, but fairly good removal at pH 5.[4]

Mercury

Mercury poisoning may be acute or chronic. Sublethal dosages result in several disorders, including a cerebral-palsy–like disease. The toxicity of mercury varies considerably with its chemical form. Monovalent mercury is nontoxic because its salts have a low solubility. Inorganic mercury compounds, which form water-soluble inorganic salts, are among the least toxic forms of mercury occurring in water. Divalent, organic, and elemental mercury, on the other hand, are highly toxic. Normal background levels of mercury in natural waters range from less than 0.5 to 5.0 µg/l. The risk from mercury in fish is much greater than from mercury in water.

Elemental mercury is soluble in oxygenated waters and may form mercuric oxide salts. These salts, in turn, may be adsorbed by minerals and bottom sediments of lakes and streams, a property that tends to prevent natural waters from carrying excessive concentrations of mercury except under unusual conditions.[8]

The principal industrial use and discharge of mercury occurs in chloralkali manufacturing.[9,16] Other major sources of mercury wastewaters include the electrical and electronics industry, explosives manufacturers, the photographic industry, chemical processing, the agricultural industry, the pulp and paper industries, the pharmaceutical industry, and paint manufacturers.

Water Treatment. There are several treatment methods available for the removal of mercury from water, but the form of mercury must be determined before a proper treatment method can be selected. In general, the organic form of mercury is more difficult to remove than the inorganic form. Inorganic mercury can be removed from drinking water by ion exchange, chemical clarification, lime softening, and activated carbon adsorption.

Chemical clarification using alum or ferric sulfate is effective in removing inorganic mercury. Research by Logsdon and Symons demonstrated that mercury removal percentages were relatively constant for varying concentrations of mercury in the range of 3 to 16 μg/l.[17] The research further indicated that inorganic mercury removals using alum steadily increased as turbidity increased, while coagulation with ferric sulfate was independent of turbidity. Typical removal percentages using alum and ferric sulfate are as follows:[1]

pH	COAGULANT	PERCENT REMOVAL INORGANIC MERCURY
7	Ferric sulfate	66
7	Alum	47
8	Ferric sulfate	97
8	Alum	38

For alum clarification, inorganic mercury removal increased from 10 percent on a water with turbidity of 2 NTU to 60 percent on water with 100 NTU. Lime softening in the pH range of 10.6 to 11.0 can remove 60 to 80 percent inorganic mercury, but at a lower pH (9.4) removal drops to only 30 percent.

Several studies of powdered activated carbon indicate that removal of inorganic mercury increases as carbon dosage and contact time are increased. Research by Thiem et al. showed that mercury removal using activated carbon is enhanced by the presence of calcium ion, or by the addition of a chelating agent such as EDTA or tannic acid.[5] These chemicals increased mercury removal 10 to 30 percent over removals obtained by carbon alone. Logsdon and Symons found that mercury removal by alum alone was only 40 percent, but the addition of activated carbon increased removals to greater than 70 percent.[17] Other studies show that 80 percent removal of 20 to 29 μg/l of inorganic mercury is possible with a 3.5-minute contact time with granular activated carbon beds.[4,8]

Ion exchange and activated carbon adsorption are effective methods for organic mercury removal. Removals of 98 percent have been achieved by ion exchange, and 80 percent using granular activated carbon. Research by Sigworth and Smith indicates that removals of 90 to over 99 percent are possible in the low concentration range using beds of granular activated carbon.[4]

Nitrate

Nitrate in drinking water can cause methemoglobinemia in infants less than three months old. Serious and sometimes fatal poisonings have occurred in infants consuming water containing more than 10 mg/l nitrate nitrogen as N. Older children and adults are not affected. Nitrate (NO_3^-) is the principal form of nitrogen found in most natural waters. Nitrate also is the most highly oxidized form of nitrogen, and usually is very stable in both ground and surface waters when there is a sufficient quantity of oxygen available.

High concentrations of nitrate normally are not found in most surface waters, unless such supplies have been contaminated from sewage or irrigation return flows. In surface waters, nitrate concentrations seldom exceed 3 mg/l, and often are less than 1 mg/l. In groundwater, nitrate concentrations may range from near zero to 1,000 mg/l. In some cases, the high concentrations can be traced to barnyard pollution or water percolating through soil that has been repeatedly fertilized.

Small quantities of nitrogen contained in igneous rock provide a minor source for the nitrate found in water supplies. More significant sources of nitrate in both surface and groundwater include nitrogen in plant debris, animal excrement, and fertilizers. Also, ammonia nitrogen present in wastewater effluent may be oxidized to nitrate when discharged to receiving waters.

Water Treatment. The methods currently available for removing nitrate from water are ion exchange and reverse osmosis. A full-scale plant in Long Island, New York, using a strong anion exchange resin reduces nitrate levels from 20 to 30 mg/l in the groundwater to 0.5 mg/l, achieving a removal efficiency of 97.5 to 98.3 percent.[1] Nitrate can be reduced 60 to 80 percent by reverse osmosis.

Selenium

There are few documented cases of poisoning from selenium in drinking water. Moreover, the maximum allowable concentration of selenium is complicated by several considerations: (1) selenium may be an essential element in nutrition; and (2) the chemical form of selenium and the protein, vitamin E, and trace-element content of the diet all apparently determine whether selenium has beneficial or adverse effects. The MCL of 0.01 mg/l is based on the total selenium content, and provides a factor of safety to prevent even minor toxic effects in humans. Selenium concentrations in rivers and streams generally range from less than 0.001 mg/l to 0.060 mg/l. Typical values normally are about 0.001 mg/l, as compared to the drinking water MCL of 0.01 mg/l.

Selenium is chemically similar to sulfur and often occurs with sulfur in mineral veins of rocks. In water, selenium may occur as selenite (SeO_3^{-2}) or, in its most oxidized state, as selenate (SeO_4^{-2}). In the quadrivalent form (valence +4), selenium may appear in groundwater, while in the hexavalent form (valence +6), it may occur naturally in both ground and surface waters. Both forms are quite stable and appear to act independently of one another when present in the same water. Selenium may be sorbed on hydroxide precipitates such as $Fe(OH)_3$ or on sediments, so that concentrations found in natural waters normally are very low.[15]

Industries that may produce selenium wastewaters include paint, pigment, and dye producers, electronics, glass manufacturers, and insecticide industries.

Water Treatment. Chemical clarification with lime, ferric sulfate, or alum and activated carbon adsorption are ineffective in removing Se^{+6}, and only moderately effective in removing Se^{+4} from water. Of these methods, clarification with ferric sulfate at low pH (below 7) is the most effective in removing the quadrivalent form.[10] In one study, an initial selenium concentration of 0.1 mg/l was reduced by approximately 85 percent using 100 mg/l of ferric sulfate at a pH of 5.5. Other research shows that the same removal efficiency is possible using 30 mg/l of coagulant for Se^{+4} concentrations of 0.03 mg/l. Limited laboratory research also indicates that both Se^{+4} and Se^{+6} may be removed by ion exchange and reverse osmosis, although no actual full-scale or pilot plant data are available.[1]

Silver

The primary effect on humans of ingesting silver is a permanent blue-gray discoloration of the skin, eyes, and mucous membranes. There are no known toxic effects. The MCL of 0.05 mg/l was established because of the permanent and irreversible nature of the silver discoloration effect.

Although it appears to be relatively uncommon for silver to occur in natural waters, silver in excess of 0.3 mg/l has been found in some U.S. surface waters. Because many silver salts such as AgCl and Ag_2S are highly insoluble, low levels of soluble silver should be expected in most waters. Silver nitrate, the most common soluble silver salt, could be a significant industrial source of silver in wastewater. This compound is used in porcelain, photographic, electroplating, and ink manufacturing industries. Of these sources, the photographic and electroplating industries are the major contributors of soluble silver wastes.

Water Treatment. Laboratory tests show that chemical clarification with ferric sulfate or alum in the 6 to 8 pH range can achieve greater than 70 percent removal with an initial silver concentration of 0.15 mg/l.[1] Settling alone, without the addition of a coagulant, achieved 50 percent removal in river water with 30 NTU and 0.15 mg/l of silver. Lime clarification will also remove silver; removal efficiency increases from 70 to 90 percent when the pH is increased from about 9 to 11.5.

Reverse osmosis and adsorption by activated carbon also have been reported to remove silver from water supplies. Laboratory studies indicate that there is a high potential for good adsorbability of silver on activated carbon.[4]

Zinc

The chemical behavior of zinc is very similar to that of cadmium, with both metals occurring in water in the +2 oxidation state. Zinc chloride and zinc sulfate are very soluble in water, but hydrolyze in solution to reduce the pH. With an excess of carbon dioxide present in solution, it is unlikely that the proper pH will occur to allow zinc to precipitate as zinc hydroxide.

Because zinc is readily adsorbed on sediments and soils, only trace amounts of it normally are found in natural ground and surface waters. Significant quantities of zinc, however, may be discharged in industrial waste streams such as plating and metal processing, stainless steel and silver tableware manufacturing, yarn and fabric production, pulp and paper production, and the pigment industry. High zinc concentrations also are known to occur in acid mine drainage water.

Water Treatment. In a study of 12 conventional water treatment plants, the average zinc removal efficiency was 48 percent.[2] Results of the study indicated that actual removals varied significantly, from a low of near zero to a high of 90 percent.

REFERENCES

1. *Manual of Treatment Techniques for Meeting the Interim Primary Drinking Water Regulations,* EPA 600/8–77–005, May, 1977.
2. Zemansky, G. M., "Removal of Trace Metals during Conventional Water Treatment," *J.AWWA,* pp. 606–609, October, 1974.
3. Smith, S.B., "Trace Metals Removal by Activated Carbon," in *Traces of Heavy Metals in Water Removal Processes and Monitoring,* EPA 902/9–74–001, pp. 55–70, November, 1973.
4. Sigworth, E. A., and Smith, S. B., "Adsorption of Inorganic Compounds by Activated Carbon," *J.AWWA,* pp. 386–391, July, 1972.
5. Thiem, L., et al., "Removal of Mercury from Drinking Water Using Activated Carbon," *J.AWWA,* pp. 447–451, August, 1976.
6. Ferguson, J. F., and Anderson, M. Q., "Chemical Forms of Arsenic in Water Supplies and Their Removal," pp. 137–158, in *Chemistry of Water Supply Treatment and Distribution,* by A. J. Rubin, Ann Arbor Science, Ann Arbor, Mich., 1974.
7. Shen, Y. S., "Study of Arsenic Removal from Drinking Water," *J.AWWA,* pp. 543–598, August, 1974.
8. Hem, J. D., "Study and Interpretation of the Chemical Characteristics of Natural Water," Geological Survey Water-Supply Paper 1473, 1959.
9. Sittig, M., *Toxic Metals, Pollution Control and Worker Protection,* Noyes Data Corporation, Park Ridge, N.J., 1976.
10. Logsdon, G. S., and Symons, J. M., "Removal of Heavy Metals by Conventional Treatment," in *Traces of Heavy Metals in Water Removal Processes and Monitoring,* EPA 902/9–74–001, pp. 225–256, November, 1973.
11. Furukawa, D. H., "Removal of Heavy Metals from Water Using Reverse Osmosis," in *Traces of Heavy Metals in Water Removal Processes and Monitoring,* EPA 902/9–74–001, pp. 179–188, November, 1973.

12. Parker, C. L., and Fong, C. C., "Fluoride Removal: Technology and Cost Estimates," *Industrial Wastes,* pp. 23–27, 1975.
13. Harmon, J. A., and Kalichman, S. B., "Defluoridation of Drinking Water in Southern California," *J.AWWA,* pp. 245–254, 1965.
14. Sollo, F. W., Jr., et al., *Fluoride Removal from Potable Water Supplies,* Research Report No. 136, Illinois State Water Survey, Urbana, Ill., September, 1978.
15. Hem, J. D., and Durum, W. H., "Solubility and Occurrence of Lead in Surface Water, *J.AWWA,* pp. 562–568, August, 1973.
16. Humenick, M. J., and Schnoor, J. L., "Improving Mercury (II) Removal by Activated Carbon," *J.AWWA,* pp. 1249–1262, December, 1974.
17. Logsdon, G. S., and Symons, J. M., "Mercury Removal by Conventional Water Treatment Techniques," *J.AWWA,* pp.554–562, August, 1973.

Chapter 3
Source Development

ALTERNATIVE SOURCES

The alternative sources of water are surface water, groundwater, and the conjunctive use of surface water and groundwater. The source selected is a key factor in determining the nature of the required purification, transmission, and storage facilities. The supply must provide a reliable quantity of water for the long-term needs of the community, and preferably will have a quality that minimizes the amount of treatment required. A detailed evaluation of all alternative sources should be made to compare yield, reliability, quality, treatment, collection, and distribution costs. Some of the general advantages and disadvantages of the alternative sources are discussed below.

Surface Water Supplies

Surface waters usually can be obtained without the need for pumping groundwaters (or with less costly pumping than groundwaters require). It is also usually possible to define the quantity and quality of a surface supply at a lower cost than that associated with exploration of a groundwater supply.

The quality of lake water is not so consistent as the quality of groundwater, but is more consistent than river water. The turbidity of river water may change rapidly during a heavy rainstorm, or from runoff due to melting snows. Lake water quality may change from wind-generated currents or when thermal stratification occurs. If the lake freezes over, the dissolved oxygen may be depleted in portions of the lake, with the result that the bottom deposits become anaerobic and many compounds become soluble. Many lakes and reservoirs have substantial quantities of iron, and occasionally manganese, in their bottom deposits. Normally they are of no particular importance to water engineers because they are oxidized and precipitated, or chelated, with organic compounds. However, when the bottom deposits in the relatively shallow areas of the lake become suspended because of wind-induced currents, these deposits can enter the intakes.

Deep lakes are affected by changes in density which cause the water at the bottom of the lake to "turn over" and rise toward the surface. Because water is at its greatest density at 39.2°F (4°C), this effect can occur in both spring and fall. In the spring, surface water heats up to 39.2°F (4°C) and then sinks to the bottom, forcing water from the bottom to the surface. As winter approaches, the surface water cools, increasing in density and displacing the warmer water immediately underneath it. Significant changes in water quality can occur during these periods.

The majority of the water served in the United States is derived from surface supplies. While groundwater often can be used with a minimum of treatment, often requiring only disinfection, most surface waters require chemical coagulation, sedimentation, filtration, and disinfection to make them suitable for use as public water supplies. A combination of treatment methods will, if properly carried out, convert a moderately polluted water into a safe drinking water. It is increasingly difficult, because of easier access and greater recreational use of streams, lakes, and watersheds, and urban and industrial development, for unfiltered surface water from protected watersheds to meet the requirements of the National Interim Primary Drinking Water Regulations, as can be inferred from Fig. 3–1A.

Fig. 3–1A. Increased recreational use of streams, lakes, and watersheds makes it increasingly difficult for unfiltered surface water to meet the requirements of the National Interim Primary Drinking Water Regulations.

Groundwater Supplies

About one-fifth of the fresh water withdrawals in the United States are from groundwater resources.[1] When available in sufficient quantity, groundwater is often the preferred source. Most is clear, cool, colorless, and quite uniform in character. Underground supplies are generally of better bacterial quality and contain less organic material than surface water, but may be more highly mineralized. Adequate natural protection of groundwater involves purification of water by infiltration into the soil, by percolation through underlying material, and by storage below the groundwater table. Also, groundwater is usually more uniform in temperature than surface water.

Groundwater sources also have some disadvantages. The cost of pumping groundwater may be greater than the cost of pumping surface water. Unless there are good geological data on the area, exploration to define the quality and quantity of groundwater could be expensive and speculative. Some groundwaters are highly mineralized, and contain large quantities of iron, manganese, sulfates, chlorides, calcium, magnesium, and other elements that are expensive to remove. Some groundwaters are high in color. Elements such as iron and manganese are held in solution at low pH values in the aquifers, because of the presence of carbon dioxide. Once the water is pumped to the surface, the free carbon dioxide is liberated, and the ferrous and manganous ions precipitate out of solution. If the groundwater is overpumped, water quality can change significantly as water from other formations flows into the system.

Conjunctive Use

When multiple sources of water with different characteristics are available—as is the case with groundwater and surface water systems—it may be possible to develop an operating strategy that capitalizes on the best features of the sources. This strategy is known as the conjunctive management of groundwater and surface water, or "conjunctive use."

Some advantages offered by conjunctive use include the use of the water storage capacity of aquifers, improved water conservation (less evaporation from surface storage), and more uniform availability of water.

Physical, social, legal, and economic factors determine the operation of conjunctive ground–surface water systems. Several models have been developed to optimize the conjunctive use of surface and groundwater sources.[3]

Surface water physical variables include availability, quality, losses (especially from reservoirs), and possible transfer. Groundwater aquifer physical variables include type of aquifer, storage capacity and hydraulic characteristics, losses, recharge features, and quality of groundwater.

Legal constraints for surface water consist of low flow requirements, surface water transfer, operation of reservoirs, navigation requirements, and allocation rights of users. Groundwater constraints include interaquifer water transfers, allocation rights of users, quality of recharge waters, and land subsidence.

Economic and financial variables include kinds of demands for water (agriculture, industry, municipal, hydroelectric, recreation, and waterborne commerce), return from economic activities, cost functions for technological activities, and project financing.

By proper balancing of the above variables, it may be possible to develop a conjunctive use system that is superior to reliance on either a ground or a surface supply.

DESIGN PERIODS FOR WATER SOURCES

The quantity of water from a source(s) should be adequate to supply the total water demand of a community, as well as a reasonable surplus for anticipated growth. An analysis of the elements making up the total water demand of a community should be conducted, including but not limited to the following items: location, climate, population growth, type and character of community, fire protection, air conditioning, metering practice, cost of water, water quality, and pressure on mains.[4]

Surface water collection works either use a source of water that is continuously adequate in quantity to satisfy present and reasonable future demands, or they convert an intermittently inadequate source into a continuously adequate supply by storing surplus water for use during periods of insufficiency.

In the case of multipurpose reservoir projects, the various demands for water should be carefully integrated. For an impounded source, allowances should be made for required water releases, evaporation, seepage, and losses due to siltation. For major costly projects such as impoundments, which require difficult planning, property acquisition, and financing, the design period should be about 50 years.

If the source of supply is located some distance from the point of use, a long supply pipeline is required. A study should be made of the economic size of the pipeline, taking into consideration cost of construction, expected future growth, and cost of operation based on power costs and other factors. Parts of the supply works that can be expanded without difficulty or excessive costs may be designed for a shorter period such as 20 years. Examples include intakes, pumping stations, or certain pipelines.

Major projects such as a centralized well field and long transmission lines may not be suited to construction in phases without incurring excessive extra costs, or they may involve difficult planning, property acquisition, or financing. For such projects a design period on the order of 50 years may be appropriate.

For development of scattered wells feeding individually into the distribution system through short pipelines, and for other projects that are readily adaptable to construction in stages, shorter design periods of 10 to 20 years are appropriate.

SURFACE WATER SUPPLIES

Safe Yield

Flowing Streams. When no storage is provided, the minimum available (considering any competing water rights) stream flow of record must exceed the estimated future water demand on the maximum day. The best hydrologic data on minimum stream flows are those recorded on the specific watershed in question. In the absence of such data or with a short period of record, it may be necessary to use estimating methods. For such estimates, stream flow and weather records of contributing or adjacent watersheds should be used. Empirical formulas and ratios contained in published literature vary widely for watersheds of the same size; therefore, they should not be considered satisfactory criteria for judging the adequacy of a source unless they are supported by hydrologic data obtained from the specific watershed.

In making estimates, a careful study should be made of all factors that affect and determine the safe yield of a proposed surface water source, including such data as: geographical location; storm paths; prevailing winds; type and intensity of precipitation; topography and size of basin; orientation of basin; types of soil; types of vegetation; condition of ground surface; type and extent of artificial drainage; extent of surface storage in lakes and swamps; condition and slope of stream channel; average slope of basin; character of drainage net; evaporation, infiltration, and other losses.

There are many methods of estimating runoff. The accuracy of the various methods depends upon the ability and the experienced judgment of the estimator in finding and supplying the correlating factors that will produce a realistic synthetic record. Consultation with the U.S. Weather Bureau and the U.S. Geological Survey is recommended.

Reservoirs and Lakes. When the demand for water is greater than the minimum rate of flow in the stream from which the water is to be taken, an impounding reservoir may be required. The development of reservoir sites is discussed later in this chapter. In general, the ideal topographic conditions for a reservoir are a narrow gorge in which a dam may be built at minimum expense, and an expanding valley immediately above the gorge that will afford a large amount of storage per unit of surface area. This minimizes evaporation loss and the growth of algae and aquatic vegetation.

A reservoir will yield only part of the long-term average runoff of the watershed that it controls. The rest either will go over the spillway in times of flooding or will be lost to evaporation, bank storage, seepage, or siltation.

Up to a certain point, increasing the amount of storage space in the reservoir will cause additional water to be on hand in drought times, and will allow the project to produce a greater dependable yield. However, the volume of water lost to lake surface evaporation also increases with reservoir size. Therefore, there is progressively less benefit from each increment of storage. For any specific site, there is a maximum practical capacity beyond which further enlargement of storage will not increase the yield, even though it might allow retention of more water during wet years.

The necessary reservoir capacity is dependent upon the quantity of water required to supply the community, the amount of water lost by evaporation from the reservoir surface, the loss of volume due to siltation and by seepage through and around the dam, and the minimum flow of the stream. The minimum stream flow can be determined accurately only by actual measurement over an extended period of time. Even where such flow records are available, there is always the possibility of a more severe drought than any previously recorded. However, most requirements will be met if provision is made for droughts that occur less than once in 50 to 100 years.

The dependable yield of a reservoir is defined as the amount of water that can be provided on a continuous basis, without deficit, under the full range of hydrologic conditions that might reasonably be expected to occur during the life of the project.[5] It is normally expressed in terms of acre-feet per year. Yields vary widely and are influenced by a large number of interrelated factors. Geographic location, rainfall, runoff, evaporation, reservoir storage capacity, the area-versus-capacity relationship, drainage area size, sedimentation, minimum drawdown limitations, and various other considerations will affect a reservoir's performance and effectiveness.

A key factor is the probable critical drought condition. As applied to reservoir yields, the severity of a drought should be considered in two ways—in terms of how much below average the runoff is likely to be, and also how long the deficient runoff conditions may be expected to last. Both aspects are important. The runoff deficiency must be made up by taking stored water out of the lake. A long drought means more loss to evaporation and less water available for supplementing the low runoff. A 2-year drought when the average runoff rate is 20 percent below normal will be more critical than a 1-year drought when the runoff is 40 percent below normal.

Potential critical drought conditions are usually derived from actual past records of runoff, rainfall, and evaporation on the stream in question or on other similar watersheds in the same general area. In most places such records have only been collected during the past 30 to 50 years, but they constitute the most realistic estimate of what might happen in the future and the approximate duration of droughts for a given area. There is no assurance that there will not be even worse conditions in the future. Because of this uncertainty, it is frequently desirable to include a factor of safety in the yield estimates.

One way to do this is to assume some of the storage capacity to remain unused at the low point of the historical critical drought, as a reserve allowance. An alternative approach is to decrease the estimated yield by some percentage of the average runoff experienced during the historical critical drought, thus assuming a potential drought with that much less runoff than occurred in the past. It is desirable to have a comfortable margin between supply and demand.

The amount of storage required to carry the community through any of the recorded past droughts at the estimated rate of consumption may be determined by preparation of a mass diagram based on the best hydrologic data available. (Preparation of mass diagrams is described in the next section of this chapter.) Mass curves of runoff should cover a period of several years of minimum rainfall.

Approval of the safety features of any planned structures should be obtained from the appropriate agency. Meeting safety requirements in the design of a dam for a reservoir requires that the spillway design flood be determined and provision made for adequate spillway capacity.

DEVELOPMENT OF RESERVOIRS

Volume Requirements

As noted above, the amount of storage required can be determined with a mass diagram. As shown in Fig. 3–1, two curves are plotted on a single

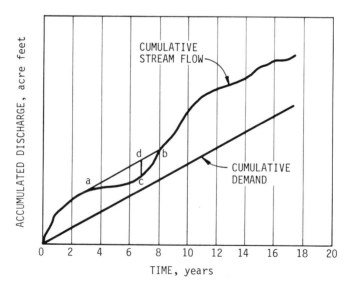

Fig. 3–1. Typical mass diagram.[6] Reproduced by permission of McGraw-Hill Book Co.

graph—one a plot of the cumulative stream flow over a period of years, and the other a plot of the cumulative demand over the same period of time. The stream flow curve is examined to determine the beginning of the longest dry period. This period will be evidenced by the portion of the stream flow curve with the flattest slope (begins at point a in Fig. 3–1). A line is then drawn from the start of the longest dry period (point a) parallel to the cumulative demand curve (line ab in Fig. 3–1). This line must intersect the cumulative stream flow curve if the reservoir is to refill (it does so at point b in this example). The ordinate value, dc, represents the volume of storage required to maintain the flow rate represented by the slope of the demand line.

The mass curve of water utilization need not be a straight line. Figure 3–2 shows a curve of irregular demand plotted with a curve of supply for the design dry period. The lower curve shows the total flow from October 1 to May 7 to be 110,000 acre-feet (135.7 Mm³). On December 15 the total was 15,000 acre-feet (18.5 Mm³); on April 1, it was 60,000 acre-feet (74.01 Mm³). Likewise the total flow from October 1 to other dates may be read from the curve. The maximum vertical distance between the curves is the storage required to meet the needs of the project. If the worst period of record is selected for the stream flow mass curve, the maximum storage requirements are obtained from the mass curve. In the illustration, it was assumed that the reservoir was full on October 1, the beginning of the period. The greatest amount of storage has been used on January 31, 74,000 acre-feet (91.28 Mm³). The reservoir was full again on May 4, when the two curves intersected.

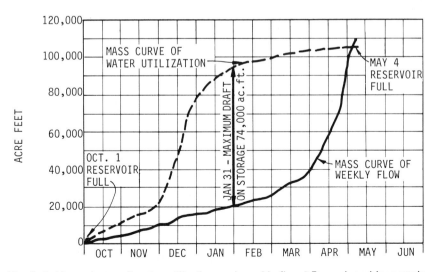

Fig. 3–2. Mass curves of water utilization and weekly flow.[6] Reproduced by permission of McGraw-Hill Book Co.

The mass curve is plotted from the hydrograph of reservoir inflow. In most cases, this is taken as the hydrograph at the dam site.

Area–Volume Relationships

The shape of the reservoir site—the width of the valley floor and the steepness of the adjoining hills—will determine how much storage can be provided and how much reservoir surface area will be created for a given volume of storage. Through analysis of the contour lines on topographic maps, it is possible to develop the relationships between reservoir surface elevation, reservoir area, and volume of storage for a given site. These relationships are normally presented in the form of tables, but can also be presented graphically as in Fig. 3–3. The arrangement shown in Fig. 3–3, with two lines plotted in opposite directions, is used so that the curves do not overlap and can be easily read.

It is usually desirable that the reservoir have as little surface area as possible in relation to the volume of storage because land costs and natural evaporation losses will be proportional to the surface acreage. Alternative sites can be

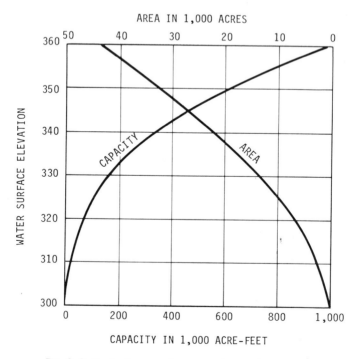

Fig. 3–3. Typical reservoir area and capacity curves.[5]

compared by plotting their respective area-versus-capacity curves on the same graph and selecting the one that offers the most storage per unit of area.

SITE SELECTION

The selection of a suitable site for a reservoir depends upon a number of interrelated factors that establish the adequacy, economy, safety, and palatability of the supply. The surface topography should create a high ratio of water storage to dam volume; a broad and branching valley for the reservoir should impinge upon a narrow gorge for the dam. The topography should also present a favorable site for an adequate spillway to pass the flood flow and a suitable route for an aqueduct or pipeline.

The subsurface geology should provide useful materials for the construction of the dam and appurtenant structures, safe foundations for the dam and spillway, and tightness against seepage of the impounded waters beneath the dam and through its abutments.

The reservoir area that is to be flooded should be sparsely inhabited, not heavily wooded, and not traversed by important roads or railroads, pipelines, or other facilities. It should contain little marshland. The area should constitute a reservoir of such shape as not to favor short-circuiting of the incoming waters to the intake, and of such depth, especially around its margins, as not to create large shallow areas. Purification of water by storage is an important asset of impounding reservoirs. Narrow reservoirs with their major axis in the direction of prevailing winds are especially subject to short-circuiting. Shallow areas often support a heavy growth of aquatic vegetation, when they are submerged, and of land plants, when they are uncovered by the lowering of the water surface. Decaying vegetation imparts odors and tastes to the water, supports algal growths, and liberates color.

The reservoir should interfere as little as possible with existing water rights; the intake should be as close as possible to the community it is to serve; and the development should preferably be at such elevation as to supply its waters by gravity.

The character of the soil and rock of the drainage basin influences the kind and amount of mineral matter in the water. These are important considerations in the selection of a site source. A geologist, or aquatic biologist with knowledge of soils and their effect on the growth of taste- and odor-producing organisms, may need to be consulted on the selection of a reservoir site. Consideration should be given to seepage of poor-quality groundwater into proposed impoundment areas.

Precipitation, temperature, sunshine, evaporation, and air movements all influence the quality of surface waters, and should be evaluated. For example, if heavy rains and floods, with their resulting turbidities, should all occur during winter months, while in the summer rains are light and the water becomes clear, algal growth will be much greater than if the conditions were reversed. The vertical circulation of water in a lake or reservoir also is related closely to climatic conditions, and should be considered.

The extent of soil erosion on the drainage basin should be evaluated, as this will determine the quantity of sand and silt reaching the reservoir. Soil erosion also affects the biological productivity of a stream, lake, or reservoir, which in turn influences the quality of the water. Under severe soil erosion conditions, subsequent siltation may seriously limit the effective life of the reservoir as a source of water supply.

The uses of the land located on potential drainage basins should be investigated because they may affect the water quality. For example, reservoirs that are fed by water draining from highly cultivated farming areas often produce extensive algal growths, the products of which are difficult to remove. Fertilization for field crops results in the addition of nitrates and phosphates that may stimulate aquatic growths. Large algal blooms have been found in small reservoirs under these conditions, and have created difficult water treatment problems. Pesticide use should be investigated and evaluated.

Peat bogs, mucky areas, swamps, and marshes on a drainage basin contribute a great amount of organic material to the waters draining from them, which may cause foul odors, undesirable tastes, acid conditions, and high color. These areas should be avoided, or reduced to a minimum by artificial drainage.

Narrow, deep bodies of water, having a minimum of shoreline, are better sources of water generally than shallow ones having extensive flats that are exposed by periodic low water levels. These features also determine the period of detention, and the probable flow time from sources of pollution through the reservoir to intake, and so should be evaluated.

The extent of industrial development on the drainage area should be considered, as industrial wastes may contain products that are difficult or impossible to remove by water treatment processes.

The extent and character of present and future recreational activities on the drainage basin should be known, as they may affect the sanitary quality of the water. The health laws, rules, and regulations of the local health departments should be consulted on this matter.

The death and decay of vegetation can injure the quality of the water by increasing the organic matter and producing food material for microorganisms. The various classes of vegetation existing on the drainage basin should be cataloged, and their effect on the quality of water should be evaluated.

Vegetation should be removed from the area to be flooded to prevent its decay and subsequent impairment of the physical character of the stored water. Even then, the organic content and color of the stored water will likely increase for a period of 3 to 5 years. The ideal reservoir site is one where the slopes are relatively steep, thus ensuring deep water over most portions of the reservoir and minimizing the growth of algae, which are most prolific in shallow portions of reservoirs. The periodic exposure of shallow areas during periods of low water also leads to the growth of semiaquatic vegetation, which will die and decay when the reservoir fills again. Shallow areas, therefore, should be filled, or isolated by dikes when economically feasible.

RESERVOIR LOSSES

Water will be lost from a reservoir by bank storage, seepage, evaporation, and siltation, and overflows when the inflow exceeds the storage capacity of the reservoir. As the water level rises and falls, water will move into and out of the surrounding reservoir banks. Generally, as the reservoir level rises, water is lost into the surrounding soils.

Where a reservoir is underlain by porous strata that have ample outlets beneath the surrounding hills, or under the dam, seepage may amount to several hundred cubic feet per second (cfs). In the usual case, a maximum loss of 10 cfs (0.28 m³/s) would be considered large.[6] The basin of a reservoir should be studied carefully, and if porous conditions are present, experts in geology and soil mechanics should be called in to analyze potential seepage losses.

Evaporation is a direct function of reservoir surface area, and is usually expressed in inches. The rate of evaporation varies directly with temperature and wind velocity and inversely with humidity. Reservoirs at high elevations generally show lower rates of evaporation, due to lower temperatures. Typically about 75 percent of the annual evaporation occurs in the 6 months from April to September, and about 20 percent occurs in the maximum month. The volume of water lost to evaporation (annual evaporation in inches × reservoir surface area) is offset to some degree by the gain of rainfall on the surface area of the reservoir (annual rainfall in inches × reservoir surface area). A negative value for evaporation–rainfall indicates a net loss; a positive value, a net gain of water.

Sedimentation (siltation) in the reservoir can cause a significant loss of volume over time. The sediments can come from erosion of the streambed itself or from erosion of the watershed.

In general, the problem is more serious in arid regions where the ground does not support a good vegetative cover. As a result, heavy rains cause excessive sheet erosion and a high percentage of silt in the streams. In humid climates, the effect of vegetative cover reduces erosion. The type of land use affects the siltation of reservoirs. One square mile of urban development has been reported to produce about the same amount of sediment as 100 square miles (259 km²) of rural land.[8] Where reservoir capacity is small relative to annual stream flow, silting may be an important consideration.

Only in rare instances is the removal of silt from a reservoir by mechanical means justified economically. Once silt has been deposited, its removal has never been particularly effective. However, where large gates can be installed and reservoir operation permits the passing of floodwaters, a large portion of the silt can be passed through the reservoir.

Sometimes certain small areas in a watershed are major silt contributors. In such cases, there may be economic justification for special check dams and debris barriers. However, reduction of soil erosion is generally a long-range undertaking. Factors involved are: proper farming methods, such as contour plowing; terracing of hillsides; reforestation or afforestation; cultivation of permanent pastures; prevention of gully formation through construction of check dams or debris barriers, and revetment of stream banks.[2]

In the design of impounding reservoirs for silt-bearing streams, suitable allowance must be made for loss of capacity by silting. Understandably, deposition is most severe in reservoirs that are large in volume relative to inflow; especially, in impoundments serving "flashy" and therefore strongly erosive streams. The proportion of sediment retained is called its "trap efficiency." Nearly 100 percent of the sediment transported by influent streams may be retained in reservoirs storing a full year's tributary flow. Trap efficiency drops to a point between 65 and 85 percent when the storage ratio is reduced to 0.5 (half a year's inflow) and to 30 to 60 percent when the storage ratio is lowered to 0.1 (5 weeks' inflow). Silting is often fast when reservoirs are first placed in service, and proceeds toward a steady state as time goes on. The typical unit weight of silt is 70 lb/cu ft (1,135 kg/m³). Thus, 1,500 tons (1,361 metric tons) of silt will occupy about 1 acre-foot (1,234 m³) of reservoir volume. A silt content of 250 mg/l is equivalent to 1 ton per million gallons (0.24 metric tons/m³) of reservoir influent flow.

INTAKES

The purpose of the plant intake is to withdraw adequate quantities of the best available grade of raw water continuously. In selecting the intake location, the lake or river bottom character, currents, and potential sources of pollution must be considered. To provide for the variability of environmental influences, the intake structure should be designed and built to permit raw water withdrawal at various levels, or locations, or both. The intake capacity, including pumping facilities, should provide sufficient raw water for the treatment plant at all times. The quantity of finished water in storage provides a buffer, and is a factor in determining the necessary intake capacity. In reservoirs, intake capacity generally equals the average rate of demand on the maximum day. Dual facilities should be provided for mechanical equipment. Pump priming must not create a cross connection between the finished and raw water supplies.

Intake facilities also should be constructed to ensure continuous raw water flow despite floods, icing, plugging with debris or sand, high winds, power failure, damage by boats, or any other occurrences. They should be inaccessible to trespass, contain adequate toilet facilities located and installed to prevent chance contamination of the raw water supply, and contain an immediate warning system for the treatment plant operator in case of failure of automatic or semiautomatic pumping operations. Because many intakes will be constructed on permeable material, the design of the structure must consider underflow and hydrostatic uplift pressure.

Intakes in large rivers should be located so that the ports are submerged at all stages of the river to a sufficient depth to avoid trouble with ice cakes or floating debris and to preclude the entraining of air. The ports should also be several feet above the bottom of the stream so that sand and gravel being transported on the bottom will not be drawn into the intake. In order to meet these requirements, it is usually necessary to locate the intake in the deepest part of the stream and away from the shore, particularly if the river is subject to large fluctuations in stage. Under these conditions, the water must be conveyed from the intake to the shore through a pipe laid upon the river bottom or through a tunnel constructed under the bottom.[6]

If the river stage does not fluctuate significantly, it may be possible to construct the intake on the shore of the river. The necessity for a tunnel or subaqueous pipeline is avoided, and the design of the intake structure simplified.

Intakes in small streams frequently require the construction of small diversion dams for the dual purpose of providing a sufficient depth of water at all flows to divert water into the intake port and a settling period in order to reduce the turbidity of the water. A small period of quiescent flow will also permit suspended leaves and wood either to rise to the surface or to sink if they have become waterlogged, and it will favor the formation of sheet ice in cold weather and thus reduce the difficulties of anchor and frazil ice.

Both bar racks and mesh screens are frequently used on the openings into the intake structure. Bar racks with spacings of 2 to 4 inches (51 to 101 mm) protect the intake from large floating objects. If hand raking is required, the racks should have a slope of 2 to 4 inches (51 to 101 mm) horizontal to 12 inches (305 mm) vertical, and a suitable raking platform should be provided above high water.[6]

Screens are used to protect against floating materials such as leaves. They should have not less than 2 and sometimes have as many as 8 meshes to the inch (79 to 315 per meter), depending upon the character of the floating matter in the water. Screens should be of corrosion-resistant metal and easily removable; and, if hand cleaning is necessary, provisions should be made for washing them with a hose stream. Screens should have a velocity of not more than 3½ inches/sec (8.89 mm/s). Low velocities and small openings are necessary to prevent the entrance of fish.

Intakes in large lakes or reservoirs should be located in the deepest water economically available. In large impounding reservoirs, the most desirable location for an intake is usually at or near the dam, where depths are greatest.

In cold climates, ice troubles are reduced in frequency and intensity if intake ports lie as much as 25 feet (7.62 m) below the water surface and entrance velocities are less than 3 to 4 inches/sec (76.2 to 101.6 mm/s). At such low velocities, frazil ice, leaves, and debris are not entrained in the flowing water, and fish are able to escape from the intake current. Also, the use of fiberglass-reinforced plastic (FRP) for intake bells has been reported to be effective in reducing frazil ice adherence.[7] Because the fiberglass material has a low thermal conductivity, the heat of fusion released by ice formation is not rapidly conducted away, and the tendency of ice to adhere to the surface of the intake structure is reduced. Also, the smooth, noncrystalline surface of the FRP intake bells does not provide points of nucleation for ice formation and growth.

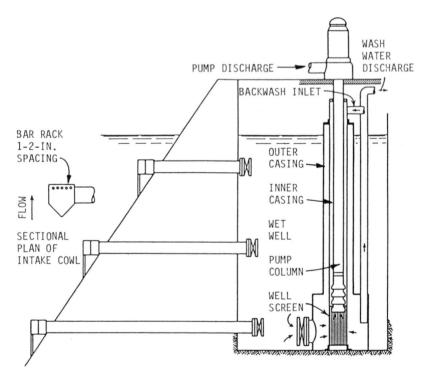

Fig. 3-4. Intake with vertical pump and backwashed well-type screen.[2] Reprinted by permission of John Wiley & Sons, Inc.

Bottom sediments may be kept out of intakes by raising entrance ports 4 to 6 feet (1.22 to 1.82 m) above the lake or reservoir floor. Ports controlled at numerous depths permit water-quality selection and optimization. A vertical interval of 15 feet (4.57 m) is common.[2] Submerged gratings are given openings of 2 to 3 inches (51 to 76.2 mm). Specifications for screens commonly call for 2 to 8 meshes to the inch (79 to 315 per meter) and face (approach) velocities of 3 or 4 inches/sec (76.2 to 102 mm/s). Typical intakes are shown in Figs. 3–4 and 3–5.

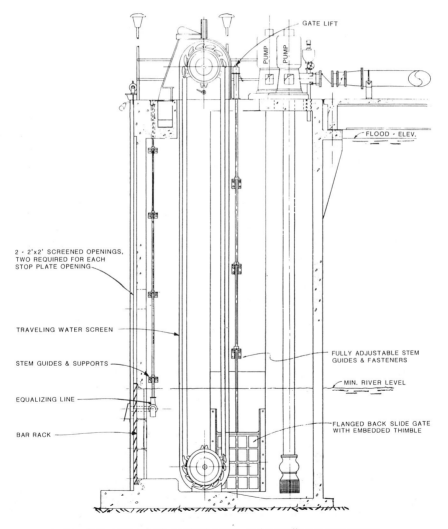

Fig. 3–5. Intake with vertical pump and traveling water screen.

RESERVOIR OUTLETS

Spillways

All natural streams are subject to periodic flooding. The storage capacity of a reservoir will usually be less than the potential volume of runoff that can come from its watershed during a large flood, and there must be a provision to pass excess floodwaters through a spillway. The spillway is vital to the safety of the structure, and it is also generally one of the more costly parts of the project.

In many cases, there will be two spillways, one known as the service spillway and one as the emergency spillway. The service spillway is usually built of reinforced concrete and passes small and medium flows. The emergency spillway has the necessary additional capacity to handle very high flows, which seldom occur but must be included in the design. The emergency structure often will simply be a channel cut through one of the abutments to discharge into the streambed below the dam. Repair of the emergency channel will, at times, be required after water flows through it, but this is acceptable because emergency bypass rarely happens. The adequacy of the combined spillway system is usually evaluated in terms of the probable maximum flood from the watershed.

Normal practice is to build the dam high enough to allow some freeboard above the maximum high water level that could occur during the probable maximum flood. This is done to keep waves from breaking over the top of the dam at the height of a storm. The amount of freeboard will vary, depending on reservoir location, depth, size, and shape. Characteristically, it will be 3 to 6 feet (0.91 to 1.83 m).[5]

Service Outlet

A requirement for most reservoirs is that they can release water and lower the lake level if desired. This provision, known as the service outlet, uses a conduit, controlled by gates or valves, passing through an abutment or under the dam at a level relatively near the bottom of the reservoir. If the outlet passes beneath the dam, it is desirable to place the control mechanism at the upstream end, so that there is no water pressure in the conduit when it is shut off. A typical configuration involves an intake tower, standing in the water at the upstream toe of the dam, with several gated ports for entry of water from the lake. The service outlet conduit connects the intake tower to an outlet channel on the downstream side of the dam.

GROUNDWATER SUPPLIES

Subsurface Distribution of Water

In the United States, groundwater storage exceeds by many times the capacity of all surface reservoirs and lakes, including the Great Lakes. It has been estimated that the total usable groundwater in storage is equivalent to the total precipitation for 10 years, or to the total surface runoff to streams and lakes for 35 years.[9] All water within the groundwater reservoirs, however, is not available for practical use because of such limiting factors as accessibility, dependability, quality, and cost of development.

Not all of the water that infiltrates the soil becomes groundwater. Three things may happen to this water. First, it may be pulled back to the surface by capillary force and be evaporated into the atmosphere. Second, it may be absorbed by plant roots growing in the soil and then reenter the atmosphere by the process of transpiration. Third, water that has infiltrated the soil deeply enough may be pulled downward by gravity until it reaches the level of the zone of saturation—the groundwater reservoir that supplies water to wells. The subsurface distribution of water is illustrated in Fig. 3–6.

The upper stratum, where the openings are only partly filled with water, is called the zone of aeration. Immediately below this, where all the openings are completely filled with water, is the zone of saturation.

The zone of aeration is divided into three belts: the belt of soil water, the intermediate belt, and the capillary fringe. The belts vary in depth, and their limits are not sharply defined by physical differences in the earth materials. A gradual transition exists from one belt to another.

The belt of soil water is of particular importance to agriculture because it furnishes the water supply for plant growth. Water passing downward from this belt escapes the reach of the roots of most plants. The depth of the belt of soil water varies with the types of soil and vegetation, and may extend from a few feet to 20 feet (6.1 m) or more below the surface.

The roots of some plants reach into the capillary fringe or the water table where this area is relatively close to the surface. This occurs mainly along stream courses. Such plants, called phreatophytes, grow without dependence upon the belt of soil water.

Water is held in the belt of soil water by molecular attraction and capillary action against the force of gravity. Molecular attraction tends to hold some water in a thin film on the surface of each soil particle, while capillary action holds water in the very small spaces between the soil particles. Only when sufficient water has entered this belt to more than satisfy the water-holding capacity of the capillary forces, does water start to percolate downward under the force of gravity.

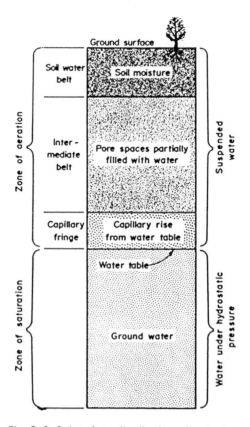

Fig. 3-6. Subsurface distribution of water.[9]

Water that does pass through the belt of soil water enters the intermediate belt, and continues its movement downward by gravitational action. Like the belt of soil water, the intermediate belt holds suspended water by molecular attraction and capillarity, the latter being the more important of the two forces. The suspended water in this belt is dead storage because it cannot be recovered for use. This belt serves only to provide a passage for water from the belt of soil water to the capillary fringe and to the zone of saturation below it. The thickness of the intermediate belt varies greatly, and has a significant effect on the time it takes water to pass through this belt to recharge the zone of saturation.

The capillary fringe lies immediately below the intermediate belt and above the zone of saturation. It holds water above the zone of saturation by capillary force acting against the force of gravity. The thickness and the amount of water held in the capillary fringe depend on the grain size of the material.

The capillary fringe in silt and clay materials is sometimes as much as 8 feet (2.43 m) thick. In coarse sand or gravel, it may be a fraction of an inch.

Water in the zone of saturation is the only part of all subsurface water that is properly referred to as groundwater. An exception to the above description of groundwater is ancient seawater found entrapped in some sedimentary formations. Groundwater of this origin is called connate water. Fresh water from precipitation percolating downward may slowly replace the salt water, but in many cases displacement of all the original seawater is not yet complete. Therefore, connate water remains in some formations in the zone of saturation.

Zone storage capacity is the total volume of the pores or openings in the rocks that are filled with water. The thickness of the zone of saturation varies from a few feet to many hundreds of feet. Factors that determine its thickness are: the local geology, the availability of pores or openings in the formations, the recharge, and the movement of water within the zone from areas of recharge toward points or areas of discharge.

Formations or strata within the saturated zone from which groundwater can be obtained for beneficial use are called aquifers. To qualify as an aquifer, a geologic formation must contain pores or open spaces that are filled with water, and these openings must be large enough to permit water to move through them toward wells and springs at a perceptible rate. Individual pores in a fine-grained material such as clay are extremely small, but the combined volume of the pores in such a formation is usually large. While a clay formation has a large water-holding capacity, water cannot move readily through the tiny open spaces. This means that a clay formation will not yield significant quantities of water to wells, and therefore it is not an aquifer even though it may be water-saturated.

A coarser material such as sand contains larger open spaces through which water can move fairly easily. A saturated sand formation is an aquifer because it can hold water, and it can transmit water at a perceptible rate when pressure differences occur.

The upper surface of the zone of saturation is called the water table. The shape of the water table is controlled partly by the topography of the land and tends, typically, to follow the shape of the land surface.

GROUNDWATER CONDITIONS

The various conditions under which groundwater may be found are shown in Figs. 3–7 and 3–8. When the upper limit of the aquifer is defined by the water table, the aquifer is referred to as a water-table aquifer or an unconfined aquifer, or free groundwater. When a well is drilled in a water-table aquifer, the static water level in the well stands at the same elevation as the water table.

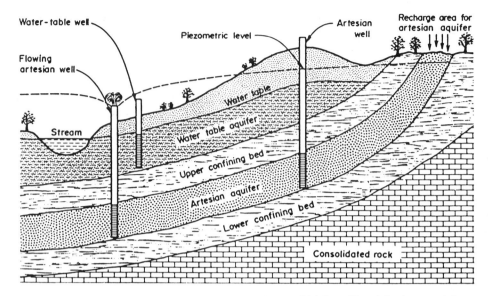

Fig. 3-7. Types of aquifers. (Courtesy of Johnson Division, Signal Environmental Systems Inc.)[9]

In some cases, a local zone of saturation may exist at some level above the main water table. This situation can occur where an impervious stratum within the zone of aeration interrupts percolation and causes groundwater to accumulate in a limited area above that stratum. The upper surface of the groundwater in such a case is called a perched water table.

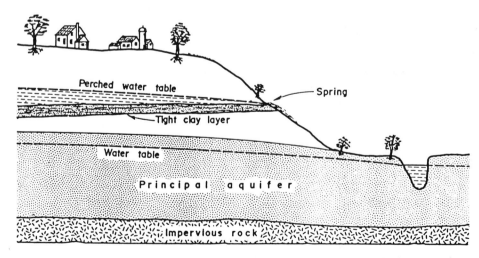

Fig. 3-8. Perched water table. (Courtesy of Johnson Division, Signal Environmental Systems Inc.)[9]

The water table periodically moves up and down—rising when more water is added to the saturated zone by vertical percolation, and dropping during dry periods when previously stored water flows out toward springs, streams, wells, and other points of groundwater discharge.

The zone of saturation may include both permeable and impermeable layers. The permeable layers are aquifers (see above). Where an aquifer is found between impermeable layers above and below it, the aquifer and the water it contains are said to be confined. Because of the upper impermeable layer, the water of the aquifer is confined at pressures greater than atmospheric. Groundwater in such a situation is said to occur under artesian conditions, and the aquifer is called an artesian aquifer. The terms confined aquifer and confined groundwater are also used.

When a well is drilled through the upper confining layer and into an artesian aquifer, water rises in the well to a level above the top of the aquifer. The water level in the well represents the artesian pressure in the aquifer. The hydraulic head, expressed in depth of water, at any point within the aquifer equals the vertical distance from this level down to the point in question.

The elevation to which the water level rises in an artesian well is referred to as the piezometric level. An imaginary surface representing the artesian pressure throughout all or part of an artesian aquifer is called the piezometric surface. This imaginary surface is analogous to the real water surface, the water table, in a water-table aquifer.

The hydrostatic pressure within an artesian aquifer is sometimes great enough to cause the water to rise in a well above the land surface. A flowing artesian well results. Nonflowing artesian wells have the advantage that the water is higher in the casing and there is less pumping head required to bring the water to the surface than in an ordinary well. In some instances it may be possible to use shorter pump columns than would be required with an ordinary well.

If the well is a flowing artesian, special precautions in drilling and completing the well must be observed. Controlling the flow, particularly if the water pressure at the surface is appreciable, is sometimes difficult, and the flow can be dangerous if not competently controlled.

Aquifer Functions and Properties

An aquifer performs two functions: (1) storage like a reservoir and (2) transmission like a pipeline. Aquifer properties related to the storage functions are porosity and specific yield. The property related to the transmission function is permeability. These properties are discussed below.

The *porosity* of a water-bearing formation is that part of its volume which consists of openings or pores not occupied by solid material. Porosity is a measure of how much groundwater can be stored in the saturated material, and is expressed as a percentage of the bulk volume of the material. For example, if 1 cubic foot (0.28 m³) of sand contains 0.30 cubic feet (0.0084 m³) of open spaces or pores, its porosity is 30 percent.

While porosity represents the amount of water an aquifer will hold, it does not indicate how much water the porous material will yield. When water is drained from a saturated material by gravity, only part of the total volume stored in its pores is released. The quantity of water that a unit volume of the material will give up when drained by gravity is called its *specific yield.*

The water that is not removed by gravity drainage is held against the force of gravity by molecular attraction and capillarity. The quantity that a unit volume retains when subjected to gravity drainage is called its *specific retention.* Both specific yield and specific retention are expressed as decimal fractions or percentages. Specific yield equals porosity minus specific retention. For example, if 0.10 cubic feet (0.0028 m³) of water is drained from 1 cubic foot (0.028 m³) of saturated sand, the specific yield of sand is 0.10, or 10 percent. Assuming that the porosity of the sand is 30 percent, its specific retention is 0.20 or 20 percent. These factors can be applied to an aquifer to determine its potential yield, as shown in the following example.[9]

Example to Determine Potential Yield. A water-table aquifer extending over an area of 20 square miles (52 km²), with an average thickness of 40 feet (12.2 m), occupies a total volume of 22.3 billion cubic feet (624.4 Mm³). If the porosity were 25 percent, this groundwater reservoir would store 5.6 billion cubic feet (156.8 Mm³) of groundwater. If the specific yield of the material were 10 percent and the upper 5 feet (1.52 m) of the aquifer were drained by lowering the water table 5 feet (1.52 m), then the total yield would be about 280 million cubic feet (7.84 Mm³) of water, or about 2.1 billion gallons (7.95 Gm³).

This quantity would supply four wells pumping 700 gpm (44.2 l/s) continuously, 12 hours each day, for 1,042 days or almost 3 years. This pumping would be sustained by the groundwater stored in the upper 5 feet (1.52 m) of the aquifer in the absence of any replenishment to the aquifer during the 3-year period.

The above example illustrates how effectively groundwater aquifers can serve as reservoirs. Their enormous capacity often makes them more effective than surface reservoirs.

The property of a water-bearing formation related to its transmission function is called its *permeability,* which is the capacity of a porous medium for transmitting water. Movement of water from one point to another in the material takes place whenever a difference in pressure or head occurs between two points. Permeability may be measured in the laboratory by noting the amount of water that will flow through a sample of sand in a certain time and under a given difference in head.

This relationship has been quantified in Darcy's law:

$$V = P \frac{h_1 - h_2}{l} \qquad (3\text{--}1)$$

where:

V = velocity of flow, ft/day

P = the coefficient of permeability, a constant that depends on the characteristics of the porous material through which the water flows, gpd/sq ft (m/h)

l = distance along the flow path between points 1 and 2, ft (cm)

h_1 = hydraulic head measured at point 1, ft (m)

h_2 = hydraulic head measured at point 2, ft (m)

$\dfrac{h - h_2}{l} =$ hydraulic gradient (headloss per unit of travel) usually denoted as I, ft/ft (m/m)

The quantity of water moving through the aquifer can be determined by multiplying the above velocity by the cross-sectional area of the aquifer through which the water is moving.

The coefficient of permeability, often simply called the permeability, depends on the size and arrangement of the particles in an unconsolidated formation and on the size and character of the surfaces of the crevices, fractures, or solution openings in a consolidated formation. It may change with any variation in these characteristics. The coefficient of permeability is the quantity of water that will flow through a unit cross-sectional area of a porous material per unit of time under a hydraulic gradient of 1.00 (100 percent) at a specified temperature.

For convenient use in well problems, P is expressed as the flow in gallons per day (m³/d) through a cross section of 1 square foot (1 m²) of a water-bearing material under a hydraulic gradient of 1.00 and at a temperature of 60°F (15.6°C). The permeability unit with these dimensions is called a Meinzer unit.

The slope of the water table or the slope of the piezometric surface is the hydraulic gradient under which groundwater movement takes place. The total flow through any vertical section of an aquifer can be calculated from the thickness of the aquifer, its width, its average coefficient of permeability, and the hydraulic gradient at the section in question. The flow q per unit width of the aquifer is:

$$q = PmIc \qquad (3\text{--}2)$$

where:

$q =$ flow, gpd/ft of aquifer width (m³/d/m of aquifer width)
$P =$ average coefficient of permeability of the material from top to bottom of the aquifer, gpd/sq ft (m/d)
$m =$ thickness of the aquifer, ft (m)
$I =$ hydraulic gradient, ft/ft (m/m)
$c = 1.0$ for English units listed (80.5 for metric units listed)

The product of P and m is often used as a single term to represent the water-transmitting capability of the entire thickness of an aquifer, and is called the coefficient of transmissibility. This coefficient is the rate of flow in gallons per day (m³/d) through a vertical section of an aquifer whose height is the thickness of the aquifer and whose width is 1 foot (1 m), when the hydraulic gradient is 1.00. The temperature assumed in this definition of the coefficient is the temperature of the groundwater in the aquifer.

The coefficient of transmissibility can be determined from aquifer pumping tests, as described later in this chapter. Determination using field test data overcomes the problem of getting reliable values of the average coefficient of permeability from laboratory tests. Inaccuracies are always present in the laboratory results because the samples are never entirely representative of the natural state of the formation from which they are taken.

Formations made up entirely of coarse, unconsolidated materials such as gravel give relatively high yields. Because of the large particle sizes in such a formation, the pores or voids are large. The large pores offer less resistance (higher permeability) to flow than the smaller pores in finer sands, so more water will flow through a unit area of the coarser material under any given pressure difference.

WELLS

Characteristics of Wells

Before we discuss the types and design, some key characteristics of wells need to be defined (refer to Figs. 3–9 and 3–10). The height to which water will rise in a well without pumping is known as the static level. When pumping starts, the water level drops, and this decline below the static level is known as the drawdown. When a water table is pumped, the water-bearing beds around the well are unwatered for some distance, which is known as the radius of influence. This depression in the formerly saturated material is called the cone of depression. Under artesian conditions, similar results occur

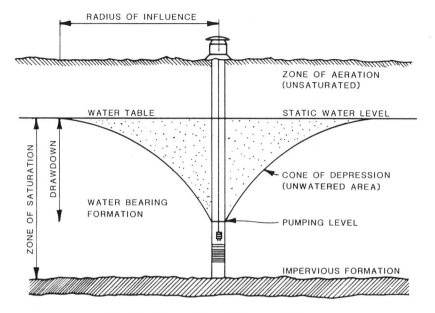

Fig. 3–9. Water-table well characteristics.[10]

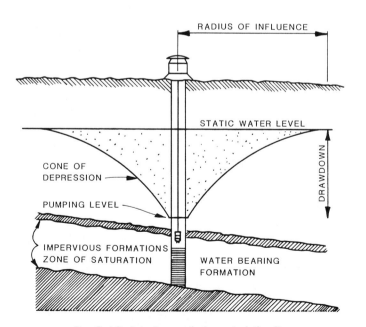

Fig. 3–10. Artesian well characteristics.[10]

with pumping, but the beds are not unwatered. It is only the pressure in the water-bearing stratum that is reduced outward to the edge of the circle of influence. When pumping from a water table is stopped, it may require days or even weeks for the water to rise to the original static level because the water must percolate back into the unwatered section. On the other hand, under artesian conditions, only a few minutes may be required for the pressure to be reestablished to, or almost to, the static level. The total lift of the pump is the drawdown, plus the friction head in the pump and pump discharge pipe from the drawdown point to the ground surface, plus any additional head required to lift the water to and above the ground.

The location of wells may have an important effect on their yield. Where a large amount of water from a formation is to be obtained by means of a number of wells, it is usually advantageous to locate them in a line perpendicular to the direction of the underflow.

The spacing of wells is important. If they are too closely spaced, the cones of depression of two or more wells will overlap and the interference between wells will cause mutual loss of head. Thus, the total lift required to raise the water to the surface will be increased, as will the pumping cost. If practicable, heavily pumped wells that draw on the same water-bearing strata should be spaced at least one-fourth of a mile apart—one-half mile is better—to avoid excessive interference.[10] The radii of the cones of depression and drawdown in the wells are related to the amount of water pumped, the permeability and thickness of the water-bearing material penetrated by the well, and the ability of the material to yield water from storage. Increasing the diameter of the well has only a small effect upon the yield.

Types of Wells

Dug Wells. Dug wells, usually excavated by hand, are shallow in depth and vary from a few to many feet in diameter. They should not be used unless geological conditions prevent the use of a drilled well. Because dug wells are especially subject to contamination by surface water, they must be carefully protected. This can be done by the use of metal casing, concrete walls, vitrified tile pipe, concrete pipe, or double brick walls. Large and deep dug wells are often constructed by sinking their liners as excavation proceeds. The casing should be seated securely in an impervious formation whenever possible. Where an impervious formation does not occur, protection of the supply is increased by extending the casing as far as is practicable below the water table—if possible, at least 10 feet (3.05 m) below the lowest level of the water table. The casing should extend several inches above the pump room floor, and the floor should be elevated some distance above normal ground level on mounded earth or, better still, on a brick or concrete foundation. A concrete apron or ring should extend around the outside of the casing at least 12 inches (0.3 m) in order to divert surface water from the well.

A tight cover over the top of the well is necessary. The pump should be placed on the well cover in such a manner that wastewater from the pump will not return to the well. The pump base should also be sealed so that rodents, insects, and rainwater cannot enter the well. After the well has been completed, it should be disinfected with chlorine. This type of well is not generally used for public supplies except where small quantities of water are required.

Driven Wells. Wells are often driven when the formation involved is a relatively shallow sand layer. These wells usually are used when adequate quantities of water are found at depths of 20 to 70 feet (6.1 to 21.3 m). They are made simply by driving a metal pipe into the water-bearing stratum (see Fig. 3–11); the pipe acts as the permanent casing. Well-points are usually driven by hand when depths are 30 feet (9.1 m) or less. For greater depths, driving tools are often suspended from a tripod or derricks. With this type

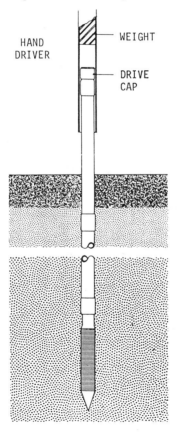

HAND DRIVER

WEIGHT

DRIVE CAP

Fig. 3–11. Typical tool used for driving well-points. (Courtesy of Johnson Division, Signal Environmental Systems Inc.)[9]

of well, special precaution should be taken because the strata penetrated are likely to show contamination. The well should be protected from surface waters by use of a concrete well top with an apron, the well pump should be properly sealed, and the well surroundings should be banked and tamped to divert surface waters. The well should have a solid casing that extends from a minimum of 12 inches (0.31 m) above the ground surface to at least 10 feet (3.05 m) below the groundwater surface. A typical driven well is shown in Fig. 3–12.

Drilled Wells. There are a large number of methods for drilling wells. Those most commonly used are percussion, rotary, or reverse-circulation drilling. The percussion method consists of lifting and dropping a heavy string of tools in the borehole. The drill bit breaks or crushes hard rock into small fragments. In soft, unconsolidated rocks, the drill bit loosens the material. Water in the borehole (added if necessary) mixes with the crushed or loosened rock to form a slurry, which is removed with a sand pump or boiler. The drill bit is of a size that will permit the casing to be introduced into the well after drilling is completed.

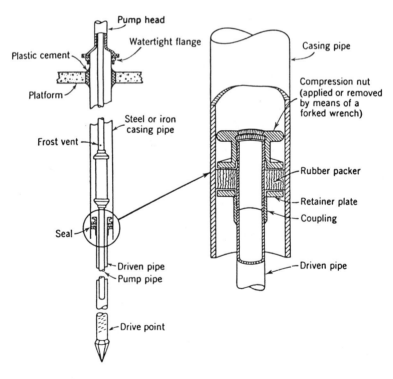

Fig. 3–12. Typical driven well.[2] Reprinted by permission of John Wiley & Sons, Inc.

In rotary drilling, a cutting bit is attached to a hollow drill rod rotated rapidly by an engine-driven rotary table. Either water or a suspension of colloidal clay is pumped down the drill pipe, flows through openings in the bit, and transports the loosened material to the surface. The clay suspensions are designed to reduce loss of drilling fluid into permeable formations, lubricate the rotating drill pipe, bind the wall against caving, and suspend the cuttings. In drilling for water, the thick drilling clay may be forced into the aquifer and reduce the flow into the well; but new methods of reaming and flushing have largely overcome such difficulties.

Reverse-circulation, rotary drilling is done with the flow of drilling fluid reversed with respect to the system used in the conventional rotary method. The drilling fluid and its load of cuttings move upward inside the drill pipe and are discharged by the pump into a settling pit. The fluid returns to the borehole by gravity flow. It moves down the annular space around the drill pipe to the bottom of the hole, picks up cuttings, and reenters the drill pipe through ports in the drill bit.

Boreholes with diameters up to 60 inches (1.53 m) can be drilled. To maintain a low velocity for the descending fluid, the diameter of the hole must be large in relation to the drill pipe. Descending velocities on the order of 1 foot/sec (0.305 m/s) or less are the rule.

Reverse-circulation offers the least expensive method for drilling large-diameter holes in soft, unconsolidated formations. Where geologic conditions are favorable, the cost per foot of borehole increases little with increase in diameter. Drilling cost for a 36- or 40-inch (0.92- or 1.02-m) hole is only moderately greater than for a 24-inch (0.61-m) hole.

Gravel-Packed Wells. The effective diameter of a well can be increased by packing gravel between the well screen and the outer limits of the borehole. A well hole is first drilled and reamed to a diameter of 24 inches (0.61 m) or more. An outer casing is then cemented in place, and the aquifer is cleaned before a smaller inner casing carrying the well screen is inserted. Gravel is then packed into the annulus between the two casings. Gravel-packed wells are favored in fine uniform sands and in loosely cemented sandstones. In fine sands, the use of gravel permits the use of larger slot openings in the well screens. In loose sandstone, the gravel can prevent sloughing sandstone from entering the well without the need to resort to very small well screen openings. Another system consists of several standard vertical wells in a circular pattern. The individual wells are generally smaller in diameter than normal production wells, and when pumped simultaneously, all the wells produce as much as an extremely large single well. Ring wells are chosen for groundwater development where thin aquifers of fine sands are encountered that would require the use of very large-diameter well screens to obtain the desired capacity in a single well. Often it is more economical to drill several 6- to 10-inch (0.15- to 0.25-m)-diameter wells than one very large-diameter well.

Horizontal Collector Wells. These wells, which are shown in Fig. 3–13, are typically used to withdraw water from nearby rivers. They are constructed by lowering a concrete caisson into unconsolidated material, sealing the bottom with a concrete plug, and jacking slotted pipes radially out from the bottom of the caisson. Water enters the main shaft or caisson through these horizontal laterals, and is pumped to the water system by vertical turbine pumps hung in the caisson. The hydraulic head differential created between the river water surface and the water level in the caisson draws water through the river bed material to the slotted pipes of the collector well. The advantage of this system is that it provides a very large intake area through which to draw water, much more than can be obtained from

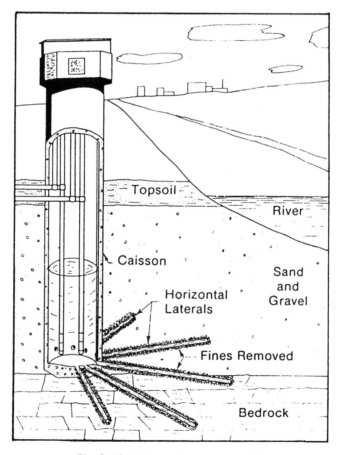

Fig. 3–13. Horizontal collector well.[11]

a single vertical well. However, collector wells are more expensive than a single vertical well.

Well Yield

The specific yield of a well is the discharge per foot of drawdown. The safe yield is the capacity of the aquifer to supply water without a continuous lowering of the water table or piezometric surface.

When water is pumped from a well, the quantity discharged initially is derived from aquifer storage immediately surrounding the well. As pumping continues, more water must be derived from storage at greater and greater distances from the well. The circular-shaped cone of depression must expand so that water can move from greater distances toward the well. The radius of influence of the well increases as the cone continues to expand. The drawdown also increases as the cone deepens to provide the additional head required to move the water from a greater distance. Over time, the cone expands and deepens at a decreasing rate because with each additional foot of horizontal expansion, a larger volume of stored water is available than from the preceding one. The cone will continue to enlarge until aquifer recharge equals the pumpage.

When the cone has stopped expanding for one or more of the above reasons, a condition of equilibrium exists. There is no further increase in drawdown with increase in time of pumping. In some wells, equilibrium occurs within a few hours after pumping begins; in others, it does not occur even though the length of the pumping period may be extended for years.

Well discharge formulas for equilibrium conditions are well established.[9] There are two basic formulas, one for artesian conditions and the other for water-table conditions. Both assume recharge at the periphery of the cone of depression. Figure 3–14 shows a vertical section of a well constructed in a water-table aquifer. The formula for the water-table well is:

$$Q = \frac{P(H^2 - h^2)}{1055 \log R/r} \tag{3–3}$$

where:

Q = well yield or pumping rate, gpm (m³/d)
P = permeability of the water-bearing sand, gpd/sq ft (m/d)
H = saturated thickness of the aquifer before pumping, ft (m)
h = depth of water in the well while pumping, ft (m)
R = radius of influence, ft (m)
r = radius of the well, ft (m)
1055 = constant for English units listed (3.993 = constant for metric units listed)

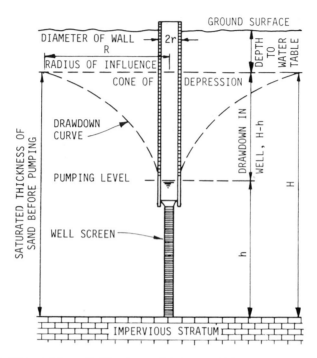

Fig. 3-14. Water-table well. (Courtesy of Johnson Division, Signal Environmental Systems Inc.)[9]

Figure 3–15 is a vertical section of a well pumping from an artesian aquifer. The formula for a well operating under artesian conditions is:

$$Q = \frac{Pm(H - h)}{528 \log R/r} \qquad (3\text{--}4)$$

where

m = thickness of aquifer, ft (m)
H = static head at bottom of aquifer, ft (m)
528 = constant for English units listed (1.9984 = constant for metric units listed)

All other terms are as defined in equation (3–3) for a water-table well.
The above formulas show how the well diameter affects yield:

$$Q = f \frac{1}{\log R/r} \qquad (3\text{--}5)$$

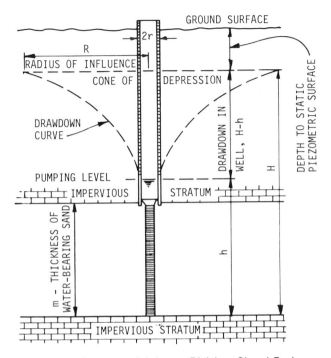

Fig. 3–15. Artesian well. (Courtesy of Johnson Division, Signal Environmental Systems Inc.)[9]

In other words, well yield is not a direct function of well diameter. For example, if a 6-inch (0.15-m) well will yield 100 gpm (6.3 1/s) with a certain drawdown, a 12-inch (0.31-m) well constructed at the same spot will yield 110 gpm (6.94 m) with the same drawdown, and an 18-inch (0.46-m) well will yield 117 gpm (7.38 1/s). A 48-inch (1.22 m) well will yield 137 gpm (8.65 1/s), or 37 percent more water than a 6-inch (0.15-m) well at the same drawdown.

Derivations of the above formulas are based on the following simplifying assumptions:[9]

1. The water-bearing materials are of uniform permeability within the radius of influence of the well.
2. The aquifer is not stratified.
3. For a water-table aquifer, the saturated thickness is constant before pumping starts; for an artesian aquifer, the aquifer thickness is constant.
4. The pumping well is 100 percent efficient.
5. The pumping well penetrates to the bottom of the aquifer.
6. Neither the water table nor the piezometric surface has any slope; both are horizontal surfaces.

7. Laminar flow exists throughout the aquifer and within the radius of influence of the well.
8. The cone of depression has reached equilibrium so that both the drawdown and the radius of influence of the well do not change with continued time of pumping at a given rate.

Uniform permeability is seldom found in a real aquifer, but the average permeability as determined from aquifer pumping tests has proved reliable for predicting well performance. For artesian wells where most of the aquifer thickness is penetrated and screened, the assumption of no stratification is not an important limitation. For water-table aquifers, where drawdown reduces the saturated thickness considerably, the situation can be handled when the stratification is known and taken into account in applying the formula.

Assumption of constant thickness is not a serious limitation because variation in aquifer thickness within the cone of depression in most situations is relatively small. Where changes in thickness are important, they can be taken into account.

Assumption that the well is 100 percent efficient can cause the calculated well yield to be seriously in error if the real well is inefficient. Therefore, the actual efficiency must be taken into account.

The water table or the piezometric surface is never truly horizontal; however, the slope is usually very flat, and its effect on calculation of well yield is negligible in most cases. The slope of the water table or the piezometric surface does cause distortion of the cone of depression, making it elliptical rather than circular.

The above formulas can be used to calculate well yield if P, H, m, and R are known. The well log provides values of H and m; R is usually estimated; P must be determined from laboratory or field tests. For a water-table aquifer, the formula for calculating P is:

$$P = \frac{1055 \ Q \ \log r_2/r_1}{(h_2{}^2 - h_1{}^2)} \qquad (3\text{–}6)$$

where:

P = permeability, gpd/sq ft (m/d)
Q = pumping rate, gpm (m³/d)
r_1 = distance to the nearest observation well, ft (m)
r_2 = distance to the farthest observation well, ft (m)
h_2 = saturated thickness at the site of the farthest observation well, ft (m)
h_1 = saturated thickness at the site of the nearest observation well, ft (m)
1055 = constant for English units listed (2405.6 = constant for metric units listed)

For artesian conditions, the formula for determining the permeability is:

$$P = \frac{528 \, Q \, \log \, r_2/r_1}{m \, (h_2 - h_1)} \tag{3-7}$$

where:

m = thickness of the aquifer, ft (m)

h_2 = head at the site of the farthest observation well, measured from bottom of aquifer, ft (m)

h_1 = head at the site of the nearest observation well, measured from bottom of aquifer, ft (m)

528 = constant for English units listed (1203.9 = constant for metric units listed)

The remaining terms are the same as for the water-table aquifer.

GROUNDWATER QUALITY CONSIDERATIONS

When water seeps downward through overlying material to the water table, particles held in suspension, including microorganisms, may be removed. The extent of removal depends on the depth and character of the overlying material. The bacterial quality of the water also generally improves during storage in the aquifer because time and storage conditions are usually unfavorable for bacterial multiplication or survival. Of course, the clarity of groundwater does not guarantee safe drinking water, and only adequate disinfection can guarantee the absence of pathogenic organisms.

After a well has been completed and is to be tested for yield, or any time after a permanent pump is installed or repaired, the well should be disinfected. The procedure is outlined in detail in AWWA A100–66, American Water Works Association Standard for Deep Wells. This standard recommends the use of hypochlorites or fresh chlorinated lime in a concentration of 50 mg/l of chlorine.

Four alternative procedures are given: (1) preparation of the disinfecting solution at the ground surface to a concentration of 50 mg/l and to a quantity at least twice the volume of the well, after which the solution is discharged rapidly into the well interior; (2) preparation of a stock solution containing 15,000 mg/l of chlorine and diluting it to the 50 mg/l concentration in the well by feeding it into a continuous flow of water; (3) adding the stock solution directly to the well and agitating the well contents with a baler or bit; and (4) placing the calculated amount of dry hypochlorite in a perforated pipe section capped at both ends and moving the pipe up and down within the well casing. The disinfection period should be at least 2 hours, and the procedure should not be attempted until after the well has been thoroughly cleansed of oil, grease, and foreign matter of any kind. After the well has been pumped until all of the chlorine solution has been removed, a sample should be collected for bacteriological examination.

All groundwater withdrawal points should be located a safe distance from sources of pollution. Sources of pollution include septic tanks and other individual or semipublic sewage disposal facilities, sewers and sewage treatment plants, industrial waste discharges, land drainage, farm animals, fertilizers, and pesticides. Where water resources are severely limited, groundwater aquifers subject to contamination may be used for water supply if adequate treatment is provided.

Because many factors affect the determination of safe distances between groundwater supplies and pollution sources, it is impractical to set fixed distances. Where insufficient information is available to determine the safe distance, the distance should be the maximum that economics, land ownership, geology, and topography will permit. If possible, a well site would be located at an elevation higher than that of any potential source of contamination. The direction of groundwater flow does not always follow the slope of the land surface; so the slope of the water table should be determined from observation wells.

Groundwater quality in aquifer systems varies spatially and may range from good to unacceptable. Water-quality problems vary from high concentrations of dissolved solids to small amounts of trace elements, organics, and pathogens that exceed drinking water standards. In some cases water-quality problems become apparent only after a water well is drilled, constructed, developed, and tested. However, test wells may be used to predict the quality of water that will be obtained from permanent, high-capacity wells.[12]

Other chapters of this book address the significance of various contaminants and techniques for their removal. A few contaminants are more frequently of concern in groundwater than in surface water, including iron, manganese, fluoride, and nitrates.

Iron can cause staining of plumbing fixtures and clothes and may encrust well screens and pipes. Concentrations greater than 0.5 mg/l are usually troublesome.

Water may pick up iron from contact with the well casing, pump parts, and piping. The more corrosive the water, the more metal it will dissolve from the iron surfaces with which it comes in contact. Water standing in a well that has been idle will have a higher iron content than the water in a water-bearing formation.

Upon contact with air, dissolved ferrous iron changes to the ferric state and precipitates. The resulting iron hydroxide and iron oxide are commonly called rust.

Well water containing iron in appreciable amounts may be completely clear and colorless when first pumped. If it stands for a time, oxygen from the air oxidizes the dissolved iron. The water grows cloudy and will yield a deposit of a rust-colored material.

Iron-bearing waters also favor the growth of iron bacteria, such as crenothrix. These growths can form so abundantly in water mains, recirculating systems, and other places, that they exert a marked clogging action.

Manganese resembles iron in its chemical behavior. Because manganese is less abundant in rock materials than iron, its occurrence in water is less common than that of iron. Manganese occurs in groundwater as soluble manganous bicarbonate, which changes to insoluble manganese hydroxide when it reacts with oxygen of the air. The stains caused by manganese are more annoying and harder to remove than those caused by iron.

Fluoride in groundwater may be derived from fluorite, the principal fluoride mineral of igneous rocks, or from any of a considerable number of complex fluoride-bearing minerals. Volcanic or fumarolic gases may also contain fluoride, and may be the source of fluoride in water. Too much fluoride in the water has been shown to be associated with the dental defect known as mottled enamel. This may appear on the teeth of children who drink water containing too much fluoride during the period when permanent teeth are formed. Conversely, small concentrations of fluoride are beneficial and help to prevent tooth decay. The desirable fluoride content varies with air temperature, as shown in Chapter 20.

Nitrate in concentrations greater than 45 mg/l (as NO_3) is undesirable in water used for domestic purposes because of the possible toxic effect that it may have on young infants. Nitrates are transported through soil without significant degradation or adsorption. Groundwater under heavily fertilized areas may have very high (more than 100 mg/l) nitrate concentrations. Other major sources may include liquid wastes and bacterial fixation of atmospheric nitrogen.[13]

REFERENCES

1. Walker, R., *Water Supply, Treatment, and Distribution,* Prentice Hall, Englewood Cliffs, N.J., 1978.
2. Fair, G. M., Geyer, J. C., and Okun, D. A., *Water and Wastewater Engineering,* John Wiley & Sons, Inc., New York, © 1966.
3. Maknoon, R., and Burges, S. J., "Conjunctive Use of Ground and Surface Water," *J.AWWA,* p. 419, August, 1978.
4. Culp, Wesner, Culp, *Technical Guidelines for Public Water Systems,* U.S. Environmental Protection Agency, Contract 68–01–2971, 1975.
5. Gooch, R. S., "Surface Water Supplies," *Manual of Water Utility Operations,* Texas Water Utilities Association, Distributed through Texas State Dept. of Health, Austin, TX. 1975.
6. Davis, C. V., *Handbook of Applied Hydraulics,* McGraw-Hill, New York, 1942.
7. Hutcheon, B. C., and Smith, D. W., "Raw Water Studies Determine Siting of New Intake," *J.AWWA,* p. 16, 1981.

8. "Water Supply Reservoirs: How Much Storage Space Remains?" Committee Report, *J.AWWA*, p. 674, 1975.
9. *Ground Water and Wells*, Johnson Division, Universal Oil Products Co., St. Paul, Minn., 1974.
10. Harvill, C. R., and Billings, C. H., "Ground Water Supplies," *Manual of Water Utility Operations*, Texas Water Utilities Association, Distributed through Texas State Dept. of Health, Austin TX. 1975.
11. Willis, R. F., "Groundwater Exploration and Development Techniques," *J.AWWA*, p. 556, 1979.
12. Fetler, C. W., Jr., "Use of Test Wells as Water Quality Predictors," *J.AWWA*, p. 516, 1975.
13. Kaufman, W. J., "Chemical Pollution of Ground Waters," *J.AWWA*, p. 152, 1974.

Chapter 4
Wastewater in Supply Sources

INTRODUCTION

It has long been a common practice for one community to discharge its treated wastewater into a stream or lake ultimately used by another community as a water supply source. One study estimated that 1 gallon (3.785 l) out of every 30 gallons (113.6 l) used for water supply had passed through the wastewater system of an upstream community, based on 155 cities studied.[1] Another showed that 90 percent of 1,246 municipal water supplies studied contained wastewaters.[2] Several utilities were forced to use water from a source when low flow was less than the combined upstream wastewater discharge flows. Water supplies drawn near the bottom of large river basins were found to contain wastewater from several thousand dischargers. About 15 million people were estimated to be served by supplies containing at least 10 percent wastewater at low flow conditions, and 4 million were served by supplies containing 100 percent wastewater.

The practice of using water supplies containing wastewaters has resulted in extensive, although often unintentional and unplanned, water reuse. Many municipalities have no alternative but to use such supplies. Careful consideration of wastewater constituents in water supplies and their fate in receiving streams can contribute to choosing intake locations that will improve supply quality.

TYPES OF CONCERNS

The general types of risks fall into four main categories: toxicological, microbiological, aesthetics, and possible chronic effects of trace organics. The degree of risk with and without wastewater in supply sources, and the effects of separation between discharge and intake for these four risk categories are discussed below.

1. Toxicological (pesticides, residues, mine drainage, accidental spills of chemicals in transport; municipal, industrial, and agricultural wastewaters; urban and rural runoff).
Many surface waters throughout the country are subject to contamination from accidental spills of toxic materials in transport, and some are so contaminated from urban

and agricultural runoff and other nonpoint pollution sources that their health risks are only slightly different from those involved in the use of river water receiving treated municipal wastewater. Increasing the distance between points of discharge and intake provides more time for detection of toxic materials that have not been removed in treatment. The removal of toxic materials by stream self-purification is not so important or effective as that in water or wastewater treatment. Advanced wastewater treatment (AWT) processes can remove nearly all the dissolved and suspended contaminants in wastewater, and can provide a high degree of protection against possible toxic contaminants.

2. Microbiological (bacteria, viruses, fungi, and other organisms from human wastes). Raw municipal wastewater is the principal source of the organisms that transmit waterborne disease. Increasing the distance between wastewater outfall and water intake provides more time for natural die-off of the pathogens. However, in cases of limited separation, any necessary removal up to complete disinfection can be obtained prior to effluent discharge. The minimum separation required then depends upon the degree and reliability of pathogen removal provided by the wastewater prior to discharge. Because microbiological tests require several days for completion, time of travel or storage in the stream is not a monitoring advantage unless 4 days or more are available.

3. Aesthetics (acceptance, taste and odor, and color). Water from streams that do not receive treated wastewater is usually more aesthetically pleasing than water obtained from streams receiving wastewater. Increasing the separation between outfall and intake, up to a point at least, probably provides greater public acceptance of the water. Physical problems of taste, odor, and color may arise from the discharge of wastewaters to water sources, although these problems can be avoided if proper water and wastewater treatment are provided.

4. Possible Chronic Effects of Trace Organics (carcinogens, mutagens, tetratogens). Many water supply sources contain trace organics. Either unreacted or in the form of their reaction products with chlorine, ozone, or other oxidants, these substances may have adverse health effects following their long-term ingestion in trace amounts. These organics may originate in nature or be present in wastewater, which also may contain synthetic organics not found in nature. Until more is known, it is assumed that there may be more risks involved in the wider variety of organics found in wastewater, compared to those of natural origin. Even if the concentration of total organics is reduced to 1 mg/l or less by adsorption on stream sediments or activated carbon, there may still be a health hazard. The degree of the hazard probably is not affected to any practical extent by the time of stream travel or storage. Removal of organics by AWT or upgraded water treatment is more important than separation distance between outfall and intake.

A review of the quality of secondary effluent from municipal treatment plants that receive sewage from residential areas (no industrial wastes) shows that it does not typically contain specific inorganic or organic chemicals in concentrations that exceed the National Interim Primary or Secondary Drinking Water Regulations.[3] However, secondary effluent does exceed the regulations for turbidity, coliform bacteria, and unspecified organics measured as foaming agents, color, and odor. Although there are no drinking water regulations for the general organic content as measured by chemical oxygen demand (COD) or total organic carbon (TOC), there is concern about the health effects of these organics. This concern is particularly justified when the organics are contributed by wastewater.

The next two sections address the concerns related to organic compounds and pathogenic organisms.

ORGANICS IN WASTEWATERS

The organic material in municipal wastewater is a mixture of many compounds that are only partially known. The three broad classes of organics in municipal wastewater are fats, carbohydrates, and proteins. They are usually considered removable by primary and secondary biological treatment, although protein is somewhat less readily removed than fats and carbohydrates. A properly operating biological treatment plant treating residential wastewater is capable of producing secondary effluent with a soluble COD of 30 to 50 mg/l and a soluble biochemical oxygen demand (BOD) of 1 to 2 mg/l. The data on TOC levels in wastewater are scarce, but the available information indicates that secondary effluent concentrations are in the 30 to 50 mg/l range. Concentration of foaming agents in secondary effluent, as indicated by tests for MBAS, is typically in the 2 to 4 mg/l range.

Trickling filter effluent from the Haifa, Israel municipal wastewater treatment plant was analyzed for total organic content, and the results were reported as percent of total COD.[4] The total COD was about 180 mg/l, and the organics were classified as: 40 to 50 percent humic substances (humic, fulvic, and hymathomelanic acids); 8.3 percent ether extractables; 13.9 percent anionic detergents; 11.5 percent carbohydrates; 22.5 percent proteins; and 1.7 percent tannins. The results of a later, more extensive study including activated sludge effluents are summarized in Tables 4–1 and 4–2.[5]

Another study found 60 percent of the organics in secondary effluent had molecular weights less than 700, and 25 percent had apparent molecular weights greater than 5,000.[6]

Table 4–1. Distribution of Organic Groupings.

	PERCENT OF TOTAL COD		
ORGANIC GROUPINGS AND FRACTIONS	MUNICIPAL WASTEWATER HIGH RATE TRICKLING FILTER	MUNICIPAL WASTEWATER STABILIZATION POND	DOMESTIC WASTEWATER EXTENDED AERATION ACTIVATED SLUDGE
Proteins	21.6	21.1	23.1
Carbohydrates	5.9	7.8	4.6
Tannins and lignins	1.3	2.1	1.0
Anionic detergents	16.6	12.2	16.0
Ether extractables	13.4	11.9	16.3
Fulvic acid	25.4	26.6	24.0
Humic acid	12.5	14.7	6.1
Hymathomelanic acid	7.7	6.7	4.8

In one study of municipal wastewater, 77 organic compounds were detected in the primary effluent, and 38 in the secondary effluent. Several compounds found in the secondary effluent were not present in the primary effluent.[7] The concentrations of individual compounds in the secondary effluent were estimated to be less than 20 μg/l. It was found in another study that soluble organics are produced in biological treatment that are more refractory to further treatments than are the organics in raw sewage.[8] It was hypothesized by these same investigators that the residual organics not removed by activated carbon are intermediate breakdown products of protein, and that these are most likely proteins that originate from the cell walls of microorganisms present in biological treatment processes.

The fate of organics during chlorination is of particular concern. One study found that chlorine-containing stable organic constituents are present

Table 4–2. Molecular Weight Distribution of Humic Substances.

	PERCENT OF HUMIC COMPOUND PRESENT		
MOLECULAR WEIGHT RANGE	FULVIC ACID	HUMIC ACID	HYMATHOMELANIC ACID
<500	27.5	17.9	4.5
500–1000	7.8	6.2	12.2
1000–5000	35.7	29.4	48.0
5000–10,000	15.3	7.8	28.0
10,000–50,000	9.4	36.7	7.3
>50,000	4.3	2.0	0

after chlorination of effluents from domestic sewage treatment plants.[9] Over 50 chlorine-containing organic compounds were separated from chlorinated secondary effluents, and 17 of these compounds were tentatively identified and quantified at the 0.5- to 4.3-μg/l level. The chlorination yield, which is the portion of the chlorine dose associated with chlorine-containing stable organic compounds at the end of the chlorination reaction period, was approximately 1.0 percent for secondary effluents that had been chlorinated with 3.2 mg/l chlorine to a residual of 1 mg/l at contact times of 15 to 45 minutes. The chlorination yield was approximately constant with respect to chlorine dosage in the range studied, but increased with increasing reaction time. Chlorination yields were approximately the same for both primary and secondary effluents. Essentially the same effects were obtained by chlorination with either chlorine gas or hypochlorite solution. In addition to the 17 chlorine-containing compounds that were identified, 32 stable organic constituents were identified, and 23 of these were quantified at 2- to 190-μg/l levels in the effluents from domestic primary sewage treatment plants. Nine stable organic constituents were identified, and eight of these were quantified at 5- to 90-μg/l levels in the effluents from domestic secondary sewage treatment plants.

Some 30 chlorinated compounds, primarily aromatic derivatives, have been attributed to chlorination of secondary effluent with chlorine dosages of 1,500 mg/l.[10] Preliminary data indicated total organic-bound chlorine to be 3,000 to 4,000 μg/l with these large chlorine dosages.

Another study found that when chlorine was added to activated sludge effluent in excess of the ratio necessary to achieve breakpoint chlorination, chloroform was formed in less than 2 minutes, and its concentration increased with reaction time.[11] For example, at the end of 2 minutes reaction time, the chloroform concentration was 30 μg/l, and it increased to 262 μg/l at the end of 24 hours. When chlorine was added at or below the ratio required to achieve breakpoint, only small amounts of chloroform were formed, with little or no increase in concentration at a longer reaction time. The chloroform concentration increased from 6.0 μg/l after 2 minutes reaction time to 10.8 μg/l after 24 hours.

In addition to the presence of organic compounds contributed directly by wastewaters, there is also the potential that inorganic nutrients, such as nitrogen and phosphorus, can contribute to organics of concern by stimulating algal growths. There is evidence that chlorophyll, algal biomass, and algal extracellular products can serve as trihalomethane precursors.[12] Both green algae and blue-green algae produce extracellular products that were found upon chlorination to yield at least as much chloroform per unit of organic carbon as has been reported from studies of humic and fulvic acids.[12]

Chapters 12, 14, and 15 describe treatment processes for the removal of organic compounds.

PATHOGENIC ORGANISMS IN WASTEWATER

Domestic sewage contains agents of human disease. Thus the location of drinking water intakes downstream from the discharge of wastewater, treated or untreated, is of concern to health authorities because of the potential for disease transmission.

In considering the transmission of disease, one must take the following factors into account: (1) the presence of agents of disease, (2) the concentration of the agent or the dose, (3) the dose–response, and (4) the host contact.

A number of bacterial diseases have been associated with the consumption of sewage-contaminated water, including typhoid fever, salmonellosis, shigellosis, cholera, and infections due to enterocytopathic *Escherichia coli* and *Yersina entercolitica.* During the period 1969–74, the majority of waterborne outbreaks of bacterial disease in the United States were due to *Shigella sp.,* followed by *Salmonella sp.,* typhoid fever, and pathogenic *Escherichia coli.* Less than 10 percent of the cases of these diseases in the United States were waterborne.[13]

The two waterborne parasitical diseases commonly associated with contaminated water are amoebic dysentery and giardiasis, both protozoan diseases. Less than 1.5 percent of waterborne disease outbreaks are due to these parasites.

There are at least 101 types of viruses that may find their way into water via fecal contamination. Of these, the most serious threat to the public health (in terms of disease severity) is the virus of infectious hepatitis (Hepatitis A). Less than 1 percent of the reported cases of hepatitis in the United States are attributable to contaminated drinking water. A large proportion of cases were associated with the consumption of contaminated shellfish.

The low incidence of infectious disease transmission via finished drinking water is a testimony to present drinking water sanitation technology. However, the potential presence of these agents in waters receiving waste discharge must be assumed, as must their presence at the intake of downstream water treatment plants.

The second factor to consider in the transmission of disease via water is the concentration of disease agents present. If the nature of an agent is sufficiently understood, then methods can be developed to quantitatively determine its presence in water. In certain instances, because of the technical difficulty of measuring an agent or agents, it may be possible to monitor a surrogate parameter rather than the actual agents. This is particularly true when a large variety of agents may be present, and the measurement of each would present a monumental task.

Infectious disease agents have traditionally been monitored in water using a surrogate parameter, the coliform–fecal coliform test. The presence of these bacteria in water is indicative of the presence of fecal material and thus the potential presence of pathogenic enteric organisms. Through the years this test has been effectively used in the quality control of finished drinking water. More recently some shadow of doubt has been cast upon the efficiency of the coliform test, as there is evidence that certain of the enteric viruses are more resistant to chlorination than are the coliform bacteria. Thus the absence of the latter organisms in finished drinking water may not guarantee the absence of enteric viruses. However, epidemiologically there is very little evidence that coliform standards for drinking water have been ineffective in protecting the public from infectious disease, including that of viral origin. To date no simple method has been devised to test routinely for the presence of animal viruses in drinking water.

The manner in which the exposed population comes into contact with a disease agent is a most important factor in evaluating the health risk from such exposure. In the case of drinking water, exposure via ingestion is more intimate and implies a greater risk to the population than a less personal exposure such as boating or fishing. Thus, in considering the public health implications of the siting of a drinking water intake below wastewater discharges, it must be assumed that the host contact with any agent that might be present is a critical one.

Another major criterion is dose–response; that is, what concentration of agent is required to bring about illness in the exposed population? This is one of the most important aspects in considering the impact of wastewater discharge upstream from community drinking water intakes. What concentrations of agents can be present without unduly affecting the consumer? How will the treatment of wastewater, the hydraulic characteristics of the intervening water course, and the treatment of the drinking water prior to distribution affect the concentration of disease agents present in the finished drinking water?

The most fundamental concept in toxicology is the dose–response relationship. Figure 4–1 is a generalized dose–response curve.

This curve assumes a normal distribution in the frequency of response, and indicates that there is some level of dose at which there is no measurable response, or threshold concentration.

Ideally the concentration of disease agents present in drinking water delivered to the consumer should be below the threshold dose. Unfortunately, the dose–response characteristics of the disease agents that might be present in water are frequently unknown. Total elimination of the potential for pathogenic organisms at the water intake is impossible. Thus, the assessment of the risk of disease in exposure to various concentrations of these agents is essential.

Some information exists concerning the dose–response relationships between humans and certain bacterial and parasitical diseases, but there is a notable deficiency in dose–response data involving enteric viruses. In the

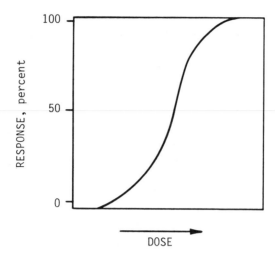

Fig. 4-1. Dose–response curve.

case of bacteria and certain parasites, it is evident that there is a threshold dose below which few cases of clinical disease occur. In order to elicit a response in 25 to 75 percent of humans challenged, the following threshold dosages were required: 1×10^2 to 1×10^3 *Shigella*; 1×10^6 to 1×10^9 *Salmonella*; 1×10^4 to 1×10^7 typhoid bacilli; 1×10^6 to 1×10^{10} *Escherichia coli*; and 1×10^3 to 1×10^8 cholera vibrios, per dose.[14]

The two enteric pathogens *Entamoeba histolytica* (amoebic dysentery) and *Giardia sp.* evidence similar dose–response relationships. It should be pointed out that, in a number of instances, infections without clinical diseases were recorded at lower exposure doses. Data for enteric virus dose–response are not well established. Some proponents believe that one unit of virus is an infective dose, whereas others think that a threshold dose exists.

The dose of infectious agents present in water is routinely measured using the coliform and fecal coliform index, and there is little epidemiological evidence to suggest that the procedures are ineffective. That is, when coliform levels that meet presently established drinking water standards are obtained, enteric infectious disease transmission is low or nondetectable.

FATE OF CONTAMINANTS IN STREAMS

There are many factors that affect contaminants in streams. These factors are discussed in numerous books and reports, including the following: *Environmental Chemistry,* by Manahan; *Aquatic Chemistry,* by Stumm and Morgan; *Trace Metals and Metal-Organic Interactions in Natural Waters,* edited by Singer; and *Aqueous Environmental Chemistry of Metals,* edited by Rubin.[15-18]

Contaminants may be contributed to streams from a number of sources, including: (1) soil and rock, (2) waste discharges, (3) storm runoff, (4) precipitation and atmospheric fallout, and (5) biological organisms. Flowing streams are dynamic systems, whose chemical and biochemical reaction rates are such that equilibrium conditions are only slowly, if ever, attained. The organic and inorganic chemicals of concern in water supply interact with one another and with other materials and organisms in various chemical and biochemical reactions. Reactions that can take place in streams to decrease, or in some cases increase, the concentration of contaminants include: (1) precipitation, (2) complexation, (3) oxidation–reduction, (4) ion exchange, (5) adsorption and/or absorption, often termed sorption, (6) flocculation, and (7) biological uptake and/or release.

Contaminants and other material are present in water in suspended, colloidal, or dissolved form. Most chemical reactions affecting natural water systems occur at the solution–solid interface or within the cells of bacteria or algae. Because of their large surface-to-volume ratio, colloidal particles suspended in water are particularly important in chemical reactions. Such particles are capable of changing contaminant concentrations through sorption, exchange, and flocculation actions. Clay minerals are important colloids in many streams.

Physical Factors

The important stream physical characteristics include: velocity, depth, turbulence, degree of mixing, temperature, turbidity, changes in cross-section, and bottom characteristics such as slope, type of material, and sorptive capacity. The effects of some of these parameters are so interrelated that their relative importance is difficult to determine. For example, the bottom slope affects the depth and the velocity of the stream, and turbulence is affected by all three characteristics.

The degree of mixing in a stream is an important parameter, and is determined by the physical characteristics. Mixing in a stream can occur vertically, laterally, and/or longitudinally. Vertical mixing normally occurs within a few tenths of a mile. Density gradients from temperature differences in the stream tend to overcome vertical mixing, but this rarely occurs. Lateral mixing normally is complete within miles and increases with the number of relatively sharp reverse bends in the stream reach. Longitudinal mixing, caused by variations in cross section and changes in direction that permit areas of quiet water and eddy currents, may require many miles. Typically, the ratio of longitudinal-to-lateral-to-vertical mixing is about 100:10:1.

Other parameters determined by the physical characteristics of the stream are the reaeration rate and solids deposits. The reaeration rate is a function of the turbulence of the river and can be determined from the velocity, depth, and temperature of the stream; it increases with increasing velocities and decreases with increasing temperature. Sedimentation of suspended particles is dependent on the degree of turbulence in the stream; settled solids can be scoured off the bottom and resuspended during periods of increased flow. This raises the stream turbidity and increases oxygen demand.

Precipitation and Complexation

In general, all the constituents of a natural chemical system, including trace metals, ions, and inorganic and organic compounds, are related to each other through complex formation and solid precipitation reactions. Variation of any constituent will give rise to variations in at least some of the other constituents. The organic component of fresh natural water is poorly known analytically, and the complexing properties of organics are only hypothesized in many cases.

The complexation of metal ions with either naturally occurring organic material or organics in waste discharges may be important reactions. Complex formation is the combination of metal cations with molecules or anions to form a coordination or complex compound. There is some indirect evidence that complexing agents play a significant role in the form of metal concentration in treated wastewater. It is possible that pollutant complexing agents in flowing streams play an important part in transporting heavy metals and preventing their removal by conventional water treatment.

Many of the metal ions found in natural waters, particularly those found at trace levels, form strong complexes with a variety of chemical species. The formation of complex compounds may have several effects: (1) the formation of insoluble compounds may remove metal ions from solution, (2) complexation also may solubilize metal ions from otherwise insoluble metal compounds, and (3) strong complexation may shift oxidation–reduction potentials. Little has been reported about levels of complexing agents and stable metal complexes in natural waters. Complexing agents are not normally determined in water analyses.

Humic and fulvic acids probably are the most important naturally occurring complexing agents. These acids are rather loosely defined; they refer to a family of compounds, similar in structure and chemical properties, formed during the decomposition of vegetation. They can strongly bind metal ions, and they are found in both water and soil. Synthetic complexing agents such as sodium tripolyphosphate, sodium ethylenediaminetetraacetate (EDTA), sodium citrate, and sodium nitrilotriacetate (NTA) are produced in large quantities and almost certainly find their way into streams through waste discharges. NTA may also solubilize heavy metals from sediments on stream bottoms, depending on pH, bicarbonate concentration, calcium concentration, and the nature of the sediments.

In addition to their phosphorus content, polyphosphate fertilizers are valued for their capacity to act as complexing agents, thus solubilizing micronutrient ions from soils and making essential metals more available to plants. Therefore, it is possible that polyphosphates present in runoff from fertilized soils may act as complexing agents in some waters prior to hydrolysis of the polyphosphate to orthophosphate.

Effects of Clay Minerals and Other Suspended Material

Clay minerals are among the most common suspended matter found in natural waters. In many streams, clays might be considered the most important mineral solids present in colloidal suspension or as sediments, for these reasons:

1. Clay minerals can fix dissolved chemicals in water and thus exert a purifying action. The ability of clays to exchange cations is an important phenomenon having an impact on the availability of trace-level metal nutrients in water.
2. Because of their high surface area and other properties, clays also may sorb organic compounds such as pesticides and herbicides, and are important in the transport and removal of organic pollutants in streams.
3. It is also believed that some microbiological processes occur at clay mineral surfaces, so that clays may participate in the degradation of organic materials.

The sorption of organic compounds by montmorillonite has been investigated in some detail; the sorption of the herbicide 2,4-D on montmorillonite, kaolinite, and illite has also been studied. The sorption process was found to be relatively slow, requiring several hundred hours to reach equilibrium. It was concluded that the overall process must be relatively complicated, involving sorption onto the clay surface and subsequent diffusion of the herbicide into the clay.

One study found that there are regional differences in the concentrations of chromium, silver, molybdenum, nickel, cobalt, manganese, and suspended sediments in streams.[19] The sediments of the Mississippi and the rivers west of it draining into the Gulf of Mexico resemble average shale in composition, while the rivers east of the Mississippi are considerably higher in metals concentration. The suspended material in the central rivers is rich in montmorillonite, whereas eastern rivers carry more organic matter, illite, and kaolinite.

The cation exchange capacity (CEC) of suspended clays in eastern streams is generally in the range of 14 to 28 milliequivalents/100 grams. Although the central streams have the higher CEC, they carry a relatively lower load of most trace metals than do the eastern streams. Therefore, the degree of trace metal transport is not strictly a function of CEC, but perhaps is due to a greater amount of trace-element-rich soil and industrial discharges. The correlation between suspended organic material and suspended trace metal levels was found to be quite low. It appears that suspended trace metals are carried primarily by suspended mineral materials.

FACTORS IN THE LOCATION OF WATER INTAKES

There are a great many factors to be considered regarding intake locations with respect to municipal wastewater discharges. Many of these factors also apply even in the absence of upstream discharges of municipal wastewater. Raw water quality may vary greatly from stream to stream; it is assumed that a preliminary selection among various streams has been made on the basis of best available raw water quality from a public health standpoint, as well as adequate quantity. The problem then becomes one of locating the intake along a given stream.

For the moment, the question of upstream wastewater discharge is set aside, and the other factors are considered. This is done in order to help place the potential hazards of wastewater contaminants in perspective with the risks involved in all water systems, even when there is no pollution from wastewater.

Important items in the location of intakes, which take into account reliability, safety, and cost, include:

- Adequacy of supply.
- Channel changes, shoal and bar formation, and silting.
- Availability of water to intake ports at all river stages and at all stream flows.
- Accessibility to intake for maintenance at all river stages and at all seasons.
- Location of the intake with respect to the city to be served.
- Navigation requirements.
- 100-year flood level.
- Need for storage dam, either in-channel or off-channel, and detention provided.
- Foundation conditions.
- Structural stability and safety of dams.
- Protection from rapid currents, wind, ice, boats, floating material, waves, and bottom sediment.
- Water depth, and ability to draw water from different depths.
- Distance from service roads and a source of electric power.
- Protection from vandalism.

Even in the absence of upstream discharges of municipal wastewater, there are public health factors to be considered in intake location. Storm runoff that makes up stream flow may contain almost any of the contaminants included in the National Interim Primary and Secondary Drinking Water Regulations. The concentrations of these substances can, in many cases, be restricted by proper location and operation of intakes. Runoff from urban areas should be avoided when possible because it is likely to be higher in contaminants than runoff from rural areas. Ability to draw water from a stream at different depths may give some control over the amounts of turbidity, color, iron, manganese, algae, and other substances in the raw water. Rivers and streams fed by rural and urban runoff can contain many substances of public health significance.

Many streams and watersheds are subject to accidental spills of agricultural chemicals and hazardous materials being transported by truck, rail, or air, whether or not there are planned municipal wastewater discharges. Metals that are picked up from distribution and service piping as a result of corrosion are also independent of water supply source and the presence of wastewater content.

Additional hazardous materials that may be contributed to raw water in streams by municipal wastewater discharges are principally organics (especially synthetic organics) and some heavy metals that have their only origin of consequence in wastewater. Also, many substances present in streams that receive no wastewater discharges may be present in higher concentrations in streams that receive municipal wastewater, particularly pathogenic bacteria, viruses, and other organisms originating in the intestinal tract of man. In the absence of wastewater discharge, control of treated water supplied to the public depends upon the monitoring of raw and finished water quality and upon the degree and reliability of water treatment provided.

With the discharge of municipal wastewater upstream of a water intake, another control must be provided in the form of wastewater treatment prior to discharge and the monitoring of effluent quality. The resulting effluent quality depends upon the degree and reliability of wastewater treatment provided. With wastewater discharge to a water source, there are three primary forces at work that affect finished water quality: (1) wastewater treatment, (2) purification in the stream, and (3) water treatment. The end product of these three forces must be water of sufficient quality for drinking. The intermediate force, stream purification, may play an important role, relative to the other two, where the stream provides great dilution, high rates of reaeration, long travel distance, and/or prolonged storage. However, the situation of greatest interest is the limiting condition, which, in a particular stream with given flow, dilution, and reaeration characteristics, becomes the minimum allowable distance between the points of waste discharge and water supply withdrawal under prevailing conditions.

This minimum separation is not a fixed distance for all situations, but, rather, one that varies widely depending on many local factors. It is a distance that must be decided on a case-by-case basis by qualified authorities. At present there is an important unresolved question that has a vital impact on whether certain impurities contributed by wastewater are to be removed at their source, in the wastewater treatment plant, or later in the water purification process. There is uncertainty about the minimum raw water quality allowable for streams that are supply sources receiving municipal wastewater, and the maximum safe load that can be placed on the water plant. The National Interim Primary and Secondary Drinking Water Regulations are based on accumulated experience treating waters from relatively unpolluted sources. There are no minimum stream quality standards for water at supply system intakes.

This lack of standards for minimum-quality raw water from polluted sources places a considerable restraint on water reuse. Water and wastewater treatment techniques for production of high-quality water are quite similar—actually almost identical in many cases. In general, it is more economical overall to apply the required treatment process in the water treatment plant rather than the wastewater plant.[3] However, uncertainty regarding the safe load that may be placed on water treatment processes may dictate the use of additional wastewater treatment rather than (or in addition to) additional water treatment.

In making case-by-case judgments of the minimum distance between the points of wastewater discharge and water supply intake, some of the items to be considered include:

- Stream flow and quality.
- Quantity and quality of treated wastewater to be discharged.
- Potential water quality improvement by stream purification processes, including dilution, reaeration, adsorption, sedimentation, and biological die-off.
- Pollution from sources other than municipal wastewater, including industries, storm runoff, agriculture, and miscellaneous nonpoint sources.
- Raw water quality at intake under most adverse conditions.
- Water treatment provided.
- Relationship between wastewater effluent characteristics and safe drinking water requirements.
- Risk assessment of intake siting options.

Historically, a city's water supply intake is located upstream from its own wastewater discharge. This principle has been followed almost without exception. However, there remains the question of how far downstream a city's water intake should be from a neighboring city's wastewater discharge.

The *USPHS Manual of Recommended Water Sanitation Practice*, 1946, states that:

Waters containing more than 20,000 coliform bacteria per 100 ml, are unsuitable for use as a source of water supply, unless they can be brought into conformance (coliforms reduced to less than 20,000/100 ml) by means of prolonged preliminary storage, or some other means of equal permanence and reliability.[20]

Many cities located on the Ohio, Mississippi, Missouri, Kansas, and other rivers have successfully produced biologically safe water from raw river water containing from 50,000 to 2,000,000 coliforms/100 ml. This has been accomplished by use of prechlorination, presedimentation, storage, double coagulation–sedimentation, and other means. A later USPHS study, reported by Walton in 1956, concluded that the routine successful treatment of high-coliform raw waters indicated that:

The current USPHS recommendations with respect to the density of coliform organisms in raw water acceptable for treatment in modern well operated plants are no longer applicable.[21]

By effective disinfection of water free from suspended material in a modern water treatment plant, river water can be disinfected with certainty. In 1969, the "AWWA Committee Report on Viruses in Water" concluded that:

There is no doubt that water can be treated so that it is always free from infectious microorganisms—it will be biologically safe.[22]

Another matter of concern is trace organics. Trihalomethane production during chlorination may be high, even with relatively low total organic concentrations. Further, enough of these organics will survive natural purification, regardless of time and distance between outfall and intake, to remain a potential problem. For trace organics, the methods of treatment in the wastewater and water treatment plants, particularly in the latter, appear to be much more important than intake location. For wastewater contributed to water supply streams, there is also concern about the organics that have yet to be identified and evaluated for possible health effects. This is true even in cases where the total quantity of all the unidentified organics is low (less than 0.5 mg/l). In this instance, the question is not distance from the point of wastewater introduction, but the complete exclusion of organics originating in municipal wastewater from water supply sources until this potential hazard is better evaluated.

The location of water intakes immediately below a municipal wastewater plant outfall should not be considered. Similarly, outfalls should not be allowed to discharge immediately above an existing waterworks intake. Either surface or underground storage of treated wastewater prior to withdrawal for subsequent water treatment should be provided. It would be foolish not to take advantage of natural purification processes in any rational plan of either direct or indirect water reuse. The natural purification processes of dilution, time of travel, separation by distance, sedimentation, bacterial die-off, adsorption, storage, and loss-of-identity have many advantages, the most important of which is the time afforded to learn the results of water quality monitoring, to detect accidental spills, and to correct treatment plant malfunction.

If zero distance between sewer outfall and water intake is not enough, what is adequate? This decision is a matter of judgment, and must be made by local public health experts and other informed professionals on a case-by-case basis supported by an adequate background of facts and circumstances concerning local conditions. However, in all cases, reaction time to emergencies is a major consideration. From this standpoint it would appear that 24 hours of combined time in travel and storage between wastewater discharge and water supply intake should be a minimum. This, then, is one criterion for intake location.

The minimum combined time may be the governing criterion, if it is accepted by health authorities as being sufficient to serve the intended purpose of reaction time. Greater separation in order to obtain more treatment through natural purification would have to be justified on an economic basis. That is, the cost of additional pipeline or storage facilities would have to be less than the cost of providing equal removal of the affected contaminants in water treatment. In most situations, it is unlikely that a longer outfall pipe will be more economical than providing additional treatment.

Once the intake location is selected, the question remains of whether removals of certain substances are to be provided in wastewater or water treatment. Items to be considered include total costs, local costs, equity of who pays the costs, and incidental benefits of treatment in the upstream location. It has been found that it is cheaper in most cases to minimize wastewater treatment beyond secondary levels and to maximize water purification.[3] However, it seems fairest for the city that produces the wastewater to pay for the costs of removing substances that have adverse effects on downstream water supplies. In addition, greater overall protection of the environment, especially of the public health aspects of recreational use of the stream, is provided by removal of undesirable materials nearest their source, in the wastewater treatment plant.

CASE STUDIES

The purpose of this section is to discuss cases that involve the indirect reuse of wastewater through location of wastewater treatment plant discharges into streams above water supply intakes. The facilities described are:

1. Occoquan, Virginia. This project discharges highly treated wastewater to a storage reservoir on Bull Run that serves as a water supply source for about 500,000 people in the Washington, D.C. area.

2. Passaic Basin, New Jersey. The Passaic River is an existing situation that involves a long history of indirect wastewater reuse.
3. Windhoek, South Africa. This project takes reclaimed wastewater directly from holding ponds to the public water supply system.

Occoquan, Virginia[31]

The Occoquan watershed drains some 600 square miles (1,555 km²) southwest of Washington, D.C., and is the source of drinking water for about 500,000 people in the Washington, D.C. area. The general location of the watershed is shown in Fig. 4–2. The first storage of runoff from the watershed for water supply occurred in 1950 when the Alexandria Water Company built a small dam near the town of Occoquan. The current 1,700-acre, 9.8 billion-gallon (37.1 Bl) reservoir was created in 1958 by the construction of the Upper or High Dam. Concurrent with construction of the dam were rapidly increasing pressures for development in the watershed. From 1940 to 1960, the population in the Washington, D.C. area nearly doubled. Most of the growth since the late 1950s has occurred in Northern Virginia and suburban Maryland.

The development in the watershed caused increasing concern of the Virginia State Health Department and State Water Resources Control Board (SWRCB) over the potential effects of wastewater discharges and urban runoff on the reservoir water quality. In December, 1959, requirements of 90 percent BOD removal and 15 days detention in holding ponds were established for conventional wastewater treatment plants in the watershed. In early 1960, the State Health Department added effluent limits of 20 mg/l for BOD and suspended solids in discharges to the reservoir. Although the concerns of the Health Department continued to increase, by 1970 there were 25 wastewater treatment plants in the Occoquan basin with a certified capacity of 5.95 mgd (22.5 Ml/d). Population was projected to increase tenfold, to 800,000, by the year 2000.

The SWRCB commissioned studies of the Occoquan Reservoir quality and the basin pollution problems in 1969. The reservoir was found to be highly eutrophic in summer and early fall, with heavy growths of blue-green algae and anaerobic conditions below a depth of 10 feet (3.05 m). It was concluded that sewage effluents were a major cause of the highly eutrophic conditions; that natural runoff from the watershed contained, at times, enough inorganic phosphorus to contribute the entire allowable load to the reservoir; and that treatment plant effluents should not be discharged to the reservoir during the algal growth season. The solution proposed was construction of secondary plants, with export of the secondary effluents to a tributary to the Potomac River.

Public hearings were held on the export proposal in the fall of 1970 with

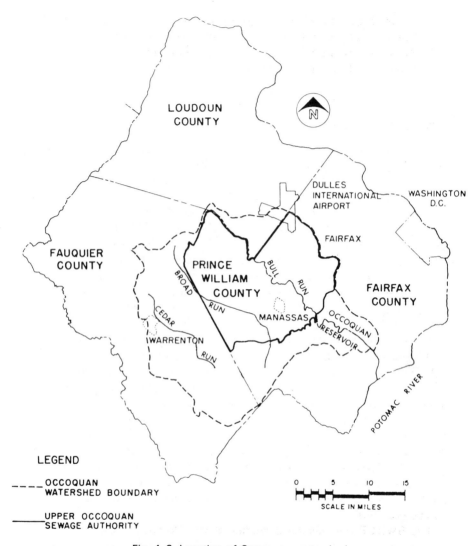

Fig. 4–2. Location of Occoquan watershed.

several citizen groups expressing opposition based upon concerns about the cost of the proposed treatment and export system, the potential impacts on growth in previously underdeveloped areas, and the need for a solution that would not merely transport the problem to the Potomac River. A special bond election was held on September 22, 1970 with the result that funds for a treatment and export system were defeated by a 2:1 margin. The portion of the sewer bonds associated with the export system ($9,000,000) was deleted from the total sewer bonds proposed ($39,000,000, which included $30,000,000 for upgrading existing plants in the basin) and subjected to a subsequent vote in the November, 1970 election. The revised $30,000,000 bond proposal (without export) passed by a 4:1 margin.

A significant development in mid-1970 was the appointment of Noman Cole as chairman of the SWRCB. Cole, a nuclear engineer, perceived the Occoquan situation as a major environmental issue. Taking actions somewhat unique in the pollution control field, Cole personally made an independent analysis of alternative approaches to handling the wastewaters in the Occoquan basin, including a trip to California to evaluate advanced wastewater treatment (AWT) at the South Lake Tahoe Public Utility District. Cole drafted a report that compared AWT and discharge to the Occoquan Reservoir with the export alternative, and reviewed his report in December, 1970, with the four jurisdictions involved (Fairfax County, Prince William County, Manassas Park, and Manassas).

The report, became final in January, 1971, first reviewed conventional wastewater treatment, the performance of the South Lake Tahoe AWT plant, existing discharges to the Occoquan Reservoir, and the export alternative. A chart similar to Fig. 4–3 was presented, which compared the quantities of BOD, phosphorus, and nitrogen being discharged to the reservoir at the then-existing flow of 4.3 mgd (16.3 Ml/d) of secondary effluent, to those quantities that would be discharged at flows of 10 and 40 mgd (37.9 and 151.4 Ml/d) of AWT effluent. The 40-mgd (151.4-Ml/d) flow would represent 12 percent of the annual inflow to the reservoir and 46 percent of the July low flow to the reservoir. This analysis concluded that even if flows should eventually reach ten times the existing wastewater flows, the quantities of pollutants discharged to the reservoir would be substantially reduced by AWT.

Although the costs of the export system were a prime factor motivating the evaluation of the AWT alternative, the SWRCB report also concluded:

Hopefully, this will help kill the archaic and technically out-of-date philosophy that the "solution to the control of water pollution is export and dilution"—or more simply and clearly stated, "Do not dump it in my yard—but it is all right to dump it in my neighbors."

We would have a valuable asset (e.g., a 10 to 40 mgd [37.9 to 151.4 Ml/d] clean water source between now and 1985). This clean water source will help stretch our water supply as well as protect our valuable park land investments on the Occoquan.

In summary, we will have preserved the Occoquan Reservoir water quality as well as provided a major source of its water supply. Further, in this approach, we will have spent our efforts, energy, talent, and financial resources for treatment of water, rather than for expensive export pipes which do not clean up any water but just move the problem elsewhere!

This approach of thoroughly cleaning up waste is equivalent to developing a new water supply resource.

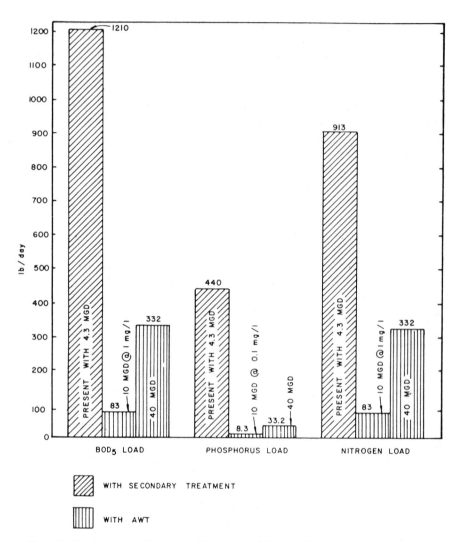

Fig.4–3. Discharges to Occoquan Reservoir with secondary treatment and AWT.

The report also presented an outline of administrative considerations that would be a part of the in-basin reuse concept. These considerations later were the nucleus of a formal State of Virginia policy for the Occoquan watershed. As reflected in newspaper articles, the initial response of the jurisdictions involved was positive:

It is kind of refreshing to see the State Water Control Board take a positive position of leadership in the field of pollution . . . . (Chairman Prince William, County Board of Supervisors, *Potomac News,* 12/2/70)

We feel that this concept is a breakthrough. (Warrenton Town Manager Edward Brower, *The Fauquier Democrat,* 12/10/70)

The whole thing was a breath of fresh air. (Jason Paige, Chairman, the Fauquier Water and Sanitation Authority, *The Fauquier Democrat,* 12/10/70)

If you're wondering where the gale-force sigh of relief came from, it's just us, pleased to have something to be positive about regarding the Occoquan Reservoir.

And positive we are about the report State Water Control Board Chairman Noman M. Cole, Jr. released Wednesday in which he concludes it would be both better and cheaper to treat sewage in the Occoquan Watershed a la Lake Tahoe than to export the effluent from the basin. (Editorial, *Northern Virginia Sun,* 12/4/70)

A preliminary state policy for the Occoquan watershed based on the in-basin reuse concept was issued in late January, 1971, and a public hearing was scheduled for March 31, 1971. Concurrently, the four jurisdictions involved formed the Upper Occoquan Sewage Authority (UOSA) to implement a regional wastewater system.

At the public hearing on the reuse concept, the Virginia State Health Department and the downstream water supply utility (the Fairfax County Water Authority—FCWA) attacked the proposed plan on the basis that viruses discharged in the AWT effluent present a health threat. Both agencies urged that the wastewaters be exported from the basin. The SWRCB conducted a further evaluation of the virus question in April, 1971 and issued a report on this issue. This evaluation concluded that:

1. The 25 existing discharges of marginally treated secondary effluent were discharging significant quantities of virus to the Occoquan Reservoir.
2. The existing discharges constituted 2 percent of the annual water supply and 10 percent of low flow months.
3. There were no apparent virus problems related to the water supply.
4. The lack of virus problems was due to 20 to 25 miles (32.2 to 40.3 km) of travel between the points of wastewater discharge and water intakes and the 100 to 190 days travel time between these points, and that the FCWA downstream water treatment processes provided the conditions needed to assure kill of any surviving virus—turbidity of 0.1 to 1 NTU and effective chlorination.
5. There were cases in the world where water supplies contained a higher percentage of wastewater than would be the case for Occoquan and yet no water supply-related virus health problems were apparent.
6. The discharge of AWT effluent should substantially reduce the existing level of discharges of virus to the Occoquan Reservoir.
7. The virus issue was not one of substance based on available evidence.

After further review of the issue, the FCWA concluded " . . . that the upgrading and replacement of the present treatment plants with an advanced waste treatment facility will not result in any increase in risk and will most probably reduce the current level of pollutants."

The U.S. EPA wrote in response to the controversy:

... AWT processes utilized provide a substantial improvement in virus removal over conventional secondary processes and additionally allow more effective chlorination of the final effluent. There is no doubt that the provision of a Tahoe-type AWT treatment facility in the Upper Occoquan Watershed, even with a discharge twice that of the present secondary plants, will provide for an improvement of water quality, including public health aspects.

The Virginia Department of Health stated:

While this Department still has reservations about the proposed policy the above cited circumstances or evidence indicates the risk is a minimal or accepted one provided all the safety features are included in the design and a meaningful management and surveillance program is included. Therefore, this Department will cease its opposition to the policy and will work with the Water Control Board in establishing the requirements for management and surveillance program.

With the virus issue resolved, the SWRCB adopted a formal policy for the Occoquan watershed on July 26, 1971.

The SWRCB policy called for AWT of all wastewaters in the basin with discharge of a high-quality effluent into the Occoquan Reservoir. The SWRCB stated that this approach offered several advantages over that of export, including: (1) solution of the sewage-related Occoquan water quality problems, while preventing further degradation of the Potomac estuary; (2) elimination of a very costly export system; (3) preservation of the water supply resources of the watershed; and (4) minimization of the possibility of a major siltation problem from the construction of the export line. The policy also established a total yearly wastewater flow allocation for the watershed through 1985 to 1990, as well as suggesting allotments to each jurisdiction. It also called for establishment of a monitoring program to provide data on plant operation and the effect of discharge on the quality of the reservoir and monitoring of siltation problems.

Until the regional AWT plant was available, the SWRCB policy required the removal of "maximum amounts" of BOD, COD, suspended solids, and phosphorus at existing plants. "Modest" expansion of existing plants was permitted, provided they were abandoned when the regional facility became available, and if no increased loading to the stream resulted.

The policy presented detailed requirements for the AWT facility and collection system. No more than three, and preferably only two, plants were specified for the watershed area. All plants were required to be at least 15 miles (24.2 km)—preferably 20 miles (32.2 km)—above the Fairfax County Water Authority's water intake.

The major treatment processes specified by the policy to meet these standards were:

- Primary settling.
- Activated sludge.
- Lime coagulation and settling for phosphorus removal.
- Two-stage recarbonation with intermediate settling.
- Mixed-media filtration.
- Nitrogen removal. (Selective ion exchange was later selected by the designer as the process to be used for nitrogen removal.)
- Granular carbon adsorption.
- Chlorination.
- Storage, on site.

The policy specified the following effluent quality as measured on a weekly average basis:

BOD: 1.0 mg/l
COD: 10.0 mg/l
Suspended solids: 0 mg/l
Nitrogen: 1 mg/l
Phosphorus: 0.1 mg/l as P
Methylene blue active substances (MBAS): 0.1 mg/l
Turbidity: 0.4 NTU
Coliform bacteria: less than 2 per 100 ml

The initial wastewater flow allocation was established as 10 mgd (37.5 Ml/d), with 5 mgd incremental increases in flow being licensed only if the monitoring program showed that the AWT effluent was not creating a water quality or public health hazard in the reservoir. A maximum watershed allotment of 39.3 mgd (148.8 Ml/d) was specified for 1985 to 1990.

The policy required that the initial plant must have 100 percent backup capacity, but after the success of the concept was established, the ratio of on-line treatment to standby capacity could be as high as 4:1. The plant was to be designed so that the failure of a single component would not interrupt plant operations necessary to meet the final effluent requirements. In a similar vein, the policy also specified two independent sources of outside power and one on-site power supply to minimize the risk of treatment interruption due to power failure. Changes from the plant design requirements were permitted only with written approval from the SWRCB, which reserved the right to hold public hearings on any proposed design changes. The plant must be staffed 24 hours per day, 7 days per week, with a minimum of five people per shift.

In addition to the design requirements for the plant, the policy also presented specific requirements concerning the design of pumping stations. These dealt with minimum standards for standby pumps, on-site power supply, dual off-site power sources designed so that failure of a single component would not degrade pumping capacity, flow measuring and recording, and provision of retention basins of a minimum 1-day capacity.

In August, 1972, the Regional Council of Governments (COG) reviewed the facility plan for the Occoquan project. The review addressed only two points related to reuse: nutrient control and virus removal. The COG review of the nutrient question concluded that the AWT process proposed would "go a long way toward abating the pollution problem in the Occoquan Reservoir." Regarding viruses, the COG review concluded that the travel distance of 20 miles (32.2 km), travel time of 100 to 190 days, and effective water treatment downstream were an adequate combination to provide virus control. COG approved the project, with the reuse aspect receiving relatively minor consideration among factors such as effects on jurisdictions' ability to control growth, conformance to regional plans, air pollution, and so on.

In accordance with the schedule approved by the state and EPA, the plans and specifications for the total project with an initial rated capacity of 10.9 mgd (41.3 Ml/d) were submitted in May, 1973. Subsequent to submission of the plans, the state and EPA advised UOSA that funding limitations would not permit award of one contract for the total project. The project had to be split for funding in two fiscal years, and EPA required that each split must result in an operable unit. Instead of splitting the project into logical construction phases, such as earthwork, concrete, buildings, and so forth, UOSA had to award contracts for total operable unit processes in the treatment plant. (For example, the first plant contract was for construction of the filtration units and carbon adsorption system.) This change, plus delays in plan review and approvals, delays from protests over bid awards, delays in EPA approvals due to new grant regulations that required all UOSA jurisdictions to comply with infiltration/inflow, user charge, industrial cost regulations, and unprecedented rates of inflation, caused the total project costs to increase. The capital costs totaled $82,000,000, and operation costs (in 1979) were estimated at $0.90/1,000 gallons ($0.24/1,000 l).[23] The operating costs were reported to be comparable to those of the multiple small plants replaced by the regional plant. The performance of the AWT has been excellent. Figures 4–4 through 4–7 summarize data on the AWT effluent concentrations of phosphorus and COD in fiscal years 1982 and 1983.

Passaic Basin, New Jersey[3]

The Passaic River, although one of the most polluted rivers in the East, serves as the major water supply source for northeastern New Jersey. A

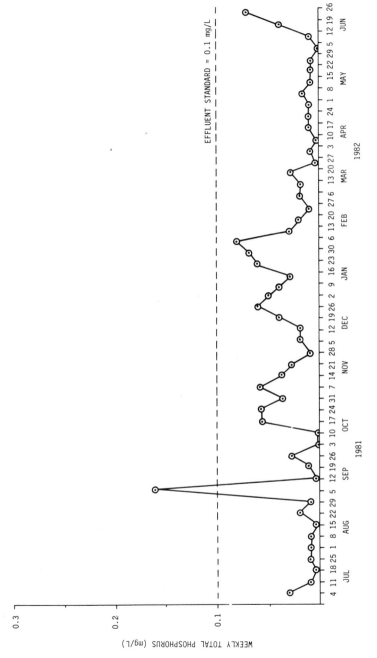

Fig. 4–4. Final effluent phosphorus concentration at the Upper Occoquan AWT plant, fiscal year 1982.[31]

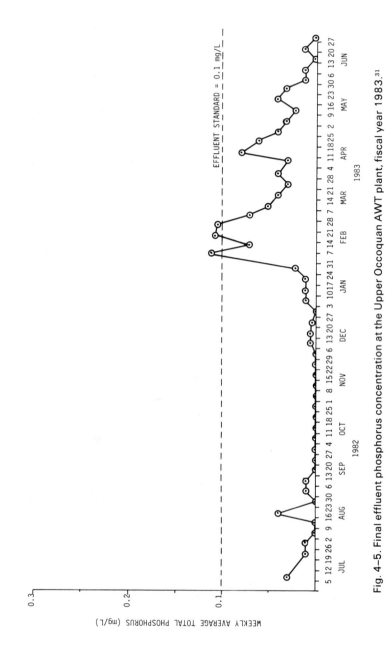

Fig. 4–5. Final effluent phosphorus concentration at the Upper Occoquan AWT plant, fiscal year 1983.[31]

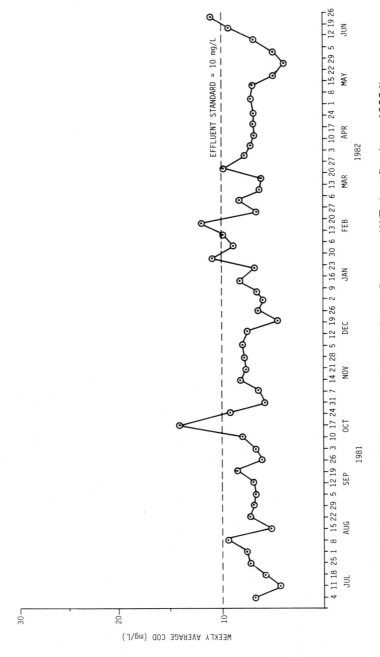

Fig. 4–6. Final effluent COD concentration at the Upper Occoquan AWT plant, fiscal year 1982.[31]

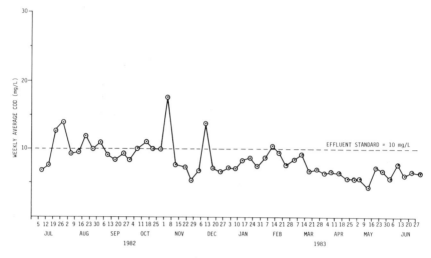

Fig. 4–7. Final effluent COD concentration at the Upper Occoquan AWT plant, fiscal year 1983.[31]

backbone for industrial development since the 1790s, the Passaic River has long been a source for water power, recreation, and potable water supply, and a means for disposing of municipal and industrial wastes. Over the years there have been many controversies over water rights for uses such as power, flow maintenance in canals, and private and public water supplies.

Beginning in the 1850s and continuing through 1890, the river served without any treatment as a direct source of potable water. By 1890 the river downstream from the City of Paterson was considered too polluted to use, because of wastes discharged from the industrial centers. In 1902, the first rapid sand filtration plant was built at Little Falls, New Jersey, where it is still in use today.

The Passaic Valley Water Commission serves over 600,000 people and is the chief water supply agency for 15 northern New Jersey communities, including Paterson, Passaic, and Clifton. There are about 115 municipal and industrial wastewater treatment plants located within the 485-square-mile (1,257 km²) area of watershed above the Little Falls water intake (see Fig. 4–8). The total wastewater flow discharged in the watershed is about 55 mgd (208.2 Ml/d), which represents up to 60 percent of the total flow in the river at Little Falls during drought years. In the past, the wastewater treatment plants have achieved between 75 and 95 percent removals of BOD, with an average of about 85 percent removal. The BOD in the river at Little Falls has averaged from 2 to 5 mg/l, but is as high as 10 mg/l during periods of low flow.

The treatment processes reported to be used at the 100 mgd (378.5 Ml/

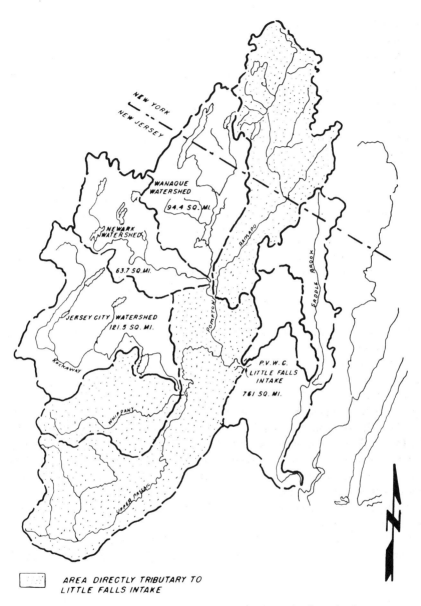

AREA DIRECTLY TRIBUTARY TO
LITTLE FALLS INTAKE

Fig. 4–8. Watersheds of the Passaic River and tributaries.[3]

d) Little Falls water treatment plant include: two-stage screening, addition of powdered activated carbon for taste and odor control, free residual or breakpoint chlorination with rapid mixing, alum coagulation, flocculation, sedimentation, dechlorination, filtration, pH adjustment with caustic soda, and post-chlorination.[3] Powdered activated carbon is used for poor water quality at the intake due to high organic concentrations resulting from waste spills upstream.

In 1971 John Wilford, then chief of the Bureau of Water Supply in the State of New Jersey, Department of Environmental Protection, wrote:

. . . . Notwithstanding the poor quality of the water in the river, the Passaic Valley Water Commission has been consistently able to produce water from this source which meets the requirements of the New Jersey Potable Water Standards. . . . No studies have been conducted to determine the incidence of viruses in either the river water or the treated water, primarily because of the impracticability of conducting viral determinations on a routine basis. It is reasonable to assume, however, that the water treatment process must effectively remove any viruses which may be present; otherwise there would have been epidemics of infectious hepatitis, etc. among the consumers. . . .

That the situation constitutes indirect reuse of the wastewaters for potable water supply is widely understood and accepted, as indicated in the following statement made in 1973 by Wendell R. Inhoffer, the general superintendent and chief engineer of the Passaic Valley Water Commission: "The Commission, with its rights to 75 mgd (283.9 Ml/d) at Little Falls admittedly depends upon this wastewater discharge to meet its requirements."

When the demand for potable water was increasing during World War II, the quality of the water in the Passaic River had so deteriorated that there were many complaints received about the drinking water treated by the 1902 filtration plant. According to Inhoffer: "During this period, it was not unusual for the Commission to receive 800 complaints a day of phenolic and chemical taste and odors, primarily due to the increase in industrial pollution created by wartime industry."

In 1948, the Commission added free residual, or breakpoint, chlorination to control taste- and odor-causing materials, particularly free ammonia. Inhoffer reported:

It can be stated that without the use of breakpoint chlorination, the Passaic River could not be safely used as a source of potable water supply. However, the use of breakpoint chlorination is costly. . . . It [is] apparent that the influence of sewage treatment plant discharges resulting in free ammonia in the raw river water significantly affects the cost of treatment. . . . Commission experience indicates that as long as one maintains a free chlorine residual of at least 50 percent after prechlorination, taste and odor standards will be maintained.

Average chlorine doses, which depended mostly upon the ammonia concentration in the raw water, ranged from 15 to 25 mg/l. The ammonia concentration, typically varied from 0.5 to 5 mg/l, with an average of 1 to 2 mg/l.

In the mid-1960s, as commercial and industrial development continued in the watershed, and with the realization that the ultimate wastewater flow could approach 100 mgd (378.5 Ml/d), the Commission became concerned about possible contaminants not normally identified in their regular testing program. In 1968, after concluding that their program which included powdered activated carbon treatment was not adequate, the Commission undertook a pilot study to investigate possible means of treatment for removal of soluble organics. In the study, a filter containing granular activated carbon was found to remove substantial portions of dissolved organics.

The similarity between the Passaic Valley Water Commission water treatment plant and an advanced wastewater treatment plant was noted by Wendell Inhoffer in 1973, when he wrote: "The 100 mgd [378.5 Ml/d] treatment plant . . . has sometimes been characterized as the tertiary sewage treatment plant for the Passaic Basin. . . ."

Windhoek, South Africa

In this project, the subject of several published papers, reclaimed water is taken directly from maturation (or holding) ponds to the public water system, and comprises as much as one-third of the total municipal water supply.[24-29] The secondary effluent is reclaimed for reuse by chemical (lime) treatment, flotation, ammonia stripping, recarbonation, filtration, foam fractionation, and granular carbon adsorption, as shown in Fig. 4–9.

Tables 4–3 and 4–4 summarize the results. The water supply sources consist of an impoundment reservoir (dam water in Table 4–4), 36 wells (borehole water), and the reclaimed wastewater. The dam water is treated in a conventional plant, while the borehole water is pumped directly to the distribution reservoirs. The reclaimed water is mixed with the purified dam water in the clearwell. The admixed streams are chlorinated to a free residual of 0.2 mg/l and then pumped to the service reservoir, where they are mixed with the borehole water before final distribution. Although enteroviruses and reoviruses entered the reclamation plant at levels as high as the $TCID_{50}$ (50 percent tissue culture infective dose) per liter, no virus could be recovered from the reclaimed wastewater.[28]

WASTEWATER AND GROUNDWATER SUPPLIES

Many of the considerations discussed for surface supplies are applicable to groundwater supplies. The effects of the receiving stream are replaced by the effects on the wastewater of passage through soil above the groundwater.

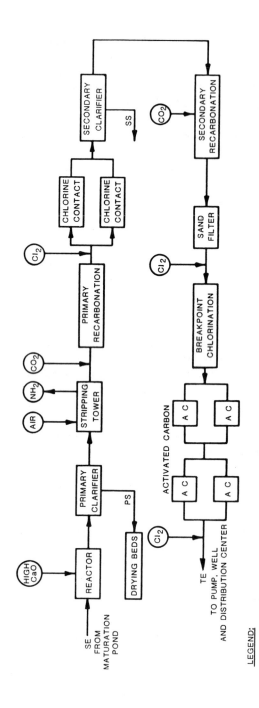

Fig. 4–9. Flow diagram of modified Windhoek water reclamation plant. SE = secondary effluent; TE = tertiary effluent; PS = primary sludge; SS = secondary sludge.[29]

LEGEND:

SE = SECONDARY EFFLUENT   PS = PRIMARY SLUDGE
TE = TERTIARY EFFLUENT   SS = SECONDARY SLUDGE

Table 4-3. Water Quality at Various Stages of Wastewater Treatment at Windhoek.*

DETERMINATION	RAW WATER	PRIMARY FLOTATION	LIME, CHLORINE, AND SEDIMENTATION	SAND FILTRATION	ACTIVATED CARBON FILTRATION
Total N (mg/l)	35	32	15	14	13
Organic N (mg/l)	3.2	1.3	0.9	0.7	Nil
Ammonia N (mg/l), as N or Na_3^-	14.9	14.0	0.2	0.3	0.1
Oxides of N (mg/l)	17	17	14	13	13
Phosphates, as PO_4 (mg/l)	10	Nil	Nil	Nil	Nil
ABS (mg/l)	8	7	4	4	0.7
BOD_5 (mg/l)	30	4	1	1	0.3
Sulfate, as SO_4 (mg/l)	108	228	220	220	220
pH	8.5	7.1	8.0	8.0	8.0

* Reference 27.

A comprehensive review of the subject of groundwater recharge with reclaimed wastewaters has been published, and offers the reader detailed guidance on such systems.[30]

Table 4-4 reflects the quality of blended water actually consumed for potable purposes (recirculated water after integration). The total dissolved solids (TDS) of the reclaimed water increased from ± 740 to 810 mg/l, mainly the result of chemical dosing. Owing to changes in alkalinity and algal density, the aluminum sulfate dosing requirements vary between 130 and 220 mg/l. Depending on the ratio of admixture of reclaimed water with purified raw water (± 200 mg/l), the TDS of the blended water is controlled well below 600 mg/l.

The COD of the maturation pond effluent is reduced by 70 percent up to the sand filter stage, with a substantial further reduction by carbon adsorption. After optimum operation conditions were established, a final-quality reclaimed water could be produced with COD less than 10 mg/l and ABS of less than 0.2 mg/l before blending with purified raw dam water. The carbon chloroform extract (CCE) of the reclaimed water is about 0.13 mg/l. The level of organics present in the recycled water has not caused any customer complaints. In fact, the public complained of the "brackish" taste of the water during a temporary shutdown of the reclamation plant.[24] It is obviously premature to judge the long-term effects of this direct recycling project, but its initial reception by the public has been favorable.

The city is situated in an arid environment. The nearest perennial river is about 500 miles (805 km) away, and obvious alternate sources of water

Table 4-4. Composition of Waters from Individual and Combined Sources at Windhoek.

CONSTITUENT, mg/l	TREATED DAM WATER A	RECLAIMED WATER B	BOREHOLE WATER C	DAM AND RECLAIMED WATER MIXED IN RATIO 1:1 = (A + B)	RETICULATED WATER BEFORE INTEGRATION = (A + C)	RETICULATED WATER AFTER INTEGRATION = (A + B + C)
Total dissolved solids (TDS)	186	740–810	498	525–609	375	475
Phosphate as PO_4	0.1	0.5–1.2	0.03	0.4	0.4	0.6
Nitrate as N	0.3	18.0	0.0	9.0	0.9	3.3
Synthetic detergents as monoxol O.T. (ABS)	—	0.05–0.40	Nil	0.10	—	0.1
Chemical oxygen demand (COD)	18	7.5–28	7.0	7.5–34	8–12	11.0–14.5

are completely absent. These factors no doubt contribute to general public awareness of water and inspire an appreciation of the availability of reclaimed wastewater.[29]

REFERENCES

1. Haney, P. D., "Water Reuse for Public Supply," *J.AWWA*, p. 73, 1969.
2. Swayne, M. D., et al., *Wastewater in Receiving Waters at Water Supply Abstraction Points,* EPA-600/2–80–044, July, 1980.
3. Culp, Wesner, Culp, *Guidance for Planning the Location of Water Supply Intakes Downstream from Municipal Wastewater Treatment Facilities,* U.S. EPA 68–01–4473, April, 1978.
4. Rebhun, M., and Manka, J., "Classification of Organics in Secondary Effluents," *Environmental Science and Technology,* p. 606, 1971.
5. Manka, J., et al., "Characterization of Organics in Secondary Effluents," *Environmental Science and Technology,* p. 1017, November, 1974.
6. Sachdev, D. R., et al., "Apparent Molecular Weights of Organics in Secondary Effluents," *J.WPCF,* p. 570, 1976.
7. Katz, S., et al., "The Determination of Stable Organic Compounds in Waste Effluents at Microgram per Liter Levels by Automatic High Resolution Ion Exchange Chromatography," *Water Research* 6:1029–1037, 1972.
8. Helfgott, T., et al., "Analytic and Process Classification of Effluents," *Proc. ASCE* SA3:779–803, June, 1970.
9. Jolley, R. L., "Chlorination Effects on Organic Constituents in Effluents from Domestic Sanitary Sewage Treatment Plants," Oak Ridge National Lab., Publication No. 565, ORNL-TM-4290, October, 1973.
10. Glaze, W. H., and Henderson, J. E., "Formation of Organochlorine Compounds from the Chlorination of Municipal Secondary Effluent," *J.WPCF,* p. 2511, 1975.
11. "Interim Report on Montgomery Simulation—Study of Formation and Removal of Volatile Chlorinated Organics," unpublished EPA report, July 8, 1975.
12. Hoehn, R. C., et al., "Algae as Sources of Trihalomethane Precursors," *J.AWWA,* p. 344, June, 1980.
13. Craun, G. F., et al., "Water Borne Disease Outbreaks in the U.S., 1971–74," *J.AWWA,* pp. 420–424, August, 1976.
14. Bryan, F. L., "Disease Transmitted by Foods Contaminated with Wastewater," in EPA Document No. 660/2–74–041, pp. 16–45, June, 1974.
15. Manahan, S. E., *Environmental Chemistry,* Willard Grant Press, Boston, Mass., 1972.
16. Stumm, W., and Morgan, J. J., *Aquatic Chemistry,* Wiley-Interscience, New York, 1970.
17. *Trace Metals and Metal–Organic Interactions in Natural Waters,* P. C. Singer, editor, Ann Arbor Science, Ann Arbor, Mich., 1973.
18. *Aqueous Environmental Chemistry of Metals,* A. J. Rubin, editor, Ann Arbor Science, Ann Arbor, Mich., 1974.
19. Turekein, K. K., and Scott, M. R., "Concentrations of Cr, Ag, Mo, Ni, Co, and Mn and Suspended Material in Streams," *Environmental Science and Technology,* pp. 940–942, November, 1967.
20. United States Public Health Service, *Manual of Recommended Water Sanitation Practice,* 1946.
21. Walton, G., "Relation of Treatment Methods to Limits for Coliform Organisms in Raw Waters," *J.AWWA,* pp. 1281–1289, October, 1956.
22. Clarke, N. A., et al., "AWWA Committee Report on Viruses in Water," *J.AWWA,* pp. 491–494, October, 1969.

23. Dale, J. T., "Wastewater Reuse: Exploring the U.S. Potential," *J. WPCF,* p. 440, March, 1979.
24. Van Vuuren, L. R. J., et al., "The Full-Scale Reclamation of Purified Sewage Effluent for the Augmentation of the Domestic Supplies of the City of Windhoek," Fifth International Water Pollution Research Conference, July, 1970.
25. Van Vuuren, L. R. J., et al., "The Flotation of Algae in Water Reclamation," *International Journal Air and Water Pollution,* p. 823, 1965.
26. Cillie, G. G., et al., "The Reclamation of Sewage Effluents for Domestic Use," Third International Conference on Water Pollution Research, 1966.
27. Stander, G. J., et al., "The Reclamation of Wastewater for Domestic and Industrial Use," Biennial Health Congress of the Institute of Public Health, October, 1966.
28. Nupen, E. M., "Virus Studies on the Windhoek Wastewater Reclamation Plant (South-West Africa)," *Water Research,* p. 661, 1970.
29. Van Vuuren, L. R. J., Clayton, A. J., and Van der Post, D. C., "Current Status of Water Reclamation at Windhoek," *J. WPCF,* p. 661, April, 1980.
30. *Health Aspects of Wastewater Recharge, A State-of-the-Art Review,* State of California, Water Information Center, Huntington, N.Y., 1978.
31. Upper Occoquan Sewage Authority (UOSA) Wastewater Treatment Plant, Summaries of Operating Data.

Chapter 5
Water Chemistry

INTRODUCTION

This chapter provides basic information to assist the engineer in understanding the role of water chemistry in determining the cause and effect relationships of water treatment processes. Natural waters are never completely pure. During precipitation and passage over or through the ground, water acquires many kinds of dissolved and suspended impurities. The concentrations of these substances are seldom large, in most instances amounting to a few hundredths of 1 percent or less. Yet they may profoundly modify the chemical properties of water and its usefulness.

The public expects water engineers to produce finished waters that are free of color, turbidity, taste, odor, and harmful metals and organics. In addition, the public desires water that is low in hardness and total solids, is noncorrosive, and is non-scale-forming. To meet these expectations, the engineer must understand and be able to apply water chemistry. Subsequent chapters describe the principles of chemistry associated with various treatment processes and the application of the basic terms and concepts presented in this chapter.

The chemical substances of interest range from dissolved gases through inorganic salts to complex organic materials, both natural and synthetic. This chapter sets forth the principles that form a common basis for the specific chemistry of these materials. Some examples of the general relations in water chemistry are presented.

ELEMENTS AND COMPOUNDS

Elements are substances that cannot be subdivided into simpler substances, and constitute the fundamental constituents of matter. Compounds represent substances formed by the union of one or more elements, the compounds having properties unlike the elements from which they are formed.

Each element differs from any other in weight, size, and chemical properties. Elements common to water engineering are given in Table 5–1, along with their symbols, atomic weights, and common valences.

Atomic weight is the weight of an element relative to that of carbon, which has an atomic weight of exactly 12. The atomic weight expressed in

Table 5-1. Elements Common in Water Engineering.[4]

NAME	SYMBOL	ATOMIC WEIGHT	COMMON VALENCE*	COMBINING WEIGHTS** $\frac{\text{ATOMIC WEIGHT}}{\text{VALENCE}} = $ COMBINING WEIGHT
Aluminum	Al	27.0	3+	9.0
Arsenic	As	74.9	3+	25.0
Barium	Ba	137.3	2+	68.7
Boron	B	10.8	3+	3.6
Bromine	Br	79.9	1−	79.9
Cadmium	Cd	112.4	2+	56.2
Calcium	Ca	40.1	2+	20.0
Carbon	C	12.0	4−	
Chlorine	Cl	35.5	1−	35.5
Chromium	Cr	52.0	3+	17.3
			6+	
Copper	Cu	63.5	2+	31.8
Fluorine	F	19.0	1−	19.0
Hydrogen	H	1.0	1+	1.0
Iodine	I	126.9	1−	126.9
Iron	Fe	55.8	2+	27.9
			3+	
Lead	Pb	207.2	2+	103.6
Magnesium	Mg	24.3	2+	12.2
Manganese	Mn	54.9	2+	27.5
			4+	
			7+	
Mercury	Hg	200.6	2+	100.3
Nickel	Ni	58.7	2+	29.4
Nitrogen	N	14.0	3−	
			5+	
Oxygen	O	16.0	2−	8.0
Phosphorus	P	31.0	5+	6.0
Potassium	K	39.1	1+	39.1
Selenium	Se	79.0	6+	13.1
Silicon	Si	28.1	4+	6.5
Silver	Ag	107.9	1+	107.9
Sodium	Na	23.0	1+	23.0
Sulfur	S	32.1	2−	16.0
Zinc	Zn	65.4	2+	32.7

* Valence is based upon the combining value of the hydrogen atom, which is assigned the unit value of 1. The valence of any other atom represents the number of atoms of hydrogen that will be replaced by it, or that are equivalent to it in combining value. Thus, an atom with a valence of 2 will replace two hydrogen atoms, or react with two hydrogen atoms.
** In ordinary chemical work, the atomic and combining weights are expressed as whole numbers, rather than using a decimal place as shown in this table for fundamental reference. Thus the atomic weight of cadmium used in ordinary chemical work would be 112, and the combining weight would be 56.

grams is called 1 *gram atomic weight* of the element. For example, 1 gram atomic weight of chlorine is 35.5 grams.

Elements can combine to form molecules. The molecular weight of these compounds equals the sum of the weights of the combined elements. It is termed the *gram molecular weight,* also referred to as a *mole.* The molecular weight of sodium chloride (NaCl) is 58.5 grams (23.0 + 35.5), while 1 mole of ammonia (NH_3) is 17.0 grams (14.0 + 3 × 1).

The chief significance of a mole is in the preparation of *molar* or *molal* solutions. A molar solution consists of 1 gram molecular weight dissolved in enough water to make 1 liter of solution. A molal solution consists of 1 gram molecular weight dissolved in 1 liter of water. The molal solution will have a volume slightly in excess of 1 liter.

Valence

Atoms combine with other atoms in definite ratios. For example, if hydrogen gas is burned in an atmosphere of gaseous oxygen, two atoms of hydrogen (no more or less) will combine with one atom of oxygen to form one molecule of water vapor, as indicated by the following simple relationship: HH + O = H_2O. This will occur irrespective of whether hydrogen or oxygen is present in quantities greater than required to react with the other substance. The combining ratios of the various elements or atoms are simple values, such as 1:1, 1:2, 1:3, and so on.

One element, hydrogen, has been selected as a reference substance. The combining or reaction value of hydrogen has been given a value of 1. Any atom that will combine with one atom of hydrogen has the same value, 1. The atom of chlorine has a combining value of 1 because it combines with one atom of hydrogen to make one molecule of hydrochloric acid (H + Cl = HCl).

One atom of oxygen will combine with two atoms of hydrogen; hence the oxygen atom has a combining value of 2. Furthermore, one atom of nitrogen will combine with three atoms of hydrogen to make a molecule of ammonia (N + 3H = NH_3); hence nitrogen has a combining value of 3. These combining values have been designated the valence, which can simply be thought of as the reaction or combining value of the atoms.

Unfortunately, some atoms have a property of combining with other atoms in more than one definite ratio; so these atoms have more than one valence. Carbon has valences of 2 and 4: one carbon atom will react with one oxygen atom to form carbon monoxide (C + O = CO), or with two oxygen atoms to form carbon dioxide (C + 2O = CO_2). Likewise, iron has valences of 2 and 3, as indicated by the two compounds, ferrous chloride ($FeCl_2$) and ferric chloride ($FeCl_3$), or two other compounds, ferrous carbonate ($FeCO_3$) and ferric carbonate ($Fe_2(CO_3)_3$).

Combining Weights

Valence determines the manner in which atoms will combine to form a molecule. The weight of the molecule equals the sum of the weights of the atoms constituting the molecule. For example, the molecular weight of sodium chloride equals the sum of the atomic weights of sodium and chlorine (23 + 35 = 58). Thus, 23 pounds of sodium will react with 35 pounds of chlorine to produce 58 pounds of salt. If, however, atoms of higher valence than 1 are involved, such as calcium with a valence of 2, the combining weights are not equal to the atomic weights, but are equal to the atomic weights divided by the valence. Thus, when calcium (atomic weight 40) combines with chlorine (atomic weight 35) to form calcium chloride ($CaCl_2$, molecular weight 110), the ratio of the weights of the two constituents is not 40:35, but 40:70 (2 × 35). This ratio is equivalent to 20:35. The value 20 for calcium equals the atomic weight of 40 divided by the valence of 2 of calcium, whereas the value of 35 for chlorine equals the atomic weight of 35 divided by the valence of 1 of chlorine. The general rule, therefore, is to divide the atomic weight of an element by its valence to secure the combining or equivalent weight.

Ions

Many substances can be dissolved in liquids, such as sugar, salt, or air in water. The molecules of the dissolved substances become dispersed in the water or other solvents in a manner characteristic of the various substances.

The dissolved substances may be subdivided into two distinct classes, electrolytes and nonelectrolytes. Electrolytes are substances that when dissolved in water become electrically charged, whereas nonelectrolytes do not exhibit this property.[5] For example, when salt is dissolved in water, the molecules dissociate into positively charged sodium atoms and negatively charged chlorine atoms, the electric charges of which are equal and thus neutralize each other ($NaCl = Na^+ + Cl^-$). However, when sugar is dissolved in water, no dissociation into electrically charged atoms occurs, but the sugar molecule remains unchanged. Sugar, therefore, is a nonelectrolyte. The electrically charged atoms formed by the dissociation of dissolved substances are called ions, and the breaking down of compounds into their constituent ions is known as ionization.

The electric charge of the ions is important because it determines the combining characteristics. Because like electric charges repel each other, positive ions will not combine with positive ions, but only with negative ions. Thus positive sodium ions will not combine with positive calcium ions but will combine with negative chlorine ions.

Ions also may consist of groups of atoms having an unbalanced electrical charge. For example, sulfuric acid, when added to water, ionizes into the positive hydrogen ion H and the negative sulfate group SO_4 ($H_2SO_4 = 2H^{+2} + SO_4^{-2}$).

Ions formed by groups of atoms, such as SO_4 or OH, have an electrical charge because one atom of the group has a charge greater than that of the other atom or atoms, with the resulting charge of the group equaling this unbalanced charge.

For example, the OH group has an unbalanced, negative electrical charge of 1 because the twofold negative charge of oxygen is only partly neutralized to one-half its value by the single positive charge of hydrogen ($O^{-2} + H^+ = OH^-$). If, however, the single negative charge of the OH group is neutralized by a single positive charge of another hydrogen ion, the opposing charges are balanced, and neutral water is formed ($H + OH = HOH = H_2O$). Table 5–2 lists several of the ions commonly encountered in water treatment and their characteristics.

Neutralization

When an acid is added to a base, neutralization occurs. The neutralization reaction results in the positive metal ion of the base combining with a negative ion of the acid to form a neutral salt, and the hydrogen ion of the acid combining with the OH ion of the base to form water. The neutralization reaction is illustrated by the reaction of sodium hydroxide with sulfuric acid to form a salt, sodium sulfate, and water:

Table 5–2. Common Ions Encountered in Water.

NAME	FORMULA	MOLECULAR WEIGHT	COMMON VALENCE	COMBINING WEIGHT
Ammonium	NH_4^+	18.0	1+	18.0
Hydroxyl	OH^-	17.0	1−	17.0
Bicarbonate	HCO_3^-	61.0	1−	61.0
Carbonate	CO_3^{-2}	60.0	2−	30.0
Orthophosphate	PO_4^{-2}	95.0	3−	31.7
Orthophosphate, mono-hydrogen	HPO_4^{-2}	96.0	2−	48.0
Orthophosphate, di-hydrogen	$H_2PO_4^-$	97.0	1−	97.0
Bisulfate	HSO_4^-	97.0	1−	97.0
Sulfate	SO_4^{-2}	96.0	2−	48.0
Bisulfite	HSO_3^-	81.0	1−	81.0
Sulfite	SO_3^{-2}	80.0	2−	40.0
Nitrite	NO_2^-	46.0	1−	46.0
Nitrate	NO_3^-	62.0	1−	62.0
Hypochlorite	OCl^-	51.5	1−	51.5

$$2NaOH + H_2SO_4 = Na_2SO_4 + 2H_2O \qquad (5-1)$$

The salts resulting from the neutralization of an acid by a base are neutral when a strong acid is neutralized by a strong base. If, however, a strong base, such as caustic soda, is neutralized by a weak acid, such as carbonic acid, the resulting salt, sodium carbonate, has an alkaline or basic reaction. If a weak base is neutralized by a strong acid, the resulting salt has an acid reaction. Thus, ferric chloride and ferrous sulfate, when used as coagulants for water, have an acid reaction.

Hydrolysis

The effect of water on certain substances is the disintegration of the substance through hydrolysis. For example, when sodium bicarbonate is dissolved in water, it hydrolyzes to sodium hydroxide and carbonic acid. Sodium hydroxide is a strong base, while carbonic acid is a weak acid; so the basic properties predominate:

$$NaHCO_3 + HOH \rightleftarrows HaOH + H_2CO_3 \qquad (5-2)$$

If, however, ferric chloride is dissolved in water, it is hydrolyzed to ferric hydroxide and hydrochloric acid, the acid in this case being stronger than the base so that the acid characteristic predominates:

$$FeCl_3 + 3HOH \rightleftarrows Fe(OH)_3 + 3HCl \qquad (5-3)$$

Hydrolysis of carbonates by water is responsible for the alkalinity of natural waters, a subject discussed later in this chapter.

Normality

A *normal* solution contains 1 gram equivalent weight of a substance, either an ion or a compound, per liter of solution. Equivalent weight is the weight of a compound that contains 1 gram atom of available hydrogen or its chemical equivalent. For all practical purposes, it can be calculated by dividing the molecular weight of the compound by the valence number of its ions.

For example, the molecular weight of hydrochloric acid is 36.5, the sum of the atomic weights of hydrogen (1) and chlorine (35.5). The gram equivalent weight of hydrochloric acid, then, is 36.5 grams. A normal solution of hydrochloric acid consists of 36.5 grams of the concentrated acid diluted to 1 liter with distilled water.

If, however, a substance has two replaceable hydrogen atoms (valence of 2), such as sulfuric acid, the gram equivalent weight is one-half of its molecular weight, which in the case of sulfuric acid is one-half of 98, or 49 [½ (2 × 1 + 32 + 4 × 16)].

Standard solutions formed by dissolving a gram equivalent weight of various substances in a liter of distilled water are all equal in their reacting value, volume for volume. That is, 10 ml of normal hydrochloric acid solution will react with, and completely neutralize, 10 ml of normal sodium hydroxide solution. Normality is important in the analytical determination of water constituents. In water analysis, it is desirable to report results in terms of milligrams per liter of some particular ion, element, or compound. It is convenient to have the titrating agent of such strength that 1 ml is equal to 1 mg of the material being measured.

A related quantity is the milliequivalent (meq), equal to 1/1,000 the gram-equivalent weight. The normality of a solution multiplied by its volume in milliliters gives the number of milliequivalents of solute. Concentrations in milliequivalents per liter (meq/l) are obtained simply by multiplying the normality by 10^3.

SOLUBILITY

Gases, liquids, and solids may all dissolve in water to form true solutions. The amount of solute present may vary continuously below a certain limit, called the "solubility." Solubility is the concentration of solute present when the solution is in a state of equilibrium with an excess of the pure solute, and is a fixed number for a given temperature, pressure, and solvent. This equilibrium and its application are important in water engineering.

Liquid and Solid Solubility

In general, the solubility of solids and liquids depends only slightly on pressure. In most water engineering situations, solubility may be considered as a function of temperature alone. Exceptions are situations where extreme pressures exist, such as deep underground aquifers, great ocean depths, or certain industrial applications. The solubilities of most inorganic salts increase with temperature, but a number of calcium compounds, such as $CaCO_3$, $CaSO_4$, and $Ca(OH)_2$, decrease in solubility with increasing temperature. The activity of a solid in regard to solubility is constant at a given temperature, and this activity is stated mathematically by the "solubility product" (K_{sp}), which is expressed as $K_{sp} = [A^+] [B^-]$ where $[A^+][B^-]$ represents the product of the molar concentrations of the ions making up the substance. The solute concentrations (i.e., $[A^+]$) are usually expressed in moles per liter:

$$\text{molarity} = \frac{\text{milligrams per liter}}{\text{gram-molecular weight} \times 10^3}$$

Solubility products may be obtained by reference to chemical textbooks or handbooks. Typical K_{sp} constants of interest in water engineering are listed in Tables 5–3 and 5–4. In an unsaturated solution, the ion product ($[A^+][B^-]$) is less than K_{sp}. If the ion product is greater than K_{sp}, the solution is supersaturated and will tend to form a precipitate.

Other ions present can affect the solubility of a substance through either the common-ion effect or the secondary salt effect. For example, in a solution containing 100 ppm of carbonate alkalinity, $CaCO_3$ has a solubility of only 0.5 ppm, although its solubility in pure water is about 13 ppm. This repression of solubility in the presence of an excess of one of the ions concerned in the solubility expression is known as the common-ion effect. The solubility of slightly soluble salts is increased when other salts that do not have an ion in common with the slightly soluble substance are present. The increased ionic strength of the solution resulting from the foreign salt causes a decrease in the activity coefficients of the slightly soluble substance. For example,

Table 5–3. Typical Solubility-Product Constants.*

EQUILIBRIUM EQUATION	K_{sp} AT 25°C**	SIGNIFICANCE IN WATER ENGINEERING
$MgCO_3 \rightleftharpoons Mg^{+2} + CO_3^{-2}$	4×10^{-5}	Hardness removal, scaling
$Mg(OH)_2 \rightleftharpoons Mg^{+2} + 2OH^-$	9×10^{-12}	Hardness removal, scaling
$CaCO_3 \rightleftharpoons Ca^{+2} + CO_3^{-2}$	5×10^{-9}	Hardness removal, scaling
$Ca(OH)_2 \rightleftharpoons Ca^{+2} + 2OH^-$	8×10^{-6}	Hardness removal
$Cu(OH)_2 \rightleftharpoons Cu^{+2} + 2OH^-$	2×10^{-19}	Heavy metal removal
$Zn(OH)_2 \rightleftharpoons Zn^{+2} + 2OH^-$	3×10^{-17}	Heavy metal removal
$Zn(OH)_2 \rightleftharpoons 2H^+ + ZnO_2^{-2}$	1×10^{-29}	Heavy metal removal
$Ni(OH)_2 \rightleftharpoons Ni^{+2} + 2OH^-$	2×10^{-16}	Heavy metal removal
$Cr(OH)_3 \rightleftharpoons Cr^{+3} + 3OH^-$	6×10^{-31}	Heavy metal removal
$Cr(OH)_3 \rightleftharpoons CrO_2^- + H^+ + H_2O$	9×10^{-17}	Heavy metal removal
$Al(OH)_3 \rightleftharpoons Al^{+3} + 3OH^-$	1×10^{-32}	Coagulation
$Al(OH)_3 \rightleftharpoons H^+ + AlO_2^- + H_2O$	4×10^{-13}	Coagulation
$Fe(OH)_3 \rightleftharpoons Fe^{+3} + 3OH^-$	6×10^{-38}	Coagulation, iron removal, corrosion
$Fe(OH)_2 \rightleftharpoons Fe^{+2} + 2OH^-$	5×10^{-15}	Coagulation, iron removal, corrosion
$Mn(OH)_3 \rightleftharpoons Mn^{+3} + 3OH^-$	1×10^{-36}	Manganese removal
$Mn(OH)_2 \rightleftharpoons Mn^{+2} + 2OH^-$	8×10^{-14}	Manganese removal
$Ca_3(PO_4)_2 \rightleftharpoons 3Ca^{+2} + 2PO_4^{-3}$	1×10^{-27}	Phosphate removal
$CaHPO_4 \rightleftharpoons Ca^{+2} + HPO_4^{-2}$	3×10^{-7}	Phosphate removal
$CaF_2 \rightleftharpoons Ca^{+2} + 2F^-$	3×10^{-11}	Fluoridation
$AgCl \rightleftharpoons Ag^+ + Cl^-$	3×10^{-10}	Chloride analysis
$BaSO_4 \rightleftharpoons Ba^{+2} + SO_4^{-2}$	1×10^{-10}	Sulfate analysis

* Reference 2. Reproduced by permission of McGraw-Hill Book Co.
** Conversion factors may be found in Appendix A.

Table 5–4. Additional Solubility-Product Constants.*

COMPOUND	K_{sp} AT $25°C$**
Carbonates	
$NiCO_3$	1.4×10^{-7}
$MnCO_3$	4.0×10^{-10}
$CuCO_3$	2.5×10^{-10}
$FeCO_3$	2.0×10^{-11}
$ZnCO_3$	3.0×10^{-11}
$CdCO_3$	5.2×10^{-12}
$PbCO_3$	1.5×10^{-13}
Hydroxides	
$Cd(OH)_2$	2.0×10^{-14}
$Pb(OH)_2$	4.2×10^{-15}
Phosphates	
$AlPO_4$	6.3×10^{-19}
$Mn_3(PO_4)_2$	1.0×10^{-22}
$Mg_3(PO_4)_2$	1.0×10^{-32}
$Pb_3(PO_4)_2$	1.0×10^{-32}
Sulfates	
$CaSO_4$	2.5×10^{-5}
$PbSO_4$	1.3×10^{-8}
Sulfides	
MnS	7.0×10^{-16}
FeS	4.0×10^{-19}
Fe_2S_3	1.0×10^{-88}

* Reference 3.
** Conversion factors may be found in Appendix A.

calcium carbonate is several times as soluble in seawater as in fresh water as a result of the secondary salt effect.

Table 5–5 (p. 149) summarizes solubility information for very soluble compounds.

Gas Solubility

The principal gases of concern for the water engineer are nitrogen, oxygen, and carbon dioxide. Nitrogen and oxygen gas do not react with water, and their concentrations can be determined by Henry's law. Carbon dioxide does react with water to form carbonic acid (H_2CO_3), which ionizes to product H^+, bicarbonate (HCO_3^-), and carbonate (CO_3^{-2}) ions. Consequently, carbon dioxide solubility is greater than that determined by Henry's law. Carbon dioxide solubility in water is discussed later.

Henry's law states that the weight of any gas that will dissolve in a given volume of liquid, at constant temperature, is directly proportional to the pressure that the gas exerts above the liquid. In equation form:

$$C_{eq} = HP_{gas} \tag{5-4}$$

where:

C_{eq} = the equilibrium gas solubility, ml/l
P_{gas} = the partial pressure of the gas above the liquid
H = the Henry's law constant (or absorption coefficient) for the gas at the given temperature, ml/l

Table 5–6 presents absorption coefficients (k_s) for several gases important in water treatment. The coefficient in Table 5–6 can be used, for example, to determine the concentration of oxygen in pure water exposed to air at 0°C and a barometric pressure of 760 mm.[3]

From Table 5–6, k_s = 49.3 ml/l. Because dry air normally includes 20.99 percent of oxygen by volume, and it must be assumed that the air in contact with the water is saturated with water vapor, the partial pressure of the oxygen is 0.2099 (760 − 4.58) = 159 mm Hg. The value of 4.58 is the vapor pressure of water at 0°C, from Table C–3. The volume concentration of oxygen, therefore, is 49.3 × 159/760 = 10.3 ml/l; and because 1 ml of oxygen weighs 2 × 16 × 10^{-3}/22.412 = 1.43 mg, the weight concentration of oxygen is 1.43 × 10.3 = 14.7 mg/l.

Table 5–7 presents solubility data for oxygen under the same conditions but at other temperatures. The significant effect of temperature on gas solubility is evident from Table 5–7. A more recent study yielded the following equation for oxygen solubility:[8]

$$C_p = C_s (P - p) \tag{5-5}$$

where:

C_p = oxygen saturation concentration in water at a given barometric pressure, mg/l
C_s = 14.652 + 10.53 [exp(−0.03896 T) − 1]
= saturation dissolved oxygen concentration at atmospheric pressure, mg/l
T = water temperature, °C
P = barometric pressure, atmospheres
p = partial pressure of water vapor in the air, atmospheres

pH

The pH is an expression of the intensity of the acid or base condition of a solution. Acids are substances that dissociate to yield hydrogen ions (H^+), and bases dissociate to yield hydroxyl ions (OH^-). pH is a way of expressing the hydrogen ion activity.

Pure water dissociates as follows:

$$H_2O = H^+ + OH^- \qquad (5\text{--}6)$$

and yields a hydrogen ion concentration equal to approximately 10^{-7} g/l. For convenience, pH is defined as:

Table 5–5. Solubility of Soluble Compounds.

COMPOUND	SOLUBILITY AT 20°C* PERCENT BY WEIGHT
Bicarbonate	
$NaHCO_3$	8.8
$Ca(HCO_3)_2$	14.2
$KHCO_3$	24.9
Carbonate	
$Na_2CO_3 \cdot 10H_2O$	17.7
$K_2CO_3 \cdot 2H_2O$	52.5
Chloride	
HCl	41.9
$NaCl$	26.5
KCl	25.4
$MgCl_2 \cdot 6H_2O$	35.3
$CaCl_2 \cdot 6H_2O$	42.7
$FeCl_2 \cdot 4H_2O$	40.7
$FeCl_3$	47.9
$BaCl_2$	37.5
Hydroxide	
$NaOH \cdot 1H_2O$	52.2
$KOH \cdot 2H_2O$	52.8
$Ca(OH)_2$	0.17
Sulfate	
H_2SO_4	Infinitely soluble
$Na_2SO_4 \cdot 7H_2O$	30.6
K_2SO_4	10.0
$MgSO_4 \cdot 7H_2O$	30.8
$CaSO_4 \cdot 2H_2O$	0.20
$FeSO_4 \cdot 7H_2O$	20.9
$Al_2(SO_4)_3 \cdot 18H_2O$	26.7

* Conversion factors may be found in Appendix A.

Table 5–6. Absorption Coefficients (k_s) of Common Gases in Water*
(milliliters of gas, reduced to 0°C and 760 mm Hg, per liter of water when partial pressure of gas is 760 mm Hg).

GAS	MOLECULAR WEIGHT	WEIGHT AT 0°C AND 760 mm HG g/l	k_s TEMPERATURE IN °C**				BOILING POINT, °C
			0°C	10°C	20°C	30°C	
Hydrogen, H_2	2.016	0.09004	21.5	19.6	18.2	17.0	−253
Methane, CH_4	16.03	0.7160	55.6	41.8	33.1	27.6	−165
Nitrogen, N_2	28.02	1.251	23.0	18.5	15.5	13.6	−195
Oxygen, O_2	32.00	1.429	49.3	38.4	31.4	26.7	−183
Ammonia, NH_3	17.03	0.7706	1,300	910	711	—	−38.5
Hydrogen sulfide, H_2S	34.08	1.523	4,690	3,520	2,670	—	−60
Carbon dioxide, CO_2	44.00	1.977	1,710	1,190	878	665	−80
Ozone, O_3	48.00	2.144	641	520	368	233	−119
Sulfur dioxide, SO_2	64.06	2.927	79,800	56,600	39,700	27,200	−10
Chlorine, Cl_2	70.91	3.167	4,610	3,100	2,260	1,770	−33.6
Air†	—	1.2928	28.8	22.6	18.7	16.1	—

* Reference 3.
** Conversion factors may be found in Appendix A.
† At sea level dry air contains 78.03% N_2, 20.99% O_2, 0.94% A, 0.03% CO_2, and 0.01% other gases by volume. For ordinary purposes it is assumed to be composed of 79% N_2 and 21% O_2.

$$\text{pH value} = \text{logarithm} \ \frac{1}{\text{hydrogen ion concentration}} \qquad (5\text{--}7)$$

Thus, for pure, neutral water, the pH is the logarithm of $1/10^{-7}$ or 7.

The pH scale ranges from 0 (extreme acidity) to 14 (extreme basic condi-

Table 5–7. Solubility of Oxygen in Fresh Water at the Temperatures Noted When Exposed to the Atmosphere Containing 20.9% Oxygen under a Barometric Pressure of 760 mm Hg.*

TEMPERATURE OF WATER, °C**	mg/l DISSOLVED OXYGEN
0	14.62
1	14.23
2	13.84
3	13.48
4	13.13
5	12.80
6	12.48
7	12.17
8	11.87
9	11.59
10	11.33
11	11.08
12	10.83
13	10.60
14	10.37
15	10.15
16	9.95
17	9.74
18	9.54
19	9.35
20	9.17
21	8.99
22	8.83
23	8.68
24	8.53
25	8.38
26	8.22
27	8.07
28	7.92
29	7.77
30	7.63

* Reference 1.
** Conversion factors may be found in Appendix A.

tions), with 7 representing neutrality. Values less than pH 7.0, therefore, represent acid characteristics, while values greater than pH 7.0 represent basic or alkaline characteristics. It should be borne in mind that decreasing pH values represents increasing concentrations of hydrogen ions. A water with a pH of 6.0 has ten times the hydrogen ion concentration of neutral water, with a pH of 7.0. Natural waters, however, seldom have pH values less than 5.5 to 6.0. Most soft waters have pH values between 6.0 and 7.0, with the values below 7.0 due to the presence of carbon dioxide and other acid constituents.

The more alkaline waters have pH values between 7.0 and 8.0. For instance, the waters of the Great Lakes have pH values of about 7.8. As evident in later discussions, pH values can be elevated or decreased significantly by treatment processes.

ALKALINITY AND ACIDITY

Alkalinity

Alkalinity, the measure of the capacity of a water to neutralize acids, is due primarily to the presence of the carbonates of potassium, sodium, calcium, and magnesium. A small portion of the alkalinity of some waters may be due to silicates, borates, and phosphates, although the latter are of little practical significance, except in highly mineralized waters. Ordinarily the alkalinity of water is due to calcium and magnesium bicarbonates alone, so that the alkalinity usually equals the carbonate hardness. The exception occurs when potassium and sodium bicarbonates contribute to the alkalinity without influencing the hardness, in which case the alkalinity will exceed the hardness.

The alkalinity produced by carbonates is due to their hydrolysis into the hydroxide and carbonic acid, the alkalinity of the hydroxide overbalancing the acidity of the weak carbonic acid. The following reaction indicates the hydrolysis of sodium carbonate to form caustic soda or sodium hydroxide and carbonic acid:

$$Na_2CO_3 + 2HOH \rightleftharpoons 2NaOH + H_2CO_3 \qquad (5–8)$$

Other carbonates hydrolyze in the same manner.

The alkalinity of a water is measured by titration of a sample with a strong acid solution, whereas acidity (see below) is measured by titration of a sample with a strong basic solution containing carbonic acid species. When acid is added to a strong basic solution, the majority of the hydrogen ions from the acid combine with the hydroxide ions of the base until the pH is about 10. This is the *carbonate equivalence point.* Then excess hydrogen ions lower the pH gradually until at pH 8.3 all carbonate ions have been converted to bicarbonates. This is the *bicarbonate equivalence point.* Additional hydrogen ions reduce the bicarbonates to carbonic acid below pH 4.5, which is the *carbonic acid equivalence point.*

Each of these equivalence points can be measured by the use of either a pH meter or indicators. The indicators commonly used are phenolphthalein, which turns from pink to colorless at pH 8.3, and methyl orange, which changes from orange to pink around pH 4.3. These equivalence points are used to determine three types of alkalinity:

1. *Caustic alkalinity:* The initial solution pH lies above the CO_3^{-2} equivalence point. The equivalents of acid required to lower the pH to the CO_3^{-2} equilvalence (pH = 10) are a measure of the caustic alkalinity.
2. *Phenolphthalein alkalinity:* The initial solution pH lies above the HCO_3^- equivalence point. The equivalents of acid required to lower the pH to the HCO_3^- equivalence point (pH = 8.3) are a measure of the phenolphthalein alkalinity.
3. *Total alkalinity:* The initial solution pH lies above the H_2CO_3 equivalence point. The equivalents of acids required to lower the pH to the H_2CO_3 equivalence point (pH = 4.5) are a measure of the total alkalinity.

Acidity

The three different types of acidity are determined at the same equivalence points as the alkalinities. However, acidity is a measure of the resistance to a pH increase.

1. *Mineral acidity:* The initial solution pH lies below the H_2CO_3 equivalence point. The equivalents of base required to raise the pH to the H_2CO_3 equivalence point (pH = 4.5) are a measure of the mineral acidity.
2. *CO_2 acidity:* The initial solution pH lies below the HCO_3^- equivalence point. The equivalents of base required to raise the pH to the HCO_3^- equivalence (pH = 8.3) are a measure of the CO_2 acidity.
3. *Total acidity:* The initial solution pH lies below the CO_3^{-2} equivalence point. The equivalents of base required to raise the pH to the CO_3^{-2} equivalence point (pH = 10) are a measure of the total acidity.

OXIDATION—REDUCTION REACTIONS

A chemical substance is oxidized when it loses electrons to a second substance that is reduced as it acquires the transferred electrons. Oxidation indicates an increase in valence of the substance being oxidized because of the loss of negatively charged electrons; reduction involves a decrease in valence. For example, ferric carbonate is reduced to ferrous carbonate through the lowering of the valence of the iron. The suffix "ous," as in ferrous, indicates the reduced state of a compound, whereas the suffix "ic," as in ferric, indicates the oxidized state of a compound.

Ferrous iron is oxidized when it is changed to ferric iron according to the equation:

$$Fe^{+2} \rightarrow Fe^{+3} + e^- \qquad (5-9)$$

where e^- = electron.

Similarly, the chloride ion is oxidized when it is changed to chlorine gas, as in the equation:

$$2Cl^- \rightarrow Cl_2 + 2e^- \qquad (5-10)$$

Equations such as (5-9) and (5-10) are called half-equations or partial equations because each represents only half of the chemical process that takes place. These partial equations are also called half-cell equations because they represent the reaction taking place in one-half of a suitable electrolytic cell while a complementary reaction is taking place in the other half of the cell at the same time. The above partial equations show that the iron and the chlorine have become more positive (less negative) as a result of a loss of electrons; that is, they have been oxidized. Reduction means simply a reversal of the change indicated, Fe^{+3} going to Fe^{+2}, and Cl_2 to Cl^-.

An oxidizing agent is an element or compound that has the power to increase the positive valence number of some other element. It contains some atom that can gain electrons. Oxidizing agents used in water treatment include:[7]

- Oxygen or air
- Ozone
- Hydrogen peroxide (very limited use)
- Potassium permanganate
- Chlorine (or hypochlorites)
- Chlorine dioxide

Specific applications of these oxidants are discussed in other chapters of this text. Selection of the most effective oxidant depends on economics, availability, ease of handling, and its ability to oxidize the constituent of concern, preferably at a high rate. The oxidizing ability of a reagent is generally expressed in terms of the reversible oxidation (or reduction) potential. The standard electrode potential, E^0, is defined as the potential of a half cell at 25°C in which all the ions are at unit activity and all gases are at partial pressures of one atmosphere.[7] Standard electrode potentials for a variety of redox reactions of interest in water and wastewater treatment are given in Table 5-8. It should be noted that the potentials are listed as oxidation

Table 5–8. Standard Oxidation Potentials at 25°C.*

	ACIDIC	E^0 (V)	BASIC
Oxygen-Related Couples:			
$2H_2O = O_2 + 4H^+ + 4e^-$	−1.229		
$4OH^- = O_2 + 2H_2O + 4e^-$			−0.401
$2H_2O = O_2 + 4H^+$ (pH 7) $+ 4e^-$		(−0.815)	
$H_2O = O(g) + 2H^+ + 2e^-$	−2.42		
$2OH^- = O(g) + H_2O + 2e^-$			−1.59
$H_2O_2 = O_2 + 2H^+ + 2e^-$	−0.682		
$OH^- + HO_2^- = O_2 + H_2O + 2e^-$			+0.076
$3OH^- = HO_2^- + H_2O + 2e^-$			−0.88
$2H_2O = H_2O_2 + 2H^+ + 2e^-$	−1.77		
$O_2 + H_2O = O_3 + 2H^+ + 2e^-$	−2.07		
$O_2 + 2OH^- = O_3 + H_2O + 2e^-$			−1.24
$H_2O_2 = HO_2 + H^+ + e^-$	−1.50		
$OH^- + HO_2^- = O_2^- + H_2O + e^-$			−0.4
$2H_2O = HO_2 + 3H^+ + 3e^-$	−1.70		
$4OH^- = O_2^- + 2H_2O + 3e^-$			−0.7
$HO_2 = O_2 + H^+ + e^-$	+0.13		
$O_2^- = O_2 + e^-$			+0.56
$H_2O = OH + H^+ + e^-$	−2.80		
$OH^- = OH + e^-$			−2.0
$OH + H_2O = H_2O_2 + H^+ + e^-$	−0.72		
$OH + 2OH^- = HO_2^- + H_2O + e^-$			+0.24
Chlorine-Related Couples:			
$2Cl^- = Cl_2 + 2e^-$	−1.36		
$Cl^- + 2OH^- = ClO^- + H_2O + 2e^-$			−0.89
$1/2Cl_2 + H_2O = HClO + H^+ + e^-$	−1.63		
$Cl^- + H_2O = HClO + H^+ + 2e^-$	−1.49		
$HClO + H_2O = HClO_2 + 2H^+ + 2e^-$	−1.64		
$ClO^- + 2OH^- = ClO_2^- + H_2O + 2e^-$			−0.66
$HClO_2 = ClO_2 + H^+ + e^-$	−1.275		
$ClO_2^- = ClO_2 + e^-$			−1.16
$ClO_2 + H_2O = ClO_3^- + 2H^+ + e^-$	−1.15		
$ClO_2 + 2OH^- = ClO_3^- + H_2O + e^-$			+0.50
Manganese-Related Couples:			
$Mn^{2+} = Mn^{3+} + e^-$	~+1.5		
$Mn^{2+} + 2H_2O = MnO_2(s) + 4H^+ + 2e^-$	−1.23		
$Mn(OH)_2 + 2OH^- = MnO_2(s) + 2H_2O + 2e^-$			+0.05
$MnO_2(s) + 4OH^- = MnO_4^- + 2H_2O + 3e^-$			−0.588
$MnO_2(s) + 2H_2O = MnO_4^- + 4H^+ + 3e^-$	−1.695		
$MnO_4^{2-} = MnO_4^- + e^-$	−0.564		
$MnO_2(s) + 2H_2O = MnO_4^{2-} + 4H^+ + 2e^-$	−2.26		
$MnO_2(s) + 4OH^- = MnO_4^{2-} + 2H_2O + 2e^-$			−0.60
$4H_2O + Mn^{2+} = MnO_4^- + 8H^+ + 5e^-$	−1.51		

* Reference 7. Reprinted by permission of John Wiley & Sons, Inc.

potentials; that is, the half-cell reactions are formulated as oxidation reactions. In this form, the stronger the oxidizing agent, the more negative is the electrode potential. The signs of these potentials change if the equations are written as reduction reactions.

Some considerations relative to oxidation–reduction reactions can be illustrated by considering reactions associated with an oxidant commonly used in water treatment, potassium permanganate ($KMnO_4$). In potassium permanganate, manganese has a valence of positive 7 if oxygen is negative 2, as indicated by the formula: $K^+Mn^{+7}(O_4^{-2})$.

Potassium permanganate acts as an oxidizing agent in one of two ways:

1. In neutral or alkaline solution it changes to MnO_2, and the valence of the manganese is positive 4. The decrease in valence for the manganese in one molecule of $KMnO_4$ in neutral solution is, therefore, from 7 to 4, a decrease of 3. Potassium permanganate in neutral or alkaline solution is, consequently, a trivalent oxidizing agent.

2. In acid solution, potassium permanganate, acting as an oxidizing agent, changes to Mn^{+2} (as, for example, in $MnSO_4$). The valence of the manganese is positive 2. The decrease in valence for manganese in one molecule of $KMnO_4$ in acid solution is from 7 to 2, a decrease of 5. Potassium permanganate in acid solution is a pentavalent oxidizing agent, and its equivalent weight as an oxidizer is one-fifth of its molecular weight. When it reacts with ferrous iron, the proportions are $1 KMnO_4$ to $5Fe^{+2}$.

In acid solution, the partial equation for the permanganate ion is:

$$MnO_4^- + 8H^+ + 5e^- \rightarrow Mn^{+2} + 4H_2O \qquad (5\text{--}11)$$

The stoichometry of the reaction can be represented as:[7]

$$MnO_4^- + 3Fe^{2+} + 2H_2O + 5OH^- \rightleftharpoons MnO_2 + 3Fe(OH)_3 \quad (5\text{--}12)$$

CORROSION

Corrosion is the tendency of all metals, when exposed to the elements, to revert to the original sources or more stable forms in which they are found in the earth. It is a process or processes by which metals wear away. The products of corrosion are usually in the form of oxides, carbonates, and sulfides. The phenomena of corrosion are the same for all metals and alloys, differing only in degree, but not in kind. A water that exhibits corrosiveness can cause problems in distribution pipelines and home plumbing systems. These problems can include health, aesthetic, and economic effects.

Electrochemical corrosion can be viewed in terms of oxidation and reduction reactions. Two reactions are possible for the corrosion of iron pipe:[6]

$$Fe + 2H_2O \rightarrow Fe(OH)_2 + 2H^+ + 2e^- \qquad (5\text{-}13)$$

$$Fe + HCO_3 \rightarrow FeCO_3 + H^+ + 2e^- \qquad (5\text{-}14)$$

The former reaction occurs when the carbonate concentration is low, whereas the latter occurs when the carbonate concentration is high.

Corrosion can be inhibited when the oxidation products form a stable solid, such as an oxide, hydroxide, or carbonate, that adheres to the pipe surface and prevents or slows direct contact between the metal and the corrosive water, thus providing a barrier to corrosion reactions. This inhibition of corrosion is known as passivation. The compounds that form when iron corrodes depend on several factors, including pH and buffer capacity. Siderite ($FeCO_3$) may form and slow the corrosion rate. If conditions favor the formation of iron hydroxides, such as $Fe(OH)_2$ and $Fe(OH)_3$, soft and porous coatings may not inhibit corrosion.

The factors favoring corrosion include a low pH and a high concentration of oxidizing substances, such as dissolved oxygen and free residual chlorine. Movement of hydrogen gas away from the cathode area promotes continued corrosion. Factors that affect the corrosivity of drinking water are listed in Table 5–9.

All types of pipes (iron, galvanized, copper, and lead) commonly found in potable water systems are subject to corrosion. The reactions for corrosion of iron pipe were presented above.

Deterioration of galvanized pipe occurs in two stages. Initially, only the galvanized or zinc layer corrodes until iron is exposed. Corrosion of the galvanized layer depends on pH, carbonate concentration, and flow. Once the galvanized layer is penetrated and iron is exposed, the galvanized pipe begins to perform as an iron pipe.

In copper pipe, in the presence of dissolved oxygen, a thin film of cuprous oxide is formed in most of the metal's surface. This film promotes a constant corrosion rate that is normally only a fraction of the corrosion rate of iron or galvanized pipe.[6] However, the passivation film can be disturbed by high-velocity water flow or dissolved by either carbonic acid or organic acids that are found in some freshwaters. Chlorides tend to increase the porosity of the passivation film. Chlorine increases the oxidation of copper and the protective film of cuprous oxide.

The corrosion of lead pipe depends on pH and alkalinity. At low alkalinities, lead is soluble throughout the pH range of drinking water. In water containing carbonate alkalinity, an insoluble film of basic lead carbonate forms in the

Table 5-9. Factors Affecting the Corrosivity of Drinking Water.*

FACTOR	EFFECT ON CORROSIVITY
pH	Low pHs generally accelerate corrosion.
Dissolved oxygen	Dissolved oxygen in water induces active corrosion, particularly of ferrous materials.
Free chlorine residual	The presence of free chlorine in water promotes corrosion of ferrous metals and copper.
Low buffering capacity	There is insufficient alkalinity to provide buffering capacity to the water and provide protective films.
High halogen and sulfate–alkalinity ratio	A molar ratio of strong mineral acids much above 0.5 results in conditions favorable to pitting corrosion.
Carbon dioxide	Carbon dioxide is particularly corrosive to copper piping.
Total dissolved solids	Higher concentrations of dissolved salts increase conductivity and may increase corrosiveness. Conductivity measurements may be used to estimate total dissolved solids.
Calcium	Calcium generally reduces corrosion by forming protective films with dissolved carbonate.
Tannins	Tannins form protective organic films over metals.
Flow rates	Turbulence at high flow rates allows oxygen or carbon dioxide to reach the surface more easily, removes protective films, and causes higher corrosion rates.
Metal ions	Certain ions, such as copper, can aggravate corrosion of downstream materials. For example, copper ions can increase the corrosion of galvanized pipe.
Temperature	High temperature increases corrosion reaction rates. High temperature also lowers the solubility of calcium carbonate and calcium sulfate and thus may cause scale formation in hot-water heaters and pipes.

* Reference 6.

intermediate pH region. The film of basic lead carbonate performs two functions: (1) by adhering to the metal surface, the film forms a physical barrier between the metal and the water; and (2) the basic carbonate or carbonate solid phase limits the lead solubility and, therefore, reduces the amount of lead that can be leached into the drinking water.[6] However, even in systems containing high pH values, corrosion can occur at very low and high alkalinities.

WATER STABILITY

The carbonic acid system is important in estimating the stability of a water. A stable water is one that exhibits neither scale-forming nor corrosion properties. In water engineering, a stable water neither dissolves nor precipitates calcium carbonate ($CaCO_3$). A thin layer of $CaCO_3$ protects pipes against corrosion. However, excessive buildup will clog the pipe and drastically reduce its carrying capacity.

A relationship called the Langelier Saturation Index (LI), based on pH values, is used to predict if a water will or will not precipitate $CaCO_3$. The index is defined as:

$$LI = pH - pH_s \qquad (5\text{--}15)$$

where:

pH = measured pH of the water
pH_s = calculated $CaCO_3$ saturation pH of the water

The value of pH_s can be computed using the following equilibrium expressions:

$$CaCO_3(s) \rightleftarrows Ca^{+2} + CO_3^{-2} \qquad (5\text{--}16)$$

and:

$$k_s = (Ca^{+2})(CO_3^{-2}) \qquad (5\text{--}17)$$

as well as:

$$HCO_3^- \rightleftarrows H^+ + CO_3^{-2} \qquad (5\text{--}18)$$

and:

$$k_2 = \frac{(H^+)(CO_3^{-2})}{(HCO_3^-)} \qquad (5\text{-}19)$$

Dividing equation (5-17) by (5-19) and rearranging yields:

$$(H^+) = \frac{k_2}{k_s}(Ca^{+2})(HCO_3^-) \qquad (5\text{-}20)$$

or:

$$pH_s = p(Ca^{+2}) + p(HCO_3^-) + p(k_2/k_s) \qquad (5\text{-}21)$$

The value of pH_s can be determined from a simplified equation, as follows:

$$pH_s = A + B - \log (Ca^{+2}) - \log (\text{alkalinity}) \qquad (5\text{-}22)$$

The values for A and B are given in Tables 5-10 and 5-11, and the log values for Ca^{+2} and the alkalinity in Table 5-12.

Table 5-10. Constant
A as Function of
Water Temperature.*

WATER TEMPERATURE °C**	A†
0	2.60
4	2.50
8	2.40
12	2.30
16	2.20
20	2.10
25	2.00
30	1.90
40	1.70
50	1.55
60	1.40
70	1.25
80	1.15

* Reference 9.
** Conversion factors may be found in Appendix A.
† Calculated from k_2 as reported by Harned and Scholes and k_2 as reported by Larson and Buswell. Values above 40°C involve extrapolation.

Table 5–11. Constant B as Function of Total Filterable Residue.*

TOTAL FILTERABLE RESIDUE mg/l	B
0	9.70
100	9.77
200	9.83
400	9.86
800	9.89
1,000	9.90

* Reference 9.

Table 5–12. Logarithms of Calcium Ion and Alkalinity Concentrations.*

CA^{+2} OR ALKALINITY mg/l CaCO$_3$	LOG
10	1.00
20	1.30
30	1.48
40	1.60
50	1.70
60	1.78
70	1.84
80	1.90
100	2.00
200	2.30
300	2.48
400	2.60
500	2.70
600	2.78
700	2.84
800	2.90
900	2.95
1,000	3.00

* Reference 9.

When LI = 0, the water is in equilibrium. When LI is positive, the water is oversaturated, and it will tend to deport $CaCO_3$. A negative LI indicates that the water is undersaturated and will tend to dissolve existing deposits of $CaCO_3$ or corrode the pipeline.

Generally, in the United States, an attempt is made in practice to keep the saturation index slightly positive so that a thin, protective coat of $CaCO_3$ is deposited within the pipe. It should be emphasized that the Langelier Index only tells if precipitation or dissolution can occur. It does not determine how much $CaCO_3$ will be deposited.

REFERENCES

1. Cox, Charles R., *Laboratory Control of Water Purification,* Case-Sheppard-Mann Publishing Corp., New York, 1946.
2. Sawyer, C. N., and McCarty, P. L., *Chemistry for Sanitary Engineers,* McGraw-Hill Book Co., New York, 1967.
3. Fair, G. M., and Geyer, J. C., *Water Supply and Waste-Water Disposal,* John Wiley and Sons, Inc., New York, 1954.
4. *Manual for Water Works Operators,* Texas Water Works and Sewerage Short School, 1951. Distributed through the cooperation of the Texas Department of Health, Austin, Tx.
5. Moeller, T., and O'Conner, R., *Ions in Aqueous Systems,* McGraw-Hill, New York, 1972.
6. Kirmeyer, G. J., and Logsdon, G. S., "Principles of Internal Corrosion and Corrosion Monitoring," *J.AWWA,* p. 78, February, 1983.
7. Weber, W. J., Jr., *Physicochemical Processes for Water Quality Control,* Wiley-Interscience, New York, © 1972.
8. Hunter, John S., and Ward, John C. "The Effects of Water Temperature and Elevation Upon Aeration," presented at the Symposium on Wastewater Treatment in Cold Climates on August 22, 1973, at the University of Saskatchewan, Saskatoon, Saskatchewan, Canada.
9. *Federal Register,* Vol. 45, No. 168, August 27, 1980, Rules and Regulations.

Chapter 6
Pretreatment of Surface Water Supplies

INTRODUCTION

Surface water supplies, whether from rivers or reservoirs, often contain impurities or characteristics that may best be removed prior to conventional water treatment. Removal or control of these impurities is described in this chapter.

Reservoirs contain algae and other microscopic organisms that can cause taste and odor problems. Anaerobic conditions may develop near the bottom of reservoirs causing solubilization of objectionable metals. Rivers may carry heavy silt loads that would overtax or complicate operation of coagulation or softening unit processes. Debris in either reservoirs or rivers may foul water transport equipment. This chapter will address pretreatment provisions that may be considered for:

- Screening of surface water for removal of general debris.
- Treatment of reservoirs for destratification.
- Treatment of reservoirs for control of algae and plant growths.
- Pretreatment of surface water for control of taste and odors.
- Presedimentation of river waters for removal of suspended silt load.

SCREENING

In rivers or lakes where the water transports debris such as leaves, branches, logs, and similar objects, it is necessary to provide screening facilities before water is withdrawn for treatment. When substantial amounts of debris are present, mechanically cleaned screens should be considered. Most water intakes include protective screens, or bar racks, that are not mechanically cleaned, and only remove the largest pieces of debris.

Screens should be located at the intake structure. Typically the screenings, whether manually or mechanically removed, will be returned to the water source in such a way that they will not return to the screen and accumulate.

Traveling water screens are operated intermittently unless the debris is particularly heavy. These screens are installed in a channel as shown in Fig. 6–1, where they move in a slot, upward to the surface. Water sprays clean the screens and sluice the debris to a channel or pipe that carries it

'Fig. 6-1. Traveling water screen. (Courtesy of Envirex Inc., a Rexnord Company)

back to the river or lake. Screen openings are normally ⅛ to ½ inch (3.18 to 12.7 mm) wide.

Protective bar racks are often placed upstream of traveling water screens as barriers to logs, large fish, and other similar objects that could cause damage or are not suitable for removal by the screen. These bars may be spaced at 3 to 4 inches (76.2 to 101.6 mm), and inlet velocities are less than 0.25 ft/sec (76.2 mm/s). They may be fixed in place with no provision for cleaning because they are not intended to collect material, but are merely to prevent entry of large material. Where they will collect debris, provision should be made to allow hoisting them to the surface to allow its removal.

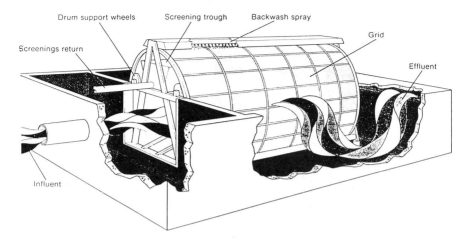

Fig. 6–2. Microscreen.

Microscreens, shown in Fig. 6–2, are also used as a pretreatment process, to remove finer material such as filamentous algae. Screening media are generally stainless steel or polyester. Media openings can be as small as 1 micron but are typically 20 to 30 microns. Microscreens typically are located at the treatment plant, and are used only when the water source is very turbid. Figure 6–3 shows a typical installation.

DESTRATIFICATION OF RESERVOIRS

The thermal stratification of lakes and reservoirs is a seasonal phenomenon. In the spring as the weather warms, heat is transferred to the surface of lakes and reservoirs. As the water is warmed, its density decreases, and the warmer, lighter water rises to the top of the reservoir and floats on a heavier cooler mass of water, causing two distinct water layers. The upper zone, or epilimnion, is characteristically well mixed and aerobic. It is separated from the lower zone, or hypolimnion, by a transition zone known as the thermocline, where the temperature of the water changes rapidly with depth between the two zones. The hypolimnion is characteristically cool and unmixed. A schematic diagram of a reservoir showing these two zones is included in Fig. 6–4.

As the spring weather begins, the surface water warms, and the stratification of the reservoir is initiated. The wind action on the water causes a zone of mixing, and a layer of water at the surface increases uniformly in temperature. The stability of the thermocline, or temperature difference, between the epilimnion and hypolimnion increases as the average air temperature increases.

Fig. 6–3. Microscreen installation. (Courtesy of Envirex Inc., a Rexnord Company.)

The thermocline may initially be 6 to 10 feet (1.82 to 3.05 m) below the surface of the water. If wind velocity increases, the mixing of the epilimnion increases, and the thermocline is found farther below the surface.

The hypolimnion is isolated from a source of oxygen. In aged reservoirs, or where a reservoir has an input of organic materials, the demand for oxygen

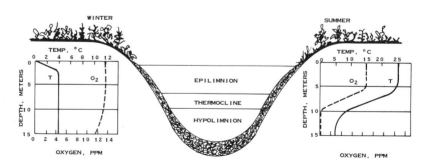

Fig. 6–4. Distribution of oxygen and temperature with lake depth.

may cause a total depletion of oxygen in all or part of the hypolimnion. The lack of oxygen may cause anaerobic conditions, and chemical constituents in the hypolimnion and benthal deposits will be chemically reduced. Hydrogen sulfide may be formed, and iron and manganese dissolved. Hydrogen sulfide imparts taste and odor to the water, and iron and manganese form precipitates (when water is subsequently aerated) and cause the water to be turbid and colored. Iron and manganese will also stain plumbing fixtures and laundry, and if present in sufficiently large concentrations, will cause taste problems.

In addition, anaerobic conditions in the hypolimnion may cause reduction in pH and an increased concentration of CO_2, which in turn may impair the capability of the treatment facilities.

The problem can be resolved by not drawing water from the hypolimnion, adding more intensive water treatment, or eliminating the problem in the reservoir. In some cases, it may be feasible to eliminate thermal stratification so as to oxygenate the hypolimnion, or to add oxygen to the hypolimnion without thermal destratification. These approaches may not be practical for very large reservoirs.

To destratify the reservoir, the cooler, more dense water from the hypolimnion is raised to mix with the less dense, warmer water of the epilimnion. If sufficient energy is added, complete dispersion of the thermocline occurs. If the oxygen resources and natural reservoir reaeration are sufficient, an aerobic reservoir may be maintained.

There are potential problems associated with destratification. In an advanced eutrophic reservoir, the accumulation of a sufficiently high organic content in the hypolimnion may, if mixed with the epilimnion, cause insufficient dissolved oxygen throughout the reservoir. In such a case, thermal destratification could do more harm than good.

In eutrophic reservoirs, dissolved nutrients in the hypolimnion, if mixed into the epilimnion, may stimulate aquatic growths, which can trigger taste and odor problems in the water supply. Also, mixing of water from the cool hypolimnion water with that of the warm epilimnion will result in an increase in water temperature at a hypolimnitic intake. The warmer water may not be desirable for a water supply.

Most of these problems appear to be of consequence only in small lakes. The studies done with thermal destratification in large reservoirs have not shown evidence of either oxygen depletion or increased phosphorus concentration in the epilimnion. Therefore, for small lakes or reservoirs, consideration should be given to avoiding thermal destratification and any associated mixing of the hypolimnion and epilimnion. Alternatively, it may be desirable to aerate the hypolimnion without mixing it with the epilimnion. Techniques to accomplish this seemingly contradictory goal have been developed by several investigations.[2,4] The most common method is to install a tube the full depth of the reservoir with the upper lip of the tube above the reservoir water surface. Air is injected near the bottom of the tube, and an airlift effect is produced, raising the water in the tube and concurrently aerating the water in transit. A slot or opening is provided in the tube below the thermocline to provide a means for the water to escape before it reaches the epilimnion, and the air continues up the tube to the atmosphere.

The two methods for in-reservoir oxygenation are (1) thermal destratification to co-mix the epilimnion and hypolimnion to provide a uniform dissolved oxygen concentration and temperature, and (2) artificial oxygenation and mixing of the hypolimnion to maintain a layered reservoir.

The energy levels required to accomplish thermal destratification have been calculated by Symons et al., who related energy efficiency of destratification to "oxygenation capacity" (OC) and "destratification efficiency" (DE).[3]

The oxygenation capacity is determined by calculating the mass input of dissolved oxygen per unit of time divided by the energy input per unit of time. The dissolved oxygen input per unit of time is the summation of increased concentration in dissolved oxygen per unit of time and the steady-state demand rate for oxygen per unit of time. In reservoirs, the demand rate for oxygen is low and often ignored as insignificant. The OC is expressed in units of pounds of oxygen transferred per kilowatt-hour.

The DE is calculated by the net change of "stability" over a set time frame divided by the total energy input over the same time frame. Stability is defined as the minimum energy needed to mix the lake, and is calculated by multiplying the weight of water in the lake by the vertical distance between the center of gravity of the lake, taking into account the density due to the thermal gradient in the lake, and the center of gravity of the lake if there were no thermal gradient. The results of this calculation are in foot-pounds, which can be converted to kilowatt-hours over the time frame used. The change in stability divided by the power input yields the DE.

As the depth of the epilimnion increases and/or as the temperature differential between the epilimnion and hypolimnion increases, the location of the center of gravity in the reservoir moves downward. The farther the center of gravity moves from the isothermal center of gravity, the more stable is the stratification, and the more energy is required to overcome the stratification.

The use of either the OC or DE in design of lake destratification is precarious without prior experience. Table 6–1 shows experienced power requirements and time required to acquire destratification, using pumps and diffused air mechanical devices.

CHEMICAL TREATMENT OF RESERVOIRS

When sufficient nutrients are available and warm, sunny conditions prevail, algal growths occur in reservoirs, and can cause taste and odor problems.

Table 6–1. Experienced Power Requirements for Destratification.*,**

LAKE	VOLUME (ACRE FT)	SURFACE AREA (ACRES)	EPILIMNION DEPTH (FT)	MECHANICAL MIXING DEVICE	TIME RQ'D FOR DESTRATIFICATION (HR)	POWER (HP)
Stewart Hollow	120	8	11	Pump	20	12
Caldwell	100	10	10	Pump	8	12
Vesuvius	1260	105	10	Pump	208	12
Bolt	2380	56	15	Pump	912	21
Indian Brook	316	18	—	Diffused Air	160	8
Wohl Ford	2510	130	—	Diffused Air	78	50
Blelham Tarn	515	27	—	Diffused Air	335	1

* Reference 1.
** Conversion factors may be found in Appendix A.

Copper sulfate and/or potassium permanganate is used as an algicide in reservoirs to reduce the number of organisms. Chlorine has also been used in some situations, but its usage must be carefully controlled because it combines with certain odor-forming compounds to cause a more intense odor and may result in trihalomethane production.

Copper sulfate is considered to be effective in controlling algal growths. There is some difference of opinion over whether copper sulfate is algistatic or algicidal; however, continuing programs in which the copper sulfate is added regularly to keep growths from occurring are more effective than programs depending on copper sulfate to destroy an existing growth. In a typical situation, copper sulfate would be added to a reservoir in a concentration of about 0.1 to 0.5 mg/l. Many times it is only necessary to add the copper sulfate around the shoreline where the water is quiescent and the algal growths are heaviest. The copper will eventually precipitate from solution. The alkalinity of the water is important in determining the dosage. If the methyl orange alkalinity is less than 50 mg/l, copper sulfate has been shown to be effective at an application rate of 0.33 mg/l. If the methyl orange alkalinity is greater than 50 mg/l, the dosage rate will be 5.4 pounds/acre (606 kg/Mm²). Depth is not a factor in high alkalinity water because precipitation of copper will occur rapidly.

Care must be exercised to prevent overdosing of copper sulfate and killing fish. Safe dosages for most fish are about 0.5 mg/l, but for trout the safe dosage is 0.14 mg/l.

Potassium permanganate has proved to be algicidal at dosages of approximately 0.5 to 2.0 mg/l, but it is used to a lesser extent than copper sulfate in controlling algal growths in reservoirs.[5] The higher dosages and cost of potassium permanganate have restricted its application for reservoir algal control.

If blue-green algae are predominant—a condition associated with low oxygen conditions—alternative control methods may be advantageous. The addition of aeration or destratification techniques may prove to be simpler and more direct than use of an algicide. However, blue-greens are more susceptible to copper sulfate than green algae.

AERATION OR PREOXIDATION

As an alternative to providing reservoir destratification for control of hydrogen sulfide, iron and manganese, and other problems, preaeration at the treatment plant may be used. Preaeration will strip hydrogen sulfide and other volatiles, and will oxidize the iron and manganese to their insoluble forms, which can then be coagulated and precipitated.

Aeration is not considered an efficient method for removing or controlling taste and odor problems other than those associated with iron, manganese, and hydrogen sulfide. Most taste- and odor-causing substances are not sufficiently volatile to be effectively removed.

Aeration may be accomplished by several methods. For example, in cascade-type aerators consisting of multiple trays with water distributed over the trays, air, either by natural or forced means, passes through the trays as the water percolates from tray to tray. The water is oxygenated, and volatile gases are removed. The cascade or step aerator is also used to carry oxygen to the water, but it lacks the stripping effect of the multiple tray aerator. The step aerator is designed to cause the water to flow down a series of steps in a thin sheet. Also, the packed towers (described in Chapter 15) used for removal of volatile organic compounds will provide aeration. Another technique is to use spray nozzles to provide intimate contact of water and air. Generally, spray aeration is accomplished in a concrete or earthen basin. The energy requirements for spray aeration are significantly greater than those for tray or step aerators.

The above methods typically are used for waterworks aeration. Diffused air or mechanical surface agitators are also potential means of aeration, but are seldom used in water treatment plants, except for removal of volatile organics (see Chapter 15).

Aeration is accomplished in a basin that includes either diffused air or mechanical surface aerators to aerate the water. The diffused air method requires an air source, and perforated pipes or diffusers to distribute the compressed air in the basin. Surface aerators are either mounted on fixed platforms or floated in the basin, and agitate the surface of the water to oxygenate the water. If the aerator is platform-mounted, care is needed in maintaining the surface water level because the surface aerator is especially sensitive to the submergence level.

The transfer of oxygen to water is a rate diffusion reaction, which can be expressed as follows:

$$dC/dt = K_L A \ (C_s - C) - r \qquad (6\text{--}1)$$

where:

dC/dt = change in oxygen concentration per unit time, mg/l/hour
$K_L A_{20}$ = mass transfer constant for particular basin configuration and aeration device, hour^{-1} at 20°C
C_s = saturated oxygen concentration, mg/l
C = ambient oxygen concentration, mg/l
r = oxygen demand of water, mg/l/hour
t = time, hour

The above equation can be integrated to determine $K_L A_{20}$, assuming that r is small and can be deleted:

$$K_L A_{20} = \frac{2.3 \, \log_{10} \left[\dfrac{C_s - C_1}{C_s - C_2} \right]}{t_2 - t_1} \qquad (6\text{--}2)$$

The value of $K_L A_{20}$ can be converted to other operating temperatures by the equation:

$$K_L A_T = K_L A_{20} \times 1.024^{(T - 20)} \qquad (6\text{--}3)$$

By measuring the incoming and outgoing oxygen concentrations, and knowing the time within the aeration device, the calculation of the $K_L A$ value will permit determination of the capability of an aeration device.

PRESEDIMENTATION

Sedimentation, as a unit process, is discussed in detail in Chapter 9. Presedimentation is a pretreatment process for control of silt load on subsequent treatment units. It is unnecessary to use presedimentation for reservoir supplies because the reservoir accumulates the silt. The silt load to the plant should be minimal, if an effective intake structure is provided. Therefore, presedimentation is a unit process almost solely associated with river supplies, and, more specifically, only for those river supplies that carry heavy silt loads.

Presedimentation basins require additional capital expenditures that are not justifiable at all plants. The silt load on most rivers is controlled by upstream reservoirs, and only during high river flow rates does the silt load become a problem. If presedimentation is not provided, the heavy turbidities will require increased chemical dosages. If the high turbidities are associated with short-duration, seasonal periods, it may not be justifiable to provide presedimentation.

Silt settles rapidly, and surface overflow rates of 1,600 to 2,000 gpd/sq ft (2.71 to 3.39 m/h) have been used. Detention times of 2 to 3 hours at peak flow rates are frequently used. Longer detention times are required where there is no sludge collection equipment.

REFERENCES

1. Knoppert, P. L., et al., "Destratification Experiments at Rotterdam," *J. AWWA* 62:449, July, 1970.
2. Irwin, W. H., et al., "Impoundment Destratification by Mechanical Pumping," *JSED, ASCE* 92:SA6:21, December, 1966.

3. Symons, J. M., et al., "Impoundment Water Quality Changes Caused by Mixing," *JSED, ASCE* 93:SA2:1, April, 1967.
4. Symons, J. M., et al., "Management and Measurement of DO in Impoundments," *JSED, ASCE* 93:SA6:181, December, 1967.
5. Ficek, K. J., and Emanual, A. G., "Surface and Well Water Treatment with Potassium Permanganate," *J. Missouri Water and Sewerage Conference,* p. 25, 1965.

Chapter 7
Coagulation and Flocculation

INTRODUCTION

Impurities in water often cause the water to appear turbid or be colored. These impurities include suspended and colloidal materials and soluble substances. Particle size ranges are shown in Fig. 7–1.[1] Because the density of many of these particles is only slightly greater than the density of water, agglomeration or aggregation of particles into a larger floc is a necessary step for their removal by sedimentation. The process that combines the particles into larger flocs is called coagulation.

Impurities that can be removed by coagulation include turbidity, bacteria, algae, color, organic compounds, oxidized iron and manganese, calcium carbonate, and clay particles. Clays are a large part of natural turbidity in raw waters, but are not directly responsible for harmful effects to humans. However, there is some evidence that clays affect human health indirectly through adsorption, transport, and release of inorganic and organic toxic constituents, viruses, and bacteria. Removal efficiencies of clay particles in water treatment are not normally monitored, although several laboratory studies have demonstrated the effectiveness of alum coagulation on clay suspensions.[2,3]

Coagulation has been shown to be effective for the removal of color and other organic constituents in water. Laboratory studies have demonstrated the removal of these constituents using iron and aluminum salts.[4-6] Humic acids were readily removed using these salts, although a large fraction of the fulvic acids was not removed. Because humic acids react with chlorine in the formation of halomethanes, their removal in the coagulation process is an important step in limiting the production of potential carcinogens.

THEORIES OF COAGULATION

The aggregation of colloidal particles takes place in two separate and distinct phases.[7,8] First, the repulsion force between particles must be overcome, a step that requires that the particles be destabilized; and, second, contact between the destabilized particles must be induced so that aggregation can occur. The destabilization step typically is achieved through the addition of chemicals, followed by thorough blending in rapid mix tanks. The aggrega-

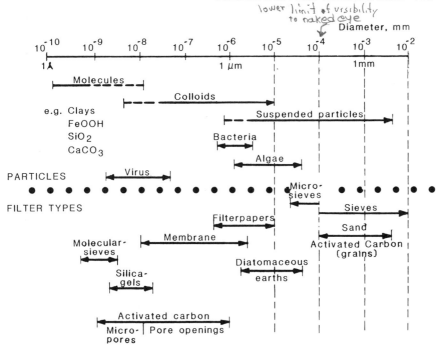

Fig. 7-1. Size spectrum of waterborne particles and filter pores.[1]

tion step is accomplished through gentle stirring in flocculation tanks. A representation of the coagulation process is shown in Fig. 7-2.[8]

It is useful to define the terms used in connection with the coagulation process.[9] Coagulation is defined as the process that causes a reduction of repulsion forces between particles or the neutralization of the charges on particles. Flocculation is defined as the aggregation of particles into larger elements. The coagulation (destabilization) step is virtually instantaneous following addition of the coagulant, while the flocculation (transport) step requires more time for development of large flocs.

Historically, two theories have been advanced to explain the coagulation process of colloidal systems:

- Chemical theory
- Physical or double-layer theory

The former theory presumes that colloids acquire electrical charges on their surfaces by ionization of chemicals present at the surface, and that

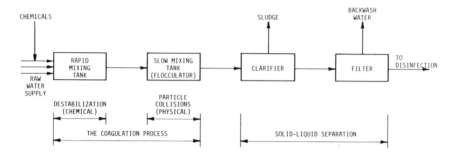

Fig. 7–2. Schematic diagram of a coagulation process.

coagulation, or destabilization, occurs through specific chemical interactions between the coagulant and the colloids. Accordingly, this theory presumes that the coagulation of colloids is a result of the precipitation of insoluble complexes that are formed by chemical reactions.

The second theory is based on the presence of physical factors, such as electrical double layers surrounding the colloidal particles in the solution and counterion adsorption. Destabilization requires a reduction in the electric potential between the fixed layer of counterions and the bulk of the liquid. This electric, or zeta, potential can be estimated by observing the movement of microscopically visible particles in an electric field. This theory is presented in more detail below. These two mechanisms are not mutually exclusive. Both theories are used to explain the process of coagulation in treatment systems containing a heterogeneous mixture of colloids.

The Electrical Double Layer

When a colloidal particle is immersed in a solution, electrical charges develop at the particle–water interface. However, a colloidal dispersion does not have a net electrical charge. For electroneutrality to exist, the charges on the colloids must be counterbalanced by ions of opposite charge (counterions) in the solution. The ions involved in establishing the electroneutrality are arranged in an electrical double layer. The concept of the electrical double layer was proposed initially by Helmholtz and later modified and improved by Gouy, Chapman, and Stern.[10-12]

The Stern-Gouy diffuse, double-layer model, which is illustrated in Fig. 7–3, can be used to describe the electrical potential in the vicinity of a colloid particle. A portion of the counterions remain in a compact ("Stern") layer on the colloid surface. The remainder of the counterions extend into the bulk of the solution, and constitute the diffuse ("Gouy-Chapman") layer. The effective thickness of the double layer is influenced significantly by the ionic concentration of the solution, but relatively little by the size of the colloid.

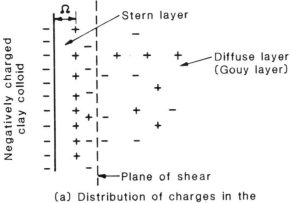

(a) Distribution of charges in the
vicinity of a colloidal particle

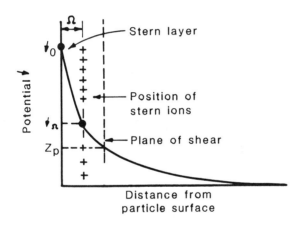

(b) Distribution of potential in the
electrical double-layer

Fig. 7–3. Stern's model for the electrical double layer.[16] Reprinted by permission of John Wiley & Sons, Inc.

The electrical potential created by the surface charges will attract counterions toward the colloidal particles. The closest approach of the counterions to the particle is limited by the size of the ions. Stern proposed that the center of the closest counterions is separated from the surface charge by a layer of thickness Ω, which represents the Stern layer. The electrical potential drops linearly across this layer. Beyond the Stern layer, in the diffuse layer, the electrical potential decreases exponentially with distance from the particle.

The magnitude of the charge on a colloid cannot be measured directly, but the value of the potential at some distance from the colloid can be computed. This potential, termed the zeta potential, can be computed by several techniques, such as electrophoresis, electroosmosis and streaming potential. Most often, the electrophoretic mobility of the colloidal particles is used to compute the zeta potential, by observing the particle mobility through a microscope. The zeta potential is:

$$Zp = \frac{4\,\pi\nu\mu}{D} \tag{7-1}$$

where:

Zp = zeta potential, millivolts
μ = electrophoretic velocity, cm/s
ν = viscosity, poise
D = dielectric constant of the liquid
π = constant
   = 3.14159

The magnitude of the zeta potential is an approximate measure of colloidal particle stability. Low zeta potentials indicate relatively unstable systems (easily coagulated), while a high zeta potential represents strong forces of separation and a stable system (difficult to coagulate).

The computed value for the zeta potential may vary significantly from the true value because of the uncertainty of values for the various constants in the equation. As a result, many researchers report experimental results in terms of electrophoretic mobility, instead of zeta potential. The electrophoretic mobility may be computed from the equation:

$$\Omega = \frac{dx}{tIR_s} \tag{7-2}$$

where:

Ω = electrophoretic mobility, $\mu/s/V/cm$
d = distance traveled μ/s
x = cross-sectional area of cell, cm²
t = time, seconds
I = current density, amperes
R_s = specific resistance of suspension, ohm-cm

Mobilities are positive (+) for those particles that migrate to the negative pole of the cell, and are negative (−) for those that migrate toward the positive pole. The point at which there is no particular migration (zero mobility) is considered to be the isoelectric point.

Coagulation Mechanisms

Coagulation can be accomplished through any of four different mechanisms:[14,15]

- Double layer compression
- Adsorption and charge neutralization
- Enmeshment by a precipitate (sweep-floc coagulation)
- Adsorption and interparticle bridging

Brief descriptions of these theories are included here.

Double Layer Compression. This mechanism relies on compressing the diffuse layer surrounding a colloid. This is accomplished by increasing the ionic strength of the solution through the addition of an indifferent electrolyte (neutral salt). The explanation for this phenomenon lies in the Schulze-Hardy rule for anions, which was based on Schulze's work on the coagulating power of cations. He noted that the coagulating power increases in the ratio of 1:10:1000 as the valency increases from 1 to 2 to 3. The Schulze-Hardy rule for anions is:

The coagulating power of a salt is determined by the valency of one of its ions. The prepotent ion is either the negative or positive ion, according to whether the colloidal particles move down or up the potential gradient. The coagulating ion is always of the opposite electrical sign to the particle.

This rule is valid for indifferent electrolytes, which are those that do not react with the solution. If such an electrolyte is added to a colloidal dispersion, the particle's surface charge will remain the same, but the added electrolyte will increase the charge density in the diffuse layer. This results in a smaller diffuse-layer volume being required to neutralize the surface charge. In other words, the diffuse layer is "compressed" toward the particle surface, reducing the thickness of the layer. At high electrolyte concentrations, particle aggregation can occur rapidly.

Adsorption and Charge Neutralization. The energy involved in an electrostatic interaction having a 100-millivolt potential difference across the diffuse layer between a colloidal particle and a monovalent coagulant ion is only about 2.3 kcal/mole. This compares with covalent bond energies in the range of 50 to 100 kcal/mole. Based on these facts, it is apparent that some coagulants can overwhelm the electrostatic effects and can be adsorbed on the surface of the colloid. If the coagulant carries a charge opposite to that of the colloid, a reduction in the zeta potential will occur, resulting in destabilization of the colloid. This process is quite different from the double-layer compression mechanism described above.

The hydrolyzed species of Al(III) and Fe(III) can cause coagulation by adsorption. However, at higher doses of Al(III) coagulation is caused by enmeshment of the colloidal particles in a precipitate of aluminum hydroxide. This aspect is discussed in the next section. Destablization by adsorption is stoichiometric. Therefore, the required coagulant dosage increases with increasing concentrations of colloids in the solution.

Enmeshment by a Precipitate (Sweep-Floc Coagulation). The addition of certain metal salts, oxides, or hydroxides to water in high enough dosages results in the rapid formation of precipitates. These precipitates enmesh the suspended colloidal particles as they settle.[3] Coagulants such as aluminum sulfate ($Al_2(SO_4)_3$), ferric chloride ($FeCl_3$), and lime (CaO or $Ca(OH)_2$) are frequently used as coagulants to form the precipitates of $Al(OH)_3(s)$, $Fe(OH)_3(s)$ and $CaCO_3(s)$. The removal of colloids by this method has been termed sweep-floc coagulation.

This process can be enhanced when the colloidal particles themselves serve as nuclei for the formation of the precipitate. Therefore, the rate of precipitation increases with an increasing concentration of colloidal particles (turbidity) in the solution. Packham reported the inverse relationship between the optimum coagulant dose and the concentration of the colloids to be removed.[3] Benefield explained this phenomenon as follows:[14]

At low colloidal concentrations a large excess of coagulant is required to produce a large amount of precipitate that will enmesh the relatively few colloidal particles as it settles. At high colloidal concentrations, coagulation will occur at a lower chemical dosage because the colloids serve as nuclei to enhance precipitate formation.

This method of coagulation does not depend upon charge neutralization; so an optimum coagulant dose does not necessarily correspond to minimum zeta potential. However, an optimum pH does exist for each coagulant.

Destabilization by Interparticle Bridging. Synthetic polymeric compounds have been shown to be effective coagulants for the destabilization of colloids in water. These coagulants can be characterized as having large molecular sizes, and multiple electrical charges along a molecular chain of carbon atoms. Both positive (cationic) and negative (anionic) polymers are capable of destabilizing negatively charged colloidal particles. Surprisingly, the most economical destabilization process is often obtained using anionic polymers.[15] The mechanisms already described cannot be used to describe this phenomenon, although a generally accepted chemical bridging theory/ model has been developed that can explain the unusual reaction associated with synthetic polymer compounds.[9,17]

The chemical bridging theory, shown schematically in Fig. 7–4, may be

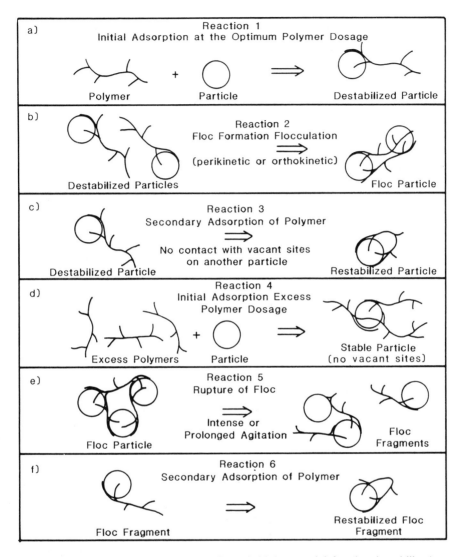

Fig. 7-4. Schematic representation of the bridging model for the destabilization of colloids by polymers.[13,15] Reprinted by permission of John Wiley & Sons, Inc.

explained as follows. The simplest form of bridging, shown in Fig. 7–4a, proposes that a polymer molecule will attach to a colloidal particle at one or more sites. Colloidal attachment is postulated to occur as a result of coulombic attraction if the charges are of opposite charge, or from ion exchange, hydrogen bonding, or van der Waal's forces.[13] The second reaction is shown in Fig. 7–4b, where the remaining length of the polymer molecule from the first reaction extends out into the bulk of the solution. If a second particle having some vacant adsorption sites contacts the extended polymer, attachment can occur to form a chemical bridge. The polymer, then, serves as the bridge. However, if the extended polymer molecule does not contact another particle, it can fold back on itself and adsorb on the remaining sites of the original particle, as shown in Fig. 7–4c. In this event, the polymer is no longer capable of serving as a bridge, and in fact, it restabilizes the original particle.

Colloidal restabilization can occur from an overdose of polymer to the sol or from extended or intense agitation. If polymer is added in excess quantities, the polymer segments may saturate the colloidal surfaces to the extent that no sites are available for interparticle bridging. This reaction, shown in Fig. 7–4d, results in restabilization of the particles. Intense or extended agitation can result in restabilization due to the destruction of previously formed polymer-surface bonds, or bridges. These reactions are depicted in Figs. 7–4e and 7–4f.

COAGULATION IN WATER TREATMENT

The coagulation process in water treatment is normally accomplished by using the adsorption or enmeshment mechanisms. The coagulants used for the adsorption process typically are synthetic organic polymers or activated silica. The coagulants used in the enmeshment principle are either metal hydroxide [Al(III) or Fe(III)] or carbonate precipitates.

The coagulation process depends on the appropriate design of the rapid mixing and flocculation processes, which are described later in this section. Only the more popular coagulants are discussed in this text.

Coagulation Using Al(III) and Fe(III)

Al(III) and Fe(III) are the salts used most frequently to coagulate colloidal material in water treatment. These salts are hydrolyzing metal ions, and a brief discussion of the chemistry of these ions is useful to understanding their ability to destabilize and coagulate colloidal particles. More detailed discussions of the chemistry of these salts are presented elsewhere.[15,18,19]

When added to water, salts of Al(III) and Fe(III) dissociate to their respective trivalent ions, Al^{+3} and Fe^{+3}, and then hydrate to form aquometal complexes. Their respective aquometal complexes are $Al(H_2O)_6^{+3}$ and $Fe(H_2O)_6^{+3}$. These complexes react with the water by replacing the H_2O molecules in the aquometal complex with OH^- ions. These subsequent reactions are called hydrolytic reactions. Hydrolytic reactions result in the formation of several soluble species, including monomeric, dimeric, and polymeric hydroxometal complexes. Typical reactions include:[8]

$$Fe(H_2O)_6^{+3} + H_2O = Fe(H_2O)_5(OH)^{+2} + H_3O^+ \qquad (7\text{-}3)$$
$$Al(H_2O)_6^{+3} + H_2O = Al(H_2O)_5(OH)^{+2} + H_3O^+ \qquad (7\text{-}4)$$
$$Al^{+3} + 4H_2O = Al(OH)_4^- + 4H^+ \qquad (7\text{-}5)$$
$$FeOH^{+2} + H_2O = Fe(OH)_2^+ + H^+ \qquad (7\text{-}6)$$
$$2Fe^{+3} + 2H_2O = Fe_2(OH)_2^{+4} + 2H^+ \qquad (7\text{-}7)$$

The last three equations shown above are presented without the H_2O ligands for simplicity.

When Al(III) or Fe(III) salts are added to water in quantities less than the solubility limit of the hydroxide, then the hydrolysis products will form, and will adsorb on the colloidal particles. Adsorption of the hydrolysis products will cause destabilization by charge neutralization. However, when the amount of Al(III) or Fe(III) added to the water exceeds the solubility limit of the hydroxide, the hydrolysis products will form as kinetic intermediates in the eventual precipitation of metal hydroxides. In this case, charge neutralization and enmeshment in the precipitate both act to destabilize and coagulate the colloids.

Typically, the amount of Al(III) or Fe(III) added in a conventional water coagulation process is sufficient to exceed the solubility limit of the respective metal hydroxides. The solubility of $Al(OH)_3(s)$ and $Fe(OH)_3(s)$ is a minimum at a specific pH, and increases as the pH increases or decreases from that point. Precipitation of amorphous metal hydroxides is necessary for sweep-floc coagulation.

The pH must be controlled to establish optimum conditions for coagulation. For pH's below the isoelectric point of the metal hydroxide, positively charged polymers (kinetic intermediates) will be formed. Adsorption of these positive polymers can destabilize negatively charged colloids by charge neutralization. Above the isoelectric point, negative polymers will predominate, and destabilization is achieved by bridge formation. Control of the coagulation process is complicated by the release of hydrogen ions, as shown by equations (7-3) through (7-7). The hydrogen ions liberated will react with the alkalinity in the water to yield:

$$Al_2(SO_4)_3 \cdot 14H_2O + 3Ca(HCO_3)_2$$
$$= 2Al(OH)_3 + 3CaSO_4 + 14H_2O + 6CO_2 \quad (7\text{-}8)$$

Equation (7-8) predicts that each mg/l of alum will consume 0.50 mg/l (as $CaCO_3$) of alkalinity. If the alkalinity is not sufficient to react with the alum and buffer the pH, then it is necessary to add alkalinity to the water in the form of lime, sodium bicarbonate, soda ash, or some other similar chemical. The following are the stoichiometric reactions:

$$Al_2(SO_4)_3 \cdot 14H_2O + 3Ca(OH)_2 = 2Al(OH)_3 + 3CaSO_4 + 14H_2O \qquad (7\text{-}9)$$

$$Al_2(SO_4)_3 \cdot 14H_2O + 3Na_2CO_3 + 3H_2O = 2Al(OH)_3 \\ + 3Na_2SO_4 + 3CO_2 + 14H_2O \quad (7\text{-}10)$$

As the amount of Al(III) or Fe(III) added to water is gradually increased from zero, the following phenomena occur. Low coagulant doses can be insufficient to destabilize the colloidal particles. Increasing the coagulant dose results in destabilization, and then rapid aggregation occurs. Increasing the coagulant dose further can cause restabilization of the dispersion at some pH's. Finally, if a sufficient quantity of coagulant is added, large amounts of metal hydroxide are precipitated that enmesh the colloidal particles, and sweep-floc coagulation occurs.

A knowledge of the interrelationships between optimum coagulant dosage, pH, and colloid concentrations, combined with an understanding of the two modes of destabilization that are caused through the addition of Al(III) or Fe(III) salts, is useful in the operation of a coagulation process. O'Melia describes four types of suspension, as follows:[8]

1. *High colloid concentration, low alkalinity.* This is the easiest system to treat, in that only one chemical parameter must be determined—the optimum coagulant dosage. Destabilization is achieved by adsorption of positively charged hydroxometal polymers; these are produced at acidic pH levels (pH 4 to 6, depending on the coagulant).

2. *High colloid concentration, high alkalinity.* In this case destabilization is again achieved by adsorption and charge neutralization at neutral and acid pH levels. Here the engineer can elect to use a high coagulant dosage. (Because of the high alkalinity, the pH will generally remain in the neutral region where the hydroxometal polymers are not highly charged so that charge neutralization is more difficult.) Alternatively, it is possible to use elutriation facilities, remove alkalinity by washing, and destabilize with a lower coagulant dosage at a lower pH.

3. *Low colloid concentration, high alkalinity.* Coagulation is readily accomplished here with a relatively high coagulant dosage by enmeshment of colloidal particles in a sweep floc. Alternatively, a coagulant aid may be added to increase the colloid concentration and increase the rate of interparticle contacts. Destabilization by adsorption and charge neutralization may then be effected at a lower dosage of primary coagulant.

4. *Low colloid concentration, low alkalinity.* Coagulation is most difficult in such systems. Al(III) and Fe(III) salts will be ineffective if used alone because the pH will be depressed too low to permit the rapid formation of a sweep floc, and the rate of interparticle contacts is presumably too slow to utilize destabilization by charge neutralization. Additional alkalinity, additional colloidal particles, or both must be added to provide effective coagulation.

Coagulation with Polymers

Synthetic organic polymers have been shown to be effective coagulants or coagulant aids. Polymers are long-chain molecules comprised of many subunits called monomers. A polymer that is comprised of only one type of monomer is termed a homopolymer, and those comprised of different monomers are termed copolymers. The number and type of subunits, or monomers, can be varied to yield a wide range of polymers having different characteristics and molecular weights.

A polymer is called a polyelectrolyte if its monomers consist of ionizable groups. Polyelectrolytes having a positive charge upon ionization are referred to as cationic polymers. Negatively charged polyelectrolytes are termed anionic polymers. Finally, polymers that do not contain ionizable groups are called nonionic polymers.

Cationic polymers can be effective in coagulating negatively charged clay particles.[17] It has been hypothesized that electrostatic forces or ion exchange is the process by which the polymers become attached to the clay particles, which is then followed by bridging. Cationic polymers do not require a large molecular weight to be effective in destabilization.

Anionic particles generally are ineffective coagulants for negatively charged clay particles.[17] Anionic polymers of large molecular weight or size are required for these molecules to bridge the energy barrier between two negatively charged particles. The minimum polymer size depends on several factors, but limited data indicate that the minimum size is on the order of a molecular weight of one million.[8] When anionic polymers are used in conjunction with an electrolyte such as NaCl or $CaCl_2$, or another coagulant such as alum, their coagulating effectiveness is increased. There is strong evidence that divalent metal ions are necessary for anionic polymers to flocculate negative colloids.

Dosages of only 0.5 to 1.5 mg/l of cationic polymer frequently are sufficient to achieve coagulation. In contrast, 5 to 150 mg/l of alum is often needed to obtain similar results. Other important differences between the use of polymers and metal ions are sludge quantities and dosage control. The use of alum or ferric chloride can result in copious volumes of sludge that must be handled, whereas the additional sludge quantity is negligible when a polymer is used. A narrow band exists for optimum polymer dosage. Overdosing or underdosing from this optimum will result in restabilization of the colloids. The control method for polymer feed systems must be precise and reliable to give satisfactory performance.

Because polymers do not affect the pH of water, their use offers a decided advantage for treating low-alkalinity waters. This is particularly true if the low-alkalinity waters are high in turbidity. Such waters would require considerable quantities of alum, which would necessitate the addition of soda ash or lime to replace the buffering capacity of the water and maintain neutral pH.

Coagulation with Magnesium

Magnesium precipitated as magnesium hydroxide ($Mg(OH)_2$) has been shown to be an effective coagulant for the removal of color and turbidity from waters.[20,21] Coagulation is achieved by enmeshing the colloidal particles in the gelatinous hydroxide precipitate. Similar findings have been shown in lime treatment of wastewaters.

Waters that are high in naturally occurring magnesium can be coagulated by raising the pH to the point at which $Mg(OH)_2$ is precipitated. Normally, pH elevation is accomplished through the addition of lime. For waters that are low in magnesium, it is necessary to add a suitable magnesium salt. In using lime to raise the pH, one advantage is that both lime and magnesium can be recovered from the sludge and reused. The recovery process takes the form of recalcination in a furnace, typically a multiple hearth furnace. The sludge disposal problems are reduced by recovery of the lime. However, the sludge handling processes are complex. The economics of treatment must be carefully evaluated before recalcination is implemented.

The chemistry of magnesium coagulation with lime addition is based on water softening and coagulation theories. The reactions involved in $Mg(OH)_2$ precipitation are:

$$CO_2 + Ca(OH)_2 \rightarrow CaCO_3 + H_2O \qquad (7\text{--}11)$$
$$Ca(HCO_3)_2 + Ca(OH)_2 \rightarrow 2CaCO_3 + 2H_2O \qquad (7\text{--}12)$$
$$Mg(HCO_3)_2 + Ca(OH)_2 \rightarrow MgCO_3 + CaCO_3 + 2H_2O \qquad (7\text{--}13)$$
$$MgCO_3 + Ca(OH)_2 \rightarrow Mg(OH)_2 + CaCO_3 \qquad (7\text{--}14)$$

Figure 7–5, a solubility diagram for magnesium, shows that precipitation is enhanced at high pH values. In practice, sufficient lime is added to raise the pH to at least 10.7, and to sustain it at that level during the clarification process.

Recovery of magnesium is achieved by recarbonation to lower the pH to the point at which $Mg(HCO_3)_2$ is formed and removed. The magnesium can then be returned to the coagulation process. Alternatively, the magnesium can be held with the lime sludge for recalcination for lime recovery. The recovered lime is then used to raise the pH as previously explained. Lime–soda softening and recarbonation are described in Chapter 21.

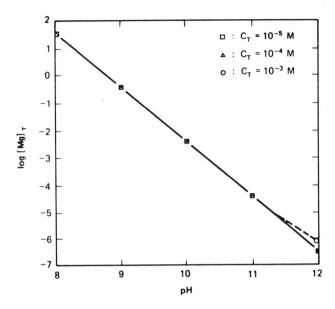

Fig. 7-5. Relationship between total soluble magnesium, pH and the final equilibrium total carbonic species concentration.[14] (Reprinted by permission of Prentice-Hall, Englewood Cliffs, New Jersey)

Coagulant Aids

Ideally, flocculated colloidal particles should settle rapidly and be strong enough to resist shearing forces. Often, the flocs do not possess these characteristics, and a coagulant aid is then added to improve floc properties. Coagulant aids that have been used include clays, activated silica, and polymers.

Bentonite clays have been used as coagulant aids for low-turbidity waters. The use of clay may reduce the amount of coagulant and improve the floc settleability. The reduction in required coagulant dose is caused by greater contact opportunities (increased colloid concentration) with subsequent charge neutralization.

The other advantage of using clays is that the floc particles are weighted by the clay particles which cause the floc to settle more rapidly than alum flocs. Bentonite doses in the range of 10 to 50 mg/l are generally sufficient for improved coagulation efficiency. However, the actual dosage should be determined by laboratory testing.

Activated silica has been used as a coagulant aid in water treatment plants. When used in conjunction with alum, it increases the rate of flocculation, improves floc toughness, and increases settleability.

The mechanisms by which activated silica are formed are described elsewhere.[22] Activated silica normally is used with alum, with the dose typically in the range of 7 to 11 percent of the alum dose. However, use of excess silica can be detrimental to coagulation. Activated silica has been successfully added both before and after alum addition, although the latter method is the more widely used. Jar tests should be used to identify the optimum combination of chemicals to use. Activated silica has been found to be an effective filter aid because it strengthens the flocs.[25]

The use of polymers has been discussed elsewhere; only its use as a coagulant aid is presented here. Both anionic and nonionic polymers have proved effective as coagulant aids. The polymers help to promote large floc particles by a bridging mechanism, after the colloidal particles have been destabilized by a coagulant such as alum. Nonionic polymers are more effective with increasing concentrations of divalent metal ions (Ca^{+2}, Mg^{+2}, etc.).

Anionic polymer doses in the range of 0.1 to 0.5 mg/l in association with alum improve floc settleability and toughness, compared to alum alone. However, overdosing the solution can inhibit coagulation and should be avoided.[24] A side benefit of using a combination of alum and polymer is the fact that frequently the alum dosage can be reduced, and less sludge is produced.

RAPID MIXING

The rapid mixer is used to provide complete and thorough mixing of the coagulant and the raw water. Typically, the rapid mix unit is the initial process in a water treatment plant. The mixing, or blending, is required to force multiple contacts between each of the colloidal particles and the products of the coagulation process. The overall process is controlled by hydrodynamic parameters, geometry, molecular properties of the water, and kinetics of the coagulation reactions. Proper design of the rapid mixer can result in lower coagulant doses and improved aggregation during flocculation.

Many devices have been used for rapid mixing, including baffled chambers, hydraulic jumps, mechanically mixed tanks, and in-line jet blending. The mechanically mixed tanks, which are typically termed "completely mixed or back-mixed units," have been the most commonly used devices. These tanks have been designed for detention times of 10 to 30 sec and velocity gradients (G) in the range of 200 to 1,000 sec^{-1}.

The principal mechanisms for alum coagulation, as discussed earlier, are adsorption and destabilization and sweep-floc coagulation.

The reactions are extremely rapid, occurring in microseconds without the formation of polymers, and less than 1 sec if polymers are formed.[15,27] However, sweep-floc coagulation is slower, occurring in the range of 1 to 7 sec.[28]

The coagulants should be dispersed as rapidly as possible in the raw water stream for adsorption–destabilization. The suggested dispersion time is 0.1 sec. This will allow the hydrolysis products that develop in 0.001 to 1 sec to cause destabilization of the colloid. The rapid mixer most suited to this requirement is an in-line blender.[29-31]

However, if sweep coagulation is used, extremely short dispersion times are not required because coagulation is due to colloidal entrapment by the hydroxide precipitate. For these conditions, a typical back-mix reactor would be suitable. Rapid formation of precipitates occurs when the solution is oversaturated by two orders of magnitude.[32]

In typical water treatment operations, only 0.001 mg/l alum (as $Al_2(SO_4)_3 \cdot 14.3H_2O$) is required to achieve that degree of oversaturation. When compared to frequently used coagulation doses of about 10 mg/l, the alum is oversaturated by nearly 10^4 times the amount needed for rapid precipitation, and can easily produce sweep floc.

Many studies do not make a distinction, regarding the time needed for rapid mixing, between the predominant modes of coagulation. Failure to recognize these differences has led to conflicting recommendations in the literature. Certain state standards require a "minimum" detention time of 30 sec in the rapid mixer.[33] This requirement is inappropriate for the adsorption–destabilization reactions and may not be long enough for sweep-floc coagulation and start of flocculation.

DESIGN EQUATIONS FOR RAPID MIXING

The most commonly used method for designing the rapid mix hardware is based on the velocity gradient or G-value, as developed by Camp.[34] The power requirements are computed from the equation:

$$G = \left(\frac{P}{\mu V} \right)^{1/2} \tag{7-15}$$

where:

G = velocity gradient, sec^{-1}
P = power input, ft-lb/sec ($N \cdot m/s$)
μ = dynamic viscosity lb-sec/sq ft ($N \cdot s/m^2$)
V = volume of water receiving input, cu ft (m^3)

Several studies have demonstrated that the velocity gradient does not completely describe the mixing process; but until a more complete understanding of the mixing process is developed, the G-value will continue to be used for designing rapid mix units.

Experience has shown that:

- For the same amount of turbidity removal, a higher G-value produces a denser floc.
- For the same turbidity removal, a higher Gt produces a denser floc.

There is a limitation to the premise that a higher Gt produces a denser floc, because high shear rates break up previously formed flocs. Therefore, the goal is to set G as high as possible without shearing the floc. Typical values for G and t are shown in Table 7–1.

Rushton developed a relationship for the power requirements for turbulent conditions:[36]

$$P = \frac{k}{g} \rho N^3 D^5$$

$$P = 1.36 \frac{k}{g} \rho N^3 D^5 \text{ (metric)}$$

(7–16)

where:

P = power, ft-lb/sec (N·m/s)
k = constant based on type of impeller
ρ = density of fluid, lb/cu ft (kg/m³)
N = impeller rotational speed, revolutions/sec
D = impeller diameter, feet (m)
g = gravitational acceleration, ft/sec² (m/s²)
 = 32.174 ft/sec² (9.81 m/s²)

The values of the constant range from 1.0 for a three-bladed impeller to 6.30 for a turbine with six flat blades.

The foregoing mathematical relations can be used to design a mechanical rapid mixer. The equations can be used to size the equipment and power requirements. Examples of their use are presented later.

Table 7–1. Typical Values for G and t.

DESCRIPTION	G, SEC^{-1}	Gt
Camp[50]	20 to 74	2×10^4 to 2×10^5
Turb/color removal; no solids recirculation[45]	50 to 100	1×10^5 to 1.5×10^5
Turb/color removal; solids contact reactors[45]	75 to 175	1.25×10^5 to 1.5×10^5
Softening; solids contact reactors[45]	130 to 200	2×10^5 to 2.5×10^5
Softening; ultrahigh solids contact[45]	250 to 400	3×10^5 to 4×10^5

RAPID MIX SYSTEMS

There are several different types of rapid mix units, some of which are described briefly in the following paragraphs.

Mechanical Mixing

Mechanical mixing is the most commonly used system for rapid mixers. This system is effective, has little headloss, and is unaffected by the volume of flows or flow variations. Typical design practice provides a contact time in the range of 10 to 30 sec and a G value in the range of 700 to 1,000 sec^{-1}.

Gemmell recommends a 20-sec or less detention time and a power input of 1 to 2 hp/cu ft/sec (26.36 to 52.72 kw/m³s) of flow.[37] He further states there are specific guidelines for determining detention time or power dissipation required to disperse chemicals. Examples of typical rapid mix units are shown in Figs. 7–6 and 7–7.

Camp conducted studies on Boston tap water using ferric sulfate.[38] These studies indicated that a maximum floc volume was obtained with a velocity gradient of 700 to 1,000 sec^{-1}, a detention time of 2 to 2½ minutes, and a flat-bladed mixer. The detention time is significant because precipitation of ferric hydroxide occurred within 8 sec. The dosage of 15 mg/l created sweep-floc coagulation.

Another study states that the G-value "does not provide a complete characterization of the mixing in the rapid-mix operation."[28] This study found that the optimum period of rapid mix was 2½ minutes at a G-value of 1,000 sec^{-1}.

TeKippe and Ham demonstrated that a rapid mixer design, which has a detention time of long enough duration to allow floc particles to reach near-equilibrium sizes, provides for optimum sedimentation.[39] The rapid mixer should be followed by a tapered flocculation velocity gradient. Visible floc formation ranged in time from 2 to 6 minutes.

Finally, Vrale and Jorden tested five different types of rapid mix units, and concluded that the "backmix reactor is very inefficient for rapid mixing."[29] They further conclude that "a tubular reactor appears to be the most efficient type."

Amirtharajah summarized recommended guidelines for designing a mechanical rapid mix unit:[40]

- A square vessel is superior in performance to a cylindrical vessel.
- Stator baffles are advantageous.
- A flat-bladed impeller performs better than a fan or propeller impeller.
- Chemicals introduced at the agitator blade level enhance coagulation.

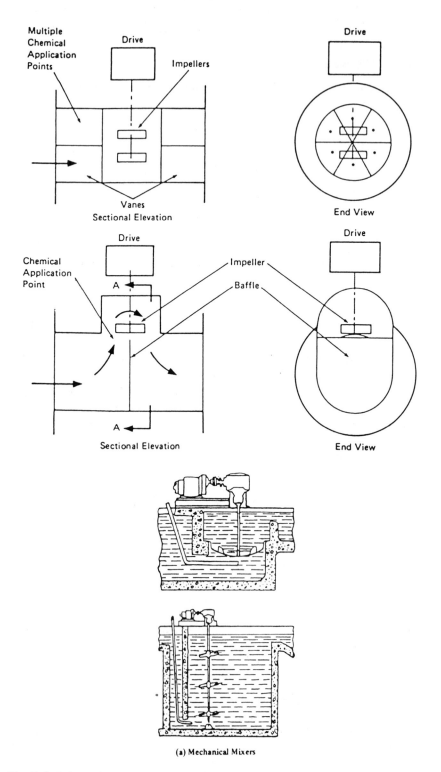

Multiple Chemical Application Points

Drive

Impellers

Vanes

Sectional Elevation

Drive

End View

Chemical Application Point

Drive

A

Impeller

Baffle

Sectional Elevation

A

Drive

Impeller

Baffle

End View

(a) Mechanical Mixers

Fig. 7–6. Schematic representations of in-line blenders. Upper unit: Walker Process Corp. Lower unit: Philadelphia Mixer Corp.

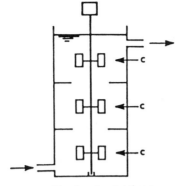

a. Mechanical Mixer

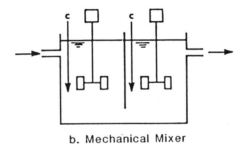

b. Mechanical Mixer

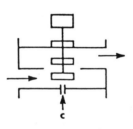

c. In-line Blender

Fig. 7–7. Representative types of mechanical agitation devices used for rapid mixing.[45]

The mechanical mixer has been used in numerous water treatment plants. Because the mixer speed can be changed by including a variable speed drive, it is amenable to operational changes due to changing conditions. Lower speeds result in lower G-values, which are used with the addition of polyelectrolytes.

In-Line Blenders

The adsorption–destabilization reactions are very rapid, and in-line blenders were developed to approach instantaneous mixing of chemicals. The G-value suggested is in the range of 3,000 to 5,000 sec^{-1}. Hudson recommended using in-line blenders with a residence time of 0.5 sec and a water hp of 0.5 hp/mgd (0.99 kW/Ml/d) of flow.[41]

Kawamura prefers the use of in-line blenders, and gives the following reasons:[42,43]

- In-line blenders provide virtually instantaneous mixing with a minimum of short-circuiting.
- There is no need to consider headlosses.
- In-line systems are less expensive than more conventional rapid mix units.

Jet Injection Blending

A study by Vrale and Jorden showed that a jet injection device, which introduced the coagulant through six holes [0.028 inch (0.71 mm) diameter], was superior to a typical backmix reactor.[29] The jet injection system required a G-value of 1,000 sec^{-1} to achieve the maximum particle aggregation rate. Other units tested required G-values of 6,000 to 9,000 sec^{-1} to achieve the same aggregation rate. Chao and Stone developed a typical design for a jet injection system.[30] The result was a G-value $\simeq$ 1,000 sec $^{-1}$ and a detention time of 0.55 sec.

A unit used for full-scale applications is shown in Fig. 7–8. The utilization of this type of unit is limited in practice, although it has been shown to have potential advantages. Disadvantages include plugging of the orifices and the fact that mixing intensity cannot be varied.

The unit shown in Fig. 7–8 has design criteria of:[43]

- G = 750 to 1,000 sec^{-1}
- Dilution ratio at maximum alum dose = 100:1
- Flow velocity at injection nozzle = 20 to 25 ft/sec (6.1 to 7.6 m/s)
- Mixing time = 1 sec

The power input, P (ft-lb/sec or watts), for this type of flash mixer can be computed from:

$$P = q(\triangle H)\rho \qquad (7\text{–}17)$$
$$P = 1.356 \, q(\triangle H)\rho \text{ (metric)}$$

where:

P = power input, ft-lb/sec (N·m/s)
q = flow rate from orifice hole—jet discharge, cu ft/sec (m³/s)
   $= v \cdot a = C_d a v$
ΔH = total jet energy loss, feet (m)
   $= v^2/2g$
ρ = density of water, lb/cu ft (kg/m³)
a = area of orifice, sq ft (m²)
v = jet velocity, ft/sec (m/s)
C_d = discharge coefficient (± 0.75)

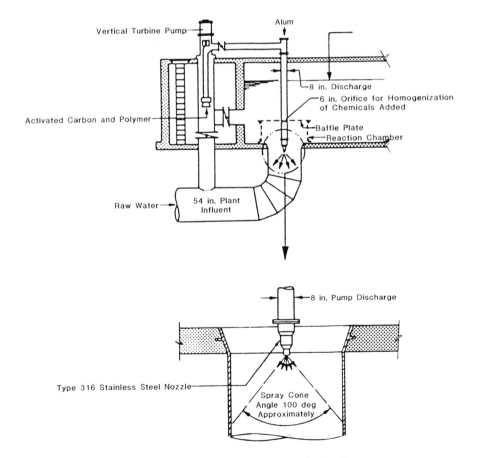

Fig. 7-8. Typical flash mixing facility.[43]

Substituting known values in equation (7–17) yields:

$$P \simeq 0.97 \ C_d \cdot a \cdot v^3 \qquad (7\text{–}18)$$

A valve on the chemical pump discharge line allows flexibility in changing the power input. This system can be adjusted to accommodate changing raw water conditions.

Hydraulic Mixing

Hydraulic jumps have been used for mixing chemicals. Frequently, plant flows are measured by a Parshall flume, or other similar device, which incorporates a hydraulic jump downstream by including an abrupt drop in the channel. The coagulants are introduced immediately upstream of the flume. Typical residence times are about 2 sec with a G-value of about 800 sec^{-1}.

Chow presents the mathematical equations required to compute the G-values.[44] The headloss through the flume varies with the flow rate and can be computed or obtained from discharge tables. The principal advantages of this unit are:

- No mechanical equipment to operate and maintain.
- Lower cost because there is no separate rapid mix unit.

DESIGN EXAMPLES FOR RAPID MIX SYSTEMS

In-Line Blender

If jar tests indicate that adsorption–destabilization is the predominant mechanism, an in-line blender is the appropriate mixer. Assume the following plant sizing and criteria:

- Average day flow = 2.0 mgd (7.57 Ml/d)
- Maximum day flow = 3.0 mgd (11.36 Ml/d)
- Temperature range = 46.4 to 68°F (8 to 20°C)

Select a typical commercial mixer at a water horsepower of 0.5 hp/mgd (0.99 kW/Ml/d) of flow. For maximum day flows, the hp requirement is:

$$\text{hp} = 0.5 \ \text{hp/mgd} \times 3.0 \ \text{mgd} = 1.5 \ \text{water hp} \ (1.12 \ \text{N} \cdot \text{m/s})$$

Assume a detention time of 0.5 sec in the mixer. This should be checked for the commercial blender being considered. The lowest temperature gives the highest water viscosity, and therefore the lowest G-value.

Power input $= 1.5 \times 550$ ft-lb/sec hp
$= 825$ ft-lb/sec $(1.12$ N $\cdot$ m/s$)$

Volume of mixer $= 1.5$ mgd $\times 1.547$ cu ft/sec/mgd $\times 0.5$ sec
$= 1.160$ cu ft $(0.033$ m$^3)$

Viscosity at $46.4°F$ $(8°C)$:
$\mu = 1.387$ centipoises $\times 2.088 \times 10^{-5}$ lb sec/sq ft/centipoise
$= 2.90 \times 10^5$ lb sec/sq ft $(1.379 \times 10^{-3}$ N $\cdot$ s/m$^2)$

Velocity gradient G:

$$G = \left[\frac{P}{\mu V}\right]^{1/2} = \left[\frac{825}{2.90 \times 10^{-5} \times 1.160}\right]^{1/2} = 4{,}900 \text{ sec}^{-1}$$

At a temperature of $68°F$ $(20°C)$:

$$G = \left[\frac{825}{2.092 \times 10^{-5} \times 1.16}\right]^{1/2} = 5{,}800 \text{ sec}^{-1}$$

The G-value for the higher temperature is above the recommended maximum value of 5,000 sec^{-1}. The design could be adjusted to reduce the G-value, but that is not considered necessary.

Backmix Reactor

Assuming the jar test results indicate sweep-floc coagulation as the predominant mechanism, a backmix reactor is suitable. The same plant size and criteria as above will be used, with an assumed mean alum dose of 25 mg/l. Using the equation, $Gt_{opt} C^{1.46} = 5.9 \times 10^6$, the value of Gt can be estimated:

$$Gt_{opt} = \frac{5.9 \times 10^{-6}}{25^{1.46}} \simeq 53{,}700$$

Although this equation was developed for only one type of colloid suspension, it does provide a starting point, and the results should be checked against normally accepted design practice.

For $G = 1{,}000$ sec^{-1}, $t_{opt} = 53.7$ sec.

For $G = 700$ sec^{-1}, $t_{opt} = 76.7$ sec.

Based on *Water Treatment Plant Design*,[45] for mixing times greater than 40 sec, a G-value of 700 sec^{-1} should be used. Use a detention time of 60 sec and a G-value of 700 sec^{-1}.

The volume of the rapid mix chamber is calculated as:

$V = 3.0$ mgd $\times 1.547$ cu ft/sec/mgd $\times 60$ sec
$= 278.4$ cu ft $(7.88$ m$^3)$

Design for at least two units, and make the dimensions of each unit compatible with the raw water influent channel or pipe.

Using two units:

$$Volume = \frac{278.4}{2} = 139.2 \text{ cu ft}$$

Use a square chamber with 5-foot side dimensions:

$$Depth = \frac{139.2}{5 \times 5} = 5.568 \text{ ft}$$

Design the rapid mix chamber for 6-foot depth. Use a turbine mixer with incline blades. The actual mixer dimensions and recommendations should be obtained directly from the manufacturer.

Compute motor horsepower by rearranging equation (7–15) as follows:

$$P = G^2 \, \mu V$$
$$= 700^2 \times 2.9 \times 10^{-5} \times 150$$
$$= 2131.5 \text{ ft-lb/sec}$$

$$Water \text{ hp} = \frac{2131.5}{550} = 3.9 \text{ hp}$$

Using an overall efficiency for motor and drive of 80 percent:

$$Motor \text{ hp} = \frac{3.9}{0.8} = 4.8 \text{ hp (Use 5 hp motor)}$$

Jet Injection

Chao and Stone present a typical design application for the jet mixing system.[30] Reference should be made to Chao and Stone and to manufacturer's literature for a more thorough understanding of this rapid mix unit.[30]

TEMPERATURE EFFECTS

Morris and Knocke studied the effect of temperature on coagulation.[46] Specifically, the study:

- Determined the impact of low temperature conditions on the efficiency of metal ion coagulants for turbidity removal from surface waters.
- Investigated fundamental parameters such as reaction rate kinetics and particle size characteristics to aid in describing the observed results.

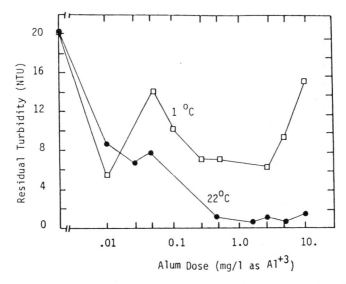

Fig. 7–9. Temperature effects on turbidity removal using aluminum sulfate.[46]

This study showed that low temperatures did not affect the rate of metal ion precipitation in the pH range 6.0 to 8.0. However, in the evaluation of turbidity removal, a decrease in water temperature was accompanied by a decrease in turbidity removal, as shown in Figs. 7–9 and 7–10 for aluminum sulfate and ferric chloride.

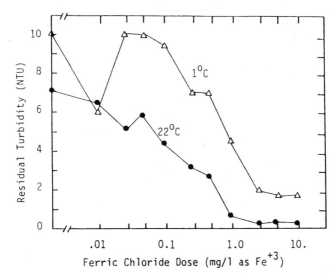

Fig. 7–10. Temperature effects on turbidity removal using ferric chloride.[46]

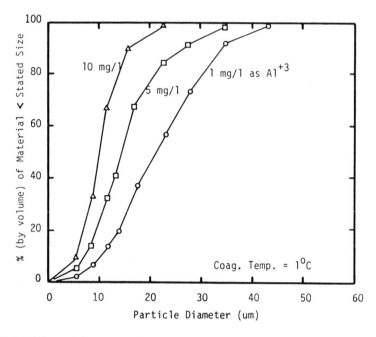

Fig. 7–11. Effect of alum coagulant dose on particle size distribution produced after flocculation.[46]

Regarding particle size distribution, an increased alum dose resulted in the production of a smaller size floc, which would settle poorly. This is shown in Fig. 7–11. The practical implication of this phenomenon is that an increase in alum dosage during cold temperatures offers no improved turbidity removal.

FLOCCULATION SYSTEMS

The flocculation process aggregates destabilized particles into larger and more easily settleable flocs. This process typically follows rapid mixing. Destabilization results from chemical reactions between the coagulant and the colloidal suspension, while flocculation is the transport step that causes the necessary collisions between the destabilized particles.

There are two major mechanisms of flocculation:

• Perikinetic, which is the aggregation of particles as a result of random thermal motion of fluid molecules. This is significant for particles that are 1 to 2 μ in size.
• Orthokinetic, which is the aggregation of particles by induced velocity gradients in the fluid.

Orthokinetic flocculation is the predominant mechanism in water treatment. In addition, sludge blanket or solids contact clarifiers cause differential and fluctuating velocities, which lead to particle collisions and aggregation. An understanding and knowledge of the kinetics created by orthokinetic flocculation is required to optimize the design of flocculators. The parameters that define the rate of aggregation also define the physical dimensions of the process equipment and basin.

This section presents information on kinetic models that can be used to design flocculation systems. Typical design examples are presented.

Orthokinetic Flocculation

In systems that are mixed (velocity gradients are induced), the velocity of the fluid varies both spatially (from point to point) and temporally (from time to time). The spatial changes in velocity are termed the velocity gradient, G, sec^{-1}. In water treatment plants, mean velocity gradients of 10 to 100/sec are typical. Flocculation tanks are ineffectual until the colloidal particles reach a size of 1 μ, through contacts produced by Brownian motion. For example, flocculation tanks cannot aggregate viruses, which are 0.1 μ in size or smaller, until they are adsorbed or enmeshed in larger flocs or particles.

Camp developed the following equations for computing the velocity gradients for mixing chambers and pipes:[34]

Mechanical mixing:

$$G = \left[\frac{P}{\mu V}\right]^{1/2} \tag{7-19}$$

$$G = 425 \left[\frac{h p_w}{t}\right]^{1/2} \tag{7-20}$$

Ports and conduits:

$$G = \left[\frac{P}{\mu V}\right]^{1/2} \cdot V^{3/2} \tag{7-21}$$

$$G = 172 \left[\frac{f}{D}\right]^{1/2} \cdot V^{3/2} \tag{7-22}$$

Baffled chambers:

$$G = 178 \left[\frac{H}{t}\right]^{1/2} \tag{7-23}$$

where:

P = power input, ft lb/sec (N·m/s)
V = velocity, ft/sec (m/s)
μ = dynamic viscosity, lb sec/sq ft (kg·s/m²)
f = Darcy-Weisbach friction factor based on roughness height of 0.00085 ft (0.2591 mm)
D = conduit diameter, ft (m)
H = headloss, ft (m)
t = theoretical detention time, minutes

Camp analyzed several flocculation basins and determined that flocculation basins having values in the range of 2×10^4 and 2×10^5 for the nondimensional parameter Gt performed satisfactorily.[50] G-values for satisfactory performance were found to be in the range of 20 to 74 sec^{-1}.

For paddle-type mechanical flocculators, the power dissipated in the liquid, P, can be determined from:

$$P = \tfrac{1}{2}\, C_d A \rho v^3 \qquad (7\text{--}24)$$

where:

P = power dissipated, ft lb/sec (N·m/s)
C_d = coeffcient of drag
    = 1.8 for flat plates
A = area of paddle, sq ft (m²)
ρ = density of liquid, slugs/cu ft, (kg/m³)
v = velocity of paddles relative to liquid
    ≈ 0.5 to 0.75 × velocity of paddle, ft/sec (m/s)

Equation (7–24) should be integrated for an elemental area if velocity changes occur along the length of a paddle due to distance from the shaft.[51]

There are two important conclusions that can be drawn from past experience:[54,55]

- There is a minimum time below which no flocculation occurs, whatever the value of G (i.e., aggregations and break-up are equal).
- The use of reactors in series can significantly reduce the overall detention time for the same degree of treatment.

The second conclusion has been identified and confirmed by other investigators, and the recommended minimum number of compartments is three.[31,50]

A recent study indicates that optimum values exist for the flocculation time, T, and the velocity gradient, G.[56] Bench-scale studies with alum as the coagulant and kaolin clay indicated that flocculation time should be in the range of 20 to 30 minutes. Increases in time did not improve flocculation significantly. The optimum value of G is computed from the equation:

$$(G^*)^{2 \cdot 8} \cdot T = K \qquad (7\text{--}25)$$

where:

$G^* =$ optimum velocity gradient, sec^{-1}
$T =$ flocculation time, minutes
$K =$ constant
$\quad = 4.9 \times 10^5$ for alum concentration of 10 mg/l
$\quad = 1.9 \times 10^5$ for alum concentration of 25 mg/l
$\quad = 0.7 \times 10^5$ for alum concentration of 50 mg/l

Equation (7–25) can be simplified by combining the empirical results into a single, approximate equation, as follows:[48]

$$(G^*)^{2.8} \cdot T = \frac{44 \times 10^5}{C} \qquad (7\text{--}26)$$

where $C =$ alum concentration, mg/l, in the range of 0 to 50 mg/l.

The optimum value of G was defined as the velocity gradient that minimizes residual turbidity by flocculation and settling. The range of optimum G was 20 to 50 sec^{-1}.

The expression $(G^*)^{2.8}\,CT$ is similar to the expression GCT suggested by other investigators as the design parameter.[37,57]

Studies have shown tapered flocculation with a diminishing velocity gradient to be more efficient than uniform velocity gradient flocculation.[43,50,58] The recommended overdesign factor, in order to handle variation in flow rates, is approximately 1.4.[59] This gives an objective value for providing operational adjustments in the design of flocculation systems.

FLOCCULATION ALTERNATIVES

The following types of mechanical mixing devices are typically used in water treatment flocculation:

- Paddle or reel-type devices
- Reciprocating units (walking beam flocculator)
- Flat blade turbines
- Axial flow propellers or turbines

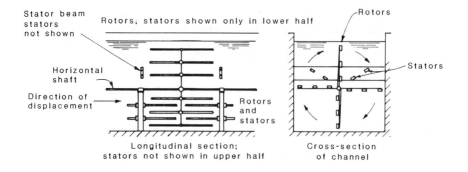

Stator beam stators not shown

Rotors; stators shown only in lower half

Rotors

Horizontal shaft

Direction of displacement

Rotors and stators

Stators

Longitudinal section; stators not shown in upper half

Cross-section of channel

(a) Paddle type with rotors and stators.

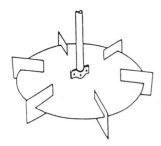

(b) Plate turbine type.

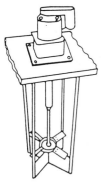

(c) Axial flow propeller type with straightening vanes.

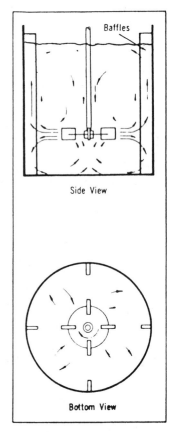

Baffles

Side View

Bottom View

(d) Schematic of radial flow pattern in baffled tank.

Fig. 7-12. Typical flocculation units.[48]

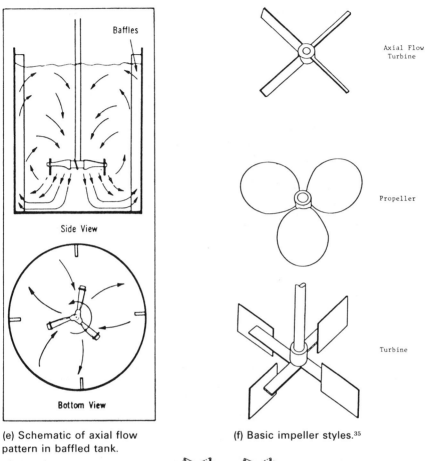

(e) Schematic of axial flow pattern in baffled tank.

(f) Basic impeller styles.[35]

(g) Walking beam flocculator.

Fig. 7-12 (continued)

Typical units are shown in Fig. 7-12. The spatial distributions of velocity gradients that the units produce are shown in Fig. 7-13 for some of the mixers.

The paddle or reel-type devices are mounted horizontally or vertically

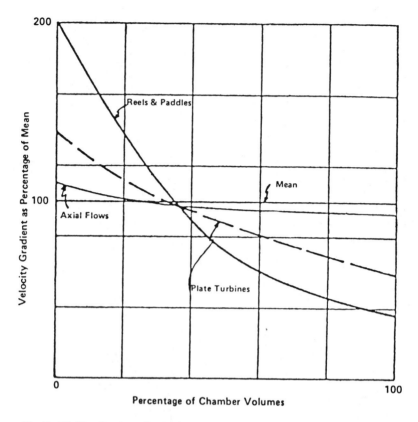

Fig. 7-13. Distribution of velocity gradients in mechanical flocculators.[41,69]

and rotate at low speeds (2 to 15 rpm). Design is based on limiting the tip speed of the paddle farthest from the center axis to 1 to 2 ft/sec. Argaman and Kaufman used a stake and stator device for their studies.[54] This unit is similar to reel-type units, but is mounted vertically, and was found to be superior to a turbine.

Walking beam flocculators are driven in a vertical direction, in a reciprocating fashion. The unit contains a series of cone-shaped devices on a vertical rod. The cone devices impart energy to the water as they move up and down, thereby creating velocity gradients. The manufacturer's literature should be consulted to design this unit.

Turbines are flat-bladed units connected to a disc or shaft. The flat blades are in the same plane as the drive shaft. The blades can be mounted vertically or horizontally, and typically operate at 10 to 15 rpm. Walker found that plate turbines are effective up to a G-value of 40 sec^{-1}, but produced high-velocity currents at G-values greater than 45 sec^{-1}.[60] His suggested design criterion is to limit the maximum peripheral velocity to 2 ft/sec for weak floc and 4 ft/sec for strong floc. Other investigators have found that turbines are the least effective units for flocculation.[41,54,60]

The axial flow unit "pumps" liquid because the impeller has pitched blades. This unit may be installed vertically or horizontally. Typically, these units are high-energy flocculation devices operating at 150 to 1,500 rpm, and there is no limitation on the tip speed. Hudson and Walker favor these units because they are simple to install and maintain, and produce uniform turbulence in the flocculator.[41,60]

DESIGN EXAMPLES, FLOCCULATION

The design examples presented will be continuations of the examples presented for the rapid mix design.

Example 1 (In-Line Blender)

Determine the optimum velocity gradient, based on alum doses of 15 to 25 mg/l. The advantages of compartmentalization have been shown before. Assume a four-compartment flocculator. Using equation (7–26):

$$(G^*)^{2.8}T = \frac{44 \times 10^5}{15} \text{ to } \frac{44 \times 10^5}{25}$$
$$= 2.933 \times 10^5 \text{ to } 1.76 \text{ to } 10^5$$

Using a flocculation time of 30 minutes because longer times do not increase efficiency significantly:

$$(G^*)^{2.8} = \frac{2.933 \times 10^5}{30} \text{ to } \frac{1.76 \times 10^5}{30}$$
$$= 9776.6 \text{ to } 5866.6$$
$$G^* = 26.6 \text{ to } 22.1 \text{ sec}^{-1}$$

Check G^* values for a detention time of $T = 20$ minutes. The corresponding range of G^* values is 30.7 to 25.6 sec^{-1}.

Use an optimal G^* value of 25 sec^{-1}, and a detention time of 20 minutes. The product Gt is 3.0×10^4, which is within the recommended guideline of 10^4 to 10^5.

To incorporate the advantages of tapered flocculation, the optimum design G^* of 25 sec^{-1} would be used in the third compartment. The G-values are tapered on either side to yield $G_1 = 45$ sec^{-1}, $G_2 = 35$ sec^{-1}, $G_3 = 25$ sec^{-1}, and $G_4 = 20$ sec^{-1} in the four compartments.

The detention time of each compartment is 5 minutes.

Design flow rate $= 2.0 \times 1.547 = 3.09$ cu ft/sec (0.087 m³/s)
Volume in each compartment $= 3.09 \times 5 \times 60 = 927$ cu ft (26.23 m³)

Assume square-shaped compartments, and a water depth of 10 feet. Then:

$$\text{Side dimension} = \left[\frac{927}{10}\right]^{1/2} = 9.62 \text{ ft. Use 10 ft (305 m)}$$

Each compartment would be 10 ft $\times$ 10 ft $\times$ 12 ft ($3.05 \times 3.05 \times 3.65$ m) deep, which provides a 2 ft (0.61 m) freeboard allowance.

For simplicity, an axial flow impeller will be used for this example. For the first compartment, $G = 40$ sec^{-1}. Provision should be made to operate the unit at variable speeds to allow adjustment of the G-value. Assume G-values in the range of 90 to 20 sec^{-1}. Use the highest viscosity with the highest probable G-value to obtain the power requirements:

$$P = G^2 \mu V$$

Viscosity at 46.4°F (8°C), $\mu = 1.3872$ centipoise $\times 2.088 \times 10^{-5}$ lb sec/sq ft centipoise

$$= 2.896 \times 10^{-5} \text{ lb sec/sq ft}$$
$$(1.377 \times 10^{-3} \text{ N} \cdot \text{s/m}^2)$$
$$P = (90)^2 \times 2.896 \times 10^{-5} \times (10 \times 10 \times 10)$$
$$= 234.6 \text{ ft-lb/sec (0.318 kW)}$$

Assume overall efficiency of the motor and drive as 75 percent. Then:

$$\text{motor horsepower} = \frac{234.6}{550 \times 0.75} = 0.57 \text{ hp (0.43 kW)}$$

Use a standard ¾ hp (0.56 kW) motor. Now check the rotational speed and G-values.

$$P = \frac{k}{g} \rho N^3 D^5$$

Assume $D = 2$ ft (0.61 m); then each blade is 1 ft (0.305 m) long.

$$(¾ \times 550 \times 0.75) = \frac{1.0}{32.2} \times 62.4 \times N^3 \times 2^5$$

$$N = 1.709 \text{ revolutions/sec}$$
$$= 102.5 \text{ rpm}$$

Use a 4 to 1 variable speed drive, and check the value for G at the lowest speed.

$$P = \frac{1.0}{32.2} \times 62.4 \times \left(\frac{25}{60}\right)^3 \times 2^5$$

$$= 4.49 \text{ ft-lb/sec } (6.09 \text{ N} \cdot \text{s/m}^2)$$

and:

$$G = \left[\frac{4.49}{2.896 \times 10^{-5} \times (10 \times 10 \times 10)}\right]^{1/2} = 12.4 \text{ sec}^{-1}$$

The range in G-values is satisfactory. For standardization, put the same unit in all four compartments. It may be possible to reduce the horsepower in the third and fourth compartments because of the lower G-values required.

Example 2 (Paddle-Type Unit)

Assume the plant flow rate is 20 mgd (75.7 Ml/d), and the lowest temperature is 53.6°F (12°C). The alum doses, based on jar tests, ranged from 20 to 30 mg/l. From equation (7–26), determine the optimum G-value.

$G^* = 24 \text{ sec}^{-1}$ and 20.8 sec^{-1}, respectively for $T = 30$ minutes.

Use a G-value of 24 sec^{-1} and a detention time of 30 minutes. The product Gt is 4.32×10^4, which is within the recommended range.

Assume a three-compartment system with the optimum G-value in the middle. The G-values on either side would be 30 sec^{-1} and 15 sec^{-1}.

Assume a rectangular shaped flocculator, and a 12-foot sidewater depth:

$$\text{Volume} = 20 \times 1.547 \times 30 \times 60$$

$$= 55,692 \text{ cu ft } (1576 \text{ m}^3)$$

Assume width of basin, L, is one-third of the length. Then:

$$L \times 3L = \frac{55,692}{12}$$

$$L = 39 \text{ ft. Use } 40 \text{ ft } (12.2 \text{ m})$$

Length of basin $= 120$ ft (31.1 m)

Width $= 40$ ft (12.2 m)

Water depth $= 12$ ft (3.66 m)

Now:

$$P = G^2 \mu V$$
$$= 24^2 \times 2.896 \times 10^{-5} \times (120 \times 40 \times 12)$$
$$= 960.8 \text{ ft lb/sec } (1.3 \text{ kW})$$

$$\text{Motor horsepower} = \frac{960.8}{550 \times 0.75} = 2.33 \text{ hp. Use 3 hp } (2.24 \text{ kW})$$

Limit the tip speed (V_p) to 1.5 ft/sec (0.46 m/s).
Compute the area of the paddles:

$$P = \tfrac{1}{2}\ C_d A \rho v^3$$

$$960.8 = \tfrac{1}{2} \times 1.8 \times A \times \frac{62.4}{32.2} \times (0.75\ V_p)^3$$

$$A = 386.9 \text{ sq ft } (36.0 \text{ m}^2)$$

The diameter of the farthest paddle is the depth of water less a clearance allowance. Use an 11-foot (2.13-m) diameter.

$$1.5 \text{ ft/sec} = 2\pi rn$$

$$= 2\pi\ 11\tfrac{1}{2} \times \frac{n}{60}$$

$$n = 2.6 \text{ rpm}$$

Assume eight paddles per shaft and three shafts. Each paddle is 36 feet (10.97 m) long.

$$\text{Width} = \frac{500}{8 \times 36 \times 3} = 0.58 \text{ ft} = 7 \text{ inches } (0.18\text{m})$$

The outer paddle is positioned at a radius of 5½ feet (1.68 m) to the outside edge and the inner paddle at a radius of 3¾ feet (1.14 m) to the outside edge.

COAGULATION CONTROL AND MONITORING

Techniques Available

An excellent physical treatment plant design and good equipment selection are worthless unless the chemical coagulation of the raw water being treated is properly carried out. Inadequate doses of coagulant will result in an excessively turbid finished water. Excessive coagulant doses may also cause this result and are wasteful of the public's funds. Excessive turbidity makes the water aesthetically unacceptable besides increasing the probability of microorganisms escaping subsequent disinfection processes. Public acceptance of the entire water supply program depends on proper coagulation.

Since achieving proper coagulation has been a universal problem of water treatment operators for many years, a wide variety of techniques have been developed for controlling the coagulation process. Most of these involve laboratory tests, the results of which then must be manually transferred to the full-scale plant operation by the plant operator.

Whether or not the coagulant dose that provides the optimum result in the laboratory tests will also provide optimum results on a plant scale depends on whether the same efficiency of mixing is achieved in both cases (which is unlikely). Also, the fact that the procedure is a batch procedure results in an inherent time lag in responding to changes in raw water conditions. This lag may be only an hour if the operator is on duty and alert to raw water conditions, but it may be several hours if the operator is off duty or involved in another task, such as equipment maintenance. This method may give satisfactory—even if not optimum—results when applied to a raw water of relatively uniform quality that contains only a moderate amount of organic turbidity. For low-turbidity waters and waters containing large amounts of organic material, no sharp distinction between coagulant doses near optimum may exist; so it is difficult to determine from the appearance of the sample whether adequate plant-scale results will be obtained. Although adequate control can be achieved under certain conditions by visual monitoring of jar tests, at other times it becomes a guessing game, with each observer seeing characteristics that may or may not affect plant performance.

Available coagulation control techniques fall into three general categories: conventional and modified jar tests where visual observations of supernatant, floc formation time, floc density, and so on are used; techniques based on particle charge; and techniques based upon filtering the coagulated water and measuring the filtrate turbidity.

Jar Tests

A standard jar test procedure is contained within AWWA manual M-12, entitled *Simplified Procedures for Water Examination.* A standard testing plan is required to provide a means for the comparison of coagulants or for coagulation aids. The procedure basically consists of adding varying coagulant dosages to several samples of the water in beakers, mixing them simultaneously with a gang mixer, allowing them to settle, and observing the results.

Manual M-12 states that:

. . . the jar tests are designed to show the nature and extent of the chemical treatment which will prove effective in the full-scale plant. Many of the chemicals added to a water supply can be evaluated on a laboratory scale by means of jar tests. Among the most important of these chemicals are coagulants, coagulant aids, alkaline compounds, softening chemicals, and activated carbon for taste and odor removal. The test also permits the appraisal of the relative merits of the aluminum and iron coagulants, or in conjunction with such coagulant aids as activated silica, polyelectrolytes, clays, stone dust, activated carbon, settled sludge, lime, and soda ash.

The manual also notes several considerations that may affect the outcome of jar testing, commenting that:

. . . even the smallest detail may have an important influence on the result of a jar test. Therefore, all samples in a series of tests should be handled as nearly alike as possible. The purpose of the test will determine such experimental conditions as flocculation and settling intervals. Since temperature plays an important role in coagulation, the raw water samples should be collected and measured only after all other preparations have been made, in order to reduce the effect that room temperature might have on the sample.

The importance of simulating plant conditions as closely as possible cannot be overemphasized. The manual states that "dosing solutions or suspensions should be prepared from the stock materials actually used in plant treatment. Distilled water used for the preparation of lime suspensions should be boiled for 15 minutes to expel the carbon dioxide and then cooled to room temperature before the lime is added." Details are presented in AWWA manual M-12 on the preparation of coagulant dosing solutions and suspensions.

In testing polymeric coagulants or coagulant aids supplied by cooperating polymer manufacturers, solutions of these products should be prepared in accordance with manufacturer specifications.

Techniques Based on Particle Charge

Zeta Potential.   The first technique of this type to receive attention as a possible means of coagulation control was the zeta potential technique. It has received a great deal of published attention, and a few references will provide the reader with a more extensive bibliography as well as a more detailed introduction to the subject if she or he so desires.[63-66]

Zeta potential is a measure, in millivolts, of the electrical potential between the bulk liquid and the layer of counterions surrounding the colloids, as described earlier. Because like charges repel, colloids in water—which are almost always negatively charged—resist coagulation. When this negative zeta potential is reduced, the repulsive forces are likewise reduced, and if agitated gently, the colloids will flocculate. In the treatment process, the reduction of zeta potential is accomplished by the addition of a positively charged ion or complex from such coagulants as aluminum sulfate, the iron salts, and cationic polyelectrolytes.

Mobility values are expressed as positive (+) for those particles that migrate toward the negative pole of the cell, and as negative (−) for those that migrate toward the positive pole. The point at which there is no particular migration (zero mobility) is considered to be the isoelectric point.

The use of this technique is also a batch procedure, as is the jar test. The general procedure involves varying the coagulant dosage and measuring the resulting zeta potential. However, each water requires comparative tests to determine the correlation between zeta potential and finished water turbidity in the plant. Organic colloids such as those constituting organic color generally require zeta potentials near zero, while clay-related turbidity is best removed at somewhat negative zeta potentials. Typical values for optimum coagulation range from +5 to −10 millivolts, depending upon the nature of the material to be removed.

The procedure for measuring zeta potential is somewhat tedious for routine use in the average water treatment plant. Black has presented a procedure for using the Briggs cell in conjunction with a microscope for zeta potential measurement.[63]

Although zeta potential measurements are useful as a research tool, it is generally agreed that this method is not easily adapted to the typical treatment plant because of the considerable degree of skill and patience required to make the measurements, and the amount of interpretation required to make the data useful. It has been the authors' experience in visiting plants that have purchased zeta meter equipment for routine operation that generally it will be found stored and inoperative. Other observers also have reported a recent decline in the popularity of zeta potential as a control technique.[61] It is subject to the same shortcomings as any other batch test, in that sudden changes in raw water conditions may not be detected until a poor-quality finished water is produced in the full-scale plant.

Streaming Current Detector. A streaming current detector may be used to provide a continuous measure of the relative charges. This technique involves placing a sample in a special cylinder containing electronic sensing electrodes at the top and the bottom. A loose-fitting piston is then partly submerged into the sample, and is reciprocated along its axis to produce an alternating current between the electrodes when the cylinder contains moving charges. A synchronous motor drives the piston and a synchronous rectifier switch, by means of which the alternating current generated by the alternating fluid motion is made to register on a dc meter (Fig. 7–14). An amplifier with adjustable negative feedback is used to provide an output proportional to the current collected by the electrodes. Readout may be by

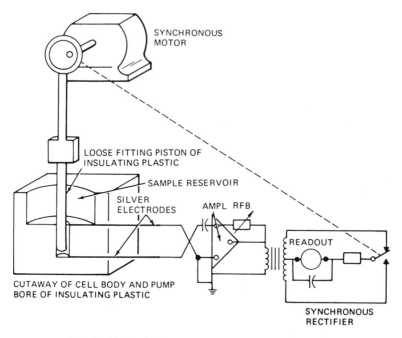

Fig. 7–14. Simplified diagram of SCD instrument.[70]

a microammeter, with calibration in arbitrary units. The alternating current is analyzed and related to zeta potential, which can be used to control coagulant dosage. A flow-through cylinder and recording ammeter may be built into the system to permit continuous monitoring. Almost all of the successful applications of the instrument have involved the titration of charge-influencing materials, either batchwise or continuously. In these applications, uncertainty as to what is measured is of little consequence as long as the addition of anionic and cationic materials changes the reading in a reproducible manner. Titrations are carried out in the same fashion as acid–base titrations with a pH meter. Continuous readings are obtained while one material is added to the other.

The use of this approach for continuous control of coagulation is subject to the other limitations of the zeta potential approach, including definite limitations when used to evaluate synthetic polymer coagulants, such as when the polymer is anionic and is used to coagulate negatively charged particles. Because the technique involves only the aspects of electronic charge, it may lead to erroneous conclusions when it is used to study coagulants not following the electrokinetic theory. As with zeta potential, the data must be correlated with the usual indices of plant performance, as there may not be any consistent relationship between charge and filtered water clarity even at a given plant for various seasons of the year. However, the continuous nature of the streaming current detector may make it attractive in some instances where cationic coagulants are used in a water that shows little variation from season to season.

Colloid Titration. A simple titration technique using an indicator that changes colors when a solution is titrated to electric neutrality has been developed.[67,68] An excess amount of positively charged polymer is added to the naturally negatively charged water. It is then back-titrated with a standard negatively charged polymer. An empirical correlation is used to relate the volume of titrant to the proper coagulant dosage, in a manner analogous to zeta potential measurement.

Like the streaming current detector, colloid titration has the advantage of using a larger, more representative sample than is used in zeta potential measurements. Since its conclusions are based on the electrokinetic properties of a suspension, its results are subject to the same limitations given for the streaming current detector and the zeta potential devices.

Continuous Filtration Techniques

Pilot Filters. The goal of the water treatment plant should be to produce the minimum possible filter effluent turbidity at the minimum chemical cost. The best measure of the efficiency of the coagulation–filtration steps would be the direct continuous measurement of the turbidity of coagulated water that has passed directly through a pilot granular filter. The application of continuous turbidity monitoring equipment to the effluent of the plant-scale filters should be a must for monitoring plant performance, but it has limited value as a control technique because of the substantial lag time between the point of chemical coagulant addition and the point of filtrate turbidity monitoring. For example, improper coagulation of the incoming raw water will result in a clarifier (with typically 1½ to 3 hours of detention time) being filled with improperly coagulated water before the results become fully apparent at the discharge of the plant filters.

The pilot filter technique is applied by sampling plant-treated coagulated water from the discharge of the plant rapid mix basin. This sample stream is then passed through a small (usually 4½-inch-diameter) pilot filter to determine whether the coagulant dose is proper, by continuously monitoring the pilot filter effluent turbidity. This technique provides a continuous, direct measurement of the turbidity, which is achieved by filtration of water that has been coagulated in the actual plant. Thus, no extrapolation from small-scale laboratory coagulation experiments is required. The only purpose of this test is to determine the proper coagulant dose; it is not to predict the length of filter run, nor to determine the optimum filter aid dose, nor to predict the rate of headloss buildup. Pilot filter columns have been used for these other purposes, but the technique of interest here is that used for monitoring the coagulation process.

The pilot filter technique has the advantages of offering a continuous monitoring of the plant-scale coagulation process with a minimum lag time. Filtering the water through the pilot filter yields immediate information about the adequacy of the coagulant dose. In a typical situation, correctness of coagulant dose is determined within 10 to 15 minutes after the raw water enters the plant. Experience at many locations shows that the pilot filter effluent turbidity is a very accurate prediction of the turbidity that will be produced by the plant-scale filter when the corresponding water reaches the plant-scale filter.

The mixed-media filter bed design described in Chapter 10 is often used in the pilot filter. The mixed-media design has the ability to accept the high solids load associated with most unsettled, coagulated waters without excessive headloss buildup. The turbidity of the pilot filter is monitored continuously and recorded. High turbidity in the filter effluent could result from either an improper coagulant dose or a breakthrough of floc from a properly coagulated water. To ensure that breakthrough does not occur, supplemental doses of polymers are injected into the pilot filter influent line. These large polymer doses may shorten the pilot filter run times, but, to prevent breakthrough, it is desirable to backwash the filter every 1 to 3 hours in any case. To provide a continuous monitoring of the coagulation process, two pilot filters are used in parallel. The system is equipped so that the filters backwash automatically on a high headloss signal. When one pilot filter enters the backwash cycle (which requires only about 10 minutes), the other pilot filter is automatically placed in service. Where very high raw water turbidities can be expected, a miniature flocculator-tube settler device can be installed ahead of the pilot filters to ensure reasonable filter run times even during periods of high turbidity. Typically, the pilot filters are contained in a console unit or in the plant control panel, which may also house turbidimeters for monitoring the plant-scale filter effluent turbidity.

At least two manufacturers (Neptune Microfloc and Turbitrol Co.) offer commercial pilot filter systems, both based on the same principle but differing somewhat in mechanical aspects.

The pilot filter system can be used to accurately indicate coagulation conditions with a minimum lag time so that the operator can make the necessary adjustments in the plant chemical feed. Alternatively, the coagulant feed can be controlled automatically.

Design for Automated Coagulant Control. This system is used in conjunction with a pilot filter system and automatically varies the plant coagulant dosage to maintain the effluent turbidity from the pilot filter at the desired set point value regardless of variations in raw water quality and other related factors. Other chemicals can be varied automatically by this system in direct relation to the coagulant dosage. The coagulant control system normally consists of a pilot filter system, an automatic control unit, and switches and controls for the chemical feed equipment. The output signal from the automatic control unit to the chemical feeders can be a time duration, current, pneumatic, or other standard instrumentation control signal. In general, the time duration signal is found to be reliable and economical and is used quite extensively.

Basically, this automated control is accomplished by comparing a 0 to 10 NTU signal from the pilot filter turbidimeter with the desired turbidity value. If the turbidity signal is greater than the desired turbidity, the alum dosage will be increased. If the input turbidity is less than the set point turbidity, the alum dosage is automatically decreased. The plant operator establishes the desired quality through a set point potentiometer, and the unit adjusts the coagulant dosage to maintain the set point value. This unit continuously optimizes the alum dosage for a given water treatment condition and consequently is constantly changing to establish the precise dosage.

The system may incorporate provisions for automatic plant flow pacing. To allow for flow pacing, a potentiometer must be connected into the plant raw water flow measurement equipment. This potentiometer is integrated into the plant flow measurement system so that the potentiometer wiper is at zero resistance with no flow and at maximum resistance at maximum plant flow. When the flow pacing feature is incorporated, the dosage output will be automatically adjusted for changes in plant flow. At the same time, changes in coagulant feed dosage requirements will be maintained by the control unit. The system may be provided with an override means so that the plant coagulant feed system may be operated in several modes.

The chemical characteristics of the raw water will have an effect upon the adequacy of alum coagulation. Thus, all other variables that could affect the pilot filter turbidity must be controlled. The most important variable is pH. If the pH of the coagulated water is not maintained in the proper range for optimum filtration, the unit cannot function properly.

Waters with low natural alkalinity will generally require addition of artificial alkalinity to maintain the coagulated water pH at an optimum value for filtration. In most cases, the alkalinity requirement is a direct function of the coagulant dosage, and the chemical feed pumps for both solutions can be proportioned over the entire range of expected coagulant dosages, so that supplementary artificial alkalinity feed is not required.

Filter Aid Control by Interface Monitoring. This technique has been developed in conjunction with dual-media filters. A sample of water is removed from the filter at a point near the interface between the coarse anthracite coal and the fine sand. The turbidity of this sample is continuously monitored and recorded, as is the filter effluent turbidity. Figure 7–15 shows the interface sampling system. The sample at the interface is obtained with a device that is a specially constructed well screen with slits small enough that the coal and sand cannot pass through. The interface screen is placed 2 inches above the fine sand prior to anthracite placement. The turbidimeter is placed as close as possible to the filter. Samples flow by gravity to the turbidimeters. This prevents air bubbles, which frequently occur when turbidimeter samples are pumped, and also provides for the fastest possible response time.

The interface turbidity sample is designed as a tool to aid the treatment plant operator in obtaining optimum performance from a dual-media filter. The turbidity value obtained at the interface indicates where floc removal is occurring within the filter bed. For instance, a high interface turbidity and a low filter effluent turbidity indicate that the proper coagulant dosage is being used but that significant amounts of floc are penetrating through the anthracite layer and must be removed by the fine sand. If the sand is forced to carry too much of the load, the headloss buildup becomes excessive, or breakthrough of turbidity could occur into the effluent. When floc removal does not occur in the anthracite layer when the proper coagulant dose is applied, the floc strength is inadequate and should be increased by the feed of polymers.

This method works with any material that can be successfully used as a filter aid or conditioner. The filter aid must be started at a known underdose

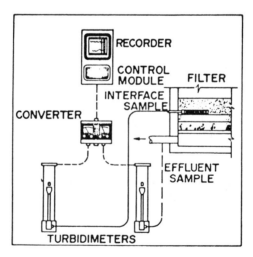

Fig. 7–15. Schematic diagram of an interface monitoring system. (Courtesy of Turbitrol Co.)

and incrementally increased until the desired floc removal is obtained in the coal layer. If an overdose is applied, it is usually impossible to use the interface concept until the filter has been washed.

It may also be desirable to adjust the filter aid dosage during the filter run, beginning with no aid or a small amount and increasing the dosage as the run progresses.

Turbidity Monitoring

With the currently available, low-cost turbidimeters, which are accurate and require little maintenance, every municipal water plant in the United States, regardless of size, should provide a continuous record of the quality of its final product. Turbidity is an important parameter in that it:

- Reflects the efficiency of the coagulation process and the overall treatment provided.
- Is related to probability of escape of pathogenic organisms from the treatment plant.
- Is a sensitive indicator of the aesthetic acceptability of the product to the consumer.

An appalling number of plants provide no continuous monitoring of plant effluent turbidity. Visual estimates based on seeing the clearwell floor, as well as plant log entries of "zero" turbidity at all times, have been witnessed by the authors at many plants. There are few production-oriented industries that do not maintain better quality control procedures than the water treatment industry, even though its performance is directly related to the public health of the nation.

For the measurement of very small amounts of turbidity, the principle of light scattering (nephelometry) is used. In this process, light is reflected at right angles to the light beam by the particles of turbidity in the liquid, as the beam of light passes through the liquid. The amount of reflected (scattered) light depends directly upon the amount of turbidity present in the liquid. This is similar to the common phenomenon of sunlight streaming through a window being reflected by otherwise invisible dust particles in the air.

If the liquid is entirely free of particles of turbidity, no scattered light will reach the photocells, and the indicating meter will read zero; thus increasing turbidity gives an increase in the meter reading.

The advantages of this system of turbidity measurement are:

- A very strong light source can be used, permitting a high degree of sensitivity; thus very small amounts of turbidity can be measured accurately.

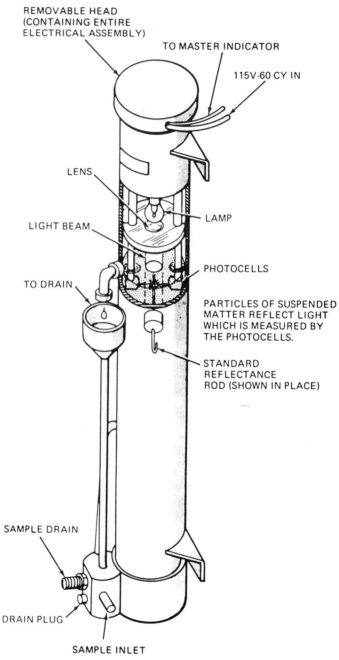

REMOVABLE HEAD
(CONTAINING ENTIRE
ELECTRICAL ASSEMBLY)

TO MASTER INDICATOR

115V-60 CY IN

LENS

LIGHT BEAM

LAMP

TO DRAIN

PHOTOCELLS

PARTICLES OF SUSPENDED
MATTER REFLECT LIGHT
WHICH IS MEASURED BY
THE PHOTOCELLS.

STANDARD
REFLECTANCE
ROD (SHOWN IN PLACE)

SAMPLE DRAIN

DRAIN PLUG

SAMPLE INLET

Fig. 7-16. Cross section of continuous-flow, low-range turbidimeter. (Courtesy of HACH Co.)

- The meter reading or output from the instrument is zero when the turbidity is zero.

Units of this type (such as manufactured by HACH Company) are available with ranges of 0 to 0.2, 0 to 1, 0 to 3, and 0 to 30 NTU. A water sample of approximately 0.5 gpm at a head of 6 inches of water is required. Figure 7–16 illustrates this type of equipment, which is best suited for measuring filter effluent turbidity or for interface turbidity monitoring.

The principle of nephelometry is the best system for turbidity measurement—although it was once limited to low and medium-range measurement because in the high ranges, with the conventional instrument, the light beam itself becomes absorbed by the high turbidity so that the instrument is unresponsive. In order to extend the light scattering principle to high-range turbidities, a design employing "Surface Scatter" was developed, as illustrated in Fig. 7–17. A very narrow beam is directed onto the surface of the liquid at an angle of 15 degrees. Part of the beam is reflected by the water surface and escapes to a light trap. The remaining portion enters the water at approxi-

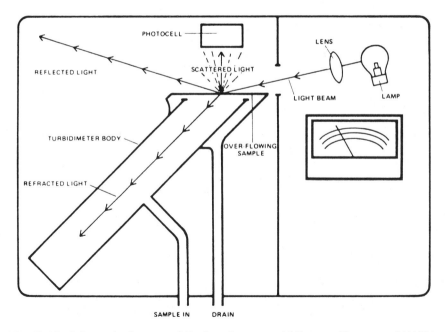

Fig. 7–17. Schematic diagram of Surface Scatter turbidimeter. (Courtesy of HACH Co.)

mately a 45-degree angle. If particles of turbidity are present, light scattering will take place, and some of the scattered light will reach the photocell. With this "Surface Scatter" design, there is no upper limit to the amount of turbidity it is possible to measure.

This unit is calibrated by using a Jackson candle turbidimeter to measure the turbidity of the same water that is flowing through the instrument. The instrument is adjusted so that its output reading corresponds to the Jackson candle reading. Ranges as high as 0 to 5000 are available, and this instrument is used for measuring the turbidity of raw water or the settling basin effluent. A 0.25 to 0.50 gpm (0.016 to 0.032 l/s) sample stream is required.

REFERENCES

1. Stumm, W., "Chemical Interactions in Particle Separation," in *Chemistry of Wastewater Technology,* A. J. Rubin, editor, Ann Arbor Science Publishers/Technomic Publishing Co., Inc., Lancaster, Pa., 1978.
2. Black, A. P., and Hannah, S. A., "Electrophoretic Studies of Turbidity Removal by Coagulation with Aluminum Sulfate," *J.AWWA,* Vol. 53, No. 4, p. 438, 1961.
3. Packham, R. F., "Some Studies of the Coagulation of Dispersed Clays with Hydrolyzing Salts," *J. Coll. Science,* Vol. 20, 1965.
4. Rook, J. J., "Haloforms in Drinking Water," *J.AWWA,* Vol. 68, No. 3, p. 168, 1976.
5. Black, A. P., and Willems, D. G., "Electrophoretic Studies of Coagulation for Removal of Organic Color," *J.AWWA,* Vol. 53, No. 5, p. 589, 1961.
6. Hall, E. S., and Packham, R. F., "Coagulation of Organic Color with Hydrolyzing Coagulants," *J.AWWA,* Vol. 57, No. 9, p. 1149, 1965.
7. Stumm, W., and O'Melia, C. R., "Stoichiometry of Coagulation," *J.AWWA,* Vol. 60, No. 5, p. 514, 1968.
8. O'Melia, Charles R., "Coagulation," in *Water Treatment Plant Design,* Robert L. Sanks, editor, Ann Arbor Science, Ann Arbor, Mich., 1978.
9. LaMer, V. K., and Healy, T. W., "Adsorption–Flocculation Reactions of Macromolecules at the Solid–Liquid Interface," *Rev. Pure App. Chem.* 13:112, 1963.
10. Gouy, G., "Sur la Constitution de la Charge Electrique a la Surface d'un Electrolyte," *Ann. Phys.* (Paris), Serie 4, 9:457, 1910.
11. Chapman, D. L., "A Contribution to the Theory of Electrocapillarity," *Phil. Mag.* 25:475, 1913.
12. Stern, O., "Zur Theorie der Elektrolytischen Doppelschicht," *A. Electrochem.* 30:508, 1924.
13. O'Melia, Charles R., "A Review of the Coagulation Process," *Public Works,* p. 82, May, 1969.
14. Benefield, Larry D., Judkins, J. F., and Weand, Barron, L., *Process Chemistry for Water and Wastewater Treatment,* Prentice-Hall, Englewood Cliffs, N.J., 1982.
15. O'Melia, C. R., "Coagulation and Flocculation," in *Physicochemical Processes for Water Quality Control,* by Walter J. Weber, Jr., Wiley-Interscience, New York, © 1972.
16. van Olphen, H., "An Introduction to Clay Colloid Chemistry," 2nd ed., Wiley-Interscience, New York, © 1977.
17. Ruehrwein, R. A., and Ward, D. W., "Mechanisms of Clay Aggregation by Polyelectrolytes," *Soil Science* 673:485, 1952.

18. Stumm, W., and Morgan, J. J. "Chemical Aspects of Coagulation," *J.AWWA* 5:971, 1962.
19. Stumm, W., "Metal Ions in Aqueous Solutions," in *Principles and Applications of Water Chemistry*, S. D. Faust and J. V. Hunter, editors, John Wiley and Sons, Inc., New York, 1967.
20. Thompson, C. G., Singley, J. E., and Black, A. P., "Magnesium Carbonate—A Recycled Coagulant," *J.AWWA*, Vol. 64, No. 1, p. 11, 1972.
21. Thompson, C. G., Singley, J. E., and Black, A. P., "Magnesium Carbonate—A Recycled Coagulant," *J.AWWA*, Vol. 64, No. 2, p. 93, 1972.
22. Stumm, W., Huper, H., and Champlin, R. L., "Formation of Polysilicates as Determined by Coagulation Effects," *Environmental Science and Technology*, Vol. 1, No. 3, p. 221, 1967.
23. Weber, W. J., Jr., *Physicochemical Processes for Water Quality Control*, John Wiley & Sons, Inc., New York, 1972.
24. Committee on Coagulation, "State of the Art of Coagulation," *J.AWWA*, Vol. 63, No. 2, p. 99, 1971.
25. Cohen, Jesse M., and Hannah, Sidney A., "Coagulation and Flocculation," in *Physicochemical Processes for Water Quality Control*, W. J. Weber, Jr., editor, John Wiley & Sons, Inc., New York, 1972.
26. Amirtharajah, A., and Mills, Kirk M., "Rapid-Mix Design for Mechanisms of Alum Coagulation," *J.AWWA*, Vol. 74, No. 4, pp. 210–216, 1982.
27. Hahn, H. H., and Stumm, W., "Kinetics of Coagulation With Hydrolyzed Al(III)," *J. Colloid Interface Sci.* 28:134–144, 1968.
28. Letterman, R. D., Orron, J. E., and Gemmell, R. S., "Influence of Rapid-Mix Parameters on Flocculation," *J.AWWA* 67:716–722, 1973.
29. Vrale, Lassi, and Jorden, Roger M., "Rapid Mixing in Water Treatment," *J.AWWA* 63:52–58, 1971.
30. Chao, Junn-Ling, and Stone, Brian G., "Initial Mixing by Jet Injection Blending," *J.AWWA* 71:570–573, 1979.
31. Hudson, H. E., and Wolfner, J. P., "Design of Mixing and Flocculation Basins," *J.AWWA*, 59:1257–1267, 1967.
32. Stumm, W., and Morgan, J. J., *Aquatic Chemistry*, Wiley-Interscience, New York, 1970.
33. *Recommended Standards for Water Works* (Ten States Standards), Health Education Service, Albany, New York, 1972.
34. Camp, T. R., and Stein, P. C., "Velocity Gradients and Internal Work in Fluid Motion," *J. Boston Soc. Civil Eng.* 30:219–237, 1943.
35. Cornwell, David A., and Bishop, Mark M., "Determining Velocity Gradients in Laboratory and Full-Scale Systems," *J.AWWA*, Vol. 75, No. 9, pp. 470–475, 1983.
36. Rushton, J. H., "Mixing of Liquids in Chemical Processing," *Ind. Eng. Chem.* 44:2931–2936, 1952.
37. Gemmell, R. S., "Mixing and Sedimentation," in *Water Quality and Treatment*, American Water Works Association, Inc., McGraw-Hill Book Co., New York, 1971.
38. Camp, T. R., "Floc Volume Concentration," *J.AWWA* 60:656–673, 1968.
39. TeKippe, R. J., and Ham, R. K., "Velocity-Gradient Paths in Coagulation," *J.AWWA* 63:439–448, July, 1971.
40. Amirtharajah, A., "Design of Rapid Mix Units," in *Water Treatment Plant Design*, Robert Sanks, editor, Ann Arbor Science, Ann Arbor, Mich., 1978.
41. Hudson, H. E., Jr., "Dynamics of Mixing and Flocculation," *Proceedings* of the 18th Annual Public Water Supply Conference, University of Illinois, Dept. of Civil Engineering.
42. Kawamura, Susumu, "Coagulation Considerations," *J.AWWA*, Vol. 65, No. 6, pp. 417–423, 1973.

43. Kawamura, S., "Considerations on Improving Flocculation," *J.AWWA,* Vol. 68, pp. 328–336, 1976.
44. Chow, V. T., *Open Channel Hydraulics,* McGraw-Hill Book Co., New York, 1959.
45. ASCE, AWWA, CSSE, *Water Treatment Plant Design,* American Water Works Association, New York, 1969.
46. Morris, Juli K., and Knocke, William R., "Temperature Effects on the Use of Metal Ion Coagulants for Water Treatment," presented at AWWA Conference, Las Vegas, Nev., 1983.
47. Swift, D. L., and Friedlander, S. K., "The Coagulation of Hydrosols by Brownian Motion and Laminar Shear Flow," *J. Colloidal Sci.* 19:621–647, 1964.
48. Amirtharajah, A., "Design of Flocculation Systems," in *Water Treatment Plant Design,* Roberts Sanks, editor, Butterworth Publishers, Stoneham, Mass., 1978.
49. Hudson, H. E., "Physical Aspects of Flocculation," *J.AWWA* 57:885–892, 1965.
50. Camp, T. R., "Flocculation and Flocculation Basins," *Trans. Am. Soc. Civil Eng.* 120:1–16, 1955.
51. Fair, G. M., Geyer, J. C., and Okun, D. A., *Water and Wastewater Engineering,* Vol. 2, John Wiley and Sons, Inc., New York, 1968.
52. Harris, H. S., Kaufman, W. J., and Krone, R. B., "Orthokinetic Flocculation in Water Purification," *J. Environ. Eng. Div.,* ASCE 92:95–111, 1966.
53. Fair, G. M., and Gemmell, R. S., "A Mathematical Model of Coagulation," *J. Colloid. Sci.* 19:360–372, 1964.
54. Argaman, Y., and Kaufman, W. J., "Turbulence and Flocculation," *J. Environ. Div., ASCE* 96:223–241, 1970.
55. Argaman, Y. A., "Pilot-Plant Studies of Flocculation," *J.AWWA* 63:775–777, December, 1971.
56. Andreu-Villegas, R., and Letterman, R. D., "Optimizing Flocculator Power Input," *J. Environ. Eng. Div., ASCE* 102:251–264, 1976.
57. Ives, K. J., "Theory of Operation of Sludge Blanket Clarifiers," *Proc. Inst. Civil Eng.* 39:243–260, 1968.
58. Ives, K. J., and Bhole, A. G., "Theory of Flocculation for Continuous Flow Systems," *J. Environ. Eng. Div., ASCE* 99:17–34, 1973.
59. Cockerham, P. W., and Himmelbau, D. M., "Stochastic Analysis of Orthokinetic Flocculation," *J. Environ. Eng. Div., ASCE* 100:279–294, 1974.
60. Walker, J. D., "High Energy Flocculation," *J.AWWA* 60:1271–1279, 1968.
61. TeKippe, R. J., and Ham, R. K., "Coagulation Testing, A Comparison of Techniques," *J.AWWA,* Vol. 62, Part I, p. 594; Part II, p. 620, 1970.
62. *Information Resource: Water Pollution Control in the Water Utility Industry,* American Water Works Association Research Foundation, EPA Report 12120 EUR 11/71, November, 1971.
63. Black, A. P., and Smith, A. L., "Determination of the Mobility of Colloidal Particles by Microelectrophoresis," *J.AWWA* 54:926, 1962.
64. Williams, R. L., "Microelectrophoretic Studies of Coagulation With Aluminum Sulfate," *J.AWWA* 57:801, 1965.
65. Riddick, T. M., "Zeta Potential and Its Application to Difficult Waters," *J.AWWA* 53:1007, August, 1961.
66. Shull, K. E., "Filtrability Techniques for Improving Water Clarification," *J.AWWA* 59:1164, 1967.
67. Kawamura, S., and Tanaka, Y., "Applying Colloid Titration Techniques to Coagulant Dosage Control," *Water and Sewage Works* 113:348, 1966.
68. Kawamura, S., Sanna, G. P., Jr., and Shumate, K. S., "Application of Colloid Titration Technique to Flocculation Control," *J.AWWA* 59:1003, 1967.

69. Hudson, Herbert E., *Water Clarification Processes, Practical Design and Evaluation,* Van Nostrand Reinhold, New York, 1981.
70. Culp, Gordon L., and Culp, Russell L., *New Concepts in Water Purification,* Van Nostrand Reinhold, New York, 1974.

Chapter 8
Chemical Storage and Feeding Systems

INTRODUCTION

The importance of the design of chemical systems in a water treatment plant cannot be overemphasized. Although the chemical systems only account for a relatively small percentage of the capital costs of a water treatment plant, they usually account for a large percentage of the annual operation and maintenance costs. The proper design of these systems can reduce operation and maintenance costs and improve treatment efficiency. This chapter describes specific design considerations for liquid, solid, and gaseous chemical systems. The groups of chemicals used for particular water treatment functions are discussed, and the specific chemicals in each group are listed, as well as their pertinent physical data and characteristics. The rationale for selecting a particular chemical is presented and explained. Important design considerations for each type of chemical system are discussed, including the selection of chemical form, delivery, storage, feeding, conveyance methods, and safety.

CHEMICAL SELECTION

Function of Chemicals

In selecting a chemical for use in a water treatment process, a primary consideration is the intended function of that chemical. Nearly all chemicals used in water treatment can be classified into functional groups. There are chemicals used for coagulants and coagulant aids, softening, taste and odor control, disinfection, dechlorination, fluoridation, fluoride adjustment, pH adjustment, and removal of certain constituents such as iron, manganese, heavy metals, and corrosion control. Each one of these classifications contains a variety of chemicals that can be used for the same purpose, and in some cases the chemicals serve more than one purpose. Table 8–1 describes the chemicals within these various functional groups.

Coagulants. In most instances, surface waters require the use of a coagulant, while well waters do not. Aluminum sulfate (alum) is the most widely used coagulant. It is available in lump, ground, or liquid form and can be shipped in bulk or in 100-pound (45.4-kg) bags. However, it may be appropri-

Table 8-1. Chemicals Used in Water Treatment.*,**

Coagulants
Aluminum ammonium sulfate
Aluminum potassium sulfate
Aluminum sulfate (alum)
Calcium hydroxide (hydrated lime)
Calcium oxide (quicklime)
Ferric chloride
Ferric sulfate
Ferrous sulfate
Sodium aluminate

Coagulant aids
Bentonite
Calcium carbonate
Organic coagulant aids
Sodium silicate

Disinfection agents
Ammonia, anhydrous
Ammonium hydroxide
Ammonium sulfate
Bromine
Calcium hypochlorite
Chlorine
Chlorine dioxide
Chlorinated lime
Iodine
Ozone
Sodium chlorite
Sodium hypochlorite

Dechlorination agents
Activated carbon
Aluminum ammonium sulfate
Ion-exchange resins†
Sodium bisulfite (sodium
  pyrosulfite)
Sodium sulfite
Sulfur dioxide

Iron and manganese removal
Chlorine
Chlorine dioxide
Ozone
Potassium permanganate
Sodium hexametaphosphate

Other uses—Algae control
Copper sulfate

Softening
Calcium hydroxide (hydrated lime)
Calcium oxide (quicklime)
Carbon dioxide
Ion-exchange resins†
Sodium carbonate (soda ash)
Sodium chloride
Sulfuric acid

Taste and odor control
Activated carbon
Bentonite
Chlorine
Chlorine dioxide
Ozone
Potassium permanganate

Fluoridation
Ammonium silicofluoride
Calcium fluoride
Hydrofluoric acid
Hydrofluosilicic acid
Sodium fluoride
Sodium silicofluoride

Fluoride adjustment
Activated alumina
Ion-exchange resins†
Magnesium oxide

pH adjustment
Calcium carbonate
Calcium hydroxide (hydrated lime)
Calcium oxide (quicklime)
Carbon dioxide
Hydrochloric acid
Sodium carbonate (soda ash)
Sodium hydroxide
Sulfuric acid

Corrosion control
Calcium hydroxide (hydrated lime)
Calcium oxide (quicklime)
Sodium carbonate (soda ash)
Sodium hexametaphosphate
Sodium hydroxide
Sodium tripolyphosphate

* Reference 1.
** See Appendix B for more information on specific chemicals.
† Consult manufacturer of particular resin for more information.

ate to use a coagulant other than aluminum sulfate, such as those listed in Table 8–1.

For instance, a raw water with high magnesium content may dictate the use of lime to allow for coagulation as well as precipitation of magnesium hydroxide. Heavy metals are removed more effectively at the high pH values resulting from the use of lime. Other coagulants include aluminum ammonium sulfate, aluminum potassium sulfate, ferric chloride, ferric sulfate, ferrous sulfate, and sodium aluminate.

Coagulant Aids. In some cases, a coagulant aid is used to improve the efficiency of the coagulation process.[2] Bentonite has been used in low-turbidity waters and in up-flow clarifiers. The bentonite clay provides a nucleus for floc formation and helps produce a heavy sediment blanket. Calcium carbonate also has been used to provide a nucleus for aluminum and iron hydroxide floc formation, and to add weight to the floc to aid in settling. Organic coagulant aids or polymers are widely used. There are both natural and synthetic types. Some are suitable for water treatment, while others are not. As discussed in Chapter 7, there are different types of polymers which can perform many functions. Polymers are supplied in solid or liquid form, and in bulk or 50-pound (22.7-kg) bags. Sodium silicate plus activated silica has been used as a coagulant aid to toughen the floc through ionic and electronic bond formation.

Softening. Softening can be accomplished by either ion exchange or the lime–soda softening process. Each process requires different chemicals. In ion exchange, the exchange resin typically is regenerated with sodium chloride, as described in Chapter 17. The lime–soda process requires calcium hydroxide or calcium oxide and sodium carbonate. Carbon dioxide or sulfuric acid is required for subsequent pH adjustment. These processes are described in Chapter 21.

Taste and Odors. The most common chemicals used for removal of tastes and odors are powdered or granular activated carbon. For intermittent or occasional taste and odor problems, oxidizing agents such as potassium permanganate, ozone, chlorine, and chlorine dioxide may be used. See Chapter 18 for a detailed review of taste and odor control.

Disinfection Agents. Chlorine is the principal disinfecting agent, although ultraviolet light, ozone, chloramines, and chlorine dioxide are also used. Bromine, iodine, and potassium permanganate have been used in specific circumstances. More detail is presented in Chapter 11.

Dechlorination Agents. The materials that have been used, or proposed, for the dechlorination of waters include granular and powdered activated carbon, hydrogen peroxide (H_2O_2), ammonia (NH_3), sodium thiosulfate ($Na_2S_2O_3$), and the sulfur (IV) species, which include sulfur dioxide (SO_2), sodium bisulfite ($NaHSO_3$), sodium sulfite (Na_2SO_3), and sodium metasulfite ($Na_2S_2O_5$). Sulfur dioxide comes in gaseous form and uses the same type of storage and feeding equipment as chlorine.

Fluoridation. Fluoride is added to the drinking water of some communities to reduce dental decay among children. Correctly proportioned, fluoridation is effective, but overdoses can be detrimental to teeth. The optimum fluoride concentration varies with air temperature, as shown in Table 2–2, because during warmer weather people drink larger quantities of water. Fluospar and calcium fluoride are the commercial fluoride compounds usually used in water treatment. Sodium silicofluoride is the most commonly used compound for fluoridation of municipal water supplies, whereas sodium fluoride is used less frequently because of higher costs. Ammonium silicofluoride and hydrofluosilicic acid also can be used for fluoridation. More detail is presented in Chapter 20.

Fluoride Reduction. Because fluoride compounds are present in natural environments (13th rank among the elements), a raw water supply may contain too much fluoride. Excess doses of fluoride may cause blackening or mottling of teeth. Several chemicals can be used for fluoride reduction, including commercially produced products such as fluorex and fluo-carbon, as well as magnesium oxide and activated alumina. The defluoridation process is discussed in Chapter 20.

pH Adjustment. There may be several points in the water treatment process where the pH must be adjusted. Several acids and bases can be used, depending on the final pH desired. Those typically used are listed in Table 8–1.

Iron and Manganese Removal. Iron and manganese concentrations above 0.3 mg/l total should be removed from water. Although not harmful to health, they cause staining and taste problems. These constituents are removed through oxidation, by settling and filtration, using chlorine, chlorine dioxide, ozone, or potassium permanganate. Small amounts (below 1 mg/l) can be sequestered by sodium hexametaphosphate. More detail is presented in Chapter 19.

Corrosion Control. At the completion of the water treatment process and prior to entering the distribution system, the water must not be corrosive.

Corrosive water will cause costly problems in the distribution and storage systems as well as for the individual consumer. Lime, soda ash, sodium hydroxide, sodium hexametaphosphate, and sodium tripolyphosphate are all used to prevent corrosive waters.

Algal Control.   Algal control in reservoirs is usually accomplished by addition of copper sulfate to the water. This topic is discussed in Chapter 6.

Raw Water Quality

Many items need to be considered in selecting the specific chemical to be used in a water treatment process, one important item being the raw water quality. The raw water must first be analyzed to determine the level and type of treatment required. It is helpful to consider the effectiveness of each chemical under consideration, based on experience and operating results from full-scale plants at other locations with similar raw water quality. However, caution should be exercised in using data from other plants. The designer should not assume that a particular chemical used at one plant will produce identical results at a different plant. The effectiveness of a chemical varies in different applications, depending upon operating conditions. Laboratory and/or pilot tests can be used to predict the effectiveness of a chemical for a given application. The objectives of laboratory studies usually are: (1) to determine what chemical dosages are needed to obtain the desired results, and (2) to obtain data for the design and operation of a pilot or full-scale facility. After the chemicals that perform unsatisfactorily are eliminated, a preliminary cost comparison can be made for the remaining chemicals to determine which should receive further consideration. Also, it should be recognized that laboratory and pilot tests do not always accurately predict plant-scale dosages and performance.

Availability and Cost

Other important considerations when selecting a chemical are its availability in a particular region, reliability of supply, and cost. There is little advantage in selecting a chemical that satisfies all the requirements of a water treatment process if the chemical is not readily available. Capital costs for handling and feeding various chemicals vary considerably, depending on the characteristics of the chemical to be fed, the form (liquid, solid, gas) in which the chemical is purchased, and the form in which the chemical ultimately is used in the treatment process.

Transportation is a significant cost for some locations. The cost at the point of origin usually is quoted by the manufacturer in cents or dollars per pound, per 100 pounds, or per ton, and varies according to the size of the order. It may be a price "f.o.b. cars" at the point of manufacture or at a regional stock point. When small lots are purchased, the f.o.b. point is important because the manufacturer ships to the regional stock point in bulk and at lower rates in order to give the customer the benefit of this savings. The point of shipment origin should always be clearly stated because transportation costs on some chemicals may be more than the cost of the chemical, especially if long hauls are involved.

Many manufacturers quote prices "f.o.b." from a distribution point but also will give the customer information on the expected cost of transportation by rail or truck to the point of usage. Sometimes manufacturers will also quote "freight allowed," which means that they will assume the freight charge on the shipment. In some shipments, it is important to compare the cost of shipment by truck to the cost of shipment by rail. While the truck rate may be higher than the rail rate, the material will be taken from the manufacturer's plant and delivered to the door of the plant by truck at no extra cost. On the other hand, by rail, even though the price given is "f.o.b. your nearest freight station," there will be extra costs for handling and hauling of the material from the freight station to the plant. The overall delivered cost calculation must consider these factors.

To avoid or minimize potential problems in obtaining chemicals, suppliers of the specific chemical should be contacted for details on chemical availability before the chemical is selected for use in the treatment plant. It is advisable to consider market trends for water treatment chemicals to anticipate possible chemical shortages or large cost increases.

Storage Life

The length of time a chemical will retain its full potency limits the amount of chemical to be purchased and delivered at any one time. If the chemical retains its full potency for 6 months, then it would not be economical to purchase it in quantities that last much longer. If potency will last for 1 year, it may be more economical to purchase the chemical in quantities sufficient to last the longer period, because of discounts for large-quantity purchases. When quantities are ordered to last over long periods, there must be adequate storage facilities. Depending on the characteristics of the selected chemical, the cost of such storage facilities may be a significant factor in determining the optimum amount of chemical purchase.

A sufficient supply of chemicals should be on hand to cover the daily operation requirements, plus an additional amount to cover the time between placement of the reorder and receipt of the material.

Compatibility with Existing System

In expanding an existing water treatment plant, important considerations are the type of chemicals currently in use, the type of equipment currently in use, the type of equipment to be used in the expansion, and compatibility with the other treatment processes. The water treatment plant under consideration may have several processes, such as softening or iron and manganese removal in addition to coagulation and filtration. The selection of the chemical may aid one or more of the other processes while still achieving its primary function. Another consideration is the type of control system. The control method may favor one chemical over another.

Labor Requirements

The level of operation and maintenance labor required to store and feed chemicals is a consideration. It is important to recognize that the labor depends on both the characteristics of the chemical and the form in which it is purchased, stored, and fed. For example, less labor is usually required for a chemical that is purchased and fed in the same form, in contrast to a chemical that is purchased in dry form, dissolved for storage, and later diluted for feeding. Table 8–2 illustrates the variation in operation and maintenance labor for several chemicals commonly used as coagulants and coagulant aids. These labor requirements include unloading, storing, and feeding operations. Unslaked lime requires relatively high labor for slaking and feeding equipment. As Table 8–2 illustrates, there are differences in labor requirements, and

Table 8–2. Typical Chemical Treatment Labor Requirements.

	CAPACITY		OPERATION AND MAINTENANCE*
CHEMICAL	LB/HR	(KG/H)	LABOR, HR/YR
Alum	10	(4.54)	150
	50	(22.7)	210
	100	(45.4)	300
	500	(227)	800
Lime (slaked)	100	(45.4)	1,800
	500	(227)	1,850
	1,000	(454)	2,100
Lime (unslaked)	100	(45.4)	2,400
	500	(227)	2,400
	1,000	(454)	2,900
Polymer (dry)	.5	(0.23)	500
	1.0	(0.45)	580
	5.0	(2.27)	750
	10.0	(4.54)	850
Polymer (liquid)	.5	(0.23)	390
	1.0	(0.45)	400
	5.0	(2.27)	420
	10.0	(4.54)	440

* Labor for operation and maintenance of unloading, storing, and feeding facilities.

these differences should be considered in selecting chemicals. Water plant treatment costs are included in Chapter 30, and include chemical feeding systems.

DESIGN CONSIDERATIONS

General Considerations

The design of chemical handling systems must take into account the type of chemical to be fed and the form of the chemical. It must also consider the methods of chemical delivery, storage, handling, mixing, and feeding, and the conveyance of the chemicals to the final feed points. The design must result in efficient handling of the chemicals from delivery and storage to the application point. The operation should be convenient and easy without placing an excess burden on operating personnel.

The chemical feeding and handling design should be versatile and allow maximum operator flexibility. A sufficient number of feed points should be included in the treatment plant to allow the operator to vary the chemical dosages to these points. The chemical feed equipment must have enough capacity to provide for an adequate range of chemical addition. There must be enough feeders that the final water quality will protect the health of the consumer even with mechanical failures. The chemical feeders and pumps should operate no lower than 20 percent of the feed capacity. The chemical feed and handling design also should take into account future expansions of the plant.

The materials chosen for conveyance, storage, and measurement of a chemical must be compatible with the properties of that chemical. Corrosive chemicals must be handled in such a manner as to minimize potential for corrosion.

The conveyance of chemicals from unloading to the application point must be done in separate conduits for each chemical. Slurry-type chemicals, especially lime, should be fed by gravity where practical. Every effort must be made to locate pneumatic conveyance tubing for chemicals vertically because horizontal tubing frequently plugs.

Liquid, Solid, or Gaseous Chemical Form

Once a chemical has been selected, it is necessary to decide whether to feed it in a liquid, solid, or gaseous form. Some chemicals are only available in one form; however, many of the chemicals, as shown in Table 8–1, are available in a variety of forms.

There are certain advantages to using the liquid form of chemicals. Liquid chemicals are generally easier to handle during loading, unloading, and feeding. The liquid form also eliminates the dust problem associated with the use of solids. Because the liquid form can be directly fed to the process, the feeding equipment is simplified, as mixers and dissolvers associated with dry chemicals are not needed. Some of the disadvantages of the liquid form are the additional cost of hauling the extra water and the additional storage space required.

The solid form of chemicals generally requires less storage space, may be less expensive, and provides a greater selection of chemicals. The disadvantages of using a solid chemical include: dusting problems, caking and lumping problems during storage and feeding, and the need for more equipment to dissolve and feed the chemical.

Chemicals in the gaseous form are not so widely available as the liquid or solid form. The most common gaseous chemicals are chlorine, sulfur dioxide, carbon dioxide, oxygen, and anhydrous ammonia. Ozone and chlorine dioxide are fed as gases but must be generated on-site. When chemicals are available in the gaseous form, that may simplify the feeding and control of the chemical as well as its storage and handling.

The selection of the chemical form will be dictated by such factors as cost, local availability, method of feed or control anticipated, compatibility with the existing facilities, and the quantity of the chemical required. The larger the quantity used, the more favorable the solid form of most chemicals becomes. The solid form may be the more economical when large quantities are used because of the high cost of liquid storage. Also, the cost of transporting the water associated with the liquid form becomes a more significant portion of the costs. However, although a liquid chemical may require a greater storage volume and have higher freight costs than the solid, it may still be more economical when used in small quantities, because the feeding equipment is less complex.

Chemical Delivery

Following selection of the chemical form, it is necessary to determine how that chemical will be delivered to the plant. Generally, bulk chemical deliveries are made by truck or rail; however, for some plants located adjacent to waterways, chemicals have been delivered by barges.

Design considerations for truck delivery include: sufficiently wide access roads to the plant to allow adequate traffic movement; the turning radius for a particular type truck for delivery and unloading of the chemical; the height clearances of monorails, underpasses, and so on; the grade of the access road; the type of pavement for the access road; and weight limits on roads and bridges. Figure 8–1 shows the minimum turning radius for delivery trucks, general height clearances required, and other design considerations.[3]

In large plants, it may be feasible to obtain chemicals through rail service if the plant is located near rail lines that could be extended to the plant economically. The capital cost required to obtain service may be offset by the savings in operation and maintenance costs, including the cost of chemicals.

The facility design for unloading the chemical depends on the form of the chemical. For chemicals in the liquid form, trucks are usually unloaded by pumping or gravity. The storage tank and unloading lines are designed to accommodate the rate at which the transporter would unload the material. The transporter should be contacted for this information.

Chemicals in the solid form generally are delivered in bulk or bag by truck or rail. Bagged chemicals typically are delivered on pallets. Design considerations for bag unloading include the location of the bag storage with respect to the unloading facility and day tanks, the manner in which the bags are to be conveyed, and the type of facility for unloading the truck. The most common way to unload bag deliveries is to use a loading dock designed so that the entire truck or rail car can be unloaded with a forklift truck, bag conveyor, or hoist. In small plants, hand trucks are sometimes used for individual bag unloading.

Bulk chemicals delivered by truck or rail can be unloaded in a variety of ways. The most common method of unloading bulk carriers is by pneumatic conveyance. The storage tanks must be designed to allow for pneumatic conveyance when this type of delivery is anticipated. Figure 8–2 illustrates a typical pneumatic conveying system. Other bulk carriers can be unloaded by gravity means, such as a chute. Other methods use mechanical means, such as bucket elevators and belt or screw conveyors. Each type of conveyance requires particular design considerations for the unloading and storage facilities. The bulk carrier should be consulted prior to design to determine all pertinent parameters.

Chemicals in the gaseous form usually come in ton cylinders, 150-pound (68.1-kg) cylinders, or large bulk truck or rail carriers. A hoist with a monorail is used to unload cylinder-type containers. Special facilities must be designed for bulk truck gas tanks and rail gas tanks so the gas can be transferred from the tanker. Figure 8–3 illustrates a bulk chlorine storage facility.[4]

Chemical Storage

Location. Chemical storage facilities should be designed to make operations as convenient and efficient as possible. The location of the storage relative to its unloading point, the feeding equipment, and the application point is important. For example, the use of hand and forklift trucks should be maximized, and the need to carry bag materials upstairs or lift heavy loads by operators should be eliminated. There should be adequate access around all storage space. Storage of stacked chemicals in bags and pallets should

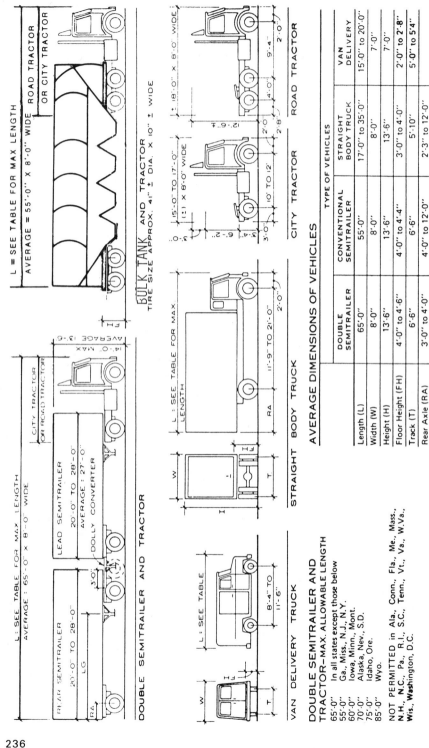

L = SEE TABLE FOR MAX LENGTH
AVERAGE = 55'-0" X 8'-0" WIDE

ROAD TRACTOR
OR CITY TRACTOR

BULK TANK AND TRACTOR
TIRE SIZE APPROX. 41" ± DIA. X 10" ± WIDE

‡ 18'-0" X 8'-0" WIDE

15'-0" TO 17'-0"
(±) X 8'-0" WIDE

ROAD TRACTOR

CITY TRACTOR

L = SEE TABLE FOR MAX LENGTH

RA

STRAIGHT BODY TRUCK

LEAD SEMITRAILER
20'-0" TO 28'-0"
AVERAGE = 27'-0"

DOLLY CONVERTER

REAR SEMITRAILER
20'-0" TO 28'-0"

LG

RA

L = SEE TABLE FOR MAX LENGTH
AVERAGE = 65'-0" X 8'-0" WIDE

CITY TRACTOR
OR ROAD TRACTOR

DOUBLE SEMITRAILER AND TRACTOR

L = SEE TABLE

FH

W

T

8'-4" TO 11'-6"

VAN DELIVERY TRUCK

AVERAGE DIMENSIONS OF VEHICLES

	TYPE OF VEHICLES			
	DOUBLE SEMITRAILER	CONVENTIONAL SEMITRAILER	STRAIGHT BODY TRUCK	VAN DELIVERY
Length (L)	65'-0"	55'-0"	17'-0" to 35'-0"	15'-0" to 20'-0"
Width (W)	8'-0"	8'-0"	8'-0"	7'-0"
Height (H)	13'-6"	13'-6"	13'-6"	7'-0"
Floor Height (FH)	4'-0" to 4'-6"	4'-0" to 4'-4"	3'-0" to 4'-0"	2'-0" to 2'-8"
Track (T)	6'-6"	6'-6"	5'-10"	5'-0" to 5'-4"
Rear Axle (RA)	3'-0" to 4'-0"	4'-0" to 12'-0"	2'-3" to 12'-0"	

DOUBLE SEMITRAILER AND TRACTOR—MAX. ALLOWABLE LENGTH

65'-0"   In all states except those below
55'-0"   Ga., Miss., N.J., N.Y.
60'-0"   Iowa, Minn., Mont.
70'-0"   Alaska, Nev., S.D.
75'-0"   Idaho, Ore.
85'-0"   Wyo.

NOT PERMITTED in Ala., Conn., Fla., Me., Mass.,
N.H., N.C., Pa., R.I., S.C., Tenn., Vt., Va., W.Va.,
Wis., Washington, D.C.

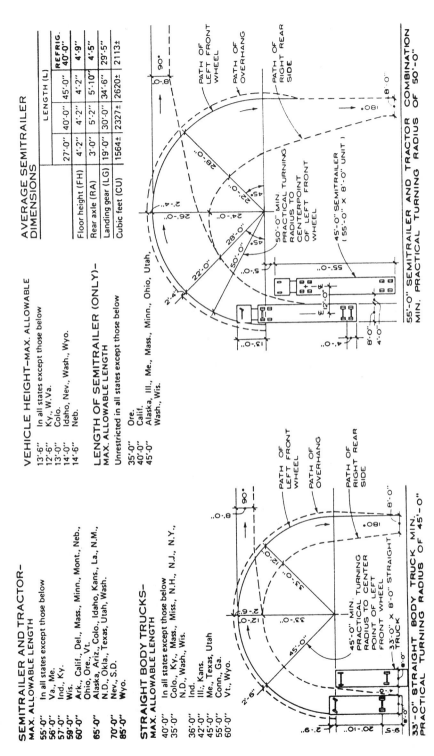

AVERAGE SEMITRAILER DIMENSIONS

	LENGTH (L)			
	27'-0"	40'-0"	45'-0"	REFRIG. 40'-0"
Floor height (FH)	4'-2"	4'-2"	4'-2"	4'-9"
Rear axle (RA)	3'-0"	5'-2"	5'-10"	4'-5"
Landing gear (LG)	19'-0"	30'-0"	34'-6"	29'-5"
Cubic feet (CU)	1564±	2327±	2620±	2113±

VEHICLE HEIGHT—MAX. ALLOWABLE

In all states except those below
13'-6" Ky., W.Va.
12'-6" Colo.
13'-0" Idaho, Nev., Wash., Wyo.
14'-0" Neb.
14'-6"

LENGTH OF SEMITRAILER (ONLY)— MAX. ALLOWABLE LENGTH

Unrestricted in all states except those below
35'-0" Ore.
40'-0" Calif.
45'-0" Alaska, Ill., Me., Mass., Minn., Ohio, Utah, Wash., Wis.

SEMITRAILER AND TRACTOR— MAX. ALLOWABLE LENGTH

55'-0" In all states except those below
56'-0" Va., Me.
57'-0" Ind., Ky.
59'-0" Wis.
60'-0" Ark., Calif., Del., Mass., Minn., Mont., Neb., Ohio, Ore., Vt.
65'-0" Alaska, Ariz., Colo., Idaho, Kans., La., N.M., N.D., Okla., Texas, Utah, Wash.
70'-0" Nev., S.D.
85'-0" Wyo.

STRAIGHT BODY TRUCKS— MAX. ALLOWABLE LENGTH

40'-0" In all states except those below
35'-0" Colo., Ky., Mass., Miss., N.H., N.J., N.Y., N.D., Wash., Wis.
36'-0" Ind.
42'-0" Ill., Kans.
45'-0" Me., Texas, Utah
55'-0" Conn., Ga.
60'-0" Vt., Wyo.

Fig. 8-1. Truck and trailer sizes. (Adapted from Reference 3.) Reprinted by permission of John Wiley & Sons, Inc.

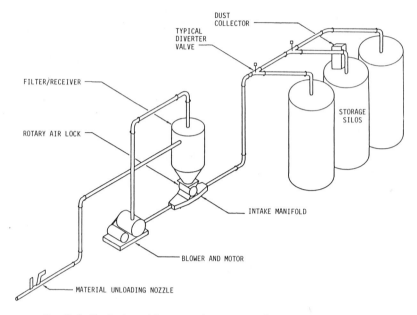

Fig. 8-2. Typical positive–negative pneumatic conveying system.

be placed in such a way that pallet trucks or hand trucks can be maneuvered very easily and efficiently.

The location of chemical storage must also take into consideration the effects of thermal and moisture changes. Liquid chemicals should be stored in an area of adequate temperature to prevent crystallization. Temperature of the space for dry chemical storage is also important to prevent condensation and to allow for proper working conditions in handling the material. Dry chemical forms should be protected from moisture. Consideration of dust control must also be given to storage of dry chemicals. It is desirable to keep dust from accumulating in the air for several reasons: to prevent dust contamination in the atmosphere; to prevent dust in the working area from being introduced into the chemical; and to prevent accumulations well below explosive levels.

Chemicals in the gaseous form should be located in areas where thermal protection is provided. Because withdrawal of gas from containers is a function of temperature, heated space is required to ensure an adequate supply of gaseous chemical.

All storage facilities for chemicals should be adequately ventilated, not only for workers' safety but also to maintain the proper thermal and moisture protection for the chemicals. Consideration also must be given to the density of the gases in work with gases and volatile chemicals. The density of the

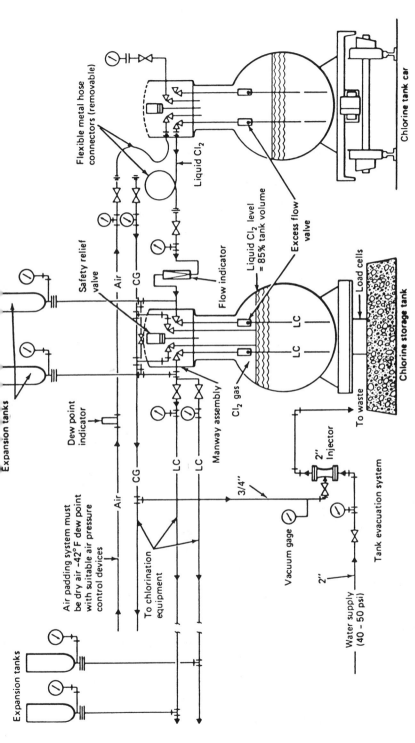

Fig. 8-3. Bulk chlorine storage and tank car unloading system. LC refers to liquid chlorine; CG refers to chlorine gas. The liquid chlorine header is in duplicate. A separate gas header is provided so that any car that arrives with too high a pressure (125 psi) may be relieved to a lower pressure. (Dip tubes must extend far enough into the tank so that liquid contents cannot exceed 85 percent of the total volume.)[4]

chemical gases in relation to air will dictate the location of ventilation facilities. For example, chlorine gas is heavier than air, and so the ventilation intake is close to the floor of the room housing the chemical.

Size. The size of chemical storage facilities should be based on cost-effectiveness, amount of chemical to be used per day, and ease of operation. Generally, storage for chemicals should contain a minimum of a 30-day supply to allow for chemical delivery, ordering, and any contingencies, or one and one-half times the bulk transport capacity, whichever is greater. The economics of bulk chemical delivery may also dictate storage size, although the quantity to be used on a daily basis may be small. The economics of sizing considerations includes: labor requirements for accepting delivery of the chemical, the actual cost of the physical storage space for the chemical, the cost of bulk versus bag delivery for a dry chemical, and the reliability of delivery. With larger plants it may be desirable to have a large amount of storage for the primary chemicals, with the ability to transfer the chemicals to day tanks.

Miscellaneous Considerations. Storage sizing should be based on the useful life of the chemical, as discussed earlier. Appurtenances that are included in the design of storage facilities include: mixing, vibration, or aeration to prevent arching and aid flow of dry chemicals; high and low level indicators; access hatches; and pressure and vacuum control devices. Mixing can be accomplished with pumping or conventional mixers. Level indicators can be in the form of sight devices, tank level gauges, or electronic devices.

Safety Considerations. Workers' safety must be considered in designing facilities for handling and storing chemicals. They should be stored in such a manner that chemicals accidentally spilled or leaked do not interact with other chemicals. Areas in which acids are stored should be separated from storage areas for bases, and liquids should be separated from dry forms of chemicals. Should a spill or accident occur, the contamination should be contained by use of containment walls or separate storage rooms. The accumulation of dusts and gases must also be controlled. Because of its hazardous nature and widespread use, the safety of chlorine has received a great deal of attention.[4,6,7]

Chemical handling also requires consideration of safety. Worker contact with chemicals should be minimized by use of machinery. If it is necessary for the workers to handle the chemicals, they should be protected from any contact with them. Consideration should be given to special clothing, emergency showers, and eyewashes. In the design of facilities for bag loading and unloading, the size of the bags and the manner in which they must be handled should be considered. Lifting of 100-pound (45.4-kg) bags higher than the waist should be avoided.

Bulk Storage. A typical bulk storage tank or bin for dry or solid chemicals is shown in Fig. 8–4, and one manufacturer's silo is shown in Fig. 8–4A. Dust collectors should be provided on manually and pneumatically filled bins. The material of construction and the required slope on the bin outlet vary with the type of chemical stored. In addition, some dry chemicals such as lime must be stored in airtight bins to keep moisture out.

Bulk storage bins should have a discharge bin gate so feeding equipment can be isolated for servicing. The bin gate should be followed by a flexible connection and a transition chute or hopper which acts as a conditioning chamber over the feeder.

Liquid storage tanks should be sized according to maximum feed rate, shipping time required, and quantity of shipment. The total storage capacity should be one and one-half times the largest anticipated shipment, and should provide at least a 30-day supply of the chemical at the design average dose. Storage tanks for most liquid chemicals may be located inside or outside. However, outdoor tanks usually must be insulated and/or heated to prevent crystallization. Storage tanks for some liquids, such as liquid caustic soda, should be provided with an air vent for gravity flow. Recirculation pumping systems frequently are used to prevent crystallization.

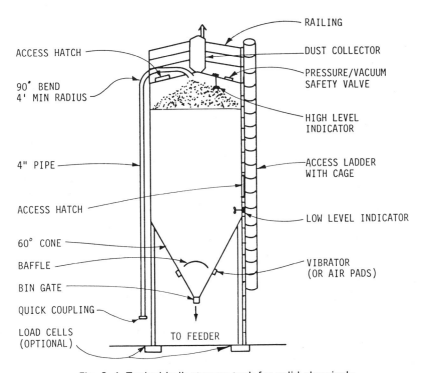

Fig. 8–4. Typical bulk storage tank for solid chemicals.

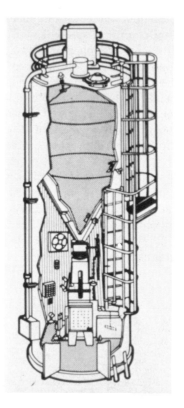

Fig. 8–4A. Bulk storage silo. (Courtesy of BIF® a Unit of General Signal)

Liquid storage tanks can be located either at ground level or above ground level, depending upon whether gravity feed or pressure feed is desired at the point of application. Figure 8–5 shows two common liquid feed systems, one with overhead storage and one with ground storage. The rotodip-type feeder or rotameter often is used for gravity feed and the metering pump for pressure feed systems. A picture of a rotodip feeder is shown in Fig. 8–5c. Overhead storage can be used to gravity-feed the rotodip, as shown in Fig. 8–5a. A centrifugal transfer pump may also be used, but needs an excess recirculation line to the storage tank, as shown in Fig. 8–5b.

Bag and Drum Storage. In general, bags or drums should be stored in a dry, cool, low-humidity area and used in proper rotation (i.e., first in, first out). Bag- or drum-loaded hoppers should have storage capacity for 8 hours at the nominal maximum feed rate so personnel are not required to fill or change the hopper more than once a shift.

Cylinder and Ton Container Storage. Whether in storage or in use, cylinders should not be permitted to stand unsupported. They should be chained to a fixed wall or support, and in such a manner as to permit ready access and removal. Ton containers should be stored horizontally, slightly elevated from ground or floor level, and blocked to prevent rolling. A convenient storage rack is obtained by supporting both ends of containers on rails of I-beams. Ton containers should not be stacked or racked more than one high unless special provision is made for easy access and removal. Chlorine cylinders and containers should be protected from impact, and handling should be kept to a minimum. Full and empty cylinders and ton containers

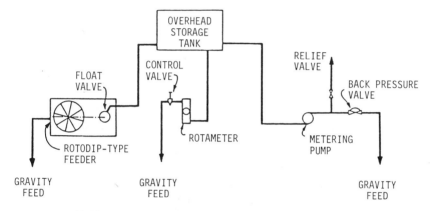

(a) Alternative liquid feed systems for overhead storage.

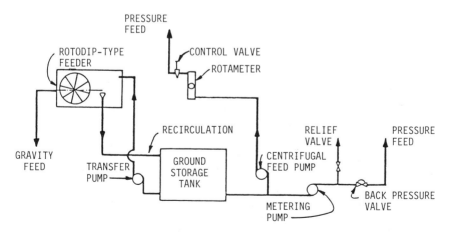

(b) Alternative liquid feed systems for ground storage.

Fig. 8–5. Liquid storage and feed systems.

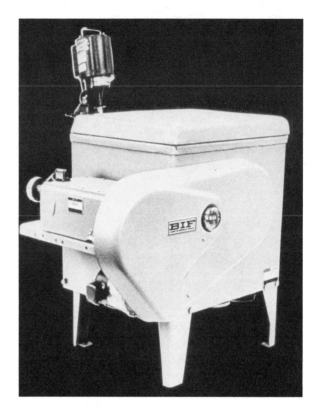

(c) Rotodip feeder. (Courtesy of BIF® a Unit of General Signal)

Fig. 8–5. (*Continued*)

should be stored separately. Figure 8–6 illustrates chlorine storage and handling for ton containers.[4]

Storage areas should be clean, cool, well ventilated, and protected from corrosive vapors and dampness. Cylinders and ton containers stored indoors should be in a fire-resistant building, away from heat sources (such as radiators, steam pipes, etc.), flammable substances, and other compressed gases. Subsurface storage areas should be avoided, especially for chlorine and sulfur dioxide. If natural ventilation is inadequate, as would be the case for chlorine or sulfur dioxide, storage and use areas should be equipped with suitable mechanical ventilators. Cylinders and ton containers stored outdoors should be shielded from direct sunlight and protected from accumulations of rain, ice, and snow.

All storage, handling, and use areas should be of such design that personnel

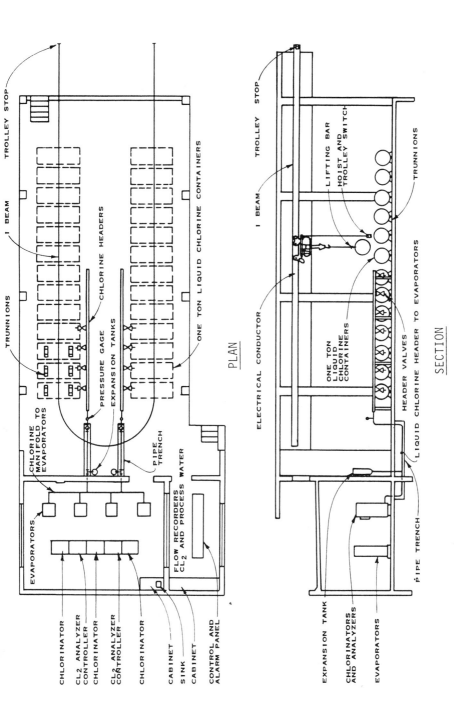

Fig. 8–6. Chlorine storage and handling facilities.

can quickly escape in emergencies. It is generally desirable to provide at least two ways to exit. Doors should open out and lead to outside galleries or platforms, fire escapes, or other unobstructed areas. Storage areas for ton cylinders should include a means for moving the cylinders to active use, and loading and unloading the delivery vehicle. Typically this is accomplished using a monorail system with a hoist. For larger facilities, a bridge crane may be more appropriate.

Feeding Systems

Feeding systems for chemicals involve conveying the chemicals from storage to the application point(s), and include pumps, conveyors, dry or liquid chemical feeders, eductors, and vacuum and pressure gas systems. This section discusses liquid chemical feeding, solid chemical feeding and dry chemical feeders, and gas feeding systems.

Liquid Chemical Feeding.   The feeding systems for liquid chemicals generally are simple, requiring only one or more metering pumps. Liquid chemicals also may be diluted and then pumped, or used with eductors or other hydraulically controlled devices. A typical solution feed system is shown in Fig. 8–7, and consists of a storage tank, transfer pump(s), a day tank for dilution, and liquid feeder(s). Some liquid chemicals can be fed directly without dilution, so that the day tank is not needed. Dilution water usually still is added to the solution feed pump discharge line after the chemical is metered, to prevent plugging of the chemical line due to crystallization, to reduce delivery time, and to help mix the chemical with the water being treated.

Liquid feed systems typically are recommended for use:

- When low chemical quantities are required.
- With less stable chemicals.
- With chemicals that are fed more easily as a liquid.
- Where handling of dusty chemicals or dangerous chemicals is undesirable.
- With chemicals available only as liquids.

Liquid feeders usually are metering pumps or orifices. These metering pumps may be positive displacement, plunger, or diaphragm-type pumps. Examples of plunger and diaphragm pumps are given in Fig. 8–8, and pictures of two styles of pump are included in Figs. 8–8A and 8–8B. Positive displacement pumps can be set to feed over a wide range by adjusting the pump stroke length. In some cases, control valves and rotameters may be all that is needed, while in other cases the rotating dipper-type feeder may be satisfac-

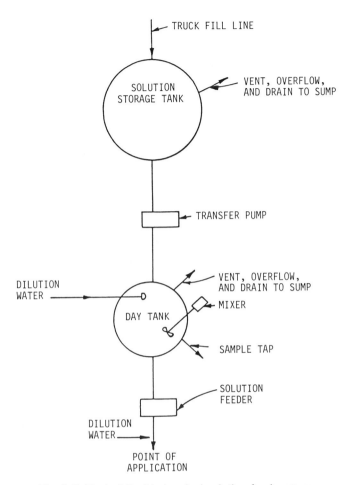

Fig. 8-7. Typical liquid chemical solution feed system.

tory. For uses such as lime slurry feeding, however, centrifugal pumps with open impellers are employed. The type of liquid feeder used depends on the viscosity, corrosivity, solubility, suction and discharge heads, and internal pressure relief requirements of the chemical.

Solid Chemical Feeding. There are a number of dry or solid chemical feeding systems available that involve direct feed of solutions or slurries. The solutions or slurries generally can be fed using pumping equipment, eductors, and other hydraulic control devices as described for liquid systems. The solutions and slurries typically are not so easy to handle as the liquid forms because they tend to form scales and precipitates. Proper solution

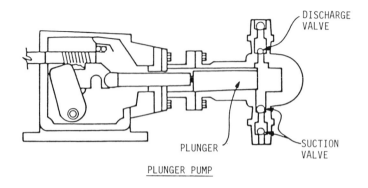

PLUNGER PUMP

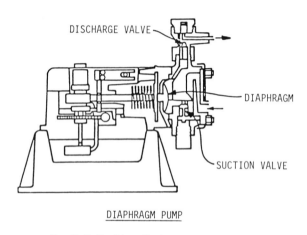

DIAPHRAGM PUMP

Fig. 8–8. Positive displacement pumps.

strength and proper mixing of the dry material must be achieved prior to pumping. The proper solution strength is attained through accurate measurement of the chemical by chemical feeders. Chemical feeders must accommodate the minimum and maximum feeding rates required. Manually controlled feeders have a common range of 10:1, but this range can be increased to about 20:1 or 30:1 with dual control systems.

Chemical feeder control can be manual, automatically proportioned to flow, dependent on some form of process feedback, or a combination of any two of these methods. More sophisticated control systems are feasible if proper sensors are available. If manual control systems are specified with the possibility of future automation, the feeders selected should be able to be converted with a minimum of expense. An example would be a feeder with an external motor that could easily be replaced with a variable speed motor or drive when automation is installed.

Fig. 8–8A. Single head solution metering pump—diaphragm type. (Courtesy of Wallace & Tiernan, division of Pennwalt Corporation.)

Standby or backup units should be included for each type of feeder used. Points of chemical addition and piping to them should be capable of handling all possible changes in dosing patterns in order to have proper flexibility of operation. Designed flexibility in hoppers, tanks, chemical feeders, and solution lines is the key to maximum benefits at least cost. More than one feed point should be provided for each chemical so that the operator can try different combinations of chemicals at the various feed points to optimize the chemical dosages.

Solids characteristics vary considerably, and the selection of a feeder must be made carefully, particularly in a smaller-sized facility where a single feeder may be used for more than one chemical. In general, provisions should be made to keep all dry chemicals cool and dry. Dryness is important, as hygroscopic (water-absorbing) chemicals may become lumpy, viscous, or even rock hard; other chemicals that absorb water less readily may become sticky from moisture on the particulate surfaces, causing increased arching in hoppers. In either case, moisture will affect the density of the chemical and may result in incorrect feed rates. Also, the effectiveness of dry chemicals, particularly polymers, may be reduced. Dust-removal equipment should be used

Fig. 8-8B. Envirotube metering pump. (Courtesy of Wallace & Tiernan, division of Pennwalt Corporation.)

at shoveling locations, bucket elevators, hoppers, and feeders for neatness, corrosion prevention, and safety reasons. In general, only limited quantities of chemical solutions should be made from dry chemicals because the shelf life of mixed chemicals (especially polymers) may be short.

Dry Chemical Feeders. The simplest method of feeding dry or solid chemicals to a mixing tank is by hand. Solid chemicals may be preweighed and added or poured by the bagful into a dissolving tank. This method is generally limited to very small operations, however, and dry chemical feed equipment is required in larger installations.

A dry feed installation is shown schematically in Fig. 8-9, and consists of a storage bin and/or hopper, a feeder, and a dissolver tank. Dry feeders are either of the volumetric or the gravimetric type. Volumetric feeders usually are used only where low feed rates are required. These feeders deliver a constant, preset amount of chemical and do not recognize changes in material density. This type of feeder must be calibrated by trial and error at the outset, and then readjusted periodically if the material changes in density.

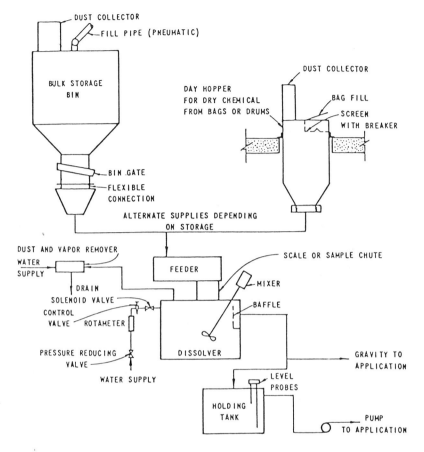

Fig. 8-9. Typical solid or dry chemical feed system.

The gravimetric feeder delivers chemicals based on the required weight per unit volume. Typically, the volumetric feeders are less expensive than the gravimetric units.

Most types of volumetric feeders generally fall into the positive displacement category. All designs of this type use some form of moving cavity of a specific or variable size. In operation, the chemical falls by gravity into the cavity, and it is almost fully enclosed and separated from the hopper's feed. The rate at which the cavity moves and is discharged, together with the cavity size, governs the amount of chemical fed. Positive displacement feeders often use air injection to enhance flowability of the material.

Rotary Paddle Feeder.   A rotary paddle feeder is especially effective for fine materials that tend to flood. The paddle or vane is located beneath the hopper discharge, with the feed being varied by means of a sliding gate and/or variable speed drive. The feed rate can be varied easily by adjusting the variable speed drive on the vane shaft. A variant of the rotary paddle feeder is the pocket feeder, also called the star or revolving door feeder, in which the paddle is tightly housed to permit delivery against vacuum or pressure. Figure 8-10 illustrates this feeder type.

Oscillating Hopper Feeder. Another type of volumetric feeder is the oscillating hopper, or oscillating throat feeder. This feeder consists of a main hopper and an oscillating hopper that swivels on the end of the main hopper. The material completely fills both hoppers and rests on the tray beneath. As the oscillating hopper moves back and forth, the scraper, which rests on the fixed tray below, is moved first to the left and then to the right. As it moves, it pushes a ribbon-like layer of dry chemical off the tray. The capacity is fixed by the length of the stroke, which may be varied by means of a micrometer screw. Further adjustment is possible by changing the clearance between the hopper and the fixed tray, which may be raised or lowered. This type of feeder is one of the most widely used in small water plants.

Oscillating Plate Feeder. In the oscillating plate feeder, a plate is mounted below the bottom spout in the storage hopper so that the chemical spills out onto the plate as it comes out of the storage hopper. A leveling bar is mounted above the plate on each of the two ends. The plate is mechanically linked to the drive motor and slowly oscillates from side to side as the feeder operates. The magnitude of oscillation can be adjusted with the mechanical

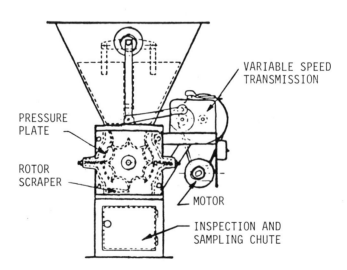

Fig. 8-10. Positive displacement rotary feeder.[5]

linkage. This provides a dosage adjustment. Each time the plate oscillates from one side to the other, a measured amount of chemical drops off it into the solution tank. The rake bar above each end of the plate helps to regulate the repeatability of the feed rate.

Grooved Disc Feeder. This feeder consists of a grooved horizontal disc that meters the chemical addition. The disc rotates, and the grooves are filled as they pass under the storage hopper. The dry chemical is struck level (excess material removed), and then a stationary plow removes the material from the groove for metering. The feed rate is varied by changing the speed of disc rotation or by changing the groove size. Typically, these feeders are used for applications requiring small feed rates of dry materials. A grooved disc feeder is illustrated in Figs. 8–11A and 8–11B.

Vibrating Feeder. The vibrating feeder is also a volumetric feeder. With it, motion is obtained by means of an electromagnet anchored to the feeding

(A)

Fig. 8–11. Grooved disk feeder. A) Inside view of feeder. B) Complete unit. (Courtesy of BIF® a Unit of General Signal)

(B)

Fig. 8–11. (*Continued*)

trough, which in turn is mounted on flexible leaf springs. The magnet, energized by pulsating current, pulls the trough sharply down and back; then the leaf springs return it up and forward to its original position. This action is repeated 3,600 times per minute (when operating on 60-cycle, ac), producing a smooth, steady flow of material.

Volumetric Belt Feeder.   The volumetric belt feeder uses a continuous belt of specific width moving from under the hopper to the dissolving tank. The material falls on the feed belt from the hopper and passes beneath a vertical gate. For a given belt speed, the position of the gate determines the volume of material passing through the feeder. A volumetric belt feeder is illustrated in Fig. 8–12.

Screw-Type Feeder.   The volumetric screw-type feeder employs a screw or helix at the bottom of the hopper to transfer dry chemical to the solution

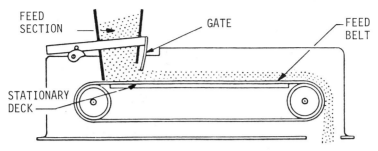

Fig. 8-12. Volumetric belt type feeder.[5]

chamber, as illustrated in Fig. 8–13, and pictured in Figs. 8–13A and 8–13B.

The basic drawback of the volumetric feeder is that it cannot compensate for changes in the density of materials and therefore is not so accurate as two other types of dry feeders: the gravimetric and loss-in-weight types. For these feeders, the volumetric design is modified to include a gravimetric or loss-in-weight controller, which allows for weighing of the material as it is fed. Both gravimetric and volumetric feeders can be used to feed in proportion to the flow of wastewater.

Belt-Type Gravimetric Feeder. Belt-type gravimetric feeders have a wide capacity range and usually can be sized for any use in a treatment plant.

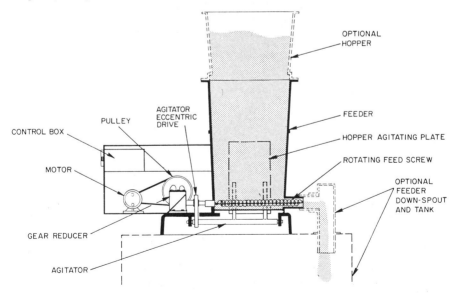

Fig. 8-13. Typical helix or screw-type volumetric feeder. (Courtesy of Wallace & Tiernan, division of Pennwalt Corporation.)[5]

Fig. 8–13A. Volumetric feeder. (Courtesy of Wallace & Tiernan, division of Pennwalt Corporation.)

Belt-type gravimetric feeders use a basic belt feeder with a weighing and control system. Feed rates can be changed by adjusting the weight per foot of belt, the belt speed, or both. Two types of gravimetric belt-type feeders are available: the pivoted belt-type and the rigid belt-type. The pivoted belt feeder consists of a feed hopper, an endless traveling belt mounted on a pivoted frame, an adjustable weight that counterbalances the load on the belt, and a means of continuously and automatically adjusting the feed of material to the belt. Dry chemical flow to the feeder can be controlled by a gate placed between the feed hopper and the belt or by controlling the amplitude of vibration in a vibrating deck placed between the feed hopper and the belt. Figure 8–14 shows a schematic of this feeder, and Fig. 8–14A presents a picture of one unit.

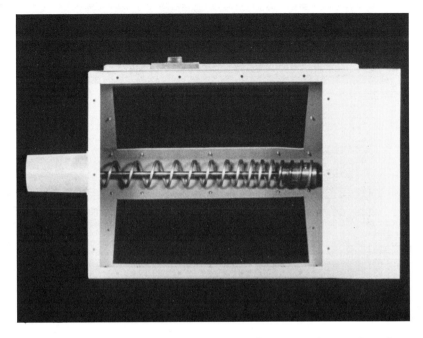

Fig. 8–13B. Helical feeder. (Courtesy of BIF® a Unit of General Signal)

The rigid belt feeder is similar to the pivoted belt feeder except for the chemical feed rate adjustment method. The pivoted belt filter is adjusted through action of the belt tilting up and down, while the rigid belt adjustment occurs through action of the scale beam dependent only on the weight of the belt. See Fig. 8–15.

Good housekeeping and the need for accurate feed rates dictate that the gravimetric feeder be shut down and thoroughly cleaned on a regular basis.

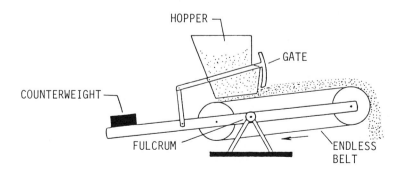

Fig. 8–14. Pivoted belt gravimetric feeder.[5]

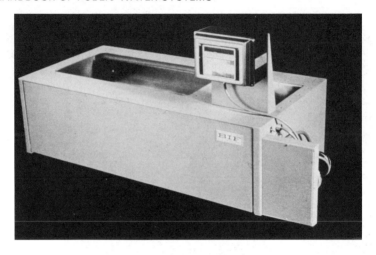

Fig. 8–14A. Gravimetric feeder. (Courtesy of BIF® a Unit of General Signal)

Chemical buildup can affect accuracy and can even jam the equipment in some cases.

Loss-in-Weight Feeder. The loss-in-weight feeder should be used where the greatest accuracy or more economical use of chemical is important. This feeder only works for feed rates up to 4,000 lb/hour (1,815 kg/h). The loss-in-weight feeder has a material hopper and feeder set on enclosed scales. The feed rate controller is used to deliver the dry chemical at the desired rate. See Fig. 8–16.

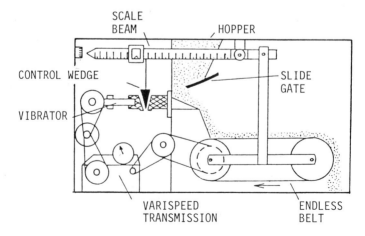

Fig. 8–15. Rigid belt gravimetric feeder.[5]

Fig. 8–16. Loss-in-weight feeder. (Courtesy of BIF® a Unit of General Signal)

Dissolvers. Dissolvers are also important to dry feed systems because any metered chemical must be wetted and mixed with water to provide a chemical solution free of lumps and undissolved particles. Most feeders, regardless of type, discharge their material to a small dissolving tank that is equipped with a nozzle system and/or mechanical agitator, depending on the solubility of the chemical being fed. It is important that the surface of each particle become completely wetted before entering the feed tank to ensure complete dispersal and to avoid clumping, settling, or floating.

A dissolver for a dry chemical feeder is unlike a chemical feeder, which by simple adjustment and change of speed can vary its output tenfold. The dissolver must be designed for the job to be done. A dissolver suitable for a rate of 10 lb/hour (4.54 kg/h) may not be suitable for dissolving at a rate of 100 lb/hour (45.4 kg/h).

The capacity of a dissolver is based on detention time, which is directly related to the wettability or rate of solution of the chemical. Therefore, the dissolver must be large enough to provide the necessary detention for both the chemical and the water at the maximum rate of feed.

It is sometimes preferable to plan the feeding equipment for the first few years of flow in the plant and replace or add on equipment in later years. This will allow a more accurate feed range to be selected. The capacity of the feeding equipment must meet the maximum dosage required on a maximum day demand, and it must be able to feed that dosage while maintaining reserve units. It is generally accepted practice to install 50 percent more than the maximum dose. For chemicals that are of primary importance in the plant, such as coagulants and disinfectants, backup units or reserve units should be provided for periods of time when the feeding equipment may be out of service. The feeding equipment's materials of construction should be compatible with the chemicals that may be used.

Feeding systems should also be designed so that an accurate inventory of chemicals can be maintained at all times. The feed rate should be checked often, as well as the total amount of chemical fed. The change in tank level should be recorded daily and possibily at more frequent intervals, depending upon feed rate and the cost of the chemical.

Gas Feeding Systems.   Gas feeding systems are mechanically and operationally simple. Gases can be fed through pressure systems, vacuum systems, or direct feed. Also they may be diluted and fed as a solution. Therefore, gas feeders can be classified as solution feed or direct feed. Solution feed vacuum-type feeders are commonly used in chlorination and in dechlorination with sulfur dioxide. In chlorination, chlorine gas is metered under vacuum, and it is mixed with water in an injector to produce a chlorine solution. The flow of chlorine gas is automatically shut off on loss of vacuum, stoppage of the solution discharge line, or loss of operating water pressure.

Direct feed or "dry feed" equipment is infrequently used, and only when either water or electricity or both is unavailable at a site. This type of equipment is nearly the same as the solution feed type except that there is no device for making and injecting an aqueous solution. The gas itself is piped directly into the water to be treated. The same equipment is used to control, withdraw, and meter the gas from the containers.

Feed System Requirements for Common Chemicals

Alum.   Dry alum must be made into a solution before being fed to the plant. Dissolving tanks must be made of a noncorrosive material, and dissolvers should be the right size to obtain the desired solution strength. The most common solution strength is 0.5 pound (0.23 kg) of alum to 1 gallon (0.06 kg/l) of water, or a 6 percent solution. The dissolving tank should be designed for a minimum detention time of 5 minutes at the maximum feed rate. Dissolvers should have water meters and mixers so that the water/alum mixture can be controlled. Most liquid alum is fed as it is delivered in a standard 50 percent solution.

Alum is usually fed by positive displacement metering pumps. Normally dilution water is added to an alum feed pump discharge line to prevent line plugging, to reduce delivery time to the point of application, and to help mix the alum with the water being treated. The output of the pumps can be controlled automatically in proportion to plant flow. This is done by setting the alum dosage for the maximum flow rate. The controls are then set to automatically adjust the off–on cycle and the amount of alum pumped to the actual flow.

Carbon Dioxide. Stack gases from on-site furnaces, such as recalcination or incineration, have been used as a source of carbon dioxide. Feeding systems for the stack gases include simple valving arrangements for admitting varying quantities of makeup gas to the suction side of constant-volume compressors. Venting of excess gas from the compressor discharge can be valved to the suction side of the compressors. The compressors deliver the stack gases to the source of use.

Pressure generators and submerged gas burners are regulated by valving arrangements on the fuel and air supply. Generation of carbon dioxide is accomplished by the combustion of a fuel (natural gas)–air mixture under water. This system is more difficult to control, requires operator attention, and demands considerable maintenance over the life of the equipment, when compared to liquid CO_2 systems.

Commercial liquid carbon dioxide is used more often because of its high purity, the simplicity and range of feeding equipment, ease of control, and smaller, less expensive piping systems. After vaporization, carbon dioxide with suitable metering and pressure reduction may be fed directly to the point of application as a gas. Metering of directly fed pressurized gas is difficult owing to the high adiabatic expansion characteristics of the gas. Also, direct feed requires extremely fine bubbles to ensure that the gas goes into solution; this in turn can lead to scaling problems. Hence, vacuum-operated, solution-type gas feeders are preferred. Such feeders generally include safety devices and operating controls in a compact panel housing, with materials of construction suitable for carbon dioxide service. Absorption of carbon dioxide in the injector water supply approaches 100 percent when a ratio of 1.0 pound (2.2 kg) of gas to 60 gallons (0.002 kg/l) of water is maintained.

Chlorine. Elemental chlorine is a poisonous yellow-green gas at ordinary temperature and pressure. The gas is stored as a moisture-free liquid under pressure in specially constructed steel containers, and is vaporized from the liquid form either directly or with heated vaporizers. Chlorine gas feeders may be classified into two types, direct feed or solution feed.

Direct or dry feed gas feeders deliver chlorine gas under pressure directly to the point of application. Direct feed gas chlorinators are less safe than solution feed chlorinators, and are used when there is no adequate water supply available for ejector operation. In solution feed vacuum-type feeders, chlorine gas is maintained under vacuum throughout the apparatus. Vacuum is created by water flow through an injector to move the chlorine from the supply system through the chlorine gas metering devices to the injector. Chlorine gas is mixed with water in the injector, and the chlorine solution is moved to the point of application. In this feeder, the vacuum controls the operation of the chlorine inlet valve so that the chlorine will not feed unless sufficient vacuum is induced through the apparatus. This type of feeder is most common because its safe operation is assured. It employs direct indication metering, and the flow of chlorine is automatically shut off on loss of vacuum, stoppage of the discharge line, or loss of operating water pressure.

In some cases it may be necessary to provide more than one point of injection from a single injector, and this requires manifolding. Valves for manifolding generally are either the ball-type or diaphragm-type.

Chlorine Dioxide.   Chlorine dioxide is a greenish-yellow gas that is quite unstable and, under certain conditions, explosive. It cannot be shipped in containers because of its explosive nature; it must be generated at the point of use and applied immediately.

Although readily soluble in water, ClO_2 does not react with water as does chlorine. Chlorine dioxide is easily expelled from aqueous solution by blowing a small amount of air through the solution. Aqueous solutions of ClO_2 are also subject to some photodecomposition.

Chlorine dioxide is generated by oxidizing sodium chlorite with chlorine (either chlorine gas or hypochlorite) at a pH of 4 or less. This means that the injector system of the chlorination assembly must be capable of delivering a chlorine solution strength greater than about 500 mg/l. Because the upper limit of this solution strength should not exceed 3,500 mg/l to prevent breakout of molecular chlorine at the point of application, the effective range of chlorine dioxide production is about 7:1. Chemical feed devices can handle ranges up to 20:1 on a flow proportional basis and 200:1 on a compound loop control system.

Ferric Chloride.   Ferric chloride is always fed as a liquid, and is normally obtained in liquid form containing 20 to 45 percent $FeCl_3$. When iron salts such as ferric chloride are used for water coagulation in soft waters, a small amount of base (such as sodium hydroxide or lime) is needed to neutralize the acidity of these strong acid salts.

Dilution of ferric chloride solution from its shipping concentration to a weaker feed solution should be avoided, because of a potential to hydrolyze.

Ferric chloride solutions may be transferred from underground storage to day tanks with rubber-lined self-priming centrifugal pumps having Teflon rotary and stationary seals. Because liquid ferric chloride can stain or leave deposits, glass-tube rotameters are not used for metering. Instead, rotodip feeders and diaphragm metering pumps made of rubber-lined steel and plastic are often used for feeding ferric chloride.

Ferric Sulfate.   Feed solutions are usually made up at a water-to-chemical ratio of 2:1 to 8:1 (on a weight basis). The usual ratio is 4:1, and the feed solution is made up in a 20-minute detention tank. Care must be taken not to dilute ferric sulfate solutions to less than 1 percent, in order to prevent hydrolysis and deposition of ferric hydroxide.

Dry feeding requirements are similar to those for dry alum except that belt type feeders are rarely used because of their open type of construction. Closed construction, as found in the volumetric and loss-in-weight type feeders, generally exposes a minimum of operating components to the vapor, and thereby minimizes maintenance. A water-jet vapor remover should be provided at the dissolver to protect both the machinery and the operator.

Ferrous Sulfate.   The granular form of ferrous sulfate has the best feeding characteristics, and gravimetric or volumetric feeding equipment may be used. The optimum chemical-to-water ratio for continuous dissolving is 0.5 lb/ gallon or 6 percent with a detention time of 5 minutes in the dissolvers. Mechanical agitation should be provided in the dissolver to ensure complete solution.

Lime.   Although lime comes in many forms, quicklime and hydrated lime are used most often for water coagulation or softening. Quicklime is almost all calcium oxide (70 to 96 percent CaO). High-calcium quicklime contains more than 88 percent CaO and less than 5 percent magnesium oxide (MgO), while dolomitic lime may contain up to 40 percent MgO.

Quicklime (unslaked lime) is almost all CaO and first must be converted to the hydrated form ($Ca(OH)_2$). Hydrated or slaked lime is a powder obtained by adding enough water to quicklime to satisfy its affinity for water. Hydrated lime needs only enough water to form milk of lime. Wetting or dissolving tanks usually are designed for 5 minutes' detention with 0.5 lb/gallon (0.06 kg/l) of water or 6 percent slurry at the highest feed rate. Hydrated lime often is used where maximum feed rates are less than 250 lb/hour (113.3 kg/h).

Dilution is not too important in lime feeding; therefore, it is not necessary to control the amount of water used in feeding. Hydraulic jets may be used for mixing in the wetting chamber of the feeder, but the jets should be the right size for the water supply pressure.

Lime is never fed as a solution because of its low solubility in water. Also, quicklime and hydrated lime usually are not applied dry directly to water for the following reasons:

- They are transported more easily as a slurry.
- A lime slurry mixes better with the water than dry lime.
- Prewetting the lime in the feeder with rapid mixing helps to ensure that all particles are wet and that none settles out in the treatment basin.

Major components of a lime feed system (illustrated in Fig. 8–17) include a storage bin, dry lime feeder, lime slaker, slurry holding tank, and lime slurry feeder. The slurry holding tank is usually needed only when the point of application is at a remote location. Quicklime feeders usually must be the belt or loss-in-weight gravimetric types because bulk density changes so much. Feed equipment usually has an adjustable feed range of at least 20:1 to match the operating range of the slaker. Lime slakers are shown in Fig. 8–17A and 8–17B.

There are two basic types of lime slakers, the paste or "pug mill" type (Fig. 8–18) and the detention-type slaker. The paste-type slaker (Fig. 8–19) adds water as required to maintain a desired mixing viscosity, so that the viscosity sets the operating retention time of the slaker. The detention-type slaker adds water to maintain a desired ratio with the lime, so that the reduction time is set by the lime feed rate. The detention slaker produces a lime slurry of about 10 percent $Ca(OH)_2$, while the paste type produces a

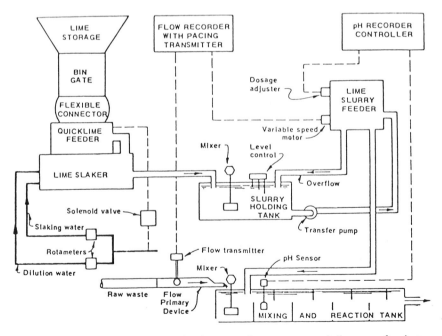

Fig. 8–17. Illustrative lime feed system for water coagulation or softening.

Fig. 8-17A. Paste-type lime slaker. (Courtesy of Wallace & Tiernan, division of Pennwalt Corporation.)

paste of about 36 percent $Ca(OH)_2$. Other differences between the two slakers are that the detention-type slaker operates with a higher water-to-lime ratio, a lower temperature, and a longer retention time. For either slaker type, vapor removers are required for feeder protection because lime slaking produces heat in hydrating the CaO to $Ca(OH)_2$.

The required slaking time varies with the source of lime. Fast-slaking limes will complete the reaction in 3 to 5 minutes, but poor-quality limes may require up to 60 minutes and an external source of heat, such as hot water or steam. Before selecting a slaker, it is advisable to determine the slaking time, best initial water temperature, and optimum water:lime ratio for the lime to be used. Procedures for slaking tests have been recommended by the American Water Works Association. More information about lime storage, handling, and use can be found in Reference 8.

Ozone. Ozone is produced commercially by the reaction of an oxygen-containing feed gas in an electrical discharge. The feed gas, which may be air, pure oxygen, or oxygen-enriched air, is passed between electrodes

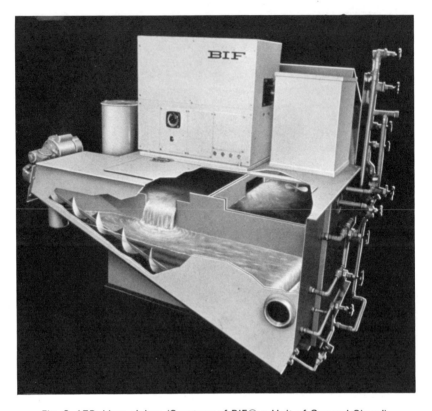

Fig. 8–17B. Lime slaker. (Courtesy of BIF® a Unit of General Signal)

separated by an insulating material. A high voltage of up to 20,000 volts is applied to the high-tension electrode. The ozone molecule, made up of three oxygen atoms, is highly unstable, and is one of the most powerful oxidizing agents known.

Ozone is a pale blue gas with a distinct pungent odor. In concentrations usually employed, the odor is detectable, but the color is not visible. It is both toxic and corrosive. Because of its unstable nature, it must be generated at the point of use. Consequently, safety aspects of transportation and storage are eliminated.

There are two basic ways to generate and use ozone in water treatment: (1) generation from air, and (2) generation from oxygen with recycling of oxygen from the ozone generation system. Figure 8–20 shows these different methods.

The once-through air approach uses conventional air drying techniques such as compression and refrigeration, followed by desiccant drying. Ozone

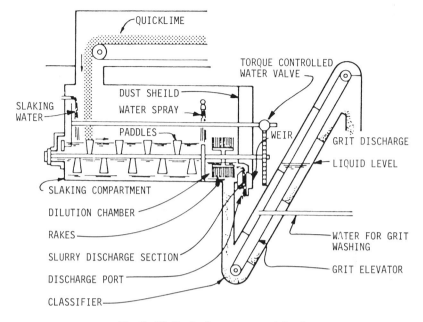

Fig. 8–18. Typical paste-type slaker.[5]

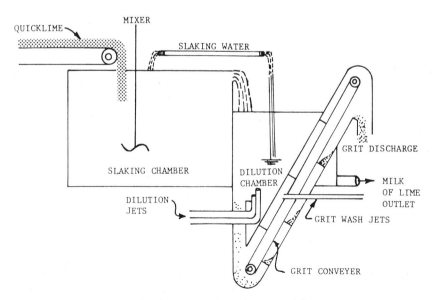

Fig. 8–19. Typical detention-type slaker.[5]

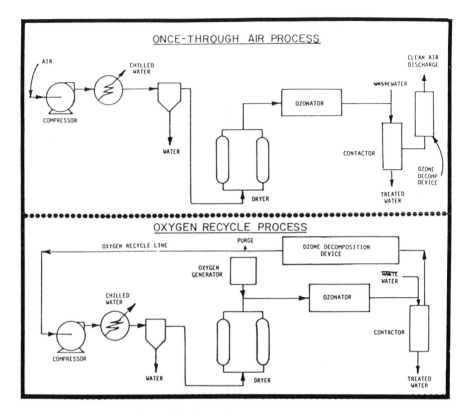

Fig. 8-20. Alternative ozonation systems.

generated from air may be 0.5 to 2.0 percent of the air by weight, but is usually produced at 1.0 percent by weight. Ozone typically is mixed with water in a contact basin as shown in Fig. 8-21. Fine bubble diffusers are used to feed the gas into the basin. Packed beds have also been used as contactors. Following treatment in the covered ozone contactor, the gas is decomposed to prevent high concentrations of ozone from being released to the atmosphere.

The oxygen recycle approach may be used to recover valuable oxygen-rich off-gas from the contactor when very pure oxygen is fed to the ozone generator. High-purity oxygen gas sometimes is used, since it enables the generator to produce two to three times as much ozone per unit time, and uses only about half as much generator power per pound compared to ozone produced from air.

There are three basic types of commercially available ozone generators: the Otto plate, the tube, and the Lowther plate. More information about the types of ozone generators available can be found in Reference 9.

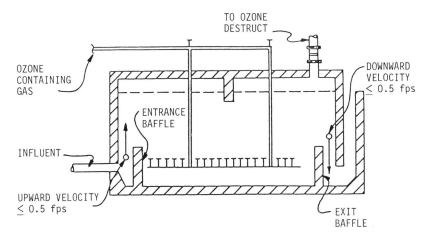

Fig. 8–21. Typical ozone contact basin using porous diffusers.

Polymers. When dry polymers are used, the polymer and water must be blended and mixed to obtain the desired solution. Initially, complete wetting of the polymer is necessary, using a funnel-type aspirator. After wetting, warm water should be added, with gentle mixing for about 1 hour. Polymer feed solution strengths are usually in the range of 0.1 to 0.75 percent. Stronger solutions are often too viscous to feed. Often the metered solution is diluted just prior to injection to the process to obtain better dispersion at the point of application.

The solution preparation system can include either a manual or an automatic blending system, with the polymer dispensed by hand or by a dry feeder to a wetting jet and then to a mixing aging tank at a controlled rate. The aged polymer solution is transported to a holding tank where metering pumps or rotodip feeders feed the polymer to the process. A schematic of a manual dry polymer feed system is shown in Fig. 8–22.

The solution preparation system may be an automatic batching system, as shown in Fig. 8–23, and pictured in Fig. 8–23A. These systems fill the holding tank with aged polymer, as required by level probes. Such a system is usually provided only at large plants.

Polymer solutions above 1 percent in strength should be avoided because they are very viscous, and difficult to handle. Most powdered polymers are stable when dry, but even in cool, dry conditions, they should not be stored as powders in unopened bags for more than 1 year. Once polymers are dissolved, they may become unstable within 2 to 3 days.

Liquid polymers need no aging, and simple dilution is the only requirement for feeding. The dosage of liquid polymers may be accurately controlled by metering pumps or rotodip feeders.

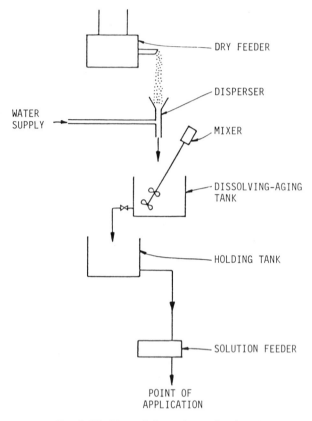

Fig. 8-22. Manual dry polymer feed system.

Because polymers can cause slippery conditions in a treatment plant, spills should be cleaned up immediately. Other safety precautions should also be observed, as specified by the manufacturer.

Powdered Activated Carbon. When powdered activated carbon is used, it is mixed directly with the water, fed as a slurry, and removed by coagulation and settling. The carbon slurry is transported by pumping the mixture at a high velocity to keep the particles from settling and collecting along the bottom of the pipe. The velocity of the slurry should be kept between 3 and 5 feet/sec. At velocities less than 3 feet/sec (0.91 m/s), carbon will settle out in the pipeline; and at velocities greater than 10 feet/sec (3.05 m/s), excessive carbon abrasion and pipe erosion will occur. At most plants, carbon slurries are fed at a concentration around 10.7 percent or 1 pound carbon/gallon (0.12 kg/l) water; typically, at this concentration, either cen-

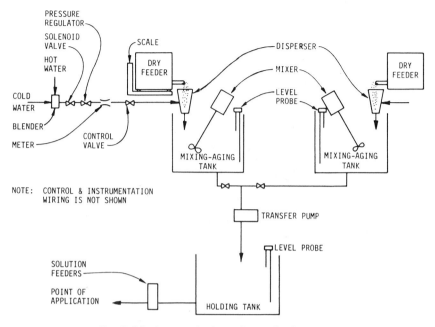

Fig. 8–23. Automatic dry polymer feed system.

trifugal pumps or a combination of centrifugal pumps and eductors is used to transport the carbon slurry. Diaphragm slurry pumps or double-acting positive displacement pumps are used for transporting higher concentrations. Another transport method used is a pressure pot system in which carbon is loaded into a pressure tank and forced out by pressurizing the vessel. The carbon slurry may be fed using a rotodip feeder.

Activated carbon is a dusty respiratory irritant, which smolders if ignited. It should be isolated from inflammable materials such as rags, chlorine compounds, and all oxidizing agents.

Soda Ash. Dense soda ash is generally used in municipal applications because of its superior handling characteristics. It has little dust and good flow characteristics, and will not arch in the bin or flood the feeder. It is relatively hard to dissolve; so ample dissolver capacity must be provided. Normal practice calls for 0.5 pound (0.23 kg) of dense soda ash per gallon of water, or a 6 percent solution retained for 20 minutes in the dissolver. Dissolving of soda ash may be hastened by the use of warm dissolving water. Mechanical or hydraulic jet mixing should be provided in the dissolver.

Sodium Aluminate. Dry sodium aluminate is not available in bulk quantities; therefore, small day-type hoppers with manual filling arrangements

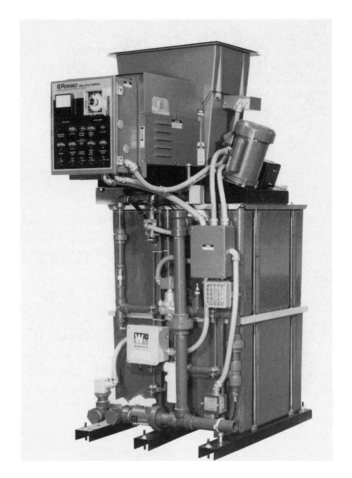

Fig. 8–23A. Polyelectrolyte batching system. (Courtesy of Wallace & Tiernan, division of Pennwalt Corporation.)

are used. Dissolvers for the free-flowing grade of sodium aluminate are normally sized for 0.5 pound/gallon (0.06 kg/l), or 6 percent solution strength with a dissolver detention time of 5 minutes at the maximum feed rate. After it is dissolved, agitation should be minimized or eliminated to prevent deterioration of the solution. Solution tanks must be covered to prevent carbonation of the solution.

Liquid sodium aluminate may be fed at shipping strength or diluted to a stable 5 to 10 percent solution. Stable solutions are prepared by direct addition of low-hardness water and mild agitation. Air agitation is not recommended.

Sodium Chlorite. Sodium chlorite ($NaClO_2$) for the generation of chlorine dioxide is available as an orange powder or as a solution.

The sodium chlorite pump is sized so as not to exceed a solution strength of 20 percent by weight or 1.66 pound/gallon (0.20 kg/l). Diaphragm pumps rather than piston pumps normally are used for handling the solutions. The solution container for sodium chlorite is sized for at least 1 day's operation.

Sodium chlorite will withstand rough handling if it is free from organic matter. However, in contact with organic materials (clothing, sawdust, brooms), it may ignite. It is sensitive to heat, friction, and impact. These problems are minimized with sodium chlorite solutions.

Sodium Hydroxide. Liquid sodium hydroxide, or caustic soda, usually is delivered in bulk shipments, and must then be transferred to storage. The caustic soda often is heated, and is fed by metering pumps as a concentrated solution. Dilution water usually is added after feeding to the pump discharge line. Feeding systems for caustic soda are about the same as for liquid alum except for materials of construction.

Caustic soda is poisonous and dangerous if handled improperly. To avoid accidental spills, all pumps, valves, and lines should be checked regularly for leaks. Operators should be properly instructed in the precautions needed for the safe handling of this chemical. Emergency eyewashes and showers should be provided close to the caustic soda storage and feed area to protect personnel from accidental spills.

Piping and Pumping of Chemicals

Piping and accessories for transporting and feeding various chemicals should be provided only after specific chemicals have been selected for use in the treatment plant. Care must be taken to use only materials that are compatible with each chemical. For example, many chemical handling systems require special materials of construction and special types of piping and transport channels, pumps, valves, and gaskets. Table 8–3 provides a summary of acceptable materials for piping and accessories for several commonly used chemicals.[5] Because sodium hydroxide (caustic soda) is dangerous to handle, a separate table is included for it; Table 8–4 provides detailed information on materials for feeding and handling equipment and piping for caustic soda systems.[5]

Manual and Low Flow Pacing for Chemical Feed Systems

There are three commonly used chemical feed systems; (1) dry feeders, (2) solution feeders, and (3) gas feeders. Each type of feeder should have one

Table 8-3. Compatible Materials for Piping and Accessories of Commonly Used Chemicals.

CHEMICAL	FRP*	PLASTICS	316SS	304SS	STEEL	RUBBER-LINED STEEL	RUBBER-LINED IRON	IRON	LEAD	COPPER	RUBBER	CONCRETE	RESIN	TEFLON	SARAN	GLASS	BRONZE
ALUM PIPES	•	•	•						•								
VALVES		•	•			•	•										
SODIUM ALUMINATE PIPES		•		•	•			•				•					
FERRIC CHLORIDE PIPES	•	•			•						•				•		
VALVES		•									•		•	•	•		
FERRIC SULFATE PIPES	•	•	•								•					•	
LIME PIPES AND CHANNELS		•			•			•			•	•					
PUMP IMPELLER LINING											•						•
SODIUM CARBONATE PIPING		•			•			•			•						
SODIUM HYDROXIDE PIPING					•						•						
CARBON DIOXIDE (COOL AND DRY)					•												
CARBON DIOXIDE (HOT AND MOIST GAS)			•	•													
CARBON DIOXIDE (SOLUTION)	•	•															
POLYMER (DRY) PIPING			•	•													

* FRP—Fiber-reinforced plastic.

Table 8–4. Materials Suitable for Caustic Soda Systems.

COMPONENTS	RECOMMENDED MATERIALS FOR USE WITH 50% NaOH UP TO 140°F
Rigid pipe	Standard-weight black iron
Flexible connections	Rigid pipe with ells or swing joints, stainless steel or rubber hose
Diluting tees	Type 304 stainless steel
Fittings	Steel
Permanent joints	Welded or screwed fittings
Unions	Screwed steel
Valves—nonleaking (plug)	
Body	Steel
Plug	Type 304 stainless steel
Pumps (centrifugal)	
Body	Steel
Impeller	Ni-resist
Packing	Blue asbestos
Storage tanks	Steel

or more means to adjust the chemical feed rate (dosage) easily. The adjustment(s) may be manual or by flow pacing, and must be accurate, repeatable, and easy to change. Also they should provide the broadest possible adjustment span from minimum to maximum feed rate. Typical ranges are 10:1 to 20:1, but greater ranges are possible.

Dry Feeders. Most dry feeders are of the belt, grooved disc, screw, or oscillating plate type. The feeding device (belt, screw, disc, etc.) is usually driven by an electric motor. Many belt feeders, particularly gravimetric-type feeders, also contain a material flow control device such as a movable gate or rotary inlet device for metering or controlling the flow of chemical to the feed belt.

Volumetric Dry Feeders. Most volumetric dry feeders are of the rotating screw or disc type, but the belt and rotary star valve types also are used. Generally, the screw or disc type is driven by an electric motor through a gear reducer drive. In some cases the drive assembly (excluding the motor) contains a variable speed or a linkage adjustment that allows feed rate changes. Otherwise, the feed rate adjustment must be made directly to the motor drive.

Manual Feed.   Manual dosage adjustment of volumetric dry feeders is accomplished by one or more of the following means:

- Drive motor for feed screw, belt, disc, or rotary valve.
  - Manual variable speed.
  - Percentage time control motor that operates on a run–stop repeat cycle with run time set by percentage timer and adjustable from 0 to 100 percent of the total cycle. Typical total run–stop cycle times are 15, 30, or 60 sec. For a system set up on a 60-sec basis, the following is typical operation:

Timer setting, percentage	Run time, sec	Off time, sec
100	60	0
75	45	15
50	30	30
25	15	45
0	0	60

  - Manually adjustable speed reducer or adjustable linkage on drive assembly.
- Control gate for belt feeder.
  - Manual setting of gate position to allow more or less material on belt.

Flow Pacing.   Automatic proportioning of volumetric dry feeders to flow (commonly called flow pacing) is readily accomplished for a variety of flow signals. The more common flow and control signals are:

- Pulse duration (on–off), with frequently used cycle times of 15, 30, and 60 cycles.
- 3 to 15 psi (20.7 to 103.5 kPa) pneumatic.
- 4 to 20 or 10 to 50 milliamperes (mA), as shown in Fig. 8–24.
- 1 to 5 volts dc.

Most modern feeders can accept one or more of the flow and control signals. If the feeder will not accept the particular signal available, signal converters are readily available to convert the signal to one that the feeder will accept. For example, a 3 to 15 psi (20.7 to 103.4 kPa) pneumatic signal can be converted to a 4 to 20 mA signal. Signal converters are relatively inexpensive and reliable.

If a volumetric feeder is automatically flow-proportioned, a means still must be available for setting the feed dosage. Typically, this is accomplished by one of two methods:

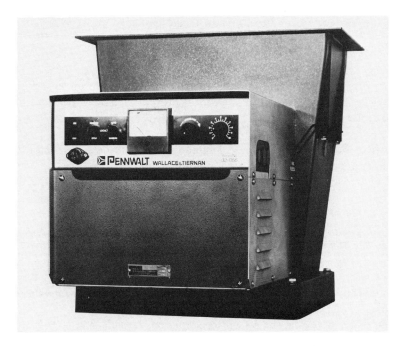

Fig. 8-24. Volumetric feeder—automatic milliampere control. (Courtesy of Wallace & Tiernan, division of Pennwalt Corporation.)

- A "manual feed rate control" built into the feeder that modifies the automatic proportioning signal within the feeder control system.
- A separate manual adjustment such as a mechanical linkage or feed gate adjustment, or a manual speed adjustment on the gear reducer drive system.

Gravimetric Dry Feeders. A gravimetric dry feeder has a built-in control system that ensures a constant feed weight, rather than volume, for any given dosage setting.

Manual Feed. For manual feed systems, an operator dosage adjustment is provided as part of this gravimetric control system. Continuous weighing of the feed belt establishes automatic internal control of the gate position (or rotary inlet valve speed), thereby maintaining a constant belt weight for a given dosage setting.

Flow Pacing. The simplest way to obtain automatic proportioning control is to provide a variable speed drive for the belt. The automatic proportioning control can be used to vary the belt drive motor speed.

If the feeders are to be shut down automatically, provisions should be made for shutdown and startup of system components such as the feeder, storage bin vibrators, water supply to dissolvers, mixers, and solution transfer pumps.

The dissolver water supply and mixer should operate on an adjustable time delay, after the feeder is stopped to prevent chemicals from settling in the dissolver.

Solution Feeders.   The most common solution feeders are motor-driven diaphragm and plunger-type feed pumps. Generally, these pumps have a built-in stroke adjustment mechanism that permits variation of the output feed rate.

Manual Feed.   For manual dosage control, the motor operates at a constant speed, and the operator adjusts the dosage with the stroke adjustment.

Flow Pacing.   Automatic proportioning control can be accomplished readily by modification of the drive motor to provide variable speed or on–off proportioning, or installation of an automatic stroke adjuster.

The drive motor can be set up to operate at variable speeds proportional to pneumatic or electric signals. On–off pulse duration signals can be applied directly to the motor starter so that the motor operates on and off in proportion to the signal. This involves many start–stop cycles for the motor, and it is recommended that only three-phase motors be used for this type of duty. When the automatic proportioning is accomplished with the drive motor, the manual stroke adjustment is still available for operator-adjusted manual dosage changes.

Automatic stroke adjusters can be installed in place of the manual adjuster on most feeders. The automatic adjusters will accept various analog control signals, including pneumatic and electric signals. These automatic stroke adjusters can be used for automatic proportioning, but in most cases there are not manual settings available for manual dosage adjustments. In some cases, a manual dosage adjustment can be incorporated into the automatic stroke adjustment mechanism, or a manual variable speed motor drive can be provided for operator manual dosage adjustment.

Probably the most satisfactory arrangement is to proportion automatically with a variable speed or pulse duration motor drive, and to leave the manual stroke adjuster for operator dosage adjustments. This approach minimizes the amount of automation, while still providing fully flow-paced operation.

If the solution feed system is to be started and stopped with plant operation, or on some other basis, consideration must be given to starting and stopping auxiliary systems such as the solution feeder, dilution water, transfer pumps, mixers, and similar equipment.

There are a number of solution batching and feed systems on the market, particularly for polymers. These systems automatically produce a feedable polymer solution from dry polymer. Typical systems include a dry polymer storage hopper, dry feed, wetting, a dissolver tank with mixing, a feed tank, and solution feeders. Most of these systems are sold as a pre-engineered package and can be supplied for manual control, automatic proportioning, and automatic start and stop operation. Generally, these systems use solution feed pumps for metering, and the previous comments concerning solution feeders are applicable.

Gas Feeders. Most modern gas feeders are vacuum-operated. The gas is accurately metered through an orifice with a fixed pressure drop across the orifice. For adjustment of the gas feed rate, the orifice size can be changed manually or automatically.

Manual Feed. Most small, inexpensive gas feeders have provisions for manual adjustment of the orifice size to change the gas-flow rate (dosage). This adjustment is made with a knob on the front of the feeder. Normally the gas-flow rate is indicated by a visual flow indicator calibrated in pounds per day.

Flow Pacing. Gas feeders may be automatically proportioned in a number of ways. Two common methods are variable vacuum control and automatic positioning of the orifice control.

Variable Vacuum Control. Variable vacuum control systems are an economical method of automatically proportioning gas feeders. The vacuum control system consists of a vacuum-producing device such as the gas injector, a restricting orifice, and an intermediate vacuum transmitter. The primary flow signal is converted to a proportional vacuum signal which is applied to the vacuum-regulating valve on the downstream side of the gas feeder orifice. The pressure ahead of the orifice is maintained at a constant value by the inlet gas pressure regulating valve. The gas feed rate is varied automatically as this proportioning vacuum signal changes with the flow.

Changes in the control vacuum signal cause comparable changes in the pressure downstream of the orifice and, therefore, in the differential across the orifice. Because the square of gas flow is proportional to the differential pressure across the orifice, the gas feed rate will vary in accordance with the control vacuum signal.

Automatic Positioning. Automatic positioning of the gas-flow control orifice is accomplished with a power positioner, which can be selected to operate from a number of input signals such as:

- 3 to 15 psi (20.7 to 103.4 kPa) pneumatic.
- 4 to 20 or 1 to 50 mA.
- 1 to 5 volts dc electric.
- Potentiometer position.
- Pulse duration.
- Pulse frequency.
- Others by special application.

When an automatic proportioning positioner is used, a manual dosage control knob is provided on the chlorinator so the operator can make manual

Fig. 8–25. Flow-paced chlorinator. (Courtesy of Wallace & Tiernan, division of Pennwalt Corporation.)

adjustments of dosage. This adjustment modifies the automatic proportioning over a wide range. A small flow-paced system is shown in Fig. 8–25.

Almost all gas feeder control schemes are based on the use of variable vacuum or the use of an orifice-proportioning positioner. In most cases, the manual dosage adjustment is retained for operator use.

Gas feeders can be started and stopped simply by starting or stopping water flow through the injector. Usually this is all that is necessary to start or stop a typical gas feed system such as a chlorinator or sulfonator.

Automatic Control for Chemical Feed Systems

Various automatic control schemes are possible for chemical feed systems beyond the automatic flow proportioning discussed for each type of feeder. For example, automatic pH control is possible using a pH sensor, controller, and pH adjustment chemical feeder for sodium hydroxide with an automatic stroke positioner.

Automatic chlorine residual control is possible using a chlorine residual analyzer, controller, and chlorinator with an automatic orifice positioner.

Fig. 8–26. Automatic chlorine control system. (Courtesy of Wallace & Tiernan, division of Pennwalt Corporation.)

These are "feed back" systems where the final control parameter, such as pH, is controlled by a previous feed of chemical. "Feed forward" systems are also possible where a parameter concentration prior to chemical feed is determined and related to the chemical feed for automatic control. An automatic control system is pictured in Fig. 8–26.

Various types of "feed forward" or "feed back" systems or combinations can be devised in theory. The problem is that there are substantial delay times in such systems that most analog controllers are not designed to handle. The delays result from the time required to change a chemical feeder setting and get the change to the injection point, as well as process delays to the sample point, delay in the sample lines, and delay in the analyzer. The total delay from the time a feed rate is changed until it is read out by an analyzer can be 5 minutes or longer. The control results can be unstable and can lead to wide, cyclic variations of the controlled variable. There are ways of overcoming these problems with proper design and equipment selection, but proper design of such systems is a very specialized skill.

Automatically controlled systems must be arranged so that auxiliary systems such as mixers, dilution water, and slakers are started and stopped as needed. Such automated systems can be designed and applied; however, their complexity and maintenance requirements are such that they should not be used unless their benefits clearly outweigh their operational disadvantages.

REFERENCES

1. BIF, Unit of General Signal, *Engineering Data*, West Warwick, R.I., 1977.
2. American Water Works Association, *Water Quality and Treatment*, 3rd ed., McGraw-Hill, New York, 1971.
3. The American Institute of Architects, *Architectural Graphic Standards*, 6th ed., John Wiley & Sons, Inc., New York, © 1970.
4. White, George Clifford, *Handbook of Chlorination*, Van Nostrand Reinhold Co., New York, 1972.
5. Heim, Nancy E., and Burris, Bruce E., Culp/Wesner/Culp, *Chemical Aids Manual for Wastewater Treatment Facilities*, USEPA, MO-25, 430/9-79-018, December, 1979.
6. AWWA, Manual M20, *Water Chlorination Principles and Practices*, 1973.
7. *Chlorine Manual*, 4th ed., The Chlorine Institute, New York, New York, 1969.
8. *Lime Handling, Application and Storage*, Bulletin 213, 2nd ed., National Lime Association, Washington, D.C., May, 1971.
9. White, G. C., *Disinfection of Wastewater and Water for Reuse*, Van Nostrand Reinhold Co., New York, 1978.

Chapter 9
Sedimentation

THEORY OF SEDIMENTATION

Sedimentation processes generally are classed in four categories:

1. *Settling of nonflocculent particles:* settling of dilute suspensions of particles that have no, or a limited, tendency to flocculate.
2. *Settling of flocculent particles:* settling of dilute suspensions of flocculent particles.
3. *Zone settling or hindered settling:* as settling particles come close together, settling as a large mass rather than as discrete particles.
4. *Compression:* accumulation of settling particles at the bottom of a settling basin, in which the particles contact each other and are supported by their compacting mass. The structure of the mass restricts further consolidation.

Settling of Nonflocculent Particles

The settling of discrete, nonflocculent particles is determined solely by the properties of the fluid and the characteristics of the particles. The key factors are the particle density, the particle volume, and the density of the fluid.

The terminal settling velocity can be calculated as:

$$V_t = \frac{2g(\rho_s - \rho_l)}{C_D \, \rho_l} \cdot \frac{V_p}{A_p} \qquad (9\text{--}1)$$

where:

V_t = terminal settling velocity, ft/sec (m/s)
g = acceleration of gravity
  = 32.2 ft/sec/sec (9.81 m/s²)
ρ_s = particle density, lbs$_m$/cu ft (kg/m³)
ρ_l = liquid density, lbs$_m$/cu ft (kg/m³)
C_D = drag coefficient, dimensionless

Vp = particle volume, cu ft (m³)
Ap = projected particle area in the direction of flow, sq ft (m²)

Equation (9–1) identifies several properties that affect sedimentation: particle density, liquid density, and the size and shape of the particles. The settling velocity of a particle varies inversely with liquid density and liquid kinematic viscosity, which is related to water temperature as shown in Table 9–1. The ratio of the kinematic viscosity at 50°F (10°C) to the kinematic viscosity at any other temperature is directly related to the change of the settling velocity of a particle. For example, changing the water temperature from 50°F (10°C) (Table 9–1) to 86°F (30°C) would increase the settling velocity of a particle by 1.310/0.804, or 1.63 times. Similarly, reducing the temperature from 50°F (10°C) to 32°F (0°C) would reduce the settling velocity by 1.310/1.792, or 0.73. Water temperature has an important effect on sedimentation tank design. With cold waters, tank overflow rates should be lower than with warmer waters.

Settling velocity also is a function of the specific gravity and the size of the particles. Table 9–2 presents some relative settling velocities that dramatically illustrate this point.

The behavior of nonflocculent particles is often used to describe ideal settling. An ideal settling basin is divided into four zones, as illustrated in Fig. 9–1: inlet, outlet, settling, and sludge zones.

Ideal settling theory results in the following equation for surface loading or overflow rate:

$$V_0 = \frac{Q}{A} \qquad (9\text{--}2)$$

where:

V_0 = settling velocity of particle that settles the depth of the basin in detention time t_0, ft/sec (m/s)
Q = rate of flow through the basin, cu ft/sec (m³/s), or gallons per day (m³/d) with appropriate conversion factors
A = surface area of the basin, sq ft (m²)

Also:

$$t_0 = \frac{C}{Q} \qquad (9\text{--}3)$$

where C = volume of settling zone, cu ft (m³).

Table 9–1. Relationship of Kinematic Viscosity to Water Temperature.*

TEMPERATURE		KINEMATIC VISCOSITY, ν	
°F	°C	$(10)^{-6}$ FT²/SEC	(10^{-6}) M²/S
32	0	1.792	1.66
41	5	1.519	1.41
50	10	1.310	1.21
59	15	1.146	1.06
68	20	1.011	0.94
86	30	0.804	0.75

* Reference 4.

All particles with a settling velocity greater than V_0 are removed. Particles
with settling velocities (V_i) less than V_0 are removed only if they enter the
basin within a vertical striking distance $h_i = V_i t_0$ from the sludge zone.
Assuming uniform distribution of particles in the inlet zone, these particles
are removed in the ratio of V_i/V_0. If the term f_i is fraction of particles
with a settling velocity of V_0 or less, the removal efficiency E for the tank
is:

$$E = (1 - f_i) + \int_0^{f_i} \frac{V_i}{V_0}\, df \qquad (9\text{--}4)$$

Table 9–2. Velocities at Which Particles of Sand and Silt Subside
in Still Water.*,**

DIAMETER OF PARTICLE, MM	CLASSIFICATION	HYDRAULIC SUBSIDING RATE, MM/S	COMPARABLE OVERFLOW RATE	
			GPM/FT²	M/H
10.0	gravel	1,000.0	1,475.0	3,599.0
1.0		100.0	148.0	361.0
0.6		63.0	93.0	227.0
0.4	coarse sand	42.0	62.0	151.0
0.2		21.0	31.0	76.0
0.1		8.0	11.8	29.0
0.06		3.8	5.6	13.7
0.04	fine sand	2.1	3.1	7.6
0.02		0.62	0.91	2.22
0.01		0.154	0.227	0.554
0.004	silt	0.0247	0.036	0.087

* Temperature, 50°F (10°C); specific gravity of sand and silt particles, 2.65; values for 10-mm to 0.1-mm
particles from Hazen's experiments; values for 0.02-mm to 0.004-mm particles from Wiley's formula; intermediate values interpolated from connecting curve.
** Reference 4.

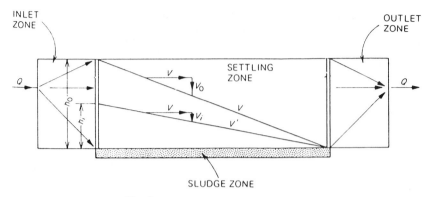

Fig. 9-1. Ideal sedimentation basin.

The overflow, Q/A (ft/sec or gpd/sq ft), should preferably be somewhat less than the settling rate of the critical particles. Equation (9–2) (using the terminal settling velocity of the critical particles) equates terminal velocity to overflow rate. Sedimentation basin loadings (Q/A) are often expressed as gpd/sq ft (m/d). Thus, under ideal settling conditions, sedimentation is independent of basin depth and detention time, and depends only on the flow rate, basin surface area, and properties of the particle and liquid.

Settling of Flocculent Particles

Ideal settling, discussed above, assumes the following:

- All particles settle discretely.
- Particles that strike the bottom remain in the sludge zone.
- Water and particulates are distributed uniformly over a vertical plane.
- Incremental volumes of water move from inlet to outlet without changing shape.
- Inlet and outlet conditions do not affect settling.

Sedimentation basins used in water treatment will not perform in accordance with ideal settling theory for the following reasons:

- Discrete particle settling is not obtained. Flocculation of particles occurs during settling. The greater the tank depth, the greater the opportunity for contact among particles.
- Inlet and outlet conditions are not ideal and do affect settling.
- Basin currents cause nonuniform flow and bottom scour.

Therefore, in the clarification of water, removal is dependent on the basin depth as well as on flow, basin surface area, and properties of the particle.

The performance of the sedimentation basin on a suspension of discrete particles can be calculated, but it is not possible to calculate sedimentation basin performance for a suspension of flocculating particles, such as a water, because settling velocities change continually. Settling analyses, however, may be performed to predict sedimentation basin performance. Settling tests, described by Camp in 1946 and by Zanoni and Blomquist and Weber more recently, can be used to approximate the design criteria to achieve a specified removal of flocculent particles.[1-3] However, there are so many factors that affect the performance of full-scale settling basins that data from the operation of other full-scale settling basins are the best source of information for design.

Zone Settling

When the settling particles come close together, they settle as a mass with a clear interface between the settling particles and the clarified liquid. For a single particle settling in a large vessel, the upward movement of the displaced water is insignificant; but if the concentration of particles is sufficiently great, the upward velocity of the water displaced by the particles becomes sizable in comparison with the settling velocity. Because the settling velocity is relative to the water, the upward flow of displaced water acts to reduce the velocity at which the particle approaches the bottom of the basin. This zone or hindered settling phenomenon is significant in dealing with large quantities of floc. It is of more import when dealing with wastewater treatment sludges than in most water treatment applications. In most of the latter, the effects of hindered settling are so negligible that usually they may be ignored.[5] For some specialized water treatment applications where the floc volume is large, as in solids contact units designed to maintain a high floc concentration (discussed later in this chapter) or in basins used to gravity-thicken sludges, zone settling must be considered.

The interface between the sludge and the clarified supernatant can be observed in a batch settling test to evaluate zone settling.[3] Initially, all the suspension is at a uniform concentration, and the height of the interface is h_0, as illustrated in Fig. 9-2.

Hindered settling of the particle–liquid interface occurs from A to B at a constant rate. Deceleration occurs in a transition zone from B to C. Consolidation of the sludge blanket in the region represents the compression zone from C to D. The solids are supported mechanically by those beneath them in the zone from C to D.

The clarification capacity of the system can be estimated from the initial rate at which the interface subsides. The area required for clarification may be calculated from:

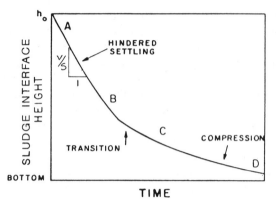

Fig. 9-2. Height of interface in zone settling.

$$A = \frac{Q}{V_s}$$ (9-5)

where:

V_s = subsidence velocity for hindered settings, ft/sec (m/s)
A = surface area, sq ft (m²)
Q = rate of flow through basin, cu ft/sec (m³/s)

The value of V_s may be computed from the slope of the hindered settling portion of the interface height versus time curve illustrated in Fig. 9-2.

The thickening capacity can be determined by consideration of the batch sedimentation characteristics of a thick suspension.[3] Adequate thickening can be accomplished when the area is calculated according to:

$$A = \frac{Qt_u}{h_0}$$ (9-6)

where:

A = thickener area, sq ft (m²)
t_u = time required to reach desired sludge concentration, seconds
h_0 = initial height of interface in batch test, ft (m)
Q = flow, cu ft/sec (m³/s)

Compression

Compression occurs when the subsiding particles accumulate at the bottom of the sedimentation basin. The weight of the particles is supported by the compacting mass. Consolidation is a relatively slow process, as illustrated by Fig. 9–2.

APPLICATIONS OF SEDIMENTATION

The two principal applications of sedimentation in water treatment are plain sedimentation and sedimentation of coagulated and flocculated waters. Plain sedimentation is used to remove solids that are present in surface waters, and that settle without chemical treatment, such as gravel, sand, silt, and so on. It is used as a preliminary process to reduce the sediment loads in the remainder of the treatment plant, and is referred to as "presedimentation." Sedimentation also is used downstream of the coagulation and flocculation processes to remove solids that have been rendered more settleable by these processes. Chemical coagulation may be geared toward removal of turbidity, color, or hardness.

Presedimentation

The purpose of presedimentation is to reduce the load of sand, silt, turbidity, bacteria, or other substances applied to subsequent treatment processes so that the subsequent processes may function more efficiently. When the raw water has exceptionally high concentrations of these substances, good removals are often obtained by plain settling and without the use of chemicals. Waters containing gravel, sand, silt, or turbidity in excess of 1,000 NTU may require pretreatment.[6]

Presedimentation basins should have hopper bottoms and/or be equipped with continuous mechanical sludge removal apparatus especially selected or designed to remove heavy silt or sand. Sludge is not removed continuously, and allowance must be made for sludge accumulation between cleanings. In manually cleaned basins, settled matter is often allowed to accumulate until it tends to impair the settled-water quality, at which time the sludge is flushed out.

Sedimentation basins that are not equipped for mechanical removal of sludge should have sloping bottoms, so that they can be rapidly drained, allowing most of the sediment to flow out with the water. Because the bulk of the material settles near the inlet end, the slope should be greatest at this point. In some plants, sluice gates are arranged to deliver raw water to the sedimentation basin to flush out the sludge. The balance is generally washed out with a fire hose. At least two basins are needed so one can remain in service while the other is cleaned.

The time between cleanings varies from a few weeks, in plants that have short periods of settling and handle very turbid water, to a year or more, where the basin capacity is large and the water is not very turbid.[5]

Because the particles to be removed are more readily settleable than chemical flocs, the detention time may be shorter and the surface overflow and weir rates may be greater for presedimentation basins than for primary or secondary settling basins. Detention times of at least 2 hours, maximum overflow rates of 3,500 gpd/sq ft (5.9 m/h), maximum weir rates of 19,000 gpd/sq ft (32.2 m/h), and minimum water depths of 3 feet (0.91 m) have been recommended as design standards for presedimentation basins.[7] When mechanical sludge removal is not provided, detention times of 2 to 3 days are often used to allow for sludge storage. Facilities for chlorination of the presedimentation basin influent are provided in many cases.

Presedimentation in itself can provide reductions in coliform organisms. Reported removals are:[8]

- Ninety percent or more for heavy coliform loadings (20,000 to 60,000 Most Probable Number (MPN) per 100 ml) and long presedimentation periods (5 to 10 days).
- About 20 percent for light coliform loadings (5,000 to 20,000 MPN per 100 ml) and shorter detention times (3 to 7 hours).

The relatively long storage periods provided by presedimentation are also valuable in providing time for the natural die-away of virus that are in an unfriendly environment.[9]

Sedimentation Following Coagulation and Flocculation

A common practice is to follow flocculation with sedimentation in order to reduce the load of solids applied to filters. A major portion of the solids can be removed from the water by gravity settling following coagulation and flocculation. As discussed in Chapter 10, development of coarse-to-fine filters, through the use of mixed-media or dual-media filters, means that much heavier loads of suspended solids can be filtered at reasonable head losses than could be handled in the past using single-media, fine-to-coarse filters. This makes it possible to consider direct filtration (without settling following flocculation) in an increasing number of situations. However, there remains a need for sedimentation in waters that are too turbid for direct filtration or that require other chemical treatment (i.e., softening) prior to filtration.

HORIZONTAL FLOW SEDIMENTATION BASINS

Horizontal flow basins are typically rectangular, square, or circular in shape. In some circular basins, the flocculator may be internal (see Fig. 9–3).

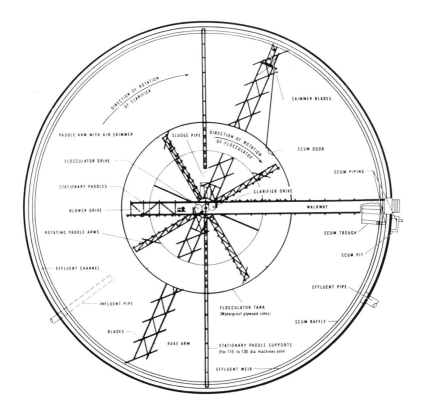

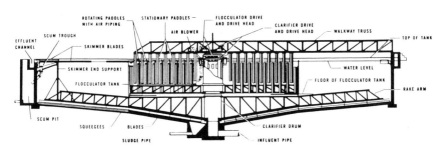

Fig. 9–3. Circular, horizontal flow clarifier with internal flocculator. (Courtesy of Dorr-Oliver, Inc.)

Despite the use of sedimentation for centuries and many attempts at mathematical formulation, good settling basin design is based more on experience and judgment than on the results of calculations. The key design considerations are overflow rates, inlet and outlet conditions, sludge removal, and basin geometry.

Overflow Rates

If ideal conditions could be achieved in a sedimentation basin, the overflow rate could equal the settling velocity of the particles to be removed. Unfortunately, it has not proved possible to achieve ideal settling conditions, and it is necessary to reduce overflow rates to less than theoretical values. Two factors are predominant in the departure of basin performance from the ideal: currents and particle interactions.[5] The most common types of currents are: surface currents, induced by the wind blowing over uncovered basins; convection currents, arising from temperature differences; density currents, due to the inflowing water having a different density (due to temperature or the suspension load) from the water in the basin; and eddy currents, produced by the incoming water. These currents distort the flow pattern from ideal conditions. As a result, safety factors are applied to the theoretical overflow rate based solely on settling velocities. Typical overflow rates for alum floc are 350 to 550 gpd/sq ft (0.59 to 0.93 m/h) and for lime floc are 1,400 to 2,100 gpd/sq ft (2.37 to 3.56 m/h).[4] The higher rates are used in warm waters, the lower rates in cold waters.

Inlet and Outlet Conditions

Poor sedimentation basin performance can result from uneven influent flow distribution, inadequate dissipation of inlet energy, nonuniform collection of effluent flow, and the associated hydraulic short-circuiting within a basin.

An ideal inlet would distribute the water uniformly over the full cross section of the tank. The inlet conditions have been found to be more critical than the outlet conditions in regard to density and inertial currents.[4]

Because of the relatively fragile nature of chemical floc, the velocity in the influent channels or pipelines to the sedimentation basin must be kept low [0.5 to 2 ft/sec (0.15 to 0.61 m/s)], to prevent destruction of the floc. Also, low velocities are needed through any inlet ports to minimize inertial currents. In rectangular basins, the flow may enter through an inlet channel across the head end of the tank (Fig. 9–4). The water passes through a number of inlet ports across the tank, usually with entrance velocities less than 2 ft/sec (0.61 m/s), distributing the flow across the full width of the tank. A perforated baffle wall located across the tank can provide excellent flow distribution (Fig. 9–5). The best location for such a baffle has been reported to be 6.5 to 8 feet (2 to 2.5 m) downstream of the basin inlet wall.[10] In the design of perforated baffles, Hudson states that four requirements must be met:[23]

1. The headloss through the ports should be about four times higher than the kinetic energy of any approaching velocities in order to equalize flow distribution both horizontally and vertically.

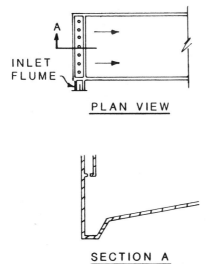

PLAN VIEW

SECTION A

Fig. 9-4. Inlet channel for rectangular basin.

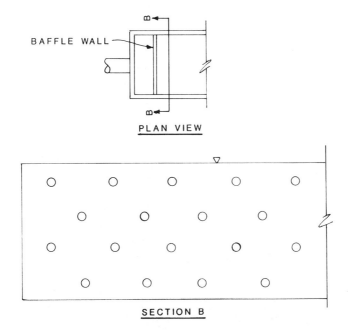

PLAN VIEW

SECTION B

Fig. 9-5. Inlet baffle wall.

2. To avoid breaking up floc, the velocity gradient through inlet conduits and ports should be held down to a value close to or little higher than that in the last compartment of the flocculators.
3. The maximum feasible number of ports should be provided in order to minimize the length of the turbulent entry zone produced by the diffusion of the submerged jets from the ports in the perforated-baffle inlet.
4. The port configuration should be such as to assure that the discharge jets will direct the flow toward the basin outlet.

It is desirable for the port diameter to be no more than the thickness of the permeable-baffle wall, so that the hydraulic behavior will cause the jets to emerge in the proper direction. Hudson presents a design procedure for determining the size and spacing of the ports.[23] He reports that it is safe to use a velocity through the perforated baffles of about 0.65 to 1 ft/sec (0.2 to 0.3 m/s). Uniformly distributed 5-inch (0.13-m) orifices with an opening ratio of 6 to 8 percent with a headloss of 0.08 to 0.12 inches (2 to 3 mm) have been found effective.[10] In cases where extreme wind currents, density flows, or large flow-rate variations occur, an intermediate diffuser wall at the basin midpoint will improve basin efficiency. Unfortunately, an intermediate diffuser wall limits the type of sludge collection system that can be used.

In circular, center-feed tanks, the water is introduced into a center influent well by an inlet pipe brought up from underneath the basin or suspended from the clarifier walkway. The velocity in the inlet pipe should not exceed 1 ft/sec (0.31 m/s). Discharging the pipe horizontally into the feed well can introduce undesirable currents, and it is better instead to discharge vertically upward from the pipe into the influent well (Fig. 9–6). The influent well may be perforated to assist in dissipating inlet velocities. One approach is to use an inlet diffusion well for energy dissipation that utilizes ports to develop tangential flow in an outer zone formed by a skirt.[22] This approach uses inlet energy to distribute flow evenly across the tank. The diameter of the outer skirt is a minimum of 25 percent of the tank diameter, representing 6 percent of the tank area. Larger skirts are used in tanks with flocculator center wells.[22]

The basin outlet system should collect the water uniformly across the width of the basin to prevent localized, high velocities from carrying floc out of the basin. Submerged weirs, or effluent ports, sometimes are used to avoid the breakup of floc that can occur with a freely discharging weir. Perforated launders with ports, commonly submerged 1 to 2 feet (0.31 to 0.61 m) below the surface, are useful in minimizing problems of floating trash passing to the filters.[23] They are also useful when it is desired to vary the water level in the basin during operation that cannot be done with weirs. Sometimes it helps operation to use the storage in the basins to permit some

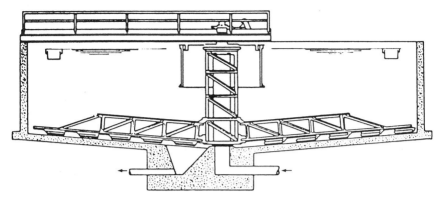

Fig. 9–6. Circular clarifier, center-fed with vertical inlet discharge.

temporary differences between the inflow to the plant and the discharge from the plant.

In northern climates, launders should be placed at such a depth as to avoid problems with icing. Fluctuating levels also may minimize ice attachment to basin walls.

An adequate length is necessary for effluent weirs to avoid excessive velocity currents. In rectangular basins, adequate weir length usually cannot be obtained with a single weir across the end of the tank, and weirs are provided in the outlet quarter or third of the tank. The weirs or launders may be aligned either parallel or transverse to the direction of flow. In center-feed circular clarifiers, weirs or launders around the periphery of the tank usually provide sufficient weir length. In some cases, an inboard annular trough located about 10 percent of the radius in from the periphery is also used to provide low weir overflow rates for very light flocs. Commonly used weir overflow rates are shown in Table 9–3.

The use of inboard launders to provide added weir length has been reported to adversely affect circular clarifier performance.[21] As shown in Fig. 9–7a, typical circular clarifier designs establish a current that moves across the basin floor, up the wall, and into the launders. A proposed approach to overcome the effects of such currents is shown in Fig. 9–7b.[21] Plant scale

Table 9–3. Weir Overflow Rates.

TYPE OF FLOC	WEIR OVERFLOW RATE GPM/FT	(M^2/H)
Alum floc, low-turbidity water	8–10	5.9–7.4
Alum floc, high-turbidity water	10–15	7.4–11.2
Lime softening floc	15–18	11.2–13.4

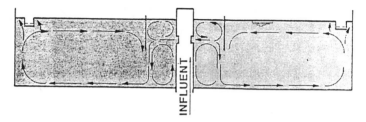

(a) Flow pattern in conventional circular clarifier.

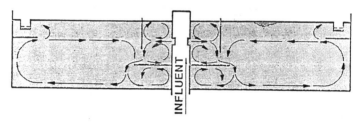

(b) Flow pattern in baffled circular clarifier.

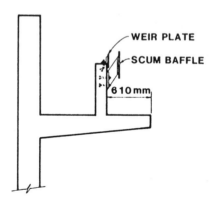

(c) Clarifier baffle at weir trough.

Fig. 9–7. Circular clarifier inlet and outlet modifications.[21]

tests using a 24-inch (0.61-m) baffle beneath a simple peripheral weir showed improved clarifier performance.[21] It was also found that the larger-diameter, design inlet wells used in flocculating clarifiers (see Fig. 9–3) provide better flow distribution and energy dissipation than the inlet wells in standard clarifiers.

Sludge Removal

Modern sedimentation basins are equipped with sludge collection and removal mechanisms to eliminate the need to shut down for cleaning. In rectangular tanks, the bottom is usually sloped gently about 5 percent from the effluent end of the tank to a sludge hopper located at the tank inlet. Two types of sludge-removal mechanisms are used in rectangular basins. One consists of a series of chains and sprockets, with the latter mounted at the top and bottom of the tank (Fig. 9–8). Wood, plastic, or steel cross flights are fitted between the chains. The flights scrape along the tank bottom, dragging the sludge to the hopper. The second type of mechanism is mounted on a carriage that travels along the top of the wall of the tank. The sludge scrapers move slowly [less than 1 ft/min (0.31 m/min)] to avoid resuspending the settled sludge.

The bottom slope of circular clarifiers is typically about 8 percent from the outer wall to a central sludge hopper, although steeper slopes up to 15 percent may be used with very high turbidities or heavy sludges, for example, lime softening sludges. Usually, a rotating sludge collector pivots around the center of the tank and scrapes the sludges toward the center hopper (see Figs. 9–3 and 9–6).

Basin Geometry

Rectangular basin widths vary from 5 to 24 feet (1.5 to 7.3 m) with a single sludge collector, and are greater with parallel sludge collectors. Length-to-width ratios of 3:1 to 5:1 are typical, as are depths of 8 to 10 feet (2.4 to 3.1 m).[4] Circular clarifier sludge removal mechanisms are typically available in 1-foot (0.31-m)-diameter increments between 10 and 30 feet (3.1 and 9.1 m), in 2-foot (0.61-m)-diameter increments between 30 and 50 feet (9.1 and

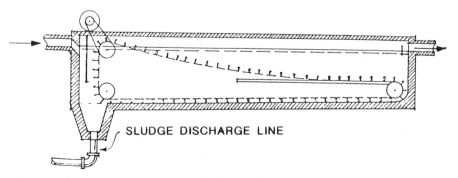

SLUDGE DISCHARGE LINE

Fig. 9–8. Rectangular sedimentation basin with flight and chain sludge removal system.

15.2 m), and in 5-foot (1.5-m)-diameter increments above 50-foot (15.2-m)-diameter units.

SOLIDS CONTACT UNITS

Solids contact units combine coagulation, flocculation, and sedimentation in a single basin. Also, settled sludge may be recirculated internally or externally. The chief advantage of the solids contact units is the easy internal recirculation and suspension of precipitated solids, particularly the recirculation of calcium carbonate sludge in treatment of hard waters. The chemical and physical reactions of lime–soda softening are accelerated by contact with previously precipitated calcium carbonate. In the treatment of surface waters containing turbidity or organic matter, internal sludge recirculation may be a disadvantage rather than an advantage. It may be desirable to purge the system of organics as quickly and completely as possible to avoid production of tastes and odors and to avoid interference with coagulation and settling by organics. The best application of solids contact basins is in the softening of well waters of constant quality at a uniform basin throughput rate. Solids contact reactors, when used in clarification operations, are usually sized for a 1 gpm/sq ft (2.44 m/h) loading rate, although they are known to work well at 2 gpm/sq ft (4.88 m/h) in certain cases. When they are used in softening, it is customary to size them for a settling velocity of 1.5 to 2 gpm/sq ft (3.66 to 4.88 m/h), and they may be operated at even higher rates. There are two types of solids contact units: sludge blanket units and slurry recirculation units.

Sludge Blanket Units

Figure 9–9 illustrates a sludge blanket unit. After passing through a mixing zone in the center, the water passes up through a sludge blanket. Sludge removal in sludge blanket units is usually by means of a concentrating chamber into which the sludge blanket overflows. Sludge drawoff is regulated by a timer-controlled valve. It is usually necessary to use activated silica or other coagulant aids together with the addition of clay to weight the floc in order to obtain results in sludge blanket units comparable or superior to those given by slurry recirculation units.[23] Sludge blanket units are more sensitive than recirculation units to increases in flow rate, or rises in water temperature, which tend to upset the sludge blanket.

Slurry Recirculation Units

Figure 9–10 illustrates a slurry recirculation unit. Substantial quantities of slurry are recycled through the central mixing zone. Often there is no sludge

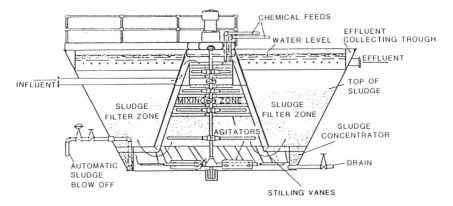

Fig. 9–9. Sludge blanket clarifier. (Courtesy of Permutit Co.)

bed formed, as the slurry can be seen falling out of the water beneath the conical hood.[23] The slurry concentration is controlled by the amount of sludge drawn from the sludge sumps, and is usually maintained at 6 to 20 percent by volume (higher in softening applications). The recirculation rate is usually maintained at the highest rate consistent with the production of clear settled water. In some instances, too high a rate may cause floc to be discharged into the effluent of the unit.

SHALLOW DEPTH SEDIMENTATION

The theoretical basis of the tube settling concept is discussed by McMichael and Yao.[15,16] There is general agreement that the tube settler concept offers a theoretically sound basis for operating clarifiers at surface loading rates two to four times higher than in deep, conventional basins.

Theory

The earlier description of the settling paths of discrete particles in an ideal, rectangular basin is useful in understanding the benefits of shallow depth sedimentation. Referring to Fig. 9–1, particles with V_i less than V_o could be completely removed if false bottoms or trays were inserted at intervals, h. Without such trays, a basin with a much greater length would be required to capture these particles. As the interval h is reduced, the size of basin required to remove a given percentage of the incoming settleable material decreases.

Theory indicates that the removal of settleable material is a function of the overflow rate and basin depth and is independent of the detention period. This theory was proposed in 1904 by Hazen when he pointed out that the

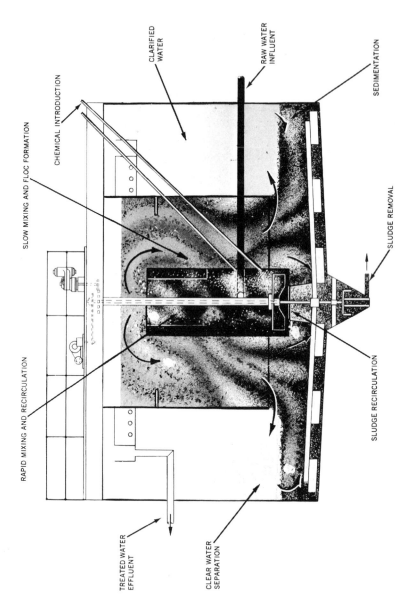

Fig. 9-10. Slurry recirculation clarifier. (Courtesy of Graver Water Conditioning Co.)

proportion of sediment removed in a settling basin is primarily a function of the surface area of the basin and is independent of the detention time.[11] He noted that doubling the surface area by inserting one horizontal tray would double the capacity of the basin. He thought that trays spaced at intervals as low as 1 inch (25.4 mm) would be very desirable if the problems of sludge removal could be resolved.

Camp presented a design for a settling basin that would capitalize on these advantages.[12] It had horizontal trays spaced at 6 inches (0.15 m), the minimum distance he thought was permissible for mechanical sludge removal. The basin had a detention time of 10.8 minutes, a velocity of 9.3 ft/min (2.83 m/min), and an overflow rate of 667 gpd/sq ft (1.13 m/h). Outlet orifices were used to distribute the flow over the width of the trays. In discussing this article, Eliassen noted that although tray tanks had been used for many years in the chemical and metallurgical industries, they had been used in only a few water or sewage treatment systems.[13] Camp believed the lack of acceptance was due to the reluctance of design engineers to depart from previously accepted practice in the size and shape of basins.[14]

Theoretically, the use of very shallow settling basins enables the detention time of the settling process to be reduced to only a few minutes, in contrast to conventional settling basin designs that use 1- to 4-hour detention. Application of this theory offers tremendous potential for minimizing the size and cost of water treatment facilities. This section describes the techniques that are now available for successful application of the principle.

Basic Systems Available

The two basic shallow depth settling systems now commercially available are illustrated in Fig. 9–11.[17,18] These configurations are (a) essentially horizontal and (b) steeply inclined.

Essentially Horizontal. The operation of the essentially horizontal tube settlers is coordinated with that of the filter following the tube settler. The tubes essentially fill with sludge before any significant amount of floc escapes. Solids leaving the tubes are captured by the filter. Each time the filter backwashes, the settler is completely drained. The tubes are inclined only slightly in the direction of normal flow (5 degrees) to promote the drainage of sludge during the backwash cycle. The rapidly filling water surface scours the sludge deposits from the tubes and carries them to waste. The water drained from the tubes is replaced with the last portion of the filter backwash water so that no additional water is lost in the tube draining procedure. This tube configuration is applicable primarily to small plants (1 mgd or less in capacity) and is often used in package plant systems (described in Chapter 22).

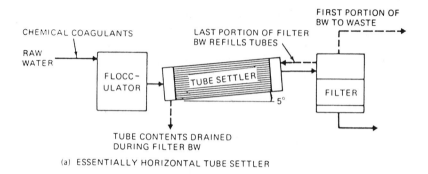

(a) ESSENTIALLY HORIZONTAL TUBE SETTLER

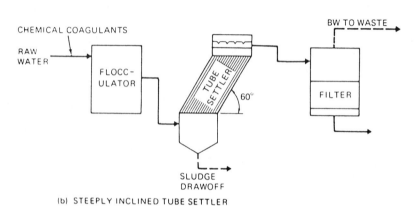

(b) STEEPLY INCLINED TUBE SETTLER

Fig. 9-11. Basic tube settler configurations shown schematically.[18]

Steeply Inclined. Sediment in tubes inclined at angles in excess of 45 degrees does not accumulate but moves down the tube to eventually exit the tubes into the plenum below. A flow pattern is established in which the settling solids are trapped in a downward-flowing stream of concentrated solids, shown schematically in Fig. 9-12. The continuous sludge removal achieved in the steeply inclined tubes eliminates the need for drainage or back-flushing of the tubes for sludge removal. The advantage of shallow settling depth coupled with that of continuous sludge removal extends the range of application of this principle to installations with capacities of many millions of gallons per day.

Various manufacturers have developed alternative approaches for incorporating steeply inclined tubes into a modular form that can be built economically and can be easily supported and installed in a sedimentation basin. One modular construction is shown in Fig. 9-13, in which the material of construction is normally PVC and ABS plastic. Extruded ABS channels are installed at a 60-degree inclination between thin sheets of PVC. By inclining

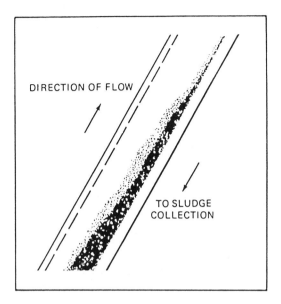

Fig. 9-12. The flow of liquid and sludge quickly establishes a countercurrent flow pattern in the steeply inclined tubes.

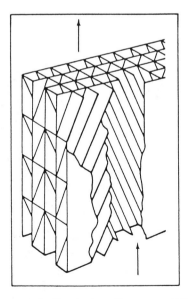

Fig. 9-13. A module of steeply inclined tubes. (Courtesy of Microfloc Products Group, Johnson Division, UOP)

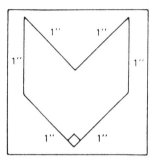

Fig. 9–14. Cross section of a chevron-shaped tube. (Courtesy of Permutit Co.)

the tube passageways rather than inclining the entire module, the rectangular module can be readily installed in either rectangular or circular basins. By alternating the direction of inclination of each row of the channels forming the tube passageways, the module becomes a self-supporting beam that needs support only at its ends. The rectangular, tubular passageways are 2 inches (50.8 mm) by 2 inches square in cross section and normally are 24 inches (0.61 m) long.

Another manufacturer, Permutit, has adopted a chevron-shaped tube cross section, as shown in Figs. 9–14 and 9–15. According to the manufacturer, the V-groove in the bottom of the tube enhances the counterflow characteristics of the sludge. The Reynolds number for water at 70°F (21.1°C) and a loading of 2,000 gpd/sq ft (3.39 m/h) on the 2-inch by 1-inch (50.8 by

Fig. 9–15. Modules of chevron-shaped tubes installed. (Courtesy of Permutit Co.)

25.4 mm) square tube is about 43. Similar conditions for the chevron configuration shown in Fig. 9–14 give a Reynolds number of 19. Both of these values indicate that flow through the tubes is well within the laminar range, and it is doubtful whether the difference is significant for hydraulic conditions. A third manufacturer, Graver Water Conditioning Company, utilizes rectangular channels similar to the construction shown in Fig. 9–13 but with all of the channels inclined in the same direction.

All of the above systems are used in configurations in which the influent is introduced beneath the tubes with the flow passing up through the tubes; but in the Lamella separator, introduced by the Parkson Corporation, the influent enters at the top of the clarification basin and is then directed downward through a series of parallel plates, as illustrated in Fig. 9–16. The

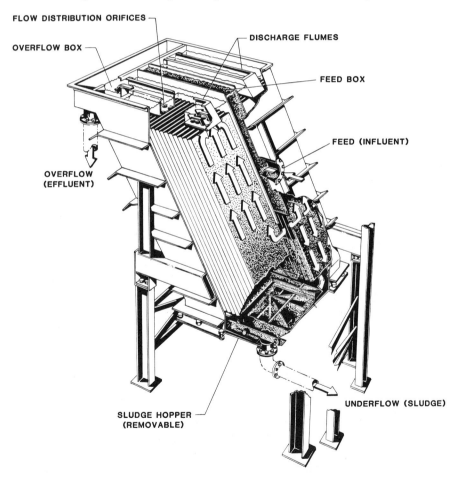

Fig. 9–16. Lamella® separator. (Courtesy of Parkson Corp.)

sludge is collected at the bottom of the basin with the sludge–water flow being in the same direction rather than countercurrent as in the above systems. The clarified water is conveyed to the top of the clarifier by returning tubes, as shown in Fig. 9–16. The plates are typically 5 feet wide by 8 feet long (1.5 by 2.4 m), spaced 1½ inches (3.8 mm) apart, inclined at 25 to 45 degrees to the horizontal, and usually constructed of PVC.

General Design Considerations

Steeply inclined tubes can be used in either upflow solids contact clarifiers or horizontal flow basins to improve performance and/or increase capacity of existing clarifiers. Of course, they can also be incorporated into the design of new facilities to reduce their size and cost. Capacities of existing basins can usually be increased by 50 to 150 percent with similar or improved effluent quality. The overflow rate at which tubes can be operated is dependent upon the design and type of clarification equipment, character of the water being treated, and the desired effluent quality. The following sections describe the most important design and operational variables that affect tube installations in existing clarifiers.

Type of Clarifier.

Horizontal Flow. The nature of the existing clarification equipment determines to some extent the allowable tube rate and the physical arrangement of modules in a basin. Ideal flow patterns are rarely experienced in practice in clarification basins. Velocities in rectangular, horizontal flow basins vary throughout the basin. Flow lines diverge at the inlet and converge at the outlet. The velocity gradient across the basin does not remain uniform because of basin drag, density currents, inlet turbulence, temperature currents, and so on. In radial flow circular basins, the flow cannot be introduced to impart velocity components in the horizontal direction only. The use of a center feedwell imparts downward currents that cause turbulence and produce a general rolling motion of the contents in an outward and upward direction.

When tube modules are installed in horizontal flow basins, it is best not to locate them too near entrance areas where possible turbulence could reduce the effectiveness of the tubes as clarification devices. For example, in a horizontal basin often as much as one-third of the basin length at the inlet end may be left uncovered by the tubes so that it may be used as a zone for the stilling of hydraulic currents. This is permissible in most basins because the required quantity of tubes to achieve a significant increase in capacity will cover only a portion of the basin. In radial flow basins, the required quantity of modules can be placed in a ring around the basin periphery, leaving an inner-ring open area between the modules and the centerwell to dissipate inlet turbulence.

Upflow Clarifiers with Solids Contact. The flow paths in solids contact basins of the upflow type are in a vertical direction through a layer or blanket of flocculated material, which is held at a certain level and maintained at a certain concentration by the controlled removal of sludge. The clarification rate is governed by the settling velocity of this blanket. The purpose of maintaining the blanket is to entrap slowly settling, small particles that otherwise would escape the basin. When the flow is increased, the level of the blanket will rise. The efficiency of the tubes is dependent upon both the overflow rate and the concentration of incoming solids. The allowable loading rate on the tubes in this situation is dependent upon the average settling velocity of the blanket, the ability of the clarifier to concentrate solids, and the capacity of the sludge removal system to maintain an equilibrium solids concentration. If sludge is not withdrawn quickly enough or if the upward velocity exceeds the average settling velocity of the blanket, the unit can become solids critical, with the result that the blanket will pass through the tubes with excessive carryover of solids into the effluent.

In expanding the capacity of an upflow, solids contact clarifier, the ability to handle increased solids may be the limiting factor. The solids loading of the basin establishes its maximum capacity. The amount of increased capacity is often limited to 50 to 100 percent of the original capacity.

Basin Geometry. The shape of a basin determines how the tube modules can be most efficiently arranged to utilize the available space. The best arrangement may be determined strictly by basin geometry once the tube quantity is established. Of course, other factors must also be considered. For example, it is desirable to locate the tubes as far as possible from areas of known turbulence or locate them to take advantage of an existing effluent launder system.

In circular basins, the tube modules are often placed in pie-shaped segments, as can be seen in Fig. 9–17. This approach is used where the entire clarification area is covered by tube modules or where the tube modules are placed in a ring around the basin outer wall. Where total coverage is required, the modules are supported by radial members that extend from an inner cone or ring to the outer wall of the basin. Where partial coverage is used and if the ring width does not exceed 10 to 12 feet (3.1 to 3.7 m), the support members may be cantilevered from the exterior walls. In any application where the entire area is not covered, a baffle wall must be installed at the inner perimeter of the modules to ensure that all flow passes through the modules. The maximum width of a pie-shaped segment at the basin perimeter is limited

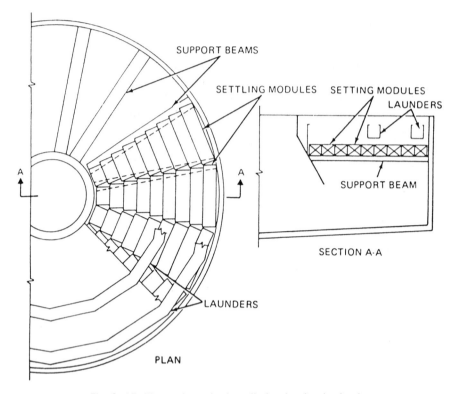

Fig. 9–17. Illustrative tube installation in circular basin.

by the maximum module length, which varies from manufacturer to manufacturer but is usually on the order of 10 to 12 feet (3.1 to 3.7 m).

In basins that have radial effluent launders, it is often possible to suspend the modules from the launders, as illustrated in Fig. 9–18. In rectangular basins, tubes are simply oriented with the long axis parallel to the sidewalls of the basin, with the support beams spanning the width of the basin, as shown in Fig. 9–19.

Tube Support Requirements. The tube support system must be able to support the weight of the tube modules when the basin is drained, as well as to make some allowance for the possibility of a workman standing on the modules. One manufacturer, Neptune Microfloc, recommends a surface loading of 7.5 lb/sq ft (36.6 kg/m²). The bearing surface width of a support member should be more than 1 inch (25.4 mm) to prevent possible shear failure of the module at the points of contact under extreme loading conditions. On the other hand, it should be as narrow as possible so as to block a minimum number of tube openings. The support members should be located

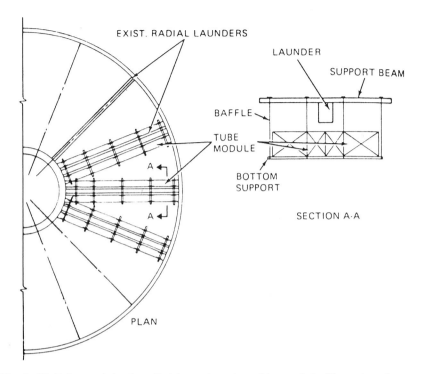

EXIST. RADIAL LAUNDERS

LAUNDER

SUPPORT BEAM

BAFFLE

TUBE MODULE

BOTTOM SUPPORT

SECTION A-A

PLAN

Fig. 9–18. Tube modules installed in conjunction with a radial effluent launder system.

a minimum of 6 inches (0.15 m) [preferably 1 foot (0.31 m)] in from the module end to develop the maximum support capability of the module.

Flocculation. In expanding a plant capacity with settling tubes, the existing flocculation facilities must be closely studied to ensure that their capacity will not be overtaxed. The tubes are settling devices and cannot remove material that has not been flocculated to the point that it is settleable.

Some older plants may have baffle flocculators that generally perform well at only one flow rate. The existing facilities, in this case, may have to be replaced or supplemented with more efficient mechanical units. In many cases, enough reserve capacity is built into existing mechanical flocculation units that the desired capacity increase can be achieved without modifying the flocculators. The exact detention time required depends on the energy input of the mixing device. A rough rule of thumb is that flocculation facilities may need upgrading if, at the increased plant flow, they provide less than 20 minutes of detention at water temperatures below 45°F (7.2°C) or 15 minutes with warmer temperatures. These detention times assume that the

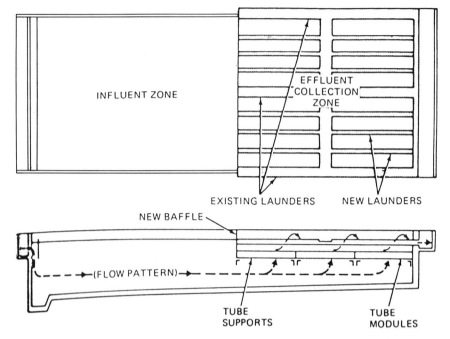

Fig. 9-19. Typical tube installation in rectangular basin.

basins are well designed so that they are free of short-circuiting and so a polymer may be fed as a coagulant aid.

Sludge Removal Facilities. Sometimes when the raw water is of very low turbidity, plants may include settling facilities designed to be manually cleaned. These basins may be cleaned infrequently, perhaps only once a year, if the sediment load is very light. In other cases, they may be cleaned every 30 days. Although it is usually desirable to have continuous mechanical sludge collectors in basins equipped with tubes, tube modules have been successfully used in manually cleaned basins. The frequency of cleaning will obviously increase if the basin's throughput is increased following the tube installation.

In general, the only adjustment required with mechanically cleaned water treatment basins will be to increase the frequency of withdrawal of the concentrated sludge from the basin. Most mechanical units have a substantial reserve capacity for handling larger quantities of sludge. In many plants, increasing the frequency of sludge withdrawal will merely involve the adjustment of a timer. In others, it may require complete modification or installation of supplementary sludge concentration facilities.

Tube Cleaning.   In certain waters, floc has a tendency to adhere to the upper edges of the tube openings. This is generally an infrequent occurrence and of no serious consequence other than detracting from the appearance of the installation. In some cases, the floc buildup eventually bridges the tube openings and results in a blanket of solids on top of the tubes that may reach 3 to 10 inches (76 to 254 mm) in depth unless some remedial action is taken. One method of removing this accumulation is to drop the water level of the basin to beneath the top of the tubes occasionally. The floc particles are dislodged and fall to the bottom of the basin. In some cases, it may not be possible to remove the basin from service to drop the level. A very gentle water current directed across the top of the tubes by a fixed-jet header as shown in Fig. 9–20 will remove floc accumulation. The general design guidelines are as follows for a water wash system:

- Provide water distribution headers at intervals not exceeding 20 feet (6.1 m).
- Provide nozzles at 1-foot (0.31-m) centers.
- Provide water supply at 60 psi (413 kPa) and about 6 gpm (0.38 l/s) per nozzle.
- Provide valving so that the cleaning system can be operated in sections in order to reduce the water supply required.

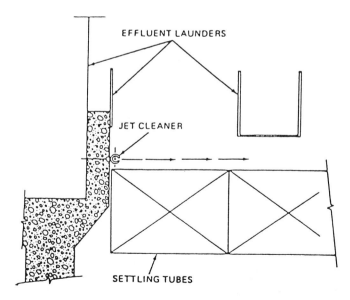

Fig. 9–20. Water wash settling tube cleaning system.

The header is operated only infrequently, typically only a few minutes per day. Another cleaning technique involves the installation of a grid of diffused air headers beneath the tubes (Fig. 9–21). General design guidelines for the air wash system are:

- Provide air at a pressure of 6 psi (41.3 kPa) and a rate of 1 cfm/sq ft (0.31 m/min) of tube surface area.
- Provide air distribution laterals 1 foot (0.31 m) below the bottom of the tubes and space at 1-foot (0.31-m) centers.
- Provide 16-inch (1.6-mm)-diameter orifices at 1-foot (0.31-m) centers, 30 degrees from bottom centerline of lateral and staggered as shown in Fig. 9–21.
- Provide valving so that the system can be operated in sections in order to reduce the air supply required.

To use this system, the influent is stopped and the air turned on and allowed to rise through the tubes, scrubbing any attached floc from them. A quiescent period of 15 to 25 minutes follows before the basin is placed back in service.

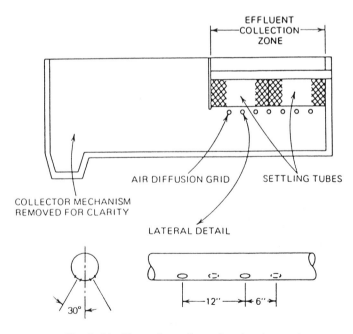

Fig. 9–21. Air wash settling tube cleaning system.

Recommended Design Criteria

Upflow Clarifiers—Loading Rates. In areas where cold water temperatures [less than 40°F (4.4°C)] occur frequently, the guidelines in Table 9–4 apply.[19,20] In warm water areas in which temperatures are nearly always above 50°F (10°C), the guidelines in Table 9–5 apply.

Of course, these guidelines are based on the assumption that both the chemical coagulation and flocculation steps have been carried out properly. Also, the sludge removal equipment has been assumed adequate.

Horizontal Flow Basins—Loading Rates. As indicated in the following tables, the raw water turbidity has a direct influence on allowable tube overflow rates, as does the raw water temperature. In cold water areas where temperatures are frequently 40°F (4.4°C) or less, the guidelines in Table 9–6 apply. In warm water areas [temperatures nearly always above 50°F (10°C)], the guidelines in Table 9–7 apply.

Table 9–4. Upflow Clarifier Loading Rates for Cold Water.*

OVERFLOW RATE, BASED ON TOTAL CLARIFIER AREA		OVERFLOW RATE, PORTION OF BASIN COVERED BY TUBES		PROBABLE EFFLUENT TURBIDITY, NTU
GPM/SQ FT	(M/H)	GPM/SQ FT	(M/H)	
1.5	3.66	2.0	4.88	1–3
1.5	3.66	3.0	7.32	1–5
1.5	3.66	4.0	9.76	3–7
2.0	4.88	2.0	4.88	1–5
2.0	4.88	3.0	4.88	3–7

* Water temperatures less than 40°F (4.4°C).

Table 9–5. Upflow Clarifier Loading Rates for Warm Water.*

OVERFLOW RATE, BASED ON TOTAL CLARIFIER AREA		OVERFLOW RATE, PORTION OF BASIN COVERED BY TUBES		PROBABLE EFFLUENT TURBIDITY, NTU
GPM/SQ FT	(M/H)	GPM/SQ FT	(M/H)	
2.0	4.88	2.0	4.88	1–3
2.0	4.88	3.0	7.32	1–5
2.0	4.88	4.0	9.76	3–7
2.5	6.10	2.5	6.10	3–7
2.5	6.10	3.0	7.32	5–10

* Water temperatures above 50°F (10°C).

Table 9–6. Loading Rates for Horizontal Flow Basins and Cold Water.*

OVERFLOW RATE, BASED ON TOTAL CLARIFIER AREA		OVERFLOW RATE, PORTION OF BASIN COVERED BY TUBES		PROBABLE EFFLUENT TURBIDITY, NTU
GPM/SQ FT	(M/H)	GPM/SQ FT	(M/H)	
Raw Water Turbidity = 0–100				
2.0	4.88	2.5	6.10	1–5
2.0	4.88	3.0	7.32	3–7
3.0	7.32	4.0	9.76	5–10
Raw Water Turbidity = 100–1,000				
2.0	4.88	2.5	6.10	3–7
2.0	4.88	3.0	7.32	5–10

* Water temperatures are frequently 40°F (4.4°C) or less.

Table 9–7. Loading Rates for Horizontal Basins and Warm Water.*

OVERFLOW RATE, BASED ON TOTAL CLARIFIER AREA		OVERFLOW RATE, PORTION OF BASIN COVERED BY TUBES		PROBABLE EFFLUENT TURBIDITY, NTU
GPM/SQ FT	(M/H)	GPM/SQ FT	(M/H)	
Raw Water Turbidity = 0–100				
2.0	4.88	2.5	6.10	1–3
2.0	4.88	3.0	7.32	1–5
2.0	4.88	4.0	9.76	3–7
3.0	7.32	3.5	8.54	1–5
3.0	7.32	4.0	9.76	3–7
Raw Water Turbidity = 100–1,000				
2.0		2.5	6.10	1–5
2.0		3.0	7.32	3–7

* Water temperatures are frequently 50°F (10°C) or above.

Effluent turbidities above 5 NTU will often obscure the tube modules through 2 feet (0.61 m) of water. This may not be aesthetically desirable to a casual observer, but such turbidities are readily treated by a mixed-media filter. If the tube clarification application is not followed by mixed-media filters, the loading rates should be selected to provide effluent turbidities in the range of 1 to 3 NTU.

Location of Tube Modules within the Basin

The tubes should be located such that they are not placed in a zone of unstable hydraulic conditions. Thus, they are frequently placed over one-half to three-quarters of the basin located nearest the effluent launders to permit the inlet portion of the clarifier to dampen out hydraulic currents. The top of the tubes should be located 2 to 4 feet (0.61 to 1.22 m) below the water surface. In general, the 2-foot (0.61-m) minimum is used in shallow basins. The submergence of 4 feet (1.22 m) would be considered only in clarifiers with a sidewater depth of 16 to 20 feet (4.9 to 6.1 m). In most basins, where sidewater depths rarely exceed 13 feet (4 m), a submergence of 2 to 3 feet (0.61 to 0.91 m) is used. The collection launders should be placed on 10- to 12-foot (3.7-m) centers over the entire area covered by tubes to ensure uniform flow distribution.

Example Applications

Horizontal Basins. An existing water treatment plant with a rated capacity of 4 mgd (15.1 Ml/d) has a horizontal, rectangular settling basin and rapid sand filters. The basin dimensions are 30 feet wide by 133 feet long (9.1 by 40.5 m). The surface overflow rate at design capacity is 1,000 gpd/sq ft (1.69 m/h). The average depth of the basin is 15 feet (4.6 m). The basin has a single overflow weir across the outlet end of the basin. The raw water is obtained from a river that has a normal maximum turbidity of 25 to 30 NTU. The water temperature rarely falls below 50°F (10°C). The settling basin is preceded by mechanical flocculation with 40 minutes of detention time at 4 mgd (15.1 Ml/d). Coagulant aids are fed during periods of high turbidity and low water temperatures to improve coagulation.

A capacity increase from 4 to 8 mgd (15.1 to 30.3 Ml/h) is desired. At 8 mgd (30.3 Ml/h), the overflow rate increases to 2,000 gpd/sq ft or 1.4 gpm/sq ft (3.39 m/h). The total basin loading of 1.4 gpm/sq ft (3.4 m/h) is below the values shown in the above guidelines for a horizontal flow basin under warm water conditions. However, at a higher basin loading of 2 gpm/sq ft (4.88 m/h) and a tube rate of 3 gpm/sq ft (7.32 m/h), the expected effluent turbidity is 1 to 5 NTU. This value is compatible with mixed-media filters, which, as discussed later in this text, can readily replace the existing sand filters. In light of the moderate raw water temperature and turbidity, a tube rate of 3 gpm/sq ft (7.32 m/h) and a basin loading rate of 1.4 gpm/sq ft (3.42 m/h) should give excellent results.

$$\text{Quantity of tubes required} = \frac{\text{Capacity, gpm}}{\text{Allowable tube rate, gpm/sq ft}} \quad (9\text{–}7)$$

$$= \frac{8 \text{ mgd} \times 700 \text{ gpm/mgd}}{3 \text{ gpm/sq ft}}$$

$$= 1,870 \text{ sq ft } (173.7 \text{ m}^2)$$

The dimensions of the area to be covered by the tube modules are determined as follows:

$$\text{Area} = \text{length} \times \text{width}$$
$$1{,}870 \text{ sq ft} = \text{length} \times 30 \text{ feet}$$
$$\text{Length} = \frac{1{,}870 \text{ sq ft}}{30 \text{ feet}} = 62.3 \text{ feet (19 m)}$$

The length of 62 feet (19 m) would be rounded off to a length readily compatible with the standard module dimensions associated with the specific modules purchased.

The modules would be installed over an area extending back from the discharge end of the basin for a distance of 62 feet (19 m). A baffle wall would be installed at the inner edge to force all flow through the modules. To improve uniform flow through the modules, three new effluent launders extending 62 feet (19 m) back from the existing end wall launder would be

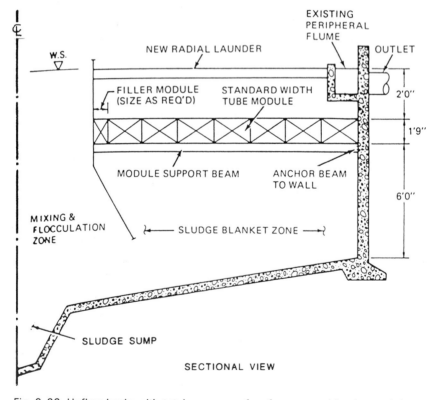

Fig. 9-22. Upflow basin with total coverage of surface area with tube modules.

required. The launders would be installed on 10-foot (3.1-m) centers, and the tubes would be submerged for a depth of 4 feet (1.22 m) because the basin is deep. The appearance of the basin would be similar to that shown in Fig. 9–22.

Upflow Basins. Assuming a plant has two 42-foot (12.8-m)-square upflow clarifiers, each designed for a flow of 3,000 gpm (16.35 Ml/d), with peripheral collection launders. The total surface area is 1,760 square feet (163.5 m²). The influent centerwell reduces the available settling area by 200 square feet (18.6 m²). The peak overflow rate now reaches 1.92 gpm/sq ft (4.68 m/h), which is high enough that the clarifier does not perform well, especially when water temperatures drop.

It is desired to increase the plant capacity to 4,000 gpm (21.8 Ml/d) per settling basin. At this flow, the loading on the total basin settling area is 2.6 gpm/sq ft (6.34 m/h). The raw water turbidity is moderate, 30 to 70 NTU, and the water temperature seldom falls below 50°F (10°C).

The guidelines for upflow basins indicate that a maximum total basin loading of 2.5 gpm/sq ft (6.1 m/h) with a corresponding tube rate of 2.5 to 3 gpm/sq ft (6.10 to 7.32 m/h) can be used. Complete coverage of the settling area would provide a tube rate of 2.6 gpm/sq ft (6.34 m/h) and would also provide a simplified support problem when compared to only partial coverage. Thus, coverage of the settling area with 1,560 square feet (144.9 m²) of tubes would be provided. Radial launders would be added to improve the flow distribution in the basin, as shown in Fig. 9–22. As the sidewater depth is only 10 feet (3.1 m), the modules would be submerged 2 feet (0.61 m).

REFERENCES

1. Camp, T. R., "Sedimentation and the Design of Settling Tanks," *Transactions American Society of Civil Engineers* 111:895, 1946.
2. Zanoni, A. E., and Blomquist, M. W., "Column Settling Tests for Flocculant Suspensions," *Proc. American Society of Civil Engineers* 101:EE3:309–318, June, 1975.
3. Weber, W. J., *Physicochemical Processes for Water Quality Control,* Wiley-Interscience, New York, 1972.
4. *Water Treatment Plant Design,* American Water Works Association, New York, 1969.
5. *Water Quality and Treatment. A Handbook of Public Water Supplies,* American Water Works Association, New York, 1971.
6. Culp, Wesner, Culp, *Technical Guidelines for Public Water Systems,* EPA 68–01–2971, June, 1975.
7. Great Lakes–Upper Mississippi River Board of State Sanitary Engineers, *Recommended Standards for Water Works,* 1968 Edition with 1972 Revisions. Published by Health Education Services, Albany, New York.
8. Walton, Graham, *Effectiveness of Water Treatment Processes,* U.S. Public Health Service, Publication No. 898, 1962.

9. AWWA Committee Report, "Viruses in Water," *J.AWWA,* p. 491, October, 1969.
10. Kawamura, S., "Hydraulic Scale-Model Simulation of the Sedimentation Process," *J.AWWA,* p. 372, July, 1981.
11. Hazen, A., "On Sedimentation," *Transactions American Society of Civil Engineers* 53:45, 1904.
12. Camp, T. R., "Sedimentation and the Design of Settling Tanks," *Transactions American Society of Civil Engineers* 111:895, 1946.
13. Eliassen, R., Discussion of (1), *Transactions of American Society of Civil Engineers* 111:947, 1946.
14. Camp, T. R., Discussion of (1), *Transactions of American Society of Civil Engineers* 111:952, 1946.
15. McMichael, F. C., "Sedimentation in Inclined Tubes and Its Application for the Design of High-Rate Sedimentation Devices," paper presented at the International Union of Theoretical and Applied Mechanics Symposium on Flow of Fluid–Solid Mixtures, University of Cambridge, England, March 24–29, 1969.
16. Yao, K. M., "Theoretical Study of High Rate Sedimentation," *J.AWWA,* p. 218, 1970.
17. Hansen, S. P., and Culp, G. L., "Applying Shallow Depth Sedimentation Theory," *J.AWWA* 59:1134, 1967.
18. Culp, G. L., Hansen, S. P., and Richardson, G. H., "High Rate Sedimentation in Water Treatment Works," *J.AWWA* 60:681, 1968.
19. Conley, W. R., and Slechta, A. F., "Recent Experiences in Plant Scale Application of the Settling Tube Concept," presented at the Water Pollution Control Federation Conference, Boston, Mass., October, 1970.
20. Culp, G. L., Hsiung, K. Y., and Conley, W. R., "Tube Clarification Process, Operating Experiences," *Journal of the Sanitary Engineering Division, ASCE,* SA5, p. 829, October, 1969.
21. Stukenberg, J., Rodman, L. C., and Touslee, V. E., "Activated Sludge Clarifier Design Improvements," presented at the 54th Annual Water Pollution Control Federation Conference, Detroit, Mich., October, 1981.
22. Parker, D. S., "Assessment of Secondary Clarification Design Concepts," presented at the California Water Pollution Control Association Conference, April, 1982.
23. Hudson, H. E., Jr., *Water Clarification Processes. Practical Design and Evaluation,* Van Nostrand Reinhold, New York, 1981.

Chapter 10
Filtration

INTRODUCTION

The term "filtration" has different connotations. Outside the waterworks profession, even in other technical disciplines, filtration is commonly thought of as a mechanical straining process. It may also have this same basic meaning in waterworks practice as applied to the passage of water through a very thin layer of porous material deposited by flow on a support septum. This type of filter has a few rather specialized applications to water treatment, as described later. Most frequently in waterworks parlance, filtration refers to the use of a relatively deep [1½ to 3 feet (0.46 to 0.91 m)] granular bed to remove particulate impurities from water. In contrast to mechanical strainers that remove only part of the coarse suspended solids, the filters used in water purification can remove all suspended solids, including virtually all colloidal particles.

Over the years the meaning of the word "filtration" as used by the waterworks industry has changed. Improved filters have been developed, and the nature of the physical and chemical processes involved in filtration has become better understood. New and improved filters are distinguished from older conventional types because they can incorporate: (1) coarse-to-fine in-depth filtration; (2) the application to the filter influent of a polymer, alum, or activated silica as a filter aid; (3) continuous monitoring of filter effluent turbidity; and (4) pilot filter control of coagulant dosage. Use of the general term "filters" overlooks important distinctions in functions and efficiency. In discussing filtration then, care must be exercised to specify the details of the particular filter under consideration.

Water filtration is a physical–chemical process for separating suspended and colloidal impurities from water by passage through a bed of granular material. There are two separate steps: (1) transport, in which suspended particles are transported to the immediate vicinity of the solid filter media; and (2) attachment, in which particles become attached to the filter media surface or to another particle previously retained in the filter. The transport step is primarily a physical process, while the attachment step is very much influenced by chemical and physical–chemical variables.

319

Filter Types

There are several ways to classify filters. They can be described according to the direction of flow through the bed, that is, downflow, upflow, biflow, radial flow, horizontal flow, fine-to-coarse, or coarse-to-fine. They may be classed according to the type of filter media used, such as sand, coal (or anthracite), coal–sand, multilayered mixed-media, or diatomaceous earth. Filters are also classed by flow rate. Slow sand filters operate at rates of 0.05 to 0.13 gpm/sq ft (0.12 to 0.32 m/h), rapid sand filters operate at rates of 1 to 2 gpm/sq ft (2.44 to 4.88 m/h), and high-rate filters operate at rates of 3 to 15 gpm/sq ft (7.32 to 36.6 m/h). Filters may also be classified by the type of system used to control the flow rate through the filter, such as constant rate, declining rate, constant level, equal loading, and constant pressure. Constant rate filtration is the most popular control system in the United States.

Another characteristic is pressure or gravity flow. Gravity filter units are usually built with an open top and constructed of concrete or steel, while pressure filters are ordinarily fabricated from steel in the form of a cylindrical tank. The available head for gravity flow usually is limited to about 8 to 12 feet (2.44 to 3.66 m), while it may be as high as 150 psi (1,033 kPa) for pressure filters. Because pressure filters have a closed top, it is not easy to inspect the filter media. Further, it is possible to disturb the media in a pressure filter by sudden changes in pressure. These two factors have tended to limit municipal applications of pressure filters to treatment of relatively unpolluted waters, such as the removal of hardness, iron, or manganese from well waters of good bacterial quality. The susceptibility to bed upset and the inability to see the media in pressure filters have been compensated for, to some extent, by the use of quick-opening manholes and by the recent development and application of recording turbidimeters for continuous monitoring of the filter effluent turbidity. The introduction of a 3-inch (76.2-mm) layer of coarse (1 mm) high-density (specific gravity 4.2) garnet or ilmenite between the fine media and the gravel supporting bed has virtually eliminated the problem of gravel upsets, which is another of the concerns about the use of pressure filters for production of potable water.

In the examination of filter sand, the terms "effective size" and "uniformity coefficient" are used. The effective size is the size of the grain, in millimeters, such that 10 percent of the particles by weight are smaller. This 10 percent by weight fraction corresponds closely in size to the median size by count, as determined by a size frequency distribution of the total number of particles in a sample.[1] The effective size is a good indicator of the hydraulic characteristics of a sand within certain limits. These limits are usually defined by means of the uniformity coefficient, which is arbitrarily taken as the ratio of the grain size with 60 percent smaller than itself to the size with 10 percent smaller than itself (effective size). This ratio thus covers the range in size of half the sand.

THE EVOLUTION OF FILTRATION

Slow Sand

In waterworks history three basic types of filters have dominated the field at different times. In the 1800s the slow sand filter was dominant. It incorporated sand with an effective size of about 0.2 mm. This very fine sand produced good-quality water from applied water of low turbidity at rates on the order of 0.05 to 0.13 gpm/sq ft (0.12 to 0.32 m/h) of bed area. The filter was cleaned by scraping a thin layer of media from the surface of the filter, washing it, and returning the washed sand to the bed. Because of the low surface rates, slow sand filters required large areas of land and were costly to install. They were also expensive to operate because of the laborious method of bed cleaning by surface scraping.

Rapid Sand

Beginning in the early 1900s under the stimulus of epidemic waterborne disease, the rapid sand filter came into general use, and it largely replaced the slow sand filter. Rapid sand filter media might vary in effective size from 0.35 to 1.0 mm, with a typical value being 0.5 mm with a uniformity coefficient of 1.3 to 1.7. This type of fine media has demonstrated the ability to handle applied turbidities of 5 through 10 NTU at rates up to 2 gpm/sq ft (4.88 m/h). The introduction of prior chemical coagulation and hydraulic backwash made possible the use of rapid rather than slow sand beds.

Rapid sand filters are cleaned by reversing the flow through the filter and backwashing the trapped particles from the bed. In backwashing single-medium sand beds, hydraulic grading of the sand grains occurs. The very finest sand accumulates at the top of the bed, and the coarser particles lie below. More than 90 percent of the particulates removed are taken out in the top few inches of the bed. Once a suspended particle has penetrated this top layer of fine sand, its chances are greatly increased for passing through the entire bed because the void spaces become larger and the opportunities for contact decrease as the particles travel downward. This is a well-recognized limitation of the rapid sand filter.

The pore openings in a rapid sand filter made up of 0.5 mm sand range from 0.1 to 0.2 mm in size. In the water applied to the filter, floc size ranges from 2 mm to less than 0.1 mm.[1] It follows from these dimensions that the larger floc particles can be removed by simple straining at the filter surface, but that much of the flocculated matter will pass into the filter and lodge within it. Unless the flocs are exceptionally strong, those that initially lodge in the filter by simple straining will subsequently break as the hydraulic gradient increases.

The flocculated material in the water passes into the filter through thousands of openings in each square foot of the filter surface and, by the end of a filter run, ordinarily lodges largely in the top 1 to 4 inches (25.4 to 101.6 mm) of the filter.

The probability of a floc particle striking the surface of the filter medium is influenced by sand size, porosity, filter rate, temperature, and density and size of floc particles. The adherence of the particles is affected by colloidal forces, including the age of the floc, temperature, coagulant concentration, type and concentration of anions, and pH.

Mixed-Media

It was not until the early 1940s, under the stimulus of a critical wartime need to produce greatly improved water quality (basically much lower turbidity water) for processing radioactive materials at Hanford, Washington, that coarse-to-fine filtration had its beginnings under the leadership and direction of Raymond Pitman and Walter Conley.[2-7] Development of the coarse-to-fine principle of filtration took place in two major steps. The first step was the development of the dual-media filter, which typically uses 24 inches (0.61 m) of anthracite coal above 6 inches (0.15 m) of silica sand. This provides a two-layer filter in which the coarse upper layer of coal acts as a roughing filter to reduce the load of particulates applied to the sand below. Because of the different specific gravities of the two materials (coal 1.4, sand 2.65), coal of the proper size in relation to the sand remains on top of the sand during backwashing. With applied turbidities of less than about 15 NTU, dual-media filters can operate under steady state conditions at 4 to 5 gpm/sq ft (9.76 to 12.2 m/h) with the production of high-quality water. Dual-media filters can retain more material removed from the water than a sand filter, but they have a low resistance to turbidity breakthrough with changing flow rates. This serious shortcoming is due to the low total surface area of media particles, which is actually less than that for a conventional sand bed. Coal–sand beds in which there is controlled mixing of the two materials near the interface perform better and wash more easily than coal–sand beds that are designed for more distinct layering (early versions).[8]

In designing a dual-media bed, it is desirable to have the coal as coarse as is consistent with solids removal to prevent surface blinding, and to have the sand as fine as possible to provide maximum solids removal. However, if the sand is too fine in relation to the coal, it will actually rise above the top of the coal during the first backwash and remain there when the filter is returned to service. For example, if 0.2 mm of sand were placed below

1.0 mm of coal, the materials would reverse during backwash, with the sand becoming the upper layer and the coal the bottom. Although the sand has a higher specific gravity, its small diameter in this case would result in its rising above the coal. The only way to enable very fine silica sand to be used in the bottom filter layer would be to use finer coal, but this would defeat the purpose of the upper filter layer because the fine coal would be susceptible to surface blinding. Experience has shown that it is not feasible to use silica sand smaller than about 0.4 mm because smaller sand would require coal small enough to result in unacceptably high headloss at rates above 3 gpm/sq ft (7.32 m/h).

To overcome the above limitation and to achieve a filter that closely approached an ideal one (Fig. 10–1c), the mixed-media concept was developed (patents held by Microfloc Products, St. Paul, Minnesota). Figure 10–1 shows the three media designs now in use. The problem of keeping a very fine medium at the bottom of the filter is overcome by using a third, very fine, heavy material (garnet, specific gravity of about 4.2, or ilmenite, specific gravity of about 4.5) beneath the coal and sand. The garnet (or ilmenite), sand, and coal particles are sized so that controlled intermixing of these materials occurs, and no discrete interface exists between them. This eliminates the stratification illustrated for the dual-media filter in Fig. 10–1b, and results in a filter that very closely approximates the ideal of a uniform decrease in pore space with increasing filter depth, as shown in Fig. 10–1c.

Actually, the term "coarse-to-fine" refers more accurately to the pore space or size rather than to the media particles themselves, as is illustrated in Fig. 10–1. By selecting the proper size distribution of each of the three media, it is possible to construct a bed that has an increasing number of particles at each successively deeper level in the filter. A typical mixed-media filter has a particle size gradation that decreases from about 2 mm at the top to about 0.15 mm at the bottom. The uniform decrease in pore space with filter depth allows the entire filter depth to be utilized for floc removal and storage. Figure 10–2 shows how particles of the different media are actually mixed throughout the bed. At all points in the bed there is some of each component, but the percentage of each changes with bed depth. There is steadily increasing efficiency of filtration in the direction of the flow. In some cases, coal of different densities can be used to further extend the coarse-to-fine concept. By using a lighter-weight coal in addition to the normal coal–sand–garnet, it is possible to increase the top grain size to 2.4 mm without changing the grain size of the rest of the filter.[9]

One of the key factors in constructing a satisfactory mixed-media bed is careful control of the size distribution of each component medium. Rarely is the size distribution of commercially available materials adequate for construction of a good mixed-media filter. The common problem is failure to remove excessive amounts of fine materials. These fines can be removed by

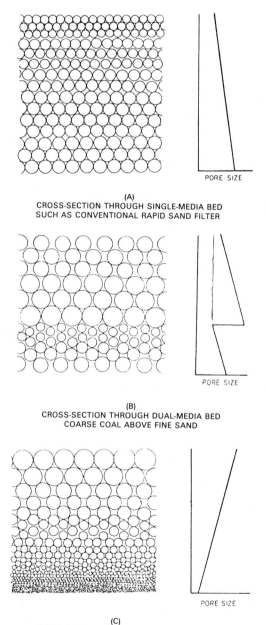

(A)
CROSS-SECTION THROUGH SINGLE-MEDIA BED
SUCH AS CONVENTIONAL RAPID SAND FILTER

(B)
CROSS-SECTION THROUGH DUAL-MEDIA BED
COARSE COAL ABOVE FINE SAND

(C)
CROSS-SECTION THROUGH IDEAL FILTER
UNIFORMLY GRADED FROM COARSE TO FINE
FROM TOP TO BOTTOM

Fig. 10–1. Graphical representation of various media designs. (Courtesy of Microfloc Products Group, Johnson Division, UOP)

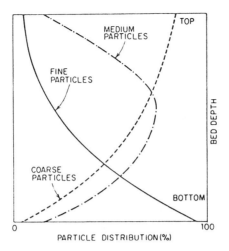

Fig. 10–2. Distribution of media in a properly designed mixed-media filter. (Courtesy of Microfloc Products Group, Johnson Division, UOP)

placing a medium in the filter, backwashing it, draining the filter, and skimming the upper surface. The procedure is repeated until the field sieve analyses indicate that an adequate particle size distribution has been obtained. The second medium is added, and the procedure is repeated; and then the third medium is added, and the entire procedure is repeated. Sometimes 20 to 30 percent of the materials may have to be skimmed and discarded to achieve the proper particle size distribution. Unfortunately, the literature is replete with comparisons of different filter media where inadequate attention was devoted to the proper sizing of the media. The use of three media rather than two assures a superior filter *only* if the three media are properly sized. The authors have run hundreds of parallel tests of single-medium, dual-media, and tri-media filters, and have found a properly designed tri-media bed to be a superior filter consistently. The reader must critically examine the literature to ascertain the basis for the design of the filter media being compared.

The vast floc storage capacity of the mixed-media bed increases the length of filter run before terminal headloss is reached. The total surface area of the grains in a mixed-media bed is greater than for a sand or dual-media bed, so that it is much more resistant to breakthrough and more tolerant to surges in flow rates. This provides a great factor of safety in filter operation. Despite the greater total surface area of grains, the initial (clean filter) headloss in the two types is comparable. At 5 gpm/sq ft (12.2 m/h) throughput, the initial headloss in either a 0.50-mm sand or a mixed-media bed is about 1½ feet (0.46 m).

Filter rates of 5 gpm/sq ft (12.2 m/h) are commonly used for design and operation of mixed-media beds, compared to 2 gpm/sq ft (4.88 m/h) for sand filters. At the same time, there is improvement in product water quality, which was the original purpose behind the development of the mixed-media bed. Along with development and acceptance of the mixed-media filter has come a recognition that the rate of filtration is only one factor (and a relatively unimportant one) affecting filter effluent quality, and that chemical dosages for optimum filtration, rather than maximum settling, as well as other variables are much more important than filtration rate to production of good water.

In the modern concept of water treatment, coagulation and filtration are inseparable. They are actually each very closely related parts of the liquid–solids separation process. Because most water plants utilize sedimentation for a preliminary gross separation of settleable solids between coagulation and filtration, the crucial direct relationship of coagulation to optimum filterability often has been overlooked.

WATER FILTER OPERATION

Mechanisms

Many publications describe filtration theory and mechanisms.[15-29] Several mechanisms are involved in particle removal by filtration, some of them physical and others chemical in nature. To explain fully the overall removal of impurities by filtration, the effects of both the physical and chemical actions occurring in a granular bed must be combined. Efficient filtration involves particle destabilization and particle transport similar to the mechanisms of coagulation; good coagulants are also efficient filter aids. The processes of coagulation and filtration are inseparable, and their interrelationships must be considered for best treatment results. One important advantage of filtration over coagulation, relative to the removal of very dilute concentrations of colloidal particles, is the much greater number of opportunities for contact afforded by the granular bed as compared to the number afforded by stirring the water. The removal efficiency of a filter bed is independent of the applied colloidal particle concentration, whereas the time of flocculation depends upon colloidal particle concentration.

There are two basic approaches to obtaining optimum filterability of water. One is to establish the primary dosage of coagulant for maximum filterability rather than for production of the most rapid-settling floc. A second approach is to add a dose of a second coagulant as a filter aid to the settled water as it enters the filter. For filter runs of practical length, filter aid to coat filter grains must be added continuously (rather than precoated). When filter aids are used, the filtration process is so effective that conventional sand filters clog very rapidly. Effective filtration without excessive headloss can be accomplished by use of a coarse-to-fine, in-depth filter.

For effective filtration, the objective of pretreatment should be to produce small, dense floc, rather than large, fluffy floc, so that the particles are small enough to penetrate the bed surface and move down into the bed. Removal of floc within a bed is accomplished primarily by contact of the floc particles with the surface of the grains or previously deposited floc, and adherence thereto. Contact is brought about principally by the convergence of stream lines at contractions in the pore channels between the grains. Of minor importance are the flocculation, sedimentation, and entrapment that occur within the pores of the bed.

Adsorption of suspended particles on the surface of the filter grains is an important factor in filter performance. Physical factors affecting adsorption include both the nature of the filter and the suspension. Adsorption is a function of the filter grain size, floc particle size, and the adhesive characteristics and shearing strength of the floc. Chemical factors affecting adsorption include the chemical characteristics of the suspended particles, the aqueous suspension medium, and the filter medium. Two of the most important chemical characteristics are the electrochemical forces and van der Waals forces (molecular cohesive forces between particles).

Filtration Efficiency

Filters are highly efficient in removing suspended and colloidal materials from water. Impurities removed by filtration included: turbidity, bacteria, algae, viruses, color, oxidized iron and manganese, radioactive particles, chemicals added in pretreatment, heavy metals, and many other substances. Because filtration is both a physical and a chemical process, there are a large number of variables that influence filter efficiency. These variables exist both in the water applied to the filter and in the filter itself. In view of the complexity of what is commonly considered by plant designers and operators to be a relatively simple process, it is surprising that the general level of filter performance is so high. Knowledge of the factors affecting filter efficiency increased quite rapidly during the 1960s. Use of this information in the design and control of filters will make possible remarkably better water quality with average or good filter operation, and will make it even more difficult to obtain unsatisfactory results with poor operation.

Filter efficiency is affected by the following properties of the applied water: temperature, filterability, and the size, nature, concentration, and adhesive qualities of suspended and colloidal particles. Cold water is notably more difficult to filter than warm water, but usually there is no control over water temperature. Filterability, which is related to the size and adhesive qualities of the suspended and colloidal impurities in the water, is the most important property. As will be discussed in more detail later, the only practical way to measure filterability is by use of a pilot filter receiving raw water plus treatment chemicals and operating continuously in parallel with plant filters. By recording pilot filter effluent turbidities, appropriate adjustments can be made in chemical treatment to obtain optimum filterability in the plant filters. The use of raw water in the pilot filter provides the necessary lead time to anticipate plant requirements. Maximum filterability is much more important to production of a water of maximum clarity (minimum turbidity) than is maximum turbidity reduction prior to filtration.

Some properties of the filter bed that affect filtration efficiency are: the size and shape of the grains, the porosity of the bed (or the hydraulic radius of the pore space), the arrangement of grains (whether from fine-to-coarse or coarse-to-fine), the depth of the bed, and the headloss through the bed. In general, filter efficiency increases with smaller grain size, lower porosity, and greater bed depth. Coarse-to-fine filters contain much more storage space for materials removed from the water than fine-to-coarse filters, and permit the practical use of much finer materials in the bottom of the bed than can be tolerated at the top of a fine-to-coarse filter.

The total surface area of filter media grains is important because it represents the total area available for adsorption of floc. Figure 10–3 summarizes the total surface area of different sizes and depths of filter media.[14] Because of the greater total surface area of the grains, smaller grain size, and lower porosity, the coarse-to-fine filter is more efficient than the fine-to-coarse filter. The much greater total grain surface area and the smaller grain size provided by mixed media as compared to dual media account for the greater resistance to breakthrough possessed by mixed media. A mixed-media filter also may permit achievement of lower turbidity at a given coagulant dosage, as shown in Fig. 10–4. A dual-media filter, which should be considered merely an intermediate step in the development of mixed media, is less resistant to breakthrough than rapid sand filters, whereas a mixed-media filter is more resistant than either.

Hydraulic throughput rate also affects filter efficiency. However, within the range of 2 to 8 gpm/sq ft (4.88 to 19.5 m/h), the rate is not nearly so significant as other variables are to effluent quality. In general, the lower the rate, the higher the efficiency. All other conditions being equal, a filter will produce a better effluent when operating at a rate of 1 gpm/sq ft (2.44 m/h) than when operating at 8 gpm/sq ft (19.5 m/h). However, it is also true that a given filter may operate entirely satisfactorily at 8 gpm/sq ft (19.5 m/h) on a properly prepared water, yet fail to produce a satisfactory effluent at 1 gpm/sq ft (2.44 m/h) when receiving an improperly pretreated water. With good filter design, the optimum throughput rate is a matter of economics rather than a question of safety.

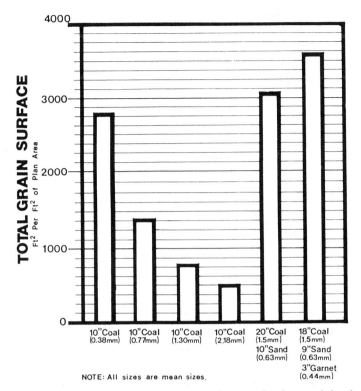

Fig. 10-3. Filter grain surface of different filter-media sizes and depths.[14]

The efficiency of filters in bacterial removal varies with the applied loading of bacteria, but with proper pretreatment should exceed 99 percent. Bacterial removal by filtration, however, should never be assumed to reach 100 percent. The water must be chlorinated for satisfactory disinfection. Coagulation, flocculation, and filtration will remove more than 98 percent of polio virus at filtration rates of 2 to 6 gpm/sq ft (2.44 to 14.6 m/h), but complete removal is dependent upon proper disinfection.[10]

The turbidity of the effluent from a properly operating filter should be less than 0.1 NTU. With proper pretreatment, filtered water should be essentially free of color, iron, and manganese. Large microorganisms, including algae, diatoms, and amoebic cysts, are readily removed from properly pretreated water by filtration.

The perfection of very sensitive, accurate, and reliable turbidimeters for the continuous monitoring of filter effluent quality has revolutionized the degree of control that can be exercised over filter performance. The turbidity of filter effluent is instantly and continuously determined, reported, and re-

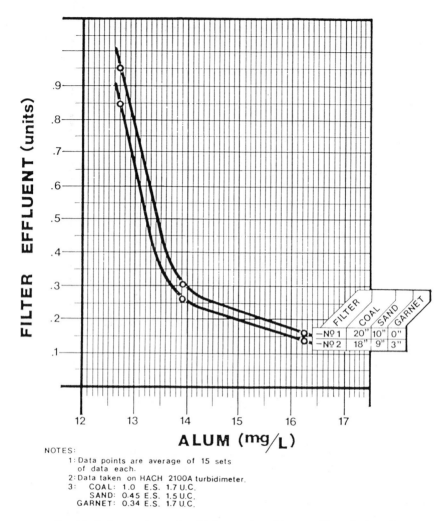

Fig. 10–4. Comparison of turbidity produced by two filter designs.[14]

NOTES:
1: Data points are average of 15 sets of data each.
2: Data taken on HACH 2100A turbidimeter.
3:    COAL: 1.0 E.S. 1.7 U.C.
      SAND: 0.45 E.S. 1.5 U.C.
      GARNET: 0.34 E.S. 1.7 U.C.

corded. The signal obtained from the instrument can be used to sound alarms, or, if necessary, to shut down an improperly operating filter unit. This greatly increases the reliability with which filters of all types may be operated, and may broaden the range of safe application for pressure filters. The significance of filter effluent turbidity is reflected in the observation that a turbidity increase of 0.1 NTU was associated with an increase in amoebic cysts by a factor of 10 to 50 when the filter had been operating at an equilibrium condition.[11]

Rating Filter Performance

There are many perspectives from which to judge filter performance: throughput rate, quantity of backwash water, length of filter runs, net water production, and finished water quality. When pilot tests are conducted to compare alternative media designs, the amount of data collected during the many runs of each medium can be overwhelming. The amount of water filtered per square foot per filter run has been suggested as a useful parameter for evaluating filter production efficiency.[12] The net water production per run can be readily calculated by subtracting the amount of backwash water required. Rating filter performance solely on volume of water processed, however, does not reflect the quality of water produced. The authors suggest the filter performance index (FPI) to express the relative performance of two or more filters operating under similar, but not necessarily identical, conditions:

$$\text{FPI} = \frac{(NTU_a - NTU_e)G}{NTU_e} \qquad (10\text{--}1)$$

where:

FPI = filter performance index
NTU_a = turbidity in units applied to filter
NTU_e = turbidity in units in filter effluent
G = gallons filtered between backwashes

Higher FPI numbers indicate better performance than low FPI numbers, with the degree of superiority represented by the percent difference in any two values.

The utility of the FPI is illustrated by the results of an extensive pilot plant test program in Sacramento, California.[13] A total of about 360 individual runs were made, consisting of 90 sets of parallel runs of a mixed tri-media filter, a sand filter, a mixed dual-media filter, and an anthracite capped sand filter. The curves and tabulations of all the filter runs made a formidably thick volume. However, the overall comparative performance can be readily grasped from the following averages:

	AVERAGE FPI
Mixed tri-media	74,000
Mixed dual-media	58,000
Anthracite capped sand	45,000
Sand	25,000

During the course of the study it was observed that the mixed tri-media filters recovered more quickly following backwash than did the other types of media. That is, only a few minutes of operation were required to produce effluent turbidities within the goal of 0.10 NTU with the mixed tri-media bed. With the other filters, periods of several hours were required for turbidities to reach the goal. This was increasingly apparent at high filter rates.

THE ROLE OF FILTRATION IN WATER TREATMENT

It is now generally acknowledged that all surface water supplies should be filtered; but a significant number of surface water supplies derived from uninhabited, "protected," watersheds still are not filtered. These unfortunate situations are attributable to economic expediency rather than the best technical judgment; but it is only a matter of time until all public water supplies obtained from surface sources will be required to provide the essential safeguard of filtration.

Some well waters meet water quality standards and goals with no other treatment than chlorination. On the other hand, many well waters require removal of iron, manganese, hardness, color, odor, turbidity, hydrogen sulfide, bacteria, viruses, or other undesirable impurities. In these cases, filtration is ordinarily a part of the overall purification process.

In water treatment, the general practice is to reduce the turbidity of water to 10 NTU before application to rapid sand filters. However, mixed-media filters can and do handle applied turbidities of up to 50 NTU on a continuous basis, and will tolerate occasional turbidity peaks of 200 NTU. This is a filter capability that should not be overlooked in selecting plant flow sheets. Several public water supplies derived from surface sources having average turbidities of 20 to 50 NTU and peaks of 100 to 200 NTU are operating quite successfully without the use of settling basins ahead of the filters.

Provisions should be made in plant design to disinfect both filter influent and effluent. Disinfectant applied ahead of a filter is a great aid in maintaining clean filters. Disinfection of the finished water makes the most effective use of the disinfectant because of the reduced disinfectant demand at this point in treatment. Provisions should also be made for application of a filter aid such as alum, activated silica, or a polymer ahead of the filters so that the water can be adjusted to optimum filterability.

Where there is no full plant operating experience with a particular water to be treated, or where the records are not adequate to determine the best treatment processes to be used, pilot plant studies are a valuable aid to producing the most efficient and economical scheme of treatment. They are highly recommended under these conditions. Pilot studies can be used to determine the kind and extent of pretreatment that may be required ahead of filtration. Only a filter itself can reveal through its operation the characteristics of floc particles and filter media that determine removal efficiency and length of filter runs. While backwash rates are well standardized, it is good to check them through pilot plant operation on a particular water to be treated. In conducting pilot plant tests, the necessity to operate under all raw-water conditions must not be overlooked. Unless the pilot plant is operated under the most severe conditions, the full-scale plant may be underdesigned and ill-equipped to produce the desired water quality at all times.

DIRECT FILTRATION

As noted above, the ability of mixed-media filters to tolerate higher applied turbidities has resulted in several applications where coagulated water is filtered directly without sedimentation.[30-45] Figure 10–5 presents approaches to direct filtration now in use.

(a) Direct filtration using alum and nonionic polymer or activated silica.

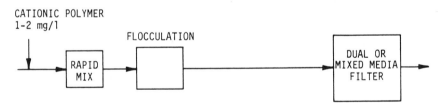

(b) Direct filtration using flocculation.

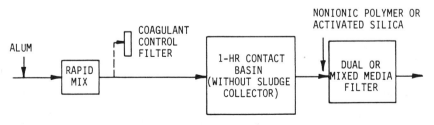

(c) Direct filtration using a contact basin.

Fig. 10–5. Approaches to direct filtration.[30]

The direct filtration arrangement shown in Fig. 10–5a consists of the addition of alum to rapid mix influent followed by the application of a nonionic polymer or activated silica to the influent of dual- or mixed-media filters. An alternative would replace the alum and nonionic polymer with the use of a cationic polymer in the raw water entering the rapid mix. In some cases, preozonation has been reported to enhance the removal of turbidity in the direct filtration process.[33] Potential mechanisms are microflocculation and the production of more polar compounds. Figure 10–5b is a direct filtration scheme utilizing a flocculation basin in which the chemical dosage alternates are the same as for Fig. 10–5a.

The Fig. 10–5c flow sheet is a direct filtration arrangement with a 1-hour contact basin between the rapid mix and the filter. In this process, the purpose of the contact basin is not to provide for settling. There is no flocculation basin, and the contact basin is not equipped with a sludge collector. The purpose of the contact basin is to increase the reliability of the process by adding a lead time of 1 hour between turbidity readings showing the results of coagulation from the coagulant-control filter and the entry of water to the filters. This helps in keeping plant operations abreast of changes in raw water quality. The coagulant options are the same as for Fig. 10–5a.

Following the proper mixing of the coagulant with raw water, a number of complex reactions take place with colloidal turbidity and color. These coagulation reactions take place in less than 1 sec. The rapid mixing process for direct filtration usually does not differ from that for conventional plants. A hydraulic jump in a Parshall flume may be used; field experience in direct filtration plants has been good with this type of mixing device. Some engineers extend the time for mechanical rapid mixing in direct filtration plants up to as much as 5 minutes, which is longer than that used in most conventional plants.

At this point in the process, the particles formed are very small, and the colloids are destabilized. When the destabilized particles collide, they stick together, with the rate of agglomeration of these microscopic destabilized particles to form visible floc depending principally upon the number of opportunities for contact they are afforded. In a still body of water, agglomeration takes place at a slow, almost imperceptible, rate; the rate can be increased by agitation or stirring of the water. In a well-designed flocculator, agglomeration of all particles might be completed in times varying from 5 to 45 minutes, when enough collisions will have occurred for the floc particles to become large enough to settle rapidly.

If settling is omitted from the plant flow sheet, as in a direct filtration plant, and if a properly designed rapid mix is provided, then usually there is no reason to include flocculation in the direct filtration process. Rather than spending money on flocculation, it may be better to improve the rapid mixing, provide finer filter media, or increase the depth of the fine filter media. The water containing the destabilized particles can be taken directly from the rapid mix basin to a granular filter where contact flocculation takes place as part of the filtration process. The flocculation rate is greatly accelerated because of the tremendous number of opportunities for contact afforded in the passage of the water through the granular bed. The floc particles become attached or absorbed to the surface of the filter grains. The smaller the filter grains, the greater are the opportunities for contact, and the more rapid is the removal of particulate matter. Small filter media also have a greater surface area per unit volume, which provides more area for attachment of floc particles to the filter grains than is available with larger grains.

This contact and the surface attachment or adsorption of particles to filter grains account for particulate removal beyond that of any simple mechanical straining action of the fine media. The pores of the filter gradually fill with floc as particles are sheared off the surfaces of filter grains. As a filter run progresses, the upper pores of the filter cannot retain any more floc, and the particles move down into the filter to find a resting place. Finally, either the headloss through the bed increases to the point where the filter must be cleaned by backwashing, or, if the floc strength is inadequate because of an underfeed of polymer, floc particles may appear in the effluent, requiring filter backwashing.

For many raw waters, direct filtration will produce water of a quality equal to that obtained by flocculation, settling, and filtration. The limitation of direct filtration is the inability to handle high concentrations of suspended solids. At some point, the amount of suspended solids will be too high for reasonable filter runs, and settling before filtration will be necessary.

The possibilities of applying direct filtration to municipal plants are good if one of the following conditions exists:

- The raw water turbidity and color are each less than 25 NTU.
- The color is low, and the maximum turbidity does not exceed 200 NTU.
- The turbidity is low, and the maximum color does not exceed 100 NTU.

The presence of paper fiber or of diatoms in excess of 1,000 areal standard units per milliliter (asu/ml) requires that settling be included in the treatment process chain. Diatom levels in excess of 200 asu/ml may require the use of special coarse coal on the top of the bed in order to extend filter runs.

The suitability of a raw water for direct filtration cannot be determined from numerical values alone; such values only provide a preliminary indication. Pilot plant tests must be performed in each case to find out whether or not direct filtration will provide satisfactory treatment under the prevailing local circumstances of raw water quality.

The chief advantage of direct filtration is the potential for capital cost savings from the elimination of sludge-collecting equipment, settling basin structures, flocculation equipment, and flocculation-basin structures. This cost reduction may make possible the provision of needed filtration for some communities that could not otherwise afford it. Operation and maintenance costs are reduced because there is less equipment to operate and maintain.

With direct filtration there may also be a savings of 10 to 30 percent in chemical costs because generally less alum is required to produce a filterable floc than to produce a settleable floc.[36] The costs for polymer may be greater than in conventional plants, but these higher costs are more than offset by the lower costs for coagulants.

Direct filtration produces less sludge than conventional treatment, and a denser sludge.[36] The collection of waste solids is simplified. Waste solids are all contained in a single stream, the waste filter-backwash water.

Pilot plant tests and plant-operating experience show that high-quality filtrate can be produced in direct filtration at filter rates of 5 to 15 gpm/sq ft (12.2 to 36.6 m/h).[36] A usual design rate is 5 gpm/sq ft (12.2 m/h). This provides for flexibility of operation and affords a margin of safety against variations in raw water quality. Filter influent and effluent piping should be designed for flows of 10 gpm/sq ft (24.4 m/h).

The design considerations presented below apply to filters used in direct filtration systems as well as to those used downstream of clarifiers.

DESIGN OF FILTER SYSTEMS

Selection of Type, Size, and Number of Units

In designing new plants, the selection of the type of unit is not influenced by existing facilities. Typically, the filter of choice would be a gravity, downflow, mixed-media bed in a concrete box with hydraulic surface wash, gravel and garnet supporting media, and Wheeler, Leopold, or pipe lateral underdrains. In small installations, the use of steel tanks or pressure filters may be indicated. These conclusions are based on information previously presented plus additional data which follow.

In medium and large plants, the minimum number of filter units is usually four. In small plants it may be two, or even one if adequate finished water storage is provided to supply service demands and backwash needs during filter shutdown for backwashing. The maximum size of a single filter unit is limited by the rate at which backwash water must be supplied, difficulties in providing uniform distribution of backwash water over large areas, reduction in filter plant capacity with one unit out of service for backwashing, and structural considerations. The largest filter units in service are about 2,100 sq ft (195 m²) in area. Units larger than about 1,000 sq ft (92.9 m²) are usually built with a central gullet so that each half of the filter can be backwashed separately. A typical gravity filter section is shown in

Fig. 10–6. Gravity filters typically do not exceed 1,000 sq ft (92.9 m²) per unit. Two filter units may be combined in one structural cell.

To filter a given quantity of water, the capital cost of piping, valves, controls, and filter structures is usually less for a minimum number of large filter units as compared to a greater number of smaller units. In expanding existing plants, it may be better not to increase the size of filter units but rather to match the existing size in order to avoid the need to increase the capacity of wash water supply and disposal facilities.

Pressure filters may be either horizontal or vertical. Horizontal filters offer much larger filter areas per unit and would normally be used when plant capacities exceed 1 to 15 mgd (3.79 to 56.78 Ml/d). A typical horizontal pressure filter vessel is shown in Fig. 10–7. Although an 8-foot (2.44-m)-diameter vessel is shown, 10-foot (0.31-m) diameters are also commonly used with lengths up to 60 feet (18.3 m). A typical vertical filter is shown in Fig. 10–8. Diameters up to 11 feet (3.35 m) are commonly used, with working pressures up to 150 psig (1033.5 kPa).

Filter Layout

One popular arrangement is to place filters side by side in two rows on opposite sides of a central pipe gallery. One end of the rows of filters should be unobstructed to allow construction of additional units in future expansion. The best pipe gallery design includes a daylight entrance. This provides good lighting, ventilation, and drainage, and improves access for operation, maintenance, and repair.

Filters should be located as close as possible to the source of influent water, the backwash water supply, the filtered water storage reservoir, and the control room. On sites that are subject to flooding, the level of the bottom of the filter boxes should be located above the maximum flood level. This arrangement permits the discharge of filter backwash water during flood periods, and avoids possible contamination of the filtered water.

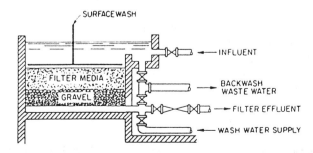

Fig. 10–6. Cross section through a typical gravity filter.

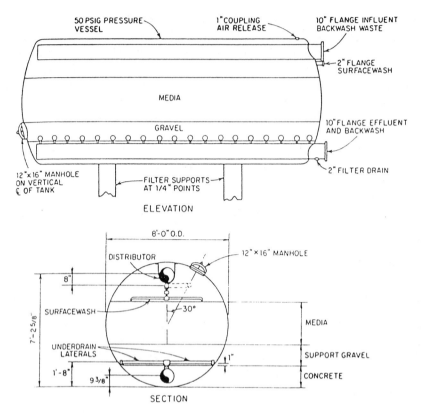

Fig. 10–7. Typical pressure filter. (Courtesy of Microfloc Products Group, Johnson Division, UOP)

Historically, the overall depth of filters from water surface to underdrains was ordinarily 8 to 10 feet (2.44 to 3.05 m). The current trend is toward deeper filter boxes to reduce the possibility for air binding and to increase the available head for filtration. Filter effluent pipes and conduits must be water-sealed to prevent the entry of air into the bottom of the filter when it is idle.

The use of common walls between filtered and unfiltered water should be avoided so as to eliminate the possibility of contamination of the finished water. For the same reason, it is better not to construct clearwell storage beneath the pipe gallery floor. Many states do not permit these types of construction.

Filter Structures

The general practice is to house and roof pipe galleries. In the past, this was also the practice for filters and filter operating galleries. Housing was

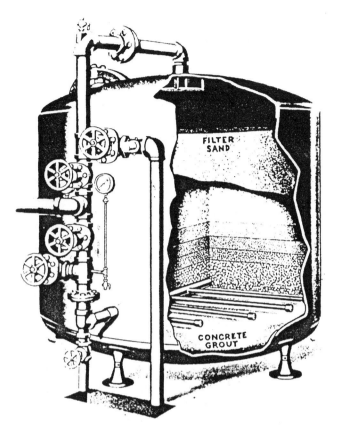

Fig. 10–8. Typical vertical pressure filter with concrete grout fill in the bottom head, pipe headers, lateral underdrains, gravel supporting bed, and filter sand. (Courtesy of Infilco Products.)

included to protect the filter controls and the water in the filters against freezing. Current practice is to locate the filter controls in a central control room and to provide perforated air lines around the periphery of each filter to control ice formation. This eliminates the need for housing the filters and filter gallery. If visual observation of filter washing without exposing the operator to the weather is considered essential, then it becomes a question of the relative costs and convenience of closed-circuit television versus housing the filters and filter gallery.

For gravity filter structures, either concrete or steel may be used. Concrete boxes may be either rectangular or circular; steel tanks are usually circular. Steel is usually preferred for small tanks and pressure filters, whereas concrete is the choice for large filters.

A number of special structural considerations involved in reinforced concrete filter boxes deserve the attention of designers. These structures must carry not only wind, dead, and live loads, but hydrostatic loads as well. To be watertight, walls should be not less than 9 inches thick. Usually, walls must be designed to withstand water pressure from either of two directions, and to resist sliding and overturning. Slabs also may be subject to hydrostatic uplift, as well as gravity loads. Shear and watertightness require special attention, particularly at keyways at the base of walls. Deflection at the top of cantilever walls and sliding at the base are common sources of trouble in water-bearing concrete structures. The concrete should be impervious to minimize the deleterious effects of freezing and thawing, and ample cover should be provided for steel reinforcement. The difference in expansion between water-cooled walls and floors and exposed slabs and roofs must be considered. The location and design of construction and expansion joints must be shown on the plans by the engineer, and not left to the judgment of the field inspector or contractor.

Filter Underdrains

Filter underdrains have a twofold purpose. The more important is to provide uniform distribution of backwash water without disturbing or upsetting the filter media above. The other is to collect the filtered water uniformly over the area of the bed. Ideally, the filter underdrain system would also serve as a direct support for the fine media. There is only one type of underdrain (porous plates) that can do this under certain special conditions, as will be described later. However, the three most widely used underdrain systems are the Wheeler, Leopold, and pipe lateral systems, all requiring an intermediate supporting bed of gravel that should be topped with a layer of coarse garnet or ilmenite to prevent upsets of the gravel layer. Figures 10–9 and 10–10 illustrate Leopold and pipe lateral underdrain systems, respectively. All of the systems accomplish the uniform distribution of wash water by introducing a controlling loss of head, usually about 3 to 15 feet (0.91 to 4.57 m), in the orifices of the underdrain system. The orifice loss must exceed the sum of the minor (manifold and lateral) headlosses in the underdrain to provide good backwash flow distribution.

The Wheeler bottom (Fig. 10–11) can be either a precast or poured-in-place false concrete bottom with a crawl space 22 inches (0.56 m) high between the structural and false bottoms. The crawl space and access manhole from the pipe gallery provide for low headloss flow distribution and a means for removal of the concrete forms. The bottom consists of a series of conical depressions at 1-foot (0.305-m) centers each way. At the bottom of each conical depression is a porcelain thimble with a ¾-inch (19.1-mm)-diameter orifice opening. Within each cone there are 14 porcelain spheres ranging in diameter from 1⅜ to 3 inches (34.9 to 76.2 mm) and arranged so as to

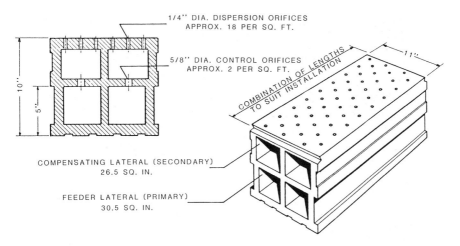

1/4'' DIA. DISPERSION ORIFICES
APPROX. 18 PER SQ. FT.

5/8'' DIA. CONTROL ORIFICES
APPROX. 2 PER SQ. FT.

COMBINATION OF LENGTHS
TO SUIT INSTALLATION

11''

10''

5''

COMPENSATING LATERAL (SECONDARY)
26.5 SQ. IN.

FEEDER LATERAL (PRIMARY)
30.5 SQ. IN.

Fig. 10–9. Leopold filter bottom. (Courtesy of F. B. Leopold Co.)

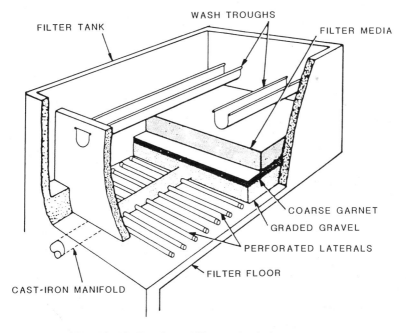

FILTER TANK

WASH TROUGHS

FILTER MEDIA

COARSE GARNET

GRADED GRAVEL

PERFORATED LATERALS

FILTER FLOOR

CAST-IRON MANIFOLD

Fig. 10–10. Pipe lateral filter underdrain system.

Fig. 10–11. Wheeler filter bottom. (Courtesy of Roberts Filter Manufacturing Co.)

dissipate the velocity head from the orifice with minimum disturbance of the gravel above. In some recent systems, inverted plastic cones with exterior fluted ridges have been used in place of the porcelain balls. Steel forms for the poured-in-place Wheeler bottom can be rented from either BIF, Providence, Rhode Island, or Roberts Filter Co., Darby, Pennsylvania.

The Leopold bottom consists of vitrified clay blocks each about 2 feet long by 11 inches wide by 10 inches high ($0.61 \times 0.28 \times 0.25$ m). The bottom half of the block contains two feeder channels about 4 inches $\times$ 4 inches (0.10×0.10 m) with two $\frac{5}{8}$-inch (15.8-mm)-diameter orifices per square foot (0.093 m³) that feed water into the two dispersion laterals about $3\frac{1}{4} \times 4$ inches (83×102 mm) above with a headloss of about 1.45 feet (0.44 m) at a flow of 15 gpm/sq ft (36.6 m/h). In the top of the block there are approximately forty-five $\frac{5}{32}$-inch (4-mm)-diameter orifices per square foot (0.093 m²) for further distribution of the wash water and for dissipation of the velocity head from the lower orifices. These blocks and installation services are available from F. B. Leopold Co., Zelienople, Pennsylvania, or Warminster Fiberglass Co., Southampton, Pennsylvania.

The perforated pipe lateral system uses a main header with several pipe laterals on both sides. The laterals have perforations on the underside so that the velocity of the jets from the perforations during backwash is dissipated against the filter floor and in the surrounding gravel. Piping materials most commonly used are steel, asbestos cement, or PVC.

Usually, orifice diameters are ¼ to ½ inch (6.4 to 12.8 mm) with spacings of 3 to 8 inches (76.2 to 203 mm). Fair and Geyer give the following guides to pipe lateral underdrain design:[19]

1. Ratio of area of orifice to area of bed served, 0.15 to 0.001.
2. Ratio of area of lateral to area of orifices served, 2:1 to 4:1.
3. Ratio of area of main to area of laterals served, 1.5:1 to 3:1.
4. Diameter of orifices, ¼ to ¾ inch (6.4 to 19.1 mm).
5. Spacing of orifices, 3 to 12 inches (76.2 to 305 mm) on centers.
6. Spacing of laterals, about the same as spacing of orifices.

Several underdrain systems have been developed with the goal of eliminating the need for the gravel support bed. These systems employ strainer-like nozzles or porous plates. Plastic or steel nozzles of various shapes are available from several manufacturers: the Eimco Corporation manufactures a plastic, conical nozzle with a screen supported between two layers of plastic; Degremont Company makes plastic nozzles that have finely slotted sides; the Edward Johnson Company manufacturers a stainless steel nozzle consisting of 4 ¾-inch (121-mm)-diameter well screen pipe with a cap welded on one end; Walker Process offers a nozzle made of thermoplastic resin having 0.25-mm-aperture vertical slots, as shown in Fig. 10–12. The nozzles can be mounted in hollow, vitrified clay blocks laid to form the filter bottom. The 9-inch (229-mm)-high hollow-core blocks are grouted in place and provide a system of ducts for filtered water flow as well as for backwash flow. The nozzles are designed with built-in orifices to meter the backwash flow evenly over the filter area. Care must be taken in selection of these nozzles to ensure that the finest filter media will not pass through the nozzle openings. The nozzles may also be mounted, either in a false bottom or in a pipe lateral.

A disadvantage of the plastic nozzles is that they are fairly fragile and can be easily broken while being installed or when the medium is placed. A potential operating problem in use of the nozzles is plugging from the inside with material found in the backwash water or with rust or other particulate matter from the filter backwash piping. Plugging frequently occurs on the first application of backwash due to the failure to remove construction debris from the system. Some nozzles made from a combination of plastic

Fig. 10–12. Filter underdrain nozzles. (Courtesy of Walker Process Corp.)

and steel have cracked, and the underdrain systems have failed because of differential contraction with temperature changes.

Porous plates made of aluminum oxide, such as those manufactured by the Carborundum Company, are also available to eliminate the need for the gravel layer. The aluminum oxide plates are usually mounted on steel or concrete frames that support the plate on at least two edges, or by studs projecting from the filter floor fastened to the corners of the plates. Mastic or caulking compounds are used to seal the joints. The disadvantages of the aluminum oxide plates are that they are brittle and easily broken during installation; also, the sealing of joints between plates may be difficult, and the plates can become plugged from either side. If plugged plates are not promptly cleaned, structural failure may occur due to decreasing differential pressures across the plates during backwash. However, the plates offer the advantages of good flow distribution and minimum filter height because the media can be placed directly on them. They may not be suitable for use in filtering water that deposits calcium carbonate or in waters high in iron or manganese, because of plugging problems under these conditions.

The decision to use a nozzle or a porous plate underdrain system should be based on personal satisfactory experience of the designer in similar applications. In general, currently available direct support systems have not yet proved to be as satisfactory as gravel support beds for fine media.

Filter Gravel

A graded gravel layer, usually 14 to 18 inches (0.36 to 0.46 m) deep, is placed over the pipe lateral system to prevent the filter media from entering

Table 10-1. Typical Gravel Bed for Pipe Underdrain System.*

		NUMBER OF LAYERS		
DESCRIPTION	BOTTOM	2	3	TOP**
Depth of layer, inches (mm)	†	3 (76)	3 (76)	4 (122)
Square mesh screen opening, inches				
Passing		¾	½	¼
Retained	¾	½	¼	⅛

* Reference 46.
** Plus coarse garnet.
† Bottom layer should extend to a point 4 inches (122 mm) above the highest outlet of wash water.

the lateral orifices and to aid in distribution of the backwash flow. A typical gravel design is shown in Table 10-1. A weakness of the gravel–pipe lateral system has been the tendency for the gravel eventually to intermix with the filter media. Such gravel "upsets" are caused by localized high velocity during backwash, introduction of air into the backwash system, or use of excessive backwash flow rates. The gravel layer can be stabilized by using 3 inches (76 mm) of garnet or ilmenite as the top layer of the gravel bed. This coarse, very heavy material will not fluidize during backwash and provides excellent stabilization for the gravel. It also prevents the fine garnet or ilmenite used in a mixed tri-media filter from mixing with the gravel support bed. The remaining major disadvantage of the gravel–pipe lateral system is the vertical space required for the gravel bed.

Gravel layers are also used with several of the commercially available underdrain systems, such as the Leopold bottoms (see Fig. 10–13) and Wheeler bottoms. Gravel depths and gradations vary for these underdrain systems. For example, Leopold recommends the gradation listed in Table 10–2, while the gravel for use with Wheeler bottoms is shown in Table 10–3.

Filtration

Gravel should consist of hard, rounded silica stones with an average specific gravity of not less than 2.5. Not more than 1 percent by weight of the material should have a specific gravity of 2.25 or less. Not more than 2 percent by weight of the gravel should consist of thin, flat, or elongated pieces (pieces in which the largest dimension exceeds five times the smallest dimension). The gravel should be free from shale, mica, clay, sand, loam, and organic impurities of any kind. The porosity of gravel in any layer should not be less than 35 percent or more than 45 percent. Gravel should be screened to proper size and uniformly graded within each layer. Not more than 8 percent by weight of any layer should be coarser or finer than the specified limit.

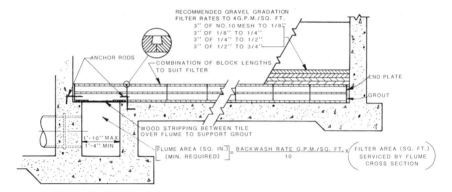

Fig. 10–13. Typical installation of Leopold filter bottom. (Courtesy of F. B. Leopold Co.)

Coarse Garnet

It is recommended that a 3-inch (76-mm) layer of high-density gravel (garnet or ilmenite) be used between the gravel bed and the fine media. This coarse, dense layer prevents disruption of the gravel. The specific gravity of the material should not be less than 4.2. The garnet or ilmenite particles in the bottom 1½-inch (38-mm) layer should be $\frac{3}{16}$ inch (4.8 mm) by 10 mesh, and in the top 1½-inch (38-mm) layer should be 1 mm in diameter. It is important to note that when the 3-inch (76-mm) coarse garnet layer is not used, another 3-inch (76-mm) layer of $\frac{3}{16}$ inch (4.8 mm) by 10 mesh silica gravel should be added as the top layer in Tables 10–1, 10–2, and 10–3. Otherwise there is apt to be migration of fine media down into the gravel supporting bed.

Table 10–2. Gravel Size and Layer Thickness for Use with Clay Tile Bottoms.

GRAVEL LAYER	LAYER THICKNESS		SIZE LIMIT	
	INCHES	(MM)	INCHES	(MM)
Bottom	2	(61)	¾ × ½	(20 × 12.7)
Second	2	(61)	½ × ¼	(12.7 × 6.4)
Top*	2	(61)	¼ × ⅛	(6.4 × 3.2)

* Plus coarse garnet.

Table 10–3. Gravel Size and Layer Thickness for Use with Wheeler Bottoms.

GRAVEL LAYER	LAYER THICKNESS, INCHES (MM)	SIZE LIMIT, INCHES (MM)
Bottom	As required to cover the underdrain systems	$1 \times 1\frac{1}{4}$ (25.4 × 31.8)
Second	3 (76)	$\frac{5}{8} \times 1$ (15.9 × 25.4)
Third	3 (76)	$\frac{3}{8} \times \frac{5}{8}$ (9.5 × 15.9)
Top*	3 (76)	$\frac{3}{16} \times \frac{3}{8}$ (4.8 × 9.5)

* Plus coarse garnet.

Gravel Placement

The filter tanks must be thoroughly cleaned before gravel is placed and kept clean throughout the placing operation. Gravel made dirty in any way should be removed and replaced with clean gravel. The bottom layer should be carefully placed by hand to avoid movement of the underdrain system and to assure free passage of water from the orifices. Each gravel layer should be completed before the next layer above is started. Workmen should not stand or walk directly on material less than ½-inch in diameter, but rather should place boards to be used as walkways. If different layers of gravel are inadvertently mixed, the mixed gravel must be removed and replaced with new material. The top of each layer should be made perfectly level by matching to a water surface at the proper level in the filter box.

Mixed Media

There is no one mixed-media design that will be optimum for all water filtration problems. Conley and Hsiung have presented techniques designed to optimize the media selection for any given filtration application.[7] Their work clearly indicates the marked effects that the quantity and quality of floc to be removed can have on media selection, as shown in Table 10–4. Pilot tests of various media designs can be more than justified by improved plant performance in most cases.

Certainly, use of tri-media does not in itself ensure superior performance, as illustrated by the experiences of Oakley and Cripps.[47] Their using the coal, sand, and garnet materials readily available to them resulted in a bed that did not have significant advantages over other filter types. Oakley reports the tri-media bed was made up of 8 inches (203 mm) of 0.7 to 0.8 mm garnet, 8 inches (203 mm) of 1.2 to 1.4 mm sand, and 8 inches (203 mm) of 1.4 to 2.4 mm coal.[48] The authors' experience indicates that this bed was too shallow and too coarse, and that a better media selection for the particular application would have been 3 inches (76 mm) of 0.4 to 0.8 mm

Table 10–4. Illustrations of Varying Media Design for Various Types of Floc Removal.

	GARNET		SILICA SAND		COAL	
TYPE OF APPLICATION	SIZE	DEPTH, INCHES (MM)	SIZE	DEPTH, INCHES (MM)	SIZE	DEPTH INCHES (MM)
Very heavy loading of fragile floc	−40 + 80*	8 (203)	−20 + 40	12 (305)	−10 + 20	22 (559)
Moderate loading of very strong floc	−20 + 40	3 (76)	−10 + 20	12 (305)	−10 + 16	15 (381)
Moderate loading of fragile floc	−20 + 80	3 (76)	−20 + 40	9 (229)	−10 + 20	8 (205)

*−40 + 80 = passing No. 4 and retained on No. 80 U.S. sieves.

garnet, 9 inches (229 mm) of 0.6 to 0.8 mm sand, and 24 inches (610 mm) of 1 to 2 mm coal.

A key factor in constructing a satisfactory mixed-media bed is careful control of the size distribution of each component medium. Rarely is the size distribution of commercially available materials adequate for construction of a good mixed-media filter. A common problem is failure to remove excessive amounts of fine materials. These fines can be removed by placing a medium in the filter, backwashing it, draining the filter, and skimming the upper surface. The procedure is repeated until field sieve analyses indicate that an adequate particle size distribution has been obtained. The second medium is added and the procedure repeated. The third medium is then added and the entire procedure repeated. Sometimes, 20 to 30 percent of the materials may have to be skimmed and discarded to achieve the proper particle size distribution.

Dual Media

As compared to mixed media, the dual-media (coal–sand) filter has less resistance to breakthrough because it is made up of coarser particles and has less total surface area of particles. The mixed-media filter is capable of producing lower finished water turbidities than the dual-media. These differences are greater and become more pronounced when the difficulty of the filtration application increases. In polishing highly pretreated waters, the differences are not so great, and some designers continue to use coal–sand media. However, it is doubtful that anyone who has observed the two filter types running side by side on the same water would select coal–sand media.

Typically, coal–sand filters consist of a coarse layer of coal about 18 inches (0.46 m) deep above a fine layer of sand about 8 inches (0.20 m) thick. Some mixing of coal and sand at their interface is desirable to avoid excessive accumulation of floc, which occurs at this point in beds graded to produce well-defined layers of sand and coal. Such intermixing reduces the void size in the lower coal layer, forcing it to remove floc that otherwise might pass through the coal layer. Typical gradations of sand and coal for use in dual-media filters are given in Table 10–5.

"Capping" Sand Filters with Coal

One inexpensive expedient to improve rapid sand filter performance is to remove about 6 inches (0.15 m) of sand from a bed and replace it with 6 inches (0.15 m) of anthracite coal. This produces a layered-type bed that has only part of the advantages of a dual-media bed that has been designed for some intermixing at the interface, but is superior in performance to a single medium. The results of this modification of sand filters at Kenosha,

Table 10-5. Typical Coal and Sand Distribution by Sieve Size in Dual-Media Bed.

COAL DISTRIBUTION BY SIEVE SIZE	
U.S. SIEVE NO.	PERCENT PASSING SIEVE
4	99–100
6	95–100
14	60–100
16	30–100
18	0–50
20	0–5

SAND DISTRIBUTION BY SIEVE SIZE	
U.S. SIEVE NO.	PERCENT PASSING SIEVE
20	96–100
30	70–90
40	0–10
50	0–5

Sheboygan, and Racine, Wisconsin, and at Evanston and Waukegan, Illinois, have been presented.[49] At Sheboygan, comparisons were made between coal-capped sand filters and sand alone under various raw water conditions, including comparisons made during the algae season. The capped filters were operated at 3 gpm/sq ft (7.32 m/h) and the sand at 2 gpm/sq ft (4.88 m/h). It was found that "the more adverse the applied water conditions, relative to algae and floc, the more dramatic are the results obtained with anthracite capped filter runs." Good water conditions will give 2 to 1. The worst water conditions may give 10 to 1 improvement in filter runs. Through the use of capped filters, short filter runs can be eliminated. At Kenosha, residual aluminum tests showed less alum passing through the capped filter than through the sand-only filter, except in cases of breakthrough. Bacteriologically, all tests showed safe water at all times. At Racine, it was found that the capped filters removed "more amorphous matter than sand alone." At Evanston, capped filter operation at 4 gpm/sq ft (9.76 m/h) used less washwater than sand filters.

These benefits—longer filter runs, reduced use of washwater, and better filter effluent quality using capped filters—could without question be further extended by replacing all of the fine media with mixed media. Then, in addition, the mixed media would eliminate the breakthrough problem mentioned, and would add greatly to the safety and reliability of filter operation. Finished water clarity could be substantially improved through use of a polymer added to the mixed media and influent as a filter aid.

Rapid Sand Filters

Virtually all new water purification plants are now being designed to use mixed-media or dual-media filters, and many existing rapid sand filter plants are being converted to use these materials. However, some single-medium (sand or coal) filters are being installed in the expansion of existing plants to match existing facilities and to avoid the cost of converting all of the old units.

For practical purposes, the size of sand grains is determined on a weight basis from sieve analysis, even though the resulting diameters may be 10 to 15 percent less than those determined by the count and weigh method, which should be used for more accurate results. The majority of rapid sand filters in use today contain sand with an effective size of 0.35 to 0.50 mm, although some have sand with an effective size as high as 0.70 mm. The uniformity coefficient is usually not less than 1.3 or more than 1.7.

Sand passing a 50-mesh (U.S. series) sieve is generally too fine for use in a rapid sand filter, as it stratifies at the surface and shortens filter runs by sealing off the top quite rapidly. Sand retained on a 16-mesh sieve is too coarse to be useful in filtration within the depths normally used in filter plants. Therefore, filter sand usually ranges in size from that passing a 16-mesh to that retained on a 50-mesh sieve. A typical sand specification is given in Table 10–6.

Sand filters should have a hydrochloric acid solubility of less than 5 percent when tested in accordance with AWWA Standard B11–53. Sand should have a specific gravity of not less than 2.5, and be clean and well graded. After placement in the filter, sand should be backwashed three times at not less than 30 percent expansion, and then the top ¼ inch (6.4 mm) of very fine material should be carefully scraped off and discarded.

Crushed anthracite coal may be used in lieu of sand as a fine granular filter medium. It has a specific gravity of 1.5, compared to 2.65 for silica

Table 10–6. Suggested Size Specifications for Filter Sand.

SIEVE NO.			RETAINED ON SIEVE, PERCENT	
SERIES TYLER $\sqrt{2}$	U.S. SERIES	OPENING, MM	MINIMUM	MAXIMUM
65	70	0.208	0	1
48	50	0.295	0	9
35	40	0.417	40	60
28	30	0.589	40	60
20	20	0.833	0	9
14	16	1.168	0	1

sand and crushed quartz. Effective sizes of coal up to 0.70 mm are used in filters. Because of the lower specific gravity of coal, only about half the backwashing velocity is needed for equal expansion (not necessarily equal washing) compared to sand. An anthracite bed of the same effective size as a sand bed has a greater bed porosity than the sand bed. Anthracite coal filter media should be clean and free of long, thin, or scaly pieces, with a hardness of 2.0 to 3.55 mm on the Moh scale and a specific gravity not less than 1.5.

Filter Control System

There are two major categories of filter control systems: (1) constant rate and (2) declining rate. The most widely used system has been the constant rate. Such filters are equipped with a means of controlling the rate of flow through each bed. By controlling either the influent or the effluent flow it is possible to divide the flow among the filters, to limit the maximum flow through any unit, and to prevent sudden flow surges. Rapid changes in rate of flow through a filter are undesirable because they literally shake the floc particles down through the bed and into the effluent.

Filter rate-of-flow controllers consist of: (1) a flow-measuring device such as a Venturi tube or a Dall tube, (2) a throttling valve such as a rubber seated butterfly valve with hydraulic or pneumatic actuator, and (3) a valve stop or positioning device. Auxiliary or overriding control of the throttling valve may be provided to maintain a predetermined minimum water level above the filter media and a maximum water level in the receiving clearwell. The flow-measuring device must be provided with straight pipe runs on either side if it is to provide the necessary accuracy. Controllers can be equipped with gauges or charts to indicate or to record rate of flow through each filter, and to summarize total flow.

Rate controllers may be set to operate at a constant fixed rate, which has several advantages from the standpoint of water quality. This means that flow through the plant must be nearly constant, or that filter units must be thrown in or out of service to meet fluctuating demands. Another method of operation is to use a variable-rate controller with a fixed-maximum setting. This permits operation of all filters at the minimum possible rate all of the time, which also has advantages.

The configuration of declining rate filters is similar to constant rate filters except that all filters usually discharge to a common effluent pressure header without the use of a flow control valve. Influent levels are essentially the same in all filters, and are nearly constant. Because the filters operate at virtually constant head, the throughput rates are higher at the start of the filter runs and gradually decline as the filter headloss increases during the filter run. Declining rate filters typically are backwashed on a selected time interval. Hudson has presented design guidance for declining filter rates.[50] The declining rate systems are mechanically simpler because they eliminate the need for mechanical flow control devices. Their mechanical simplicity must be weighed against the lack of control of filtration rate at the start of each filter run. Depending upon the ratio of dirty filters to clean filters at the start of a run, the rate through the just-backwashed filter could become very high and increase the chances of turbidity breakthrough.

Filter Backwashing

During the service cycle of filter operation, particulate matter removed from the applied water accumulates on the surface of the grains of fine media and in the pore spaces between grains. With continued operation of a filter, the materials removed from the water and stored within the bed reduce the porosity of the bed. This has two effects on filter operation: it increases the headloss through the filter, and it increases the shearing stresses on the accumulated floc. Eventually the total hydraulic headloss may approach or equal the head necessary to provide the desired flow rate through the filter, or there may be a leakage or breakthrough of floc particles into the filter effluent. Just before either of these outcomes can occur, the filter should be removed from service for cleaning. In the old slow-sand filters the arrangement of sand particles is fine to coarse in the direction of filtration (down); most of the impurities removed from the water collect on the top surface of the bed, which can be cleaned by mechanical scraping and removal of about ½ inch (12.7 mm) of sand and floc. In rapid sand filters, there is somewhat deeper penetration of particles into the bed because of the coarser media used and the higher flow rates employed. However, most of the materials are stored in the top few inches of a rapid sand filter bed. In dual-media and mixed-media beds, floc is stored throughout the bed depth to within a few inches of the bottom of the fine media.

Rapid sand, dual-media, and mixed-media filters are cleaned by hydraulic backwashing (upflow) with potable water. Thorough cleaning of the bed makes it advisable in the case of single-medium filters and mandatory in the case of dual- or mixed-media filters to use auxiliary scour or so-called surface wash devices before or during the backwash cycle. Backwash flow rates of 15 to 20 gpm/sq ft (36.6 to 48.8 m/h) should be provided. A 20 to 50 percent expansion of the filter bed is usually adequate to suspend the bottom grains. The optimum rate of washwater application is a direct function of water temperature, as expansion of the bed varies inversely with viscosity of the washwater. For example, a backwash rate of 18 gpm/sq ft (43.9 m/h) at 68°F (20°C) equates to 15.7 gpm/sq ft (38.3 m/h) at 41°F (5°C) and 20 gpm/sq ft (48.8 m/h) at 95°F (35°C). The time required for complete washing varies from 3 to 15 minutes.

Following the washing process, water should be filtered to waste until the turbidity drops to an acceptable value. Filter-to-waste outlets should be through an air-gap-to-waste drain, which may require from 2 to 20 minutes, depending on pretreatment and type of filter. This practice was discontinued for many years, but modern recording turbidimeters have shown that this operation is valuable in the production of a high-quality water. Operating the washed filter at a slow rate at the start of a filter run may accomplish the same purpose. A recording turbidimeter for continuous monitoring of the effluent from each individual filter unit is of great value in controlling this operation at the start of a run, as well as in predicting or detecting filter breakthrough at the end of a run.

As backwashing begins, the sand grains do not move apart quickly and uniformly throughout the bed. Time is required for the sand to equilibrate at its expanded spacing in the upward flow of washwater. If the backwash is turned on suddenly, it lifts the sand bed bodily above the gravel layer, forming an open space between the sand and gravel. The sand bed then breaks at one or more points, causing sand boils and subsequent upsetting of the supporting gravel layers, so that the gravel section must be rebuilt. It is essential that the backwash valve open slowly.

The time from start to full backwash flow should be at least 30 seconds and perhaps longer, and should be restricted by devices built into the plant. This is frequently done by means of an automatically regulated master wash valve, controlled hydraulically or electrically and designed so that it cannot open too fast. Alternatively, a speed controller could be installed on the operator of each washwater valve.

Filters can be seriously damaged by slugs of air introduced during filter backwashing. The supporting gravel can be overturned and mixed with the fine media, which requires removal and replacement of all media for proper repair. Air can be unintentionally introduced to the bottom of the filter in a number of ways. If a vertical pump is used for the backwash supply, air may collect in the vertical pump column between backwashings. The air can be eliminated without harm by starting the pump against a closed discharge valve and bleeding the air out from behind the valve through a pressure air release valve. The pressure air release valve must have sufficient capacity to discharge the accumulated air in a few seconds.

Also, air or dissolved oxygen, released from the water on standing and warming in the washwater supply piping, may accumulate at high points in the piping and be swept into the filter underdrains by the inrushing washwater. This can be avoided by placing a pressure air release valve at the high point in the line, and providing a ½-inch (12.7-mm) pressure water connection to the washwater supply header to keep the line full of water and to expel the air.

The entry for washwater into the filter bottom must be designed to dissipate the velocity head of the washwater in such a manner that uniform distribution of washwater is obtained. Lack of attention to this important design factor has often led to difficult and expensive alterations and corrective repairs to filters.

Horizontal centrifugal pumps may be used to supply water from the treated water storage reservoir for filter backwashing, provided they are located where positive suction head is available, or are provided with an adequate priming system. Otherwise, a vertical pump unit may be suspended in the clearwell. The pumps should be installed with an adequate air release valve, a nonslam check valve, and a throttling valve in the discharge. Single washwater supply pumps are often installed, but consideration should be given to provisions for standby service, including multiple pump units and/or standby generators.

Washwater may also be supplied by gravity flow from a storage tank located above the top of the filter boxes. Washwater supply tanks usually have a minimum capacity equal to a 7-minute wash for one filter unit, but may be larger. The bottom of the tank must be high enough above the filter wash troughs to supply water at the rate required for backwashing as determined by a hydraulic analysis of the washwater system. This distance is usually at least 10 feet (3.05 m), but more often is 25 feet (7.6 m) or greater. Washwater tanks should be equipped with an overflow line, and a vent for release and admission of air above the high-water level. Washwater tanks are often filled by means of a small electric driven pump equipped with a high-water level cutoff. They may also be filled from the high-service pump discharge line through an altitude or float valve, but this is usually wasteful of electrical energy. Washwater tanks may be constructed of steel or concrete. Provisions should be made for shutdown of the tanks for maintenance and repair.

The washwater supply line must be equipped with provisions for accurately (within ± 5 percent) measuring and regulating the rate of washwater flow. The rate of flow indicator should be visible to the operator at the point at which the rate of flow is regulated.

The use of a high pressure, above 15 psi (103.4 kPa), source of filter backwash water through a pressure-reducing valve is not advised. Numerous failures of systems using pressure-reducing valves have so thoroughly upset and mixed the supporting gravel and fine media that these materials have had to be completely removed from the filter and replaced with new media.

In some cases, pollution control regulations prohibit the once common practice of discharging backwash wastewaters directly to a stream. In these cases, the backwash wastewater must be reprocessed. The rate of backwash flow, if it is returned directly to an upstream clarifier, is usually large enough in relation to the design flow through the clarifier to cause a hydraulic overload and upset of the clarifier. In this case, the backwash wastes should be collected in a storage tank and recycled at a controlled rate. The volume of backwash wastewater is typically 2 to 5 percent of the plant throughput, and plant components must be sized to handle this recycled flow.

Washwater Troughs

To equalize the head on the underdrainage system during backwashing of the filter, and thus to aid in uniform distribution of the washwater, a system of weirs and troughs is ordinarily used at the top of the filter to collect the backwash water after it emerges from the sand and to conduct it to the washwater gullet or drain. The bottom of the trough should be above the top of the expanded sand to prevent possible loss of sand during backwashing. The clear horizontal distance between troughs should be one and one-half to two times the distance between the top of the trough and the top of the filter media during the filter cycle. The cross section of the trough depends somewhat on the material of construction, and the semicircular bottom is used frequently for this measurement. Troughs may be made of concrete, fiberglass reinforced plastic, or other structurally adequate and corrosion-resistant materials. The dimensions of a filter trough may be determined by use of the following equation:

$$Q = 2.49 \ bh^{3/2} \tag{10-2}$$

where:

Q = the rate of discharge, cu ft/sec (m³/h)
b = width of trough, ft (m)
h = maximum water depth in trough, ft (m)
2.49 = constant for English units (4951.7 is the constant for metric units shown)

Some freeboard should be allowed to prevent flooding of washwater troughs and uneven distribution of washwater.

The most popular material for construction of washwater troughs is reinforced concrete. Ordinarily concrete troughs are precast on the job in an inverted position, and then placed in position after forms have been removed. Washwater troughs may also be factory-fabricated from steel, fiberglass, or transite.

If the filter troughs are to serve their purpose properly, the weir edges must be honed to an absolutely smooth and perfectly level edge after placement. This leveling can be performed by matching the finished edges of all troughs in a single filter to a still water surface at the desired overflow elevation.

Filter Agitators

Present practice in the United States leans heavily toward installation of the essential, but misnamed, "surface wash" on all new filters. "Auxiliary scour" or "filter agitation" better describes the function of this device, as it aids in cleaning much more than the filter surface. Rotary surface washers are the most common, but fixed jets also are used successfully.

Adequate surface wash improves filter cleaning and prevents mudball formation and filter cracking. Conventional rotary surface wash equipment consists of arms on a fixed swivel supported from the washwater troughs about 2 inches (50.8 mm) above the surface of the unexpanded filter media. In dual- or mixed-media filters, dual arm devices have also been used to provide better cleaning at the coal–sand interface. In these systems, a second set of arms is located about 6 inches (152 mm) above the coal–sand interface in a dual-media filter or about 18 inches (457 mm) above the top of the gravel in a mixed-media filter. The arms are fitted with a series of nozzles, and revolve because of the water jet reaction. Water pressures of 40 to 100 psi (275.6 to 689 kPa) are required for the operation of the rotary surface wash, depending upon the diameter of the arms. The volume required is about 0.5 gpm/sq ft (1.22 m/h). Surface washers are usually started about 1 minute in advance of the normal backwash flow, and turned off a minute or so before the end of the backwash period. It is recommended that the nozzles be equipped with rubber caps which act to prevent entry of fine filter media and plugging of the nozzles.

Another concept in rotary agitator design has been developed by Roberts Filter Manufacturing Co., Darby, Pennsylvania (see Fig. 10–14). This agitator utilizes the aspirator effect of the agitator water supply to introduce a controlled amount of air into the agitator arms. The air–water jet spray provides improved scour and separation of accumulated floc. The agitators also include a rotary valve that cuts off the water supply to the end sprays during the instant that the arms are closest to the filter walls. This reduces erosion of the concrete at this point, which is sometimes a problem with conventional rotary surface washers.

Baylis has designed a fixed-jet surface wash system consisting of a grid of distributing pipes extending to within a couple of inches of the top surface of the bed. Nozzles with five ¼-inch (6.4-mm) holes are spaced at about 24- to 30-inch (0.61- to 0.76-m) centers each way. The required flow is about 2 gpm/sq ft (4.88 m/h) at a head of 20 to 60 feet (6.1 to 18.3 m).

Surface wash piping is a direct cross connection between filtered and unfiltered water. The normal practice is to bring the surface wash header into the filter over the top of the filter box and to fit it with a vacuum breaker and a check valve at the high point to prevent backsiphoning. A single vacuum breaker on the surface wash header will suffice. An alternate is to use settled

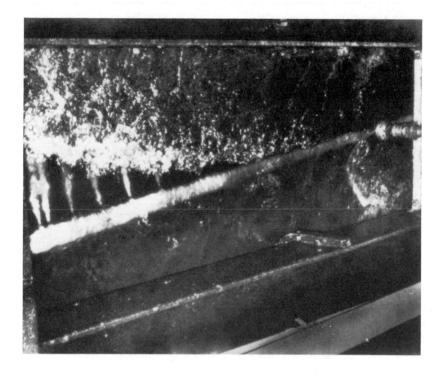

a)

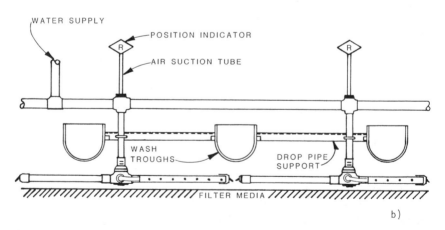

b)

Fig. 10–14. The Roberts AW rotary agitator.[51] (a) Here the sweep arms are held stationary to illustrate the violent action of the sprays using an air/water mixture. At this point the control ports in the rotary joint are still closed as the end nozzle passes the wall of the filter. (b) Elevation drawing of a typical installation showing relationship of the stator supports to wash troughs. (Courtesy of Roberts Filter Manufacturing Co.)

water through a separate surface wash pump. Valve positions for various cycles of filter operation are shown in Table 10–7.

An alternative to mechanical surface wash employment is the use of air–water backwash techniques, in which air is injected through the filter underdrain system to break up mudballs as it rises through the filter. These techniques have been used fairly extensively in Europe, but lost favor in the United States about 30 years ago because of problems encountered with upsetting the beds and a loss of media. Recent improvements in underdrain systems have led to renewed interest in them. Because of the tendency of the air to float coal out of a filter, the backwash water and the air cannot be applied at the same time with beds utilizing coal. The air–water technique applied to these beds usually consists of the following steps:

- Stop the influent and lower the water level to within 2 to 6 inches (51 to 152 mm) of the filter surface.
- Apply air alone at a rate 2 to 5 cfm/sq ft (4.88 to 12.2 m/h) for 3 to 10 minutes.
- Apply a small amount of backwash water [about 2 gpm/sq ft (4.88 m/h)], with the air continuing until the water is within 8 to 12 inches (0.20 to 0.31 m) of the washwater trough (the air is then shut off).
- Continue water backwash at 8 to 10 gpm/sq ft (19.52 to 24.4 m/h) until the filter is clean.

An important consideration is that a mixed-media bed will require 1 to 2 minutes of wash at 12 to 15 gpm/sq ft (29.3 to 36.6 m/h) at the end of the wash period to classify the media properly.

It has been the authors' experience that although lower backwash rates are used with the air–water techniques, the amount of backwash water required to achieve the same degree of cleaning as is obtained with a surface wash in conjunction with backwash is not significantly different. The air–water techniques increase the complexity of the backwash cycle, lengthen the filter downtime for backwashing, increase the risk of media loss, and

Table 10–7. Valve Positions During Various Treatment Operations.

	VALVE POSITION		FILTERING TO
VALVE	FILTERING	BACKWASHING	WASTE
Influent	Open	Closed	Open
Effluent	Open	Closed	Closed
Filter to waste	Closed	Closed	Open
Washwater supply	Closed	Open	Closed
Washwater drain	Closed	Open	Closed
Surface wash	Closed	Open	Closed

offer little, if any, advantage in the quantity of backwash water required to achieve a given degree of cleaning.

Loss of Head

The loss of head through a filter provides valuable information about the condition of the bed and its proper operation. An increase in the initial loss of head for successive runs over a period of time may indicate clogging of the underdrains or gravel, the need for auxiliary scour, or insufficient washing of the beds. The rate of headloss increase during a run yields considerable information concerning the efficiency of both the pretreatment and the filtration operation.

The determination of headloss through a filter is a very simple matter, involving only the measurement of the relative water levels on either side of the filter. The simplest form of headloss device for gravity filters is made up of two transparent tubes installed side by side in the pipe gallery with a gauge board between, graduated in feet. One tube is connected to the filter or filter influent line, the other to the effluent line. The headloss is the observed difference in water levels. More sophisticated methods for measuring the difference in water levels include float-actuated and differential-pressure-cell-actuated indicating and recording devices. The headloss may be indicated and recorded at some remote point if desired. Headloss equipment may be used in connection with control systems for automatic backwashing of filters, as one means of initiating the backwash cycle when the headloss reaches some preset maximum value.

Use of Polymers as Filtration Aids

Polymers are high-molecular-weight, water-soluble compounds that can be used as primary coagulants, settling aids, or filtration aids. They may be cationic, anionic, or nonionic in charge. Generally, the doses required as a filtration aid are less than 0.1 mg/l. Used as a filtration aid, polymer is added to increase the strength of the chemical floc and to control the depth of penetration of floc into the filter. For maximum effectiveness as a filtration aid, the polymer should be added directly to the filter influent and not in an upstream settling basin or flocculator. However, if polymers are used upstream as settling aids, it may not be necessary to add any additional polymer as a filtration aid.

Figure 10–15 illustrates the effects of polymers as filter aids. The conditions represented in part A of the figure illustrate the results of a fragile floc shearing and then penetrating the filter, causing a premature termination of its run due to breakthrough of excessively high effluent turbidity. If the polymer dose is too high (part B), the floc is too strong to permit penetration

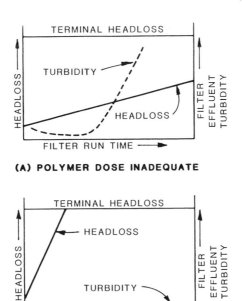

(A) POLYMER DOSE INADEQUATE

(B) POLYMER DOSE EXCESSIVE

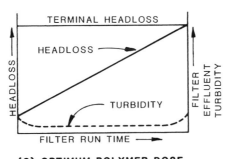

(C) OPTIMUM POLYMER DOSE

Fig. 10–15. Effects of polymers as filtration aids.

into the filter, causing a rapid buildup of headloss in the upper portion of the filter and premature termination due to excessive headloss. The optimum polymer dose will permit the terminal headloss to be reached simultaneously with the first sign of increasing filter effluent turbidity (part C).

Many polymers are delivered in a dry form. They are not easily dissolved, and special polymer mixing and feeding equipment is required. Typical equipment is shown in Figs. 10–16 and 10–17. Many polymers are biodegradable and cannot be stored in dilute solution for more than a few days without

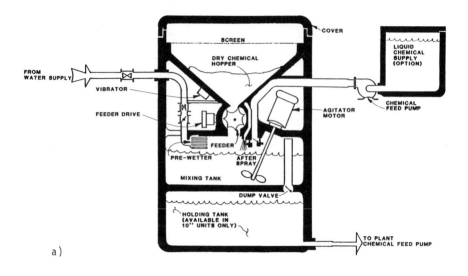

a)

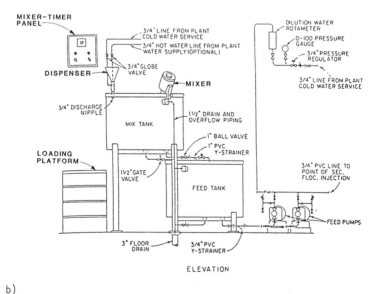

b)

Fig. 10–16. Typical polymer mixing and feed equipment. a) Automatic polymer system (Chemix ®). b) Manual polymer system. (Courtesy of Microfloc Products Group, Johnson Division, UOP)

Fig. 10-17. Typical polymer feed station. (Courtesy of Microfloc Products Group, Johnson Division, UOP)

suffering significant degradation and loss of strength. (Chapter 8 contains detailed information on feeding and handling systems for polymers.)

Monitoring Filter Effluent Turbidity

Inexpensive, reliable turbidimeters are available to monitor the quality of the filter effluent continuously. (See Chapter 7 for a description of turbidity monitoring systems.) Should the turbidity exceed the desired level, a signal from the turbidimeter can be used to sound an alarm or to initiate the back-wash program. Recording of the turbidimeter output provides a continuous record of filter performance. The importance of continuously monitoring and recording filter effluent turbidity has been summarized by the American Water Works Association:[1]

Operators accustomed to using the continuous-monitoring recording filtered-water turbidimeters quickly notice departures from normal. They recognize the effects of operations such as basin shutdowns, filtration-rate increases caused by defective control devices, taking filters out of service for washing, coagulant underfeeds or interruptions, improper alkali dosages, changes in rawwater character, changes in floc strength, etc. Initial and final filter breakthroughs also show up conspicuously. By seeing the effects of these phenomena, operators learn to anticipate, prevent, control, or correct them. This new knowledge and awareness helps to make feasible the operation of water treatment plants at increasingly higher rates.

Pilot Filters

A small filter receiving raw water dosed with the treatment chemicals is often operated in parallel with the plant filters. The pilot filter directly measures the filterability of the water under actual plant operating conditions. By fitting the pilot filter with a recording turbidimeter, chemical dosages can be adjusted to obtain the desired quality of filter effluent before water in the full-scale plant reaches the filters. (Detailed descriptions of pilot filters and their use are contained in Chapter 7.)

Control of Filter Operation

Until about 1960, most filter plants depended primarily upon pretreatment facilities to produce filter influent of low turbidity, and on low filter rates [3 gpm/sq ft (7.32 m/h) or less] to produce a satisfactory filter effluent. Composite or grab samples of filtered water were examined for turbidity or bacterial content. The results of these tests were used to adjust pretreatment and for record purposes, rather than for direct control of filter operation. Cotton plug filters were introduced by Baylis and used by others to evaluate filter performance based on long-term operation. Since 1960, three major developments have greatly increased the degree of control over the efficiency and reliability of filter operation:[1]

- The use of 0.01 to 0.05 mg/l of a polymer to control floc strength, depth of floc penetration into the bed, and effluent turbidity.
- The use of continuous recording turbidimeters having a sensitivity of 0.01 NTU.
- The use of a control bed or pilot filter to anticipate pretreatment requirements for optimum filter performance.

These three major advances allow a degree of control and direct adjustment of filter operation not previously attainable. It should be pointed out that the use of polymer as a filter aid is not suited to single-medium, fine-to-coarse filters, as they are sealed off quite rapidly at the surface by such application to the filter influent.

Filter runs are usually terminated when either the headloss reaches a predetermined value, the filter effluent turbidity exceeds the desired maximum, or a certain amount of time passes. Each of these events is adaptable to instrumentation that can be used to signal the need for backwashing or to trigger a fully automated backwash system. Automatic control of filter backwashing may be provided by an automatic sequencing circuit (step switch) which is interlocked so that the necessary prerequisites for each step are completed before the next step is begun. At the receipt of a backwash start signal, the following events will occur in the sequence listed in this illustrative program. Filter influent and effluent valves close. Any chemical feed to the filter being backwashed stops. Plant chemical feeds adjust to the new plant flow rate, to maintain proper chemical feed to the filters still in service. The waste valve starts to open. When the waste valve reaches the fully open position and actuates a limit switch, the surface wash pump starts, and the surface wash valves open. Surface wash flow to waste continues for a period of time adjustable up to 10 minutes.

At the end of the initial surface wash period, usually 1 to 2 minutes, the main backwash valve opens. The backwash and surface wash both continue for a period of time, usually 6 to 7 minutes, adjustable up to 30 minutes. The backwash flow rate is indicated on a controller and is controlled automatically to a manual set point. At the end of the combined wash periods, the surface wash valves close, and the surface wash pump stops. The backwash continues without surface wash for a time, usually 1 to 2 minutes, adjustable up to 30 minutes. At the completion of the backwash period, the backwash valve has closed, and the waste valve starts to close. When the latter valve has closed, influent and effluent waste valves open, and the bed filters to waste for a period of time, usually 3 to 7 minutes, adjustable up to 30 minutes. The backwash delay timer resets and begins a new timing cycle, adjustable up to 12 hours. The bed selector switch steps to the next filter. Chemical feed to the clean filter is reestablished. At the end of the filter-to-waste period, the effluent waste valve closes, and the effluent valve opens to restore the cleaned bed to normal filter service. Provision should be made for optional manual operation of all automatic features.

It may be desirable to set alarms for certain functions that affect filter operation, on a conveniently located annunciator panel. These alarm functions include high turbidity, high headloss, low plant flow, low backwash flow rate, and excessive length of backwash.

Filter Piping, Valves, and Conduits

Cast-iron pipe and fittings and coal-tar enamel-lined welded steel pipe and fittings are the most widely used materials for filter piping. The layout of filter piping must include consideration of the ease of valve removal for repair and easy access for maintenance. Flexible pipe joints should be provided at all structure walls to prevent pipeline breaks due to differential settlement. The use of steel pipe can reduce flexible joint requirements, and color coding of the filter piping is a valuable operating aid. The filter piping is usually designed for the flows and velocities shown in Table 10–8.

The rubber-seated, pneumatically actuated and operated butterfly valve has almost entirely replaced the hydraulically actuated and operated gate valves that were formerly used extensively as filter valves. Of the two types,

Table 10–8. Filter Piping Design Flows and Velocities.

DESCRIPTION	VELOCITY FT/SEC (M/S)		MAXIMUM FLOW, GPM/SQ FT (M/H) OF FILTER AREA
Influent	1–4	(0.305–1.22)	8–12 (19.5–29.3)
Effluent	3–6	(0.92–1.83)	8–12 (19.5–29.3)
Washwater supply	5–10	(1.52–3.05)	15–25 (36.6–61)
Backwash waste	3–8	(0.92–2.44)	15–25 (36.6–61)
Filter to waste	6–12	(1.83–3.66)	4–8 (9.8–19.5)

the butterfly valve is smaller, lighter, easier to install, and better for throttling services, and it can be installed and operated in any position. The valves should be factory-equipped with the desired valve stops, limit switches, and position indicators because field-mounting of these devices is often unsatisfactory.

Each filter unit, except split beds, should have six valves for its proper operation: influent, effluent, washwater supply, washwater drain, surface wash, and filter-to-waste. The positions of these valves during the three cycles of filter operation were given in Table 10–7. The filter influent should enter the filter box so that the velocity of the incoming water does not disrupt the surface of the fine media. This is often done by directing the influent stream against the gullet wall, thus dissipating the velocity head within the gullet. It can also be done by locating the influent pipe below the top of the filter troughs so that the water enters the filter through the troughs. A further precaution is to install an influent valve with throttling control for use in slowly refilling the beds. The filter-to-waste connection to the filter should have positive air gap protection against backsiphoning from the drain to the filter bottom. The filter-to-waste, effluent, and washwater supply lines usually are manifolded for common connection to the filter underdrain system.

In the design of pipe galleries, reinforced concrete flumes and box conduits and concrete-encased concrete pipe may be used for washwater drains or other service when located adjacent to the pipe gallery floor, but should not be installed overhead because of difficulties with cracks and leaks. Invariably, pipe galleries with overhead concrete conduits are drippy, damp, unsightly places with a humid atmosphere that discourages good housekeeping by making it difficult to maintain. Instead, pipe galleries should be provided with positive drainage, good ventilation, sufficient light, and dehumidification equipment (if required by the prevailing climate). Filter influent and effluent lines should be provided with sample taps. Figure 10–18 is a photograph of the filter pipe gallery at the Richland, Washington Water Treatment Plant.

Fig. 10–18. Filter pipe gallery at the Richland, Washington Water Treatment Plant.

Filter Problems and Their Solutions

Problems in filter operation and performance can arise from either poor design or poor operation. However, during the past decade tremendous advances in the engineering design of filters and filter controls and appurtenances have made water filtration an inherently stable, extremely efficient, and highly reliable unit treatment process. With proper design and good operation, all the problems of the past are easily solved.

Some potential filter problems are the following:

- Surface clogging and cracking.
- Short runs due to rapid increases in headloss.
- Short runs due to floc breakthrough and high effluent turbidity.
- Variations in effluent quality with changes in applied water flow rate or quality.
- Gravel displacement or mounding.
- Mudballs.
- Growth of filter grains, bed shrinkage, and media pulling away from sidewalls.
- Sand leakage.
- Loss of media.
- Negative head and air binding.
- Air leakage into the system.

Surface Clogging and Cracking. These conditions are usually caused by rapid accumulations of solids on the top surface of the fine media. This is not a problem in dual- or mixed-media filters because of the greater porosity of their top surface, compared to sand. Also, when a filter aid is used with dual- or mixed-media, the dosage can be reduced, as necessary (or eliminated), to allow particulates to penetrate deeper into the bed. In other words, regulation of the polymer dosage to the filter influent gives some control over the effective porosity of the filter, to accommodate changes in incoming floc characteristics.

Rapid Increases in Headloss. This is related to the problem just discussed. Dual- and mixed-media beds collect particulates throughout the depth of the bed, rather than mostly at the surface of the bed as with a sand or other surface-type filter, and are much less susceptible to this problem than the surface-type filters. Also the flexibility provided by use of a polymer as a filter aid allows control of the rate of headloss buildup by dosage changes.

Floc Breakthrough. Floc breakthrough can be avoided by using mixed-media filter units. As mentioned earlier, this is one important point of superiority of mixed-media over dual-media and sand filters. It arises because of the much greater surface area of the grains in a mixed-media filter compared to sand or dual-media. The finest medium is 40 to 80 mesh in a mixed-media bed (10 percent of the total bed), 40 to 50 mesh (9 percent of total) in a sand bed, and 40 to 50 mesh (5 percent of total) in a dual-media filter. The finest medium (garnet) in a mixed-media bed has the advantage not only of being finer but also of being located at the very bottom of the filter where the applied load is lightest, and where it can serve its intended purpose as a polishing agent. Floc storage depths above the finest media in various typical beds are as follows:

- Sand, 0 inches
- Coal–sand, 18 inches (0.46 m)
- Mixed-media, 27 inches (0.69 m)

A further advantage of mixed-media filters in this regard lies in the greater total number of media particles contained in an equal volume of bed. This tremendously increases the number of opportunities for contact between media and colloids in the water, which greatly enhances removal of these colloids. It is these superior properties of mixed-media in resistance to leakage of particulates and much greater removal of colloids that make the use of single- or dual-media beds debatable under any circumstances. When coupled with the proper use of polymers as a filter aid, effluent filter media design can eliminate short runs due to floc breakthrough.

Variation in Effluent Quality. Effluent quality changes with variations in applied water flow or influent quality occur much less often in mixed-media beds than with either sand or dual-media. Again this relates to the greater total surface area of the fine media, the greater total number of fine media particles, and the smaller size of pore openings at the bottom of the filter bed.

Mounding. Gravel displacement or mounding can be eliminated by use of a 3-inch (76-mm) layer of coarse garnet or ilmenite between fine media and the gravel supporting bed as previously recommended, and by limiting the total flow and head of water available for backwash. Do not draw washwater from a high pressure source through a pressure-reducing valve that could fail.

Mudball Formation. This can be eliminated by providing an adequate backwash flow rate [up to 20 gpm/sq ft (48.8 m/h)] and a properly designed system for auxiliary scour (surface wash). The successful use of beds employing filter aids and in-depth filtration is dependent on provision and operation of a good system of auxiliary scour. Small media particles enhance mudball formation. It has been shown that the size of sand particles in mudballs is much smaller than the effective size of the sand media.[25] Thus, proper removal of the fines when placing media will assist in mudball control.

Growth of Filter Grains. Growth of filter grains, bed shrinkage, and media pulling away from filter sidewalls are related problems. Again, the provision and use of adequate backwash facilities including surface wash are the keys. It is the compressibility of filter grains that are heavily coated with materials filtered, deposited, or absorbed from the water that is the root of these difficulties. These problems usually can be avoided by proper backwashing. The growth of particles refers to a macroscopic increase in size and not to the development of a microscopic film of polymer and other chemicals. The microscopic film results from proper use of a filter aid that is beneficial in adsorption and retention of particulates for the period of a single operational cycle. This microscopic layer is not nearly thick enough to create a problem by increasing the compressibility of the bed. An alum or polymer film on filter grains may actually be an aid in reducing the adherence of calcium carbonate and facilitating its removal during backwashing. However, in filtering lime-softened waters it is important to adjust the pH of the filter influent by addition of carbon dioxide or acid to a level at which calcium carbonate deposition does not occur.

Sand Leakage. This can be prevented by using the coarse garnet layer between the fine media and gravel supporting bed as recommended earlier.

The garnet layer prevents the downward migration and escape of media fines.

Loss of Media. The loss of filter media, particularly coal during backwashing, is one problem for which there is no complete solution. Losses can be reduced by increasing the distance between the top of the expanded bed during maximum backwash flows and the washwater troughs. It also can be helped by cutting off air wash or auxiliary scour 1 or 2 minutes before the end of the main backwash.

Negative Head and Air Binding. These can be avoided in most cases, but there may be a few extreme situations, usually of short duration, where they cannot be entirely eliminated. In any case it is a good idea to provide a water depth of at least 5 feet (1.5 m) above the surface of the unexpanded filter bed. The more depth the better, at least as far as negative head and air binding are concerned. The filter should not be operated to terminal headlosses that are greater than the depth of submergence of the filter media, in order to minimize the potential for air binding.

When filter influent water contains dissolved oxygen at or near saturation levels, and when the pressure is reduced by siphon action to less than atmospheric at a point below the surface of the fine media, the oxygen comes out of solution, and gas bubbles are released. They may accumulate within the bed and tremendously increase the resistance to flow, or headloss. When flow through a filter is stopped and the water level is lowered in preparation for backwashing, bed pressures are reduced, and more oxygen is released. Even further release of bubbles occurs during backwash, which may lead to loss of media in the waste backwash water when bubbles adhere to coal or sand particles and carry them into the washwater troughs. More frequent filter backwashing may alleviate the problem to some extent, as there is then less time for bubbles to accumulate. However, when the problem is acute—as it may be in the spring when surface water is warming, and oxygen solubility is decreasing at the higher temperatures—it may only be endured and not solved. Maintaining maximum water depths above the beds and frequent backwashing may help, but may not completely eliminate the difficulties.

Air. Air can leak into the system by a variety of means. If the filter rate control valve is located above the hydraulic gradient, air may enter the system through the stuffing box between the valve and the valve operator. The problem is aggravated by the aspiration effect of a control valve located in a reduced cross section of pipe. A similar problem can occur during the filtration cycle through the stuffing boxes on surface wash systems, where one of the surface wash arms is located in the media. If there is a negative head in the filter, air may be pulled through the stuffing box into the bed during the filtration cycle.[50] Such leakage can be prevented by locating the stuffing box at an elevation that is always submerged during the filter cycle.

Existing Plants: Expansion and Conversion

Because mixed-media filters operate more efficiently, safely, and reliably at 5 gpm/sq ft (12.2 m/h) than do conventional rapid sand filters at only 2 gpm/sq ft (4.88 m/h), there obviously is great potential for expanding the capacity of existing plants at least up to double with only the nominal expense of replacing sand with mixed media. This, of course, has been done in a great many instances. Because of the ability of mixed-media filters to remove and store solids from high-turbidity waters, often it is not necessary to add settling basin capacity in plant expansion. In other cases this must be done, or, as an alternative, settling tubes may be installed in existing basins because this change will allow increasing basin throughput without loss of settling efficiency. (Chapter 22 discusses this subject in detail.)

Filter Design Checklist

1. Filter media sizing and selection should be based on pilot tests. If this is not possible, data should be obtained from similar applications to determine the suitability of the media design.
2. In dual- and mixed-media filter systems, provisions should be made for the addition of polyelectrolytes directly to the filter influent.
3. The turbidity of each filter unit should be monitored continuously and recorded.
4. The flow and headloss through each filter should be monitored continuously and recorded.
5. Provisions should be made for the addition of disinfectant directly to the filter influent.
6. Pressure filters must be equipped with pressure and vacuum air release valves.
7. Provisions should be made to divert any filter effluent of unsatisfactory quality (i.e., provide a filter to waste cycle).
8. Provisions should be made for automatic initiation and completion of the filter backwash cycle. The filter controls and pipe galleries should be housed.
9. Filter piping should be color-coded.
10. The filter system layout must enable easy removal of pumps and valves for maintenance.
11. The backwash rate selected must be based upon the specific filter media used and the wastewater temperature variations expected.

12. Filter backwash supply storage should have a volume at least adequate to complete two filter backwashes.
13. Adequate surface wash or air scour facilities must be provided.
14. There should be adequate backwash and surface wash pump capacity available with the largest single pumps out of service.
15. Backwash supply lines must be equipped with air release valves.
16. A means should be provided to indicate the backwash flow rate continuously and to enable positive control of the filter backwash rate. A means should also be provided to limit the filter backwash rate positively to a preset maximum value.
17. The filter design must incorporate underdrains and backwash wastewater collection devices that ensure uniform distribution of backwash water and filter influent.
18. The filter system should be equipped with an alarm system that will indicate major malfunctions.
19. Construction details must prevent cross connections and backflow.

OTHER TYPES OF FILTERS

Diatomite Filters

Diatomite filters consist of a layer of diatomaceous earth about $\frac{1}{8}$ inch (3.2 mm) thick supported on a septum or filter element. The thin precoat layer of diatomaceous earth is subject to cracking and must be supplemented by a continuous body feed of diatomite. The problems inherent in maintaining a perfect film of diatomaceous earth between filtered and unfiltered water have restricted the use of diatomite filters for municipal purposes, except under certain favorable conditions.

The present status of the diatomite filter and its potential uses have been summarized as follows:[1]

The diatomite filter is an acceptable tool in the waterworks industry, but is only a step in the water-treatment process, the details of which must be predicated on the characteristics of the raw water and what must be done to condition the water prior to filtration. Present municipal experience relates principally to turbidity removal from surface waters, but experience is growing with installations designed for iron and manganese removal from groundwaters and with filtration of lime-soda ash-softened water and coagulated surface waters. Municipal experience with turbidity removal has been principally where the actual turbidity is relatively low and the bacteriological quality is good; thus little pretreatment is provided. There is no common agreement on the upper limits of turbidity that can be handled without pretreatment. There has been some extension of the use of diatomite filters in some of the areas (i.e., marginally, moderately, and grossly polluted surface waters) which have been of particular concern to regulatory public health personnel. No one should recommend diatomite filtration in the near future for filtration of a grossly or even moderately polluted supply. However, appropriate prior conditioning of such waters by one of several available means, including coagulation and settling or chemical treatment of the filter medium or water to improve water filtrability, will undoubtedly lead to wider application of diatomite filters. Currently, effective water conditioning coupled with filtration through carefully engineered filters, an appropriate filter aid body feed system, and with adequate disinfection should provide a substantial margin of safety for marginally polluted waters and even sometimes for moderately polluted waters.

An advantage of a diatomite filtration plant for potable water is the lower first cost of such a plant. On waters containing low suspended solids, the diatomite filter installation cost should be somewhat lower than the cost of a conventional rapid-sand filtration plant. Diatomite filters will thus find application in potable water treatment under the following conditions:

- In cases where the diatomite plant will be found to produce water at a lower total cost than any practical alternative.
- In cases where financial capacity is tightly circumscribed, when the lower first cost of a diatomite filter installation may be the major factor in the final choice of the plant.
- For emergency or standby service at locations experiencing large seasonal variations in water demand, when the lower first cost of the diatomite filter may prove to be economical.

Some of the important operating parameters of diatomite filters have been summarized:[11]

- A precoat of 2.45 to 4.89 lb/sq ft (0.5 to 1.0 kg/m^2) is applied to prepare the filter.
- A continuous feed of filter aid as body feed is necessary to prevent the cake from clogging with the particles being filtered out.
- Acceptable cleaning of the filter will maintain at least 95 percent of the septum area available for flow after 100 filter run cycles.
- Because of the precoat, DE filters do not require filter-to-waste upon startup.
- If the flow to the filter is disrupted, the filter cake drops off the septum. When the filter is restarted, clean diatomite and filtered water should be used to recoat the filter to reduce the potential for passage of pathogens.
- It may be necessary to adjust the body feed rate in proportion to the raw water turbidity to prevent short runs.
- Although filter runs can be 2 to 4 days long, decomposition of organic matter trapped in the filter cake may necessitate shorter runs to avoid taste and odor problems.

- Vacuum DE filters offer the advantages of not requiring pressure vessels and being visible during backwash; however, they have the disadvantage of an increased potential for release of gases, which can cause shorter runs.
- DE filters can provide very effective removal of cysts, algae, and asbestos. In some cases, alum coating of the DE improves performance.
- The rate of body feed and size of diatomite used are critical variables affecting the length of the run.

Upflow Filters

Upflow filtration has an obvious theoretical advantage because coarse-to-fine filtration can be achieved with a single medium such as sand with almost perfect gradation of both pore space and grain size from coarse to fine in the direction of filtration (upward). Since the bed is backwashed in the same direction but at higher flow rates, the desired relative positions of fine media are maintained or reestablished with each backwash. The inherent advantage of upflow filtration has long been recognized, and, under laboratory conditions, the anticipated high filtration efficiency has been verified by several workers.

The difficulty with upflow filtration comes when the headloss above a given level exceeds the weight of the bed above that level, at which time the bed lifts or partially fluidizes, allowing previously removed solids to escape in the effluent. In Russia, bed depths up to 6 feet (1.82 m) are used in an attempt to minimize bed lifting. In the United States, parallel plates or a metal grid is placed at the top of the fine media. The spacing of the plates or the size of the openings in the grid is such that the media grains arch across the open space to restrain the bed against expansion. These restraining bar systems have about 75 percent open area in the best designs developed to date. Figure 10–19 illustrates an upflow filter with a restraining grid system. Even with the use of a restraining grid or a deep bed, there may be problems with excessive pressures or sudden variations in pressure that break the sand bridge or cause the bed to expand and lose its filter action.

The frequency of breakthrough is rare, but the fact that it can occur at all, say with poor operation, has been sufficient to raise questions concerning public health implications and to limit the use of upflow filters for potable water applications. In areas that are free of health considerations, upflow filters have found wide application and have given excellent service. These areas include process water, wastewater treatment, deep well injection water, API separator effluent, cooling water, and other similar applications. Until a foolproof means of restraining a dirty upflow bed is developed, it appears likely that potable water applications of the upflow filter in the United States will continue to be limited.

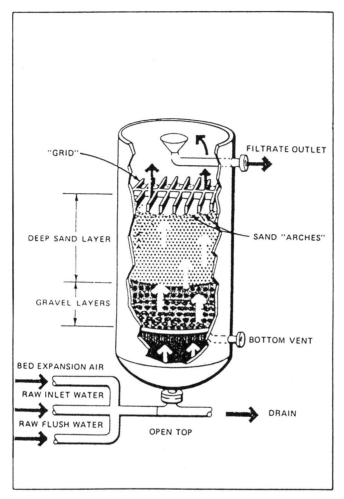

Fig. 10–19. Upflow filter with restraining grid.

Biflow Filters

Biflow filters are an outgrowth of upflow filters, in that the divided flow (downward from the top and upflow from the bottom; see Fig. 10–20) is an attempt to restrain the bottom upflow portion of the bed by placing a downflow filter above it. Biflow filters are used in Holland and Russia but not to any extent in the United States. They permit filtration in two opposite directions at the same time. Essentially the top and bottom halves are completely independent filters of equal capacity which results in some savings in structure and underdrains.

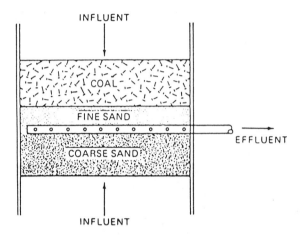

Fig. 10–20. Section through a dual-media biflow filter.

Unfortunately, the biflow filter has an inherent limitation that seems to preclude development of a unit that will produce an exceptionally high-quality effluent. First consider a single-medium biflow bed. The finest materials are at the top of the upper downflow bed. This arrangement makes the top half of the bed a rapid sand or surface-type filter, and the quality of water produced at best cannot exceed that from a rapid sand filter. The bottom half of this same filter is a coarse-to-fine filter, but unfortunately the finest material at the top outlet from the bed is somewhat coarser than the finest material that can be successfully used in a rapid sand filter. Obviously, the effluent from this bed will be of lesser quality than that from the rapid-sand downflow filter above. This situation has been recognized by researchers and revealed by pilot tests, and led to a consideration of the dual-media biflow filter, illustrated in Fig. 10–20. The idea here, and the advantage over the single-medium biflow bed, is that this arrangement places the fine sand closer to the mid-collector. It provides a dual-media (coal–sand) downflow bed above a coarse-to-fine single-medium (sand) upflow bed. Again there are practical limits on the coarseness of the sand. If the sand is finer than that ordinarily used in a rapid sand filter, as it must be to build the best upflow filter in the bottom half of the bed, then it will be so fine that excessive amounts of it will rise into the coal bed during backwashing. The gradation of a sand that is suitable for the dual-media bed in the upper half of the bed will be too coarse to provide the best possible filtration in the upflow bottom half of the bed. The quality of effluent from either half will not approach that from a mixed-media filter. This problem is so basic that there does not appear to be an easy solution to the dilemma, but perhaps one will be found.

REFERENCES

1. *Water Quality and Treatment. A Handbook of Public Water Supplies,* American Water Works Association, McGraw-Hill, New York, 1971.
2. Conley, W. R., Jr., and Pitman, R. W., "Innovations in Water Clarification," *J.AWWA,* p. 1319, 1960.
3. Conley, W. R., Jr., "Experiences with Anthracite Sand Filters," *J.AWWA,* p. 1473, 1961.
4. Rice, A. H., and Conley, W. R., "The Microfloc Process in Water Treatment," *Tappi,* p. 167A, 1961.
5. Conley, W. R., Jr., "Integration of the Clarification Process," *J.AWWA,* p. 1333, 1965.
6. Pitman, R. W., and Wells, G. W., "Activated Silica as a Filter Conditioner," *J.AWWA,* p. 1167, 1968.
7. Conley, W. R., Jr., and Hsiung, K., "Design and Application of Multimedia Filters," *J.AWWA,* p. 97, 1969.
8. Cleasby, J. L., and Sejkora, G. D., "Effect of Media Intermixing on Dual Media Filtration," *Journal of the Environmental Engineering Division, ASCE,* p. 503, August, 1975.
9. *Mixed Media,* Neptune MicroFloc Bulletin No. 4206.
10. Robeck, G. G., Clarke, N. A., and Dostal, K. A., "Effectiveness of Water Treatment Processes in Virus Removal," *J.AWWA,* p. 1275, 1962.
11. Logsdon, G. S., and Fox, K., "Getting Your Money's Worth from Filtration," *J.AWWA,* p. 249, May, 1982.
12. Trussell, R. R., et al., "Recent Developments in Filtration System Design," *J.AWWA,* p. 705, December, 1980.
13. Culp/Wesner/Culp, *Pilot Filtration Tests at the American River Water Treatment Plant,* prepared for the City of Sacramento, Calif., May, 1982.
14. Hsiung, A. K., Conley, W. R., and Hansen, S. P., "The Effect of Media Selection on Filtration Performance," presented at the Hawaii section, American Water Works Association, April, 1976.
15. American Water Works Association, *State of the Art of Water Filtration,* Committee Report, p. 662, 1972.
16. Camp, T. R., "Theory of Water Filtration," *Proceedings American Society of Civil Engineers,* p. 1, 1964.
17. Cleasby, J. L., "Approaches to a Filtrability Index for Granular Filters, *J.AWWA,* p. 372, 1969.
18. Cleasby, J. L., Williamson, M. M., and Baumann, R. E., "Effect of Filtration Rate Changes on Quality," *J.AWWA,* p. 869, 1963.
19. Fair, G. M., and Geyer, J. C., *Water Supply and Wastewater Disposal,* John Wiley and Sons, New York, 1954.
20. Diaper, E. W. J., and Ives, K. J., "Filtration through Size Graded Media," *J. San. Eng. Div., Proceedings American Society of Civil Engineers,* p. 89, June, 1963.
21. Hudson, H. E., Jr., "Physical Aspects of Filtration," *J.AWWA,* p. 33, 1969.
22. Hudson, H. E., Jr., "Theory of the Functioning of Filters," *J.AWWA,* p. 868, 1948.
23. Hudson, H. E., Jr., "Functional Design of Rapid Sand Filters," *J. San. Eng. Div., ASCE,* p. 17, 1963.
24. Ives, K. J., "Progress in Filtration," *J.AWWA,* p. 1225, 1964.
25. Kawamura, S., "Design and Operation of High-Rate Filters," *J.AWWA,* pp. 535, 653, 705, 1975.
26. Minz, D. M., *Modern Theory of Filtration,* Special Subject No. 10, International Water Supply Assoc., London, 1966.
27. Robeck, G. G., Dostal, K. A., and Woodward, R. L., "Studies of Modifications in Water Filtration," *J.AWWA,* p. 198, 1964.

28. Yao, K. M., Hubibjun, M. T., and O'Melia, C. R., "Water & Waste Filtration—Concepts and Application," *ES&T,* p. 1105, November, 1971.
29. O'Melia, C. R., and Stumm, Werner, "Theory of Water Filtration," *J.AWWA,* p. 1393, 1967.
30. Culp, R. L., "Direct Filtration," *J.AWWA,* p. 375, July, 1977.
31. Moulton, F., Bedc, C., and Guthrie, J. L., "Mixed-Media Filters (Clean) 50 MGD without Clarifiers," *Chemical Processing,* March, 1968.
32. Willis, J. F., "Direct Filtration, an Economic Answer to Water Treatment Needs," *Public Works,* p. 87, November, 1972.
33. Tate, C. H., and Trussell, R. R., "Recent Developments in Direct Filtration," *J.AWWA,* p. 165, March, 1980.
34. Foley, P. D., "Experience with Direct Filtration at Ontario's Lake Huron Treatment Plant," *J.AWWA,* p. 162, March, 1980.
35. McCormick, R. F., and King, P. H., "Factors That Affect Use of Direct Filtration in Treating Surface Waters," *J.AWWA,* p. 234, May, 1982.
36. Wagner, E. G., and Hudson, H. E., Jr., "Low Dosage High-Rate Direct Filtration," *J.AWWA,* p. 256, May, 1982.
37. "The Status of Direct Filtration," *J.AWWA,* p. 405, July, 1980.
38. Tate, C. H., Lang, J. S., and Hutchinson, H. L., "Pilot Plant Tests of Direct Filtration," *J.AWWA,* p. 379, July, 1977.
39. Logsdon, G. S., Clark, R. M., and Tate, C. H., "Costs of Direct Filtration to Meet Drinking Water Regulations," *J.AWWA,* p. 134, March, 1980.
40. Hutchinson, W., and Foley, P. D., "Operational and Experimental Results of Direct Filtration," *J.AWWA,* p. 86, February, 1974.
41. Sweeney, G. E., and Prendiville, P. W., "Direct Filtration: An Economic Answer to a City's Water Needs," *J.AWWA,* p. 65, February, 1974.
42. Hutchinson, W. R., "High-Rate Direct Filtration," *J.AWWA,* p. 292, June, 1976.
43. Spink, C. M., and Monscvitz, J. T., "Design and Operation of a 200-mgd Direct-Filtration Facility," *J.AWWA,* p. 127, February, 1974.
44. Monscvitz, J. T., et al., "Some Practical Experience in Direct Filtration," *J.AWWA,* p. 584, October, 1978.
45. Culp, R. L., "New Water Treatment Methods Serve Richland," *Public Works,* p. 86, July, 1964.
46. Culp, R. L., "Filtration," Chapter 8, *Water Treatment Plant Design Manual,* American Water Works Association, New York, 1969.
47. Oakley, H. R., and Cripps, T., "British Practice in the Tertiary Treatment of Sewage," *J.WPCF,* p. 36, 1969.
48. Oakley, H. R., personal communication, 1969.
49. Nelson, O. F., "Capping Sand Filters," *J.AWWA,* p. 539, 1969.
50. Hudson, H. E., *Water Clarification Processes. Practical Design and Evaluation,* Van Nostrand Reinhold, New York, 1981.
51. Culp, Gordon L., and Culp, Russell, L., *New Concepts in Water Purification,* Van Nostrand Reinhold, New York, 1974.

Chapter 11
Removal of Pathogenic Bacteria and Viruses

SIGNIFICANCE OF WATERBORNE DISEASES

It was not until the establishment of the germ theory of disease by Pasteur in the mid-1880s that water as a carrier of disease-producing organisms could be understood, although the epidemiological relation between water and disease had been suggested as early as 1854. In that year, London experienced the "Broad Street Well" cholera epidemic, and Dr. John Snow conducted his now famous epidemiological study. He concluded the well had become contaminated by a visitor who arrived in the vicinity with the disease. Cholera was one of the first diseases to be recognized as capable of being waterborne. Also, that probably was the first reported disease epidemic due to direct recycling of nondisinfected wastewater. Now, over 100 years later, the list of potential waterborne diseases due to microorganisms is considerably larger, and includes the bacterial, viral, and parasitic microorganisms shown in Tables 11–1, 11–2, and 11–3, respectively.[1,2,3]

During the years 1971 to 1978, forty-three states and Puerto Rico reported 224 outbreaks of waterborne diseases that affected about 48,000 people and included two deaths. Since 1970, there has been a pronounced increase in such outbreaks, with an average of 34 per year. There are two reasons for this increase:

- More active and accurate surveillance by regulatory agencies.
- Modification of the definitions of water systems by the U.S. Environmental Protection Agency in 1976 to 1978 to correspond with those in the Safe Drinking Water Act (PL 93–523).

As a result, more systems and people are under surveillance. Figure 11–1 shows the etiology of waterborne diseases for 1971 to 1978.

The number of disease outbreaks identified as waterborne in the United States dropped steadily following the introduction of chlorination of potable water. Figure 11–2 shows the annual number of reported waterborne disease outbreaks from 1920 through 1960 and the growth of U.S. water treatment plants using chlorine disinfection from 1910 through 1958. There is an obvious relationship between the drop in waterborne disease outbreaks and the number

Table 11–1. Waterborne Bacterial Diseases.*

CAUSATIVE AGENT	DISEASE	SYMPTOMS	RESERVOIR
Salmonella typhosa	Typhoid fever	Incubation period 7–14 days. Headache, nausea, loss of appetite, constipation or diarrhea, insomnia, sore throat, bronchitis, abdominal pain, nose bleeding, shivering and increasing fever. Rose spots on trunk.	Feces and urine of typhoid carrier or patient.
S. paratyphii S. schottinulleri S. hirschfeldii C.	Paratyphoid fever	General infection characterized by continued fever, diarrheal disturbances, sometimes rose spots on trunk. Incubation period 1–10 days.	Feces and urine of carrier or patient.
Shigella flexneri Sh. dysenteriae Sh. sonnei Sh. paradisinteriae	Bacillary dysentery	Acute onset with diarrhea, fever, tenesmus and stool frequently containing mucus and blood. Incubation period 1–7 days.	Bowel discharges of carriers and infected persons.
Vibrio comma	Cholera	Diarrhea, vomiting, ricewater stools, thirst, pain, coma. Incubation period a few hours to 5 days.	Bowel discharges, vomitus, carriers.
Pasteurella tularensis	Tularemia	Sudden onset with pains and fever; prostration. Incubation period 1–10 days.	Rodent, rabbit, horsefly, woodtick, dog, fox, hog.
Brucella melitensis	Brucellosis (undulant fever)	Irregular fever, sweating, chills, pain in muscles.	Tissues, blood, milk, urine, infected animal.
Pseudomonas pseudomallei	Melioidosis	Acute diarrhea, vomiting, high fever, delirium, mania.	Rats, guinea pigs, cats, rabbits, dogs, horses.
Microorganisms	Gastroenteritis	Diarrhea, nausea, vomiting, cramps, possibly fever. Incubation period 8–12 hr, average.	Probably man and animals.
Leptospira icterohaemorrhagiae (spirochete)	Leptospirosis (Weil's disease)	Fevers, rigors, headaches, nausea, muscular pains, vomiting, thirst, prostration and jaundice may occur.	Urine and feces of rats, swine, dogs, cats, mice, foxes, sheep.
Enteropathogenic E. coli	Gastroenteritis	Watery diarrhea, nausea, prostration and dehydration.	Feces of carrier.

* References 1 and 2. Reprinted by permission of John Wiley & Sons, Inc.

of water treatment plants using chlorine disinfection. Of course, improvements in other water treatment procedures (flocculation, filtration, etc.) have contributed to the decrease in waterborne disease; but disinfection is, and will remain, the most important treatment process for the prevention of such disease.

Although there has been a marked decline in waterborne disease outbreaks since 1920, there has been no decline since the mid-1950s. In fact, there has been an increase, as shown in Fig. 11–3. Table 11–4 presents a summary of the known outbreaks of waterborne diseases for the 10-year period 1961 through 1970.

Two agents never before associated with documented waterborne outbreaks in the United States appeared during the 1961 through 1970 period: enteropathogenic *E. coli* (EEC) and *Giardia lamblia*. Various serotypes of EEC have been implicated as the etiologic agent responsible for disease in newborn infants, usually the result of cross contamination in nurseries. Now, there have been several well-documented outbreaks of EEC (serotypes 0111:B4 and 0124:B27) associated with adult waterborne disease.

Giardia lamblia is a flagellated protozoan responsible for giardiasis. The three outbreaks of giardiasis (see Table 11–4) occurred in resort or recreational areas. Another protozoal infection sometimes transmitted by the waterborne route is amoebiasis or amoebic dysentery, which occurred in three outbreaks. The etiologic agent for this disease is *Entamoeba histolytica,* an amoeba.

The major cause of outbreaks in public systems is contamination of the distribution system—primarily via cross connections and back siphonage. However, when cases of illness are considered, the picture changes. Outbreaks resulting from contamination of the distribution system are usually quite contained, and few illnesses result. However, contamination of the source or a breakdown in treatment produces many illnesses. When cases of illness are considered, the major perpetrators are source contamination and treatment deficiencies.[18] About 46 percent of the outbreaks in public systems were found to have been related to these two particular causes; 92 percent of the cases of illness in the public systems occurred as a result of source and treatment deficiencies.

REMOVAL BY COAGULATION, SETTLING, FILTRATION, AND ADSORPTION PROCESSES

Although the addition of halogens, such as chlorine, is called to mind when one considers the removal of pathogenic bacteria and viruses, other unit processes also provide substantial degrees of removal.

The chemical coagulation process has been noted by several researchers as providing high degrees of virus removal. For example, alum coagulation was found to remove 95 to 99 percent of Coxsackie virus, and ferric chloride

Table 11-2. The Human Enteric Viruses That Can Be Waterborne and Known Diseases Associated With These Viruses.*

GROUP	SUBGROUP	NO. OF TYPES OR SUBTYPES	DISEASE ENTITIES ASSOCIATED WITH THESE VIRUSES	PATHOLOGIC CHANGES IN PATIENTS	ORGANS WHERE VIRUS MULTIPLIES
Enterovirus	Poliovirus	3	Muscular paralysis	Destruction of motor neurons	Intestinal mucosa, spinal cord, brain stem
			Aseptic meningitis	Inflammation of meninges from virus	Meninges
			Febrile episode	Viremia and viral multiplication	Intestinal mucosa and lymph
	Echo virus	34	Aseptic meningitis	Same as above	Same as above
			Muscular paralysis	Same as above	Same as above
			Guillain-Barre's Syndrome**	Destruction of motor neurons	Spinal cord
			Exanthem	Dilation and rupture of blood vessels	Skin
			Respiratory diseases	Viral invasion of parenchymiatous of respiratory tracts and secondary inflammatory responses	Respiratory tracts and lungs
			Diarrhea	Not well known	
			Epidemic myalgia		
			Pericarditis and myocarditis	Viral invasion of cells with secondary responses	Pericardial and myocardial tissue
			Hepatitis	Same as above	Liver parenchyma
			Herpangina†	Viral invasion of mucosa with secondary inflammatory responses	Mouth
	Coxsackie virus	>24			
	A		Acute lymphatic pharyngitis	Same as above	Lymph nodes and pharynx
			Aseptic meningitis	Same as above	Same as above
			Muscular paralysis	Same as above	Same as above

Virus	No.	Disease	Mechanism	Tissue/Cells affected
		Hand-foot-mouth disease‡	Viral invasion of cells of skin of hands and feet and mucosa of mouth	Skin of hands and feet and much of mouth
		Respiratory disease	Same as above	Same as above
		Infantile diarrhea	Viral invasion of cells of mucosa	Intestinal mucosa
		Hepatitis	Viral invasion of liver cells	Parenchyma cells of liver
B	6	Pericarditis and myocarditis	Same as above	Same as above
		Pleurodynia§	Viral invasion of muscle cells	Intercostal muscles
		Aseptic meningitis	Same as above	Same as above
		Muscular paralysis	Same as above	Same as above
		Meningoencephalitis	Viral invasional invasion of cells	Meninges and brains
		Pericarditis, endocarditis, myocarditis	Same as above	Same as above
		Respiratory diseases	Same as above	Same as above
		Hepatitis or rash	Same as above	Same as above
		Spontaneous abortion	Viral invasion of vascular cells(?)	Placenta
		Insulin-dependent diabetes	Viral invasion of insulin-producing cells	Langerhans' cells of pancreases
		Congenital heart anomalies	Viral invasion of muscle cells	Developing heart
Reo virus	6	Not well known	Not well known	Same as above
Adenovirus	31	Respiratory diseases	Same as above	Same as above
		Acute conjunctivitis	Viral invasion of cells and secondary inflammatory responses	Conjunctival cells and blood vessels
		Acute appendicitis	Viral invasion of mucosa cells	Appendia and lymph nodes
		Intussusception	Viral invasion of lymph nodes(?)	Intestinal lymph nodes(?)
		Subacute thyroiditis	Viral invasion of parenchyma cells	Thyroid
		Sarcoma in hamsters	Transformation of cells	Muscle cells
Hepatitis	>2	Infectious hepatitis	Invasion of parenchyma cells	Liver
		Serum hepatitis	Invasion of parenchyma cells	Liver
		Down's Syndrome***	Invasion of cells	Frontal lobe of brain, muscle, bones

* Reference 3. ** Ascending type of muscular paralysis † Febrie episode with sores in mouth ‡ Rash and blisters on hand-foot-mouth with fever § Pleuritis type of pain with fever
*** Mongolism

Table 11-3. Waterborne Diseases from Parasites.*

CAUSATIVE AGENT	DISEASE	SYMPTOMS
Endamoeba histolytica	Amebiasis	Diarrhea alternating with constipation, chronic dysentery with mucus and blood
Giardia lamblia	Giardiasis	Intermittent diarrhea
Naegleria gruberi	Amoebic meningocephalitis	Death
Taenia saginata (beef tapeworm)		Abdominal pain, digestive disturbance, loss of weight
Ascario lumbricoides (round worm)	Ascariasis	Vomiting, live worms in feces
Schistosoma mansoni (blood fluke)	Schistosomiasis	Liver and bladder infection

* Reference 2. Reprinted by permission of John Wiley & Sons, Inc.

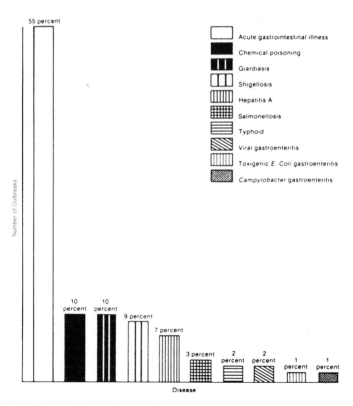

Fig. 11-1. Etiology of waterborne disease in the United States, 1971–1978. (Courtesy of AWWA.[4])

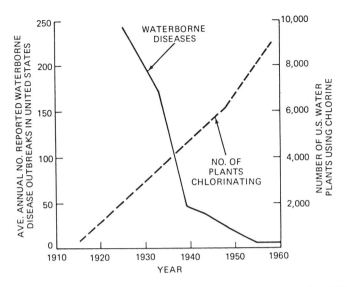

Fig. 11-2. Average annual number of waterborne disease outbreaks 1920–1960 and the number of water plants using chlorine.

was found to remove 92 to 94 percent of the same virus.[5] With both alum and $FeCl_3$, good virus removal was contingent upon good floc formation and the absence of interfering substances. The effectiveness of removal was not temperature-dependent with either coagulant. It has been postulated that aluminum coordinates with the carboxyl groups in the virus's protein coat, followed by incorporation of the complex into the precipitating hydrated aluminum oxide. Viruses were not inactivated by alum coagulation, and could

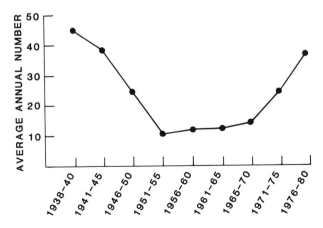

Fig. 11-3. Average annual number of waterborne outbreaks, 1938–1980.

Table 11–4. Waterborne Disease Outbreaks 1961 Through 1970 by the Type of Illness and System.

ILLNESS	PRIVATE SYSTEMS		PUBLIC SYSTEMS		TOTAL	
	OUTBREAKS	CASES	OUTBREAKS	CASES	OUTBREAKS	CASES
Gastroenteritis	25	4,498	14	22,048	39	26,546
Infectious hepatitis	22	664	8[a]	239	30[a]	903
Shigellosis	16	939	3	727	19	1,666
Typhoid	14	104			14	104
Salmonellosis	4[b]	96	5	16,610	9	16,706
Chemical poisoning	7	42	2	4	9	46
Enteropathogenic						
E. coli	4	188			4	188
Giardiasis	1	19	2	157	3	176
Amoebiasis	2	14	1	25	3	39
Total	95	6,564	35	39,810	130	46,374

[a] One gastroenteritis outbreak also included seven cases of infectious hepatitis.
[b] One gastroenteritis outbreak was preceded by outbreak of 38 cases of salmonellosis.

be partially recovered from the sludge. The presence of organic material was shown to decrease the amount of virus removed by alum or ferric chloride coagulation.

Thorup et al. conducted a laboratory study on the removal of a bacterium (bacteriophage T2 of E. coli) and a virus (Type 1 Poliovirus) from artificially seeded water.[6] The polyelectrolyte alone was ineffective for virus reduction, but when the clay mineral montmorillonite was added, up to 82 percent virus removal was effected by the clay itself. The use of either alum or ferric sulfate as the primary coagulant resulted in virus removals that were not so good as those mentioned above. However, the presence of coagulated clay increased virus removal to as much as 86 percent.

The alum coagulation of suspensions of E. coli was conducted by Hanna and Rubin, and bacterial removal, as measured by turbidity reduction, was found to be significant.[8] Ferric chloride was used by Manwaring et al. to coagulate the bacterial virus bacteriophage MS2.[9] The optimum ferric chloride dosage was 40 to 60 mg/l with the optimum virus and turbidity removals occurring at a pH value of 5.0. The highest virus removal was 99.5 percent, with an average input of 3.9×10^5 PFU/ml. Similar results were obtained with alum.

Removal of a bacterial virus (MS2 against E. coli) using a bench-scale diatomaceous earth filter was investigated by Anurhor and Engelbrecht.[10] Filtration was aided through the addition of a cationic polymer. The best removals (about 75 percent) were obtained by a polymer dose of 0.25 mg/l, while poorer removals occurred at higher polymer doses.

Sproul reviewed the published data on virus removal by coagulation processes.[7] Enteric virus removal should be in the range of 90 to 99.999 percent by coagulation of water containing poliovirus and Coxsackie virus A2.

High pH Conditions

High pH values are obtained when drinking water is softened by the addition of lime and soda ash to remove calcium (Ca^{+2}) and magnesium (Mg^{+2}) ions that cause water hardness. Calcium hydroxide ($Ca(OH)_2$) is usually used to raise the pH to values as high as 10.5, which are maintained up to 6 hours. (See Chapter 21 for a discussion of water softening.) High pH values in water also can be obtained by using sodium hydroxide (NaOH).

Disinfection is readily accomplished at a high pH, as discussed in this section, but the pH must be reduced before the water is consumed. In a large water treatment plant, this is accomplished by recarbonation (as described in Chapter 21).

Lime coagulation has been shown to be effective in the removal and inactivation of viruses. The mechanism of inactivation under alkaline conditions is probably denaturation of the protein coat and disruption of the virus. In some cases, complete loss of structural integrity of the virus may occur under high pH conditions.

The pH reached is a critical factor in determining the degree of virus inactivation. Figure 11–4 shows a marked difference in virus inactivation

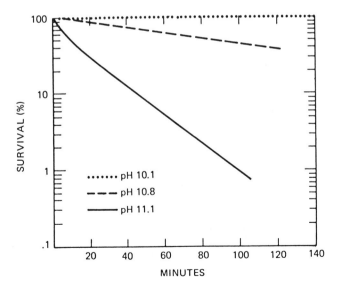

Fig. 11–4. Inactivation of Poliovirus 1 by high pH at 25°C in lime-treated (500 mg/l $Ca(OH)_2$), sand-filtered secondary effluents.

as the pH is increased from 10.1 to 10.8 and then to 11.1. These data indicate that it may be useful to increase the pH to slightly higher levels during excess lime softening that those now achieved, to obtain higher inactivations during softening. Figure 11–5 illustrates the effects of pH on *E. coli* and *S. typhosa*.

The microbial inactivation being considered is that caused solely by high pH. When the pH of water is raised, the opportunity exists for precipitation of many compounds. These precipitates provide opportunities for adsorption and coagulation of microorganisms, and will cause increased removal over that obtained from the pH alone. The following studies obviated this effect or made it insignificant through their experimental procedures.

An early study determined the times required to obtain 100 percent inactivation of several bacterial species at initial concentrations of approximately 1,500 organisms/ml.[11] The tests were conducted in 20–25°C dechlorinated tap water, using calcium hydroxide for pH adjustment. The results are shown in Table 11–5.

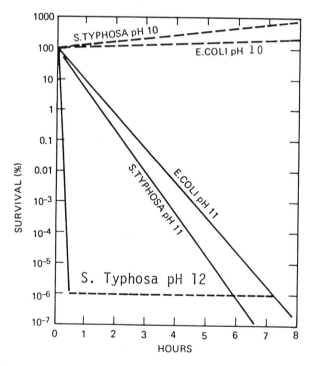

Fig. 11–5. Survival of *E. coli* at pH 11 and 12 and *S. typhosa* at pH 10, 11, and 12 at 25°C.[18]

Table 11-5. Contact Times Necessary for High pH Conditions to Effect 100 Percent Inactivation of Initial Concentrations of 1,500 Organisms/ml.*

	CONTACT TIME REQUIRED FOR 100% INACTIVATION BY HIGH pH CONDITIONS, MIN**				
pH	*Escherichia coli*	*Enterobacter aerogenes*	*Pseudomonas aeruginosa*	*Salmonella typhi*	*Shigella dysenteriae*
9.01–9.5	>540	—	—	>540	—
9.51–10.0	>600	>600	420	>540	>300
10.01–10.5	>600	>600	300	>540	>300
10.51–11.0	600	>540	240	240	180
11.01–11.5	300	>600	120	120	75

* Reproduced from Wattie and Chambers, Reference 11. (Courtesy of American Water Works Association.)
** Experiments were conducted at 20–25°C.

Riehl, et al. reported *E. coli* inactivations of 95 percent in 8 hours at pH 10.5 and 5°C, 100 percent in 2 hours at pH 10.5 and 15°C, and 100 percent in approximately 30 minutes at pH 10.5 and 25°C in distilled water.[12] Lime was used for pH adjustment. Inactivations of 50 percent and 55 percent in 10 hours were noted for *Salmonella montevideo* and *S. typhi,* respectively, at 2°C and pH 10.6 in distilled water. Initial concentrations of organisms were approximately 1,000/ml. Higher temperatures resulted in more inactivation at the same pH and contact time.

Filtration

Virus removal by filtration has also been studied.[13] Up to 98 percent of Poliovirus Type 1 was removed in a coal and sand filter at filtration rates from 2 to 6 gpm/sq ft (4.88 to 14.64 m/h) when a low dose of alum was fed ahead of the filters. Up to 99 percent was removed when the alum dose was increased, and conventional flocculation and sedimentation were carried out ahead of the filter.

Floc breakthrough of the filter, sufficient to cause a turbidity of less than 0.5 NTU, was usually accompanied by virus breakthrough. Addition of a polyelectrolyte to the filter influent decreased virus in the effluent by increasing floc strength and floc retention in the filter.

While the bulk of studies on virus removal have been laboratory studies, Walton reported a study of coliform bacteria removal in over 80 full-scale water treatment plants.[14] He found that the removal of coliform bacteria by coagulation, sedimentation, and filtration averaged about 98 percent for the several plants studied. The average coliform densities in filtered but un-chlorinated waters ranged from 2.9 to 200 per 100 ml. These data make it clear that chlorination must also be provided if adequate coliform removals are always to be achieved.

Although the coagulation and filtration processes cannot be relied upon to provide the only means of pathogen removal, these processes perform an important role in assuring the ability to completely disinfect the water. Hudson noted the importance of minimizing finished water turbidity to maximize assurance of disinfection.[15]

Hudson concluded that filtration plants operated to attain a high degree of removal of one impurity tend to accomplish high removals of other suspended materials. Examples of parallelism in removal of turbidity showed removals of manganese, microorganisms, and bacteria. Plants producing very clear water also tended to secure low bacterial counts accompanied by a low incidence of viral diseases. He also concluded that the production of high-quality water requires striving toward high goals as measured by several, not just one or two, quality criteria. These criteria include filtered water turbidity, bacteria as indicated by plate counts and by presumptive and confirmed coliform determinations, and thorough chlorination. The operating data he reviewed for plants treating polluted water indicate that low virus disease rates occur in cities where the water treatment operators aim to produce a superior product rather than a tolerable water.

The report of an AWWA Committee on viruses in water noted that viruses, because of their small size, more easily become enmeshed in a protective coating of turbidity-contributing matter than bacteria would.[16] For most effective disinfection the committee concluded, turbidities should be kept below 1 NTU; indeed, they thought it would be best to keep the turbidity as low as 0.1 NTU, as recommended by AWWA water quality goals. They noted that the limit of 5 NTU of turbidity specified in the USPHS "Drinking Water Standards," 1962, is meant to apply to protected watersheds and not to filtration plant effluents. With turbidities as low as 0.1 to 1, they concluded that a preplant chlorine feed need be only enough to have a 1 mg/l free chlorine residual after a 30-minute contact time. Post-chlorination practice would then depend upon the ability to maintain such residuals throughout the distribution system.

Summary

The disinfection of water depends upon optimizing the unit processes of chemical coagulation, sedimentation, and filtration to produce minimum water turbidities. This will assure maximum contact between any remaining pathogens and the disinfectant added.

ALTERNATIVE DISINFECTANTS

Many types of disinfectant are used for municipal water systems. Disinfection of municipal waters is intended to destroy the pathogenic and indicator organisms. Alternative disinfection processes for water include:

- Physical treatment by application of heat.
- Irradiation by ultraviolet (UV) light.
- Metal ions, for example, copper and silver.
- Strong acids and bases, that is, pH adjustment.
- Surface active agents, for example, the quaternary ammonium compounds.
- Chemical oxidants, such as chlorine (Cl_2), bromine (Br_2), iodine (I_2), ozone (O_3), chlorine dioxide (ClO_2), potassium permanganate ($KMnO_4$), ferrate (FeO_4^{-2}), and organohalogen compounds.[19]

Several of these methods are not suitable for large-scale disinfection of public drinking water supplies. Chemical oxidants, UV irradiation, certain metal ions, and insoluble contact disinfectants, will be considered in this text.

CHLORINATION

History

According to the AWWA, the earliest recorded use of chlorine directly for water disinfection was on an experimental basis, in connection with filtration studies at Louisville, Kentucky, in 1896.[17] In North America, the first continuous, municipal application of chlorine to water was in 1908, to disinfect the 40 mgd Boonton reservoir supply of the Jersey City, New Jersey, water utility.

Today the principal sources of chlorine for water disinfection are:

- The elemental form, as liquefied compressed gas
- Calcium hypochlorite
- Chlorine bleach solutions
- Chlorine dioxide

Alternate Forms of Chlorine

For purposes of disinfection of municipal supplies, chlorine is primarily used in two forms: as a gaseous element or as a solid or liquid chlorine-containing hypochlorite compound. Gaseous chlorine is generally considered the least costly form of chlorine that can be used in large facilities. Hypochlorite forms have been used primarily in small systems (fewer than 5,000 persons) or in large systems where safety concerns related to handling the gaseous form outweigh economic concerns.

Gaseous Chlorine. In the gaseous state, chlorine is greenish-yellow in color and about 2.48 times as heavy as air; the liquid is amber-colored and about 1.44 times as heavy as water. Unconfined liquid chlorine rapidly vaporizes to the gas; 1 volume of liquid yields about 450 volumes of gas. Chlorine is only slightly soluble in water, its maximum solubility being approximately 1 percent at 49.2°F.

Chlorine confined in a container may exist as a gas, a liquid, or a mixture of both. Thus, any consideration of liquid chlorine includes that of gaseous chlorine. The vapor pressure of chlorine in a container is a function of temperature, and is independent of the contained volume of chlorine; therefore, gage pressure is not an indication of container contents.

Chlorine gas is a respiratory irritant, with concentrations in air of 3 to 5 mg/l by volume readily detectable. It can cause varying degrees of irritation of the skin, mucous membranes, and respiratory system, depending on the concentration and duration of exposure. In extreme cases, death can occur from suffocation. The severely irritating effect of the gas makes it unlikely that any person would remain in a chlorine-contaminated atmosphere unless he were unconscious or trapped. Liquid chlorine may cause skin and eye burns upon contact with those tissues; when unconfined in a container, it rapidly vaporizes to the gas, producing the above effects.

Chlorine is shipped in cylinders, tank cars, tank trucks, and barges as a liquefied gas under pressure. Containers commonly employed include 100- and 150-lb cylinders, 1-ton containers, 16-, 30-, 55-, 85-, and 90-ton tank cars, 15- to 16-ton tank trucks, and barges varying from 55- to 1,110-ton capacity. Ton containers, loaded to about 2,000 lb of chlorine, have a gross weight as high as 3,700 lb. Chlorine tank barges have either four or six tanks, each with a capacity of from 85 to 185 tons.

Chlorine Compounds. Many chlorine-containing compounds that could be used for the disinfection of water are shown in Table 11–6. Selection of the appropriate compound depends on several factors:

- Size of the system
- Location of the treatment facility with respect to obtaining chlorine gas
- Safety considerations
- Requirements for chlorine residual in the distribution system

Table 11–6. Percent Available Chlorine of Various
Chlorine Materials.*

MATERIAL	AVAILABLE CHLORINE, %
Cl_2, Chlorine	100 (by definition)
Bleaching Powder (chloride of lime, etc.)	35–37
$Ca(OCl)_2$, Calcium Hypochlorite	99.2
Commercial Preparations	70–74
$NaOCl$, Sodium Hypochlorite (unstable)	95.2
Commercial Bleach (industrial)	12–15
Commercial Bleach (household)	3–5
ClO_2, Chlorine Dioxide	263.0
NH_2Cl, Monochloramine	137.9
$NHCl_2$, Dichloramine	165.0
NCl_3, Nitrogen Trichloride	176.7
$HOOCC_6H_4SO_2NCl_2$ (Halazone)	52.4
NClCONClCONCl CO, Trichloroisocyanuric Acid	91.5
CONClCONClCONH, Dichloroisocyanuric Acid	71.7
CONClCONClCON Na, Sodium Dichloroisocyanurate	64.5

* Reproduced from Laubusch.[19] (Courtesy of the McGraw-Hill Book Company.)

Gaseous chlorine is used for large volumes of municipal and industrial waters. Small-scale facilities frequently apply calcium or sodium hypochlorite.

The term "available chlorine content" (ACC) was initially used as a method of comparing the disinfection efficiencies of the various chlorinated compounds. Gaseous Cl_2 was considered to be completely "available" for disinfection reactions. The ACC is a measure of the oxidizing power of the chemical and is expressed in terms of an equivalent quantity of chlorine. It is determined analytically by titration of iodine (I_2) and the chlorine compound of interest. The ACC is expressed as the percentage by weight of the original chlorine, 70.914. The reaction of chlorine and iodine is:

$$Cl_2 + 2I^- = 2Cl^- + I_2 \qquad (11-1)$$

The ACC is computed as follows:

$$\% \text{ ACC} = \frac{n\,MW_c}{MW_a} \times 100 \qquad (11-2)$$

where:

ACC = available chlorine content
n = number of moles of I_2 released

MW_a = molecular weight of compound of interest
MW_c = molecular weight of chlorine = 70.914

Hypochlorites. Calcium hypochlorite, a dry bleach, has been used widely since the late 1920s. Present-day commercial high-test calcium hypochlorite products contain at least 70 percent available chlorine and about 3 to 5 percent lime.

Calcium hypochlorite is readily soluble in water, varying from about 21.5 g/100 ml at 0°C to 23.4 g/100 ml at 40°C. Tablet forms dissolve more slowly than the granular materials, and provide a fairly steady source of available chlorine over an 18- to 24-hour period.

Granular forms usually are shipped in 35-lb or 100-lb drums, cartons containing 3¾-lb resealable cans, or cases containing nine 5-lb resealable cans. Tablet forms are shipped in 35-lb and 100-lb drums, and in cases containing twelve 3¾-lb or 4-lb resealable plastic containers.

Commercial sodium hypochlorite usually contains 12 to 15 percent available chlorine at the time of manufacture, and is available only in liquid form. It is marketed in carboys and rubber-lined drums of up to 50-gallon volume, and in trucks. All NaOCl solutions are unstable to some degree and deteriorate more rapidly than calcium hypochlorite, as shown in Table 11–7. The effect can be minimized by care in the manufacturing processes and by controlling the alkalinity of the solution. The greatest stability is attained with a pH close to 11.0, and in the absence of heavy metal cations. Storage temperatures should not exceed about 85°F; above that level, the rate of decomposition rapidly increases. While storage in a cool, darkened area greatly limits the deterioration rate, most large manufacturers recommend a maximum shelf life of 60 to 90 days. All hypochlorite solutions are corrosive to some degree and affect the skin, eyes, and other body tissues that they contact.

On-Site Manufacture of Hypochlorite

Chlorine gas, although economical, is risky to store and use. This is true especially in populated areas where the release of a 50-ton tank, for example, could require the evacuation of a 5-square-mile area. If this were to occur in a highly populated area, it is evident that the results could be disastrous. In order to eliminate the risks inherent in chlorine storage and use, some large cities, such as New York, Chicago, and Providence, have shifted to the use of concentrated solutions of sodium hypochlorite.

Sodium hypochlorite is the form of chlorine that comes closest to being as economically feasible as liquid chlorine as a chlorinating agent. It is clearly a safer form to handle than elemental chlorine. The major reason why sodium hypochlorite is not more widely used as a chlorinating agent is economic. Any slight dosage savings does not offset the increased cost of sodium hypochlorite compared to liquid chlorine. Other hypochlorite salts, such as calcium

Table 11-7. Stability of NaOCl Solutions.*

AVAILABLE CHLORINE TRADE PERCENT	CHLORINE, g/l	HALF-LIFE DAYS, 77°F
3	30	1700
6	60	700
9	90	250
12	120	180
15	150	100
18	180	60

* Reproduced from Culp and Culp.[20]

hypochlorite, are several times more costly than sodium hypochlorite solutions.

The relative safety of sodium hypochlorite is well known, but the large quantities that must be used present a problem. For example, a water treatment plant using 6 tons of chlorine per day requires 83 tons per day of hypochlorite because it is used in maximum concentrations of 15 percent. This requires transportation and storage costs for 30,000 tons per year. Consequently, the transportation and storage costs are substantial factors in the cost of a liquid sodium hypochlorite operation. Large quantities of the hypochlorite cannot be stored economically because of the short half-life of the 15 percent concentration.

A solution to the storage and safety problem is to use a system that can produce sodium hypochlorite at the treatment plant as needed. Several commercial systems have been introduced that can produce hypochlorite electrolytically from sodium chloride.[21-24]

The electrolytic cell converts the chloride ion to hypochlorite ion as follows:

$$NaCl + H_2O \xrightarrow{e} NaOCl + H_2\uparrow \qquad (11\text{--}3)$$

The raw materials required are salt (either in a brine solution or in seawater), power, and water. Because calcium and magnesium in the brine will precipitate and foul the system, it is desirable that both the salt and water be as free as possible of these constituents. Ordinary rock salt contains sufficient hardness that use of evaporated salt, solar salt, or southern rock salt is recommended. In installations where either softened, deionized water or tap water of sufficient quality is not available, water softeners with automatic regeneration cycles should be included in the system.

Chlorine Reactions

When chlorine is dissolved in water at temperatures between 49°F and 212°F, it reacts to form hypochlorous and hydrochloric acids:

$$Cl_2 + H_2O \rightarrow HOCl + H^+ + Cl^- \qquad (11\text{-}4)$$

This reaction is essentially complete within a very few seconds.[25,26] The hypochlorous acid ionizes or dissociates practically instantaneously into hydrogen and hypochlorite ions:

$$HOCl \leftrightharpoons H^+ + OCl^- \qquad (11\text{-}5)$$

These reactions represent the basis for the use of chlorine in most sanitary applications.

The extent of chlorine hydrolysis is controlled by the $[H^+]$ in equation (11-5). At pH values > 3 and with $[Cl_2]_t < 100$ mg/l, little molecular Cl_2 is present. Superchlorination of water will reduce the natural alkalinity in substantial quantities. For example, 1 mg/l Cl_2 reacts with 1.41 mg/l of alkalinity as $CaCO_3$.

$HOCl$ and OCl^- have considerably different capabilities for the destruction of microorganisms; therefore, it is important to know the distribution of $HOCl$ and OCl^-. Figure 11-6 shows the percentage dissociation of $HOCl$ and OCl^-. The K_i values cited in the graph are those proposed by Morris.[27]

Hypochlorite chorine forms also ionize in water and yield hypochlorite ions, which establish equilibrium with hydrogen ions:

$$CaOCl_2 + 2H_2O \rightarrow 2HOCl + Ca(OH)_2 \qquad (11\text{-}6)$$

$$NaOCl + H_2O \rightarrow HOCl + NaOH \qquad (11\text{-}7)$$

Chlorine has a strong affinity for other materials, particularly reducing agents, which sometimes exert a considerable chlorine demand. In these reactions, the chlorine atom manifests its great tendency to lose electrons to form chloride ions or organic chlorides. Reacting substances typically include Fe^{+2}, Mn^{+2}, NO_2^-, N_2S, and the greater part of the organic material present. The reactions with inorganic reducing substances are generally very rapid; those with organic materials are generally slow, the extent depending on the kind and form of excess available chlorine present.

Reactions with Ammonia

In aqueous systems, chlorine reacts with ammonia (NH_3) to form the chloramines:

$$NH_3 + HOCl = \underset{\text{monochloramine}}{NH_2Cl} + H_2O \qquad (11\text{-}8)$$

$$NH_2Cl + HOCl = \underset{\text{dichloramine}}{NHCl_2} + H_2O \qquad (11\text{-}9)$$

$$NHCl_2 + HOCl = NCl_3 + H_2O \qquad (11\text{--}10)$$
$$\text{trichloramine}$$
$$\text{(nitrogen trichloride)}$$

These chloramines have many different properties from those of the HOCl and OCl⁻ forms of chlorine. They exist in various proportions, depending on the relative rates of formation of monochloramine and dichloramine, which change with the relative concentrations of chlorine and ammonia as well as with pH and temperature. Above about pH 9, monochloramines exist almost exclusively; at about pH 6.5, monochloramines and dichloramines coexist in approximately equal amounts; between pH 6.5 and pH 4.5, dichloramines predominate; while trichloramines exist below about pH 4.5.

The point where all ammonia is converted to trichloramine or oxidized to free nitrogen is referred to as the breakpoint. Chlorination below this level is combined available residual chlorination; that above this level is free

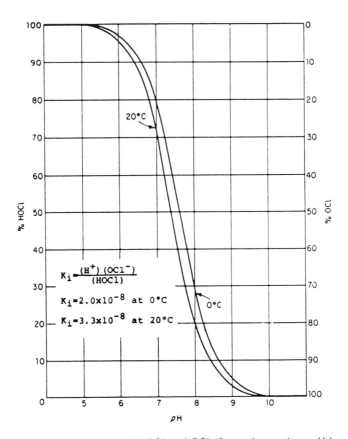

Fig. 11-6. Relative amounts of HOCl and OCl⁻ formed at various pH levels.[27]

available residual chlorination. Figure 11–7 shows a graphical illustration of the breakpoint reaction. The breakpoint curve results from the reactions shown in equations (11–8), (11–9), and (11–10).

Beginning about 1915, the influence of ammonia on the disinfecting capacity of chlorine was observed and reported by many investigators. The widespread adoption of chlorine–ammonia treatment followed recognition that the combination of chlorine and ammonia produced a more stable disinfecting residual than that produced by chlorine alone, and that the process could be applied to limit the development of objectionable tastes. Chlorine–ammonia treatment later declined in popularity, however, largely because of the realization of the superior bactericidal efficiency of hypochlorous acid. Currently, the principal application is for post-treatment, to provide long-lasting chloramine residuals in potable water distribution systems.

The rate of monochloramine formation is extremely rapid and sensitive to changes in the $[H^+]$.[29] The reaction is usually 90 percent complete in 1 minute or less. The reaction rate is at its maximum at pH 8.5. The production of $NHCl_2$ predominates in a pH range of 7 to 9 and at a $HOCl:NH_3$ ratio of 3:1 to 5:1 (by weight). At higher ratios or at lower pH values, dichloramines and trichloramines are formed. Monochloramine formation accounts for the chlorine residual peak shown in Fig. 11–7.

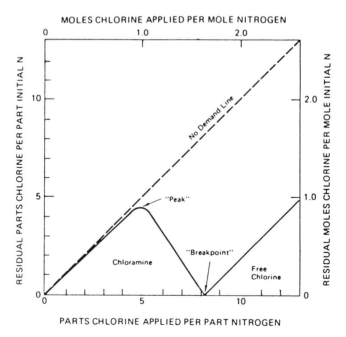

Fig. 11–7. Diagrammatic representation of completed breakpoint reaction.[28]

The rate of the ammonia reaction is:

$$V_r = k_1 [NH_3][HOCl]\gamma NH_3 \gamma HOCl/\gamma x \qquad (11\text{–}11)$$

where:

V_r = velocity of reaction
k_1 = rate constant
γ = activity coefficients (γx is for the activated complex)

At 25°C, the second-order rate constant k_1 between the uncharged molecules is 5.1×10^6 l/mol-sec.[30] The relative percentages of di- and monochloramine for equimolar Cl_2 and NH_3 and for 25 percent excess NH_3 are seen in Fig. 11–8.[31]

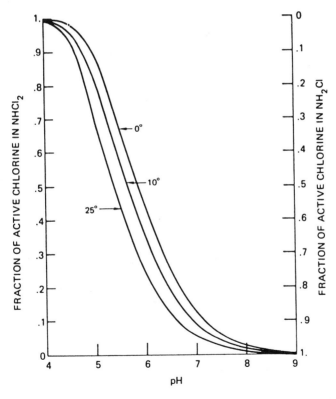

Fig. 11–8. Proportions of monochloramine (NH_2Cl) and dichloramine ($NHCl_2$) formed in water chlorination with equimolar concentrations of chlorine and ammonia at different temperatures (°C).[28]

In Fig. 11–7, the breakpoint becomes significant in the pH range of 7 to 9. When the $Cl_2:NH_3$ molar ratios are 0 to 1, NH_2Cl is formed. This creates the peak seen in Fig. 11–7. At molar ratios > 1, $NHCl_2$ is formed, but it decomposes rapidly. Therefore, when the molar ratio is between 1 and about 1.65, the chlorine residual decreases until the breakpoint occurs. At this point the NH_3 has been converted mainly to N_2 and some NO_3.[91] After the breakpoint, the chlorine residual exists as "free available" chlorine, that is, HOCl and OCl^-.

Except for the formation of NO_3, the stoichiometry of the breakpoint reaction is explained by two reactions:

$$HOCl + NH_3 = NH_2Cl + H_2O \qquad (11–12)$$

and:

$$2NH_2Cl + HOCl = N_2 + H_2O + 3H^+ + 3Cl^- \qquad (11–13)$$

Efficacy of Chlorine

It is important to know the relative resistances of viruses, bacteria, and cysts to chlorine to be certain that chlorination practice will be adequate. There are wide differences in the susceptibility of various pathogens to chlorine. It is now agreed that the waterborne pathogens in order of disinfection difficulty generally are bacteria $<$ viruses $<$ cysts, and that the difficulty increases in sewage or waters contaminated with sewage. The destruction of pathogens by chlorination is dependent upon water temperature, pH, time of contact, degree of mixing, turbidity, presence of interfering substances, and concentration of chlorine available. Chloramines formed if ammonia is present are much less effective disinfectants than free chlorine. For example, in one study, at normal pH values, approximately 40 times more chloramine than free chlorine was required to produce a near 100 percent kill of E. coli in the same time period.[32] For S. typhosa, this ratio was about 25:1. To obtain a near 100 percent kill with the same amounts of residual chloramine and free chlorine required approximately 100 times the contact period for chloramine, compared to free chlorine.

An empirical equation was developed by White to relate the required detention time to chlorine dose for a 99.6 to 100 percent kill.[18] The equation is:

$$T = aC^b \qquad (11–14)$$

where:

$T =$ contact time, minutes
$C =$ disinfectant dosage, mg/l
$a =$ proportionality constant for a given organism, pH value, and temperature
$b =$ slope of a log–log plot of T vs. C

Figure 11–9 shows plots of the destruction of microorganisms using chlorine. Note that most of the curves have a slope, or b value, of -1, and equation (11–14) can be reduced to:

$$T = \frac{a}{C} \qquad (11\text{–}15)$$

and:

$$a = CT \qquad (11\text{–}16)$$

Equation (11–16) can be used to compare disinfectants and also to establish either dose or contact time if a is computed from Fig. 11–9, or other similar plots. The percentage kill should be in the range of 99.6 to 100 percent.

Figure 11–10 illustrates the relative resistance of three viruses and *E. coli*. While the polio and Coxsackie viruses are considerably more resistant than *E. coli* HOCl, the adenovirus tested is apparently more sensitive. One-tenth milligram per liter of chlorine as HOCl destroyed 99 percent of *E. coli* in about 99 seconds. The same quantity of the adenovirus was destroyed in about one-third that time by the same amount of HOCl, but at this HOCl concentration the same amount of poliovirus required almost 85 minutes, more than five times longer, and the Coxsackie virus required over 40 minutes, more than 24 times longer than the *E. coli*.

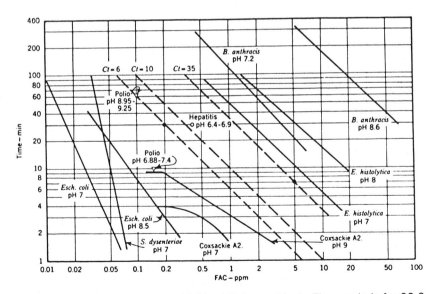

Fig. 11–9. Disinfection vs. free available chlorine residuals. Time scale is for 99.6–100 percent kill. Temperature was in the range 20–29°C, with pH as indicated. (Courtesy of AWWA.[33])

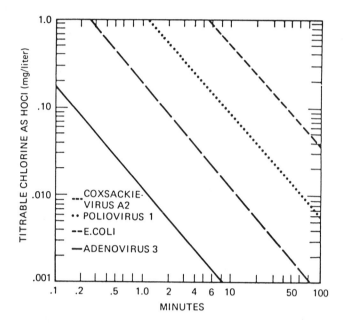

Fig. 11-10. Relationship between concentration and time for 99 percent destruction of *E. coli* and three viruses by hypochlorous acid (HOCl) at 0-6°C.

An early study by Butterfield et al. considered the percentages of inactivation as a function of time for *E. coli, Enterobacter aerogenes, Pseudomonas aeruginosa, Salmonella typhi,* and *Shigella dysenteriae.* [34] The results of Butterfield's free available chlorine research show that HOCl is more effective than OCl⁻, and the higher temperatures, 20 to 25°C, lowered the contact time and chlorine dosage required at 2 to 5°C for a 99 to 100 percent kill of *E. coli.*

Generally, the HOCl form has been considered to be a far more effective disinfectant than OCl⁻. Several investigators have reported that HOCl is 70 to 80 times as bactericidal as OCl⁻, and that increasing the pH reduces germicidal efficiency because most of the free chlorine exists as the less microbiocidal OCl⁻ at the higher pH levels. However, one study reported that one virus (Poliovirus 1) was more rapidly inactivated at pH levels (pH = 10) where the free chlorine is in the form of OCl⁻ rather than HOCl.[35] This finding is contrary to the conclusions reached by many other researchers.

Both pH and temperature have a marked effect on the rate of virus kill by chlorine. Table 11-8 summarizes data on virus inactivation rates at varying pH at 10°C in filtered secondary sewage effluent. The constants in Table 11-8 were determined from experimental results using model f2 virus that was seeded in filtered secondary sewage effluent buffered to the desired pH

Table 11-8. Viral Inactivation Rate Constants at Varying pH
of Sewage at 10°C.*

| | FLASH MIXING OF 30 mg/l CHLORINE | | |
SEWAGE pH	VIRAL INACTIVATION RATE CONSTANT (MIN^{-1})	FREE CHLORINE DURATION (SEC)	VIRAL KILL (PERCENT)
5.0	120.0	110.0	>99.9999
5.5	72.0	24.0	>99.9999
6.0	44.0	10.0	99.93
6.5	26.0	5.6	92.0
7.0	16.0	5.4	76.0
7.5	9.5	5.0	55.0
8.0	5.6	5.0	40.0
8.5	3.5	4.9	25.0[a]
9.0	2.0	4.8	15.0[a]
9.5	1.2	4.6	10.0[a]

* Reference 20.
[a] Essentially control survival.

value. Chlorine solution was flash-mixed at a dose of 30 mg/l, and the time of persistence of any free chlorine as well as virus survival was determined.[36]

Other studies show that decreasing the pH from 7.0 to 6.0 reduced the required virus inactivation time by about 50 percent, and that a rise in pH from 7.0 to 8.8 or 9.0 increased the inactivation period about six times. In the presence of low free residuals, the virus inactivation rate is markedly affected by variations of temperature and pH. Clarke's work on virus destruction by chlorine suggests that the temperature coefficient for a 10°C change is in the range of 2 to 3, indicating that the inactivation time must be increased two to three times when the temperature is lowered 10°C. Data also indicate that the chlorine concentration coefficient lies in the range 0.7 to 0.9, which means that the inactivation time is reduced a little less than half when the free chlorine concentration is doubled. Therefore, to increase virus kill there is some advantage in increasing the contact time instead of raising the chlorine content. This observation has been confirmed, in that one study found increased contact time to be more important than increased chlorine dosage in the inactivation of the T2 phage in excess of 90 percent.[37] Inactivation of poliovirus in water also has been found to be more dependent on contact time.

Laboratory studies were conducted to assess the effect of chlorine over the pH range 6 to 10 on five types of polioviruses and two types of Coxsackie viruses.[38] The results show that chlorine is at least four times as effective at pH 6 as at pH 9.

The effect of chlorine on a mixture of cysts consisting predominantly of

Entamoeba histolytica and *Entamoeba coli* has been reported.[39] The cysticidal efficiency was related to the residual concentration of halogen measured chemically after 10 minutes of contact time. It was concluded that the chemical species of free and combined halogens that prevail in low-pH waters were superior cysticides when compared to forms that predominate in waters of high pH. Cysts were not affected by the $[H^+]$ concentration alone (pH 5 to 10); however, the possibility that a low pH enhanced the ability of the halogen to penetrate the cyst wall cannot be dismissed. The most rapid-acting cysticide was found to be the hypochlorous acid (HOCl) form of free available chlorine. The results of this study show that cysts are not likely to survive where prechlorination to a free residual is practiced.

The AWWA Committee report on viruses in water concluded that:

in the prechlorination of raw water, any enteric virus so far studied would be destroyed by a free chlorine residual of about 1.0 ppm, provided this concentration could be maintained for about 30 minutes and that the virus was not embedded in particulate material. In postchlorination practices where relatively low chlorine residuals are usually maintained, and in water of about 20°C and pH values not more than 8.0 to 8.5, a free chlorine residual of 0.2 to 0.3 ppm would probably destroy in 30 minutes most viruses so far examined.[16]

The committee also states "there is no doubt that water can be treated so that it is always free from infectious microorganisms—it will be biologically safe. Adequate treatment means clarification (coagulation, sedimentation, and filtration), followed by effective disinfection. Effective disinfection can be carried out only on water free from suspended material." They caution that "positive coliform tests certainly indicate possible virus contamination, but a negative coliform test may not indicate freedom from viruses."

There is a need for improved virus detection techniques in light of the greater susceptibility of the coliform indicator organism to destruction by chlorination. In the meantime, the probability of viral survival can be minimized through low turbidity, low pH, high free chlorine residual, adequate mixing, and long contact times.

Chloramines

The bactericidal properties of the chloramines at various pH values and temperatures are summarized in Table 11–9.[40] Dichloramine was a more effective disinfectant than monochloramine because lower dosages and contact times were required at lower pH values. A 100 percent kill with the same contact time requires about 25 times as much chloramine as free chlorine. Conversely, for the same kill using equal quantities of chlorine and chloramine, approximately 100 times the contact time is required for chloramine.

Table 11–9. Effect of pH and Temperature on Contact Time for
100 Percent Kill of *E. coli* by Chloramines.*

pH	NH₂Cl (%)ᵃ	DOSAGE (mg/l) 2–5°C	DOSAGE (mg/l) 20–25°C	CONTACT TIME (MIN) 2–5°C	CONTACT TIME (MIN) 20–25°C
6.5	35.		0.3		60.
7.0	51.	0.6	0.3	180.	90.
7.8	84.		0.3		180.
8.5	98.	1.8	0.6	20.	120.
9.5	100.	1.8	0.9	90.	180.
10.5	100.		1.5		180.

* Reference 40.
ᵃ Estimated, balance is NHCl₂.

Kelly and Sanderson researched the effect of combined chlorine on poliovirus and Coxsackie virus at 25°C and a pH value of 7.0.[41] A concentration of at least 9.0 mg/l was necessary for inactivation of polioviruses with a contact time of 30 minutes and of 6 mg/l over 60 minutes; 0.5 ppm required 7 hours of contact.

Figure 11–11 shows the comparisons that were made among two enterovi-

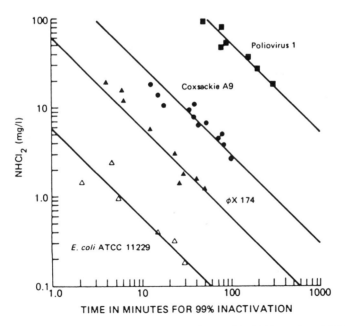

Fig. 11–11. Inactivation of various microorganisms with dichloramine (NHCl₂) at pH 4.5 and 15°C.[92]

ruses, Poliovirus 1 and Coxsackie virus A9, the bacteriophage ϕX-174, and *E. coli*. Poliovirus 1 is the most resistant organism, and *E. coli* the least resistant to $NHCl_2$. Dichloramine was observed to be a more effective disinfectant than monochloramine in these experiments.

Table 11–10 summarizes the various types of chlorine dosages required for 99 percent inactivation of *E. coli* and Poliovirus 1. Johanneson cites the individual studies in this table.[42]

Rapid Mixing

In addition to the chemical and water quality considerations discussed above, complete and uniform mixing of the chlorine with the water to be disinfected is important. Many full-scale chlorination facilities are inadequately designed, with the chlorine applied directly to the chlorine contact basin, sometimes even without a diffuser. Flows in the contact basins are usually highly stratified, resulting in very little dispersion of the chlorine. Disinfection is one process for which any measurable short-circuiting is completely ruinous to

Table 11–10. Dosages of Various Chlorine Species Required for 99 Percent Inactivation of *E. coli* and Poliovirus 1.*

TEST MICRO-ORGANISM	DISINFECTING AGENT	CONCEN-TRATION (mg/l)	CONTACT TIME (MIN)	Ct	pH	TEMPER-ATURE (°C)
E. coli	Hypochlorous acid (HOCl)	0.1	0.4	0.04	6.0	5
	Hypochlorite ion (OCl⁻)	1.0	0.92	0.92	10.0	5
	Monochloramine (NH₂Cl)	1.0	175.0	175.0	9.0	5
		1.0	64	64.0	9.0	15
		1.2	33.5	40.2	9.0	25
	Dichloramine (NHCl₂)	1.0	5.5	5.5	4.5	15
Poliovirus 1	Hypochlorous acid (HOCl)	1.0	1.0	1.0	6.0	0
		0.5	2.1	1.05	6.0	5
		1.0	2.1	2.1	6.0	5
		1.0	1.0	1.0	6.0	15
	Hypochlorite ion (OCl⁻)	0.5	21	10.5	10.0	5
		1.0	3.5	3.5	10.0	15
	Monochloramine (NH₂Cl)	10	90	900	9.0	15
		10	32	320	9.0	25
	Dichloramine (NHCl₂)	100	140	14,000	4.5	5
		100	50	5,000	4.5	15

* Reference 42.

process efficiency. Attention to tank shape mixing and to proper baffling is critical.

Studies at the Sanitary Engineering Research Laboratory of the University of California evaluated the importance of rapid initial mixing in chlorination.[43] The researchers believed that prolonged segregation of the chlorine from the bacteria results in poorer performance. They concluded that rapid initial mixing allows more lethal residuals to come in contact with the bacteria and act as killing agents.[44]

Collins also concluded that backmixing that occurs in a propeller-mixed tank may contribute to poorer disinfection.[43] It was noted that reactions may take place between the more bactericidal residuals initially formed as chlorine enters the reactor and the more complex residuals that form over a longer time. Thus, the simple and more bactericidal compounds may be "destroyed" before they can act. Collins evaluated the effects of adding a tubular reactor with a short ($t = 0.12$ minute) detention time upstream of a constant flow stirred tank reactor (CSTR) ($t = 37.5$ minutes), at both pilot and full scale.

Although the total mean residence time was essentially the same, the kill obtained when the chlorine was added at the head of the tubular reactor, rather than directly to the CSTR, was much greater. The coliform survival ratio—the parameter used to indicate bactericidal effectiveness—was about one order of magnitude lower in this case. However, the coliform survival ratio for the CSTR alone was approximately the same regardless of whether the chlorine was added directly to the CSTR or to the tubular reactor. The increase in kill was attributed solely to the more rapid mixing that occurred in the tubular reactor. Another result is that for each run the chlorine residual in the CSTR was slightly lower when the chlorine was applied to the tubular reactor. It was concluded that the various reactions involved were driven to a greater degree of completion because of the more rapid mixing.

Collins concluded that improved performance is provided by applying the chlorine to plug flow reactors designed to provide rapid initial mixing. Where mechanically stirred mixers are necessary, the chlorine solution should be applied to the stream ahead of the mixer. The two process streams (chlorine solution and water) would not be applied directly and separately to the mechanical mixer.

IODINE

Historically, iodine is better known for its use as a disinfectant for abrasions and other skin wounds than for disinfecting water systems. It is also used in medical facilities as a sanitizing compound. It has only been used in piped water systems in emergency situations, and then only in small, independent utilities.[28]

Iodine is the only common halogen that is a solid at room temperature, and it possesses the highest atomic weight, 126.91. Its formula weight is 253.84. Of the four common halogens, it is the least soluble in water, has the lowest standard oxidation potential for reduction to halide, and reacts least readily with organic compounds. Iodine is blackish-gray in color, and has a specific gravity of 4.93. Its melting point is 113.6°C, and its boiling point 184°C. Taken collectively, its characteristics mean that low iodine residuals should be more stable and persist longer in the presence of organic material than corresponding residuals of any of the other halogens.

Iodine reacts with water in the following manner:

$$I_2 + H_2O \leftrightharpoons HOI + H^+ + I^- \tag{11-17}$$

$$HOI \leftrightharpoons H^+ + OI^- \tag{11-18}$$

and therefore:

$$K_H = \frac{[HOI][H^+][I^-]}{[I_2]} \tag{11-19}$$

K_H at 25°C is 3×10^{-13} M^2/I^2.[45] Consequently, the extent of hydrolysis for I_2 is less than for Cl_2 and Br_2 and is controlled by $[H^+]$. The effect of pH on I_2 and HOI is shown in Table 11-11.

At pH values above 4.0, hypoiodous acid (HOI) undergoes dissociation, as shown in equation (11-18). At pH values over 8.0, HOI is unstable and will not form the hypoiodite ion, but decomposes according to the following equation:

$$3HOI + 2(OH^-) \leftrightharpoons HIO_3 + 2H_2O + 2I^- \tag{11-20}$$

This eventually results in the formation of iodate (IO_3^-) and iodide (I^-).

McAlpine suggested the reaction of I_2 with NH_3 to be as follows:[46]

$$I_2 + 2NH_3 = NH_2I + NH_4^+ + I^- \tag{11-21}$$

An average K_{eq} of 1.8 was determined when the molar ratio of NH_3 to I_2 was on the order of 4,000:1, which is unlikely in treated waters.

Concern has been expressed over the possible harmful effects of iodinated water on the thyroid function, as well as objectionable tastes and odors. However, field analyses of human users of water treated with iodine have failed to disclose any adverse effects on health. A study by Black indicated that few people were able to detect a concentration of 1.0 mg/l elemental iodine in water by either color, taste, or odor.[47] Black also indicated that

Table 11–11. Effect of pH on the Hydrolysis of Iodine.*,**

pH	CONTENT OF RESIDUAL (%)		
	I_2	HOI	OI^-
5	99	1	0
6	90	10	0
7	52	48	0
8	12	88	0.005

* Reproduced from Black et al.[47] (Courtesy of the American Water Works Association.)
** Total iodine residual, 0.5 ppm.

the iodine ion cannot be detected in concentrations exceeding 5.0 mg/l. The upper limit of acceptability was judged to be at a dose of 16 mg/l.

Iodine is not likely to combine with ammonia, and does not combine with many other organic compounds; therefore, the production of tastes and odors is minimized. Iodine does not produce any detectable taste or odor in water containing several ppb phenol, unlike chlorine.

The bactericidal effects of iodine show only a slight difference between pH 5.0 and pH 7.0. A pH of 9 or above results in a reduction in efficiency. Hypoiodus acid formed at higher pH values is less bactericidal than free iodine. At pH 7.0, it was found that a 99.99 percent kill of the most resistant of six bacterial strains tested could be achieved in 5 minutes with a free iodine residual of 1.0 mg/l.[48]

In general, it has been found that the bactericidal action of iodine resembles that of chlorine with respect to temperature and pH, but higher concentrations of iodine are required to produce comparable kills under similar conditions. However, considering the higher organisms and especially cysts and spores, iodine and bromine may have some advantages.[49] HOI destroys viruses at a rate considerably faster than that achieved by I_2. IO_3^- and I^- are essentially inert as viricides, and I_3 has little or no viricidal activity.

Chang and Morris determined that I_2 concentrations of 5 to 10 mg/l were effective against all types of waterborne pathogenic organisms within 10 minutes, and at room temperature.[50] Germicidal activity is maintained over a pH range 3 to 8. Chambers et al. studied the effect of $[I_2]$, pH, contact time, and temperature on 13 enteric bacteria in the laboratory.[28,51] At 2 to 5°C, some bacterial organisms needed three to four times as much I_2 for inactivation ($\approx$ 99.9 percent kill) at pH 9.0 as at pH 7.5.

The results of several studies on the efficacy of I_2 are summarized in Table 11–12. At pH 6, I_2 is nearly as effective a viricide as HOCl. However, iodine seems unlikely to become a municipal water supply disinfectant in any broad sense because of its high cost and restricted availability when compared to chlorine.

Table 11-12. Concentrations of Iodine and Contact Times
Necessary for 99 Percent Inactivation of Polio- and f2
Viruses With Flash Mixing.*

TEST MICROORGANISM	IODINE, (mg/l)	CONTACT TIME (MIN)	Ct^a	pH	TEMPERATURE (°C)
f$_2$ Virus	13	10	130	4.0	5
f$_2$ Virus	12	10	120	5.0	
f$_2$ Virus	7.5	10	75	6.0	
f$_2$ Virus	5	10	50	7.0	
f$_2$ Virus	3.3	10	33	8.0	
f$_2$ Virus	2.7	10	27	9.0	
f$_2$ Virus	2.5	10	25	10.0	
f$_2$ Virus	7.6	10	76	4.0	25–27
Poliovirus 1	30	3	90	4.0	
f$_2$ Virus	64	10	64	5.0	
f$_2$ Virus	4.0	10	40	6.0	
Poliovirus 1	1.25	39	49	6.0	
Poliovirus 1	6.35	9	57	6.0	
Poliovirus 1	12.7	5	63	6.0	
Poliovirus 1	38	1.6	60	6.0	
Poliovirus 1	30	2.0	60	6.0	
f$_2$ Virus	3.0	10	30	7.0	
Poliovirus 1	20	1.5	30	7.0	
f$_2$ Virus	2.5	10	25	8.0	
f$_2$ Virus	2.0	10	20	9.0	
f$_2$ Virus	1.5	10	15	10.0	
Poliovirus 1	30	0.5	15	10.0	

* Reference 28.
^{a} Concentration of iodine multiplied by contact time.

BROMINE

Bromine, Br_2, is a dark, brownish-red, heavy liquid.[18] It gives off a heavy, brownish-red vapor with a sharp, penetrating and suffocating odor at room temperature. Liquid bromine is corrosive and destructive to organic tissues. With an atomic number of 35, a molar weight of 159.832, a specific gravity of 3.12, and a boiling point of 58.78°C, bromine is the only nonmetallic element that is a liquid at room temperature. It is produced by the oxidation of bromine-rich brines (0.05 to 0.6 percent Br^-) with chlorine. Bromine is then stripped with air or steam and is collected as liquid Br_2.

Bromine was first applied to water as a disinfectant in the form of liquid bromine (Br_2),[28] and it can be added as bromine chloride gas (BrCl).[52] Oxidation with aqueous chlorine yields either bromine or hypobromous acid (HOBr). Both bromine and bromine chloride hydrolyze to hypobromous acid:

$$Br_2 + H_2O \rightarrow HOBr + H^+ + Br^- \tag{11-22}$$
$$BrCl + H_2O \rightarrow HOBr + H^+ + Cl^- \tag{11-23}$$

Molecular bromine is present in water at slightly acid pH and high bromide concentrations, and has an equilibrium constant of 5.8×10^{-9} at 25°C for equation (11–22). Bromine chloride has a higher hydrolysis constant than molecular bromine, and typically does not exist as the molecular form under conditions of water treatment. The ratio of molecular bromine to hypobromous acid depends on both pH and bromide concentration, as follows:

$$\log \frac{(Br_2)}{(HOBr)} = \log (Br^-) - pH + 8.24 \tag{11-24}$$

For a solution containing 10 mg/l bromide, and a pH of 6.3, 1 percent of the bromine is Br_2. At lower bromide concentrations, or higher pH, aqueous bromine occurs principally as hypobromous acid. Reaction (11–25) has a dissociation constant of 2×10^{-9} at 25°C:

$$HOBr \rightleftharpoons H^+ + OBr^- \tag{11-25}$$

Above pH 8.7 and at 25°C, the hypobromite ion (OBr^-) is the major form of bromine. Lower temperatures decrease equilibrium values, thus increasing the pH range in which hypobromous acid is the major form of bromine.

Bromine and bromine chloride also react with basic nitrogen compounds to form combined bromine or bromamines:[42,53,54]

$$HOBr + NH_3 \rightarrow NH_2Br + H_2O \tag{11-26}$$
$$NH_2Br + HOBr \rightarrow NHBr_2 + H_2O \tag{11-27}$$
$$NHBr_2 + HOBr \rightarrow NBr_3 + H_2O \tag{11-28}$$

Kinetic data have shown that at pH values of 6, 7, and 8 a breakpoint occurs for bromine at a N:Br mole ratio of 2:3.[55] NBR_3 is the most stable species when there is excess Br_2. However, $NHBr_2$ is most stable in the presence of excess NH_3. The stability of both bromines decreases at high pH.

The bromine breakpoint for ammonia (NH_3) solutions corresponds to 17 mg/l bromine for 1 mg/l of ammonia nitrogen (NH_3–N). At this point, a minimum of bromamine stability occurs as NH_3–N is oxidized to nitrogen gas, as follows:[56]

$$3HOBr + 2NH_3 \rightarrow N_2 + 3HBr + 3H_2O \tag{11-29}$$

At bromine-to-ammonia ratios higher than this and in the acid pH range, nitrogen tribromide (NBr$_3$) is stable. At lower bromine concentrations, dibromamine (NHBr$_2$) can be present but unstable. Organic bromamines are also formed, but there is little information on their forms or stability.

Kruse et al. found that 4 mg/l of bromine as hypobromous acid resulted in 5 logs or disinfection of *E. coli* in 10 minutes at 0°C.[57] At pH 4.5 in the presence of significant Br$_2$, 2 mg/l bromine at 0°C gave 4 logs of *E. coli* disinfection in 3 minutes. However, Kruse also reported that an increase in the molar concentration of bromide reduced its effectiveness.

Using 4 mg/l bromine at pH 7.5 and 0°C, 0.1 M bromide decreased *E. coli* disinfection to 2 logs of inactivation from 4.5 logs at 0.001 M bromide. This is in conflict with observations made at low pH, since high bromide should also produce bromine. The data at pH 4.5 showed 4 logs inactivation with a $C \cdot t$ of 6, compared to 5 logs inactivation from a $C \cdot t$ of 40 for hypobromous acid.

Reoviruses and polioviruses have been evaluated in bromine disinfection studies. Several investigators have studied the effect of HOBr on single and aggregated particles at pH = 7.0 and 2°C.[28] The amount of aggregation affected the inactivation rates. For example, reoviruses, as single particles, required only 1 second for a 1,000-fold decrease in their numbers at 0.46 mg/l HOBr as Br$_2$.[58] The time was doubled for aggregated particles for a 1,000-fold decrease.

Many studies, with measured, controlled residual concentrations, demonstrate that the degree of aggregation, or clumping, had a marked effect on the apparent, observed inactivation rates with reovirus and Type 1 Poliovirus.[58-61] Single particles of reoviruses required only 1 second for 3 logs of inactivation at pH 7 and 2°C at 0.46 mg/l HOBr as bromine. Aggregated samples required longer contact times to virus inactivation. At 3 logs of inactivation, the time required doubled for the same concentration condition.[61]

Floyd studied the inactivation of single poliovirus particles in buffered, distilled water.[58] Table 11–13 gives the calculated $C \cdot t$ values required to yield 99 percent inactivation for different forms of bromine near 1 mg/l expressed as bromine. Dibromamine, nitrogen tribromide, and hypobromous acid are less efficient at higher concentrations, while the hypobromite ion is more effective as the concentration is increased.[58]

Four milligrams per liter of bromine as hypobromous acid at pH 7 and 0°C resulted in 3.7 logs of inactivation of f2 *E. coli* phage virus in 10 minutes.[57] At higher bromide concentrations and lower pH, where Br$_2$ becomes the principal form of bromine, the rates of inactivation increased. A $C \cdot t$ of 40 was required for 3.7 logs inactivation at pH 7.0 and 0°C compared to a $C \cdot t$ of 1 for 4.5 logs inactivation at 0°C and pH 4.8.

In dose–response experiments in distilled water at pH levels from pH 4 to 10, bromine was found to be the most effective of the halogens.[62]

Table 11-13. Exposure $(C \cdot t)^*$ to Various Bromine Compounds Required for 99 Percent Inactivation of Poliovirus 1, Mahoney.**

CHEMICAL FORM	Ct	TEMPERATURE (°C)	pH
Dibromamine (NHBr₂)	1.2	4	7.0
Nitrogen tribromide (NBr₃)	0.19	5	7.0
Bromine (Br₂)	0.03	4	5.0
Hypobromite ion (OBr⁻)	0.01	2	10.0
Hypobromous acid (HOBr)	0.24	2	7.0
Hypobromous acid	0.21	10	7.0
Hypobromous acid	0.06	20	7.0

* Concentration of compound multiplied by contact time.
** Reproduced from Floyd et al. as reported in Reference 28.

Under conditions likely to be found during the treatment of natural water supplies, free and combined bromine appear to be practical, effective cystides, at least as far as cysts of E. histolytica are concerned.

The efficacy of BrCl and Cl₂ on poliovirus was tested in experiments conducted at pH 6 to compare HOBr and HOCl.[63] Concentrations of BrCl and Cl₂ were premixed and then inoculated with 1 ml of poliovirus (10^5 to 10^6 PFU/ml). The results were that BrCl (HOBr) is two to three times more effective than Cl₂ (HOCl) for inactivation of poliovirus.

OZONE

It has been reported that "nearly a thousand cities purify their water with ozone," with the largest such plant being 238 mgd.[64] Ozone is one of the two most potent and effective germicides used in water treatment. Only free residual chlorine can approximate it in germicidal power. The use of ozone for disinfection of municipal drinking water actually antedates chlorination.

The advantages of using ozone are its high germicidal effectiveness, which is the greatest of all known substances, even against resistant organisms such as viruses and cysts; its ability to ameliorate many problems of odor, taste, and color in water supplies; and the fact that upon decomposition, the only residual material is more dissolved oxygen. In addition, its potency is unaffected by pH or ammonia content.

Ozone has the molecular formula O_3, a molecular weight of 48 g/mol, and a density, as a gas, of 2.154 g/l at 0°C and 1 atm. Ozone is about 13 times more soluble in water than is oxygen. At saturation in water at 20°C, a 2 percent weight mixture of ozone and oxygen contains about 11 mg of ozone and 40 mg of oxygen per liter.

Ozone is a powerful oxidant that reacts rapidly with most organic and many inorganic compounds.[93] During disinfection, only minor amounts of ammonia are oxidized when ozone is used. The limited reaction of ozone with ammonia is desirable, but its fast reaction rate with most organic and many inorganic compounds further shortens its persistence in water.

Ozone must be produced at its point of usage because of a short half-life of about 40 minutes at pH 7.6. It is generated by passing dry air between two high-potential electrodes to convert oxygen into ozone. Improvements in technology of ozone production have bettered the reliability and economy of its generation. The development of dielectric, ozone-resistant materials has simplified the design and operation of ozone generators; in addition, new electronic designs have increased the efficiency of ozone production. Approximately twice the percent of ozone by weight is obtained if oxygen, rather than air, is used as the feed stream. Power requirements are about 13 to 22 kWh/kg of ozone that is generated from air, and approximately half that when oxygen is used.[94] Compressors and dryers may increase these requirements by 20 to 50 percent.

Ozone has been found to be many times more effective than chlorine in inactivating the virus of poliomyelitis. Under experimental conditions, an identical dilution of the same strain and pool of virus, when exposed to chlorine in residual amounts of 0.5 to 1.0 mg/l and to ozone in residual amounts of 0.05 to 0.45 mg/l, was inactivated within 2 minutes by ozone, whereas 1½ to 2 hours were required for inactivation by chlorine.

Ozone has also been reported to be several times faster than chlorine in its germicidal effects on *Endamoeba histolytica*. For most cases, an ozone residual of 0.1 mg/l for 5 minutes is adequate to disinfect a water low in organics and free of suspended material.[65] Organic material does exert a significant ozone demand.

Nonetheless, the use of ozone does have disadvantages, as summarized by Morris.[66] Because it must be produced electrically as it is needed and cannot be stored, it is difficult to adjust treatment to variations in load or to changes in raw water quality with regard to ozone demand. As a result, ozone historically has been found most useful for supplies with low or constant demand, such as groundwater sources. Moreover, although ozone is a highly potent oxidant, it is quite selective and by no means universal in its action. Some otherwise easily oxidized substances, such as ethanol, do not react readily with ozone.

In some instances in which river waters rather heavily polluted with organic matter have been ozonated, results have indicated a rupture of large organic molecules into fragments more easily metabolized by microorganisms. This fragmentation, coupled with the inability of ozone to maintain an active concentration in the distribution system, led to increased slime growths and consequent deterioration of water quality during distribution.

In many ways, the desirable properties of ozone and chlorine as disinfectants are complementary. Ozone provides fast-acting germicidal and viricidal potency, commonly with beneficial results regarding taste, odor, and color. Chlorine provides sustained, flexible, controllable germicidal action that continues to be beneficial during distribution. Thus, it would seem that a combination of ozonation and chlorination might provide an almost ideal form of water supply disinfection.

This procedure is being employed in a number of places, most particularly in the Netherlands at Amsterdam on Rhine River water. Ozone is used initially for phenol oxidation, viral destruction, and general improvement in physical quality; after normal water treatment, post-chlorination is used to assure hygienic quality throughout the system. It seems logical that this approach will find increasing use in the United States as improvements in ozone-generating equipment are made.

Keller et al. studied ozone inactivation of viruses by using both batch tests and a pilot plant with a 10 gpm (38 l/min) flow rate.[67] Inactivation of Poliovirus 2 and Coxsackie virus B3 was more than 99.9 percent in the batch tests with an ozone residual of 0.8 and 1.7 mg/l and a 5-minute contact time. Greater than 99.999 percent inactivation of Coxsackie virus B3 was achieved in the pilot plant with an ozone dosage of 1.45 mg/l, which provided an ozone residual of 0.28 mg/l after 1 minute in lake water.

A wide range of ozone dosages, contact times, and percentages of inactivation have been studied. Table 11–14 summarizes this information where $C \cdot t$

Table 11–14. Concentration of Ozone and Contact Time Necessary for 99 Percent Inactivation of *E. coli* and Poliovirus 1.*,**

TEST MICROORGANISMS	OZONE (mg/l)	CONTACT TIME (MINUTES)	Ct†	pH	TEMPERATURE (°C)
E. coli	0.07	0.083	0.006	7.2	1
	0.065	0.33	0.022	7.2	1
	0.04	0.50	0.02	7.2	1
	0.01	0.275	0.027	6.0	11
	0.01	0.35	0.035	6.0	11
	0.0006	1.7	0.001	7.0	12
	0.0023	1.03	0.002	7.0	12
	0.0125	0.33	0.004	7.0	12
Polio 1	<0.3	0.13	<0.04	7.2	5
	0.245	0.50	0.12	7.0	24
	0.042	10	0.42	7.0	25
	<0.03	0.16	<0.005	7.0	20

* Reference 28.
** Conversion factors may be found in Appendix A.
† Concentration of ozone multiplied by contact time.

relationships are noted. It is seen that the $C \cdot t$ product varies over a broad range from 10^{-3} to 4.2×10^{-1}.

CHLORINE DIOXIDE

Sir Humphrey Davey first prepared chlorine dioxide (ClO_2) in the early nineteenth century. By combining potassium chlorate ($KClO_3$) and hydrochloric acid (HCl), he produced a greenish-yellow gas, which he named "euchlorine." Later, this gas was found to be a mixture of chlorine dioxide and chlorine. Chlorine dioxide was first used in 1944 in the United States at the water treatment plant in Niagara Falls, New York, to control phenolic tastes and odors.[68]

Chlorine dioxide is an unstable gas, explosive at temperatures above $-40°C$.[18] As a result, it is normally generated at the point of use. ClO_2 is seldom used as a gas because concentrations above 11 percent may cause mild explosions. It explodes with changes in its environment such as rising temperatures and exposure to light, and when it contacts an organic substance. The mere transfer from one container to another can cause an explosion.

Chlorine dioxide has an odor that is similar to that of chlorine and chlorine monoxide.[18] It has a density of 2.4 (air = 1), a boiling point of 11°C, and a melting point of $-59°C$. The vapor pressure of ClO_2 is given by the following equation:[69]

$$\log P = \frac{7.74 \times 1.375}{T} \qquad (11\text{--}30)$$

where:

P = vapor pressure, torr
T = absolute temperature, °K

The equation is valid for a range in temperature of -38 to $+40°C$. Chlorine dioxide reacts with a wide variety of organic and inorganic chemicals under conditions that are usually found in water treatment systems. However, two important reactions do not occur. Chlorine dioxide per se does not react to cause the formation of trihalomethanes (THM's). Chlorine dioxide does not react with ammonia, but will react with other amines.[70] The amine structure determines reactivity.

Both sodium chlorate ($NaClO_3$) and sodium chlorite ($NaClO_2$) may be used to generate chlorine dioxide, but the use of $NaClO_3$ is the more efficient process. Commercial processes used in North America for large-scale production of chlorine dioxide are based on the following three reactions:

$$2NaClO_3 + H_2SO_4 + SO_2 \rightarrow 2ClO_2 + 2NaHSO_4 \qquad (11-31)$$

$$2NaClO_3 + CH_3OH + H_2SO_4 \rightarrow 2ClO_2$$
$$+ HCHO + Na_2SO_4 + 2H_2O \qquad (11-32)$$

$$NaClO_3 + NaCl + H_2SO_4 \rightarrow ClO_2 + \frac{1}{2}Cl_2 + Na_2SO_4 + H_2O \quad (11-33)$$

A detailed review of chlorine dioxide chemistry is given by Gall.[71]
Chlorine dioxide can be prepared from chlorine and sodium chlorite as follows:

$$Cl_2 + H_2O \rightarrow HOCl + HCl \qquad (11-34)$$

$$HOCl + HCl + 2NaClO_2 \rightarrow 2ClO_2 + 2NaCl + H_2O \qquad (11-35)$$

$$Cl_2 + 2NaClO_2 \rightarrow 2ClO_2 + 2NaCl \qquad (11-36)$$

The theoretical weight ratio of sodium chlorite to chlorine is 1.00:0.39, and with 80 percent available sodium chlorite, the weight ratio is 1:0.30. In practice, the recommended chlorite-to-chlorine ratio is 1:1.[71] The excess chlorine lowers the pH, thereby increasing the reaction rate and optimizing the yield of chlorine dioxide.

Chlorine dioxide also may be prepared from sodium hypochlorite (NaOCl) and sodium chlorite, and follows the reaction in equation (11–36).

The addition of a strong acid such as sulfuric acid (H_2SO_4) or hydrochloric acid to sodium chlorite, as shown in the following reactions, will also yield chlorine dioxide:

$$10NaClO_2 + 5H_2SO_4 \rightarrow 8ClO_2 + 5Na_2SO_4 + 2HCl + 4H_2O \quad (11-37)$$

$$5NaClO_2 + 4HCl \rightarrow 4ClO_2 + 5NaCl + 2H_2O \qquad (11-38)$$

One of the few stable nonmetallic inorganic free radicals, chlorine dioxide does not contain available chlorine in the form of hypochlorous acid or hypochlorite ion (OCl^-).[70] In terms of available chlorine, chlorine dioxide has 263 percent of, or more than 2.5 times, the oxidizing capacity of chlorine:

$$\frac{5e^- \times 35.45}{67.45} \times 100 = 263\% \qquad (11-39)$$

The weight ratio of chlorine dioxide to available chlorine is 67.45:35.45 or 1.9, but this increased oxidizing capacity is rarely realized in water treatment applications. The reduction of chlorine dioxide depends heavily on pH and the nature of the reducing agent. At neutral or alkaline pH, chlorine dioxide utilizes only 20 percent of its oxidizing capacity. At low pH, ClO_2 utilizes all its oxidizing capacity.

Ridenour and Ingols reported that chlorine dioxide was at least as effective as chlorine against *E. coli* after 30 minutes at similar residual concentrations.[72] The bactericidal activity of chlorine dioxide was not affected by pH values from 6.0 to 10.0. They also reported that the efficiency of chlorine dioxide decreased as the temperature decreased.

The bactericidal effectiveness of chlorine was compared with that of chlorine dioxide at pH 6.5 and 8.5 in a demand-free buffered system.[73] At pH 6.5, chlorine was more effective than chlorine dioxide, whereas at pH 8.5, chlorine dioxide was dramatically more effective than chlorine. Chlorine dioxide was significantly more efficient then chlorine in the presence of organic and nitrogenous material. Temperature also affects the rate of inactivation of bacteria with chlorine dioxide.[74]

Figure 11–12 shows the concentration–time relationships for 99 percent disinfection of Poliovirus 1, Coxsackie virus A9, and *E. coli.* Each curve has a slope of approximately -1. The plots in Fig. 11–12b show that ClO_2 is a more effective viricide as the pH increases.

POTASSIUM PERMANGANATE

Potassium permanganate ($KMnO_4$) is a strong oxidizing agent, which was first used as a municipal water treatment chemical by Sir Alexander Houston of the London Metropolitan Water Board in 1913. In the United States, it was first used in Rochester, New York, in 1927. It has been used widely in waterworks as an algicide, as an oxidant to control tastes and odors, to remove iron and manganese from solution, and as a disinfectant.

Reduction of the permanganate ion produces insoluble manganese oxide (Mn_3O_4) hydrates. Potassium permanganate typically is applied as pretreatment, and is followed by filtration. Addition of potassium permanganate to a finished water is unacceptable because of the pink color of the compound itself or the brown color of the oxides.

Cleasby et al. and Kemp et al. summarized the scientific references to disinfection by potassium permanganate published prior to 1960.[76,77] Cleasby et al. evaluated the efficacy of $KMnO_4$ on *E. coli,* prepared as a lactose broth culture. Bacterial inactivation was relatively ineffective, but slightly better at the higher temperatures. Increases in pH decreased the disinfection rates.

The relatively high cost, ineffective bactericidal action, and aesthetic unsuitability of maintaining a residual in the distribution system make potassium permanganate a generally unsatisfactory disinfectant for drinking water.

ULTRAVIOLET LIGHT IRRADIATION

Electromagnetic radiation, in wavelengths from 240 to 280 nm, is an effective agent for killing bacteria and other microorganisms in water.[78] A low-pressure

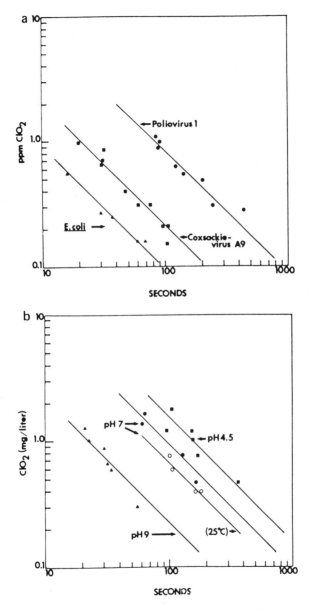

Fig. 11–12. (a) Concentration–time relationship for 99 percent destruction or inactivation of Poliovirus 1, Coxsackie virus A9, and *E. coli* by chlorine dioxide at 15°C and pH 7.0. (b) Effect of pH on inactivation of Poliovirus 1 at 21°C and pH 4.5, 7, and 9, and at 25°C and pH 7. (Courtesy of Ann Arbor Science Publishers.[75])

mercury arc enclosed in special UV-transmitting glass emits between 30 and 90 percent of its energy at a wavelength of 253.7 nm.[28]

UV disinfection systems can be either sealed from the water or open to it. Lamps are placed above the water at the apex of parabolic reflectors, which typically are aluminum. The open nature of the system can permit contamination and damage. Tubular reactors are most common in water treatment because they are sealed and operate under pressure. Multiple lamp reactors are being used to increase throughput. The units generally contain lamps that are positioned parallel to the flow of water through the unit.

Lamps that are surrounded by water include an insulation space to maintain their efficiency. The maximum efficiency of modern cold cathode lamps is near 40°C. Efficiency drops off to 50 percent output at 24°C and 65°C.

The degree of agitation and the plug flow characteristics of the contactor are important for an efficient unit. Short-circuiting or nonplug flow characteristics of contactors limit the efficiency of the process. Agitation through the contactor aids in bringing the microorganisms close to the UV source.

Dose

The dose, D, of electromagnetic radiation applied to a solution is measured as follows:

$$D = I_0 t \qquad (11\text{--}40)$$

where:

D = electromagnetic dose, $\mu W \cdot s/cm^2$
I_0 = radiant energy input, $\mu W/cm^2$
t = time of exposure, seconds

Measurements of energy input, I_0, per unit area of lamp exposed do not consider change in intensity through the depth of the exposed fluid, or the angle over which the energy interacts. In studies of wastewater disinfection, attempts have been made to correct the decrease in intensity with depth.[79,80] The usual water quality indices, such as COD, color, and turbidity, do not describe the loss of intensity through the solution adequately. However, it has been suggested that color should be a maximum of 5 units, iron a maximum of 3.7 mg/l, and turbidity a maximum of 5 units for treatment efficiency not to fall below acceptable limits.[81] These values are based on a maximum water depth of about 7.5 cm.

The biocidal dose of UV energy depends on the intensity of the absorbed UV energy and the time of interaction. This dose is a function of the following factors:

- The energy input from the UV source.
- Energy dispersion based on distance from the source.
- Depth of the fluid between the organisms and the UV source.
- Absorptivity of the fluid.
- Losses and reflection of UV light within the contactor.

All of these factors determine the actual intensity of radiation that is available to the microorganism at any one point within the contactor.[78,82] Ultraviolet radiation produces no residual. Therefore, monitoring and control of disinfection efficiency are more difficult for UV sources than for chemical disinfectants.

Biocidal Activity

Of the Gram-negative bacteria, *E. coli* consistently is more resistant to UV disinfection than are the *Salmonella* and *Shigella* species. The vegetative forms of the Gram-positive bacteria are more resistant than *E. coli.*

A dosage of 2,400 μW·s/cm² at a wavelength of 254 nm produced one lethe (1 log survival) of disinfection of *E. coli.*[78] A shipboard disinfection unit with dosages ranging from 3,000 to 11,000 μW·s/cm² and a maximum water depth of approximately 19 cm was tested. At 3,000 μW·s/cm², 0.02 to 0.04 percent of the *E. coli* survived. More than 4 logs of inactivation of polio, echo, and Coxsackie virus was achieved with a 4,000 μW·s/cm² dose. The U.S. Department of Health, Education, and Welfare (HEW) issued a policy statement of April 1, 1966, based on the work by Huff et al. stating criteria for the acceptability of UV disinfection. A minimum dosage of 16,000 μW·s/cm² with a maximum water depth of approximately 7.5 cm was required by HEW.[28]

Morris determined the the dose of 254-nm UV energy required to inactivate a variety of microorganisms.[83] The results of this work are shown in Table 11–15. Huff's values are similar in magnitude to those of Morris, partly because of the UV reflections from the contactor and aluminum surface, which approximately double the estimated dosage.[81,83]

Mechanism of Action

UV inactivation is thought to be caused by the microorganism's directly absorbing the UV energy, which causes a molecular change in the biochemical components of the organism. The major site of UV absorption in microorganisms is the purine and pyrimidine components of the nucleoproteins.

The repair of damaged nucleoprotein with light of a wavelength longer than that of the damaging radiation is commonly referred to as photoreactivation.[84] This process occurs when the visible wavelength is in the range

Table 11–15. Ultraviolet Energy
Necessary to Inactivate Various
Organisms.*

TEST MICROORGANISM	LETHAL DOSE, (μw·s/cm^2)
Escherichia coli	360
Staphylococcus aureus	210
Serratia marcescens	290
Sarcina lutea	1,250
Bacillus globiggii spores	1,300
T3 coliphage	160
Poliovirus	780
Vaccinia virus	30
Semliki Forrest virus	470
EMC virus	650

* Reference 83.

of 300 to 550 nm.[85] Also, the repair of UV-damaged nucleoproteins can occur in the dark.

UV light disinfection does not produce a residual. If a residual is required, a second disinfectant must be used. Current technology is limited to the use of UV light in small systems.

SILVER

The antibacterial action of silver and silver nitrate ($AgNO_3$) was noted first by Raulin in 1869. Since the late nineteenth century, but especially since World War II, there have been considerable efforts to exploit the use of silver as a disinfectant, particularly for individual (home) water systems and swimming pools. Silver has been used both as a salt, most commonly silver nitrate, and as metallic silver, either bound in filter beds, generated by electrolytic devices, or applied as a colloidal suspension.

However, silver and its compounds are inefficient and costly disinfectants that are unsuitable for use in municipal drinking water supplies. To achieve acceptable disinfection in a reasonable time would require concentrations exceeding the MCL of 0.05 mg/l.

HYDROGEN PEROXIDE

Hydrogen peroxide (H_2O_2) is a strong oxidizing agent that has been used as a disinfection agent for more than a century. Its instability and the difficulty of preparing concentrated solutions have limited its use. However, the development of electrochemical and other processes has resulted in the production of pure hydrogen peroxide in high concentrations. This product has been used to disinfect a variety of products, although its use in drinking water disinfection appears minimal.

Biocidal Activity

It was reported that the bactericidal activity of hydrogen peroxide required catalytic Fe^{+2} or Cu^{+2}. Yoshpe-Purer and Eylan worked with *E. coli, Salmonella typhi,* and *Staphylococcus aureus* in pure cultures, and in mixtures of bacteria, and studied the effect of concentrations (from 30 to 60 mg/l), contact time (10 to 420 minutes), and initial concentration of organisms on the disinfection of these organisms.[86] Although hydrogen peroxide residuals were not measured, researchers concluded that bacterial inactivations were relatively slow, that *E. coli* was more resistant to hydrogen peroxide than *S. typhi* or *S. aureus,* and that the required inactivation time was increased as the initial concentration of organisms was increased.

Several studies of virus inactivation by hydrogen peroxide have been reported. A 99 percent inactivation of poliovirus was obtained in about 6 hours with 0.3 percent (3,000 mg/l) hydrogen peroxide.[87] Another study used rhinovirus (types 1A, 1B, and 7), and showed that a 1.5 percent (15,000 mg/l) concentration required approximately 24 minutes for 99 percent inactivation, whereas a concentration of 3 percent (30,000 mg/l) took about 4 minutes.[88]

INSOLUBLE CONTACT DISINFECTANTS (ICD)

Janauer introduced the concept of disinfecting water by passing it over a solid, germicidal "surface." Disinfection is achieved *without* the release of biocidal moieties into the bulk phase containing the microorganisms.[89] Janauer listed the properties of an ideal ICD, which are reproduced here as Table 11–16.[89] There are other subcategories of ICD's such as those proposed by Janauer and Isquith et al.[89,90] The latter relies on creating ICD's by chemically

Table 11–16. Properties of An (Ideal) Insoluble Disinfectant (ICD).*

1. Is a completely insoluble solid (immiscible if a liquid).
2. Kills/inactivates target microorganisms on short contact.
3. Is effective against a wide spectrum of harmful microorganisms.
4. Has easily accessible germicidal functions/sites.
5. "Attracts" target microorganisms (charge; special interactions).
6. Is not irreversibly "poisoned" while damaging cells.
7. Has a high disinfectant capacity (which can be regenerated).
8. Is resistant to "fouling" by organic and inorganic traces.
9. Is insensitive to variations in temperature and pH.
10. Does not release harmful substances during/after water treatment.
11. Is chemically, mechanically, thermally stable (shelf life).
12. Is cost-effective.

* Reference 89.

Table 11-17. Summary of Major Possible Disinfection Methods for Drinking Water.[28]

DISINFECTION AGENT[a]	TECHNOLOGICAL STATUS	EFFICACY IN DEMAND-FREE SYSTEMS[b]			PERSISTENCE OF RESIDUAL IN DISTRIBUTION SYSTEM
		BACTERIA	VIRUSES	PROTOZOAN CYSTS	
Chlorine[c]					
As hypochlorous acid (HOCl)	Widespread use in U.S. drinking water	++++	++++	++	Good
As hypochlorite ion (OCl⁻)		+++	++	NDR[d]	
Ozone[c]	Widespread use in drinking water outside United States, particularly in France, Switzerland, and the province of Quebec	++++	++++	++++	No residual possible
Chlorine dioxide[c,e]	Widespread use of disinfection (both primary and for distribution system residual) in Europe; limited use in United States to counteract taste and odor problems and to disinfect drinking water	++++	++++	NDR[d]	Fair to good (but possible health effects)

					Good (but possible health effects)
Iodine As diatomic iodine (I$_2$)	No reports of large-scale use in drinking water	++++	+++	+++	Good (but possible health effects)
As hypoiodous acid (HOI)		++++	++++	+	
Bromine	Limited use for disinfection of drinking water	++++[f]	++++[f]	+++[f]	Fair
Chloramines	Limited present use on a large scale in U.S. drinking water	++	+	+	Excellent

[a] The sequence in which these agents are listed does not constitute a ranking.

[b] Ratings: ++++, excellent biocidal activity; +++, good biocidal activity; ++, moderate biocidal activity; +, low biocidal activity.

[c] By-product production and disinfectant demand are reduced by removal of organics from raw water prior to disinfection.

[d] Either no data reported or only available data were not free from confounding factors, thus rendering them not amenable to comparison with other data.

[e] MCL 1.0 mg/l because of health effects (Symons et al., 1977).

[f] Poor in the presence of organic material.

Table 11-18. Summary of Minor Possible Disinfection Methods for Drinking Water.[28]

DISINFECTION AGENT[a]	TECHNOLOGICAL STATUS	EFFICACY IN DEMAND-FREE SYSTEMS[b]			PERSISTENCE OF RESIDUAL IN DISTRIBUTION SYSTEM
		BACTERIA	VIRUSES	PROTOZOAN CYSTS	
Ferrate	No reports of use in drinking water	++	+++	NDR[c]	Poor
High pH conditions (pH 12–12.5)	No reports of large-scale use in drinking water	+++	+++	NDR[c]	Feasibility restricted since consumption of high-pH water not recommended
Hydrogen peroxide	No reports of large-scale use in drinking water	±	±	NDR[c]	Poor
Ionizing radiation	No reports of use in drinking water	++	++	NDR[c]	No residual possible
Potassium permanganate	Limited use for disinfection	±	NDR[c]	NDR[c]	Good, but aesthetically undesirable
Silver[d]	No reports of large-scale use in drinking water	+	NDR[c]	+	Good, but possible health effects
UV light	Use limited to small systems	+++	+++	NDR[c]	No residual possible

[a] The sequence in which these agents are listed does not constitute a ranking.
[b] Ratings: ++++, excellent biocidal activity; +++, good biocidal activity; ++, moderate biocidal activity; +, low biocidal activity; ±, of little or questionable value.
[c] Either no data reported or only available data were not free from confounding factors, thus rendering them not amenable to comparison with other data.
[d] MCL 0.05 mg/l because of health effects (Symons et al., 1977).

Table 11–19. Status of Possible Methods for Drinking Water Disinfection.[28]

DISINFECTION AGENT	SUITABILITY AS INACTIVATING AGENT	LIMITATIONS	SUITABILITY FOR DRINKING WATER DISINFECTION[a]
Chlorine	Yes	Efficacy decreases with increasing pH; affected by ammonia or organic nitrogen	Yes
Ozone	Yes	On-site generation required; no residual; other disinfectant needed for residual	Yes
Chlorine dioxide	Yes	On-site generation required; interim MCL 1.0 mg/liter	Yes
Iodine	Yes	Biocidal activity sensitive to pH	No
Bromine	Yes	Lack of technological experience; activity may be pH-sensitive	No
Chloramines	No	Mediocre bactericide, poor virucide	No[b]
Ferrate	Yes	Moderate bactericide; good virucide; residual unstable; lack of technological experience	No
High pH conditions	No	Poor biocide	No
Hydrogen peroxide	No	Poor biocide	No
Ionizing radiation	Yes	Lack of technological experience	No
Potassium permanganate	No	Poor biocide	No
Silver	No	Poor biocide; MCL 0.05 mg/liter	No
UV light	Yes	Adequate biocide; no residual; use limited by equipment maintenance considerations	No

[a] This evaluation relates solely to the suitability for controlling infectious disease transmission.
[b] Chloramines may have use as a secondary disinfectant in the distribution system in view of their persistence.

bonding a layer of an antimicrobial to the substrate. The substrate could be glass, sand, cellulose, and so on.

Studies of the efficacy of insoluble polymeric contact disinfectant (IPCD) resins showed that after only 5 minutes of contact, E. coli was reduced to zero from as high as 3.0×10^6 CFU/ml. IPCD resins also have been shown to be effective against polio and rotaviruses.

Janauer noted that the use of IPCD resins for potable water disinfection would avoid the introduction of chlorinated hydrocarbons, as well as the taste of dissolved chlorine. Janauer estimated that the cost of IPCD resins might be about twice the cost of conventional resins used for anion exchange resins. Because of the need for maintaining a residual disinfectant in the distribution system, IPCD's would not be used for central systems.

The IPCD resins may be suitable for small point-of-use water treatment systems, however. These could include road stops, state and national parks, rural residences, and so on. A small cartridge of an IPCD resin can disinfect large volumes of water (prefiltered) with little or no maintenance.

SUMMARY

Tables 11–17 and 11–18 summarize information on potential disinfection techniques.[28] Table 11–19 summarizes the general status of alternative disinfectants.

REFERENCES

1. Salvato, J. A., Jr., *Environmental Engineering and Sanitation,* 2nd ed., John Wiley & Sons, New York, © 1972.
2. Geldreich, E. E., in *Water Pollution Microbiology,* R. Mitchell, editor, John Wiley & Sons, New York, © 1972.
3. Taylor, F. B., "Viruses—What is Their Significance in Water Supplies," *J.AWWA* 66:306, 1974.
4. Craun, G. F., "Outbreaks of Waterborne Disease in the United States: 1971–1978," *J.AWWA* 73:360, 1981.
5. Berg, G., "Removal of Viruses from Water and Wastewater," Dept. of Civil Engineering, University of Illinois, 1971.
6. Thorup, R. T., et al., "Virus Removal by Coagulation with Polyelectrolytes," *J.AWWA* 62:97, 1970.
7. Sproul, O. T., *Critical Review of Virus Removal by Coagulation Processes and pH Modifications,* EPA Report 600/2–80–004, U.S. EPA, Cincinnati, Ohio, 1980.
8. Hanna, G. P., and Rubin, A. J., "Effect of Sulfate and Other Ions on Coagulation with Aluminum (III)," *J.AWWA* 62:315, 1970.
9. Manwaring, J. F., et al., "Removal of Viruses by Coagulation and Flocculation," *J.AWWA* 63:298, 1971.
10. Amirhor, P., and Engelbrecht, R. S., "Virus Removal by Polyelectrolyte-Aided Filtration," *J.AWWA* 67:187, 1975.
11. Wattie, E., and Chambers, C. W., "Relative Resistance of Coliform Organisms and Certain Enteric Pathogens to Excess-Lime Treatment," *J.AWWA* 35:709–720, 1943.

12. Riehl, M. L., Weiser, H. H., and Rheins, B. T., "Effect of Lime-Treated Water upon Survival of Bacteria," *J.AWWA* 44:466–470, 1952.
13. Robeck, G. G., Clarke, N. A., and Dostal, J. A., "Effectiveness of Water Treatment Processes in Virus Removal," *J.AWWA*, p. 1275, October, 1962.
14. Walton, G., *Effectiveness of Water Treatment Processes as Measured by Coliform Reduction,* U.S. Public Health Service Publication No. 898, 1962.
15. Hudson, H. E., "High Quality Water Production and Viral Disease," *J.AWWA*, p. 1265, October, 1962.
16. Committee Report, "Viruses in Water," *J.AWWA*, p. 491, October, 1969.
17. "Water Chlorination—Principles and Practices," *American Water Works Association, Manual of Water Supply Practices, M–20,* 1973.
18. White, G. C., *Handbook of Chlorination,* Van Nostrand Reinhold, New York, 1972.
19. Laubusch, E. J., in *Water Quality and Treatment,* 3rd ed., McGraw-Hill Book Company, New York, 1971.
20. Culp, Gordon L., and Culp, Russell L., *New Concepts in Water Purification,* Van Nostrand Reinhold, New York, 1974.
21. "Chloropac," Engelhard Industries brochure.
22. "Sanilec Systems," Diamond Shamrock brochure.
23. "Cloromat," Ionics, Inc., brochure.
24. "Pep-Clor Systems," Pacific Engineering and Production Company of Nevada brochure.
25. Morris, J. C., *J. Am. Chem. Soc.* 68:1692, 1946.
26. Shilov, E. A., and Solodushenkov, S. N., *Comp. Rend. Acad. Sci.* l'URSS 3(1):17, 1936.
27. Morris, J. C., *J. Phys. Chem.* 70:3798, 1966.
28. *Drinking Water and Health,* Vol. 2, National Academy Press, Washington, D.C., 1980.
29. Weil, I., and Morris, J. C., *J. Am. Chem. Soc.* 71:1664, 1949.
30. Morris, J. C., in *Principles and Applications of Water Chemistry,* S. D. Faust and J. V. Hunter, editors, John Wiley and Sons, New York, 1967.
31. Faust, Samuel D., and Aly, Osman M., *Chemistry of Water Treatment,* Butterworth Publishers, Woburn, Mass., Ann Arbor Science, Ann Arbor, Mich., 1983.
32. Kabler, P. W., Chang, S. L., Clarke, N. A., and Clark, H. F., "Pathogenic Bacteria and Viruses in Water Supplies," 5th Sanitary Engineering Conference, University of Illinois, January, 1963.
33. Baumann, E. R., and Ludwig, D. D., "Free Available Chlorine Residuals for Small Nonpublic Water Supplies," *J.AWWA* 54:1379, 1962.
34. Butterfield, C. T., et al., *Public Health Rep.* 58:1837, 1943.
35. Scarpino, P. V., Berg, G., Chang, S. L., Dahling, D., and Lucas, M., "A Comparative Study of the Inactivation of Viruses in Water by Chlorine," *Water Research,* p. 959, 1972.
36. Kruse, C. W., Olivieri, V. P., and Kawata, K., "The Enhancement of Viral Inactivation by Halogens," *Water and Sewage Works,* p. 187, June, 1971.
37. Burns, R. W., and Sproul, O. J., "Vericidal Effects of Chlorine in Wastewater," *J.WPCF,* p. 1834, November, 1967.
38. Kelly, S., and Sanderson, W. W., *Am. J. Public Health* 48:1323, 1958.
39. Stringer, R., and Kruse, C. W. "Amoebic Cysticidal Properties of Halogens," *Proceedings, National Specialty Conference on Disinfection,* American Society of Civil Engineers, New York, 1970.
40. Butterfield, C. I., and Wattie, C., *Public Health Rep.* 61:157, 1946.
41. Kelly, S. M., and Sanderson, W. W., *Am. J. Public Health* 50:14, 1960.
42. Johanneson, J. K., *J. Am. Public Health Assoc.* 50:1731, 1960.
43. Collins, H. G., "Chlorination—How to Use It," *Proceedings,* Fifth Annual Sanitary Engineering Symposium, California Department of Public Health, May, 1970.
44. Stenquist, R. J., and Kaufman, W. J., *Initial Mixing in Coagulation Processes,* EPA Report EPA–R2–72–053, November, 1972.

45. Chang, S. L., *J. American Pharmaceutical Assoc.* 47:417, 1958.

46. McAlpine, R. K., *J.AWWA* 74:725, 1952.

47. Black, A. P., et al., "Use of Iodine for Disinfection," *J.AWWA* 57:1401, 1965.

48. Karalekas, P. C., Jr., Kusminski, L. N., and Feng. T. H., "Recent Developments in the Use of Iodine for Water Disinfection," *J. New England Water Works Assoc.,* p. 152, June, 1970.

49. McKee, J. E., Brokaw, C. J., and McLaughlin, R. T., "Chemical and Colicidal Effects of Halogens in Sewage," *J.WPCF,* p. 795, August, 1960.

50. Chang, S. L., and Morris, J. C., *Ind. Eng. Chem.* 45:1009, 1953.

51. Chambers, C. W., et al., *Soap. Sanit. Chem.* 28:149, 1952.

52. Mills, J. F. "Interhalogens and Halogen Mixtures as Disinfectants," pp. 113–143 in *Disinfection: Water and Wastewater,* J. D. Johnson, editor, Ann Arbor Science Publishers, Ann Arbor, Mich., 1975.

53. Galal-Gorchev, H., and Morris, J. C., "Formation and Stability of Bromamide, Bromimide, and Nitrogen Tribromide in Aqueous Solution," *Inorg. Chem.* 4:899–905, 1965.

54. Johnson, J. D., and Overby, R., "Bromine and Bromamine Disinfection Chemistry," *J. San. Eng. Div. Am. Soc. Civ. Eng.* 97:617–628, 1971.

55. Morris, J. C., "Aspects of the Quantitative Assessment of Germicidal Efficiency," pp. 1–10 in *Disinfection: Water and Wastewater,* J. D. Johnson, editor, Ann Arbor Science Publishers, Ann Arbor, Mich., 1975.

56. Inmann, G. W., Jr., LaPointe, T. F., and Johnson, J. D., "Kinetics of Nitrogen Tribromide Decomposition in Aqueous Solution," *Inorg. Chem.* 15:3037–3042, 1976.

57. Kruse, C. W., Hsu, Y., Griffiths, A. C., and Stringer, R., "Halogen Action on Bacteria, Viruses, and Protozoa," pp. 113–136 in *Proceedings,* National Specialty Conference on Disinfection, American Society of Civil Engineers, New York, 1970.

58. Floyd, R., Sharp, D. G., and Johnson, J. D., "Inactivation of Single Poliovirus Particles in Water by Hypobromite Ion, Molecular Bromine, Dibromamine, and Tribromamine," *Environ. Sci. Tech.* 12:1031–1035, 1978.

59. Floyd, R., Johnson, J. D., and Sharp, D. G. "Inactivation by Bromine of Single Poliovirus Particles in Water," *Appl. Environ. Microbiol.* 31:298–303, 1976.

60. Sharp, D. G., Floyd, R., and Johnson, J. D., "Nature of the Surviving Plague Forming Unit of Reovirus in Water Containing Bromine," *Appl. Environ. Microbiol.* 29:94–101, 1975.

61. Sharp, D. G., Floyd, R., and Johnson, J. D., "Initial Fast Reaction of Bromine on Reovirus in Turbulent Flowing Water," *Appl. Environ. Microbiol.* 31:173–181, 1975.

62. Stringer, R. P., Cramer, W. N., and Kruse, C. W., "Comparison of Bromine, Chlorine, and Iodine as Disinfectants for Amoebic Cysts," pp. 193–209 in *Disinfection: Water and Wastewater,* J. D. Johnson, editor, Ann Arbor Science Publishers, Ann Arbor, Mich., 1975.

63. Keswick, B. H., et al., "Comparative Disinfection Efficiency of Bromine Chloride and Chlorine for Poliovirus," *J.AWWA* 70:573, 1978.

64. Bendes, R. J., "Ozonation, Next Step to Water Purification," *Power,* August, 1970.

65. Hann, V. A., "Disinfection of Drinking Water with Ozone," *J.AWWA,* p. 1316, October, 1956.

66. Morris, J., "Chlorination and Disinfection—State of the Art," *J.AWWA,* p. 769, December, 1971.

67. Keller, J. W., Morin, R. A., and Schaffernoth, T. J., "Ozone Disinfection Pilot Plant Studies at Laconia, New Hampshire," *J.AWWA,* 66:730, 1974.

68. Synan, J. F., MacMahon, J. D., and Vincent, G. P., "Chlorine Dioxide in Potable Water Treatment," *Water Eng.,* Vol. 48, 1945.

69. Masschelein, W. J., in *Chlorine Dioxide: Chemistry and Environmental Impact of Oxychlorine Compounds,* Rip G. Rice, editor, Ann Arbor Science Publishers, Ann Arbor, Mich., 1979.

70. Rosenblatt, D. H., "Chlorine Dioxide: Chemical and Physical Properties," pp. 332–343 in *Ozone/Chlorine Dioxide Oxidation Products of Organic Materials,* R. G. Rice and J. A.

Cotruvo, editors, proceedings of a conference held in Cincinnati, Ohio, November 17–19, 1976, sponsored by the International Ozone Institute and the U.S. Environmental Protection Agency; Ozone Press International, Cleveland, Ohio, 1978.

71. Gall, R. J., "Chlorine Dioxide—An Overview of Its Preparation, Properties, and Uses," pp. 356–382 in *Ozone/Chlorine Dioxide Products of Organic Materials,* R. G. Rice and J. A. Cotruvo, editors, proceedings of a conference held in Cincinnati, Ohio, November 17–19, 1976, sponsored by the International Ozone Institute and the U.S. Environmental Protection Agency; Ozone Press International, Cleveland, Ohio, 1978.

72. Ridenour, G. M., and Ingols, R. S., "Bactericidal Properties of Chlorine Dioxide," *J.AWWA,* Vol. 39, 1947.

73. Bernarde, M. A., Israel, B. M., Olivieri, V. P., and Granstrom, M. L., "Efficiency of Chlorine Dioxide as a Bactericide," *Appl. Microbiol.,* Vol. 13, 1965.

74. Bernarde, M. A., Snow, W. A., and Olivieri, V. P., "Chlorine Dioxide Disinfection Temperature Effects," *J. Applied Bacteriol.,* Vol. 30, 1967.

75. Cronier, S., et al., in *Water Chlorination,* Vol. 2, R. L. Jolley et al., editors, Ann Arbor Science Publishers, Ann Arbor, Mich., 1978.

76. Cleasby, J. L., Bauman, E. R., and Black, C. D., "Effectiveness of Potassium Permanganate for Disinfection," *J.AWWA,* Vol. 56, p. 466, 1964.

77. Kemp, H. T., Fuller, R. G., and Davidson, R. S., "Potassium Permanganate as an Algicide," *J.AWWA,* Vol. 58, p. 255, 1966.

78. Luckiesh, M., and Holladay, L. L., "Disinfecting Water by Means of Germicidal Lamps," *Gen. Electr. Rev.,* Vol. 47, 1944.

79. Roeber, J. A., and Hoot, F. M., *Ultraviolet Disinfections of Activated Sludge Effluent Discharging to Shellfish Waters,* EPA 600/2-75/060, Municipal Environmental Research Laboratory, Office of Research and Development, U.S. EPA, Cincinnati, Ohio, 1975.

80. Severin, B. F., "Disinfection of Municipal Wastewater Effluents with Ultraviolet Light," paper presented at WPCF Conference, Anaheim, Calif., 1978.

81. Huff, C. B., Smith, H. F., Boring, W. D., and Clarke, N. A., "Study of Ultraviolet Light Disinfection of Water and Factors in Treatment Efficiency," *Public Health Rep.* 80:695–705, 1965.

82. Luckiesh, M., Taylor, A. H., and Kerr, G. P., "Germicidal Energy," *Gen. Electr. Rev.* 47(9):7–9, 1944.

83. Morris, E. J., "The Practical Use of Ultraviolet Radiation for Disinfection Purposes," *Med. Lab. Technol.* 29:41–47, 1972.

84. Witkin, E. M., "Ultraviolet Mutagenesis and Inducible DNA Repair in *Escherichia coli,* " *Bacteriol. Rev.* 40:869–907, 1976.

85. Jagger, J., "Photoreactivation," *Radiat. Res. Suppl.* 2:75–90, 1960.

86. Yoshe-Purer, Y., and Eylan, E., "Disinfection of Water by Hydrogen Peroxide," *Health Lab. Sci.* 5:233–238, 1968.

87. Lund, E., "Significance of Oxidation in Chemical Inactivation of Poliovirus," *Arch. Gesamite Virusforsch.* 12:648–660, 1963, as reported in Reference 48.

88. Mentel, R., and Schmidt, J., "Investigations on Rhinovirus Inactivation by Hydrogen Peroxide," *Acta Virol.* 17:351–354, 1973, as reported in Reference 48.

89. *Progress in Chemical Disinfection: New Concepts and Materials,* proceedings of a SUNY Conversation in the Disciplines, March 26–27, 1982, Gilbert E. Janauer, Conference Chairman, Dept. of Chemistry, SUNY-Binghamton; co-edited by Dr. William C. Ghiorse, Dept. of Microbiology, Cornell University.

90. Isquith, A. J., Abbott, E. A., and Walters, P. A., "Surface-Bonded Antimicrobial Activity of an Organosilicon Quaternary Ammonium Chloride," *Appl. Microbiol.* 24(6):859–863, 1972.

91. Wei, I. W., and Morris, J. C., in *Chemistry of Water Supply Treatment and Distribution,* A. J. Rubin, editor, Ann Arbor Science Publishers, Ann Arbor, Mich., 1974.

92. Esposito, P., et al., "Destruction by Dichloramine of Viruses and Bacteria in Water," Abstract No. G99, American Society of Microbiologists, Washington, D.C., 1974.
93. Singer, P. C., and Zilli, W. B., "Ozonation of Ammonia: Application to Wastewater Treatment," in *Proceedings First International Symposium on Ozone for Water and Wastewater Treatment,* R. G. Rice and M. E. Browning, editors, International Ozone Institute, Waterbury, Conn., 1975.
94. Rosen, A. M., "Ozone Generation and Its Economical Application to Wastewater Treatment," *Water and Sewage Works,* Vol. 119, 1972.

Chapter 12
Activated Carbon Treatment

INTRODUCTION

Activated carbon adsorption is the most effective and reliable water treatment process available for the removal of a broad spectrum of organic substances dissolved in water.[1] In surface waters about 94 percent of the organics commonly present are adsorbed on carbon.[2] In groundwater supply sources, both natural organics and man-made organic contaminants are also generally amenable to carbon treatment. Because of this wide range of applicability and the effectiveness of this method of treatment, the use of activated carbon adsorption will be increasingly important in the future. When organics that are not readily adsorbed constitute a water quality problem, other treatment may be used in place of carbon, or carbon usage may be supplemented by other processes such as air-stripping. (More detail is provided in References 3 through 7.)

PAST PRACTICES

In the past, the principal reason for the use of activated carbon treatment was to control taste- and odor-causing organics. This was mainly for aesthetic rather than public health purposes. To control tastes and odors in water, the carbon dosage requirements are low and the necessary contact times short. This set of circumstances permitted the application of powdered activated carbon (PAC) rather than granular activated carbon (GAC). The use of PAC was advantageous because in most cases it could be used with no changes or additions to existing treatment facilities other than installation of the powdered carbon storage and feed equipment. Contact time was provided in existing settling basins, and the spent carbon was removed in existing rapid sand filters and disposed of along with the settling basin sludges.

RECENT DEVELOPMENTS

Over the past 15 years there has been a tremendous proliferation in the number, variety, and quantity of complex organic chemicals used for agricultural, industrial, and domestic purposes. Many of these substances eventually find their way into sources of public water supply. Even in very low concentrations many of these compounds have toxic, carcinogenic, mutagenic, or terato-genic properties that may produce long-term insidious health effects in water consumers. (These chemicals are discussed in greater detail in References 8 and 9.)

Development of this extensive, bewildering array of new synthetic organic chemicals has been paralled by the development of new sophisticated and extremely sensitive equipment that can detect and measure very minute concentrations of these synthetic organics in water. Monitoring capability has progressed rapidly from parts per million to parts per billion, parts per trillion, and beyond. Some chemists predict that the day may not be far away when it will become possible to detect individual molecules of substances in a water sample.

One surprising result of this new-found ability to detect trace quantities of organics in water is that examination of well water supply sources has revealed the universal presence of naturally occurring organics such as aldehydes, ketones, terpenes, humic compounds, and other substances. Some of these materials, as well as some organic pollutants in well water, may react with chlorine to form trihalomethanes and other undesirable chlorinated hydrocarbons in drinking water. This potentially hazardous situation has been the subject of much recent concern for public health officials and water consumers. (Chapter 14 discusses trihalomethane formation and control measures.)

CURRENT TREATMENT TRENDS

As already discussed, both natural and synthetic organics in water are for the most part adsorbable on activated carbon. However, compared to requirements for taste and odor removal, dosage and contact times are much greater. This difference in carbon removal efficiency dictates the use of granular rather than powdered activated carbon. When granular activated carbon is used for taste and odor control, it is possible to add a shallow bed (1 foot or less) of GAC on top of an existing sand filter, or to substitute a properly sized and graded bed (24 to 36 inches) of GAC in lieu of the fine media in a rapid sand filter, with satisfactory results. However, experience has demonstrated that such shallow beds of GAC generally are not suitable for removal of natural or synthetic organics for other control purposes. Deeper beds and longer contact times are necessary for economic, efficient removal of these trace materials. The majority of new installations will be separate deep-bed GAC contactors. The contactors may be located in the treatment process train, either ahead of or following plant filters. When GAC contactors are located after the filters, they will be either downflow or an upflow–downflow series configuration in order to avoid the leakage of carbon fines that is common to all upflow carbon beds. (Treatment trends also are discussed in References 10 through 14.)

When GAC contactors are located ahead of filters, they may be either upflow or downflow. The use of upflow GAC contactors makes it possible to take advantage of countercurrent flow principles that may make more efficient use of the carbon. Generally speaking, GAC empty bed contact times of from 15 to 45 minutes will be required for most installations. The time to be provided for a particular supply depends upon the characteristics and concentrations of the organics to be removed and the other properties of the water to be treated.

Most water treatment plant installations of GAC will involve the construction and use of on-site reactivation facilities. One exception would be small plants using less than 200 pounds of carbon per day (91 kg/d), which could economically use carbon on a once-through, throwaway basis. Another would be plants that use between 200 and 1,500 pounds per day (91 and 681 kg/d) of GAC. Such plants might consider central off-site carbon reactivation if the cost of hauling would permit economical operation.[15]

Trihalomethanes

GAC may be used in either of two ways for trihalomethane (THM) control. It can be used directly for removal of trihalomethanes, or it can be used indirectly to remove the precursors. In either case, good to very good removal is technically feasible. When the GAC is fresh, removal is nearly complete, but toward exhaustion, breakthrough begins. Loading is proportional to influent concentration. THM's containing bromine are adsorbed better than chloroform. (Removal of trihalomethanes is described in Chapter 14 and in References 16 through 18.)

Alternate to GAC for THM control.   It is generally more economical to replace the use of free chlorine with an alternate disinfectant (such as ozone, chlorine dioxide, or chloramines) rather than to use GAC to remove THM's or their precursor organics.[19] The major disadvantage of doing this is the lack of any precursor removal. Also, the alternate disinfectants may produce undesirable oxidation products other than THM's in the water. Further, ozone does not produce a residual for distribution system protection; chloramine is a weaker disinfectant than free chlorine; and chlorine dioxide has chlorite and chlorate as inorganic by-products, anionic species whose health effects are currently unknown.[20]

Volatile Organics

Volatile organic chemicals occur in both untreated and treated drinking water. Significant concentrations are more likely to be found in well waters than in surface waters. The potential health effects and acceptable limits of these substances in drinking water are of current concern, but these issues may not be resolved for some time. In the meantime, research is under way to determine what degree of volatile organic removal is achievable so that water utilities can better assess the options of providing treatment or developing a new source. Volatile organics can be removed by aeration or adsorption on GAC or synthetic resins, or combinations of these processes, as described in detail in Chapter 15. Data are being developed on the effects of strong oxidants such as ozone and the reverse osmosis process in removing volatile organics. Boiling tap water for 5 minutes can also be effective for removing most of these organic compounds. In regard to the removal of volatile organics by aeration, the question has been raised of whether or not the exhaust gases create a problem. The evidence gathered to date on this subject indicates that the likelihood of creating an air pollution problem by aerating solvent-contaminated drinking water is remote.[21]

PRINCIPLES OF CARBON ADSORPTION

Activated carbon removes organic contaminants from water by a process of adsorption that results from the attraction and accumulation of one substance on the surface of another.[22-25] In general, the chemical nature of the carbon surface is of relatively minor significance in the adsorption of organics from water and is secondary to the magnitude of the surface area of carbon available. Hence a high surface area is the prime consideration in adsorption. Granular activated carbons typically have surface areas of 2.44 to 6.84 million sq ft/lb (500 to 1,400 m²/g). Activated carbon has a preference for organic compounds and, because of this selectivity, is particularly effective in removing organic compounds that may cause taste and odor problems in water supplies. In addition to the taste and odor problems, trace concentrations of organics in drinking water supplies may have adverse health effects over the long term, as discussed in more detail in Chapter 15.

Activated carbon can be made from a variety of materials such as coal, wood, nut shells, and pulping waste (see below). Activated carbon made from wood is normally termed charcoal and was the first type of granular carbon to find its way into municipal water treatment.

Granular carbons made from coal are hard and dense, and can be pumped in a water slurry without appreciable deterioration. Hydraulic handling of carbon allows dust-free loading and unloading of filters. Coal-derived granular carbons are well suited to water treatment because the carbon wets rapidly and does not float, but conforms to a neatly packed bed with acceptable pressure drop characteristics. Also, the carbon is quite dense, and thus makes available more adsorption capacity in any given filter volume. Much of the surface area available for adsorption in granular carbon particles is found in the pores within the granular carbon particles created during the activation process.

The major contribution to surface area is located in pores of molecular dimensions. A molecule will not readily penetrate into a pore smaller than a certain critical diameter and will be excluded from pores smaller than this; thus, molecules are "screened out" by pores smaller than a minimum diameter that is a characteristic of the adsorbate and is related to molecular size. Furthermore, for any molecule the effective surface area for adsorption can exist only in pores that the molecule can enter.

Figure 12–1 illustrates this concept for the case in which two adsorbate molecules in a solvent (not shown) compete with each other for adsorbent surface. Because of the irregular shape of both pores and molecules and also by virtue of constant molecular motion, the fine pores are not blocked by the large molecules but are still free for entry by small molecules. As a contributing factor, the greater mobility of the smaller molecule should permit it to diffuse ahead of the larger molecule and penetrate the fine pores first. The forces of attraction between the carbon and adsorbing molecules are known to be greater, the more similar the adsorbing molecules are in size

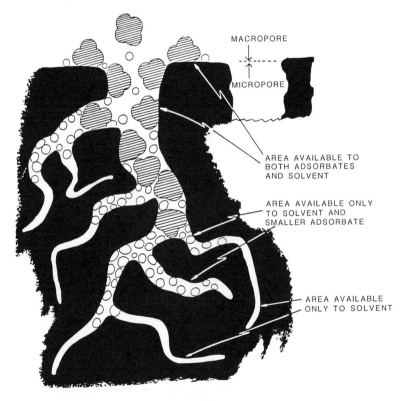

MACROPORE

MICROPORE

AREA AVAILABLE TO
BOTH ADSORBATES
AND SOLVENT

AREA AVAILABLE ONLY
TO SOLVENT AND
SMALLER ADSORBATE

AREA AVAILABLE
ONLY TO SOLVENT

Fig. 12–1. Concept of molecular screening in micropores. (Courtesy of Calgon Carbon Corp.)

to the pores. The most tenacious adsorption takes place when the pores are barely large enough to admit the adsorbing molecules. The smaller the pores with respect to the molecules, the greater the forces of attraction. The pores cannot be so small, however, that the adsorbing molecules find it difficult to enter, or the adsorption for those molecules will be greatly reduced. The pore structure of activated carbons is extremely important in determining their adsorptive properties.

MANUFACTURE OF GRANULAR CARBONS

Activated carbons can be made from a variety of carbonaceous raw materials. Generally, the process consists of selection of an organic raw material, dehydration and carbonization, and, finally, activation to produce a highly porous structure. Raw materials that have been used to produce activated carbons include wood, lignin, nut shells, coal, lignite, peat, bagasse, sawdust, and petroleum residues. Granular carbons prepared from coal generally have the best physical characteristics (specifically hardness and density) for use in water treatment filters or contactors. The better grades of powdered carbons are made from lignin and lignite. The dehydration and carbonization step is usually accomplished by the slow heating of the raw material in the absence of air. Sometimes a dehydrating agent such as zinc chloride or phosphoric acid is used. The activation step may be carried out with chemicals in the solid or liquid state, or by using mixtures of gases. The more modern techniques use mixtures of oxidizing gases such as steam, air, and carbon dioxide.

Activated carbons are usually classified according to their physical state (e.g., powdered or granular) and their use (e.g., water grade, decolorizing, liquid phase, or gas phase). Granular carbons are those materials that are greater in particle size than approximately 150 mesh, whereas powdered carbons are those that are smaller in particle size.

Characterization of Granular Carbons

The important characteristics of a carbon must be expressed in terms of both adsorptive characteristics and physical properties. The adsorptive capacity of a carbon can be measured to a fair degree by experimentally determining the adsorption isotherm in the system under consideration. Simple capacity tests such as the Iodine Number or the Molasses Decolorizing Index also may be used as a measure of adsorptive capacity. (Further details are presented in References 26 through 30.)

The adsorption isotherm is the relationship, at a given temperature, between the amount of a substance adsorbed and its concentration in the surrounding solution. If a color adsorption isotherm is taken as an example, the adsorption isotherm would consist of a curve plotted with residual color in the water as the abscissa, and the color adsorbed per gram of carbon as the ordinate. A reading taken at any point on the isotherm gives the amount of color adsorbed (at equilibrium conditions) per unit weight of carbon, which is the carbon adsorptive capacity at a particular color concentration and water temperature. In very dilute solutions, such as water supplies, a logarithmic isotherm plotting usually gives a straight line.

In this connection, a useful formula is the Freundlich equation, which relates the amount of impurity in the solution to that adsorbed, as follows:

$$\frac{x}{m} = kC^{1/n} \qquad\qquad (12\text{--}1)$$

where:

$x =$ amount of color adsorbed
$m =$ weight of carbon

$\dfrac{x}{m} =$ amount of color adsorbed per unit weight of carbon

k and n are constants
$C =$ unadsorbed concentration of color left in solution

In logarithmic form:

$$\log \frac{x}{m} = \log k + \frac{1}{n} \log C \qquad\qquad (12\text{--}2)$$

in which $1/n$ represents the slope of the straight line isotherm. Detailed procedures for establishing the experimental conditions and conducting and interpreting isotherm adsorption tests are presented elsewhere.[31,32]

From an isotherm test, it can be determined whether or not a particular purification can be effected. The test will also show the approximate capacity of the carbon for the application, and provide a rough estimate of the carbon dosage required. Isotherm tests also afford a convenient means of studying the effects of pH and temperature on adsorption. Isotherms put a large amount of data into concise from for ready evaluation and interpretation. Isotherms obtained under identical conditions using the same test solutions for two test carbons can be quickly and conveniently compared to reveal the relative merits of the carbons.

For some applications, it is not necessary to know the complete adsorption isotherm to determine the best carbon for a particular application or to specify the most appropriate carbon. In these cases, it is possible to use simple capacity

Table 12-1. Typical Granular Carbon Specifications.

ITEM	UNITS	VALUE
Total surface area	sq ft/lb (m²/g)	(860–1500)
Bulk density	lb/cu ft (kg/m³)	26 (421.5)
Particle density, wetted in water	(g/cc)	1.3–1.4
Effective size	(mm)	0.8–0.9
Uniformity coefficient		1.9 or less
Mean particle diameter	(mm)	1.5–1.7
Iodine Number		850 Min
Abrasion Number		70 Min
Ash	%	8 Max
Moisture	%	2 Max

tests such as the Iodine Number or Molasses Decolorizing Index as an appropriate measure of adsorptive capacity.

The Iodine Number is the number of milligrams of iodine adsorbed per gram of carbon when in equilibrium, under specified conditions, with a solution of 0.02 N iodine concentration.[31,32] The Iodine Number is an approximate measure of the adsorptive capacity of a carbon for small molecules such as iodine. The Molasses Decolorizing Index is, roughly, a measure of the adsorptive capacity of the carbon for color bodies in a specified molasses solution, compared to a standard carbon. The Molasses Decolorizing Index is a measure, therefore, of the adsorptive capacity for large molecules such as color bodies.

The physical properties of granular carbons that are important in performance include resistance to breakage, particle size, and degree of dustiness. For measuring resistance to breakage there are empirical tests such as the

Table 12-2. Comparison of Coal and Granular Carbon As Filtration Media.

		FILTRASORB*	
	VALUES FOR HARD COAL MEDIA	8 × 30 MESH**	14 × 40 MESH**
Real density, g/cc	1.5–1.6	2.1	2.1
Particle density in water, g/cc†	—	1.5–1.6	1.5–1.6
Effective size			
Single media filters	0.5 mm	—	0.5 mm
Multi-media filters	0.8 mm	0.8 mm	—
Uniformity coefficient	Less than 1.75	1.9 or less	1.7 or less

* Coal-based granular activated carbon, products of Calgon Corporation.
** U.S. Sieve Series.
† The pores of the activated carbon filled with water.

Table 12-3. Typical Size Distribution of
Granular Activated Carbon.

	8 × 30 MESH	14 × 40 MESH
Sieve Size U.S. Standard Series		
Larger than No. 8—max %	3	—
Larger than No. 14—max %	—	3
Smaller than No. 30—max %	1	—
Smaller than No. 40—max %	—	1
Mean particle diameter (mm)	1.6	0.9

Abrasion Number and the Hardness Number. Particle size is determined by a screen analysis, from which the mean particle diameter and effective size can be calculated. Density is simply the weight per unit volume of the carbon. Typical specifications of a granular carbon suitable for water treatment applications are given in Table 12-1.

For incorporation in filter and contactor design, carbon particle size distribution is important. If the carbon is to replace anthracite coal in a dual-media filter, it should have similar filtration characteristics and similar back-washing characteristics to the coal. Fortunately, as shown in Table 12-2, commercial carbons are available with characteristics that are similar to anthracite coals used as filter media. The typical size distribution of a commercial carbon is shown in Table 12-3.

The headloss through granular carbon is a function of the carbon size, the depth of the carbon layer, the hydraulic throughput rate, and the water temperature. Figure 12-2 presents the headloss data for some commercial carbons in downflow service as filter media following backwashing. The back-wash characteristics are shown in Fig. 12-3.

DEVELOPING DESIGN BASIS FOR WATER TREATMENT WITH ACTIVATED CARBON

Historical Use of Activated Carbon in Water Purification

Powdered Carbon. Powdered activated carbon (PAC) has been used successfully for more than 50 years to remove taste and odor from public drinking water supplies.[33,34] Except for particle size, this PAC is often identical to the GAC used in wastewater treatment. In 1930, Hassler reported that PAC was being used in more than 150 municipal water treatment plants.[35] In this type of use, PAC dosages usually are in the range of 1 to 5 mg/l, although dosages as high as 20 to 30 mg/l have been used in some places for short periods of time when taste and odor problems were severe. PAC

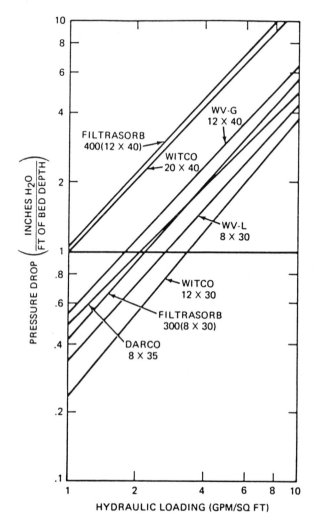

Fig. 12-2. Pressure drop vs. hydraulic loading.

is commonly used on a one-time, throwaway basis, with no attempt at recovery or reuse.

The principal benefit of PAC in water treatment is to remove taste and odor. In some waters, PAC may also remove color or organics that otherwise would interfere with coagulation or filtration. During its widespread use by waterworks for more than 50 years, no harmful effects have been reported.

Granular Carbon. Altogether, 46 water treatment plants in the United States are now using GAC (AWWA, 1979). In these plants, GAC is used

CARBON: 12 X 40, 8 X 30
LIQUID: WATER AT 22°C

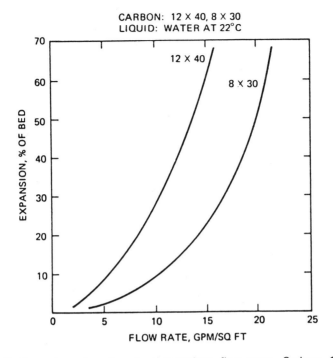

Fig. 12–3. Expansion of carbon bed at various flow rates. Carbon: 12 × 40, 8 × 30; liquid: water at 22°C.

to control taste and odor in drinking water. At the plants in Hopewell, Virginia and Davenport, Iowa, however, other benefits have been observed by the American Water Works Service Co., especially the reduction of trihalomethanes in finished water. Historical uses of GAC are discussed in References 36 through 42 for specific cases.

Because there has been no need for on-site reactivation of carbon, few useful data are available for estimating the costs involved in organics removal.

Recently, Manchester, New Hampshire installed a fluidized bed in its water plant for on-site GAC reactivation. Capital costs are available from EPA and the city, but operating and maintenance costs will not be forthcoming until more data are accumulated.

Cincinnati, Ohio has completed pilot testing of GAC adsorption and on-site regeneration. At this writing, the associated consulting firms of Malcolm Pirnie and Culp/Wesner/Culp are designing a 175 mgd (662.4 Ml/d) post-filtration GAC contacting and regeneration facility for the City Water Works in Cincinnati. This is the subject of a special discussion that follows later in this chapter.

Use of GAC in Food and Drink Processing and in Other Industries

There are more than 50 carbon reactivation furnace installations in the United States, serving many of the largest GAC users in the country. Included are seven large sugar refiners and nine large corn syrup refiners that use GAC for decolorizing their food products. The list also includes 15 industries that use GAC to remove organics from wastewater discharges and 19 cities that use GAC in the advanced treatment of their wastewaters.

GAC has also been used for many years in industrial water purification and in the production of soft drinks, pharmaceuticals, fats and oils, and alcoholic beverages. From 1935 to 1960, the utilization of decolorizing grades of GAC increased from 11,000 to 60,000 tons per year (9,977 to 54,420 metric tons/y) (API, 1969).

Extrapolating GAC Experience in Municipal Wastewater Reclamation to Water Treatment

South Lake Tahoe. The first plant-scale use of GAC in a municipal wastewater treatment plant was at South Lake Tahoe, California, in 1965. This plant has operated continuously since that time and now has 19 years of operating experience with GAC; the GAC system has processed more than 14 billion gallons (53 Gl) of pretreated municipal wastewater. The reclaimed water COD ranges from 10 to 30 mg/l. The South Tahoe installation was a United States Environmental Protection Agency (EPA) demonstration plant. For 3 years, EPA funded the collection of very detailed and complete plant operating data and cost information.[43]

Except for a few months in 1977–1978 when some carbon was reactivated off-site during an emergency, all carbon has been reactivated at the plant with the use of a multiple-hearth furnace. Carbon losses in reactivation have averaged about 8 percent. This means that all of the original carbon has now been replaced by makeup carbon. More significantly, plant experience has now verified the results of bench-scale tests of reactivation (made in 1963)—tests indicating that the GAC could be maintained at or near full adsorptive capacity by means of thermal reactivation until the carbon was eventually all replaced by making up losses from attrition and burning. It has been demonstrated that the GAC has a service life of at least 12 cycles of reactivation under actual full-scale plant operating conditions.

At South Tahoe, the carbon furnace refractories were all replaced after 10 years of service. In 1977–1978, owing to maladjustment of the burners in the reactivation furnace, excess oxygen was present during reactivation of a considerable volume of GAC. This carbon was overburned and suffered a dramatic loss (over 50 percent) in adsorptive capacity. Fortunately, it was possible to fully restore the adsorptive capacity of the carbon in one pass through the furnace, once the burners were readjusted properly. During this emergency, some off-site reactivation of GAC was necessary. This is the only major difficulty encountered in 18 years of operation of the GAC system. Corrosion of the exhaust gas scrubber has been something of a problem, especially when breakpoint chlorination has been practiced before GAC adsorption. Exhaust gas scrubbers should be fabricated from 316 SS (stainless steel) where such corrosive conditions prevail. Also the 50 psi (344.5 kPa), 304 SS carbon column inlet and outlet screens are being replaced by 100 psi (689 kPa), 316 SS over a period of years because of corrosion and mechanical distortion. The average service life of the original screens will probably exceed 15 years by the time all are replaced.

Other Wastewater Installations. Water reclamation plants constructed at the Orange County Water District (Water Factory 21), California;[44,45] the Upper Occoquan Sewer Authority, Virginia; and the Tahoe–Truckee Sanitation Agency, California, have the same configuration as the South Tahoe Plant both in respect to the type of GAC facilities provided and the high degree of pretreatment afforded. All of these plants operate successfully with few GAC system problems. As might be expected with second- and third-generation designs, these later plants embody some minor improvements over the original South Tahoe installation, although there are no major changes.

Although there are many other successful applications of GAC in advanced waste treatment (AWT) plants, the experience with GAC in AWT in some places has, unfortunately, been poor. These failures in AWT applications have not stemmed from deficiencies in the basic GAC processes or in organics adsorption and thermal reactivation, but, rather, from mechanical problems.

Operational Problems. In discussing the operational problems encountered with GAC systems, those problems associated specifically with sewage treatment must be distinguished from general problems that might be encountered with any type of GAC system. Many problems with GAC in wastewater treatment will not occur in water purification. For example, in water treatment, few or no problems could be expected with excessive slime growths, hydrogen sulfide gas production, or corrosion from adsorbed organics released during carbon regeneration. Other operational problems include GAC bed losses and desorption of adsorbed compounds.[46,47]

Some of the types of problems encountered with GAC systems in wastewater treatment include:

- Inadequate carbon transfer and feed equipment.
- Undersized slurry and transfer lines.

- Failure to provide for air venting from backwash lines with destruction of contactor bottoms and disruption of GAC.
- Failure to house automatic control systems or otherwise protect them from the weather.
- Inadequate means for continuous, uniform feed to the reactivation furnace, resulting in temperature fluctuations, inconsistent reactivation efficiency, and wasted energy.
- Location of furnace and auxiliary drive motors in areas of very high ambient temperature (e.g., above the top of the furnace).
- The use of nozzles in filter and carbon contactor bottoms, which produced major failures in carbon systems just as they have for many years in water filtration plants. Their use is risky.
- The use of air-assisted backwash of carbon contactors.

Common problems related to wastewater treatment that do not apply to water purification have been growth of biological organisms in the carbon contactors and development of anaerobic conditions with the production of corrosive hydrogen sulfide. Both of these problems have been successfully circumvented by providing adequate flow through the columns or frequent backwashing. Failure to provide adequate pretreatment has caused column clogging and mudballs with the need for more frequent backwashing.

Corrosion has been a problem with some of the GAC systems. The furnace system, transfer piping, and storage tanks are susceptible components. Many operations require frequent replacement of the rabble arms and teeth, and replacement of the hearths every few years. At one installation, titanium or ceramic-coated rabble teeth were no more resistant to corrosion than stainless steel teeth. In one case, the corrosion problem in the furnace was solved by eliminating the use of auxiliary steam during reactivation. In another case, corrosion was linked to fluctuating temperatures in the hearths caused by irregular feed to the furnace and frequent startup and shutdown. These problems can be partially remedied by better operation and avoided by better engineering design.

In several industrial applications, the wastewater itself has been highly corrosive. In these cases, the contactors have been subject to corrosion. At Spreckles Sugar, the epoxy linings in the columns must be replaced every 3 years; Republic Steel also replaces its column linings on a regular basis. This kind of corrosion would not be expected in water treatment plants.

By properly applying the best current engineering design knowledge and practices for GAC systems, these rather serious problems might be avoided. When waterworks engineers apply GAC to produce high-quality drinking water, they should make the most of the experiences of the consultants for industry and wastewater agencies.

Extending Wastewater Treatment Experience to Water Treatment Systems. Caution must be observed in extrapolating GAC cost data from operating industrial installations and municipal wastewater treatment plants to the design of waterworks. The purpose for using GAC in each of these types of applications is generally the same—to remove organics. There are important differences, however. In industry, the GAC serves to remove a rather narrow band of organics—color molecules—from a viscous liquid. In wastewater treatment, the GAC removes (with or without biological activity) a broad spectrum of organic substances from water, as measured by BOD, COD, and TOC. In water treatment, the objectives of GAC treatment are not completely defined at this time, but for raw waters with color or taste and odor problems, using GAC unquestionably improves the drinking water from an aesthetic standpoint. In many cases, the cost of GAC may be warranted for these purposes alone. For the great number of water systems without color or taste and odor problems, the main concerns with respect to organics are the health effects from prolonged periods of ingesting trace quantities of possible carcinogens.

Public health officials and waterworks managers still disagree as to whether potential health risks due to the presence of minute traces of organics in drinking water are sufficient to warrant the cost of GAC treatment. A major concern is that the potentially harmful organics in drinking water have not all been identified at this time, and many of those that have been tagged as suspect have widely different adsorptive characteristics. Some adsorb readily on GAC, while others do not.

GAC loading rates at exhaustion of adsorptive capacity vary widely among the different potentially hazardous organics. This affects the length of service life of GAC before carbon reactivation or replacement is necessary—a determining factor in GAC treatment costs. Similarly, the reactivation times and temperatures for thermal reactivation of GAC saturated with different organics also differ, and not all are known at this time. Again, this has an important bearing on GAC treatment costs. Because of the widely varying adsorptive and reactivation characteristics of trace organics on GAC in water supplies, pilot plant tests of both adsorption and reactivation are mandatory preludes to treatment system design.

Over a period of years, perhaps general, average design parameters will emerge from the results of pilot plant studies and demonstration projects; but this time is not yet at hand. Once the GAC design parameters for water treatment have been established from pilot tests for a particular water source, then the knowledge and experience from other GAC installations in industry and wastewater plants can be put to good use. Carbon dosages, GAC contact times, and spent carbon reactivation times and temperatures can be determined. Contactor sizes can be calculated, and furnace sizes and fuel requirements can be estimated. Transport facilities for GAC in water treatment can be the same as for other types of GAC installations, provided the differences in quantities and possible differences in the viscosities of carbon slurries due to any slime growth are taken into account. Also, with the GAC design parameters pinpointed as a result of pilot plant studies, construction costs can be accurately estimated based on the costs of existing installations in AWT and industry. The estimates cannot, however, be based on a flow basis; rather, they must be based on adsorption and reactivation data applicable to each specific installation.

Selecting the most economical number of contactors for a water system of a certain size involves the same principles as are used for other systems. Because of shipping regulations, factory-fabricated contactor vessels are generally limited to about 12 feet (3.7 m) maximum diameter. For large-capacity installations, a small number of field-erected steel vessels or poured-in-place concrete vessels may be economical.

Because upflow contactors provide all of the advantages of countercurrent operation with respect to carbon savings, they are favored for most types of service. The exception is water treatment. In this case, downflow is used because the discharge of carbon fines in the effluent (a characteristic of upflow columns) is avoided.

Cost estimates must be evaluated on the same basis as all other estimates of construction cost; there is no reason why they should be more or less accurate than estimates made for the rest of the treatment plant. Fifteen percent is generally accepted as an allowable difference between costs estimated from construction plans and the best bid received from the contractors. With good pilot plant data and with proper application of cost data from existing GAC installations, preliminary cost estimates for GAC treatment of public water supplies should be accurate enough for planning purposes. (Cost data for GAC systems are included in Chapter 30.)

The extrapolation of wastewater treatment experience with GAC to the design of water treatment systems is a task for trained, experienced, engineering professionals.

DESIGNING GAC ADSORPTION AND REACTIVATION SYSTEMS FOR WATER TREATMENT

The following discussion is devoted to some of the procedures and details for developing the design basis and costs for GAC water treatment systems. Criteria are obtained from pilot plant test results and information from full-scale applications. (Further information on the design and costs of GAC adsorption and reactivation systems may be found in References 48 through 55.)

GAC System Components

Systems utilizing GAC are rather simple. In general, they provide for the following system components:

- Carbon contactors for the water to be treated for the length of time required to obtain the necessary removal of organics.
- Reactivation or replacement of spent carbon.
- Transport of makeup or reactivated carbon into the contactors.
- Transport of spent carbon from the contactors to reactivation or hauling facilities.

These facilities are discussed in more detail later in this section.

Pilot Plant Tests

Despite the simplicity of GAC systems, laboratory and pilot plant tests are needed to select the carbon and the most economical plant design for both water and wastewater treatment projects.[56-58] Pilot column tests make it possible to:

- Determine treatability.
- Select the best carbon for the specific purpose based on performance.
- Determine the required empty bed contact time.
- Establish the required carbon dosage that, together with laboratory tests of reactivation, will determine the capacity of the carbon reactivation furnace or the necessary carbon replacement costs.
- Determine the effects of influent water quality variations on plant operation.

During pilot testing, the influence of longer carbon contact times on reactivation frequency can be measured. These measurements allow costs to be minimized through a proper balance of these two design factors.

Design of Pilot GAC Columns

Detailed information, including a list of materials, on the design and construction of pilot GAC columns is presented elsewhere, as are descriptions of the analytic methodology for monitoring pilot column tests.[59] Also included are:

- Data on the adsorbability of various organic compounds.
- Data on the performance of GAC in their removal.

- Information on the use of multiple-hearth, infrared, fluidized-bed, and rotary kiln furnaces for reactivation of spent GAC.
- Example calculations for balancing added costs of increased contact time versus savings (if any) from less frequent reactivation.

Frequency of Reactivation

One of the principal differences in costs for GAC treatment between water and wastewater is the more frequent reactivation required in water purification due to earlier breakthrough of the organics of concern. In wastewater treatment, GAC may be expected to adsorb 0.30 to 0.55 lb COD/lb carbon (0.30 to 0.55 kg/kg) before the carbon is exhausted. From the limited amount of data available from research studies and pilot plant tests (mostly unpublished), it appears that some organics of concern in water treatment may break through at carbon loadings as low as 0.05 to 0.25 lb organics/lb carbon (0.05 to 0.25 kg/kg). The actual allowable carbon loading or carbon dosage for a given case must be determined from pilot plant tests. Costs taken from wastewater cost curves, which are plots of flow in mgd versus cost (capital or operation and maintenance costs), cannot be applied directly to water treatment. Allowance must be made in the capital costs for the different reactivation capacity needed, and in the operation and maintenance costs for the actual amount of carbon to be reactivated or replaced.

Because the organics adsorbed from water generally are more volatile than those adsorbed from wastewater, the increased reactivation frequency due to lighter carbon loading may be partially offset, or more than offset, by the reduced reactivation requirements of the more volatile organics. The times and temperatures required for reactivation may be reduced, owing to both the greater volatility and the lighter loading of organics on the carbon. (Removal of VOC's is discussed in Chapter 15.)

From the limited experimental reactivations to date, it appears that reactivation temperatures *may* be less than the 1650 to 1750°F (899 to 954°C) required for wastewater carbons. The shorter reactivation times required for water purification carbons may allow the number of hearths in a multiple-hearth reactivation furnace to be reduced. Also, less fuel may be required for reactivation. These factors must be determined on a case-by-case basis.

GAC Contactors

Selection of the general type of carbon contactor to be used for a particular water treatment plant application may be based on several considerations, including economics and the judgment and experience of the engineering designer.[60,61] The choice generally would be made from three types of downflow vessels:

1. Deep-bed, factory-fabricated, steel pressure vessels of 12-foot (3.66-m) maximum diameter might be used over a range of carbon volumes from 2,000 to 50,000 cu ft (56 to 1,400 m³).

2. Shallow-bed reinforced concrete, gravity-filter-type boxes may be used for carbon volumes ranging from 1,000 to 200,000 cu ft (28 to 5,600 m³). Shallow beds probably will be used only when short contact times are sufficient or when long service cycles between carbon regenerations can be expected from pilot plant test results.

3. Deep-bed, site-fabricated, 20- to 30-foot (6.1- to 9.1-m)-diameter, open concrete or steel, gravity tanks may be used for carbon volumes ranging from 6,000 to 200,000 cu ft (168 to 5,600 m³), or larger.

These ranges overlap, and the designer may very well make the final selection based on local factors, other than total capacity, that affect efficiency and cost.

As previously mentioned, the current design trend in retrofitting existing water treatment plants with new GAC adsorption facilities is to provide separate, post-filtration, downflow contactors.[62,63] Contactor flow rates are usually 3 to 7 gpm/sq ft (7.32 to 17.08 m/h), and GAC bed depths are normally 2.5 to 15 feet (0.76 to 4.57 m). A direct linear relationship between contact time and carbon bed performance has been found in full-scale plant tests and concurrent small column tests. Carbon performance at a given contact time has been found to be unaffected by variations in hydraulic loading rates in the 2 to 10 gpm/sq ft (4.88 to 24.4 m/h) range. Thus, in terms of adsorption only, contact time is the governing criterion, and not surface loading rate within the ranges of practicality.

When the granular carbon bed is functioning as both a turbidity removal and an adsorption unit, there may be reasons to limit the bed depth and flow rate parameters to remove turbidity effectively and to backwash the filter properly. If GAC is to be effective in turbidity removal, it must be hard enough to withstand vigorous backwash agitation. At the same time, it should be dense enough to expand during the backwash cycle and to settle quickly for immediate resumption of filtration. As discussed earlier, coal-based granular carbon possesses approximately the same density and filtration characteristics as anthracite, and has found increasing use in the water field.

Particle size of the carbon, in addition to contact time, should be considered carefully as a design factor. Reduction of particle size for a given set of flow conditions is a means of increasing adsorption rates and thereby improving adsorption performance. Reduction in particle size to improve adsorption must be consistent with other significant factors such as headloss and backwash expansion. The length of the filter run in an adsorption–filtration bed would also be a problem, if too small a particle size were chosen.

Where existing rapid sand filters are being converted to adsorption–filtration units, the permissible depth of the carbon layer will be limited by the freeboard available in the existing structure. Adequate space between the carbon surface and the backwash trough bottoms should be available to permit at least 30 percent expansion of the carbon layer.

The AWT experience with GAC contactors may be applied to water purification if some differences in requirements are taken into account. The required contact time must be determined from pilot plant test results. Although contactors may be designed for a downflow or upflow mode of operation, and upflow packed beds or expanded beds provide maximum carbon efficiency through the use of countercurrent flow principles, the leakage of some carbon fines (1 to 5 mg/l) in upflow carbon column effluents makes downflow carbon beds the preferred choice in most municipal water treatment applications. At the Orange County Water Factory 21, upflow beds were converted to downflow beds to correct a problem with escaping carbon fines. This full-scale plant operating experience indicates that leakage of carbon fines is not a problem in properly operated downflow GAC contactors.

Single beds or two beds in series may be used. Open gravity beds or closed pressure vessels are permissible. Structures may be properly protected steel or reinforced concrete. In general, small plants will use steel, and large plants may use steel or reinforced concrete.

Sand in rapid filters has, in some instances, been replaced with GAC. In situations where contact times are short and GAC regeneration or replacement cycles are exceptionally long (several months or years), as may be the case in taste and odor removal, this may be a solution. However, with the short cycles anticipated for most organics, conventional concrete-box-style filter beds may not be well suited to GAC contact; deeper beds may be more economical in first cost and provide more efficient use of GAC. Beds deeper than conventional filter boxes, or contactors with greater aspect ratios of depth to area, provide much greater economy in capital costs. The contactor cost for the needed volume of carbon is much less. In water slurry, carbon can be moved from contactors with conical bottoms easily and quickly, and with virtually no labor. Flat-bottomed filters of a type that require labor to move the carbon unnecessarily add to carbon transport costs. The labor required to remove carbon from flat-bottomed beds varies considerably in existing installations from a little labor to a great deal, depending upon the design of the evacuation equipment.

For many GAC installations intended for precursor organic removal or synthetic organic removal, specially designed GAC contactors should be installed. Contactors should be equipped with flow-measuring devices. Separate GAC contactors are especially advantageous where GAC treatment is required only seasonally because the contactors then can be bypassed when they are not needed, possibly saving unnecessary exhaustion and reactivation of GAC.

Tremendous cost savings can be realized in GAC treatment of water through proper selection and design of the carbon contactors. The design of carbon contactor underdrains requires experienced, expert attention.

Carbon Contactor Underdrains

Although good proven underdrain systems, such as Johnson screens and Leopold block, are available, often they are not used, and there have been numerous underdrain failures due to poor design. Earlier underdrain designs have failed in many installations for conventional filter service, and they continue to be misapplied to GAC contactors as well as filters.

GAC Reactivation or Replacement

Spent carbon may be removed from contactors and replaced with virgin carbon, or it may be reactivated either on-site or off-site. The most economical procedure depends on the quantities of GAC involved. As already discussed, for larger volumes on-site reactivation is more economical. For small quantities of carbon, replacement or off-site reactivation will probably be most economical.

Carbon may be reactivated thermally to very near virgin activity. Carbon burning losses may, however, be excessive under these conditions.

Experience in industrial and wastewater treatment indicates that carbon losses can be minimized (held to 8 to 10 percent per cycle) if the activity of reactivated carbon (as indicated by the Iodine Number) is held at about 90 percent of the virgin activity. To remove certain organics, there may be no decrease in actual organics removal despite a 10 percent drop in Iodine Number.

Thermal Reactivation Equipment

GAC may be reactivated in a multiple-hearth furnace, a fluidized-bed furnace, a rotary kiln, or an electric infrared furnace. Spent GAC is drained dry in a screen-equipped tank (40 percent moisture content) or in a dewatering screw (40 to 50 percent moisture) before being introduced to the reactivation furnace. Dewatered carbon is usually transported by a screw conveyor. Following thermal reactivation, the GAC is cooled in a quench tank. The water–carbon slurry may then be transported by means of diaphragm slurry pumps, eductors, or a blow-tank. The reactivated carbon may contain fines produced during conveyance; these fines should be removed in a wash tank or in the contactor. Maximum furnace temperatures and retention time in the furnace are determined by the amount (lb organics/lb carbon) and nature (molecular weight or volatility) of the organics adsorbed.

Off-gases from carbon reactivation present no air pollution problems provided they are properly scrubbed. In some cases, an afterburner may also be required for odor control.

Fluidized-bed and multiple-hearth furnaces are the simplest, most reliable, and easiest to operate for GAC reactivation. The infrared and fluid-bed units still have problems to be worked out.

It is necessary with all four types of furnaces to specify top quality materials to suit the conditions of service, and to see that these materials are properly installed. Corrosion resistance is important in the furnace itself and especially in all auxiliaries to the furnace. (More information concerning thermal reactivation equipment is offered in References 64 and 65.)

Required Furnace Capacity

The principal cost differences between GAC treatment of water and of wastewater may lie in the capital cost of the furnace and in the O&M cost for carbon reactivation. Water purification carbons typically are easier to regenerate (less time in furnace and lower furnace temperatures) and more lightly loaded (greater volume of carbon to be reactivated per pound of organics removed). Accurate estimates of GAC costs require knowledge and consideration of these two factors. It is not possible to use AWT cost curves based on mgd (Ml/d) throughput or plant capacity to obtain costs for water treatment, unless the differences in reactivation requirements are taken into account.

Carbon Transport

The primary use of air or pneumatic transport of carbon is in bulk handling of makeup carbon. Once carbon is introduced into the adsorption–regeneration system, it is usually transported hydraulically in slurry form.

Handling characteristics have been experimentally studied by using water slurries of 12 × 40 mesh granular carbon in a 2-inch (50.8-mm) pipeline. The data indicate that a maximum of 3 lb carbon/gal (0.36 kg carbon/l) of water could be transported hydraulically, but that it is better to use a ratio of 1 lb carbon/gal (0.12 kg carbon/l). The velocity necessary to prevent settling of carbon is a function of pipe diameter, granule size, and liquid and particle density. The minimum linear velocity to prevent carbon settling was found to be 3.0 ft/sec (0.91 m/s). It is recommended that a linear velocity of 3.5 to 5.0 ft/sec (1.07 to 1.53 m/s) be used. Velocities of over 10 ft/sec (3.05 m/s) are objectionable because of carbon abrasion and pipe erosion. Carbon delivery rates are a function of pipe diameter, slurry concentration, and linear velocity. Data are shown in Fig. 12–4 and 12–5 for a 2-inch and a 1-inch pipe, respectively. Pressure drop data for various slurry concentrations and velocities in 2-inch (50.8-mm) pipe are shown in Fig. 12–6.

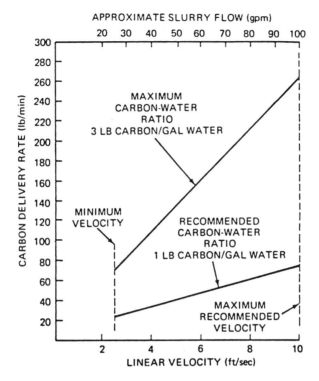

APPROXIMATE SLURRY FLOW (gpm)

Fig. 12-4. Carbon delivery rate (2-inch pipe).

Pilot plant tests indicate that after an initial higher rate, the rate of attrition for activated carbon in moving water slurries is approximately constant for any given velocity, reaching an approximate value of 0.12 percent fines generated per exhaustion–regeneration cycle. This deterioration of the carbon with cyclic operation has been reported to be independent of the velocity of the slurry, within the recommended range of 3.5 to 5 ft/sec (1.07 to 1.53 m/s). Loss of carbon by attrition in hydraulic handling apparently is not related to the type of pump (diaphragm or centrifugal) used.

Carbon slurries can be transported by using water or air pressure centrifugal pumps, eductors, or diaphragm pumps. The choice of motive power is a combination of owner preference, turndown capabilities, economics, and differential head requirements.

CARBON ADSORPTION ENHANCEMENT WITH OZONE

General

Depending upon water quality, ozone pretreatment ahead of GAC adsorption may increase GAC service life by 5 to 100 percent.[66] Under most circum-

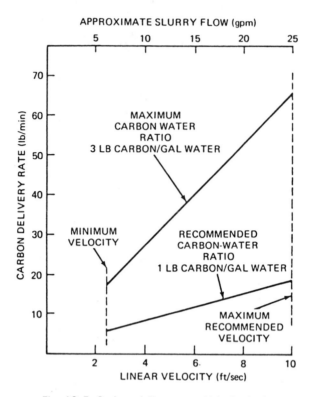

Fig. 12–5. Carbon delivery rate (1-inch pipe).

stances a 10 to 20 percent extension of carbon service life can be expected. Preozonation tests should be a part of all GAC pilot plant operations. Once the data are available on GAC savings, they should be compared to the cost of ozonation to determine the economics of providing ozone pretreatment.

Philadelphia Study

At Philadelphia, laboratory and pilot-scale studies were conducted to investigate the combined unit processes of oxidation by ozonation and adsorption by GAC.[68] In addition, the effect of chlorine and chlorine by-products on the ozone and GAC processes was investigated.

The study used Delaware River water previously treated by two coagulation/filtration plants: a full-scale plant operating with a 2 mg/l free chlorine residual and a 30,000 gpd (0.14 l/d) pilot plant operating without any disinfection. The rapid sand filter effluents from each were applied to parallel GAC and ozone/GAC systems. The GAC beds remained in service between 360 and 500 days.

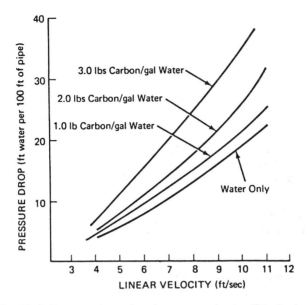

Fig. 12–6. Pressure drop of carbon–water slurries (2-inch pipe).

The removal of trace organics substances at the ng/l to µg/l level and the removal of TOC at the mg/l level were carefully monitored, along with microbial parameters of biological speciation and growth rates. System comparisons were made using estimated total costs of each unit process, as determined by the carbon regeneration rate needed to maintain various effluent criteria.

Conclusions of the Philadelphia study were:

- Enhanced TOC removal before treatment does not appear to increase the adsorptive capacity of the GAC for trace organics of health concern.
- The cost of preozonation is not sufficiently offset by the lowered GAC operating costs associated with less frequent GAC regenerations, when the removal of the halogenated, volatile organic compounds is the controlling criterion. For TOC reductions, preozonation may be cost-effective, depending on the exact criteria chosen.
- Chlorination appears to produce compounds resistant to further oxidation by ozone, and thus should be applied only after ozonation.
- Chlorination and ozonation each affect the adsorptive capacity of the specific organic compounds differently.
- After GAC effluents were effectively disinfected with chlorine, no bacterial regrowth was found after 5 days.
- Over a 2-year period, there were no operational problems associated

with maintaining a conventional treatment system on-line without any type of predisinfection.

OTHER EPA PROJECTS USING GAC FOR REMOVAL OF ORGANIC CONTAMINANTS FROM DRINKING WATER

Jefferson Parish, Louisiana

At Jefferson Parish, concern over the detection of known and suspected carcinogens in the New Orleans, Louisiana area water supplies resulted in a research effort to provide data for removal of these compounds.[70] The primary objective of the study was to examine the efficiency of using granular activated carbon (GAC) for the removal of organic contaminants in drinking water. Two full-scale systems were compared and evaluated—a post-filtration adsorption GAC filter in series with a sand filter and a combined filtration–adsorption GAC filter. Also examined were the ability of pilot GAC columns to predict the organics removal efficiency of full-scale GAC filters, the effects of varying empty-bed contact times, and variations in the organic contents of drinking water resulting from different types of water treatment processes. Correlations between certain nonspecific parameters (such as total organic carbon) and specific individual parameters were examined in the hope that one or more nonspecific parameters could be used as a surrogate monitoring parameter.

Conclusions of the Jefferson Parish study were:

- The conventional treatment process effectively removed some organic contaminants, notably the surrogate parameters: total organic carbon (TOC) (3 to 25 percent) and trihalomethane formation potential (THMFP) (22 to 40 percent) across the precipitator; and ultraviolet (UV) (50 to 55 percent), rapid fluorescence measurement (RFM) (14 to 28 percent), and emission fluorescence scan (EMF) (10 to 31 percent) following the addition of 0.5 to 1.0 mg/l potassium permanganate. These results indicate that conventional water treatment is significantly, but not totally, effective in removing general types or organic contaminants from drinking water. No significant differences in organics removal were observed between lime softening and polyelectrolyte polymer treatment. Because of the nonreactive nature of monochloramine with the THM precursor substances, the formation of trihalomethanes (THM) was not affected by the difference in pH of these treatment processes.
- The GAC systems were effective in removing organic contaminants. GAC systems operating in the adsorber mode exhibited the same relative adsorption efficiencies as those operating in the filter adsorber mode; this indicated the applicability of either type of system for organics

removal. No observable reduction in the adsorption efficiency of the filter adsorber existed because of pore blockage by turbidity.

- The GAC pilot column simulators were generally quite effective in predicting the adsorption efficiencies of their respective full-scale counterparts after taking into account the variations caused by GAC losses during backwashing in the full-scale systems. Thus, the pilot column system can be an effective and economical tool in the design of full-scale GAC facilities.

- The breakthrough profiles of the surrogate parameters through the various GAC systems did not correlate reliability with those of the individual parameters under study when simple linear correlations were used. Multiple, nonlinear-type correlations were indicated, but none was found that correlated well with the data.

- Evidence indicates that a critical GAC bed depth (which varies for different substances) is required to remove a particular organic contaminant. This critical bed depth is generally deeper for the THM's than it is for the higher-molecular-weight substances such as the chlorinated hydrocarbon insecticides, because THM substances break through the GAC beds at different rates. For example, the GAC beds with a 20-minute EBCT were saturated with THM's within 100 days, but a contactor with a 10-minute EBCT continued to remove more than 90 percent of the chlorinated hydrocarbon insecticides at the end of each phase.

- The increase in EBCT across the series contactors resulted in increased adsorption efficiency for most of the parameters under study. But the increment of the increase in efficiency across each consecutive contactor decreased as EBCT increased.

- A steady-state condition was reached for the surrogate parameters of TOC, UV, RFM, EMF, and THMFP after approximately 100 days, when a relatively constant removal of these constituents was observed. The relative removal levels for each constituent increased with EBCT up to 30 minutes; after that, no further significant removal was observed.

- A concentration gradient effect was observed for both types of GAC system. Even after saturation, large surges in influent concentration were effectively adsorbed with no apparent adverse effects in the effluent of the GAC systems. This phenomenon greatly improves the effectiveness of GAC filtration in combating chemical spills in surface water sources.

- Organic substances that had little or no adsorptive affinity for GAC at the ng/l level and did not exhibit the typical S-shaped breakthrough curve were: phthalates, n-alkanes, and substituted benzene derivatives. Some of these substances exhibited a low degree of constant removal. Also, the surrogate parameters (TOC, UV, RFM, EMF, and THMFP) showed less than 100 percent removal after 40 minutes of EBCT during the series contactor studies.

- Variations in the effectiveness of the different types of GAC used during this project were observed for the removal of the lower-molecular-weight volatile substances.
- The 12 × 40 mesh GAC medium in the filter adsorber appeared to remove turbidity as well as, if not better than, the sand medium when equal depths of the media were compared. Thus because the adsorption efficiencies of the two GAC systems (adsorber and filter adsorber) are similar, conversion from sand filtration to GAC filtration (sand replacement) would appear to be advantageous for organic removal and turbidity reduction.

Cincinnati, Ohio

A post-filtration GAC adsorption facility with a capacity of 175 mgd (662.4 Ml/d) for treatment of Cincinnati's Ohio River drinking water supply is currently under design by Malcolm Pirnie, Inc., in association with Culp/Wesner/Culp, Consulting Engineers.

The design of the full-scale plant is based upon the results of a cooperative study by the U.S. EPA and the Cincinnati Water Works (CWW). A report of this work entitled *Feasibility Study of GAC Adsorption and On-Site Regeneration* was published in October, 1983.[62] It was prepared by Richard Miller and David J. Hartman of Cincinnati. Jack DeMarco and Ben J. Lykins, Jr. were EPA project officers.

The abstract of this report follows:

This project determines whether the use of granular activated carbon (GAC) is feasible for removing certain trace organics from Ohio River water while treating it for human consumption. The study used both deep-bed contactors and conventional-depth gravity filters, with on-site GAC regeneration.

The primary objective of the study was to determine if selected organics could be reduced to a predesignated level without adversely affecting the level of treatment provided by the existing plant and at a cost acceptable to consumers. A secondary objective was to develop plant design and operating parameters for full-scale plant conversion to GAC treatment. The study was unusual because it employed full-sized filters, contactors, and carbon regeneration furnace instead of the pilot-scale components used by most water quality researchers.

In the first phase of the project, three existing rapid sand filters were converted to GAC filter adsorbers. Various GAC bed depths and types were studied to compare organic removal efficiencies, bed lives, general water quality characteristics, the need for a sand underlayer and operational problems.

The second phase involved the use of pilot-scale GAC components to investigate the effects of regeneration on the carbon's adsorptive capability and to determine the reliability of pilot columns as performance indicators for full-scale components. The relative performances of lignite and bituminous-based GAC were also studied.

The last phase of this project studied the relative performance of GAC filters to post-filtration GAC contactors, the most advantageous empty bed contact time for the contactors and the effectiveness of on-site GAC regeneration. Pilot columns were also operated in parallel with the full-sized units to assess the usefulness of pilot columns as predictors of full-scale operation. During this phase, an attempt was made to maximize the use of currently available organic analysis techniques. Additional organic analytical techniques such as acid extract GC/FID profiles, closed loop stripping analyses and carbon-adsorbable organohalides provided a broad data base. Finally, a significant aspect of this project was the development of preliminary cost estimates for full-plant conversion to GAC.

The conclusions stated below are based solely on the governing conditions and findings of this particular project. Although certain findings may well apply at other locations, particularly on the Ohio River, the conclusions are not all-encompassing and may be inappropriate under differing conditions.

1. No turbidity reduction benefit was derived from the requirement of Section 4.2.1.6 in the Recommended Standards for Water Works, commonly referred to as the "Ten State Standards," for a 12-inch (0.305-m) sand underlayer to GAC or from the requirement that replacement media be the same effective size as filter sand.[4]
2. Bacterial growth within the GAC filters and contactors was experienced. Harmful bacteria were eliminated by post-chlorination.
3. Post-chlorination would be an absolute necessity if the entire plant were converted to GAC.
4. Bituminous-based GAC outperformed lignite-based GAC with respect to service life, weight of contaminants adsorbed, and cost per weight of contaminants adsorbed.
5. Pilot columns were reasonably predictive of full-scale GAC systems for organics removal.
6. Floc removal by GAC filters had little effect on the carbon's adsorptive ability.
7. The optimum GAC empty bed contact time (EBCT) would be between 7.0 and 15 minutes during annual average conditions and greater than 15 minutes during critical summer conditions.
8. Regeneration restored the GAC to its virgin adsorptive capacity.
9. GAC regeneration losses averaged 15 percent by volume for ten contactor regenerations and 18.5 percent by volume for six GAC filter regenerations.

Costs.   Cost conclusions [for a 175 mgd (662.4 Ml/d) plant] made in the report are based on preliminary estimates using cost curves developed for general application. These cost curves were applied to site-specific design criteria. Actual costs could be considerably different when determined for this or other sites based on a detailed engineering design.

1. The preliminary cost estimates for full-scale design, construction and operation of a GAC system at the CWW treatment plant indicate that a capital investment of approximately $40 million (based on 1981 dollars) may be required to reduce total organic carbon (TOC) concentrations to a specified criterion of 1,000 μg/l using either GAC filters or contactors.
2. A GAC filter system compared to a contactor system will annually cost (in 1981 dollars) about twice as much for operating and maintenance costs ($8.0 vs. $4.0 million) and about 1.5 times as much for total costs, including capital amortization ($12.4 vs. $8.5 million).
3. The estimated increase in the unit production costs of water to reduce finished water from an average of about 2,100 μg/l to a treatment goal of 1,000 μg/l TOC will be $0.24/1,000 gal ($0.06/m^3) for GAC filters (7.5 min EBCT) and $0.165/1,000 gal ($0.04/m^3) for contactors (15.0 min EBCT), all in 1981 dollars.
4. The cost to regenerate GAC on-site over the life of the project averaged about $0.21/lb ($0.46/kg).

ACTIVATED CARBON FOR DECHLORINATION OF WATERS

Carbon has been used for over half a century to dechlorinate water, the early systems using granular coal and coke.[69] The proposed mechanism was originally considered as:

$$2Cl_2 + C + 2H_2O \rightleftarrows 4HCl + CO_2 \qquad (12\text{–}3)$$

However, this reaction describes the summation of the reactions taking place on the surface of the carbon. The actual reaction according to Kovach is:[67]

$$Cl_2 + H_2O \rightarrow 2H^+ + 2Cl^- + O \qquad (12\text{–}4)$$

The chemisorbed nascent oxygen (last term on right) decomposes in either of the following two ways:

$$C_x O_x \rightarrow C + CO \qquad (12\text{–}5)$$
$$C_x O_x \rightarrow C + CO_2 \qquad (12\text{–}6)$$

An empirically derived equation to describe the process is:

$$\log \frac{C_0}{C} = k \frac{L^{1/2}}{V} \qquad (12\text{–}7)$$

where:

C_0 = inlet chlorine concentration, mg/l
C = outlet chlorine concentration, mg/l
L = bed length, ft (m)
K = a constant
V = flow rate, mgd (Ml/d)

The process of chlorine removal from water is not a pure adsorption process but also involves a chemical reaction between the chlorine and the water. The initial step is the adsorption of chlorine in the form of hypochlorous acid on the carbon surface. The subsequent step is decomposition of the hypochlorous acid and nascent oxygen.

The hydrochloric acid is not readily adsorbed and is released into solution, while the chemisorbed oxygen builds up to form an oxide complex on the carbon surface. This surface oxide, which reduces the efficiency of the dechlorination process, is fairly stable at low reaction rates. At high inlet chlorine concentrations (1,000 mg/l), free carbon monoxide and carbon dioxide can be found in the effluent water. Removal of the surface oxide is incomplete below 752°F (400°C), making in situ regeneration uneconomical. It is postulated that even under conditions of industrial dechlorination (below 30 mg/l chlorine), some carbon dioxide is generated; however, the quantity is too low to be detectable against the background carbon dioxide content. Several investigators also postulated the formation of $HClO_3$ above 2,660 mg/l chlorine concentration, but recent work shows that the thermodynamics of this reaction, at least in static systems, is unfavorable.

The basic design parameters related to the operation of an activated carbon dechlorinator are:

- Chemical state of the free chlorine, including pH
- Inlet chlorine concentration
- Flow rate through the carbon bed
- Particle size of the carbon
- Type of carbon
- Temperature
- Manner of operation (continuous or interrupted)
- Impurities present

It has also been proposed that the dechlorination process is controlled by the slow chemical reaction on the surface, that is, decomposition of HOCl, in which case equation (12–7) above can be modified to:

$$\log \frac{C_0}{C} = k_1 \frac{L}{V} \qquad (12\text{–}8)$$

Based on current data, the use of either equation can be justified, depending on the conditions of the experiment. Because k_1 is less than k, the latter equation contains a built-in safety factor.

While the establishment of a design based on either equation can be performed easily, expected variations in process conditions are rarely taken into account. The equations show the importance of using the actual chlorine concentrations expected in the influent water; acceleration of the test by using artificially high chlorine concentrations for rapid carbon evaluation leads to faulty conclusions.

The chemical state of the chlorine strongly influences the dechlorination process. For example, chloramines considerably reduce dechlorination efficiency. Unfortunately, no quantitative evaluation was available to the authors on the effects of chloramines. In some instances where chlorine dioxide is used, the effect of pH becomes important because the $NaClO_2$ formed is more stable than $HOCl$; thus, any evaluation of activated carbon should be subjected to the same chlorine forms, or the pH should be adjusted.

Dechlorination efficiency for any activated carbon is also strongly dependent on particle size, and the smallest mean particle size should be selected for any carbon grade. Of course, the limiting factor in such a selection is the maximum allowable pressure drop. The micro- and macroporosity of the carbon also influence dechlorination efficiency. Because the dechlorination process is related to the surface area of the carbon, macropores are necessary to make the high surface areas accessible for the adsorption reaction. The raw material from which the carbon is made generally does not have any effect on dechlorination efficiency.

The presence of colloidal impurities or high organic concentrations can substantially reduce the carbon's useful life as a dechlorinator. The colloids may plug the pores of the carbon, reducing the surface area available for dechlorination. Organics can form an adsorbate film on the carbon, causing the same end result. If pretreatment steps are adequate to remove these impurities, beds of activated carbon will have a long useful life as dechlorinators.

If the experimental conditions are carefully selected so that they match the expected operating conditions, then it is fairly straightforward to evaluate activated carbon in dechlorination and to size the adsorber.

Recent tests have determined efficiency values for specific carbons available to industrial and municipal treatment operations in the United States. The results for flow rates, concentration of influent, and type of carbon are shown in Fig. 12–7. They are based on a chlorine breakpoint of 0.01 mg/l. The pH of the water was 7 and the temperature was 73.4°F (23°C). The absence of bacteria or any organic interference was assumed.

The life of the carbon in dechlorination service is extremely long. For example, with 2.5 feet (0.76 m) of carbon in a 1 mgd (3.785 Ml/d) sand

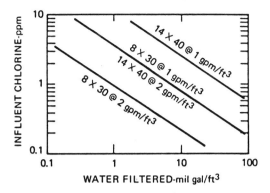

Fig. 12–7. Effect of flow rate and mesh size on dechlorination by granular carbon filters.

filter operated at 2.5 gpm/sq ft (6.1 m/h), the carbon volume is 700 cu ft (19.6 m³), and it could process 700 MG (2,650 Ml) of 4 mg/l free-chlorine influent water before a breakpoint of 0.01 mg/l chlorine would be reached. A bed processing water containing 2 mg/l chlorine under similar conditions would last about 6 years. The effect of mesh size is pronounced. As indicated in Fig. 12–7, a reduction in particle size reflected in the reduction of mesh size from 8 × 30 to 14 × 40 allows a doubling of flow rate without a sacrifice in efficiency.

Dechlorination will proceed concurrently with adsorption of organic contaminants. Long-chain organic molecules such as those of detergents will reduce dechlorination efficiency somewhat, but many common water impurities such as phenol have little apparent effect upon the dechlorination reaction.

A rise in temperature and a lowering of pH favor dechlorination. Figure 12–8 indicates the relationship of these factors as they vary from pH 7 and

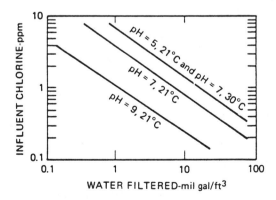

Fig. 12–8. Effect of pH and temperature on dechlorination by granular carbon (8 × 30 mesh at 1 gpm/ft³).

70°F (21°C). Mesh size was 8×30, and flow rate was 1 gpm/cu ft carbon (8.03 m³/h/m³ carbon). A breakpoint of 0.01 mg/l Cl_2 and the absence of bacteria or any organic interference were assumed. It is unlikely that a deliberate change in pH or temperature favoring dechlorination alone would be economically feasible, unless existing conditions significantly retarded the process. These data are values determined with chlorine in distilled water. Variance in hydraulic loading, suspended matter, and certain adsorbed organics, as noted above, could adversely affect dechlorination efficiency.

REFERENCES

1. Suffet, I. H., and McGuire, M. J., *Activated Carbon Adsorption of Organics from the Aqueous Phase,* 2 vols., Ann Arbor Science, Ann Arbor, Mich., 1980.
2. Chow, D. K., and David, M. M., "Compounds Resistant to Carbon Adsorption in Municipal Wastewater," *J.AWWA,* p. 5566, 1977.
3. Weber, Walter J., *Physicochemical Process for Water Quality Control,* Wiley-Interscience, New York, 1972.
4. Symons, J. M., and Robeck, G. C., "Treatment Processes for Coping with Variations in Raw Water Quality," presented at the 1973 American Water Works Association Conference, Las Vegas, Nev., May 16, 1973.
5. Hager, D. C., and Fulker, R. D., "Adsorption and Filtration with Granular Activated Carbon," *Journal of the Society for Water Treatment and Examination* 17:41, 1968.
6. Suffet, I. H., "An Evaluation of Activated Carbon for Drinking Water Treatment: A National Academy of Science Report," *J.AWWA,* p. 41, January, 1980.
7. Meijiers, A. P., et al., "Objectives and Procedures for GAC Treatment," *J.AWWA,* p. 628, November, 1979.
8. Hager, D. C., and Flentje, M. E., "Removal of Organic Contaminants by Granular Carbon Filtration," *J.AWWA* 57:1440, November, 1965.
9. Culp, G. L., and Culp, R. L., *New Concepts in Water Purification,* Van Nostrand Reinhold, New York, 1974.
10. Yohe, T. L., et al., "Specific Organics Removal by GAC Adsorption," *J.AWWA,* p. 402, August, 1981.
11. Sontheimer, H., "Applying Oxidation and Adsorption Techniques: A Summary of Progress," *J.AWWA,* p. 612, November, 1979.
12. Woods, P. R., and DeMarco, Jack, "Treatment of Groundwater with GAC," *J.AWWA,* p. 674, November, 1979.
13. Scholekamp, M., "The Use of GAC Filtration to Ensure Quality in Drinking Water from Surface Sources," *J.AWWA,* p. 638, November, 1979.
14. Schulhof, P., "An Evolutionary Approach to Activated Carbon Treatment," *J.AWWA,* p. 648, November, 1979.
15. U.S. EPA, *Process Design Manual for Carbon Adsorption,* EPA Technology Transfer, Cincinnati, Ohio, October, 1973.
16. Anderson, M. C., et al., "Controlling Trihalomethanes with PAC," *J.AWWA,* p. 432, August, 1981.
17. Kim, B. R., and Snoeyink, V. L., "The Monochloramine GAC Reaction in Adsorption Systems," *J.AWWA,* p. 488, August, 1980.
18. Quinn, J. E., and Snoeyink, V. L., "Removal of Total Organic Halogen by GAC Adsorbers," *J.AWWA,* p. 483, August, 1980.

19. Dyksen, J. E., and Hess, Alan F., "Alternatives for Controlling Organics in Groundwater Supplies," *J.A WWA*, p. 394, August, 1982.
20. Symons, J. M., et al., *Treatment Techniques for Controlling Trihalomethanes in Drinking Water*, EPA-600/2–81–156, U.S. EPA, MERL, Cincinnati, Ohio, September, 1981.
21. Love, T. O., Jr., et al., *Treatment of Volatile Organic Compounds in Drinking Water*, EPA-600/8–83–019, U.S. EPA, MERL, Cincinnati, Ohio, May, 1983.
22. Cheremisinoff, P. N., and Ellerbusch, F., *Carbon Adsorption Handbook*, Ann Arbor Science, Ann Arbor, Mich., 1978.
23. Cotruvo, J. A., "Improving Adsorption Techniques: An International Effort," *J.A WWA*, p. 610, November, 1979.
24. Weber, Walter J., Jr., and Pirabazari, M., "Adsorption of Toxic and Carcinogenic Compounds from Water," *J.A WWA*, p. 203, April, 1982.
25. Lee, M. C., et al., "Activated Carbon Adsorption of Humic Substances," *J.A WWA*, p. 440, August, 1981.
26. Randtke, S. J., and Jespen, C. P., "Effects of Salts on Activated Carbon Adsorption of Fulvic Acids," *J.A WWA*, p. 84, February, 1982.
27. Theim, L., et al., "Removal of Mercury from Drinking Water Using Activated Carbon," *J.A WWA*, p. 447, August, 1978.
28. Weber, W. J., Jr., and Van Vliet, B. M., "Synthetic Adsorbents and Activated Carbon for Water Treatment," *J.A WWA*, pp. 420, 426, August, 1981.
29. Randtke, S. J., and Snoeynik, V. L., "Evaluating GAC Adsorptive Capacity," *J.A WWA*, p. 406, August, 1983.
30. Boening, P. H., et al., "Activated Carbon vs. Resin Adsorption of Humic Substances," *J.A WWA*, p. 54, January, 1980.
31. Calgon Corporation, Pittsburgh, Pa., *Water & Waste Treatment with Filtrasorb GAC*, 1971.
32. Culp, R. L., and Culp, G. L., *Advanced Wastewater Treatment*, Van Nostrand Reinhold, New York, 1971.
33. Nielson, H. L., and Whipple, D. E., "An Improved PAC Feed Facility," *J.A WWA*, p. 324, June, 1979.
34. Hyndshaw, A. Y., "The Selection of Granular Versus Powdered Activated Carbon," *Water and Wastes Engineering*, p. 49, February, 1970.
35. Hassler, J. W., *Purification With Activated Carbon*, Chemical Publishers Company, New York, 1974.
36. Dostal, K. A., Pierson, R. C., Hager, D. G., and Robeck, C. C., "Carbon Bed Design Criteria Study at Nitro, W.Va.," *J.A WWA* 57:663, May, 1965.
37. Flentje, M. E., and Hager, D. C., "Advances in Taste and Odor Removal with Granular Carbon Filters," *Water and Sewage Works*, February, 1964.
38. Klaffke, K., "Granular Carbon System Purifes Water at Del City, Oklahoma," *Water and Wastes Engineering*, March, 1968.
39. Hansen, R. E., "Granular Carbon Filters for Taste and Odor Removal," *J.A WWA*, p. 176, March, 1972.
40. Culp/Wesner/Culp, *Granular Activated Carbon Installations*, U.S. EPA, MERL, January, 1981.
41. Culp, R. L., and Clark, R. M., "Granular Activated Carbon Installations," *J.A WWA*, p. 398, August, 1983.
42. Culp, R. L., Faisst, J. A., and Smith, C. E., *Granular Activated Carbon Installations*, Final Report, Contract No. CI-76–0288, to U.S. EPA, MERL, Cincinnati, Ohio, 1982.
43. Culp, R. L., et al., *Advanced Wastewater Treatment as Practiced at South Tahoe*, U.S. EPA Project Report 17010 ELQ 08/71, Washington, D.C., 1971.
44. McCarty, Perry L., et al., "Advanced Treatment for Wastewater Reclamation at Water Factory 21," Technical Report #267, Department of Civil Engineering, Stanford University, August, 1982.

45. McCarty, Perry L., Argo, David, and Reinhard, Martin, "Operational Experiences with GAC at Water Factory 21," *J.AWWA*, p. 683, November, 1979.
46. Ferrara, A. P., "Controlling Bed Losses of GAC Through Proper Filter Operation," *J.AWWA*, p. 60, January, 1980.
47. Thacker, W. E., et al., "Desorption of Compounds during Operation of GAC Adsorption Systems," *J.AWWA*, p. 144, March, 1983.
48. Sentheimer, J. H., "Design Criteria and Process Schemes for GAC Filters," *J.AWWA*, p. 618, November, 1979.
49. Roberts, Paul V., and Summers, R. S., "Performance of GAC for Total Organic Carbon Removal," *J.AWWA*, p. 113, February, 1982.
50. Constantine, T. A., "Advanced Water Treatment for Color and Organics Removal," *J.AWWA*, p. 310, June, 1982.
51. Randtke, S. J., and Jepsen, C. P., "Chemical Pre-Treatment for Activated Carbon Adsorption," *J.AWWA*, p. 411, August, 1981.
52. Trussell, R. R., and Trussell, A. R., "Evaluation and Treatment of Synthetic Organics in Drinking Water Supplies," *J.AWWA*, p. 458, August, 1980.
53. McCreary, J. J., and Snoeyinks, V. L., "GAC in Water Treatment," *J.AWWA*, p. 437, August, 1977.
54. Clark, R. M., and Dorsey, Paul, "The Cost of Compliance: An EPA Estimate for Organics Control," *J.AWWA*, p. 450, August, 1980.
55. Gumerman, R. C., Culp, R. L., and Clark, R. M., "The Cost of GAC Treatment in the U.S.," *J.AWWA*, p. 690, November, 1979.
56. Cairo, P. R., et al., "Evaluating Regenerated Activated Carbon Through Laboratory and Pilot Column Studies," *J.AWWA*, p. 94, February, 1982.
57. Stover, E. L., and Kincannon, D. F., "Contaminated Groundwater Treatability—A Case Study," *J.AWWA*, p. 292, June, 1983.
58. Cairo, P. R., et al., "Pilot Plant Testing of GAC Adsorption Systems," *J.AWWA*, p. 660, November, 1979.
59. U.S. Environmental Protection Agency, *Interim Treatment Guide for Controlling Organic Contaminants in Drinking Water Using Granular Activated Carbon*, James M. Symons, editor, Water Supply Research Division, MERL, Cincinnati, Ohio, January, 1978.
60. Oulman, C. S., "The Logisitic Curve as a Model for Carbon Bed Design," *J.AWWA*, p. 51, January, 1980.
61. AWWA Committee Report, "Assessing Microbial Activity on GAC," *J.AWWA*, p. 447, August, 1981.
62. Miller, Richard, and Hartman, D. J., *Feasibility Study of Granular Activated Carbon Adsorption and On-site Regeneration*, Cincinnati Water Works Cooperative Agreement No. CR 805443 with U.S. EPA, MERL, Cincinnati, Ohio, Jack Marco and B. W. Lykins, Jr., project officers, 1983.
63. Westerhoff, G. P., Hess, Alan F., and Culp, R. L., *Conceptual Development Plan for the California Water Treatment Plant Granular Activated Carbon Facilities*, Cincinnati Water Works, Malcolm Pirnie in association with Culp/Wesner/Culp, Cincinnati, Ohio, 1983.
64. Juhola, A. J., and Tepper, F., *Regeneration of Spent GAC*, FWPCA Report No. TW RC-7, Cincinnati, Ohio, February, 1969.
65. Culp, G. L., and Slechta, A., *Plant Scale Regeneration of GAC*, U.S.P.H.S. Grant 84–01, Final Progress Report, February, 1966.
66. Culp, R. L., and Hansen, S. P., "Carbon Adsorption Enhancement With Ozone," *J.WPCF*, p. 270, February, 1980.
67. Kovach, J. L., "Activated Carbon Dechlorination," *Industrial Water Engineering*, p. 30, October–November, 1971.

68. U.S. EPA, Philadelphia Water Dept., and Drexel University, *Removing Organics from Philadelphia Drinking Water by Combined Ozonation and Adsorption*, EPA, Report No. 600/52–83–048, J. Keith Carswell, EPA project officer.

69. U.S. EPA, *Feasibility Study of GAC Adsorption and On-Site Regeneration*, EPA Report No. 600/2–82–087–A, 1982.

70. U.S. EPA, *Organic Contaminant Removal in Lower Mississippi River Drinking Water by GAC Adsorption*, EPA Report No. 600/52–83–032, June, 1978.

Chapter 13
Demineralization

INTRODUCTION

This chapter describes the water demineralization processes of electrodialysis (ED) and reverse osmosis (RO), and presents performance data from test facilities and operating plants. The primary emphasis is on presenting information on contaminant-removal capabilities and operating characteristics of ED and RO systems. Detailed consideration of the theory and system configuration of these and other systems, such as distillation, freezing, and solar beds, is available in other publications.[1-4]

Demineralization processes have primarily been used to remove dissolved inorganic material from water and some wastewaters. However, both ED and RO will also remove suspended material, organic material, bacteria, and viruses to some degree. Reverse osmosis systems, in particular, are capable of removing significant amounts of material other than dissolved inorganics.

Electrodialysis and reverse osmosis use different types of membranes for the contaminant-removal process, but the basic system is similar for both processes and includes the components shown in Fig. 13–1. Removal of dissolved inorganic material is often expressed as percent salt rejection, with production expressed as percent recovery. The following example illustrates these terms:

Feed water:	500 gpm (31.55 1/s); 100 mg/l sodium or 600 lb/day (272 kg/d)
Product water:	400 gpm (25.24 1/s); 10 mg/l sodium or 48 lb/day (21.8 kg/d)
Sodium rejection:	92 percent
Water recovery:	80 percent
Waste brine flow:	100 gpm (6.31 1/s); 460 mg/l sodium or 552 lb/day (250 kg/d)

Waste brine is produced by all ED and RO systems. In this example, the waste brine flow is 100 gpm (6.31 1/s) and contains 460 mg/l sodium.

Treatment of the waste brine may be required, and brine disposal is a major consideration in these systems. Pretreatment also may be required, depending on the process and type of membrane. At a minimum, pretreatment

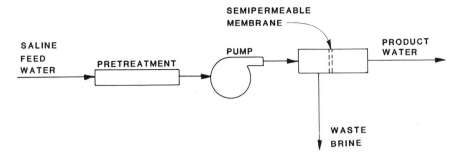

Fig. 13-1. Demineralization process flow diagram.

with cartridge filtration is used in all ED and RO plants. Fouling and scaling of the membrane surfaces can be a significant problem, and the need for pretreatment must be carefully evaluated.

Electrodialysis and reverse osmosis plants have been used primarily in small municipal systems to produce potable water from brackish groundwater, or as part of systems to produce high-quality industrial process water. There are over 80 RO and ED plants desalting brackish groundwater in Florida alone.[5] Several RO plants have recently been constructed to desalt seawater, and the process is competitive with distillation in some areas, particularly for smaller plants.[6-8]

ELECTRODIALYSIS

Electrodialysis (ED) is a well-developed process with over 20 years of operation on brackish well water supplies. A 650,000-gpd ED plant, manufactured by Ionics, Inc., Watertown, Massachusetts, began operation on well water at Buckeye, Arizona in September, 1962, and has been in continuous operation since then.

Typical removals of inorganic salts from brackish water by ED range from 25 to 40 percent of dissolved solids per stage of treatment. Higher removals require treatment by multiple stages in series. Energy required for ED is about 0.2 to 0.4 kWh/1,000 gallons (52.8 to 105.7 kWh/Ml) for each 100 mg/l dissolved solids removed, plus 2 to 3 kWh/1,000 gallons (528 to 798 kWh/Ml) for pumping feed water and brine. This corresponds to an energy requirement of 2.6 to 4.2 kWh/1,000 gallons (687 to 1,110 kWh/Ml) of product water to remove 300 mg/l of dissolved solids.

In the ED process, brackish water flows between alternating cation-permeable and anion-permeable membranes, as illustrated in Fig. 13-2. A direct electric current (DC) provides the motive force to cause ions to migrate through the membranes. Many alternating cation and anion membranes, each separated by a plastic spacer, are assembled into membrane stacks. A picture

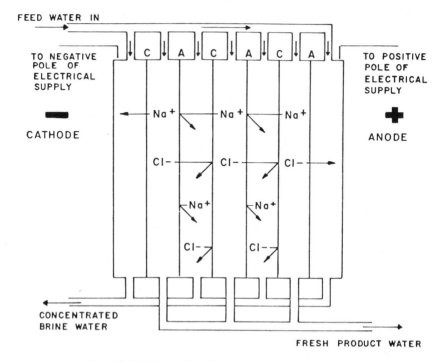

Fig. 13–2. Electrodialysis demineralization process.

of the membrane arrangement is shown in Fig. 13–3. The spacers contain the water streams within the stack and direct the flow of water through a circuitous path across the exposed face of the membranes. Several hundred membranes and their separating spacers are usually assembled between a single set of electrodes to form a membrane stack. End plates and tie rods complete the assembly.

When a membrane is placed between two salt solutions and subjected to the passage of a direct electric current, most of the current will be carried through the membrane by ions; hence, the membrane is said to be ion-selective. Typical selectivities are greater than 90 percent. When the passage of current is continued for a sufficient length of time, the solution on the side of the membrane that is furnishing the ions becomes more concentrated. These desalting and concentrating phenomena occur in thin layers of solution immediately adjacent to the membrane, resulting in the desalting of the bulk of the solution.

Passage of water between the membranes of a single stack, or stage, usually

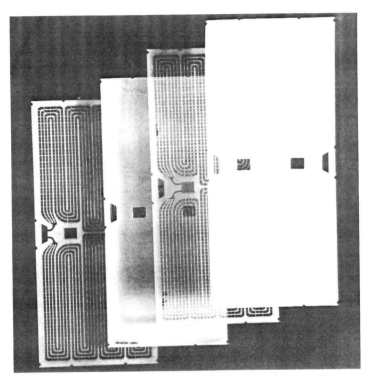

Fig. 13-3. Electrodialysis membranes and spacers. (Courtesy of Ionics, Incorporated.)

requires 10 to 20 seconds, during which time the entering minerals in the feed water are removed. The actual percentage removal that is achieved varies with water temperature, type and amounts of ions present, flow rate of the water, and stack design. Typical removals per stage range from 25 to 40 percent, systems employ one to six stages. An ED system will operate at temperatures up to 100°F (43°C), with the removal efficiency increasing with increasing temperature. Ion-selective membranes in commercial electro- dialysis equipment are commonly guaranteed for as long as 5 years, and experience has demonstrated an effective life of over 10 years.

Ionics, Inc. developed an ED unit called electrodialysis reversal (EDR) that minimizes the addition of acid or other chemicals for scale control.[9] Since 1974, virtually all ED plants constructed have been the EDR type. Until the EDR process was developed, all ED plants (and RO as well) operated with the transfer of water or salts in one direction. Membranes had a brine side and a product side. Films of scale, slime, and other deposits generally formed on one side of the membrane only.

Partially successful attempts were made to control the precipitation of insoluble salts such as calcium carbonate and calcium sulfate on the brine side of unidirectional membrane processes by the addition of acids, polyphosphates, or similar agents. These chemical additions can control calcium carbonate scale effectively, and calcium sulfate scale less effectively, when precise dosing is maintained. However, when errors, failures, or upsets occur, scale starts to form at an accelerating rate.

The symmetrical nature of ED membranes allowed the development of the EDR process. The EDR system utilizes a standard electrodialysis array of alternating cation and anion membranes, separated by alternating product and brine compartments, as shown in Fig. 13–2. The array is operated in the standard ED manner for a fixed period of time, for example, 20 minutes, and then the process is reversed by an automatic timing circuit in the following sequence:

1. The direction of the DC field is reversed by reversing the polarity of the electrodes. This polarity reversal immediately begins converting the product compartments into brine compartments, and the brine compartments into product compartments, by reversing the direction of flow of the ions.
2. Automatic valves interchange the feed to and discharge from the product and brine compartments.
3. There is a one- to two-minute period immediately following the polarity reversal when the water from both sets of compartments is of lower quality, and both streams are automatically diverted to waste. This "purge" of the brine and product compartments every 20 minutes breaks up polarization films, carries off loose scale, and reduces the tendency of deposits to build up.

In operation, the EDR process requires the addition of a timing control unit, automatic valves to interchange the product and brine streams, and relays to reverse the polarity of the DC power supply.

The EDR process has greatly reduced membrane scaling and fouling. In many plants the need for chemicals to control scaling and fouling has been eliminated through use of an EDR system.

EDR Performance

The Orange County Water District tested an EDR unit on highly treated wastewater in 1980.[10]

Feed water to the EDR unit was pretreated by activated sludge, chemical clarification with lime, air stripping, recarbonation, mixed-media filtration, and activated carbon adsorption. The treated feed water was then pumped through a 10-micron cartridge filter to the EDR unit at a pressure of about 35 psi (241.15 kPa). The unit operated over a period of 9 months at nearly steady state conditions. The product flow rate was 0.5 gpm (0.032 l/s) and did not vary during the 9-month test period by more than 0.03 gpm (0.0019 l/s). The test results are summarized in Table 13–1. The average salt rejection was 89 percent based on daily measurements of electrical conductivity at an average feed water temperature of 83°F (28.3°C). There was no decline in the rejection throughout the test period. The data in Table 13–1 show that the removal of the most common inorganic ions averaged about 90 percent. Removal of organic material, as measured by COD and TOC, was about 60 percent. The EDR unit operated throughout the test period with very few operational or maintenance problems. The membranes were cleaned in place with an acid solution about every 1,000 hours, and cartridge filters were replaced after about 340 hours of operation.

EDR plants from 15,000 to 288,000 gpd (0.0379 to 1.09 Ml/d) capacity located in the United States are summarized in Table 13–2. The characteristics and performance of several larger plants are described in a recent report by E. P. Geishecker, and the following case descriptions summarize the information on four of these installations.[11] Two EDR plants are pictured in Figs. 13–4 and 13–5.

Dell City, Texas

Dell City installed a four-stage EDR plant with four stacks and 400 membrane pairs per stack, installed in 1975 to treat groundwater for municipal supply.

Feed water:	222,000 gpd (0.84 Ml/d)
Product water:	100,000 gpd (0.379 Ml/d)
Recovery:	45 percent
Feed water temperature:	70°F (21.1°C)
Total electricity usage:	10.5 kWh/1,000 gallons (2,774 kWh/Ml) product

WATER ANALYSIS
(mg/l EXCEPT pH)

	FEED	PRODUCT	WASTE
Sodium	309	66	453
Calcium	483	42	845
Magnesium	188	20	316
Bicarbonate	204	74	306
Chloride	438	52	695
Sulfate	1,759	184	2,941
Nitrate	60	7	95
Fluoride	2.5	0.6	2.9
TDS	3,455	446	5,674
pH	7.3	7.0	7.5

Table 13-1. Electrodialysis Performance Summary, Orange County Water District.

| CONSTITUENT | AVERAGE CONCENTRATION, mg/l | | PERCENT AVERAGE REMOVAL |
	FEED	PRODUCT	
Calcium	80	7	92
Magnesium	3.2	0.4	88
Sodium	225	23	90
Potassium	17	1.0	94
Ammonium	20	1.3	94
Bicarbonate	116	15	87
Chloride	263	17	94
Sulfate	286	31	89
Barium	0.019	0.005	75
Cadmium	0.0014	0.0004	71
Iron	0.016	0.007	56
Lead	0.0075	0.0014	81
COD	10	4	60
TOC	5.0	2.1	59

Table 13-2. EDR Plant Installations 15,000 gpd to 288,000 gpd.

| LOCATION | INSTALLATION YEAR | CAPACITY | | TOTAL DISSOLVED SOLIDS, mg/l | |
		GPD	(Ml/d)	FEED	PRODUCT
New Jersey	1974	75,000	(0.284)	350	30
New Jersey	1974	75,000	(0.284)	350	30
New Jersey	1974	75,000	(0.284)	350	30
Hawaii	1975	30,000	(0.114)	1,400	250
Texas	1975	100,000	(0.379)	3,175	475
Arizona	1976	15,000	(0.057)	1,300	200
Arizona	1976	30,000	(0.114)	650	30
Washington	1977	264,000	(0.999)	800	400
Washington	1977	264,000	(0.999)	800	400
Hawaii	1979	30,000	(0.114)	1,400	250
Florida	1979	100,000	(0.379)	2,200	500
Texas	1980	100,000	(0.379)	480	80
Indiana	1980	100,000	(0.379)	460	200
Utah	1981	288,000	(1.090)	1,750	500
Utah	1981	288,000	(1.090)	1,750	500
Arizona	1981	20,800	(0.079)	2,050	100
Wyoming	1981	30,000	(0.114)	2,648	160
Bahamas	1981	30,000	(0.114)	1,638	665
Texas	1981	288,000	(1.090)	750	110
Texas	1981	288,000	(1.090)	750	110
Texas	1982	288,000	(1.090)	750	110
Texas	1982	288,000	(1.090)	750	110
Puerto Rico	1982	15,000	(0.057)	285	28

Fig. 13-4. 100,000-gpd EDR plant. (Courtesy of Ionics, Incorporated.)

Fig. 13-5. 1,150,000-gpd EDR plant. (Courtesy of Ionics, Incorporated.)

No chemical pretreatment is used,[16] but the waste brine is supersaturated with calcium sulfate.

Coupeville, Washington

Coupeville has a two-stage EDR plant with four stacks and 450 membrane pairs per stack, installed in 1977 to treat groundwater used for the municipal supply.

Feedwater:	704,000 gpd (2.66 Ml/d)
Product water:	528,000 gpd (2.00 Ml/d)
Recovery:	75 percent
Feed water temperature:	55°F (12.8°C)
Total electricity usage:	3.0 kWh/1,000 gallons (793 kWh/Ml) product (includes pumping and DC for EDR)

WATER ANALYSIS
(mg/l EXCEPT pH)

	FEED	PRODUCT	WASTE
Sodium	65	39	155
Calcium	84	33	134
Magnesium	56	22	152
Bicarbonate	527	268	893
Chloride	96	30	316
Sulfate	34	11	113
TDS	862	403	1,763
pH	8.2	7.9	7.8

No chemical pretreatment is used in this plant.

Sanibel Island, Florida

Sanibel Island has a three-stage EDR plant, with 24 stacks and 350 membrane pairs per stack, installed in 1973 to treat groundwater for municipal supply.

Feed water:	2,340,000 gpd (8.86 Ml/d)
Product water:	1,800,000 gpd (6.82 Ml/d)
Recovery:	77 percent
Feed water temperature:	80°F (26.7°C)
Total electricity usage:	8.5 kWh/1,000 gallons (2,246 kWh/Ml) product

WATER ANALYSIS
(mg/l EXCEPT pH)

	FEED	PRODUCT	WASTE
Sodium	573	105	2,954
Calcium	114	12	844
Magnesium	105	13	616
Bicarbonate	160	60	573
Chloride	949	116	5,373
Sulfate	394	65	2,173
Strontium	8.3	1.6	34.8
Fluoride	1.9	1.1	4.5
TDS	2,297	372	12,495
pH	7.4	7.1	7.6

Pretreatment consists of aeration, chlorination, and activated carbon adsorption for removal of hydrogen sulfide.

Foss Reservoir, Oklahoma

The City of Foss Reservoir installed a four-stage custom ED plant (not EDR), with four stacks and 320 membrane pairs per stack, in 1974, to treat a surface reservoir for municipal supply.

Feed water:	4,257,000 gpd (16.1 Ml/d)
Product water:	2,980,000 gpd (11.28 Ml/d)
Recovery:	70 percent
Feed water temperature:	34 to 85°F (1.1 to 29.4°C)
Total electricity usage:	5.0 kWh/1,000 gallons (1,321 kWh/Ml) product

WATER ANALYSIS
(mg/l EXCEPT pH)

	FEED	PRODUCT	WASTE
Sodium	98	49	183
Calcium	203	41	462
Magnesium	121	29	199
Bicarbonate	177	13	8
Chloride	48	7	110
Sulfate	1,000	278	2,167
TDS	1,647	417	3,129
pH	8.2	7.2	5.5

This is a standard ED plant constructed before development of EDR. Chemicals are added to the waste brine system to prevent scaling.

REVERSE OSMOSIS

The reverse osmosis process has been intensively developed during recent years, and there are many studies on the fundamental process and its use in various locations.[12-14] RO membranes remove a high percentage of almost all inorganic ions, turbidity, organic material, bacteria, and viruses. The dissolved solids in brackish water usually can be reduced to 50 to 100 mg/l in one stage. Energy requirements for commercially available RO systems were about 7 to 9 kWh/1,000 gallons (1,849 to 2,378 kWh/Ml) product water until relatively recently. Most RO systems required pressures of 450 to 600 psi (3.1 to 4.13 mPa) until the recent development of membranes that will operate at pressure as low as 200 psi (1.38 mPa).

A natural phenomenon known as osmosis occurs when solutions of two different concentrations are separated by a semipermeable membrane such as cellophane. Water tends to pass through a membrane from the more dilute side to the more concentrated side, thus producing equal TDS concentrations on both sides of the membrane. The ideal osmotic membrane permits passage of water molecules but prevents passage of ions such as sodium and chloride. For example, if a solution of sodium chloride in water is separated from pure water by means of a semipermeable membrane, water will pass more rapidly in the direction of the salt solution. At equilibrium, the quantities of water passing in both directions are equal, and the difference in pressure on opposite sides of the membrane is defined as the osmotic pressure of the solution having that particular concentration of TDS.

The magnitude of the osmotic pressure depends on the concentration of the salt solution and is related to the solution's vapor pressure and temperature. By exerting pressure on the salt solution, the osmosis process can be reversed. When the pressure on the salt solution is greater than the osmotic pressure, fresh water diffuses through the membrane in the direction opposite to that of the normal osmotic flow; hence, the name of the process, reverse osmosis. Other membrane materials have been developed and are now available in commercial systems.[15]

Many materials have been studied for possible use as membranes. One widely used membrane is modified cellulose acetate film. The modified cellulose acetate membrane currently in general use is approximately 100 microns thick (0.004 inch). Under pressures of up to 600 psi (4.13 mPa), an RO membrane acts like a filter to retain ions on the brackish water side while permitting pure, or nearly pure, water to pass through the membrane.

Operating plants utilize the RO principle in several different process designs and membrane configurations. There are three types of commercially available membrane systems that have been used in operating plants:

1. Spiral wound
2. Hollow fine fiber
3. Tubular

The spiral wound RO module was developed by Gulf Environmental Systems Company (now Fluid Systems Division, UOP) under contract to the U.S. Office of Saline Water. It was conceived as a method of obtaining a relatively high ratio of membrane area to pressure vessel volume. The membrane is supported on both sides of a backing material and sealed with glue on three of the four edges of the laminate. The laminate is also sealed to a central tube which is drilled. The membrane surfaces are separated by a screen material that acts as a brine spacer. The entire package is then rolled into a spiral configuration and wrapped in a cylindrical form. Feed flow is parallel to the central tube, while the permeate flow is through the membrane toward the central tube. A spiral wound membrane and module assembly is illustrated in Fig. 13–6. Many brackish water and wastewater desalting plants utilize this type of system.

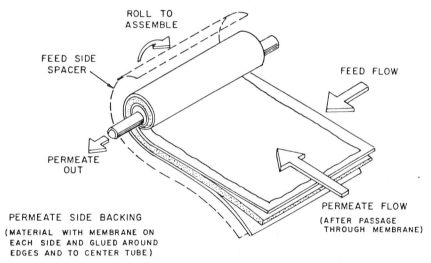

SPIRAL WOUND REVERSE OSMOSIS MODULE

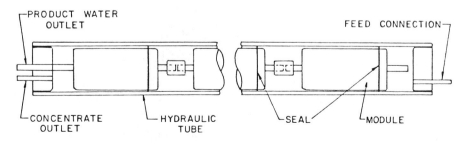

PRESSURE VESSEL ASSEMBLY

Fig. 13–6. Spiral-wound reverse osmosis module (Courtesy of Johnson Division, Signal Environmental Systems, Inc.)

The hollow fiber type was developed by DuPont and Dow Chemical. The membranes manufactured by DuPont are made of aromatic polyamide fibers about the size of human hairs. In these very small diameters, fibers can withstand high pressures. The saline water is under pressure on the outside of the fibers, which are obtained in a pressure vessel, and product water flows inside of the fiber to the open end. A DuPont module is illustrated in Fig. 13–7. For operating plants, the membrane modules are assembled in a configuration similar to the spiral-wound-type unit. Some municipal desalting plants and seawater desalting plants in the Middle East use this type of membrane.

Tubular membrane processes operate on much the same principle as the hollow fine fiber except that the tubes are much larger in diameter, on the order of 0.5 inch (13 mm). Use of this type of membrane system is normally limited to special situations such as for wastewater with high suspended

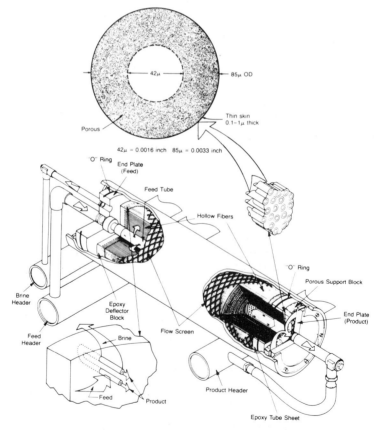

Fig. 13–7. Hollow fiber reverse osmosis module. (Permasep® Permeator as manufactured by E. I. duPont de Nemours & Co.)

solids concentration. It is not economically competitive with other available systems for treatment of most waters. A photograph of an RO module is shown in Fig. 13–8, and photographs of a 1.0-mgd RO plant and a 5-mgd RO plant are shown in Figs. 13–9 and 13–10.

Ultrafiltration, a process similar to RO, also uses a semipermeable membrane. The distinction between RO and ultrafiltration membranes is arbitrary, but the latter are usually reserved for the separation of particles in the 0.001- to 10-micron range. This is about one order of magnitude larger than material removed by RO.

RO Performance

Typical inorganic ion rejections in a 3,600 mg/l TDS feed water at 70 to 75 percent water recovery and operating pressure of 400 psi (2.76 mPa) are shown in Table 13–3.[12] All saline water contains a mixture of monovalent and divalent ions. The data in Table 13–3 illustrate that divalent ion rejections exceed TDS rejections, and monovalent ion rejections are less than TDS rejections.

A 5-mgd (18.9 ml/d) RO plant, with spiral wound membranes as illustrated

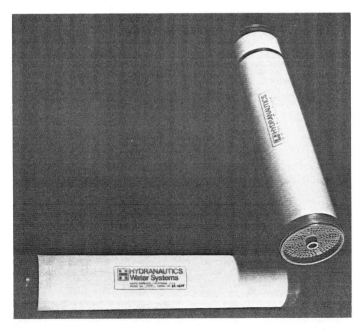

Fig. 13–8. RO modules. (Courtesy of Hydranautics Water Systems.)

Fig. 13–9. 1.0-mgd RO system at Yanbu, Saudi Arabia. Installed April, 1978. Feed water 2,950 TDS from deep wells. Plant recovery 80 percent. Above view shows four hydrablock assemblies containing pressure tubes and membranes. (Courtesy of Hydranautics Water Systems.)

Fig. 13–10. 5-mgd spiral wound reverse osmosis plant. (Courtesy of Orange County Water District.)

Table 13–3. Typical Ion
Removal by Reverse
Osmosis.

ION	PERCENT REJECTION AT pH 6.0*
Ca^{+2}	99+
Mg^{+2}	99+
Na^{+1}	96
K^{+1}	96
Fe^{+2}	99+
Mn^{+2}	99.9+
†B^{+3}	0
SO_4^{-2}	99+
Cl^{-1}	96
†HCO_3^{-1}	70–80
NO_3^{-1}	70–80
PO_4^{-3}	99+
†F^{-1}	80
SiO_2	86–90

* Expressed as percent of incoming TDS
concentration.
† Highly pH-dependent.

in Fig. 13–6, was installed in Orange County, California in 1977. Feed water
to the RO system is wastewater treated by activated sludge, chemical clarification with lime, air stripping, recarbonation, and mixed-media filtration. The
results of extensive test data on this plant were recently published by the
Orange County Water District and Stanford University.[16,17] The data on
major inorganic ions and organics, as measured by COD and TOC, are
summarized in Table 13–4. These data show that over a 15-month period,
the RO system removed 91 percent of the TDS and over 90 percent of the
organic material; it was efficient in the removal of all constituents except
boron, lead, and mercury. However, the feed water concentrations of lead
and mercury were low—substantially below the MCL of 50 $\mu g/l$ for lead
and 2 $\mu g/l$ for mercury.

EPA and the Sarasota, Florida County Health Department evaluated the
effectiveness of eight RO systems in removing radium and other inorganic
constituents.[18] The capacity of the eight systems ranged from 800 gpd to
1.0 mgd (0.003 to 3.79 Ml/d). The plant of the City of Venice, Florida
was recently expanded to 2.0 mgd (7.57 Ml/d) with new, low pressure, hollow
fiber membranes manufactured by Dow Chemical Company. Five of the
eight systems had Ra-226 concentrations above 5 pCi/l. RO treatment reduced
the Ra-226 level to below the MCL in all systems.

Seven of the eight water supplies had selenium concentrations (0.017 to

Table 13–4. Reverse Osmosis Performance Summary, Orange County Water District.

CONSTITUENT	CONCENTRATION* FEED	PRODUCT	PERCENT REJECTION
Calcium	85	1.3	98.5
Magnesium	2	0.0	100
Sodium	225	22	90.2
Potassium	14	1.1	92.1
Bicarbonate	126	14	88.9
Chloride	279	28	90.0
Sulfate	285	3	98.9
Boron	0.63	0.52	17.5
Fluoride	0.9	0.16	82.2
TDS	997	90	91.0
COD	25	0.9	96.4
TOC	11	1.0	91.0
Arsenic	<5	<5	—
Barium	33	1.1	97.0
Beryllium	0.2	<1	—
Cadmium	1.7	0.1	94.1
Cobalt	1.2	<1	—
Chromium	7	0.8	88.6
Copper	18	4	77.8
Iron	103	3	97.1
Lead	0.6	0.6	0
Manganese	6	0.1	98.3
Mercury	0.4	0.3	25.0
Nickel	25	1.5	94.0
Selenium	<5	<5	—
Silver	0.5	0.15	70.0

* Calcium through TOC in μg/l. Arsenic through silver in μg/l. Geometric means of samples collected over 15-month period in 1980–1981.

0.025 mg/l) slightly above the 0.01 mg/l MCL. Three water supplies also had flouride levels (2.0 to 2.2 mg/l) exceeding the MCL. The test data showed that the RO treatment systems lowered the concentrations of both elements below the MCL. Selenium concentrations of all product waters were below the detectable limits of 0.005 mg/l, and the flouride concentrations were 0.4 to 0.8 mg/l.

Energy Recovery from RO Plants

The waste brine stream from an RO system is under high pressure and offers the potential for energy recovery. Guy and Singh presented information on alternative energy recovery systems, including impulse turbines, reversed pump turbines, integrated hydroturbines, and so-called work exchangers.[19]

The 100-mgd (378.5-Ml/d) brackish water RO plant in Yuma, Arizona will use two impulse turbines to drive a 1,000-kW generator. Impulse turbines are also planned or in use at several smaller plants. Several seawater RO plants use reversed pump turbines for energy recovery. Flow work exchangers are the most efficient method, with efficiencies over 90 percent; however, the units are in various stages of development and are not commercially available.

COMPARISON OF PROCESSES

There are various factors to consider in selecting either reverse osmosis or electrodialysis for a particular application. Glueckstern reported an experience in Israel with ED, RO, and distillation processes; however, the ED plant began operation in 1971 and is not considered representative of modern ED technology.[20] Several RO plants have all operated satisfactorily. McNutt and Sherwood presented a methodology for matching the desalination process to product water quality requirements and site conditions.[21]

In general, ED and RO are cost-competitive for brackish waters to TDS concentrations of about 3,000 mg/l. Some factors that should be considered in evaluating or comparing the processes include:

- Energy requirements. ED is more energy efficient in lower-TDS waters. Energy requirements for RO are relatively constant for brackish waters, while ED energy requirements vary directly with the TDS concentration.
- Feed water quality and the need for pretreatment. ED may be capable of operation without chemical addition. RO may not require pretreatment for removal of organic material. Activated carbon pretreatment for feed waters containing organic material probably will be required for ED. Requirements for removal of suspended material vary and must be carefully evaluated.
- Membrane characteristics. Some membranes are more sensitive than others to pH, chlorine and other oxidizing agents, and temperature.
- Product water requirements. ED and RO are both generally efficient in the removal of most constituents, but rejections are not the same for all ions.

REFERENCES

1. Howe, E. D., *Fundamentals of Water Desalination,* Marcel Dekker, New York, 1974.
2. Buros, O. K., *The U.S.A.I.D. Desalination Manual,* U.S. Agency for International Development, Washington, D.C., August, 1980.
3. Curran, H. M., et al., *State-of-the-Art of Membrane and Ion Exchange Desalting Processes,*

U.S. Department of the Interior, Office of Water Research and Technology, November, 1976.

4. Rogers, A. N., et al., *Desalination Technology Report on the State-of-the-Art,* The Metropolitan Water District of Southern California, February, 1983.
5. Dykes, G. M., "Desalting Water in Florida," *J.AWWA,* pp. 104–107, March, 1983.
6. Brandt, D. C., "Seawater Reverse Osmosis—Three Case Histories Using DuPont Permasep B-10 Permeators," *WSIA Journal,* pp. 25–32, January, 1979.
7. Foreman, G. E., et al., "Performance of the 12,000 CM/D Seawater Reverse Osmosis Desalination Plant at Jeddah, Saudi Arabia, January 1979 through January 1981," paper presented at Second World Congress of Chemical Engineering, Montreal, Canada, October, 1981.
8. Reed, S. A., et al., "Desalting Sea Water and Brackish Waters: A Cost Update, 1979," *WSIA Journal,* pp. 17–31, January, 1980.
9. Katz, W. E., "The Electrodialysis Reversal ('EDR') Process," paper presented at the International Congress on Desalination and Water Reuse.
10. Argo, D. G., *Evaluation of Membrane Processes and Their Role in Wastewater Reclamation,* U.S. Department of Interior, Office of Water Research and Technology, November, 1980.
11. Geishecker, E. P., "Operating Summaries for Selected Electrodialysis Reversal Plants in the United States," paper presented at the WSIA Desalination Workshop, San Diego, Calif., January, 1983.
12. *Reverse Osmosis Technical Manual,* U.S. Department of the Interior, Office of Water Research and Technology, OWRT TT/80 2, July, 1979.
13. *Desalting Demonstration Plant Feasibility Study, Virginia Beach, Virginia,* U.S. Department of the Interior, Office of Water Research and Technology, Report No. 79-30-R, September, 1979.
14. *Desalting Demonstration Plant Feasibility Study, Alamogordo, New Mexico,* U.S. Department of the Interior, Office of Water Research and Technology, Report No. 79-29-R, September, 1979.
15. Riley, R. F., et al., "Recent Developments in Thin-Film Composite Reverse Osmosis Membrane Systems," paper presented at AICE Joint symposium in Water Filtration and Purification, Philadelphia, Pa., June, 1980.
16. McCarty, P. L., et al., "Advanced Treatment for Wastewater Reclamation at Water Factory 21," Department of Civil Engineering, Stanford University Technical Report No. 267, August, 1982.
17. Argo, D. G., and Ridgway, "Biological Fouling of Reverse Osmosis Membranes at Water Factory 21," paper presented at the Water Supply Improvement Association 10th Annual Conference, Honolulu, Hawaii, July, 1982.
18. Sorg, T. J., et al., "Removal of Radium-226 From Sarasota County, Fla., Drinking Water by Reverse Osmosis," *J.AWWA,* pp. 230–237, April, 1980.
19. Guy, D. B., and Singh, R., "Some Alternate Methods of Energy Recovery from Reverse Osmosis Plants," *WSIA Journal,* pp. 37–51, July, 1982.
20. Glueckstern, P., "A Water Utility's Experience Regarding the Reliability and Operating Costs of Various Desalination Technologies over the Past Decade," *WSIA Journal,* V. 10, No. 1, pp. 19–34, January, 1983.
21. McNutt, J. F., and Sherwood, K. H., "Energy and Water Quality Considerations in Selecting a Cost Effective Desalination Process," paper presented at the Water Supply Improvement Association 10th Annual Conference, Honolulu, Hawaii, July, 1982.

Chapter 14
Removal of Trihalomethanes

INTRODUCTION

Trihalomethanes (THM's) are formed by the reaction of free chlorine with certain naturally occurring organic compounds. The generalized formula for the reaction is:

$$\text{Free chlorine and/or bromine} + \text{Organic}$$
$$\text{precursors} \rightarrow \text{Trihalomethanes} + \text{By-product compounds} \qquad (14\text{--}1)$$

The naturally occurring organic precursors generally are humic substances such as humic and fulvic acids. Industrial wastes are not a factor in production of THM's.

On November 29, 1979, regulations[1] were adopted that established a maximum contaminant level (MCL) of 0.1 mg/l for the total trihalomethane (TTHM) concentration, defined as the sum of the concentration of trichloromethane or chloroform ($CHCl_3$), tribromomethane or bromoform ($CHBr_3$), bromodichloromethane ($CHBrCl_2$), and dibromochloromethane ($CHBr_2Cl$).

A key to the evaluation of THM control strategies is assurance that the selected technique will continue to result in the production of microbiologically high-quality water. This is an important consideration because several control techniques use less effective disinfectants than free chlorine and/or allow penetration of undisinfected water through the treatment works to a greater degree than occurs with conventional prechlorination. This further penetration is particularly important in surface supplies during warm water conditions and during periods of heavy runoff, which generally produce high raw water turbidity and microbiological concentrations.

DEFINITION OF THM MEASUREMENT PARAMETERS

Four THM parameters are important to an understanding of the THM regulations and THM control strategies. These parameters, shown graphically in Fig. 14–1, are discussed in the following paragraphs.

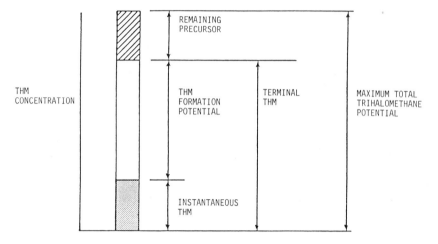

Fig. 14-1. THM measurement parameters and their relationships.

Instantaneous THM

Instantaneous THM (Inst THM) is the THM concentration at the moment of sampling. Compliance with the THM MCL is determined using Inst THM. It can be expressed in terms of the sum of the individual species, or as each of the four major species (chloroform, bromodichloromethane, dibromochloromethane, and bromoform). Although other THM species may be formed, the THM MCL includes only these four species.

Terminal THM

Terminal THM (Term THM) is a measurement of the THM present after a specified time period. In most cases, the time utilized is equivalent to the time of treatment plus the distribution system detention time. The temperature of the test is the temperature of the water in the distribution system.

THM Formation Potential

Trihalomethane Formation Potential (THMFP) is the increase in THM concentration during the storage period for the Terminal THM test. In other words, Terminal THM minus Instantaneous THM equals THMFP:

$$\text{Term THM} - \text{Inst THM} = \text{THMFP} \tag{14-2}$$

Maximum Total Trihalomethane Potential

Maximum Total Trihalomethane Potential (MTTP) is an attempt to maximize the formation of THM's, with the test results being indicative of how high the total trihalomethane (TTHM) concentration in the distribution system might become under conditions favoring TTHM formation. The test is conducted over 7 days at a temperature of 25°C (77°F). A chlorine residual of at least 0.2 mg/l must be present at the completion of the test, or the MTTP test must be rerun using an initial chlorine concentration of 5 mg/l and a solution pH buffered to 9.0–9.5 (EPA Method 510.1). Reference 2 presents additional information on conduct of the MTTP test.

FACTORS INFLUENCING THE RATE OF THM FORMATION AND TERMINAL THM CONCENTRATION

The rate of THM formation and the terminal THM concentration are influenced by a number of factors. Among the more important are:

- Temperature
- pH
- Organic precursors
- Free chlorine concentration
- Bromide concentration

Temperature

Higher water temperatures increase the rate of THM formation, and generally increase the terminal THM concentration. Figure 14–2 illustrates the influence of temperature on chloroform formation in Ohio River water. Utilities using surface water supplies generally experience highest THM concentrations during the summer months, when temperatures are highest. Although water temperature is a principal factor influencing such increases, changes in precursor concentration and type may also be influencing factors. Figure 14–3 illustrates seasonal variations in THM concentrations for Ohio River water at Cincinnati, Ohio and Evansville, Indiana. For groundwater supplies, temperature variations are less dramatic, and seasonal variation in THM's will be less pronounced. Many utilities have found that treatment techniques for THM control are not necessary during the winter months, when low water temperatures decrease the reaction rate and when organic precursor concentrations are lowest.

pH

As pH increases, the rate of THM formation increases, and often the terminal THM also increases. This impact is believed to be due to changes in the active groups on the surface of the molecule, or due to structural changes in the shape of the molecule, making the precursors more reactive. The impact of high pH is most evident in lime softening plants, which often operate at pH values up to 11.0.

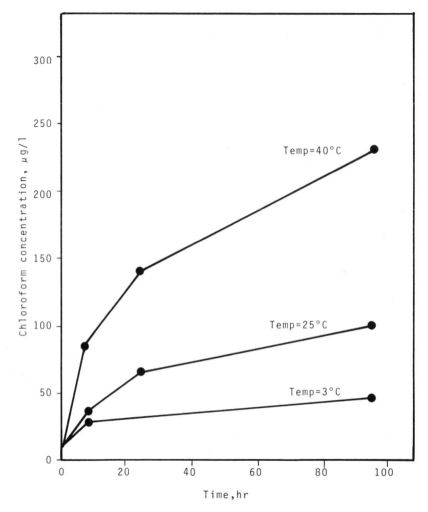

Fig. 14–2. Influence of temperature on chloroform production rate, pH = 7, chlorine dose = 10 mg/l.[2]

Organic Precursors

The type and concentration of organic precursors influence the reaction rate. Because of the diverse nature of the precursors, the best way to measure their overall impact is to conduct a THMFP test. This test will indicate the quantity of precursor available to react with free chlorine. At the completion of the Terminal THM test, all of the organic precursors will not have reacted, as shown in Fig. 14–1. Under other test conditions, such as longer

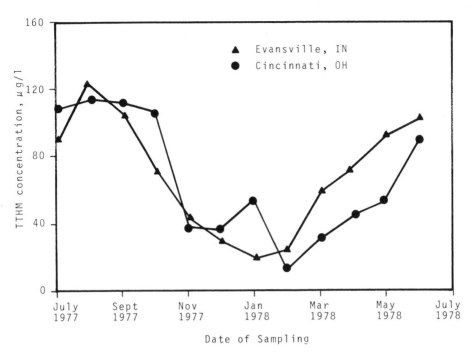

Fig. 14-3. Seasonal influence on finished water TTHM concentration, Ohio River water.[3]

time of reaction or higher temperature, a portion of this precursor material may react. Thus, it is important when measuring THMFP to select test conditions representative of the water supply system being evaluated.

Free Chlorine Concentration

The presence of free chlorine is necessary for the THM formation reaction to proceed. Without free chlorine, THM formation does not occur. Thus, it is important that chlorine residual measurements be conducted in the field simultaneously with the collection of samples for Term THM, to assure that a residual is present. However, free chlorine residuals in excess of the chlorine demand have little impact on accelerating the rate of THM formation. Initial mixing and reactor design will influence the rate of formation, even when the chlorine residual is in excess of the demand.

Bromide Concentration

Bromide ion is oxidized to bromine by free chlorine. The bromine ion can then react with organic precursors to form THM's. Three of the four trihalomethanes included in the THM MCL require the presence of bromine for formation. Importantly, the rate of formation of bromine-containing THM's is faster than the rate of chloroform (a nonbromine THM) formation because bromine competes more effectively than chlorine for the active sites on organic precursor molecules. Bromide ions occur naturally in many water supplies, while in some other surface or groundwater supplies the bromide ion is the result of seawater intrusion.

Importance of the Rate of Formation

In evaluating a water system, the time between initial chlorine application to the water and the time for the water to flow to the farthest point in the distribution system is a key factor. If the rate of THM formation is slow, the Terminal THM concentration will be low at this farthest point, compared to the Term THM concentration that would occur with a higher THM formation rate. Therefore, the rate of formation is important because it affects the Term THM concentration in the distribution system, and thus compliance with the THM MCL.

NECESSARY STEPS FOR COMPLIANCE WITH THE THM MCL

The necessary steps for compliance with the THM MCL are shown in Fig. 14–4. These steps are discussed in the following pages.

Collection of Quarterly Samples

The first step is collection of quarterly samples, for quarters ending on March 31, June 30, September 30, and December 31. These quarterly sample results must be submitted to the primacy agency within 30 days of the system's receipt of the results, unless otherwise stipulated by the primacy agency. In states that have been granted primary enforcement responsibility by the EPA, the primacy agency is the agency of the state government with jurisdiction over public water systems. In states without primary enforcement responsibility, the primacy agency is the Regional Office of the U.S. Environmental Protection Agency.

Sampling requirements and conditions for reduced sampling frequency are different for surface water and groundwater supplies. The following two sections discuss principal differences. A complete discussion of sampling requirements is included in References 1 and 4.

Quarterly Sampling for Surface Water Supplies. For surface water systems, a minimum of four samples per quarter is required for each treatment plant. Quarterly samples from the distribution system are to be collected

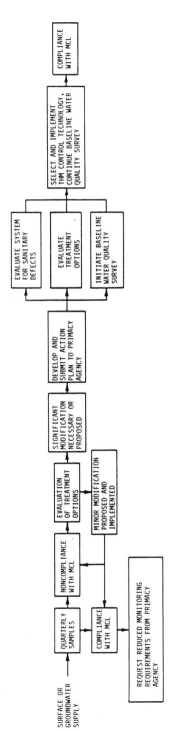

Fig. 14-4. Necessary steps for compliance with the THM MCL.

495

within a 24-hour period, and at least 25 percent are to be at locations in the distribution system that reflect the maximum residence time. The remaining 75 percent are also to be collected from the distribution system, at representative locations based on individuals served, sources of water, and treatment methods utilized.

Compliance is based upon a running average of the four quarterly averages. After 1 year of compliance monitoring, a request to decrease the monitoring frequency can be made to the primacy agency if the instantaneous TTHM concentration has never exceeded 0.1 mg/l. The decision to reduce the sampling frequency will be made on a case-by-case basis by the primacy agency. If a reduced sampling frequency of one sample per quarter is allowed, this sample should be collected at the point in the distribution system with the maximum residence time.

When there is more than one treatment plant, more than one surface supply, or a combination of surface and groundwater supplies, the number of samples can be reduced, but generally not to a frequency as low as one sample per quarter. Also, if a system is operating on a reduced monitoring frequency and there is a significant change in the source water or the type of treatment utilized, the sampling frequency must immediately be increased to four samples per quarter. A discussion of the probable number of reduced samples for such situations is presented in Reference 4.

Quarterly Sampling for Groundwater Supplies. Groundwater supplies generally are more consistent in quality and have lower precursor concentrations than surface supplies. Thus, monitoring requirements for THM's are less restrictive for groundwater supplies.

For systems using only groundwater, the monitoring frequency can be as low as one MTTP (maximum total trihalomethane potential) test per year per aquifer used as a source of raw water. This sample must be collected from the distribution system location having the longest residence time. Before the primacy agency will permit this reduced frequency, substantiation must be presented by the utility that the MTTP is less than 0.1 mg/l during the time of the year when maximum TTHM formation would occur (high temperature and pH, for example) and/or when the highest TOC concentration occurs.

Evaluation or Treatment Options

If a utility is not in compliance with the THM MCL, and an alternative water supply is not available or economically feasible, a treatment technique change should be considered. In some cases only minor modifications will be necessary to achieve compliance. Examples of minor modifications are:

- Improved precursor removal by seasonal use of PAC.
- Seasonally increased doses of coagulant or coagulant aid.

Assuming that the disinfectant practices remain unchanged, these changes are considered minor because they would not significantly impact the microbial quality of the finished water.

A treatment modification is considered significant if the impact on microbial quality of the finished water has not been established. Examples of such significant changes are:

- Change in disinfectant type, dosage, point of application, or contact time.
- Change in the source water, either partially or completely.
- Addition of granular activated carbon as a treatment process.
- Addition of an open finished water reservoir.

Action Plan

If minor modifications do not achieve compliance, and a significant treatment modification is deemed to be necessary, an action plan must be submitted to the primacy agency. The three minimum components of the action plan are:

- Evaluation of system for sanitary defects
- Baseline water quality survey
- Evaluation of treatment options

Evaluation of System for Sanitary Defects. The intent of the sanitary survey is to identify and correct any sanitary defects, unsound treatment practices, or inadequate operation and maintenance practices before any changes are implemented for the control of THM's. This survey should include the entire system: source(s), treatment plant(s), and distribution system.

There is no exact survey protocol to be followed, although Reference 4 presents the minimum recommended elements of a sanitary survey.

Baseline Water Quality Survey. The action plan must include a detailed description of a baseline water quality monitoring survey. This water quality survey should be conducted for a minimum of 6 months before a significant treatment technique change is made, and a minimum of 12 months afterward. The purpose of the survey is to detect changes in finished water quality that result from changes in the treatment technique. Specifically, the survey is aimed at detecting changes that may lead to increased public health risk due to the presence of pathogenic organisms in the finished water. To account for seasonal variations, the survey period before the treatment technique change should include the month having the highest water temperature of the year.

Selection of parameters that must be evaluated in this baseline survey is the responsibility of the utility. Consideration of the following parameters is recommended:

Microbial Parameters
- Total coliform organisms
- Standard plate count (20°C and 35°C)
- Heterotrophic plate count (at temperature of water)
- Total fecal coliform organisms
- Fecal streptococci
- Enteric virus (grossly polluted source waters only)
- Coliphage

Other Parameters
- THM
- Disinfectant residual
- pH
- Temperature
- Turbidity
- Total organic carbon

While it is essential that total coliforms be monitored, it is strongly recommended that other microbial parameters also be measured, to ensure that subtle water quality changes do not go undetected. Monitoring other microbial parameters is important because many pathogenic bacteria and virus are more resistant to environmental stress than coliform, and these pathogens may multiply in the finished water even though coliform counts remain low.

The minimum recommended sampling locations for the baseline survey are:

- Raw water.
- In the treatment plant immediately prior to final disinfection.
- Treatment plant finished water, immediately before it enters the distribution system.
- At end(s) of the distribution system.

Generally, weekly analysis of coliforms, plate count organisms, and disinfectant residuals will be satisfactory. However, during periods of abnormally poor raw water quality, daily monitoring may be justified.

The information in the water quality survey should be used in conjunction with the sanitary survey and analysis of existing treatment facilities to determine if there is a potential public health risk due to microbial penetration of the treatment barrier. When data are examined, any water quality deterioration in the plant finished water or the distribution system should be carefully evaluated. This evaluation should include consideration of seasonal changes in raw water quality and unusual weather conditions.

Evaluation of Treatment Options. The third step in the action plan is to outline an approach to be followed for evaluation of treatment techniques for THM control. In developing and presenting an approach, it is important that existing processes be evaluated and optimized for THM control. To assist in this evaluation, a THM profile (Inst THM and THMFP) through the existing plant should be prepared. This profile will indicate the extent of THM formation, as well as the rate of THM formation as water moves through each of the unit treatment processes. This knowledge is essential in the optimization of existing unit processes for THM control.

If it appears to the utility that optimization of existing unit processes will not be sufficient to achieve compliance, the action plan should present the approach that will be followed to evaluate other treatment options. In deciding which treatment options should be evaluated, factors to be considered are the ability to lower THM's, impact on disinfection efficiency and finished water quality, and cost.

A discussion of the most suitable technologies for THM control and the cost of these technologies follows.

TREATMENT METHODS FOR CONTROL OF THM FORMATION

Since the initial discovery in 1974 that THM's may be formed during the chlorination of drinking water, many treatment technologies have been evaluated for reduction of THM formation. References 5 and 6 present detailed evaluations of treatment techniques that are suitable for control of THM formation.

The February 28, 1983 amendments to the THM implementation regulations define two general categories of technology for controlling THM's.[7] These categories are defined below, and the treatment techniques included in each category are listed in Table 14–1.

Best Generally Available Treatment Methods for Reducing TTHM's

These treatment methods are the methods of choice for the control of THM formation. Each is relatively low in cost, simple in operation, and generally effective for THM control.

Table 14–1. Most Suitable Treatment Technologies for Control of TTHM.

Best Generally Available Treatment Methods

- Use chlorine dioxide as an alternate or supplement disinfectant or oxidant.
- Use chloramines as an alternate or supplemental disinfectant or oxidant.
- Improve existing clarification for THM precursor removal.
- Change point of chlorination. When necessary, use a substitute preoxidant such as chloramines or chlorine dioxide. Potassium permanganate may be useful in some situations, but not as a total substitute for chlorine or other preoxidant.
- Use powdered activated carbon on a seasonal basis to reduce precursor or THM concentrations; dosage not to exceed 10 mg/l on an average annual basis.

Other Treatment Methods

- Add off-line storage for precursor reduction.
- Add aeration for THM reduction.
- Add clarification to the treatment train.
- Consider alternative source of raw water.
- Use ozone as an alternative or supplemental disinfectant or oxidant.

Other Treatment Methods

These treatment methods, while effective for THM control, may not be technically feasible or economically reasonable. When a variance is issued, a schedule of compliance may require examination of these methods to determine whether they could significantly reduce TTHM concentrations, and whether the reductions are commensurate with costs incurred.

A summary of the advantages and disadvantages of the best generally available treatment methods is presented in Table 14–2. Also shown are the most suitable applications for each of these treatment methods.

GUIDANCE FOR THM CONTROL IN SYSTEMS ORIGINALLY USING ONLY CHLORINATION

Many systems with low turbidity and low coliform concentrations in the raw water supply use chlorination as the only form of treatment, principally for disinfection. Such treatment may be used for either groundwater or surface supplies, although it is much more predominant, and generally more appropriate, for groundwater supplies.

The choices for best generally available treatment are few, as shown in Fig. 14–5. The principal factors to be considered in the decision-making process are discussed in the following sections, with reference to Fig. 14–6, which shows bar graphs of hypothetical results for Instantaneous THM, THMFP, and Terminal THM.

Original Condition

Figure 14–6a shows the original treatment, which consists of chlorine addition to a trunk pipeline that feeds the distribution system. Following chlorine addition, Inst THM is 50 μg/l and THMFP is 90 μg/l before entering the distribution system. In the distribution system, the average Inst THM is 140 μg/l, which is a noncompliance situation.

Treatment Options

Chloramine Disinfection. There are two options for production of chloramines. One is to add ammonia concurrently with chlorine, resulting in immediate chloramine formation (Fig. 14–6b). This approach results in only minimal THM formation, during the time that the chlorine and ammonia have not reacted. The second approach (Fig. 14–6c) is to add chlorine with ammonia addition somewhat later. Figure 14–6c illustrates ammonia addition 15 minutes after chlorine addition. This approach allows free chlorine disinfection until chloramine formation. THM's are formed during the free chlorine contact period, but THM formation ceases with the addition of sufficient quantities of ammonia. This technique may be useful if the system can be operated in this manner, and if concentration of Inst THM production during the free chlorine stage does not result in a noncompliance situation.

Chlorine Dioxide. Substitution of chlorine dioxide at the original chlorination point will reduce THM's (Fig. 14–6d). As shown, only moderate THM formation occurs, due to the presence of some residual-free chlorine in the chlorine dioxide. Provided the residual chlorine is not excessive, the MCL can be achieved with this approach. If the quantity of chlorine used is carefully controlled, it is possible to generate chlorine dioxide containing very little free chlorine, and the resultant formation of THM's will be minimized.

Microbiological Considerations

For systems using only chlorination, the raw water can be assumed to have low levels of bacterial contamination. The following discussion of microbiological considerations is based upon this assumption.

The microbiological significance of any of these changes depends to a great degree upon the detention time between the point of disinfectant addition and the first service connections in the distribution system. If this time is short, immediate chloramine formation (Fig. 14–6b) may not provide adequate disinfection, because of the slow rate of chloramine disinfection. In such cases, 15 or more minutes of free chlorine contact prior to ammonia addition (Fig. 14–6c) would provide better disinfection. For systems with long

Table 14-2. Summary of Features of Best Generally Available Treatment Methods.

TREATMENT METHOD	ADVANTAGES	DISADVANTAGES	MOST SUITABLE APPLICATIONS
Chlorine dioxide	1. Easy to prepare and feed. 2. Good disinfectant. 3. Also useful for iron and manganese oxidation. 4. Destroys taste and odor-producing phenolic compounds. 5. No demand due to NH_3 in raw water.	1. If not closely controlled, free chlorine can be present along with generated chlorine dioxide. 2. Distribution system concentrations of chlorine dioxide, chlorite, and chlorate should be less than 1.0 mg/l due to potential adverse health effects. 3. High cost.	1. Pre- and post-disinfectant chlorine dioxide residual should be above 0.2 mg/l.
Chloramines	1. Easy to produce. 2. Excellent for maintaining a residual in the distribution system. 3. Does not form THM's. 4. Most effective in high-pH waters.	1. Weaker disinfectant than chlorine or chlorine dioxide. 2. Slow-reacting disinfectant. 3. Poor viricide.	1. Should not be used as a primary disinfectant. 2. Good use is as a post-disinfectant, with either chlorine or chlorine dioxide as a predisinfectant. 3. Good for use in lime softening plants, due to high pH. 4. Where residual is above 0.5 mg/l in distribution system.

Improve existing clarification for precursor removal	1. Cost is generally low. 2. Usually results in increased removal of turbidity and microorganisms. 3. Generally results in lower oxidant demand.	1. Increased sludge quantities. 2. May require baffling changes in settling basin.	1. Any plant with existing clarification facilities.
Change point of chlorination	1. Potential decrease in chlorine dosage. 2. Chlorine costs may decrease. 3. Chlorine dioxide or chloramines may be used as preoxidants. Potassium permanganate may be used in combination with other preoxidants.	1. If an alternate predisinfectant is not used, disinfection contact time is reduced, allowing microbial penetration farther into the treatment works. 2. Can lead to slime growths in settling basin if an alternate predisinfectant is not used.	1. Plants where a high percentage of precursors, or rapidly reacting precursors, are removed during coagulation/settling. 2. Plants with bromide in raw water, as THM's containing bromine are very rapidly formed. 3. Free chlorine residual in the distribution system should be maintained above 0.2 mg/l.
Use powdered activated carbon on a seasonal basis	1. Can be used intermittently during periods of poor raw water quality. 2. Can be used for removal of either precursors or THM's. 3. Also removes taste and odor.	1. May have high cost, if storage and feed facilities are not already in place. 2. Additional sludge generation. 3. Annual average not to exceed 10 mg/l.	1. Plants with TTHM's slightly above MCL, due to seasonal variations in raw water quality.

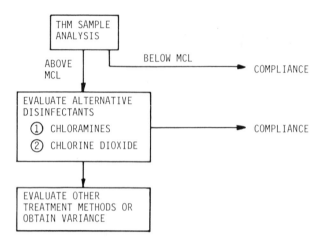

Fig. 14–5. Suggested decision format for THM compliance in systems originally using only chlorination.

detention times between chloramine formation and service connections, immediate chloramine formation may be adequate.

Chlorine dioxide addition (Fig. 14–6d) should provide good disinfection if a chlorine dioxide residual can be maintained in the distribution system. The presence of some free chlorine in the chloride dioxide will enhance disinfection, although it will result in some THM formation. EPA recommends that the combined residual of chlorine dioxide, chlorite ion, and chlorate ion should not exceed 1.0 mg/l at any point in the distribution system. Use of chlorine dioxide may be precluded in systems with water having a high oxidant demand because it may be impossible to maintain a chlorine dioxide residual in the extremities of the distribution system while complying with this recommendation.

GUIDANCE FOR SYSTEMS USING CONVENTIONAL TREATMENT

The majority of systems using surface water supplies use conventional treatment, which in this case is defined as the following unit processes: rapid mix, flocculation, settling, filtration, and clearwell storage. Chlorine addition is customarily done before rapid mix and after filtration.

The choices for best generally available treatment methods for THM control are shown in Fig. 14–7, arranged in a logical order of consideration. For a hypothetical water supply, bar graphs for Inst THM, THMFP, and Term THM at various stages of treatment and in the distribution system are shown in Fig. 14–8.

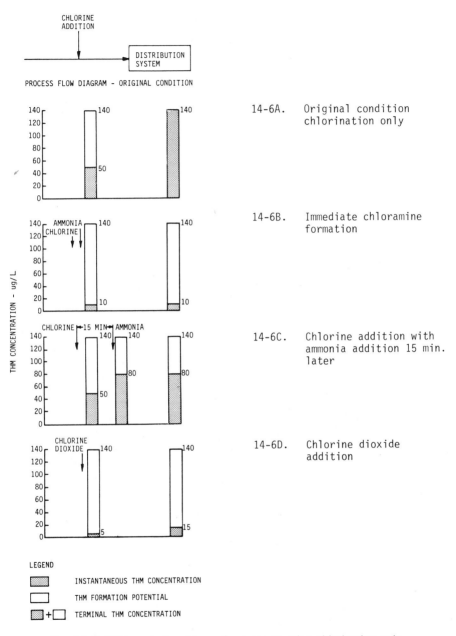

Fig. 14-6. THM control measures for systems using chlorination only.

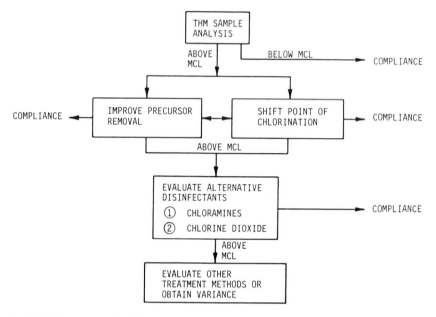

Fig. 14–7. Suggested decision format for THM compliance in conventional treatment plants.

Original Condition

Chlorine is added as a predisinfectant prior to rapid mix, and as a post-disinfectant after filtration. THM formation begins with the initial chlorine addition, and continues throughout the treatment train, assuming a residual is maintained, and in the distribution system. As shown in Fig. 14–8a, THMFP decreases during settling and filtration due to removal of organic precursors. Also shown is an increase in the Inst THM throughout the plant as the THM formation reaction proceeds. The distribution system Inst THM averages 140 μg/l, which would not be in compliance if this were the 12-month running average of the quarterly samples.

Treatment Options

Improve Precursor Removal. One of the first treatment options that should be evaluated for this noncompliant system is to improve precursor removal during clarification and filtration. This concept, which is particularly suited to highly turbid waters, may be accomplished at existing treatment plants in several ways:

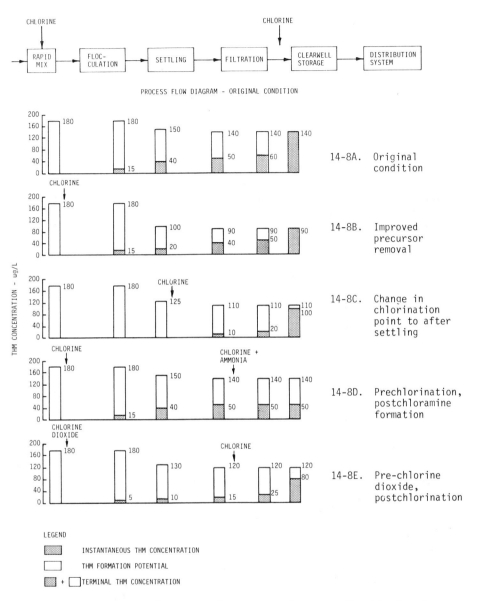

Fig. 14-8. THM control measures for systems using conventional treatment.

- Increase coagulant dose, use a different coagulant, or use a combination of coagulants.
- Use a polymer, increase polymer dose, or use a different polymer.
- Use powdered activated carbon on a seasonal basis.

Figure 14–8b shows hypothetical THM reductions when precursor removal is improved. Note that a greater reduction of THMFP occurs during settling and filtration than in the original condition (Fig. 14–8a), due to the improved precursor removal. The Inst THM formation rate is also slower, due to the enhanced removal of some of the faster-reacting precursors during settling and filtration. Since there is no change in chlorination dosage or point of chlorine addition, disinfection efficiency will not be compromised. Rather, disinfection efficiency should be enhanced because of the increased removal of turbidity, which in many cases decreases disinfection efficiency. This is based upon the assumption that the pH is not raised to improve coagulation. If pH is increased, disinfection efficiency could be decreased, due to a shift in free chlorine concentration from hypochlorous acid (HOCl) to hypochlorite ion (OCl$^-$), the latter being a weaker disinfectant.

In many, if not most, cases, the point of chlorination is shifted at the same time that increased precursor removal is implemented. The two techniques are not necessarily interdependent, however.

Shift Point of Chlorination to After Settling. When the initial application of chlorine is delayed until after settling, substantial precursor removal will have occurred during settling, and lower Term THM concentrations will result. In many cases, this technique is used in conjunction with an improvement in precursor removal (discussed above) to produce even greater THM reductions. Shifting the point of chlorination to after settling should be done with some caution because it will allow further penetration of pathogenic organisms into the treatment works. In situations where remaining THM precursors are fast-reacting and/or the distribution system has a relatively long detention time, this technique may not be adequate to achieve compliance.

Figure 14–8c shows changes in Inst THM formation rate relative to original conditions. Following settling, the THMFP is slightly greater than original conditions because not all of the precursors that form Inst THM at this point in the original condition are removed during settling. In the distribution system, Inst THM is less than original conditions because some of the precursor material has been removed during settling prior to any addition of chlorine.

Prechlorination, Post-chloramine Formation. This technique, shown in Fig. 14–8d, combines the use of prechlorination with post-disinfection by chloramines, which are capable of maintaining a long-term residual.

Inst THM formation and THMFP reduction are equivalent to the original conditions. When post-chlorine is added with ammonia to form chloramines, essentially no further production of THM's occurs. There may be some unreacted THMFP in the distribution system. However, this could only be converted to Inst THM if free chlorine were present. This unreacted THMFP illustrates the Inst THM reduction attributable to the use of chloramines as a post-disinfectant. This technique is useful when THM formation is slow.

Pre–Chlorine Dioxide, Post-chlorination. In this technique, chlorine dioxide is used as a predisinfectant and chlorine is the post-disinfectant. Low chlorine dioxide dosages must be used in order to keep the residual chlorine dioxide/chlorite/chlorate concentration below EPA's recommended level of 1.0 mg/l. If the chlorine dioxide contains some chlorine, as a result of incomplete reaction during chlorine dioxide generation, some Inst THM will be formed. This technique, shown in Fig. 14–8e, is advantageous because predisinfection is achieved with little THM formation, and precursors are removed during settling and filtration, prior to chlorine application.

Microbiological Considerations

Systems using conventional treatment for surface supplies typically have moderate, and sometimes seasonally heavy, bacterial contamination in the raw water supply. To a large extent, the disinfection efficiency of the different concepts discussed depends upon the extent of this bacterial contamination in the raw water. For moderate to heavy bacterial contamination, more than 500 coliforms/100 ml, daily sampling for standard plate count, organisms (SPC), and coliforms is recommended as a minimum. If there are significant upstream wastewater discharges, periodic virus sampling should be considered, and coliphage and fecal strep monitoring are recommended. For supplies with only light to moderate bacterial contamination, less that 500 coliforms/ 100 ml, weekly sampling for SPC and coliform is generally adequate.

A discussion of the microbiological considerations of each treatment technique previously described follows.

Improved Precursor Removal. Disinfection will probably be better than the original condition, because of increased removals of turbidity, organics, and bacteria during settling. A potential for decrease in pre- and post-chlorine dosages exists, without a change in disinfection efficiency.

Change in Chlorination Point to After Settling. This technique may be suitable for supplies with low to moderate bacterial contamination, but probably not for supplies with high bacterial concentrations, or during periods of heavy contamination. From a disinfection standpoint, the problem is the increased penetration of microorganisms into the treatment plant. This could be particularly critical in systems with a short clearwell or distribution system detention time prior to service.

Prechlorination, Post-chloramines. If adequate prechlorination is used, good overall disinfection should occur. Post-chloramines have a relatively slow rate of disinfection, but maintain a long-term residual in the distribution system. A long distribution system detention time is beneficial when post-chloramines are used.

Pre–Chlorine Dioxide, Post-chlorination. This is a highly satisfactory technique, assuming proper dosages are used.

GUIDANCE FOR SYSTEMS USING LIME SOFTENING

Lime softening plants using prechlorination generally produce high THM concentrations because of the high pH conditions achieved during softening. The typical unit operations used in lime softening include rapid mix, flocculation/settling, recarbonation, filtration, and clearwell storage.

Figure 14–9 presents best generally available treatment methods for THM control in lime softening plants. The arrangement shown in Fig. 14–9 is a logical order for evaluation of control techniques. For a hypothetical water supply, bar graphs are shown in Fig. 14–10 for Inst THM, THMFP, and Term THM for both the original condition and feasible control techniques.

The most feasible control measures are to change the point of chlorination to after recarbonation or to use a prechlorination/post-chloramine approach. If these techniques are not effective, a secondary choice would be pre–chlorine dioxide/post-chloramines. This technique is a second choice due to the high cost of constructing and operating chlorine dioxide facilities.

Original Conditions

Using prechlorination, there is normally a rapid formation of Inst THM during rapid mix, flocculation, and settling. This rapid rate of formation is to a great degree attributable to the high pH during softening. After recarbonation, when the pH is lowered, the rate of formation slows. Figure 14–10a illustrates an increase in the THMFP and Term THM following lime addition (which increases pH), and a subsequent decrease in THMFP and Term THM following settling and filtration, due to removal of precursors. Inst THM formation continues throughout the treatment train, and the system is not in compliance, as the distribution system Inst THM concentration is 140 μg/l.

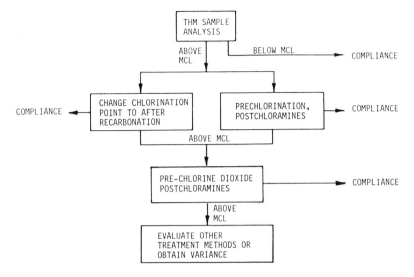

Fig. 14–9. Suggested decision format for THM compliance in lime softening plants.

Treatment Options

Change in Point of Chlorination to After Recarbonation.
Figure 14–10b shows the impact of changing the point of chlorination to after recarbonation. Such a change prevents any Inst THM formation until after recarbonation. Also, additional removal of rapidly acting precursor material occurs during flocculation/settling, reducing the THMFP to 130 μg/l. This example shows a distribution system Inst THM concentration of 90 μg/l. To achieve such an Inst THM, the precursors removed during softening and filtration would have to be predominantly rapidly reacting precursors, and/or the detention time in the distribution system would have to be short.

Prechlorination, Post-chloramines.
This approach combines use of a strong predisinfectant with a long-lasting post-disinfectant that does not produce THM's. As shown in Fig. 14–10c, Inst THM and THMFP are identical to the original condition up to the point of chlorine/ammonia addition, which results in chloramine formation. At this point, formation of Inst THM ceases as long as sufficient ammonia is added, and there is unreacted THMFP in the distribution system.

Pre–Chlorine Dioxide, Post-chloramines.
Only very low Inst THM concentrations are produced in this concept, and only if there is residual free chlorine in the chlorine dioxide. THM concentrations are shown in Fig. 14–10d. The cost of this concept is higher than that of the other suggested

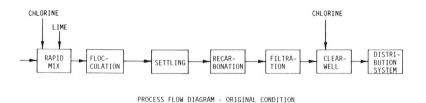

PROCESS FLOW DIAGRAM - ORIGINAL CONDITION

14-10A. Original condition

14-10B. Change in chlorination point to after recarbonation

THM CONCENTRATION - ug/L

14-10C. Prechlorination, postchloramines

14-10D. Pre-chlorine dioxide, postchloramines

LEGEND

INSTANTANEOUS THM CONCENTRATION

THM FORMATION POTENTIAL

+ TERMINAL THM CONCENTRATION

Fig. 14-10. THM control measure for systems using lime softening.

control techniques because of the high cost of the chlorine dioxide generation facilities.

Microbiological Considerations

During lime softening, the high pH conditions aid in the disinfection process. Thus, there is additional protection that is not provided by conventional treatment. Principal microbiological considerations are discussed in the following paragraphs for each suggested technique.

Change in Point of Chlorination to After Recarbonation. Increased microbial penetration into the treatment plant will be experienced. High pH will curtail this penetration to a degree, but an increase in microbiological monitoring to once a day is recommended, particularly during periods of high raw water bacterial concentration.

Prechlorination, Post-chloramines. Use of a strong predisinfectant combined with high pH should produce excellent disinfection. Then, use of long-lasting chloramines will provide additional disinfection as necessary.

Pre–Chlorine Dioxide, Post-chloramines. Results should be comparable to the previously discussed concept, as long as a chlorine dioxide residual is maintained through filtration. Overall comparability depends upon the residual of the predisinfectant.

SUMMARY OF MICROBIOLOGICAL CONCERNS

One of the principal functions of a water treatment plant is to act as a barrier to microorganisms, particularly pathogens. This barrier is especially important in surface supplies, which in many cases contain discharges from upstream wastewater treatment plants, as well as stormwater runoff and animal waste discharges.

To meet the requirements of the trihalomethane regulation, it is essential that adequate disinfection be provided. Two general THM control concepts, which can place additional stress on the microbial removal efficiency of the treatment works, are:

- Changing the initial point of chlorine addition to after settling.
- Use of alternate disinfectants (chloramines or chlorine dioxide).

The impact of these changes on the microbial population is most significant during periods of poor raw water quality and during warm water conditions. When the water is warmest, microbial activity is at or near its maximum.

Fortunately, disinfection efficiency is also enhanced during warm weather.

The objective of the baseline water quality survey is to collect adequate microbiological data before any treatment changes are made, as well as after changes are made. The baseline data collected before any changes are made for THM control should verify that adequate disinfection is being achieved. Second, the data collected will be a basis for comparison after changes are made.

Following changes, microbiological monitoring should continue for at least a year, and even longer where chloramines are used. Particularly important in this phase of the water quality survey are:

1. Increased monitoring during abnormally heavy pollution loads in the raw water.
2. Monitoring at dead-end or slow-flow portions of the distribution system.

In most situations, weekly monitoring of microbial parameters (coliforms and SPC as a minimum) is sufficient. However, during high raw water bacterial concentrations, daily sampling is recommended. Where upstream wastewater treatment plant discharges occur, periodic enteric virus or coliphage testing should be considered.

Long-term bacteriological changes are often noted first at dead-ends, or slow-flow locations. These locations are often the first to show the effect of insufficient disinfectant residuals and intermittent penetration of the treatment barrier. For systems using chloramines, it is important to monitor over even longer periods than a year, as long-term problems such as increased bacteria levels, drop-off of chloramine residual, and taste and odor complaints may occur.

COST OF THM CONTROL TECHNIQUES

Cost of control techniques is discussed in Chapter 30.

REFERENCES

1. *Federal Register,* Vol. 44, No. 231, 68628–68707, November 29, 1979, as corrected by *Federal Register,* Vol. 45, No. 49, 15542–15547, March 11, 1980.
2. Stevens, A. A., Slocum, C. J., Seeger, D. R., and Roebuck, G. G., "Chlorination of Organics in Drinking Water," *J.AWWA* 68:615–620, November, 1976.
3. Ohio River Valley Water Sanitation Commission, *Water Treatment Process Modifications for Trihalomethane Control and Organic Substances in the Ohio River,* EPA–600/2–80–028, U.S. EPA, Cincinnati, Ohio, March, 1980, NTIS Accession No. PB 81–301222.
4. *Guidance for the Sampling, Analysis, and Monitoring of Trihalomethanes in Drinking Water,* Science and Technology Branch, Criteria and Standards Division, U.S. EPA, Washington, D.C., February, 1983.

5. Symons, J. M., Stevens, A. A., Clark, R. M., Geldreich, E. E., Love, O. T., and Demarco, J., *Treatment Techniques for Controlling Trihalomethanes in Drinking Water,* EPA 600/2–81–156, U.S. EPA, Cincinnati, Ohio, September, 1981, 289 pp.
6. *Technologies and Costs for the Removal of Trihalomethanes from Drinking Water,* Science and Technology Branch, Criteria and Standards Division, Office of Drinking Water, U.S. EPA, Washington, D.C., February, 1982.
7. *Federal Register,* Vol. 48, No. 40, 8406–8414, February 28, 1983.

Chapter 15
Volatile Organic Compounds

INTRODUCTION

Synthetic volatile organic compounds (VOC's) are increasingly being detected in drinking water sources, and particularly in groundwaters once thought to be pristine. A 1982 study estimated that 15 to 20 percent of the groundwater systems across the United States contain synthetic organics.[1] Unlike trihalomethanes, these compounds are not disinfection by-products, but are pollutants entering groundwater aquifers through improper storage and handling of chemicals or wastewater disposal activities. They are named VOC's because of their distinctive common property of high volatility relative to other organic substances such as pesticides.

The presence of VOC's in groundwater poses a threat to one of the nation's most important resources. Approximately 80 percent of all public water supplies in this country rely on groundwater resources for potable water, and about 96 percent of all water used for rural domestic purposes (individual home water supplies) is obtained from groundwater.[2] Groundwater usage by public water supply systems tripled between 1950 and 1975, increasing from 3.5 BG/day (13.25 × 10⁹ l/d) to 11 BG/day (41.6 × 10⁹ l/d).[3]

OCCURRENCE OF VOC's

Over the past decade, several federal surveys, including the National Organics Reconnaissance Survey, the National Organics Monitoring Survey, and many state surveys, have identified VOC's in numerous groundwater-supplied potable water systems.[4-7] Although monitoring has been concentrated in areas of suspected problems, particularly New England and the mid-Atlantic states, it is anticipated that groundwater supplies in all areas of the country will be found to be affected to some extent by synthetic organic chemicals. A summary of the monitoring data from the state studies indicating maximum contaminant levels detected to date is shown in Table 15–1.[8,9]

The monitoring data indicate that VOC's occur at disturbingly high concentrations, often orders of magnitude higher than those found in raw or treated drinking water drawn from the most contaminated surface supplies. Generally, the concentrations of VOC's in groundwater have been several hundred micrograms per liter, with some instances of concentrations in the milligrams per liter range.

Table 15-1. Summary of VOC Occurrence Data Taken from State Surveys.*

CHEMICAL	NO. STATES	NO. WELLS TESTED**	% POSITIVE†	MAX. μg/l
Trichloroethylene	8	2,894	28	35,000
Carbon tetrachloride	4	1,659	10	379
Tetrachloroethylene	5	1,652	14	50
1,2-Dichloroethane	2	1,212	7	400
1,1,1-Trichloroethane	3	1,611	23	401,300
1,1-Dichloroethane	9	785	18	11,330
Dichloroethylenes (3)	8	781	23	860
Methylene chloride	10	1,183	2	3,600
Vinyl chloride	9	1,033	7	380

* References 8 and 9.
** Ratio of community wells to private wells unknown.
† Not a statistical value.

VOC's are seldom detected in concentrations greater than a few micrograms per liter in surface waters because the compounds do not occur naturally and are relatively volatile. However, surface water subject to wastewater discharges may contain elevated concentrations of organic solvents during periods of ice cover when volatilization of these solvents is restricted.

The monitoring data from both federal and state studies reveal interesting characteristics of affected groundwater supplies:[2]

- An affected groundwater supply typically contains several VOC's.
- Trichloroethylene (TCE), an industrial solvent and degreaser, has been detected most frequently and in the highest concentrations.
- Tetrachloroethylene (PCE) ranks second in occurrence.
- Within a specific well, at least one or two organic compounds at relatively high concentrations (100 to 500 μg/l) will likely predominate, with several other identifiable compounds present at lower concentrations (less than 50 μg/l).
- A given well field may include one well with a preponderance of one or two compounds at high concentrations, whereas in another well in the same area several different compounds may dominate.
- A groundwater system with all of its wells affected to the same extent is unlikely.

VOC's OF CONCERN

Many of the VOC's found in potable water supplies are recognized as a threat to public health that in some instances must be dealt with by removing the chemicals from water supplies through suitable treatment. Presently, several synthetic organic compounds are undergoing review for possible inclusion in the National Revised Drinking Water Regulations (Table 15–2).[10,12] The EPA has established recommended maximum contaminant levels (RMCL's) for the first nine compounds listed in Table 15–2.[11] These compounds have been detected at the greatest frequency and concentration and are the primary focus of concern at this time. The other compounds listed in Table 15–2 are not now scheduled for regulation because of insufficient data. These VOC's, as well as others, will be considered for regulation when more occurrence and toxicological data are available.

RMCL's are nonenforceable health goals that are set at levels that will result in "no known or anticipated adverse health effects with an adequate margin of safety." RMCL's for noncarcinogens are based on chronic toxicity data. Those for carcinogens are set at zero. Maximum contaminant levels (MCL's), however, are enforceable standards set as close as possible to the RMCL's. MCL's are based on health, treatment technologies, cost, and other factors. They are typically set after considerable public input has been received on the RMCL's.

Table 15–2. Volatile Organic Compounds Undergoing Review for Possible Inclusion in the National Revised Primary Drinking Water Regulations.*

COMPOUND	RECOMMENDED MAXIMUM CONTAMINANT LEVEL, mg/l
Trichloroethylene	0
Tetrachloroethylene	0
Carbon tetrachloride	0
1,1,1-Trichloroethane	0.2
1,2-Dichloroethane	0
Vinyl chloride	0
Benzene	0
1,1-Dichloroethylene	0
1,4-Dichlorobenzene	0.75
1,2-Dichlorobenzene	**
1,3-Dichlorobenzene	**
Chlorobenzene	**
Trichlorobenzene(s)	**
Methylene Chloride	**
cis-1,2-Dichloroethylene	**
trans-1,2-Dichloroethylene	**

* Reference 12.
** No recommended maximum contaminant level has been proposed to date.

Only two of the nine VOC's scheduled for regulation are noncarcinogenic: 1,1,1-trichloroethane and 1,4-dichlorobenzene. The remaining seven VOC's are carcinogenic. Vinyl chloride and benzene are known human carcinogens; the other five have produced tumors in animals.

SOURCES OF VOC's

Volatile organic chemicals are a widely used class of compounds employed in many types of industrial, commercial, agricultural, and household activities. Presently, VOC's are produced at a rate of over 20 billion pounds (9.1 × 10^9 kg) per year.[12] Both the multitude of uses and the magnitude of production of VOC's contribute to the introduction of these contaminants into the environment.

Table 15–3 lists the major uses of the six VOC's most frequently detected in water supplies. This list illustrates the vast number of pathways by which VOC's can enter the environment, and may prove useful in attempts to isolate specific causes of contamination.

It is generally perceived that most VOC contamination of groundwater is the result of improper surface or underground disposal of hazardous waste from industrial activities. Groundwater contamination may also occur as a result of activities not intended for waste disposal such as accidental spills or leaking storage tanks. The potential for groundwater contamination is revealed by Fig. 15–1, which shows existing industrial liquid waste disposal sites throughout the country. The widespread extent of these sites shows that the potential for groundwater quality degradation from these sources is great.

Occasionally, VOC's may also be introduced during the treatment, storage, and conveyance of drinking water. Carbon tetrachloride, for example, is a known contaminant of chlorine produced by the graphite-anode process.[13] Disinfection with chlorine produced by this process can be a significant source of carbon tetrachloride in treated drinking water.[14] Similarly, other products used in the production and distribution of water are sometimes sources of contaminants. Tetrachloroethylene is leached from polyvinyl-toluene-lined asbestos cement pipe.[15] Trichloroethylene is present in certain joint compounds used in reservoir liners and covers, and vinyl chloride has been found as a residual monomer in polyvinyl chloride pipe manufactured before 1977.[16]

VOC's AND THE GROUNDWATER ENVIRONMENT

VOC's can enter an aquifer and be transported great distances because they have little affinity for soils.[17-19] At present, approximately 1 percent of the nation's groundwater is thought to be contaminated by synthetic organic pollutants.[20] However, this estimate is only a rough approximation based

Table 15-3. Uses of Most Commonly Occurring VOC's.*

SUBSTANCES	USES
Trichloroethylene	Mainly as a degreasing solvent in metal industries. Common ingredient in household products such as spot removers, rug cleaners, air fresheners; dry cleaning fluids; refrigerants and inhalation anesthetics.
Tetrachloroethylene	Mainly as a dry-cleaning solvent in commercial and coin-operated systems. Used as textile scouring solvent; dried vegetable fumigant; rug and upholstery cleaner; stain, spot, lipstick, and rust remover; printing ink ingredient; heat transfer media; chemical intermediate in the production of other organic compounds; and metal degreaser.
Carbon tetrachloride	Mainly in manufacture of fluorocarbons used as refrigerants, foam-blowing agents, and solvents. Used in fumigants, although use in grain fumigation is decreasing; minor uses in metal cleaning and manufacture of paint and plastics. Banned for use in consumer goods in 1970 and as aerosol propellant in 1978.
1,1,1-Trichloroethane	Mainly in metal cleaning. Used for leather tanning; vapor depressant in aerosols and solvent for adhesives; septic tank degreasers; drain cleaners; inks; shoe polishes; cutting oils; and many other products.
1,2-Dichloroethane	Mainly as intermediate in manufacture of vinyl chloride monomers. Use as intermediate in manufacture of chlorinated solvents such as tetrachloroethylene, trichloroethylene, and 1,1,1-trichloroethane; as solvent for cleaning textiles, cleaning PVC-processing equipment, processing pharmaceutical equipment, extracting oil from oil seeds; and in manufacturing paints, coatings, and adhesives, fumigating stored grain products, and lead-scavenging additives.
Vinyl chloride	Mainly in the manufacture of plastics, polyvinyl chloride (PVC) resins, and polyvinyl chloride fabrication. Vinyl chloride and PVC used as raw materials in various industries such as rubber, glass, paper, and automotive; and in manufacture of electrical wire insulation and cables, pipe, industrial and household equipment, medical supplies, food-packaging materials, and building and construction products.

* Reference 12.

on an incomplete survey. Moreover, groundwater contamination inherently is characterized by long time lags measured in decades.[17]

Groundwater provides a unique environment for VOC's because:[21]

- It has limited contact with the atmosphere; hence, volatile compounds do not evaporate quickly.
- The surface environment below the active soil zone is relatively abiotic, allowing little biodegradation.
- The temperature of groundwater undergoes slow and limited fluctuations throughout the annual climatic cycle.

Fig. 15–1. Industrial sites containing liquid waste disposal facilities.[9]

- The groundwater moves slowly, without turbulence, in a dark environment.

When VOC's are introduced into a groundwater system, they maintain a discrete pattern, since laminar flow does not allow turbulent mixing. Dilution is accomplished by dispersion and diffusion. Thus, the characteristics of an aquifer tend to preserve, more than dissipate, VOC's.[21]

An adequate understanding of the extent of contamination within a given aquifer requires a thorough understanding of the aquifer geology and groundwater movement within that aquifer, as well as analyses from nearly every groundwater discharge within the aquifer. Contamination can be caused by a single discharge (such as a railroad accident) and remain undetected for several years. In cases of very slow groundwater movement, a contaminant may remain localized and impact only a small area. In any situation, predicting the likelihood of contamination at a particular point without a large data base is extremely difficult.[22] Often, the necessary data on aquifer contamination can only be obtained through the development of a comprehensive groundwater monitoring program. Excellent discussion of the design and installation of monitoring well networks is available from several sources.[21,23]

ALTERNATIVES FOR CONTROLLING VOC's

When a water supply is found to be contaminated with VOC's, various strategies are available to address the problem. These strategies may be classified as either management or treatment strategies, as shown in Fig. 15–2. Selection

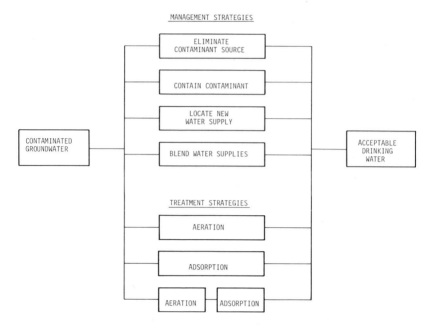

Fig. 15-2. Alternative strategies for controlling VOC concentrations in contaminated groundwater.

of the proper strategy for a specific contamination problem requires an extensive evaluation of a wide range of factors. Among the most important criteria to be considered are:

- Public opinion
- Regulatory agency acceptance
- Cost
- Long-term effectiveness
- VOC reduction efficiency
- Residual effects
- Ease of implementation
- Reliability

Most of the public water supply systems likely to be affected by VOC contamination are small, decentralized systems, serving fewer than 10,000 people. For these systems, the following additional considerations should be heavily weighed in evaluating control strategies:[24]

- The existing level of treatment in small systems typically consists only of chlorination; thus, any treatment technique will represent a new and more complicated technology.
- Small water utilities have limited resources; thus, both capital and operating costs may be significant obstacles in controlling VOC's.
- Operators of small systems typically have other duties in addition to

water plant operation and maintenance; therefore, simplicity in operation and maintenance, including monitoring, should be given special consideration.

MANAGEMENT STRATEGIES

The long-term interest of both the water purveyor and the consumer would best be served by the availability of an uncontaminated source.[25] The strategies available for accomplishing this goal are primarily management strategies. Therefore, it is essential that decision makers give priority to these options when determining a course of action.

Management strategies involve control of the water supply source to reduce or eliminate the presence of chemical compounds. These strategies include:

- Elimination of the contaminant source.
- Containment of the contaminant.
- Location of a new water supply source.
- Blending of existing water supply sources.

Management strategies have the potential advantage of low capital cost, but they are not always implementable. In addition, the technical aspects of some of the management strategies are not well defined and are difficult to assess. This is particularly true for those strategies that attempt to predict or control contaminant movement in an aquifer.

Elimination of the Contaminant Source

If the source of the VOC's can be identified and the size of the groundwater aquifer is small, elimination of the source may be an effective solution. This can be accomplished by purging the groundwater, treating the pumped water, and returning the water to replenish the aquifer. However, sources of organic contaminants are often difficult to locate because of the ease with which the chemicals migrate through the soil. Additionally, the size of the affected aquifer and the degree of infiltration may be such that many years would be required to purge the aquifer even after the source is identified and eliminated. For these reasons, eliminating the source of the compounds may be a management technique that is only practical for a limited number of affected utilities.

Containment of the Contaminant

This technique involves the use of one or more purge wells to halt further migration of the VOC's into a well field, thus protecting the remaining wells from increased contamination.[26] In order for this technique to be effective, the contaminant must be detected early, before extensive contamination of the aquifer has occurred. Also, a thorough understanding of groundwater movement is necessary to halt contaminant migration. One problem with this approach is the disposal of the contaminated water removed by the purge wells.

Locating a New Supply Source

Locating a new supply source involves developing a new well in an unaffected aquifer, tapping a surface water source, or purchasing water from a neighboring community. This approach has been utilized by several communities. However, this method may not be practical because: (1) an unaffected source of supply may not be available nearby, and the cost of developing a new source that is far removed from the service area may be prohibitive; (2) developing a new groundwater supply would not eliminate the possibility of the compound migrating to the new supply; and (3) a neighboring community's supply may not be capable of providing enough additional water to replace a large supply.[24] Consequently, this type of control may only be practical for very small systems, or for systems where only a portion of the supply has been affected.

Blending

A fourth management strategy for systems with multiple wells is to blend water from several wells to reduce the concentration of the compound by dilution. Depending on the levels of the compounds in each well, blending water from the wells prior to pumping into the distribution system could reduce the concentration of a specific compound to acceptable levels. Use of this approach may be limited due to lack of system flexibility, insufficient dilution ability, or lack of consumer acceptance.

TREATMENT STRATEGIES

Other than some incidental evaporative losses, conventional water treatment consisting of coagulation, sedimentation, filtration, and chlorination has been found to be largely ineffective for reducing the concentration of VOC's. However, because of the generally hydrophobic nature of VOC's and their tendency to partition into other phases, they can be removed by aeration, adsorption, or a combination of these processes.

Table 15–4 lists properties of selected VOC's that are pertinent to removal techniques. These techniques are examined in depth in the remainder of this chapter.

Table 15-4. Properties of Selected Volatile Organic Compounds (After Love).[*]

COMPOUND	MOLECULAR WEIGHT (g/mol)	DENSITY (g/ml)	BOILING POINT °C atm	SOLUBILITY mg/l	VAPOR PRESSURE mm Hg	HENRY'S CONSTANT[**] $(atm-m^3)\ 10^{-3}$ mol
Trichloroethylene	132	1.46	86.7	1,100	74	11.7
Tetrachloroethylene	166	1.62	121	140	18.6	28.7
cis-1,2-Dichloroethylene	97	1.29	60	3,500	206	8
trans-1,2-Dichloroethylene	97	1.26	48	6,300	271	5.2
1,1-Dichloroethylene	97	1.22	32	40	495	150
Vinyl chloride	62	0.92	-14	60	2,660	6,400
1,1,1-Trichloroethane	133	1.34	74.1	4,400	100	4.92
1,2-Dichloroethane	99	1.24	83.5	8,700	82	1.10
Carbon tetrachloride	154	1.59	76.7	800	91.3	30.2
Methylene chloride	85	1.33	40	19,400	438	3.19
Benzene	78	0.89	80	1,780	95	5.55
Chlorobenzene	113	1.11	132	448	15	3.93
1,2-Dichlorobenzene	147	1.31	180	100	1.0	1.94
1,3-Dichlorobenzene	147	1.29	173	123	2.0	1.94
1,4-Dichlorobenzene	147	1.25	147	79	1.0	1.94
1,2,4-Trichlorobenzene	182	1.45	219	30	0.29	1.42

* Reference 27.
** Obtained from Reference 28.

AERATION

Aeration is a unit process in which water and air are brought into contact with each other for the purpose of transferring volatile substances to water (gas absorption) or from water (air stripping). The latter process, air stripping, is applicable to VOC control, and has been effectively used in water treatment to remove hydrogen sulfide and carbon dioxide and also to remove certain taste- and odor-producing compounds. Although the use of air stripping solely for the purpose of controlling trace organics is a relatively new concept in the drinking water industry, for removal of many VOC's it is a cost-effective alternative to adsorption, provided the desired effluent concentrations can be achieved.

There are some limitations and concerns associated with air stripping. The process is temperature-sensitive, with poor removals occurring at low temperatures. Consequently, it may not be feasible in cold climates. In addition, questions have been raised concerning the secondary effects of aeration. These include potential air quality problems created by exhaust gases from the aerator, and the potential for water quality deterioration from airborne particulates, oxidized inorganics, instability resulting in corrosion, and biological growth in the aeration unit. Also, aeration results in increased dissolved oxygen levels in the water, that may prove detrimental to distribution system piping. These problems are not thought to be significant, but they do require consideration.

Principles of Aeration

The kinetic theory of gases states that molecules of dissolved gases can readily move between the gas and liquid phases. Consequently, if water contains a volatile contaminant in excess of its equilibrium level, the contaminant will move from the liquid phase (water) to the gas phase (air) until equilibrium is reached. Through use of this basic operating principle, the air stripping process eventually will allow all of the contaminant to be removed from solution if the air in contact with the water is continuously replenished with contaminant-free air.

The two major factors that determine the efficiency of removal of a specific organic by aeration are: (1) the ratio of concentration in the gaseous phase to the concentration in the aqueous phase at equilibrium (measured by Henry's constant) and (2) the rate at which equilibrium is obtained.[29] The departure from equilibrium conditions provides the driving force for mass transfer. The mass transport of the VOC molecules to the surface for release into the gaseous phase determines the rate of achievement of equilibrium.

Principles of Equilibrium

For dilute, nonideal solutions such as those typically encountered in potable water treatment, the basic relationship between the concentration of a volatile compound in the liquid phase to its concentration in the gaseous phase is given by Henry's law. According to Henry's law, at equilibrium, the partial pressure of the contaminant in the gaseous phase above a solution is directly related to the concentration of the contaminant in the solution. The phenomenon may be represented mathematically as:

$$P_A = H_A X_A *$$ (15–1)

where:

P_A = the partial pressure of contaminant A in the gaseous phase, atm
H_A = Henry's constant for contaminant A, atm
$X_A *$ = Mole fraction of contaminant A in liquid in equilibrium with gas phase, mol/mol

The mole fraction is defined as the molar concentration of the contamimant in the solution ($kmol/m^3$) divided by the sum of the molar concentrations of all components in the solution. Because water is the dominant component in any solution of interest in potable water treatment and because water contains 55.6 $kmol/m^3$, mole fractions may be approximated as:

$$X_A = \frac{C_A}{55.6}$$ (15–2)

where C_A = concentration of contaminant A in solution, $kmol/m^3$.

Henry's law can then be expressed as a function of the molar concentration of the contaminant, as follows:

$$P_A = \frac{C_A}{55.6} \cdot H_A$$ (15–3)

with the units of Henry's constant in atm · $m^3/kmol$.

According to Dalton's law, the total pressure of an ideal gas mixture is the sum of the partial pressures of all of its components. From this, it may be shown that the partial pressure of a particular contaminant in the air is the product of the total pressure and the volume fraction or the gas phase mole fraction of the contaminant in air.

$$P_A = Y_A P_T \tag{15-4}$$

where:

Y_A = mole fraction of contaminant A in the gas phase, mol/mol
P_T = total pressure, atm

Combining equations (15–1) and (15–4) relates the mole fraction of contaminant A in the liquid and gas phases at equilibrium:

$$Y_A = X_A * \frac{H_A}{P_T} \tag{15-5}$$

As equation (15–5) reveals, the larger Henry's constant, the greater will be the equilibrium concentration of A in the air. Thus, contaminants with a large Henry's constant are more easily removed by air stripping. The Henry's constants for several compounds are shown in Table 15–4. Henry's constant can also be estimated using the following equation:[28]

$$H = \frac{VP \times MW}{S} \tag{15-6}$$

where:

H = Henry's constant, m³ atm/mol
VP = vapor pressure of pure solute, atm
MW = molecular weight of solute, g/mol
S = solubility of solute, g/m³ or mg/l

Most volatile compounds of interest follow Henry's law quite satisfactorily in the range of concentrations experienced in potable water treatment. However, the Henry's constants listed on Table 15–4 were computed at higher concentrations than those usually encountered in water treatment practice. Data are not yet available to fully support extrapolation of Henry's constants into very low concentration regions. For these compounds, the calculated Henry's constant may be high, which would exaggerate their potential for removal by air stripping. However, it is not anticipated that the values of Henry's constant in the low concentration region will differ by more than a factor of 2.[30] Thus, the computed values may be used for feasibility studies, but experimentally derived values are preferred for design purposes.

Like all equilibrium constants, Henry's constants are directly affected by temperature. If the enthalpy caused by dissolution of the contaminant in water is considered independent of temperature, the relationship between Henry's constant and temperature can be expressed as:

$$\log H = \frac{-\Delta H^\circ}{RT} + C \qquad (15\text{--}7)$$

where:

R = universal gas constant, 1.987 kcal/kmol · °K
T = absolute temperature, °K
ΔH° = change in enthalpy due to dissolution of component A in water, kcal/kmol
C = constant, specific to particular contaminants

Table 15–5 lists values of ΔH° and C for 20 compounds of interest in water treatment. These values were developed from least squares of data. In all cases, correlation coefficients were greater than 0.99. The values show that Henry's constant, for the VOC's considered, increases about threefold for every 10°C temperature rise.[30]

Table 15–5. Henry's Constant Temperature Dependence.*

COMPOUND	FORMULA	ΔH°, KCAL/KMOL	C
Benzene	C_6H_6	3,690	8.68
Chloroform	$CHCl_3$	4,000	9.10
Carbon tetrachloride	CCl_4	4,050	10.06
Methane	CH_4	1,540	7.22
Ammonia	NH_3	3,750	6.31
Chloromethane	CH_3Cl	2,480	6.93
1,2-Dichloromethane	$C_2H_4Cl_2$	3,620	7.92
1,1,1-Trichloroethane	$C_2H_3Cl_3$	3,960	9.39
1,1-Dichloroethane	$C_2H_4Cl_2$	3,780	8.87
Trichloroethylene	C_2HCl_3	3,410	8.59
Tetrachloroethylene	C_2Cl_4	4,290	10.38
Carbon dioxide	CO_2	2,070	6.73
Hydrogen sulfide	H_2S	20	5.84
Chlorine	Cl_2	1,740	5.75
Chlorine dioxide	ClO_2	2,930	6.76
Sulfur dioxide	SO_2	2,400	5.68
Difluorochloromethane	CHF_2Cl	2,920	8.18
Oxygen	O_2	1,450	7.11
Nitrogen	N_2	1,120	6.85
Ozone	O_3	2,520	8.05

* Reference 30.

Principles of Mass Transfer

The rate of mass transfer of a volatile substance from water to air is generally proportional to the difference between the concentration of the contaminant in solution and the equilibrium concentration of the contaminant in solution at the system temperature, as defined by Henry's law. The relationship is expressed as follows:

$$M = K_L a \ (C_A * - C_A) \tag{15-8}$$

where:

M = mass of substance transferred per unit time and volume, lb/hr/cu ft (Kg/hr/m ^{3})

K_L = the overall liquid mass transfer coefficient, ft/hr (m/hr)

a = effective area for mass transfer, sq ft/cu ft (m²/m³)

$C_A *$ = liquid phase concentration in equilibrium with the gas phase concentration, lb/cu ft (kg/m³)

C_A = bulk liquid phase concentration, lb/cu ft (kg/m³)

The mass transfer coefficient, K_L, is a function of the physical and operational characteristics of the air stripping system, temperature and detention time of the liquid, and the particular compound being removed. K_L incorporates the diffusional resistance to mass transfer across both the liquid side and the gas (air) side of the gas–liquid interfacial area, and is related to the local gas and liquid phase mass transfer coefficients, K_g and K_L, respectively. For most potable drinking water applications, the overall mass transfer rate is generally controlled by liquid phase resistance. Consequently, air stripping processes should be designed to maximize the liquid mass transfer coefficient.

The effective area, a, represents the total liquid–air interfacial area created in the air stripping unit through the production of water droplets or air bubbles. The effective area is thus a function of the equipment used. Optimal air stripping units maximize the effective area.

In most applications, the mass transfer coefficient, K_L, and the effective area, a, are evaluated as a single constant, $K_L a$.

ALTERNATIVE AIR STRIPPING METHODS

Numerous types of aeration devices have been developed in which air stripping can occur. These alternatives may be classified into three general categories:

• Diffused aeration, which involves the injection of air into water.
• Spray aeration, which involves the injection of water into air.

• Waterfall aeration, which involves the cascading of water over media, forming droplets or thin films of water to contact with air.

All three types of system employ mechanical energy to create air–water interfaces across which mass transfer of the contaminant can occur. Typical examples of air stripping equipment are shown in Fig. 15–3.

Diffused Aeration

Diffused aeration involves the injection of air into water through perforated pipes, porous diffusers, or other impingement devices to produce a multitude

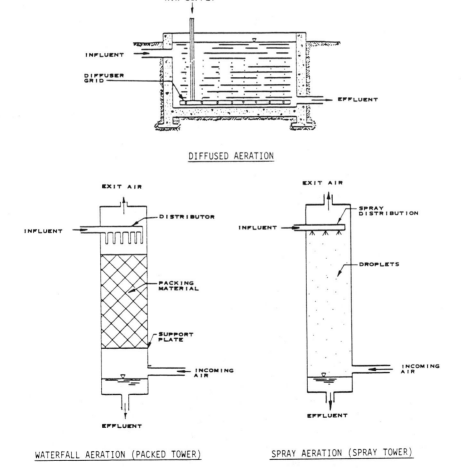

Fig. 15–3. Schematics of air stripping equipment.

of fine bubbles. As the bubbles rise, mass transfer of the contaminant occurs across the water–air interface until the bubbles reach the surface or become saturated with the contaminant.

Ideally, diffused aeration is conducted in a counterflow mode; however, most systems are operated as continuously stirred units, thereby reducing potential stripping efficiency. Although diffused aeration is generally used for gas absorption, principally absorption of oxygen, it is also effective at removing highly volatile organics. Removal efficiencies of up to 90 percent have been reported for several of the compounds listed in Table 15–4.

A variation of diffused aeration used for the removal of VOC's from well water supplies is in-well aeration.[31] Aside from the obvious advantages of this type of treatment system, several disadvantages exist. The in-well aeration method results in the dissolution of large volumes of air in the water, causing a milky appearance. Consequently, contact time with the atmosphere would have to be provided before the milky water could be pumped into the distribution system. Another disadvantage of in-well aeration is reduced efficiency.

Spray Aeration

Spray aeration involves the production of small water droplets by spraying water through nozzles. The small droplets produced expose a large interstitial surface area through which VOC's can migrate from the liquid phase to the gaseous phase. Spray systems can either be exposed to the environment or enclosed in a tower. In a tower, additional mass transfer occurs as the droplets descend through a countercurrent air flow. In both systems, the bulk of mass transfer occurs as the droplets are formed at the nozzles and as the droplets impact the water surface.

Waterfall Aeration

Waterfall aeration accomplishes the same result as diffused or spray aeration by causing the water to fall through the air and break into droplets or thin films. Although several different types of waterfall aerators exist, the most applicable unit for VOC removal is the countercurrent packed tower.

A countercurrent packed tower consists of a column 3 to 10 feet (0.9 to 3.05 m) in diameter and 15 to 30 feet (4.6 to 9.14 m) high containing packing material. This packing serves to continually disrupt the liquid flow, producing and renewing the air-to-water interface, thus improving mass transfer of the contaminant from water. Packing can be glass, ceramic, or plastic and is available in various geometrical shapes that provide high void volumes and high surface areas. The packing can be placed either randomly or in a fixed configuration within the column.

In packed towers, water flows downward by gravity, and air is forced upward. The untreated water is usually distributed at the top of the packing with sprays or distribution trays, and the air is blown through the tower in a forced or induced draft. The inside wall of the aeration column has several redistributors that force the water over the packing and prevent the water from running down the walls. The packed tower system is characterized by a high liquid–air interfacial area compared to the total volume of liquid in the column. Packed towers typically have void volumes in excess of 90 percent, which minimize air pressure drop through the column.

Packed towers have been utilized extensively in the chemical process industry, but applications for water treatment are less frequent. The most common application is the coke tray aerator used to remove H_2S and to oxidize iron and manganese. Packed towers can provide high removals of VOC's (greater than 99 percent) with relatively low gas pressures. Clogging of the packing with suspended solids can occur, however, and scaling can be a problem, particularly at low temperatures.[32]

DESIGN CONSIDERATIONS FOR AIR STRIPPING

Extensive development has occurred with air stripping processes associated with concentrated organic solutions, and the design procedures developed in these applications generally can be applied to the removal of VOC's from drinking water. The key factors affecting removal rates of VOC's by air stripping are:[31]

- Physical and chemical characteristics of the contaminant
- Temperature of the water and the air
- Air-to-water ratio
- Contact time
- Available area for mass transfer

The first two factors are determined by the contaminant to be removed and the liquid stream. The remaining factors are dependent on the aeration equipment and operating conditions. These factors can be evaluated through pilot testing.

Diffused Aeration

Design procedures for using diffused aeration to strip VOC's are not well developed. For best results, the diffused aeration should be conducted in a countercurrent mode with water entering at the top and exiting at the bottom, and exhausted air exiting at the top.

The removal efficiency of diffused aeration for stripping organics can be improved by decreasing the bubble size, increasing the water depth, increasing detention time, and increasing the volumetric air-to-water ratio. Finer bubbles are more efficient than coarse bubbles because of their greater interfacial area. However, diffusers that produce smaller bubbles are more expensive. In addition, at high air-to-water ratios, fine bubbles often coalesce into large bubbles, thereby eliminating any advantage.

Spray Aeration

Design procedures for removal of VOC's by spray aeration also have not been well developed. The principal design parameters for spray systems include the nozzle type, nozzle size, operating pressure loss, and process configuration. Depending on the nozzle type and pressure drop, droplet size can vary from 300 to 3,000 μm. Pressure drop usually ranges from 20 to 50 psi (137.8 to 344.5 kPa). Consequently, spray systems are relatively energy-intensive. Spray systems exhibit only modest mass transfer rates due to back-mixing in the tower and in the air.[33] Consequently, achievement of removal efficiencies in excess of 90 percent may not be feasible.

Packed Towers

The design of countercurrent packed towers has been well developed in the chemical process industry.[33,34] Recently, a number of researchers have developed design procedures for removing VOC's from groundwater with packed towers.[30,32,35-37] This section presents the general design considerations associated with packed towers. For a more detailed, step-by-step discussion of design procedures, the reader is referred to the literature cited.

A process schematic for a countercurrent tower with cross-sectional area A and depth of packing Z is shown in Fig. 15–4. Water containing a high level of contaminant (C_{in}) enters the top of the tower and flows downward at a superficial velocity $V_L = Q_L/A$ (Q_L = volumetric flow rate), exiting at the bottom with a low concentration (C_{out}). Correspondingly, forced air containing little or no contaminant (P_{in}) enters the bottom of the tower and travels upward at a superficial velocity $V_G = Q_G/A$, exiting the top of the tower with a higher level of contaminant or partial pressure (P_{out}).

As Kavanaugh and Trussel have shown, solution of the steady-state mass transfer equation for the case of dilute solutions (Henry's law valid) leads to a packed tower design relation stating that the depth of packing Z required to achieve a desired removal performance is the product of the number of transfer units (NTrU) and the height of a transfer unit (HTU).[30] That is:

$$Z = (NTrU) \cdot (HTU) \tag{15-9}$$

The term NTrU characterizes the difficulty in stripping a compound to a desired level, and is given by the equation:

$$\text{NTrU} = \frac{R}{R-1} \cdot \ln \text{(natural log)} \frac{C_\text{in}/C_\text{out}\,(R-1)+1}{R} \quad (15\text{–}10)$$

where:

R = stripping factor defined as $R = H_A G/L$
G = superficial molar air flow rate, lb mol/hr/sq ft
L = superficial molar liquid flow rate, lb mol/hr/sq ft

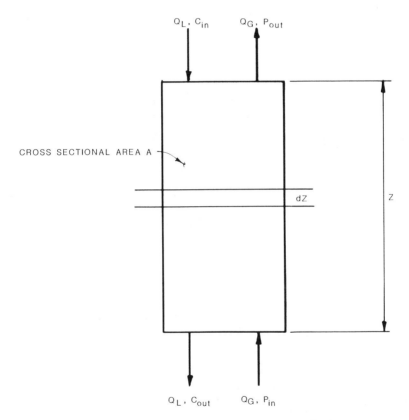

Fig. 15–4. Process schematic for countercurrent packed tower.

The NTrU is dependent on the desired removal efficiency, the air-to-water ratio, and Henry's constant. Given a specific Henry's constant and a desired removal efficiency, the NTrU can be computed for a packed column for a

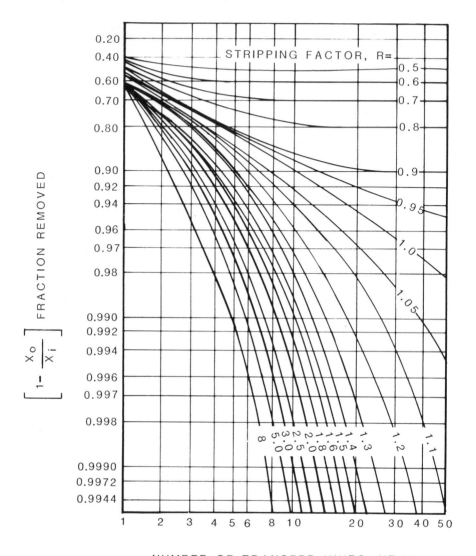

Fig. 15–5. Dependence of number of transfer units on removal efficiency and stripping factor.[32]

given stripping factor or air-to-water ratio. Such a relationship is shown in Fig. 15–5.

Optimum column designs are typically based on stripping factors between 1.2 and 5. The effect of varying R on the number of transfer units is shown in Fig. 15–6 for several removal efficiencies. For 90 percent removal, the

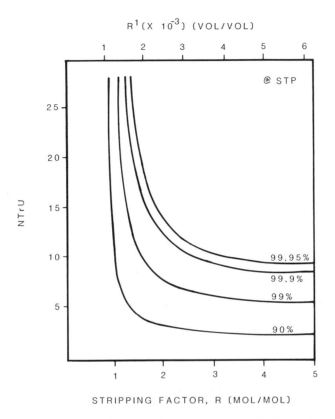

Fig. 15-6. Effect of increasing stripping factor on NTrU for various removal efficiencies.[31]

NTrU's increase rapidly for $R < 1$. Diminishing returns set in as R is increased beyond 2. For very high removals (> 90 percent), a stripping factor between 2 and 5 may provide the most economical design.[30]

The HTU characterizes the efficiency of mass transfer from water to air and is a function of the liquid loading rate and $K_L a$. HTU is defined as follows:

$$HTU = \frac{L}{K_L a \, C_0} \qquad (15\text{--}11)$$

where $C_0 =$ molar density of water, lb mol/cu ft.

Computation of the HTU requires data on the mass transfer coefficients for the system under consideration. In some cases, packing manufacturers will supply mass transfer data for air–water systems as a function of temperature and liquid flow rates. It is preferred that values for $K_L a$ be determined from pilot studies utilizing the contaminated water. However, in the absence of such data, mass transfer correlations from the literature can be used for preliminary design estimates. A typical empirical correlation used for liquid-phase mass transfer coefficients in towers containing randomly packed materials is the Sherwood-Hollaway correlation:[33]

$$\frac{K_L a}{D_A} = \alpha \left[\frac{L'}{\mu_L} \right]^{1-n} \left[\frac{\mu_L}{\rho_L D_A} \right]^{0.5} \tag{15-12}$$

where:

D_A = molecular diffusion coefficient of the compound to be removed (solute A) in water, sq ft/hr (m²/h)
α = constant
n = constant
μ_L = liquid viscosity, lb/ft·hr (kg/m·h)
ρ_L = density, lb/cu ft (kg/m³)
L' = liquid mass flux rate, lb/sq ft·hr (kg/m²·h)
$K_L a$ = mass transfer rate, hr⁻¹

The empirical constants n and α depend on the type and size of the packing. The constant n ranges from 0.2 to 0.5, and α ranges from 20 to 200. Other correlations may be found in the mass transfer literature.[33]

As stated earlier, the ability of a dissolved contaminant to be removed by air stripping is directly related to its Henry's law constant. Figure 15-7 demonstrates the effect of the type of compound on packing depth and air-to-water ratio.[31] An air-to-water ratio of about 10:1 is required to achieve 95 percent removal of trichloroethylene (TCE) with 15 feet (4.57 m) of 1-inch (2.5-cm) packing medium. For 95 percent removal of a less volatile compound such as 1,2-dichloroethane, an air-to-water ratio of about 120:1 is required for the same column design.[31]

The desired removal efficiency also affects the design of a packed column. Figure 15-8 illustrates the relationship between the air-to-water ratio and packing depth to achieve various efficiencies for TCE removal.[31] As illustrated, achieving 80 percent removal of TCE with an air-to-water ratio of 20:1 requires a column height of 6 feet (1.83 m) when 1-inch (2.5-cm) packing medium is used. By contrast, attaining 99 percent removal of TCE at the same air-to-water ratio requires a 20-foot (6.1-m) column height.

The temperature of the water supply also affects the design of a packed column. Most groundwater supplies exhibit a water temperature of about 55°F (13°C). However, some northern-latitude groundwaters may be as low as 45°F (6°C), and others may be as high as 75°F (24°C).[31] The relationship

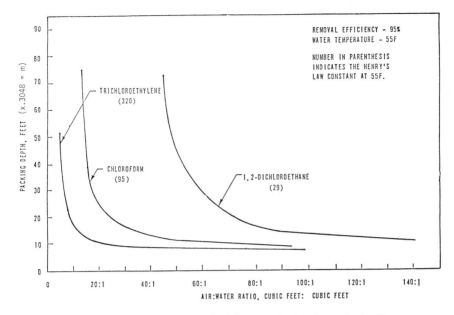

Fig. 15-7. Effect of type of VOC on packed column design.[31]

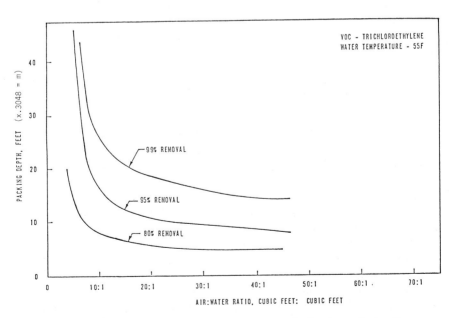

Fig. 15-8. Effect of removal efficiency on packed column design.[31]

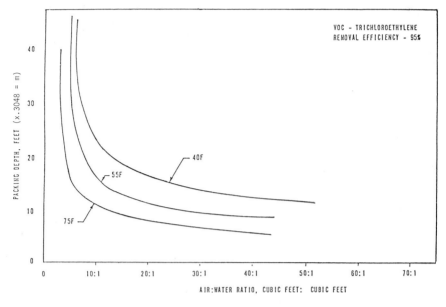

Fig. 15–9. Effect of water temperature on packed column design.[31]

of water temperature to removal efficiency is shown in Fig. 15–9, where at a 20:1 air-to-water ratio, a packing depth of about 8 feet (2.44 m) is required for 75°F (24°C), while a packing depth of about 16 feet (4.88 m) is required for 40°F (4°C).[31]

AIR STRIPPING PROCESS SELECTION

All of the aeration techniques discussed have been successfully used to reduce VOC concentrations in either full-scale or pilot-scale applications. Reported results from a few of these applications are presented in Table 15–6. Selection of the appropriate air stripping alternative depends on site considerations, the characteristics of the contaminant, the desired removal efficiency, and cost.

Figure 15–10 shows schematically the possible zones of economically feasible operation of air stripping processes. The two principal parameters controlling the selection are Henry's constant and the desired removal percentage. When removal efficiencies do not have to exceed 90 percent, both diffused aeration and spray towers may be cost-effective alternatives. Assuming that equal power is required for both spray and diffused air systems, the zone separating these two processes lies approximately at a Henry's constant of 1,000 atm.[31]

Table 15-6. VOC Removal Experience by Aeration.*

LOCATION	VOC	AIR:WATER RATIO (CFM/CFM)	CONTACT TIME (MIN)	INFLUENT CONCENTRATION (μG/L)	REMOVAL (%)
Diffused Air					
Cincinnati, OH	TCE**	4:1	10	241	78
	TCE	8:1	10	241	97
	TCE	16:1	10	241	99
Glen Cove, NY	TCE	5:1 to 20:1	5 to 20	132 to 313	73
Port Charlotte, FL					
6-ft-deep basin	THM	7.5:1	6	—	58
	THM	15:1	6	—	73
12-ft-deep basin	THM	7.5:1	6	—	73
	THM	15:1	6	—	84
	THM	22.5:1	6	—	85
Ft. Lauderdale, FL	THM	8:1 to 36:1	5 to 10	200 to 250	18 to 39
Tray Tower Aerator					
Bristol Borough, PA	TCE	—	—	80	75
	PCE	—	—	250	72
Multiple Tray Aerator					
Ft. Lauderdale, FL	THM	—	—	250 to 280	8 to 12
Camden, NJ	TCE	30:1	—	15 to 35	23 to 93
	PCE	30:1	—	6 to 20	18 to 80
Norwalk, CT	TCE	—	—	—	52
	PCE	—	—	—	40
Smyrna, DE	TCE	0	—	5 to 70	40 to 60
Packed Column					
Pelham, MA	TCEA	5:1 to 60:1	—	42 to 100	88 to 92
	TCEA	3:1 to 114:1	—	630 to 1,200	50 to 70
	TCEA	11:1 to 27:1	—	680	50 to 82
New Haven, CT	TCE	15:1	—	17 to 23	82 to 87
	TCE	25:1	—	19 to 140	85 to 90
	TCE	30:1 to 50:1	—	5 to 135	89 to 94
	TCE	8:1 to 10:1	—	8 to 10	67 to 78
	TCE	19:1 to 26:1	—	126 to 133	76 to 92
	TCE	13:1 to 19:1	—	23 to 34	75 to 87
	PCE	13:1 to 19:1	—	1.7	70+
	PCE	7:1 to 26:1	—	10 to 20	0.5 to 0.7
	PCE	8:1 to 10:1	—	2.4 to 2.7	3 to 5

* Reference 31.
** Spiked.

For removals greater than 90 percent, a packed tower probably is the only alternative. The line shown delineating the zone for packed towers assumes that the height of the transfer unit is approximately 3 feet (1 m), and that the maximum economical depth of packing is approximately 30 feet (10 m). For a ratio of local mass transfer coefficients of $k_l/k_g = 0.01$

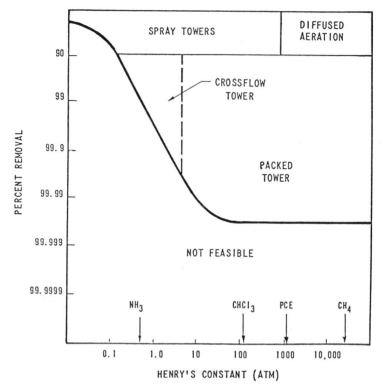

Fig. 15-10. Diagram for selection of feasible aeration process for control of volatile compounds.[31]

and a stripping factor greater than 5, it can be shown that the maximum removal efficiency over a wide range of Henry's law constants is approximately 99.5 percent.[31] As the Henry's constant decreases, the gas phase resistance (k_g) becomes important, and high gas ratios are required for system design. Under these conditions, packed towers become less economical. At some point, only a crossflow tower can handle the high gas flow rates.

ADSORPTION

Like aeration, adsorption has a spectrum of effectiveness. This process, however, is more complicated than aeration, and water quality can have a definite influence on performance. An in-depth discussion of adsorption on granular activated carbon (GAC) is presented in Chapter 12. Discussion in this chapter is limited to the application of adsorption for the removal of low-molecular-weight, volatile compounds. These compounds are more weakly adsorbed on GAC than larger-molecular-weight substances commonly found in surface waters.

Alternative Adsorbents

Adsorbents that could possibly be used to remove VOC's from groundwater include GAC, powdered activated carbon (PAC), and synthetic resins.

Granular Activated Carbon. GAC exhibits a wide range of effectiveness in adsorbing different compounds, and generally tends to adsorb high-molecular-weight compounds more readily than low-molecular-weight substances such as VOC's. Nonetheless, GAC is presently the best available adsorbent for the removal of VOC's.

Many different types of GAC are commercially produced. This allows a wide range of selection for such properties as surface area, particle size, hardness, and density. In addition, the different types of GAC can exhibit different adsorption isotherms for specific compounds. Selection of the proper type of GAC requires careful evaluation of the specific contaminant to be removed, the desired rate of adsorption, the reactor configuration and operational mode, the backwash frequency, and the type and frequency of regeneration. Detailed discussions on these topics are presented in Chapter 12.

Powdered Activated Carbon. Powdered activated carbon (PAC) has been used traditionally for removal of trace organic compounds causing taste and odor problems. The results of all studies to date indicate that PAC can be effective in removing certain higher-molecular-weight compounds; however, the lower-molecular-weight compounds such as VOC's are not satisfactorily removed except when very high dosages of PAC are used. Also, the use of PAC requires coagulation–sedimentation facilities not normally used for treating groundwater supplies.

Synthetic Resins. Synthetic resins appear to have limited application for VOC removal from groundwater.[38] The most promising of these is the carbonized resin Ambersorb® XE-340, made by Rohm and Haas. This material is designed to remove low-molecular-weight, nonpolar organics; has little affinity for competing high-molecular-weight, polar organics; and can be regenerated in situ by steam in certain applications.[39] Results obtained with this material have been encouraging.[27] Unfortunately, Rohm and Haas has decided not to make this material commercially available.[38]

XE-340 is similar to activated carbon in appearance, but retains some of the properties of the polystrene synthetic resin from which it is made by carbonation. By comparison to a typical activated carbon, XE-340 is more dense and more difficult to wet, and has a smaller total pore volume, a much smaller volume of macropores, a smaller surface area, a much lower ash content, and a smaller particle size.[39] Because of its promise for VOC removal and the possibility that some materials with similar properties may exist or become available in the future, data for XE-340 are included in this chapter (see below).

Adsorption of VOC's on Granular Activated Carbon

The two most significant properties of organic contaminants affecting their adsorption on activated carbon are solubility and affinity. The less soluble the contaminant, the better it is adsorbed. Also, the greater the specific attraction of the contaminant to the carbon surface, the greater its potential for adsorption. Increased molecular weight, decreased polarity, decreasing ionic character, and low pH for organic acids or high pH for organic bases are factors that decrease solubility of organics and, as a result, increase adsorption.[40] In general, higher-molecular-weight compounds are more strongly attracted by carbon than low-molecular-weight compounds.

The relative adsorbability of several VOC's is shown by the adsorption isotherms in Fig. 15–11. Tetrachloroethylene, one of the most commonly occurring VOC's in groundwater, is the VOC most readily adsorbed by GAC, according to isotherm data. Of the chlorinated hydrocarbon solvents most frequently detected in groundwater, unsaturated organic compounds such as ethylenes are more readily adsorbed on carbon than saturated compounds such as ethanes.[40] Thus, a treatment system to remove tetrachloroethylene would yield a much longer carbon life than would be obtained if it were necessary to remove equal amounts of 1,1,1-trichloroethane. Although adsorption isotherms are useful in developing preliminary data about the removal of organics on GAC, they do not yield sufficient data to develop design criteria for GAC systems. Adsorption isotherms are equilibrium batch tests, whereas full-scale GAC units are dynamic, continuous processes. Also, adsorption isotherms do not take into account competitive adsorption where several compounds may be present in the water supply. For these reasons, pilot studies are needed to develop specific design criteria.

Adsorption of VOC's on Synthetic Resins

Data from VOC removal studies with the synthetic resin Ambersorb® XE-340 are shown in Table 15–7. For most of these compounds, the capacity for adsorbence per gram of resin has been shown to be much greater than that of GAC. Also, the resin has been found to be particularly effective in removing the volatile, nonpolar compounds not well adsorbed by GAC. Studies have indicated that the resin is capable of treating nearly twice the volume of water per cubic foot of media that GAC can treat prior to breakthrough.[40] Empty bed contact time has typically been 5 to 7 minutes for resin columns, compared to 18 minutes for GAC columns when studies are conducted under similar operating conditions. Resin capacity may be reduced at the high pH of lime softening and where competition for adsorption sites occur.

Two important differences exist between the resins and activated carbon. First, resins have a much higher cost per pound than GAC. At present, costs for the resins are approximately $10 per pound as opposed to $0.05

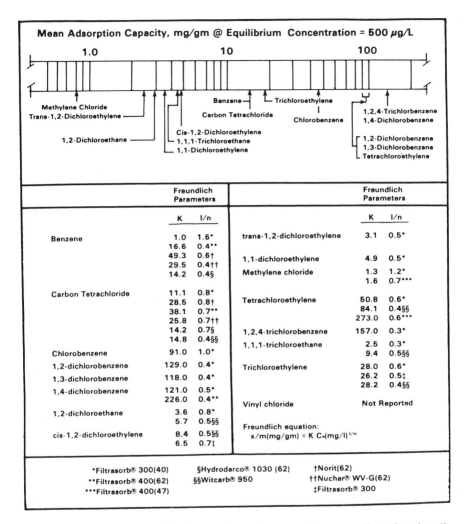

Fig. 15–11. Comparison of isotherm adsorption capacities on activated carbon.[27]

per pound for GAC.[33] Second, it appears that XE-340 can be successfully regenerated in situ with low-temperature steam. This latter factor, coupled with smaller contact tanks, shorter contact times, and higher adsorption capacity per pound of adsorbent, may make resins more cost-effective for specific installations.

Factors Affecting Adsorption of VOC's From Groundwater

Competitive Adsorption. Adsorption of trace organics can be affected by the amount of background organic carbon, generally measured as total

Table 15–7. Adsorption of Trichloroethylene and Related Solvents by Ambersorb® XE-340.*

	AVG. CONC., μg/l	BED DEPTH, FT	BED DEPTH, (m)	EMPTY BED CONTACT TIME, MINUTES	LOADING TO 0.1 μg/l BREAKTHROUGH, m³/m³**
Trichloroethylene	215	1	0.3	2	83,700†
	210	2	0.6	4	78,600†
	210	4	1.2	7.5	>53,300†
	177	2.5	0.8	9	>20,160
	4	2.5	0.8	8.5	>123,340
	3	0.8	0.2	5	>117,000
Tetrachloroethylene	41	1	0.3	2	>99,900†
	51	2	0.6	4	78,600†
	65	4	1.2	7.5	>53,300†
	70	1	0.3	2	106,000†
	94	2.5	0.8	5	112,900
	1,400	2.5	0.8	9	17,920
	3	2.5	0.8	8.5	>123,340
	2	2.5	0.8	9	>20,160
1,1,1-Trichloroethane	5	4	1.2	7.5	39,300†
	33	2.5	0.8	9	56,000
	237	0.8	0.2	5	82,600
	23	0.8	0.2	5	>100,800
	1	2.5	0.8	9	>20,160
Carbon Tetrachloride	19	2.5	0.8	5	7,560
	19	2.5	0.8	10	15,120
cis-1,2-Dichloroethylene	40	1	0.3	2	37,200†
	38	2	0.6	4	39,500†
	40	4	1.2	7.5	19,700†
	40	1	0.3	2	36,400†
	25	2.5	0.8	6	14,400
	22	2.5	0.8	6	7,200
	16	2.5	0.8	6	11,500
	6	2.5	0.8	9	>20,160
	2	2.5	0.8	8.5	>59,000 but <123,340
1,2-Dichloroethane	1	0.8	0.2	5	108,860
1,1-Dichloroethylene	122	0.8	0.2	5	80,600
	4	0.8	0.2	5	110,800

* Reference 27.
** m³ water/m³ carbon.
† Breakthrough defined by shape of wavefront curve; generally 20 to 25 μg/l of contaminant in adsorbent effluent.

organic carbon. It has been shown that a high background organic content (i.e., 1.5 to 2 mg/l or more) can result in lower than expected adsorption capacities, thus underscoring the fact that competition for adsorption sites is a necessary consideration when generalizing on adsorption efficiency.[41]

One equation that describes competitive effects is the Langmuir equation:

$$q_j = \frac{Q_{max} \cdot b_j \cdot C_j}{1 + \sum_i b_i C_i} \qquad (15\text{--}13)$$

where:

Q_{max} = amount adsorbed per unit weight of adsorbent in forming a complete monolayer on the surface, mol/g

b = constant related to the energy or net enthalpy of adsorption $[b \propto e^{-(\Delta H/RT)}]$

q_j = amount of species j adsorbed per unit weight at concentration C, mol/g

C = equilibrium concentration, mol/l

As the number of species that compete, i, and their concentration increase, the term $\Sigma_i b_i C_i$ increases, and the amount of species j adsorbed decreases.

For small amounts of adsorption $(b_j \cdot C_j \ll 1)$, the specific adsorption is proportional to the final concentration of adsorbate in solution, yielding:

$$q_j = Q_{max} \cdot b_j \cdot C_j \qquad (15\text{--}14)$$

For large amounts of adsorption, $(b_j \cdot C_j \gg 1)$ and:

$$q_j \simeq Q_{max} \qquad (15\text{--}15)$$

If background organics compete with targeted trace organics for adsorption sites, more frequent regeneration will be necessary. Competitive adsorption may also result in the displacement of previously adsorbed compounds, thus causing the effluent concentration of a contaminant to be greater than the influent.[39] This usually will occur when an adsorbent is at or near saturation with one compound and a second strongly adsorbing compound appears in the influent. If the column is properly monitored and the adsorbent is replaced or regenerated before complete saturation, effluent concentrations greater than influent caused by displacement can be avoided.

Desorption. Desorption of compounds must also be considered when adsorption is applied to groundwater treatment because the adsorption reaction is usually reversible. Desorption may occur because of competitive adsorption (i.e., is displacement by other compounds) or by a reversal of the adsorption reaction owing to a decrease in influent concentration.

Abrupt desorption has not been observed. However, periods have been observed when the effluent concentration of an organic could exceed its influent value after passing through an adsorbent.[27] It is generally agreed that the increased effluent concentration occurs when the influent concentration of the adsorbate declines, and the resulting desorption or "leakage" can be explained by the adsorption equilibrium theory.

ph and Temperature. The pH and dissolved inorganic solids in water do not directly affect adsorption of VOC's because VOC's are nonionic and do not interact strongly with inorganic ions.[38] The temperature of the water will affect both the rate and capacity of adsorption. As temperature increases, the rate of adsorption will increase, but the capacity for the adsorbate will decrease.[38]

Iron and Manganese. Iron and manganese often exist in groundwater in the reduced forms, Fe^{+2} and Mn^{+2}. During treatment, they are oxidized and precipitate as $Fe(OH)_3$ and MnO_2. Activated carbon may catalyze this oxidation. Accumulation of the precipitates in carbon beds results in rapid headloss buildup and plugging of the carbon pores.

GAC DESIGN CONSIDERATIONS

General design considerations for treatment of water by GAC are presented in Chapter 12. However, the design of a GAC system for removing VOC's from groundwater differs in several respects from the design of systems for removal of trace organic compounds from surface waters. In designing for VOC removal, the following factors must be considered:

- The species of contaminants to be removed from groundwater are generally more limited and easy to define than those encountered in surface waters. Also, the concentration of contaminant in the raw water does not vary as widely over short time periods in groundwater supplies as it does in surface waters.
- The VOC molecule is small and diffuses rapidly, compared to larger-molecular-weight compounds commonly found in surface waters. This behavior, combined with the presence of fewer and more uniform concentrations of competing substances, results in more defined and predictable breakthrough curves for VOC's.[38]
- Groundwater may contain reduced inorganic material that will precipitate in the carbon beds and adversely affect adsorption. Such circumstances will likely require pretreatment.
- Single-stage GAC systems may be more applicable to VOC removal from groundwater than from surface water because the flows requiring treatment are small, and the breakthrough curves are better defined than for surface waters. In this case, parallel operation may not be justified on a cost basis.

- Because VOC's are weakly adsorbed on GAC, regeneration of carbon saturated with VOC's could possibly be accomplished using milder conditions than those used to regenerate GAC used for surface waters. There is a possibility that steam regeneration could be successfully used in place of thermal regeneration, although efforts to date have not been particularly successful.

COMBINATION AERATION–ADSORPTION

Preceding the adsorption process with aeration can be an effective method of reducing VOC concentrations to very low levels while decreasing the frequency of adsorbent regeneration. The aeration step reduces the organic load to the adsorbent and may remove compounds competing for adsorption sites. Combination aeration–adsorption appears quite attractive when several different types of contaminants such as humic acids and VOC's need to be reduced to very low concentrations.

REFERENCES

1. Westrick, J. J., Mello, J. W., and Thomas, R. F., "The Ground Water Supply Survey Summary of Volatile Organic Contaminant Occurrence Data," Office of Drinking Water, Technical Support Division, U.S. EPA, Cincinnati, Ohio, June, 1982.
2. Hess, A. F., Dyksen, J. E., and Dunn, H. J., "Is Your Community's Groundwater Safe to Drink?," *Public Works,* October 1983.
3. Murray, C. R., and Reeves, E. B., "Estimated Use of Water in the United States in 1975," Circular 765, Geological Survey, U.S. Dept. of the Interior, Washington, D.C., 1977.
4. "National Organics Monitoring Survey (NOMS)," Office of Drinking Water, Technical Support Division, U.S. EPA, Cincinnati, Ohio, 1978.
5. Symons, J., et al., "National Organics Reconnaissance Survey for Halogenated Organics," *J. AWWA* 67(11):634, November 1975.
6. Kim, N., and Stone, D. W., *Organic Chemicals in Drinking Water,* New York State Dept. of Health, Bureau of Public Water Supply, Albany, N.Y., 1980.
7. Johnson, D., et al., *Chemical Contamination,* Commonwealth of Mass., Special Legislative Commission on Water Supply, Boston, September, 1979.
8. "The Occurrence of Volatile Organics in Drinking Water," U.S. EPA Briefing Paper, Office of Drinking Water, Cincinnati, Ohio, March 6, 1980.
9. *Contamination of Ground Water by Toxic Organic Chemicals,* Council on Environmental Quality, Washington, D.C., January, 1981.
10. "National Revised Primary Drinking Water Regulations, Volatile Synthetic Organic Chemicals in Drinking Water," *Federal Register,* Vol. 47, No. 43, p. 9350, March 4, 1982.
11. "EPA Publishes RMCL's for Nine VOC's," *AWWA Mainstream,* Vol. 28, No. 7, July, 1984.
12. DeMarco, J., "History of Treatment of Volatile Organic Chemicals in Water," *Occurrence and Removal of Volatile Organic Chemicals from Drinking Water,* AWWA Research Foundation, 1983.
13. "Control of Organic Chemical Contaminants in Drinking Water," *Federal Register,* Vol. 43, No. 28, p. 5759, February 9, 1978.

14. Cairo, P. R., Lee, R. G., Aptowicz, B. S., and Blankenship, W. M., "Is Your Chlorine Safe to Drink?," *J.AWWA* 71(8):450–453, August, 1979.
15. Larson, C. D., Love, O. T., and Reynolds, G. B., "Tetrachloroethylene Leached from Lined Asbestos Cement Pipe into Drinking Water," *J.AWWA* 75(4):184–188, April, 1983.
16. Dressman, R. C., and McFarren, E. F., "Determination of Vinyl Chloride Migration from Polyvinyl Chloride Pipe into Water Using Improved Gas Chromatographic Methodology," *J.AWWA* 70(1):29, January, 1978.
17. Roberts, P. V., "Nature of Organic Contaminants in Groundwater and Approaches to Treatment," *Proceedings,* AWWA Seminar, Organic Chemical Contaminants in Groundwater: Transport and Removal, 101st Annual National AWWA Conference, St. Louis, Mo., June, 1981.
18. Chiou, C. T., Peters, L. J., and Freed, V. H., "A Physical Concept of Soil–Water Equilibria for Nonionic Organic Compounds," *Science,* 206:831–832, November, 1979.
19. Dunlap, W., "Preliminary Laboratory Study of Transport and Fate of Selected Organics in a Soil Profile," Groundwater Research Center, U.S. EPA, Ada, Okla., 1980.
20. U.S. EPA, "Proposed Groundwater Protection Strategy," Office of Drinking Water, U.S. EPA, Washington, D.C., November, 1980.
21. Braids, O., "Volatile Organic Compounds and the Ground Water Environment," *Occurrence and Removal of Volatile Organic Chemicals from Drinking Water,* AWWA Research Foundation, 1983.
22. Hubbs, S. A., "Occurrence of Volatile Organic Chemicals," *Occurrence and Removal of Volatile Organic Chemicals from Drinking Water,* AWWA Research Foundation, 1983.
23. Fenn, D., *Procedures Manual for Ground Water Monitoring at Solid Waste Disposal Facilities,* EPA/530/SW-611, U.S. EPA, Washington, D.C., 1977.
24. Dyksen, J. E., and Hess, A. F., "Alternatives for Controlling Organics in Water Supply Sources," *J.AWWA,* August, 1982.
25. Singley, J. E., and Moser, P. H., "Evaluation of Alternatives," *Occurrence and Removal of Volatile Organic Chemicals from Drinking Water,* AWWA Research Foundation, 1983.
26. Minsley, B., "Tetrachloroethylene Contamination of Groundwater in Kalamazoo," *J.AWWA,* June, 1983.
27. Love, O. T., et al., "Treatment of Volatile Organic Compounds in Drinking Water," EPA-600/8–83–019, May, 1983.
28. Warner, P. H., Cohen, J. M., and Ireland, J. C., "Determination of Henry's Law Constants of Selected Priority Pollutants," Municipal Environmental Research Laboratory, Office of Research and Development, U.S. EPA, Cincinnati, Ohio, 1980.
29. Singley, J. E., and Williamson, D., *Aeration for the Removal of Volatile Synthetic Organic Compounds.*
30. Kavanaugh, M. C., and Trussell, R. R., "Design of Aeration Towers to Strip Volatile Contaminants from Drinking Water," *J.AWWA,* December, 1980.
31. Hess, A. F., Dyksen, J. E., and Dunn, H. J., "Control Strategy—Aeration Treatment Technique," *Occurrence and Removal of Volatile Organic Chemicals from Drinking Water,* AWWA Research Foundation, 1983.
32. Kavanaugh, M. C., and Trussell, R. R., "Air Stripping as a Treatment Process," *Proceedings,* AWWA Annual Conference, St. Louis, Mo., 1981.
33. Treybal, R. E., *Mass Transfer Operations,* 3rd ed., McGraw-Hill Book Co., New York, 1980.
34. Perry, R. H., and Chilton, C. H., editors, *Chemical Engineer's Handbook,* 5th ed., McGraw-Hill Book Co., New York, 1973.
35. Singley, J. E., et al., *Trace Organics Removal by Air Stripping,* AWWA Research Foundation, 1980.

36. Cummins, M. D., and Westrick, J. J., "Packed Column Air Stripping for Removal of Volatile Compounds," *Proceedings,* ASCE National Conference on Environmental Engineering, July, 1982.
37. Termaath, S. G., "Packed Tower Aeration for Volatile Organic Carbon Removal," paper presented at Seminar on Small Water Systems Technology, U.S. EPA, Cincinnati, Ohio, April, 1982.
38. Snoeyink, V., "Control Strategy—Adsorption Techniques," *Occurrence and Removal of Volatile Organic Chemicals from Drinking Water,* AWWA Research Foundation, 1983.
39. Snoeyink, V., "Adsorption as a Treatment Process for Organic Contaminant Removal from Groundwater," *Proceedings,* AWWA Seminar: Organic Chemical Contaminants in Groundwater: Transport and Removal, 101st Annual National AWWA Conference, St. Louis, Mo., June, 1981.
40. Dyksen, J. E., and Hess, A. F., "Alternatives for Controlling Organics in Groundwater Supplies," *J.AWWA,* August, 1982.
41. Symons, J. M., Carswell, J. K., Demarco, J., and Love, O. T., "Removal of Organic Contaminants from Drinking Water Using Techniques Other than Granular Activated Carbon Alone—A Progress Report," Drinking Water Research Division, U.S. EPA, Cincinnati, Ohio, May, 1979.

Chapter 16
Asbestos Fiber Removal

INTRODUCTION

Asbestos minerals have been used beneficially for a long time. However, the discovery of asbestos fibers in drinking waters has raised the question of health effects of ingested asbestos, and has resulted in techniques to reduce the concentration of asbestos in drinking water. The health effects of ingested asbestos are unclear, and are the subject of real and emotional public awareness. As a result of public health concern, there has been development of effective techniques to reduce asbestos concentrations in drinking water.

To date, a maximum contaminant level (MCL) has not been established for asbestos, and the American Water Works Association (AWWA) believes that regulation of asbestos is unwarranted at this time.[1] However, studies continue to be undertaken to determine whether a definite causal relationship between asbestos in drinking water and adverse health effects can be demonstrated. Should such a link be found, then the information presented in this chapter should prove useful in devising strategies to reduce asbestos levels.

DISCOVERY OF ASBESTOS IN DRINKING WATER

The occurrence of asbestos in drinking water was brought to national attention in 1973 when asbestos fibers were found in the water supply for Duluth, Minnesota and other communities on the north shore of Lake Superior.[2-4] In a significant legal case involving pollution control, the fibers were shown to have originated at a taconite processing plant at Silver Bay, Minnesota.[5]

Since 1973, asbestos fibers have been found in the water supplies of communities throughout the United States; and because of its many uses, asbestos is a nearly ubiquitous substance in our environment.[6-9] A survey of 426 water supplies in 48 states, Puerto Rico, the Virgin Islands, and the District of Columbia revealed that 71 percent of the systems reported asbestos concentrations greater than the detection level of 0.01 million fibers per liter (MFL) (Table 16–1).[10] Reported asbestos concentrations for selected communities are presented in Table 16–2.

Table 16-1. Distribution of Reported Asbestos Concentrations in Drinking Water.*,**

HIGHEST REPORTED ASBESTOS CONCENTRATION (MILLIONS OF FIBERS/LITER)	NUMBER OF SUPPLIES	PERCENTAGE
Below detectable limits	121	28.4
Less than one	221	51.9
One to ten	39	9.2
Greater than ten	45	10.6
Total	426	100.1

* Reference 10.
** Data taken from 426 supplies in 48 states, Puerto Rico, the Virgin Islands, and the District of Columbia.

Table 16-2. Reported Asbestos Concentrations for Selected Communities.

COMMUNITY	TYPE OF ASBESTOS	RANGE OF ASBESTOS CONCENTRATION, MILLIONS OF FIBERS/LITER
Los Angeles, CA	Chrysotile	BDL*-1,900
San Francisco, CA	Chrysotile	BDL-130
New Milford, CT	—	BDL
Lakeland, FL	Chrysotile	0.2-17
Chicago, IL	Chrysotile and amphibole	5-45
Duluth, MN	Amphibole	10-1,000
Socorro, NM	Chrysotile	153-2,190
Northridge, OH	—	BDL
Paint, PA	Chrysotile	4-19
Bishopville, SC	Chrysotile	0.4-500
Everett, WA	Chrysotile	35-230
Seattle, WA	Chrysotile and amphibole	BDL-25

* BDL = Below detection limit.

HEALTH CONCERNS

It is generally accepted that exposure to inhaled asbestos fibers significantly increases the risk of lung cancer and pleural mesothelioma, especially among smokers. Higher than expected incidence rates of peritoneal mesothelioma and gastric, kidney, and colon cancer among workers occupationally exposed to airborne asbestos suggest that ingested asbestos may also be a hazard because many inhaled asbestos particles are cleared from the respiratory tract and swallowed.[11] This recognition of the carcinogenic potential of ingested asbestos gives cause for concern about the presence of asbestos in drinking water.

The health effects of ingested asbestos are being investigated in animal feeding studies.[12-16] To date, these studies are inconclusive, although there is evidence that asbestos fibers can penetrate the gastrointestinal tract and migrate to other parts of the body.[17,18] Very little information exists regarding the means by which asbestos fibers may cause disease, or the ingested dosage level associated with increased cancer incidence.

Several epidemiological studies have also been conducted to determine whether a causal link exists between asbestos in drinking water and cancer.[19-23] For the most part, these studies have not found such a link, although one study found a weak but statistically significant association.[23]

NATURE OF ASBESTOS

Asbestos is a generic term that includes a number of fibrous silicate minerals that can generally be classified as belonging to either the serpentine or the amphibole group. A classification of the various asbestos minerals is contained in Table 16–3. It is important to be aware that the specific mineral names imply a chemical discreteness that does not exist, because within each mineral series the percent chemical composition can vary.[24] The relatively nonspecific definition of asbestos and geographic differences in the types present make identification and quantitation in water difficult.

The fibrous form of serpentine is chrysotile, a layered silicate that is the magnesium analog of the clay kaolinite. Chrysotile consists of parallel bundles of submicroscopic fibers generally 0.03 to 0.10 micron in diameter and less than 10 microns in length. These serpentine fibers carry a positive charge. The chrysotile fiber accounts for 90 percent of the world consumption of asbestos and is the most common form found in aquatic environments.[25]

Table 16–3. Classification of Asbestos Minerals.

GROUP	CHEMICAL FORMULA
Serpentine Group	
Chrysotile	$Mg_3Si_{12}O_5(OH)_4$
Amphibole Group	
Actinolite	$Ca_2(Mg,Fe)_5Si_8O_{22}(OH,F)_2$
Amosite	$Mg_3Fe_{11}Si_{16}O_{44}(OH)_4$
Anthophyllite	$(Mg,Fe)_7Si_8O_{22}(OH,F)_2$
Crocidolite	$Fe_5Na_2Si_8O_{22}(OH)_2$
Cummingtonite	$(Mg,Fe)_7Si_8O_{22}(OH)_2$
Glaucophane	$Na_2Mg_3Al_2Si_8O_{22}(OH)_2$
Tremolite	$Ca_2Mg_5Si_8O_{22}(OH,F)_2$

The amphibole forms of asbestos include actinolite, amosite, anthophyllite, crocidolite, cummingtonite, glaucophane, and tremolite. These fibers range from 0.1 to 100 microns in length and are negatively charged.

SOURCES OF ASBESTOS

Asbestos fibers enter surface waters from both natural and anthropogenic sources. Large deposits of chrysotile and amphibole are found throughout North America (Fig. 16–1), and leaching from asbestos mineral deposits in contact with surface water constitutes the major natural source. Studies have also found that precipitation may result in surface water contamination through scavenging of airborne asbestos.[27]

Anthropogenic sources of asbestos fibers in water include mining operations and use of products containing asbestos. Chrysotile and amphibole are commercially mined to make over 3,000 products, including filters, auto brake and clutch linings, floor covering, paper products, textiles, gasket materials, and asbestos cement pipe. However, asbestos may be facing ultimate extinction in the marketplace under a sweeping ban contemplated by the EPA.[28,29]

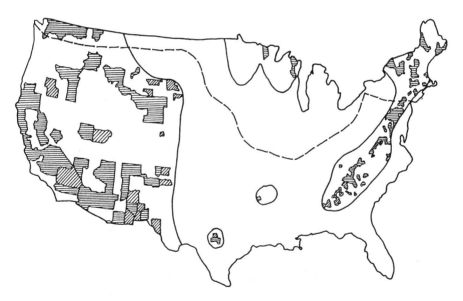

Fig. 16–1. Locations of asbestos sources in the United States. Diagonally lined areas are counties where amphibole asbestos fibers have been reported. Horizontally lined areas are counties where chrysotile or serpentine rock or both have been reported. (Where amphiboles and serpentine are found in a county, it is horizontally lined.) Solid lines surround those areas where fiber-bearing rocks might exist. The dashed line shows lowest limit of glacial activity. Rocks not native to the area can be found north of this line. (Adapted from Reference 26.)

This proposed rule would initially set production limits on asbestos-containing products and eventually phase out such products completely.

ASBESTOS CEMENT PIPE

It is estimated that there are over 200,000 miles (322,000 km) of asbestos cement (A-C) pipe in use in the United States. For this reason, A-C pipe has been the subject of numerous studies and considerable controversy over whether use of A-C for potable water distribution constitutes a health hazard. The AWWA has adopted the position that use of A-C pipe is safe. However, several reports indicate there may be reason for concern.[10,11,30] To date, there is no irrefutable evidence to support these concerns, but further research is needed.

Utilities using materials containing asbestos in their water systems should follow work practices for handling, cutting, machining, and tapping of such materials as recommended by the AWWA.[31] In addition, utilities should adopt goals for corrosivity that minimize potential fiber detachment from walls of A-C pipe. The following are conditions indicative of situations in which water is most likely to be attacking A-C pipe:[32]

- An initial aggressive index below 11. The aggressive index (AI) is given as:

$$AI = pH + (AH) \qquad (16\text{--}1)$$

where:

A = total alkalinity in mg/l as $CaCO_3$
H = calcium concentration in mg/l as $CaCO_3$

- A significant increase in pH and the concentration of calcium occurring as the water flows through the pipe.
- Water not containing iron, manganese, or similar metals.
- Significant asbestos fiber counts being consistently found in representative water samples collected at locations where:
  1. The flow is sufficient to clean the pipe of debris.
  2. Drilling and tapping of the pipe has not been performed nearby or during the sampling period.

The chloride, sulfate, and orthophosphate salts of zinc have been tested and found to provide substantial protection for A-C pipe when they are added to the water at the proper concentration and pH ranges, and when those chemicals are maintained throughout the distribution system.[33]

MONITORING TECHNIQUES

A number of techniques for monitoring the fiber content in water have been considered and tried in asbestos filtration research programs.[34] These techniques are briefly described in the following paragraphs.

Electron Microscope

The transmission electron microscope (TEM) method provides positive identification of asbestos fibers, and the EPA has proposed its use in counting asbestos fibers.[35] Electron microscopy is both slow and expensive, however. Sample analysis generally requires more than one working day, and it is not uncommon to wait for results for 3 or 4 weeks. Consequently, TEM analyses are not useful for monitoring ongoing operations. They are relegated to reviewing past performance of treatment strategies for asbestos removal.

X-Ray Diffraction

X-ray diffraction measures the concentration of asbestos mineral.[3] The method is quicker than, and costs a fraction of the electron microscope method. The technique measures the mass of the material and not the number of fibers, which is a disadvantage. Large pieces of amphibole may greatly increase the mass detected without there being a greater number of fibers. X-ray diffraction is not used to determine the mass of chrysotile in water supplies because of many technical problems.[34]

Laser-Illuminated Optical Detection

A laser-illuminated optical detector (LIOD) identifies particulates by using the differences in the scattering signatures of different types of particulates.[36] It is effective at detecting fiber levels as low as 0.05 MFL. With results obtainable in minutes, the LIOD provides a rapid estimate of the number of fibers in water. It also can identify the particle types. Particles must be present in the calibration sample in order for this method to be effective.[34]

Turbidity Measurement

Turbidity is only a gross reflection of the potential presence of asbestos, and does not count asbestos particles. A given turbidity reading cannot be related to a fiber count. Also, asbestos fibers in concentrations found in most raw waters and nearly all filtered waters are too low to cause a turbidimeter to register any light scatter.

An advantage of turbidity monitoring is that this measurement has been performed by water utilities for many years. Water utility personnel are familiar with turbidimeters and experienced in their use. Turbidimeters also are inexpensive. When continuous turbidimeters are used, changes in filtered water turbidity can be immediately detected, enabling a quick response. Responding to changes in treated water turbidity is the proper approach for the use of turbidity measurements in filtration plants operated for asbestos fiber removal.

TURBIDITY AND ASBESTOS CONCENTRATION CORRELATIONS

Turbidity and asbestos counts in raw water typically correlate poorly. However, the correlation between filtered water turbidities and asbestos counts generally is much better although exceptions to this finding have been observed.[38] Figure 16–2 shows typical observed relationships between turbidity and asbestos counts in filtered water. As this figure indicates, effective asbestos removal is usually associated with a finished water turbidity of 0.1 NTU or less. The correlation between turbidity and chrysotile fiber counts is generally poorer than that between turbidity and amphibole fiber counts.

The turbidity–fiber relationship is comparable to the coliform–pathogen relationship.[34] The presence of coliforms signals the possibility that pathogens might be present; however, there is no other connection between coliforms and pathogens. When raw water contains asbestos fibers, a rise in filtered water turbidity probably indicates an increase in filtered water fiber count. However, the fiber count cannot be estimated on the basis of the abnormally high turbidity reading. The relationship between turbidity and asbestos fiber removal efficiency is shown in Fig. 16–3, which indicates the percentage of samples of a given fiber removal efficiency having turbidity equal to or less than a certain value. Figure 16–3 clearly shows that higher filtered water turbidities are associated with lower percentages of fiber removal.

REMOVAL OF ASBESTOS FIBERS

Effective removal of asbestos fibers generally can be achieved with minor variations in conventional or direct filtration treatment techniques. These variations include optimization of pretreatment and filtration, and implementation of comprehensive process monitoring. The process sequence typically used for asbestos removal is shown in Fig. 16–4.

As a general rule, when asbestos removal is required, filter plants should be designed and operated to achieve an effluent turbidity of 0.10 NTU or lower. This is considerably lower than the 1 NTU MCL set forth in EPA's Interim Primary Drinking Water Regulations, but is similar to the AWWA's water quality goal of 0.1 NTU.[39,40] The 0.10 NTU goal is based on experience

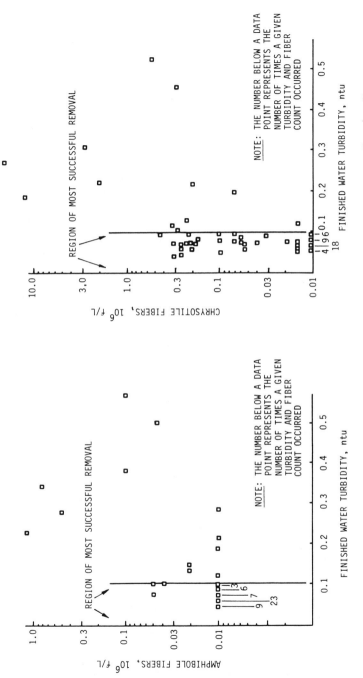

Fig. 16–2. Amphibole count and chrysotile count vs. finished water turbidity, Seattle pilot plant.[34]

559

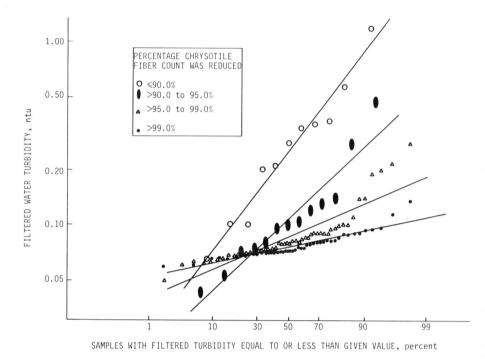

Fig. 16–3. Relationship of turbidity and chrysotile removal by granular media filtration, Seattle pilot plant.[34]

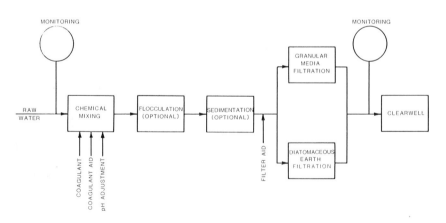

Fig. 16–4. Typical process sequence for asbestos removal.

in treating water having low turbidity—generally 1 NTU and nearly always less than 5 NTU.

When the 0.10 NTU goal has been achieved, it has generally been found that amphibole fibers can be removed consistently down to the detection limit. Chrysotile is more difficult to remove than amphibole and results are more variable.

At present, there is little experience in reducing asbestos concentrations in turbid waters. It appears that turbid waters may be easier to treat than clear waters because of the presence of a large number of particles that aid particle agglomeration. Limited data suggest that if the raw water turbidity is greater than 1 NTU, asbestos fiber removal can be accomplished when filtered water turbidity exceeds 0.10 NTU.[34] However, too few data and too much uncertainty presently exist to suggest an operating goal other than 0.10 NTU.

Both the processes that condition raw water for filtration and the filtration process itself play an important role in the removal of asbestos fibers. For a filtration plant to be consistently effective in removing asbestos fibers, each of the unit processes in the clarification process must be operated effectively. One inadequate step in the treatment train can result in higher filtered water turbidity and fiber counts. In such a system, the role of the operator cannot be overemphasized.

Removal Mechanism

Filtration can be very effective in removing asbestos fibers. The mechanism by which fibers are removed is uncertain. Straining by the granular media is unlikely because of the very small size of the fibers; sedimentation or impingement on the media is more probable. Fibers should be destabilized in order to adhere to the filter media or to fibers previously removed.[41]

In the pH range of 6 to 8, it is believed that the surface charge of amphibole fibers is negative and chrysotile fibers are positive, based upon the measurable charge of highly concentrated suspensions of fibers. This information would indicate that anionic polymers would be effective for chrysotile fiber removal, and cationic polymers would not. In fact, the use of cationic polymers or a combination of alum and cationic polymers has been effective for the removal of both chrysotile and amphibole fibers. Successful pilot plant studies have demonstrated that the use of anionic polymers is not necessary to filter out chrysotile asbestos.[41,42] Although the reason for this is not certain, it appears that chrysotile fibers attach to, or become associated with, clay particles, bacteria, and other small negatively charged particles in the raw water.[41] Consequently, conditioning and filtering water for the removal of the latter particles results in the concurrent removal of chrysotile.

PROCESS CONSIDERATIONS

General

Until the state of knowledge regarding asbestos removal is better developed, it is recommended that pilot plant studies be conducted prior to designing a filter plant in order to optimize filtration rates, coagulant dose and type, the need for flocculation and sedimentation, and the filtration mode. The plant should be designed to give the operator as much operating flexibility as possible, and continuous and extensive monitoring should be provided in order to detect changes in finished water quality as soon as they occur.

Chemical Conditioning

Proper conditioning of raw water prior to filtration is essential for effective fiber removal. Conditioning must result in complete destabilization of particles in the raw water and formation of floc that is sufficiently strong to be retained by the filter media. Alum and cationic polymers have been successfully used as coagulants for asbestos fiber removal, both alone and in combination. Polymers are often used as filter aids.

An important aspect of treatment process control is maintaining the proper dosage of treatment chemicals. Polymers generally have a range of concentration that gives optimum efficiency, and reduced effectiveness can result from either an overdose or an underdose of coagulant aid or filter aid. Inorganic coagulants must be used in doses adequate to assure that particles are destabilized prior to filtration. Available data suggest that asbestos fibers are the last particles to be destabilized by aluminum hydroxide floc formation.[42] Consequently, filtration plant operators trying to maximize asbestos fiber removal should not try to minimize application of chemical coagulant.

Another important consideration is maintaining the proper pH for optimum particle removal. Experience with alum coagulation has shown that optimum floc formation and fiber removal occur within a narrow pH range. Plants should be designed with convenient pH monitoring in mind, and means for pH control (both acid and base) should be provided so that the optimum pH value can be maintained at all times.

Jar tests are often useful for determining optimum chemical dosages. However, a floc that has good filtration characteristics is not necessarily visible to the eye. For this reason, use of a fine-sand column to filter water from the jar test is recommended to optimize chemical doses. Such a coagulation–filtration testing mechanism has been in use for many years at the Metropolitan Water District of Southern California.[43]

Chemical addition facilities must achieve thorough mixing of chemicals and must provide considerable flexibility to allow the operator to optimize treatment in the face of varying raw water quality. Use of two or three back mix or static mix units in series is a recommended design feature. This allows separate addition and dispersion of coagulant, polymer, and pH adjustment chemicals. Back mix units also allow mixing time and energy to be varied.

FLOCCULATION/SEDIMENTATION

Experience related to flocculation has been varied. The full-scale Duluth, Minnesota plant operated for the first six months without flocculation, and appeared to function well.[41] By contrast, other studies indicated that attaining low-turbidity filtered effluent was difficult when in-line filtration was used. Inclusion of flocculation in the treatment train resulted in greater water production efficiencies and a longer number of operating hours during which filtered water turbidity was 0.10 NTU or lower.[44,45]

Sedimentation has not been found necessary in the studies conducted to date. However, all these studies were conducted with clear waters that readily lent themselves to direct filtration. Treatment of more turbid waters may require sedimentation facilities.

Because the need for flocculation and sedimentation may vary at different locations, these processes should be studied in pilot plant investigations to determine whether the extra costs for flocculation will result in improved plant performance. In making the evaluation, the design engineer should be aware of the probability and magnitude of raw water quality upsets, and should provide facilities to adequately treat the water during those periods.

GRANULAR MEDIA FILTRATION

Asbestos fiber removal in excess of 99.99 percent has been achieved by granular media filtration.[41] Both dual-media and mixed-media filters have been effectively employed to remove asbestos at filtration rates of 2 to 10 gpm/sq ft (4.88 to 24.4 m/h). Media depth is typically 30 to 36 inches (0.76 to 0.91 m). Table 16–4 presents information on various types of filter media that have been successfully employed for asbestos removal. Limited data suggest that better performance is obtained from mixed-media filters than dual-media filters in the form of longer filter runs, shorter ripening times, and greater fiber removal efficiency at high filtration rates [greater than 7.5 gpm/sq ft (18.3 m/h)].[42,46] Effective asbestos fiber removal has been achieved by filters operated in both the declining rate mode and the constant rate mode.[28] An advantage of declining rate filtration is that abrupt rate changes are avoided, as such changes can result in increased finished water fiber counts, particularly when a filter is operating at high headloss.[41]

Table 16-4. Filter Media Data.*

LOCATION OF FILTERS	FILTER MATERIAL LISTED TOP TO BOTTOM	DEPTH		EFFECTIVE SIZE, mm	UNIFORMITY COEFFICIENT
		INCHES	m		
Duluth, MN	Anthracite	21	0.53	1.0–1.1	1.7
Dual Media	Sand	9	0.23	0.42–0.45	1.35–1.7
Duluth, MN	Anthracite	16.5	0.42	1.0–1.1	1.7
Mixed Media	Sand	9	0.23	0.42–0.45	1.35–1.7
	Garnet	4.5	0.11	0.21–0.32	1.8
Two Harbors, MN	Anthracite	16.5	0.42	1.0–1.1	1.7
	Sand	9	0.23	0.42–0.55	1.8
	Illmenite	4.5	0.11	0.18–0.32	1.8
Silver Bay, MN	Anthracite	15	0.38	1.16	1.43
	Sand	15	0.38	0.45	1.41
Everett, WA	Anthracite	22	0.56	0.96	1.82
	Sand	11	0.28	0.45	1.56
	Garnet	3	0.08	0.34	1.53

* Reference 41.

The release of air in filter media causes the filter structure to deteriorate, and during filter runs causes air binding and related operating problems such as shorter runs and turbidity breakthrough. It is important that the filter be designed to maintain positive pressure throughout by providing a sufficient depth of water over the top of the filter media under all operating conditions. (This subject is discussed in some detail in Chapter 10.)

Filter effluent turbidity increases may be indicative of release of captured fibers from the filter. To retain captured fibers, a rise in turbidity of 0.1 NTU from a filter that has been producing a 0.10 NTU filtered water quality is sufficient cause to backwash the filter.

DIATOMACEOUS EARTH FILTRATION

A limited amount of testing with diatomaceous earth (DE) filtration has shown this process to be effective for amphibole removal. As yet, no information is available for chrysotile removal by DE filtration.

For amphibole removal, the use of pressure filtration rather than vacuum filtration is recommended. Pressure filtration provides a large positive pressure or driving force through the filter and prevents the problem of air bubble formation encountered when vacuum filtering very cold waters.

Alum-coated diatomite has been found particularly effective for amphibole removal. Alum-coated filter media have a positive charge, and amphibole fibers in the pH range of 6 to 8 are negatively charged. This charge difference enhances the attraction of amphiboles to the diatomite filter-aid particles.[34]

Surface attachment occurs, facilitating the removal of amphibole fibers that are too small to be removed by straining.

The grade of diatomite selected should be fine enough to produce a filtered water of 0.10 NTU or lower. If the septum pores are too large to permit effective precoating with fine diatomite, a two-stage precoating process can be used, with coarse DE coated on the septum and then a second layer of fine DE placed over the first precoat layer. Only the second layer of precoat would need to be alum-coated.

MONITORING

Maintaining close control of water quality at the treatment plant is essential. Frequent or continuous monitoring of raw and treated water is strongly recommended. The most readily used quality control for filtered water is turbidity, which should be measured continuously. Effluent from each filter should be monitored so the operator can control the filters and backwash each one at the proper time. A small increase in turbidity (0.1 to 0.2 NTU) may be associated with a very large increase in asbestos fiber count in filtered water.

Experience has also shown that successful fiber removal occurs within a narrow pH range. Hence, the designer should also provide for convenient monitoring of pH and means of pH control, so that the optimum pH value can be maintained at all times.

REVIEW OF ASBESTOS REMOVAL EXPERIENCE

Duluth, Minnesota

The presence of asbestos fibers in Duluth's water supply became known in the summer of 1973. Following this discovery, Duluth and other north shore Lake Superior communities adopted a policy of reducing asbestos concentrations in their water. Extensive pilot plant studies were conducted in 1974 to determine the optimum treatment method for reducing asbestos levels. Using the information gained from these studies, a full-scale plant was designed, and became operational in 1977.[41,45]

Duluth's raw water source is Lake Superior, one of the largest bodies of fresh water in the world. Except for asbestos fibers, Lake Superior's water quality generally is excellent. Raw water turbidity at Duluth is usually less than 1 NTU and typically exceeds 10 NTU only once or twice a year when violent storms on Lake Superior stir up bottom sediments near the shore in the Duluth area.

Amphibole fiber counts in the Duluth intake typically range from 10 MFL to 100 MFL, and can exceed 1000 MFL during violent storms on the lake.

The primary source of asbestos in the lake is believed to be an iron ore beneficiation (concentration) plant in Silver Bay, Minnesota. From 1957 to 1980, this plant deposited more than 60,000 tons/day (54,420 metric tons/d) of asbestos-laden tailings into the lake.[34]

Table 16-5. Design Criteria for Duluth, Minnesota Plant.*

DESCRIPTION	CAPACITY OR CRITERIA
Design flow	30 mgd (113.6 Ml/d)
Maximum capacity	36 mgd (136.3 Ml/d)
Chemical application	9 mg/l alum
At rapid mix	0.07 mg/l 985N caustic soda
Before filter	0.006 mg/l 985N caustic soda
pH to filter	6.9 summer, 7.35 winter
Rapid mix	
Type	Turbine
Number	3 in series
Mix time at design flow	30 seconds each
Intensity, G, sec^{-1}	500
Flocculation	
Type	Horizontal paddle
Number in series	4
Floc time at design	40 minutes
Intensity, G sec^{-1}	4 to 45
Sedimentation	
Type of basin	Rectangular, horizontal flow
Theoretical detention time	140 minutes
Filtration	
Rate at design flow	4.9 gpm/sq ft (12 m/h)
Typical operating rate	3.3 gpm/sq ft (8 m/h)
Typical run, hours	
Direct filtration	22
Conventional	32
Backwash	
Rate	16.3 gpm/sq ft (39.8 m/h)
Duration, minutes	6
Surface wash rates	2.6 gpm/sq ft (6.3 m/h)
Duration	Until backwash = 13 gpm/sq ft (31.7 m/h)
Waste handling	
Discharge time	Intermittent
Holding basin volume	240,000 gallons (0.908 Ml)
Thickening time	4.8 hours
Treatment	Upflow flocculater clarifier to thicken wash water, over flow to head of plant, under flow to lagoon for natural freezing
Ultimate disposal of solids	Landfilled after freezing, thawing, and dewatering

* Reference 41.

Description of Treatment Facilities.   The unit processes incorporated into the Duluth plant allow variations of direct filtration as well as conventional treatment. The plant can be operated with rapid mixing; rapid mixing and flocculation; or rapid mixing, flocculation, and sedimentation ahead of the filters. Design criteria for the Duluth plant are presented in Table 16–5, and a schematic diagram of the plant is shown in Fig. 16–5.

In addition to the various modes of operation, the plant design allows flexibility in chemical feed application. Each of the three rapid mix sections is equipped with diffusers for polymer, coagulant, and pH adjustment chemicals. Valving is arranged so that any of the chemicals can be fed into one or more of the rapid-mix chambers. The use of three back mix units in series allows mixing time and energy to be varied. Flocculation and sedimentation are both optional processes at Duluth; both processes, or sedimentation alone, can be bypassed when water quality permits.

Three mixed-media filters and one dual-media filter were installed at Duluth to compare the media concepts. Deep filter boxes with a water depth of 7 feet (2.13 m) over the top of the filter media were designed to minimize air binding. The filters can be operated either under a constant head or under a fixed flow rate. Continuous turbidity monitoring is provided at each filter.

Wash water and sludge are stored in sludge lagoons and frozen naturally in the winter. After the spring thaw, supernatant is returned to the head of the plant, and the dewatered solids remain in the lagoons. When enough solids accumulate, they will be removed and buried in a landfill.

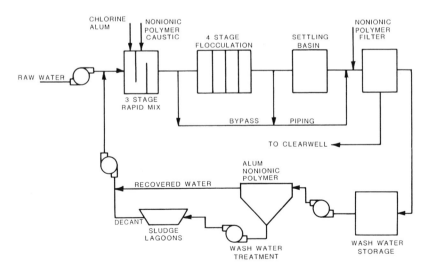

Fig. 16–5. Flow diagram for filtration plant at Duluth, showing treatment train options.

Operating Results.   The Duluth filtration plant has been operated specifically for asbestos removal since 1977. Monitoring data are shown in Fig. 16–6. Typically, filtered water has a turbidity of 0.05 NTU or lower, and amphibole fiber counts are below 0.10 MFL. Filtered water quality is not influenced greatly by short-term increases in raw water turbidity or amphibole fiber concentration.

Alum is added as the primary coagulant for the removal of turbidity, suspended solids, and asbestos fibers, and has proved to be very effective for the removal of asbestos fibers at Duluth. Optimum removal occurs in a pH range of 7.00 to 7.20 measured over the filter media. Within this range, the proper floc is formed. Dosage rates are determined by raw water turbidity; a higher influent turbidity represents higher asbestos fiber concentrations and requires an increase in the alum feed rate. A nonionic polymer is used as a coagulant and filtering aid; it strengthens the floc and decreases the filter ripening time. The dosage rate is determined by the alum dose and by filter headloss buildup. Sodium hydroxide can be added for pH control.

The filter plant was operated in the direct filtration mode for two periods totaling approximately seven months: January 1, 1977 to June 28, 1977, and November 7 through 28, 1977. From July 1977 to May 15, 1979, excluding November, 1977, the plant was operated in the conventional treatment mode. Discounting the first period when fiber counts in plant effluent were 0.15 MFL because of plant start-up problems and personnel inexperience, a comparison of the second period of direct filtration with the conventional mode results indicates no significant difference in final effluent fiber counts. Direct filtration resulted in fiber counts of 0.05 MFL, and conventional filtration resulted in counts of 0.03 MFL.

With direct filtration, the filters generally require backwashing after 20 to 24 hours of operation. A slightly longer filter run can be obtained with conventional treatment operation. The filters are washed at 8.0 feet (2.44 m) of headloss. Turbidity breakthrough under normal operating conditions does not occur before 8.0 feet (2.44 m) of headloss.

The maximum allowable effluent turbidity is 0.1 NTU. At or before this point, the filter is backwashed. Because the normal effluent turbidity is 0.04 NTU, a filter is backwashed at 0.05 NTU if an upward trend exists.

At Duluth, samples of clarified washwater and lagoon supernatant were analyzed for asbestos because these liquid streams were recycled through the plant. One lagoon decant sample contained 64 MFL. Three samples of clarifier overflow had fiber counts ranging from 27 MFL to 100 MFL. These values were similar to the amphibole fiber counts in raw lake water at that time. These data and the consistently low fiber counts in the filtered water for the first 3½ years of operation at Duluth indicated that recycling did not affect the water quality when the plant was operated properly.

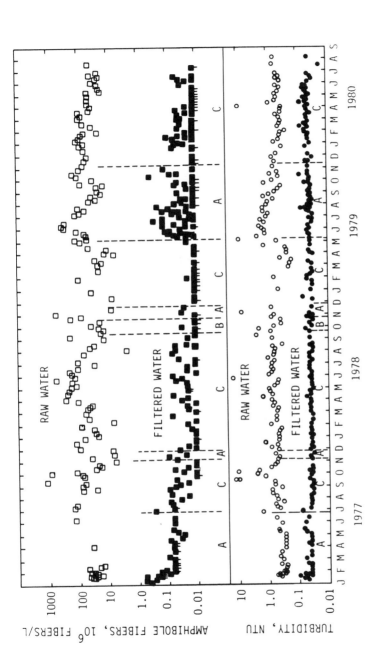

Fig. 16-6. Raw and filtered water turbidity and asbestos monitoring data at Duluth.[41]

SAMPLES BELOW DETECTION LIMIT (BDL) ARE SHOWN AT THE DETECTION LIMIT ■

TREATMENT TRAIN

A. RAPID MIX, FILTRATION
B. RAPID MIX, FLOCCULATION, FILTRATION
C. RAPID MIX, FLOCCULATION, SEDIMENTATION, FILTRATION

Seattle, Washington

In 1975, the EPA discovered naturally occuring asbestos fibers in the Tolt River, a 100-mgd (378.5-Ml/d) source of water for the City of Seattle.[41-44] Both amphibole and chrysotile fibers were discovered. Although earlier research had been conducted on amphibole fiber removal, little was known about removal methods for chrysotile. To reduce lead time should fiber removal become necessary, the Seattle Water Department conducted pilot studies to develop methods for removal of these contaminants.

The Tolt water supply is a high-quality source of water originating from rainfall and snowmelt runoff in the Cascade Mountains. It is soft (9 mg/l hardness as $CaCO_3$), has little buffering capacity (5 mg/l alkalinity as $CaCO_3$), has a low pH (6.7), and currently meets the MCL for turbidity without filtration; however, turbidity does exceed the desirable goal of 1.0 NTU on a seasonal basis. The turbidity has not exceeded the 5.0 NTU MCL since the National Interim Primary Drinking Water Regulations became effective in June, 1977.

Amphibole fiber counts range from less than 0.4 MFL to 5.7 MFL, while chrysotile fiber counts range from 1.2 MFL to 26 MFL. Both amphibole and chrysotile counts fluctuate with the season and appear to be related to raw water turbidity.

Description of Facilities. The pilot study was conducted using a modified 20-gpm (1.26-l/s) Waterboy package treatment plant. Because of the low raw water turbidities in the Tolt water supply, direct filtration techniques were applied to treat the water. Three-stage static mixers were used to blend chemicals with the raw water, and flocculation was employed prior to granular media filtration. Three chemical feed modes were employed: alum coagulation; alum and cationic polymer; and cationic polymer alone. Both dual- and mixed-media filters were employed for comparison purposes.

Operating Results. The Tolt study found that finished water turbidities could consistently be reduced to less than 0.10 NTU, and effective removal of both amphibole and chrysotile fibers could be achieved with granular media filtration preceded by any of the chemical pretreatment options presented in Table 16-6.

Amphibole fibers were consistently removed down to the detection limit of 0.01 MFL, and removal efficiencies were in excess of 98 percent. Turbidity spikes and poor destabilization were consistently associated with high amphibole counts in the finished water.

Chrysotile fibers were more difficult to remove than amphibole; nevertheless, when finished water turbidities were $\leq$ 0.10 NTU, 50 percent of the time chrysotile counts were $\leq 0.02 \times 10^6$ fibers per liter. When turbidities were $\geq$ 0.10 NTU, 50 percent of the time chrysotile was $\geq 0.27 \times 10^6$

Table 16–6. Optimal Chemical Dosages Utilized in Tolt Asbestiform Study.*

COAGULANT		ALUM AND CATIONIC POLYMER		CATIONIC POLYMER ALONE	
CHEMICAL	DOSAGE	CHEMICAL	DOSAGE	CHEMICAL	DOSAGE
Aluminum sulfate (alum)	7–10 mg/l	Alum	3–5 mg/l	Cationic polymer	3 mg/l
Lime [Ca(OH)$_2$], pH range = 6.1–6.7	1–4 mg/l	Cationic polymer	2 mg/l		
Filter aid (non-ionic or anionic polymer)	0.02–0.25 mg/l	Filter aid (non-ionic or anionic polymer)	0.1–0.3 mg/l	Filter aid (non-ionic or anionic polymer)	**

* Reference 42.
** Not reported.

fibers per liter. This indicates that there was over ten times more asbestos present in the finished water when turbidity was ≥ 0.10 NTU compared to times when turbidity was ≤ 0.10 NTU. Turbidity spikes, filter breakthrough, and poor destabilization were consistently associated with high finished water chrysotile counts.

When alum coagulation was employed, pH control was found to be very critical. For effective destabilization to occur, pH had to be maintained between 6.1 and 6.7.

The study found flocculation to be a necessary unit process, as water production efficiencies and final water quality decreased significantly when flocculation was omitted in the treatment sequence. This was particularly true when the raw water turbidity exceeded 1.5 NTU and water temperatures were cold [41 to 42.8°F (5 to 6°C)].

Filter loading rates up to 10 gpm/ft² (24.4 m/h) were found to be effective in reducing finished water turbidity to ≤ 0.10 NTU, and reducing asbestos fiber concentrations to nondetectable levels. Both dual- and mixed-media filters were able to meet water quality and process goals, and they had similar efficiencies at loading rates between 5.5 and 7.5 gpm/sq ft (13.4 and 18.3 m/h). Below 5.5 gpm/sq ft (13.4 m/h), the coarse coal dual-media filter was more efficient, and above 7.5 gpm/sq ft, (18.3 m/h) the mixed-media filter demonstrated a higher efficiency.

Metropolitan Water District of Southern California

In 1980, the Metropolitan Water District of Southern California discovered the presence of significant concentrations of chrysotile asbestos in its raw water supply.[38] This finding prompted Metropolitan to conduct a study that attempted to optimize removal of asbestos particles at five large treatment plants receiving low to high influent concentrations of chrysotile asbestos fibers.

Approximately one-half of Metropolitan's water supply [700,000 acre-feet (863,450,000 m³)] is imported Colorado River Water (CRW). The remaining supply is Northern California water delivered through the State Water Project (SWP). Metropolitan has contracted to receive more than 2 million acre-feet (2,467 Gm³) of SWP water annually.

Asbestos sampling of these water resources indicated that chrysotile levels in CRW were highly variable and ranged from below detectable limits to 13 MFL. By contrast, chrysotile fiber concentrations in SWP were found to range from 230 MFL to 1,900 MFL. The high concentrations in SWP were caused by runoff from deposits of serpentine rock throughout the state and by a massive contamination problem in the California aqueduct near Coalinga. Efforts have been made to correct the latter problem. No evidence of amphibole asbestos fibers was found in either source of supply. Studies have shown that considerable reductions in asbestos concentrations occur in the large storage reservoirs at the end of the SWP.

Description of Facilities. Metropolitan has five water filtration plants that were built over a 38-year period: Weymouth (1941), Diemer (1964), Jensen (1972), Skinner (1976), and Mills (1979). These plants treat 1.9 × 10⁹ gal (7.19 × 10⁶ m³) of water at peak flows. Two of the plants, Weymouth and Diemer, receive a blend of water from Lake Silverwood (SWP) and Lake Mathews (CRW). The Skinner plant receives a blend of SWP and CRW from Lake Skinner. The Jensen and Mills plants are supplied with SWP from Castaic and Silverwood lakes, respectively.

The five plants differ in the details of their layout and equipment but are similar in the arrangement of their unit processes. Figure 16–7 describes the general treatment scheme common to all the plants. Rapid mixing is accomplished at the Weymouth, Diemer, and Jensen plants by vertical turbine mixers, and at the Skinner and Mills plants by multijet slide gates. The Weymouth plant uses constant-rate/constant-head, dual-media filters; the Diemer and Jensen plants use constant-rate/rising-head (proportional to head-loss), dual-media filters; and the Skinner and Mills plants use constant-rate/rising-head, dual-media filters, and the Skinner and Mills plants use constant-rate/rising-head, dual-media filters. The chemical treatment used at the plants includes pre-, intermediate-, and post-chlorination; coagulation with alum and cationic polymer; and pH stabilization with caustic soda. Table 16–7 describes the basic design and operational information for the treatment plants. Because of low influent turbidities, all five plants are operated as direct filtration plants even though sedimentation basins are available for use.

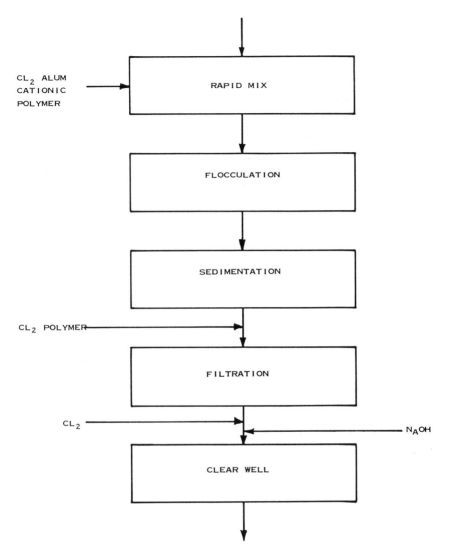

Fig. 16–7. General treatment schematic for Metropolitan plants.[38]

To optimize treatment at the five plants, bench-scale coagulation–filtration tests were run to determine optimal effluent chemical dosages. A goal of 0.1 NTU was set for turbidity during optimized treatment. Optimized treatment was then varied with standard treatment from December 10, 1980 through May 27, 1981. All of the treatment plants operated routinely through the tests with the exception of the Mills Plant, which operated only 8 hours per day, several days per week.

Table 16-7. Design and Operational Information for Metropolitan's Five Filtration Plants.*

PARAMETER	WEYMOUTH	DIEMER	JENSEN	SKINNER	MILLS
Capacity					
Design, mgd (Ml/d)	400 (1,510)	400 (1,510)	550 (2,080)	240 (900)	75 (280)
Peak, mgd (Ml/d)	520 (1,960)	520 (1,960)	550 (2,080)	290 (1,090)	75 (280)
Flocculation time at rated capacity					
Average, minutes	30	28	11	15	6
Sedimentation time at rated capacity					
Average, minutes	126	120	45	56	61
Filter media depth**					
Coal, inches (mm)	20 (500) 12 (300)	12 (300) 20 (500)	20 (500)	20 (500)	20 (500)
Sand, inches (mm)	8 (200) 9 (220)	12 (300) 8 (200)	8 (200)	8 (200)	8 (200)
Filtration rate at rated capacity					
Average, gpm/sq ft (m/h)	3 (7.32) 2 (3)	3 (7.32) 2 (3)	8 (19.52) 5 (8)	8 (19.52) 5 (8)	8 (19.52) 5 (8)
Actual filtration rate†					
Range, mm/s (gpm/sq ft)	2.6–3.1 (6.3–7.6)	3.0–3.7 (7.3–9.1)	1.7–3.7 (4.2–9.1)	3.6–7.7 (8.8–18.8)	3.7–5.2 (9.0–12.7)
Amount of SPW in blend					
Range, percent	30–55	31–54	100	17–43	100
Turbidity* range					
Influent, NTU	1.3–2.3	1.3–2.0	0.8–7.5	2.1–6.8	0.9–3.3
Effluent, NTU	0.08–0.12	0.07–0.11	0.08–0.11	0.10–0.34	0.08–0.21
Range of chemical dosages during optimized treatment					
Chlorine, mg/l	0.8–2.8	1.0–2.0	2.3–3.6	2.9–3.3	4.3–5.8
Alum, mg/l	3.8–9.9	2.5–6.6	0.0–5.0	2.6–6.4	5.5–10.5
Cationic polymer, mg/l	1.4–3.0	1.6–2.4	0.5–1.9	0.3–3.1	1.5–3.2
Caustic soda, mg/l	3.5–6.8	3.5–5.3	3.7–5.3	2.2–5.3	3.6–6.3

* Reference 38.
** Two sets of coal and sand depth are shown for Weymouth and Diemer because half of each plant was completed at a different time from a different design.
† During optimized treatment.

Operating Results. A summary of the filtration plants' performances during optimized treatment is presented in Table 16–8. In general, optimized treatment resulted in lower effluent asbestos levels than standard treatment, but the levels were highly variable and higher than expected by the investigators.

Table 16–8. Summary of Filtration Plant Performances for
Optimized Asbestos Removal.*

PARAMETER	WEYMOUTH	DIEMER	JENSEN	SKINNER	MILLS
Asbestos in influent, MFL					
Mean	690	650	5.7	1.9	770
Median	730	600	2.6	1.1	690
Maximum	1,300	1,300	45	6.4	1,900
Minimum	230	280	9.19	BDL**	230
Asbestos in effluent, MFL					
Mean	8.3	35	0.41	0.31	6.7
Median	5.2	2.5	BDL	0.13	0.86
Maximum	31	200	5.6	2.3	5.8
Minimum	0.37	0.25	BDL	BDL	0.11
Effluent samples with an asbestos level ≤ 1 MFL, percent	7	33	93	93	5.3
Mean removal of asbestos, percent	98.89	94.74			99.33
Time the turbidity was ≤ 0.10 NTU, percent	75	86	43	4	33
Median effluent turbidity for optimized treatment, NTU	0.07	0.09	0.12	0.18	0.12

* Reference 38.
** BDL = Below detection limit.

Asbestos fiber concentrations in the five treatment plant influents were different because of the different sources of supply, the presence or absence of source water reservoirs, and variations in the blends in SWP and CRW. Reservoirs were found to be effective in reducing asbestos concentrations.

The ability of each plant to remove asbestos fibers was dependent on the asbestos concentration in the influent and on the efficiencies of the individual treatment processes. In two cases (Weymouth and Diemer plants), a high percentage removal of asbestos fibers did not result in levels as low as those at a plant that was less effective in particle removal (Jensen Plant).

The 0.10 NTU goal for filter effluent turbidity was not achieved on a consistent basis by all five treatment plants. For example, the Skinner plant had significant problems in meeting the goal. The reason for this inability to meet the goal is not clear, but it is thought to have been a problem with balancing flows and with a lack of continuous effluent turbidity monitoring equipment.

Attainment of the 0.10 NTU level did not guarantee a low effluent asbestos concentration (< 1 MFL).

REFERENCES

1. American Water Works Association, "Council Approves Asbestos Policy Statement," *AWWA Mainstream,* November, 1983.
2. Peterson, D. L., "The Duluth Experience—Asbestos, Water, and the Public," *J.AWWA,* January, 1978.
3. Cook, P. M., Glass, G. E., and Tucker, J. H., "Asbestiform Amphibole Minerals; Detection and Measurement of High Concentrations in Municipal Water Supplies," *Science* 185:853–855, September 6, 1974.
4. Kay, G. H., "Asbestos in Drinking Water," *J.AWWA,* September, 1974.
5. United States of America et al. v. Reserve Mining Co. et al. (file No. 5–72, Civ. 19-D Minn.).
6. McMillan, L. M., Stout, R. G., and Willey, B. F., "Asbestos in Raw and Treated Water: An Electron Microscope Study," *Environmental Science and Technology,* April, 1977.
7. Oliver, T., and Murr, L. E., "The Electron Microscope Study of Asbestiform Fiber Concentrations in Rio Grande Valley Water Supplies," *J.AWWA,* August, 1977.
8. Logsdon, G. S., *Water Filtration for Asbestos Fiber Removal,* EPA-600/2–79–206, Cincinnati, Ohio, 1979.
9. Stewart, I., *Asbestos Fibers in Natural Runoff and Discharges from Sources Manufacturing Asbestos Products,* EPA-560/6–76–018, 1976.
10. Millette, J. R., Pansing, M. F., and Boone, R. L., "Asbestos Cement Materials Used in Water Supply," *Water Engineering and Management,* March, 1981.
11. McCabe, L. J., and Millette, J. R., "Health Effects and Prevalence of Asbestos Fibers in Drinking Water," *Proceedings,* 1979 Annual Conference, AWWA, Part 2, 1979.
12. Moore, J. A., "NIEHS Oral Asbestos Study," *Proceedings,* Workshop on Asbestos: Definitions and Measurement Methods, NBS Spec. Publ. 506, 1978.
13. Hallenbeck, W. H., and Patel-Mandlik, K., *Fate of Ingested Chrysotile Asbestos Fiber in the Newborn Baboon,* Environmental Health Effects Research Report, Office of Research and Development, EPA-600/1–78–069, 1978.
14. Barry, B., and Fisher, H., "Effects of Asbestos Fiber on Mammalian Cells in Culture," 33rd Annual *Proceedings,* Electron Microscope Society America, Las Vegas, G. W. Bailey, editor (supported by EPA grant R-801379), 1975.
15. Hart, R., et al., *Effects of Selected Asbestos Fiber on Cellular and Molecular Parameters,* Environmental Health Effects Research Report, Office of Research and Development, EPA-600/1–79–021, 1979.
16. Reiss, B., Weisburger, J. H., and Williams, G. M., *Asbestos and Gastrointestinal Cancer: Cell Culture Studies,* Environmental Health Effects Research Report, Office of Research and Development, EPA 600–1–79–023, 1979.
17. Cook, P. M., and Olson, O. F., "Ingested Mineral Fibers: Elimination in Human Urine," *Science* 204(13):195–198, 1979.
18. Cunningham, H. M., Moodie, C. A., Laurence, G. A., and Pontefract, R. D., "Chronic Effects of Ingested Asbestos in Rats," *Arch. Environmental Contamination and Toxicology* 6:507–513, 1977.
19. Craun, G. F., Millette, J. R., Woodhull, R. S., and Laiuppa, R., "Exposure to Asbestos Fibers in Water Distribution Systems," *Proceedings,* 97th AWWA Conference, May, 1977, Denver, Colo., 1977.
20. Mason, T. J., et al., "Asbestos-like fibers in Duluth Water Supply. Relation to Cancer Mortality," *Journal American Medical Association* 228:1019–1020, 1974.
21. Levy, B. S., et al., "Investigating Possible Effects of Asbestos in City Water: Surveillance of Gastrointestinal Cancer Incidence in Duluth, Minnesota," *American Journal Epidemiology,* 103:362–368, 1976.

22. Polissar, L., et al., "Cancer Incidence in Relation to Asbestos in Drinking Water in the Puget Sound Region," *American Journal Epidemiology* 116:314, 1982.
23. Cooper, R. C., Kanarek, M., Murchio, J., Conforti, P., Jackson, L., Callard, R., and Lysmer, D., *Asbestos in Domestic Water Supplies in Five California Counties,* Final Report on Grant No. R-804366-02, December, 1978.
24. Cooper, R. C., and Cooper, W. C., "Public Health Aspects of Asbestos Fibers in Drinking Water," *J.AWWA,* June, 1978.
25. Lawrence, J., Tosine, H. M., Zimmerman, H. W., and Pang, T. W. S., "Removal of Asbestos from Potable Water by Coagulation and Filtration," *Water Research* 9:397–400, 1975.
26. Olson, H. L., "Asbestos in the Potable-Water Supplies," *J. AWWA* 66:515, 1974.
27. Hesse, C. S., and Hallenbeck, W. H., "Chrysotile Asbestos in Raw Water, *Asbestos in Potable Water,* Office of Water Resources and Technology, January, 1977.
28. "Asbestos Product Ban Planned by EPA," *Engineering News Record,* October 6, 1983.
29. "EPA Stirs Health Controversy with Proposed Asbestos Ban," *AWWA Mainstream,* November, 1983.
30. Kanarek, M. S., Conforti, P. M., and Jackson, L. A., "Chrysotile Asbestos Fibers in Drinking Water from Asbestos-Cement Pipe," *Environmental Science and Technology,* August, 1981.
31. AWWA, Manual M16, *Work Practices for Asbestos-Cement Pipe.*
32. Buelow, R. W., et al., "The Behavior of Asbestos Cement Pipe under Various Water Quality Conditions: A Progress Report, Part 1, Experimental Results," *J.AWWA,* February, 1980.
33. Schock, M. N., and Buelow, R. W., "The Behavior of Asbestos Cement Pipe under Various Water Quality Conditions: Part 2, Theoretical Considerations," *J.AWWA,* December, 1981.
34. Logsdon, G. S., Symons, J. M., and Sorg, T. J., "Monitoring Water Filters for Asbestos Removal," *ASCE Journal of Environmental Engineering,* December, 1981.
35. Anderson, C. H., and Long, J. M., *Interim Method for Determining Asbestos in Water,* EPA-600/4-80-005, Athens, Georgia, 1980.
36. Kiehl, S. R., Smith, D. T., and Syom, M., *Optical Detections of Fiber Particles in Water,* EPA-600/2-79-127, Cincinnati, Ohio, 1979.
37. Lawrence, J., and Zimmerman, H. W., "Asbestos in Water: Mining and Processing Effluent Treatment," *J.WPCF,* January, 1977.
38. McGuire, M. J., Bowers, A. E., and Bowers, D. A., "Optimizing Lake-Scale Water Treatment Plants for Asbestos-Fiber Removal," *J.AWWA,* July, 1983.
39. *National Interim Primary Drinking Water Regulations,* EPA-570/9-76-003, Washington, D.C., 1976.
40. American Water Works Association Statement of Policy, "Quality Goals for Potable Water," *J.AWWA,* December, 1968.
41. Logsdon, G. S., Evavold, G. L., Patton, J. L., and Watkins, J., Jr., "Filter Plant Design for Asbestos Fiber Removal," *Journal ASCE, August, 1983.*
42. Kirmeyer, G. J., Logsdon, G. S., Courchene, J. E., and Jones, R. R., "Removal of Naturally Occurring Asbestos Fibers from Seattle's Cascade Mountains Water Source," *Proceedings,* 1979 Annual Conference, AWWA, Part 2, 1979.
43. Watkins, J., Jr., Ryder, R. A., and Persich, W. A., "Investigation of Turbidity, Asbestos Fibers, and Particle Counting Techniques as Indices of Water Treatability of a Cascade Mountain Water Source," Paper 33-3, *Proceedings,* 1978 Annual Conference, AWWA, Part 2, 1979.
44. Kirmeyer, G. J., *Seattle Tolt Water Supply Mixed Asbestiform Removal Study,* EPA-600/2-79-125, August, 1979.
45. Peterson, D. L., Schleppenbach, F. X., and Zaudtke, T. M., "Studies of Asbestos Removal by Direct Filtration of a Lake Superior Water," *J.AWWA,* March, 1980.
46. Baumann, E. R., "Diatomite Filters for Asbestiform Fiber Removal from Water," presented at the 1975 AWWA Annual Conference, Minneapolis, Minn. (printed as Appendix 1 of EPA-670/2-75-050).

Chapter 17
Ion Exchange

INTRODUCTION

Typical ion exchange applications in water treatment include softening, fluoride removal with activated alumina, and nitrate removal. Softening is by far the most common water treatment application. This chapter discusses softening and anion exchange nitrate removal; fluoride removal is discussed in Chapter 20.

The operative principal in ion exchange reactions is the transfer of an ion in solution for an ion fixed to the surface of a resin. Resin selectivity results in the preferential removal of certain ions. Selectivity varies with the type of exchange resin, the solution ionic strength, relative amounts of different ions, water temperature, and other factors. An understanding of ion exchange fundamentals is important to both the design and operation of ion exchange treatment facilities.

PILOT TESTING

Use of scale-model ion exchange reactors is recommended to establish design and operational parameters for the particular water supply being evaluated. Pilot testing will help in optimization of design parameters, and in large systems may avoid costly errors in treatment process design.

In smaller systems where the cost of pilot testing may not be economically justified, the knowledge and experience of ion exchange equipment manufacturers and resin suppliers is a valuable resource. Many manufacturers will analyze representative water samples, and will recommend design criteria and resin type. The cost for such services is usually minimal. A prudent measure is to obtain recommendations from at least two manufacturers.

FUNDAMENTAL ION EXCHANGE CONCEPTS

Operation of an ion exchange treatment plant involves four separate cycles:

- Service or operation
- Backwash
- Regeneration
- Rinsing

The service and regeneration cycles are discussed in the following paragraphs.

Service Cycle

During the service cycle, ions are removed from solution. A reaction of a cation exchange resin operating in the sodium cycle is:

$$RNa + Ca + Cl \rightarrow RCa + Na + Cl \qquad (17\text{--}1)$$

where R = exchange resin.

In this and subsequent equations, the ionic charge sign and valence are omitted for simplicity. In this reaction, the hardness-creating cation calcium is exchanged for sodium. The chloride ion or other anions that are present in the feed water are unaffected by the cation exchange reaction. However, when the resin is operated in the chloride cycle, product water chloride concentrations are higher than the raw water chloride concentrations.

A typical anion exchange reaction, with the resin (R) operating in the chloride cycle, is:

$$RCl + Na + NO_3 \rightarrow RNO_3 + Na + Cl \qquad (17\text{--}2)$$

During the service cycle, certain ions are removed selectively over other ions. The general order of selectivity for cations and anions is:

CATIONS	ANIONS
Ca	SO_4
Mg	NO_3
Na	Cl
H	HCO_3

Ion selectivity can produce situations that are at first perplexing. Nitrate removal is a good example. Because of the selectivity of sulfate over nitrate, sulfate tends to be removed before nitrate, and in a downflow operating mode produces an upper "layer" of resin exhausted by sulfate and a lower "layer" of resin exhausted by nitrate. During the service cycle, each of these layers moves progressively downward through the resin bed. When the nitrate removal capacity of the resin is exhausted, nitrate appears in the effluent. If operation continues beyond this point, the nitrate in the exchanger product water will soon exceed the raw water nitrate concentration. Thus, the nitrate removal of the resin is strongly influenced by the sulfate, nitrate, and total anion concentrations.

Regeneration

During regeneration, the ions removed by the resin during the service cycle are displaced by a strong concentration of regenerant. For operation in the sodium or chloride cycle, the regenerant is sodium chloride. Regeneration of the resins discussed previously proceeds according to the following reactions:

Exhausted cation exchange resin:

$$RCa + Na + Cl \rightarrow RNa + Ca + Cl \qquad (17-3)$$

Exhausted anion exchange resin:

$$RNO_3 + Na + Cl \rightarrow RCl + Na + NO_3 \qquad (17-4)$$

The brine solution remaining after regeneration often represents a significant disposal problem.

Softening Applications

Softening is the most commonly used ion exchange application in water treatment. The resins are strong acid, cation exchange resins in the sodium form. Typical resins are Amberlite IR-120 Plus, Duolite C-20, Dowex HCR, and Permutit Q-100. This discussion is based primarily upon the characteristics of the Amberlite IR-120 Plus resin.

Cation exchange resin can be contained in a pressure steel vessel or an open concrete basin. Pressure vessels are most commonly used in small plants, while open concrete basins are usually used in larger facilities.

Operating Conditions

Suggested conditions for each of the operating phases are shown in Table 17-1 for three different resins. Each of the four phases is discussed below.

Treatment. Treatment may be either upflow or downflow, although downflow treatment is the most common in water treatment applications. The service flow rate will vary between 2 and 5 gpm/cu ft (0.27 and 0.67 m^3/ m/m^3), depending on the type of resin used (see Table 17-1) and the raw water hardness. The length of the service cycle is a function of the hardness concentration in the raw water and the exchange capacity of the resin. Manufacturers typically express resin capacity in terms of kilograins of hardness (as $CaCO_3$) per cubic foot of resin. The conversion factors for hardness ions are listed in Table 17-2.

Table 17-1. Suggested Operating Conditions for Amberlite IR-120 Plus, Duolite C-20, and Permutit Q-100 in the Sodium Cycle.*,**

PARAMETER	UNITS OF EXPRESSION	AMBERLITE IR-120 PLUS	RESIN TYPE DUOLITE C-20	PERMUTIT Q-100
Service flow rate	gpm/cu ft	2	2–5	2–4
Exchange capacity	kilograins/cu ft of resin	17.5–34.5	18–37	41.4
Minimum bed depth	inches	24	24	24
Backwash flow rate	gpm/sq ft	6 @ 72°F	6–8 @ 50–75°F	—†
Backwash water requirement	gal/cu ft of resin	30††	10–50	—
Regenerant concentration	% NaCl	10	10–20	10–14
Regenerant flow rate	gpm/cu ft	0.5–1.0	0.2–1.0	0.5
Regenerant level	lb NaCl/cu ft of resin	5–30	5–20	5.5–16
Rinse flow rate	gpm/cu ft	1.0 initially, then 1.5	1.0 initially, then up to 5.0	—
Rinse water requirement	gal/cu ft of resin	25–75	10–50	25–50

* References 1, 2, and 3.
** Conversion factors may be found in Appendix A.
† 50 to 75 percent bed expansion required.
†† Based on 2-foot bed depth.

Table 17–2. Hardness Conversion Factors.

ION	MULTIPLIER TO CONVERT FROM mg/l AS THE ION TO GRAINS/GAL AS CaCO₃
Ca^{+2}	0.146
Mg^{+2}	0.240
Fe^{+2}	0.105
Mn^{+2}	0.106

Amberlite IR-120 Plus has an exchange capacity between 17.5 and 34.5 kilograins of hardness (as $CaCO_3$) per cubic foot (618 and 1,219 kilograins/ m^3) of resin. The actual capacity depends upon the regenerant level used. The more regenerant used per cubic foot of resin, the greater the exchange capacity of the resin is; however, the overall regenerant cost per unit of hardness removed also increases. The relationship between regenerant level and exchange capacity is shown in Table 17–3. In water treatment applications, 10 lb NaCl/cu ft of resin (161 kg NaCl/m^3 of resin) represents a reasonable compromise between regenerant usage and regeneration efficiency.

Pretreatment is required in many situations. Suspended solids should be low (zero if possible), and turbidity should be less than 1 NTU, to prevent bed plugging. Ferrous iron should be removed prior to ion exchange treatment, as it can oxidize to the ferric form within the bed if oxygen is present.

If the water to be treated is a well water containing no dissolved oxygen, the resin can effectively remove ferrous and manganous ions. Conversely, if oxygen is present, or if iron or manganese is present in the oxidized form, it should be removed prior to ion exchange treatment, to avoid resin fouling. Chlorine should not be present in the feed water.

During conventional downflow treatment, there is a pressure drop across the resin, which is a function of the service flow rate, water temperature, and amount of suspended material in the raw water. Figure 17–1 illustrates pressure drop across a clean resin bed. When suspended matter is present, it is frequently filtered out by the resin, thereby reducing the void volume in the upper portion of the bed and causing increased resistance to flow. In extreme cases, resin particles can shatter, producing fines that are lost during backwash. Although some suspended material may be tolerated, it is wise to employ a resin bed as an ion exchange medium and not as a filtering mechanism.

Organic material in the feed water can also result in resin fouling. The fouling is caused by deposition within the resin, as well as by bacterial growth within the resin bed.

Backwash. Following completion of the treatment or operation cycle, the resin bed should be backwashed for approximately 10 minutes at a rate

Table 17–3. Relationship of Regenerant Level, Resin Exchange Capacity, and Regeneration Efficiency for Amberlite IR-120 Plus.*

REGENERANT LEVEL		RESIN EXCHANGE CAPACITY (AS CaCO₃)			REGENERATION EFFICIENCIES		
lb NaCl	kg NaCl	KILOGRAINS	KILOGRAINS		lb NaCl		kg NaCl
cu ft OF RESIN	m³ OF RESIN	cu ft	m³		KILOGRAINS REMOVED		KILOGRAINS REMOVED
5	80.2	17.5	618		0.28		0.127
10	161	24.5	866		0.41		0.186
15	241	29.0	1,025		0.51		0.232
20	321	32.5	1,148		0.62		0.282
25	401	34.0	1,202		0.74		0.336
30	481	34.5	1,219		0.87		0.395

* Reference 8.

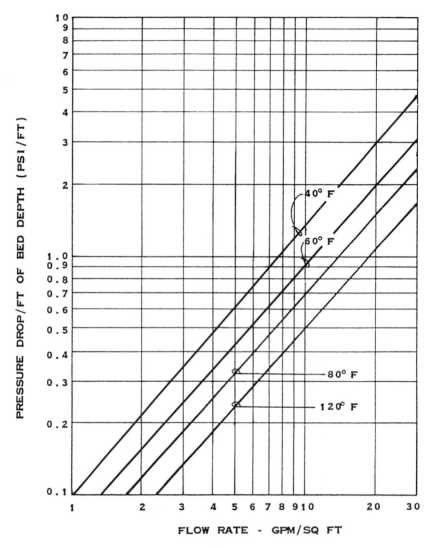

Fig. 17–1. Pressure drop during cation exchange softening.[1] (Courtesy of Rohm and Haas Co.)

that will cause a bed expansion between 50 and 75 percent. The goal of backwashing is to remove material that was filtered out by the bed in its upper layers. If this material is not removed during backwashing, it will lead to channeled water flow through the bed, premature leakage of hardness, and reduced exchange capacity. Backwash characteristics of Amberlite IR-120 Plus are shown in Fig. 17–2.

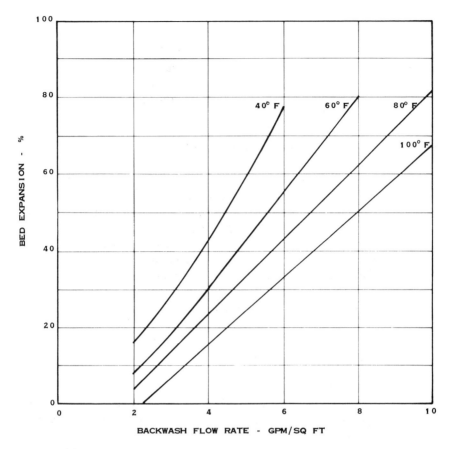

Fig. 17–2. Bed expansion vs. backwash flow rate for Amberlite IR-120 Plus cation exchange resin.[1]

Regeneration.   Regenerant brine concentration has little effect on resin exchange capacity. For example, Amberlite IR-120 Plus resin shows only a 7 percent decrease in capacity when a 4 percent sodium chloride solution is used instead of a 10 percent solution. A much more critical factor is the regenerant flow rate, as lower flow rates give longer contact time between the brine and the resin. Regenerant flow rates in excess of 1 gpm/cu ft (0.133 m³/m/m³) are not recommended, as appreciable capacity reduction is experienced at higher flow rates. Regeneration is typically in a downflow manner, with a minimum contact time of 30 minutes.

Rinsing.   After regeneration, rinsing is necessary to remove remaining brine from the resin bed. The initial rinse flow rate should be the same as that used during regeneration; rinse water is introduced while the regeneration process is concluding. After one to two bed volumes of rinse water have been added, the flow rate should be increased to 1.5 gpm/cu ft (0.2 m^3/m/ m^3). This should continue until the sodium concentration in effluent from the bed is less than 5 mg/l above the rinse water sodium concentration. These two rinse operations are referred to as the slow rinse and the fast rinse.

Design Considerations for Cation Exchangers

Exchange Vessels. Cation exchange units are similar to sand filters except that cation exchange resin replaces the sand. Small plants generally use pressure units, as this allows pumping directly through the exchanger and into the distribution system. Larger plants typically use open, gravity-flow units, as they are less costly than pressure units per unit volume.

Pressure vessels may be either vertical or horizontal, with vertical vessels preferred. Disadvantages of horizontal vessels include poor hydraulic distribution and loss of exchange capacity because resin located in the "bulges" at the ends and side of the pressure vessels is not fully utilized. The vessel interior should be coated to provide protection against the corrosive sodium chloride regenerant. All piping should be corrosion-resistant, with plastic pipe most frequently utilized.

Resin Depth. The minimum resin depth is 2 feet (0.61 m), and the maximum resin depth 5 feet (1.52 m). The maximum resin depth is established by allowable headloss through the bed and by expanded bed depth during backwashing. Expansion of the resin bed by 50 to 75 percent during backwash should be allowed in the design, without resin reaching the bottom of the wash troughs. If sufficient contactor volume is not left for resin expansion, resin will be lost during backwash.

Sizing the Contactors. Contactor size is a function of several parameters:

- Resin volume
- Maximum raw water hardness
- Time interval between regenerations
- Product water storage

Resin volume can be varied by changing the bed depth and the number of contactors. Because maximum raw water hardness is not within the control of the plant designer, the only remaining variable is regeneration frequency. If the plant is automatically operated, a minimum regeneration frequency of 8 to 12 hours is suggested. However, in manually operated plants that have one shift per day, the frequency should be 24 hours. Sufficient product water storage to "average out" the reduction in product water flow during contactor regeneration is desirable. Without storage, the treatment system must be designed to continuously meet distribution system demands even with one contactor out of service for regeneration.

The required resin volume is calculated according to the formula:

$$V = \frac{QHT\,100}{E\,24\%OP}$$

$$V = \frac{QHT}{240E\,\%OP} \quad \text{(metric)}$$

(17–5)

where:

V = required resin volume, cu ft (m³)
Q = flow, gal/day (l/d)
H = raw water hardness, kilograins/gal as $CaCO_3$ (mg/l)
T = regeneration frequency, hours (h)
E = resin exchange capacity, kilograins as $CaCO_3$/cu ft (g/m³)
$\%OP$ = average percent of time that contactors are operating

Underdrains. Similar to a sand filter, the underdrain collects water during operation, distributes backwash water, and collects brine and rinse water during and following regeneration. Good flow distribution is essential if each of the four phases of the cation exchange process is to be performed efficiently. The total area of the holes in the underdrain should be 0.16 to 0.18 percent of the surface area of the exchanger. Porous plates or lateral manifolds provide good distribution. Porous plates can only be used with relatively clear water, as suspended material will clog the plates. A number of proprietary under-drains are available.

Gravel Layer. To prevent the resin from being washed out of the under-drain system, as well as to achieve good flow distribution during backwash, a layer of graded gravel is used. Porous plate underdrains do not require a gravel layer. A total gravel depth of 15 to 18 inches (0.38 to 0.46 m) using three or more layers of gravel graded from ⅛ to 1 inch (3.2 to 25.4 mm) is typical.

Wash Troughs. Wash troughs are located at an elevation above the highest level reached by the resin during backwash. Thus, wash through elevation must account for gravel depth and media depth during backwash. Bed expansion during backwash is shown in Fig. 17–2.

Brine and Rinse Water Distribution. Assuming downflow regeneration, the brine distribution manifold is placed immediately above the softener bed. Prior to brine introduction, water in the contactor is drawn down to slightly above the top of the resin. This allows brine to be introduced with minimal disruption of the resin. Rinse water, assuming downflow rinse, is introduced either through the brine distribution system or by flooding the bed if this can be done without disrupting the resin surface.

Brine and Salt Storage. Salt storage and brine production are usually accomplished in the same basin. Salt is added to the basin in excess of the quantity that can dissolve. Thus, as concentrated brine (approximately 26 percent at saturation) is withdrawn from the bottom of the basin, fresh water introduced at the top of the basin dissolves additional salt. Undissolved salt should always be present.

Tank volume is established by the quantity of salt that must be stored, which is a function of the proximity of the plant to a reliable source of salt and the method of salt delivery. Large plants can take advantage of the lower cost of bulk delivery by either truck or rail; smaller plants use salt delivered in bags. Some large plants use unrefined rock salt, at a significant cost savings. However, unrefined rock salt contains sand and silt, which require removal. Sand precipitates during brine production, while silt removal usually is accomplished in a separate holding basin. Periodic basin cleaning is necessary to remove sand and silt.

Saturated brine is removed from the basin by either pumping or a hydraulic eductor. Maintaining the brine at a consistent concentration is essential to assure that the proper salt dose is used during each regeneration. A sampling tap or a meter should be provided.

Water addition to the brine tank should be done with an air gap to prevent any possible cross connection. Also, an overflow from the brine tank to a sewer should be provided. The density of salt is approximately 70 lb/cu ft (1,123 kg/m³).

Product Water Blending

Product water from the exchanger will have a slight hardness (about 1 percent of influent hardness), due to bed "leakage." Because a product water hardness concentration of about 100 mg/l as $CaCO_3$ is usually desirable, blending should be done to reduce operating costs. This may be accomplished by blending low-hardness product water with unsoftened water in the correct proportion to achieve the desired final water hardness, or by running an exchanger past the point where the normal background leakage concentration is exceeded. The latter technique requires sufficient storage capacity for softened water.

Brine Disposal

Potential techniques for brine disposal are discharge to sanitary sewers, evaporation ponds (lined or unlined), the ocean or an estuary, or disposal wells. In all cases, the disposal technique must be approved by regulatory agencies, and, in the case of sewer disposal, by the sewering agency. Particular attention must be given to degradation of groundwater and surface water quality. Chlorination may be necessary in some cases.

The cost of brine disposal is a key factor in the overall economic analysis that should be made to select the treatment technique.

NITRATE REMOVAL APPLICATIONS

Nitrate ions can be removed from water supplies using strong-base ion exchange resins operated in the chloride cycle. Resins that can be used are Duolite A-101D and A-104, Dowex SAR, Ionac A-550, and Amberlite IRA-900 and IRA-910. These resin types can be regenerated using sodium chloride brine.

These resins will remove a number of anions in addition to nitrate, including sulfate, nitrite, chloride, and bicarbonate. The order of affinity for these anions is:

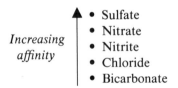

Increasing affinity

- Sulfate
- Nitrate
- Nitrite
- Chloride
- Bicarbonate

Thus, sulfate ions are preferentially removed over nitrate ions, and nitrate ions are preferentially removed over nitrite, chloride, and bicarbonate ions. As the resin becomes exhausted, bicarbonate breaks through first, followed by chloride, then nitrite, then nitrate. The sulfate ion is the most strongly removed, and during the end of the service cycle, will displace previously removed nitrate ions. This is an undesirable situation, and can result in a higher nitrate concentration in the exchanger product water than in the exchanger feed water. The service cycle must be terminated before this point.

Design and Operating Parameters

Equipment virtually identical to that described for cation exchange can also be used for anion exchange. Similarly also, sodium chloride is used for resin regeneration, and thus the salt storage and brine feed facilities are identical to cation exchange applications.

Operation is usually downflow, followed by upflow backwash, and downflow regeneration and rinsing. Suggested design parameters for Duolite A-104 resin are presented in Table 17–4.

Pretreatment. Pretreatment requirements are basically the same as for cation exchange resins, and include upper limits for iron, suspended solids, and organics.

Resin Capacity Determination. In order to size the exchangers adequately, the capacity of the resin must be determined for the particular water being treated. Resin capacity is determined primarily by the ratio of nitrate to total anions, and the ratio of sulfate to total anions.

The first step in the capacity determination is to conduct a complete water analysis. Using this analysis, all free anions should be converted to mg/l as calcium carbonate, using the multipliers in Table 17–5.

Next, the ratio of nitrate (as $CaCO_3$) to total anions (as $CaCO_3$) is calculated, and entered into Fig. 17–3 to determine nitrate removal capacity in kilograins per cubic foot of resin. Then, this resin capacity must be adjusted to account for the presence of sulfate ions. This is accomplished by calculating the ratio of sulfate (as $CaCO_3$) to total anions (as $CaCO_3$), and entering into Fig. 17–4 to determine the percent resin efficiency. This resin efficiency must be multiplied by the nitrate removal capacity from Fig. 17–3.

Table 17–4. Suggested Design Parameters for Duolite A-104 Resin.[*,**]

PARAMETER	DESIGN/OPERATING VALUE
Minimun bed depth	30 inches
Service flow rate	Up to 5 gpm/cu ft of resin
Backwash flow rate	2 to 3 gpm/sq ft
Regenerant concentration (NaCl)	10 to 12% by weight
Regenerant dosage	10 to 18 lb NaCl/cu ft of resin
Regenerant flow rate	0.5 gpm/cu ft of resin
Regenerant contact time	50 to 80 minutes
Rinse flow rate	0.5 gpm/cu ft per bed volume plus 2 gpm/cu ft of resin
Rinse volume	50 to 70 gal/cu ft of resin

[*] Reference 4.
[**] Conversion factors are included in Appendix A.

Table 17–5. Hardness Conversion Factors.

ION	MULTIPLIER TO CONVERT FROM mg/l AS THE ION TO mg/l AS CaCO₃
SO_4^{-2}	1.04
NO_3^-	0.81
Cl^-	1.41
HCO_3^-	0.82

For illustrative purposes, assume a water supply with 118 mg/l of sulfate, 105 mg/l of nitrate, 55 mg/l of chloride, and 206 mg/l of bicarbonate. Proceed by calculating nitrate/total anion and sulfate/total anion ratios.

ION	CONCENTRATION, mg/l	CONCENTRATION AS mg/l CaCO₃
Sulfate	118	122.7
Nitrate	105	85.1
Chloride	55	77.6
Bicarbonate	206	168.9
	Total anions =	454.3

$$\frac{\text{Nitrate}}{\text{Total anions}} = \frac{85.1}{454.3} = 0.181$$

$$\frac{\text{Sulfate}}{\text{Total anions}} = \frac{122.7}{454.3} = 0.27$$

Using Fig. 17–3, the uncorrected exchange capacity if 7.75 kilograins/cu ft (273.9 kilograins/m³). The correction factor for the sulfate ion concentration is 71 percent, as shown in Fig. 17–4. Thus, the resin capacity corrected for sulfate concentration is:

$$7.75 \ (0.71) = 5.50 \text{ kilograins/cu ft } (194.4 \text{ kilograins/m}^3)$$

Required Resin Volume. The volume of resin required is a function of three factors:

- Resin capacity.
- Required run time between regenerations.

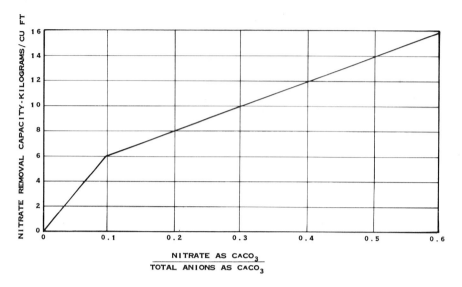

Fig. 17–3. Typical plot showing nitrate removal capacity as a function of the nitrate/total anion ratio.

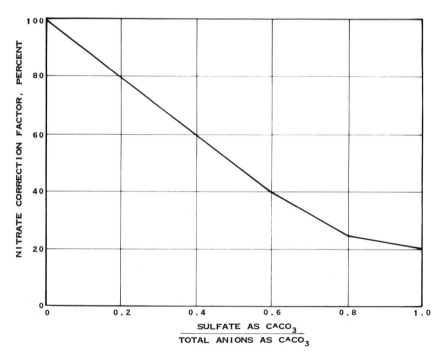

Fig. 17–4. Typical plot showing efficiency of nitrate removal as a function of the nitrate/total anion ratio.

* Resin bed depth and flow per cubic foot as recommended by the resin manufacturer.

Using the previously calculated resin capacity, a flow rate of 0.25 mgd (0.95 Ml/d), an operating time of 12 hours between regenerations, a bed depth of 3 feet (0.91 m), and a maximum service flow rate of 5 gpm/cu ft (0.67 m^3/m/m^3), the required resin volume is calculated as follows:

$$\text{Nitrate feed per 12 hours} =$$

$$(0.25 \text{ mgd}) (8.34) (105 \text{ mg/l}) \left(\frac{12 \text{ hr}}{24 \text{ hr/day}}\right) \cdot (7 \text{ kilograins/lb})$$

$$= 766 \text{ kilograins}$$

$$\text{Required resin volume} = \frac{766 \text{ kilograins}}{5 \text{ kilograins/cu ft}}$$

$$= 153 \text{ cu ft}$$

A check of the service flow rate per cubic foot of resin shows that it is 173.6 gpm/153 cu ft, or 1.13 gpm/cu ft. This is less than the maximum rate recommended by the manufacturer and is therefore acceptable. At a 3-foot (0.91-m) bed depth, the required contactor area is:

$$\frac{153 \text{ cu ft}}{3 \text{ ft bed depth}} = 51 \text{ sq ft}$$

This is equivalent to a loading rate of 3.4 gpm/sq ft (0.45 m^3/m/m^3). Consideration should be given to a deeper resin depth, which would give a higher loading. Loading rates up to 20 gpm/sq ft (0.82 m^3/m/m^3) are reported to provide satisfactory nitrate removal.

Bed Expansion During Backwash.   Contactor design must include allowance for bed expansion during backwash. As with cation exchange resins, the percentage bed expansion is a function of water temperature and backwash flow rate. Figure 17–5 presents backwash expansion curves for Amberlite IRA-900 operating in the chloride form.

Resin Regeneration.   Resin regeneration can be either complete or partial. If resin is completely regenerated, the salt requirement is high, but there is little leakage of nitrate ions during operation. Conversely, if a low amount of salt is used, regeneration is incomplete, and nitrate leakage occurs during normal operation. Run time between regenerations is lessened as the salt requirement is decreased. The overall advantage of incomplete regeneration

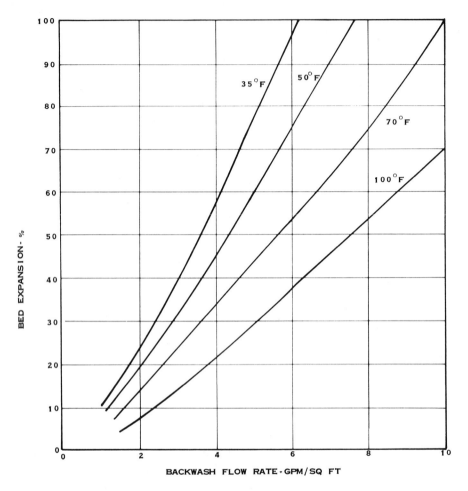

Fig. 17-5. Bed expansion vs. backwash flow rate for Duolite A-104 anion exchange resin.[4]

is a substantial decrease in salt requirements. The optimum regenerant dosage should be determined during pilot testing.

Operating Nitrate Removal Installations

Two operating installations are discussed in this section: McFarland, California and Curryville, Pennsylvania. Both installations use fixed bed anion exchange resins operated in the chloride cycle.

McFarland, California. This facility has a design capacity of 1.0 mgd (3.785 Ml/d), and treats well water. It was placed into operation in November,

1983, and in an effort to optimize operation there were plans to investigate various operational procedures during the first year.[5,6] The plant uses three contactors, each 6 feet (1.83 m) in diameter and 10 feet (3.05 m) high. The resin is Duolite A-101D, and each vessel was initially charged with 3 feet (0.91 m) of resin. After initial testing at this depth, it was planned to increase the resin depth to 5 feet (1.52 m). The brine storage tank is 12 feet (3.66 m) in diameter and 12 feet (3.66 m) high.

To reduce salt requirements for the regeneration step, only partial regeneration is performed. Salt requirements are reduced by 80 percent, with only a 23 percent reduction in the nitrate removal capacity of the resin. Salt requirements are 5 lb/cu ft (80 kg/m³) of resin. Regeneration is accomplished by passing a 6 percent brine solution downward through the bed. After this "brining" stage, the bed is slow-rinsed with product water, followed by an upflow backwash to remove suspended materials and to declassify the resin. Declassification is accomplished alternately using half of the underdrain/backwash laterals. The original design incorporated separate piping to accomplish such split backwash.

Each regeneration produces about 10 bed volumes of waste from brining, rinsing, and backwash, and this waste is discharged to the sanitary sewer system. Wastewater represents approximately 3 percent of the product water.

The three-exchanger system is designed to treat 1 mgd on a continuous basis. Each contactor can treat 0.5 mgd, and one contactor is for standby. With two units operating, one unit is placed into operation when the second unit is at the midpoint of its cycle. This avoids the need to regenerate two contactors simultaneously. Regeneration frequency is about every 9 hours, or 300 bed volumes treated.

The capital cost of the facilities at McFarland was $218,400, divided as follows:

Vessels, piping, and valves	$ 99,700
Site work (instrumentation, mechanical and electrical)	81,200
Brine tank	17,500
Building	10,000
Process controllers	10,000
	$218,400

Operating costs during the first year were projected to be:

Supplies	
Salt	$18,000/yr
Resin (10% makeup/yr)	5,800
Electricity	5,300
Repair and maintenance (5% of equipment cost)	6,900
	$36,000/yr

Electricity was calculated on the basis of added head imposed on the well pump, due to passage of the water through the anion exchange facilities. There was no labor cost, as no additional personnel were added when the new facilities were placed into operation. To some extent, this was made possible by the presence of a full-time technician paid for by an EPA grant during the first year of operation.

Raw water quality at McFarland is as follows:

Sulfate (SO$_4$)	104 mg/1
Nitrate (NO$_3$)	70.4 mg/1
Chloride (Cl)	89.2 mg/1
Bicarbonate (HCO$_3$)	91.8 mg/1

Curryville, Pennsylvania.   The facility was designed to treat up to 3,000 gpd (11.4 m³/d), but only averages about 650 gpd (2.46 m³/d).[7] Raw water nitrate concentration is only 49.5 mg/1 as NO$_3$. About 10 percent of the water is treated for nitrate removal, and this treated water is then blended with the remaining 90 percent of untreated water. The facilities were constructed in early 1979 at a cost of $30,000.

The facilities consist of a Permutit Model ED20 water softener to contain the resin, and a 24-inch-diameter brine tank. Design and operating data are shown in Table 17–6. The unit regenerates automatically after 18,000 gallons

Table 17–6. Design and Operating Data for Curryville, Pennsylvania.*

Design capacity	
Average	28 gpm
Peak	36 gpm
Bed dimensions	
Diameter	20 inches
Height	32 inches
Resin volume	5.5 cu ft
Resin type	Ionac, A-550, strongly basic
Regeneration cycle	
Time	70 minutes
Salt consumption	
Pounds	45
Pounds/cu ft of resin	8.2
Water consumption	
Backwash	200 gallons
Brine	130 gallons
Slow rinse	190 gallons
Fast rinse	150 gallons
Total	670 gallons

*Conversion factors may be found in Appendix A.

(68.1 m³) of water has been treated, and regeneration consumes about 45 pounds (20 kg) of salt. Regeneration requires about 70 minutes and consumes 130 gallons (492 l) of water. Wastewater and spent brine are discharged to a septic tank. The plant operator visits the plant twice weekly, spending about 1 hour per visit.

REFERENCES

1. *Amberlite IR-120 Plus Technical Bulletin* No. IE-169-70/76/78, © 1978, Rohm and Haas Company, Philadelphia, Pa., March, 1980.
2. Duolite C-20 Technical Sheet, Rohm and Haas Company, Philadelphia, Pa.
3. Q-100 Ion Exchange Resin Fact Sheet, Permutit, Paramus, N.J.
4. Duolite A-104 Data Sheet, Rohm and Haas Company, Philadelphia, Pa.
5. Guter, G. A., *Removal of Nitrate from Contaminated Water Supplies for Public Use,* EPA-600/52-81-029, U.S. EPA, Cincinnati, Ohio, April, 1981.
6. Gumerman, R. C., Burris, B. E., and Hansen, S. P., *Estimation of Small System Water Treatment Costs,* U.S. EPA, Cincinnati, Ohio, Contract No. 68-03-3093.
7. *Nitrate Removal Guide,* U.S. EPA, Washington, D.C.
8. *Amberlite IR-120 Plus Technical Bulletin* No. IE-123-67/78, Rohm and Haas Company, Philadelphia, Pa., May, 1980.

Chapter 18
Taste and Odor Control

INTRODUCTION

Although treated water may be free of pathogens and safe to drink, the presence of tastes or odors can cause customer concerns and complaints. It is appropriate that the water treatment industry places a high priority on producing a product that is odor- and taste-free. As discussed in this chapter, tastes and odors are frequently caused by salts and organisms that pose no problem to the safety of water supplies. However, the layperson may interpret taste and odor as an indication of a health threat and certainly an indicator of an unsatisfactory product.

This chapter discusses physiological and psychological aspects of taste and odors, causes and sources, measurement, and methods of control.

Odor is caused by volatile substances in water. Although combinations of basic odors are limitless, odors can be grouped under eight general classifications:[1]

- Aromatic (spicy)
- Balsamic (flowery)
- Chemcial
- Disagreeable
- Earthy
- Grassy
- Musty
- Vegetable

Tastes are caused by dissolved salts. Four basic tastes are recognizable:

- Salty
- Sweet
- Sour
- Bitter

Taste is a sensation transmitted by buds on the tongue. Dissolved inorganic salts of iron, manganese, zinc, copper, sodium, and potassium are detectable by taste but are odorless. Sodium chloride is detectable in concentrations of 300 to 900 mg/l as NaCl. At concentrations between 1,000 to 1,500 mg/l, it becomes objectionable. Fair and Geyer suggest that sulfates of sodium as well as chlorides and sulfates of potassium, calcium, and magnesium have similar thresholds of detection and refusal.[2] Tastes can also be produced by some algae and industrial wastes. Taste and odor are closely related and often are difficult to distinguish because taste is easily masked by olfactory sensations. Most reported studies regarding taste and odor deal only with odor. Throughout this chapter, there will be no further attempt to differentiate taste and odor.

TERMINOLOGY, CHARACTERIZATION, AND MEASUREMENT

Terminology and Characterization

Two descriptions that are commonly used to describe an odorous substance are intensity and characterization.[3] Intensity relates to relative strength; characterization describes how a substance smells. In addition to the sensations described earlier in this chapter, the Crocker-Henderson system suggests four basic odors: fragrant, acid, burnt, and caprylic (rancid). This association of an odor with a certain chemical or type of odor is commonly used. However, classification of odors is difficult due to the complex nature of odor sensation. Consequently, *Standard Methods* recommends that descriptions of odors be used only as a guideline.[4]

Two important phenomena regarding taste and odor are fatigue and adaptation.[5] Fatigue is the loss of sensitivity caused by continuous stimulation. Frequently, the result is an inability to detect increases in concentration. To prevent inaccuracies in test observations, adequate rest is required between observations. Adaptation is the adjustment of sensitivity to environmental conditions. The type of odor affects the type, rate, and degree of adaptation. Generally, an increase in concentration produces an increase in rate of adaptation. Geldard has suggested that people are partially odor-adapted most of the time.[6]

Tests

One of the most common odor evaluation methods is the threshold odor method, in which an odor value is given a threshold odor number (TON). The TON is the extent to which a sample must be diluted for the odor to be just detectable. For example, assume that various-size samples of a water are diluted to a 200-ml test volume. If the odor first becomes detectable in a sample containing 25 ml of the test water, then the TON is 8. This method is common for samples with low odor intensity such as treated water supplies. A survey reported by Sigworth indicated that 42 percent of water treatment plants use the TON method.[7] The test is typically performed at 60°C because it approximates domestic hot water temperatures. Other plants use 40°C because it approximates body temperature. A TON of 2 is nearly always palatable, as is a TON of 5 in a majority of cases.[7]

The TON is determined using a panel of several observers. Differences in individual sensitivity make it less than a precise number, however. Another method, described in *Standard Methods,* involves estimation of the odor by assigning a value expressed in Roman numerals.[4] Zero corresponds to completely palatable water; V denotes totally unpalatable water. In a survey reported by Sigworth, 31 percent of the 241 plants reporting on odor evaluation used this method.[7]

CAUSES OF TASTE AND ODOR

Sources and causes of taste and odor include:

- Microorganisms: bacteria, algae, and actinomycetes
- Decayed vegetation
- Municipal and industrial wastewater
- Agricultural runoff: animal wastes and fertilizers, chemical fertilizers, insecticides, herbicides
- Hydrogen sulfide
- Inorganic minerals

In the survey of 241 plants mentioned above, 82 percent reported algae, 67 percent decaying vegetation, 38 percent industrial wastes, 9 percent sewage, and 14 percent other wastes or causes.[7]

Microorganisms

Algae. Algae and decomposition of algae are the most common sources of taste and odors in water supplies. Decomposition is caused by action of fungi, actinomycetes, and other bacteria. Odors and tastes produced by algae are frequently described as fishy, grassy, aromatic, earthy, musty, and septic. Algal odor can be objectionable even when threshold odors are low. Algal-caused threshold odor numbers have been reported over ranges of 1 to 40.[8] A summary of odors, tastes, and tongue sensations associated with algae in water is presented in Table 18–1.[9] Taste and odor algae, organized by group, are presented in Table 18–2.[9]

Among the most potent odor producers is the group of flagellates known as *Synura,* probably the most troublesome algae in waterworks practice. Their bitter taste has been described as cucumber or muskmellon, and it leaves a dry, metallic sensation on the tongue. *Synura* cause a fishy odor when present in large numbers.[8] Odor problems are intensified by chlorine concentrations commonly used for disinfection. High concentrations of chlorine will destroy organisms and fragment cell matter. These flagellates are easily removed by filtration, but fragmented cells may not be removed. Unfortunately, filtration does not remove the *Synura* odor.

The following is a summary of algal-related taste and odors as described by White.[8]

- *Anabaena, Anacystis* (formerly known as *Microcystis, Polycystis,* and *Clathrocystis*), and *Aphanizomenon* are blue-green algae that produce "pigpen" odors. They impart a grassy to moldy odor in small concentrations. Where concentrations are higher, foul odor undoubtedly develops from decomposition products as the algae die.
- Green algae are less often associated with tastes and odors in water than blue-green algae. Their growth may help somewhat to inhibit or keep in check the blue-green algae and the diatoms, and thereby be helpful in the control of water quality.
- *Asterionella* and *Tabellaria* (500 units or more) impart aromatic geranium-like odors that change to fishy when present in large numbers.
- *Ceratium* is an armored flagellate that is abundant in California and is responsible for odors varying from fishy to septic. It can proliferate rapidly during any season.
- *Dictyosphaerium,* a green alga, will produce a grassy to nasturtium odor as well as a fishy odor in larger concentrations. Some swimming green algae, such as *Volvox,* may also produce fishy odors.
- *Dinobryon* has a fishy odor when the standard areal count reaches 30/ml.
- *Stephanodiscus* has a vegetable oily taste but is relatively odor-free.
- *Synedra* produces an earthy to musty odor, and inhibits proper floc formation.

In the past, it was generally believed that algae were direct causes of taste and odors because they secreted oils. Also, algal protein has been found to yield organic acids, alcohols, ketones, and esters, which can be odor-producing.[10] However, other investigators have shown that no chemical compounds were isolated from algal cultures that were in good physical condition.[11] They indicated that odors attributed to algae by other investigators were actually produced by Actinomycetes. They also found that peaks in taste and odor did not correspond to algae blooms; actually the highest odors were recorded weeks after peak algal growth.

Actinomycetes. Actinomycetes (ray fungi) are a major source of taste and odor in water supplies. They are considered members of the Schizomy-

Table 18–1. Odors, Tastes, and Tongue Sensations Associated with Algae in Water.*

ALGAL GENUS	ALGAL GROUP	ODOR WHEN ALGAE ARE—		TASTE	TONGUE SENSATION
		MODERATE	ABUNDANT		
Actinastrum	Green		Grassy, musty		
Anabaena	Blue-green	Grassy, nasturtium, musty	Septic		
Anabaenopsis	Blue-green		Grassy		
Anacystis	Blue-green	Grassy	Septic	Sweet	
Aphanizomenon	Blue-green	Grassy, nasturtium, musty	Septic	Sweet	Dry
Asterionella	Diatom	Geranium, spicy	Fishy		
Ceratium	Flagellate	Fishy	Septic	Bitter	
Chara	Green	Skunk, garlic	Spoiled, garlic		
Chlamydomonas	Flagellate	Musty, grassy	Fishy, septic	Sweet	Slick
Chlorella	Green		Musty		
Chrysosphaerella	Flagellate		Fishy		
Cladophora	Green		Septic		
(Clathrocystis)	See *Anacystis.*				
Closterium	Green		Grassy		
(Coelosphaerium)	See *Gomphosphaeria.*				
Cosmarium	Green		Grassy		
Cryptomonas	Flagellate	Violet	Violet	Sweet	
Cyclotella	Diatom	Geranium	Fishy		
Cylindrospermum	Blue-green	Grassy	Septic		
Diatoma	Diatom		Aromatic		
Dictyosphaerium	Green	Grassy, nasturtium	Fishy		
Dinobryon	Flagellate	Violet	Fishy		
Eudorina	Flagellate		Fishy		
Euglena	Flagellate		Fishy	Sweet	Slick
Fragilaria	Diatom	Geranium	Musty		

Glenodinium	Flagellate		Fishy		Slick
(Gloeocapsa)	See Anacystis.				
Gloeocystis	Green		Septic		
Gloeotrichia	Blue-green	Grassy	Grassy		
Gomphosphaeria	Blue-green		Grassy	Sweet	
Gonium	Flagellate		Fishy		
Hydrodictyon	Green		Septic		
Mallomonas	Flagellate	Violet	Fishy		
Melosira	Diatom	Geranium	Musty		Slick
Meridion	Diatom		Spicy		
(Microcystis)	See Anacystis.				
Nitella	Green	Grassy	Grassy, septic	Bitter	
Nostoc	Blue-green	Musty	Septic		
Oscillatoria	Blue-green	Grassy	Musty, spicy		
Pandorina	Flagellate		Fishy		
Pediastrum	Green		Grassy		
Peridinium	Flagellate	Cucumber	Fishy		
Pleurosigma	Diatom		Fishy		
Rivularia	Blue-green	Grassy	Musty		
Scenedesmus	Green		Grassy		
Spirogyra	Green		Grassy		
Staurastrum	Green		Grassy		
Stephanodiscus	Diatom	Geranium	Fishy		Slick
Synedra	Diatom	Grassy	Musty		Slick
Synura	Flagellate	Cucumber, muskmelon, spicy	Fishy	Bitter	Dry, metallic, slick
Tabellaria	Diatom	Geranium	Fishy		
Tribonema	Green		Fishy		
(Uroglena)	See Uroglenopsis.				
Uroglenopsis	Flagellate	Cucumber	Fishy		Slick
Ulothrix	Green		Grassy		
Volvox	Flagellate	Fishy	Fishy		

* Reference 9.

Table 18–2. Representative Species of Taste and Odor Algae.*

GROUP AND ALGAE

Blue-Green Algae (Myxophyceae):
  Anabaena circinalis
  Anabaena planctonica
  Anacystis cyanea
  Aphanizomenon flos-aquae
  Cylindrospermum musicola
  Gomphosphaeria lacustris, kuetzingianum type
  Oscillatoria curviceps
  Rivularia haematites
Green Algae (nonmotile Chlorophyceae, etc.):
  Chara vulgaris
  Cladophora insignis
  Cosmarium portianum
  Dictyosphaerium ehrenbergianum
  Gloeocystis planctonica
  Hydrodictyon reticulatum
  Nitella gracilis
  Pediastrum tetras
  Scenedesmus abundans
  Spirogyra majuscula
  Staurastrum paradoxum
Diatoms (Bacillariophyceae):
  Asterionella gracillima
  Cyclotella compta
  Diatoma vulgare
  Fragilaria construens
  Stephanodiscus niagarae
  Synedra ulna
  Tabellaria fenestrata
Flagellates (Chrysophyceae, Euglenophyceae, etc.):
  Ceratium hirundinella
  Chlamydomonas globosa
  Chrysosphaerella longispina
  Cryptomonas erosa
  Dinobryon divergens
  Euglena sanguinea
  Glenodinium palustre
  Mallomonas caudata
  Pandorina morum
  Peridinium cinctum
  Synura uvella
  Uroglenopsis americana
  Volvox aureus

* Reference 9.

cetes, and they impart an earthy, musty-type odor resembling freshly plowed soil.[12] They are temperature-sensitive. Growth and germination will not occur below 59°F (15°C); above that temperature, growth is uninhibited up to 100°F (38°C). Growth is poor in deep water, but shallow reservoirs can produce high concentrations in warm weather. High alkalinity enhances growth, which is poor under acidic conditions.[3]

Actinomycetes spores are not always removed by waterworks filters. After passing through filters, spores attach to mineral deposits in distribution piping where they can grow and impart taste and odor not found in water leaving the treatment plant. Problems are especially severe at dead-ends of distribution systems where scale and debris accumulate and provide nutrients for growth of microorganisms. Even where water is not stagnant, nutrients can be provided by adsorption of organic matter on encrustations and scale in the distribution system. Growth of Actinomycetes has been reported in pipes in buildings where there is low water use at night, or during weekends at temperatures of 68°F (20°C).[13]

Nitrogen requirements for this organism are high, and a symbiotic relationship with algae is essential.[14] Large colonies of Actinomycetes were reported in association with a heavy growth of the green algae *Cladophora* in reservoirs in Oklahoma.[15] Other sources of nutrition include animal remains and aquatic plants. Germination often takes place in the bottom area of a reservoir or stream where nutrients are concentrated in residues. Actinomycetes produce their most intense odors and luxuriant growth when they grow in conjunction with blue-green algae.[16]

Decayed Vegetation

Odors attributed to decaying vegetation are due to:[7]

- Decaying algae
- Grass
- Weeds
- Leaves
- Disturbances of stream and reservoir sediments due to variations in velocity or water level

The above materials are continually decomposing, and are washed into watercourses during rainy periods. Decayed vegetation is second only to algae as a cause of taste and odor.[1]

Numerous types of odors have been attributed to decaying vegetation. Table 18–3 indicates odor types identified at various plants by Sigworth, who has suggested that Actinomycetes may often be responsible for odors generally attributed to decaying vegetation.[7]

Table 18–3. Odor Types Due to Decaying Vegetation.*

ODOR	NUMBER OF PLANTS REPORTING
Musty	69
Earthy	28
Woody	20
Moldy	17
Swampy	12
Grassy	9
Fishy	8
Wet leaves	7

* Reference 7.

Municipal and Industrial Wastes

Municipal Wastes. Sewage contains organic material, sulfides, ammonia, detergents, and other substances that can produce odor in drinking water; and odors from those constituents may be intensified when water is chlorinated. Threshold odor numbers of wastewater effluents have been measured at levels between 500 and 12,500.[17] Clearly, tremendous receiving water volumes or flows are required to dilute these wastes to the level where odors are barely detectable. Fortunately, streams remove some of the organic materials by biological degradation, adsorption, and sedimentation. Nevertheless, municipal wastewater can be a significant source of taste and odor in public water supplies.

Industrial Wastes. In industrialized regions of the United States, water supplies are often drawn from watercourses that contain industrial wastes. Water supplies containing these wastes may have taste and odor, which may be accentuated by chlorination. Although industrial wastes occur less frequently than natural odor sources, they can nonetheless produce high odor levels. Odor characteristics and threshold concentrations of selected compounds are presented in Table 18–4.[18] Odors classified by chemical type are presented in Table 18–5, and threshold levels of various chemicals are presented in Table 18–6.[19,20] Many of these chemicals are manufactured in large quantities and eventually find their way into industrial effluents and ultimately into water supplies. Values shown in Table 18–5 were obtained at 104°F (40°C) with only one chemical present in each sample. When chemicals are mixed, different thresholds are possible due to antagonistic or synergistic

Table 18–4. Odor Characteristics and Threshold Concentrations.*

SUBSTANCE	FORMULA	THRESHOLD ODOR (mg/l)	REMARKS
Allyl mercaptan	$CH_2{:}CH \cdot CH_2 \cdot SH$	0.00005	Very disagreeable, garlic-like odor
Ammonia	NH_3	0.037	Sharp, pungent odor
Benzyl mercaptan	$C_6H_5CH_2 \cdot SH$	0.00019	Unpleasant odor
Chlorine	Cl_2	0.010	Pungent, irritating odor
Chlorophenol	$Cl \cdot C_6H_4 \cdot OH$	0.00018	Medicinal odor
Crotyl mercaptan	$CH_3 \cdot CH{:}CH \cdot CH_2SH$	0.000029	Skunk odor
Diphenyl sulfide	$(C_6H_5)_2S$	0.000048	Unpleasant odor
Ethyl mercaptan	$CH_3CH_2 \cdot SH$	0.00019	Odor of decayed cabbage
Ethyl sulfide	$(C_2H_5)_2S$	0.00025	Nauseating odor
Hydrogen sulfide	H_2S	0.0011	Rotten egg odor
Methyl mercaptan	CH_3SH	0.0011	Odor of decayed cabbage
Methyl sulfide	$(CH_3)_2S$	0.0011	Odor of decayed vegetables
Pyridine	C_6H_5N	0.0037	Disagreeable, irritating odor
Skatole	C_9H_9N	0.0012	Fecal odor, nauseating
Sulfur dioxide	SO_2	0.009	Pungent, irritating odor
Thiocresol	$CH_3C_6H_4 \cdot SH$	0.0001	Rancid, skunklike odor
Thiophenol	C_6H_5SH	0.000062	Putrid, nauseating odor

* Reference 18.

effects. Data by Rosen et al. imply that odor addition is a characteristic of chemical mixtures.[21] They indicate that it may not be sufficient that a receiving stream dilute an odorous substance below its odor threshold because the substance might react with other substances to produce an easily detectable odor level. Baker reported synergistic (intensification) effects when odor testing selected chemical mixtures.[22]

Some of the most potent odors are discharged in effluents related to dyes, medicinal products, coke (quench water), ammonia recovery, wood oil, phenols, creosols and petroleum products, textiles, and paper product manufacturing.[8] Sigworth reported that phenol, chemical, and petroleum wastes were most frequently mentioned as odor sources.[7] As shown in Table 18–6, phenol has a low threshold odor level.

Phenols and related compounds at concentrations of 1 μg/l may cause taste when chlorinated. Although phenol is often attributed to industrial wastes, it is also present in sewage and is produced in decay of vegetation.[23] Streams draining nonindustrial watersheds have been found to contain phenol. Hoak reported that oak leaves suspended in river water for a month developed a phenol concentration of 1,460 μg/l. When diluted and chlorinated to free residual, the solution produced medicinal odors with a TON of 500.[24] Chlorination of phenolic compounds produces an iodine or medicinal taste. These

Table 18-5. Odors Classified by Chemical Types.*

ODOR CHARACTERISTICS**

SWEETNESS	PUNGENCY	SMOKINESS	ROTTENNESS	ODOR CLASS	CHEMICAL TYPES	EXAMPLES
100	50	0–50	50	Estery	Esters, ethers, lower ketones	Lacquer, solvents, most fruits, many flowers
100	50–100	0–100	50	Alcoholic	Phenols and cresols, alcohols, hydrocarbons	Creosote, tars, smokes, alcohol, liquor, rose and spicy flowers, spices and herbs
50	50	0–50	50	Carbonyl	Aldehydes, higher ketones	Rancid fats, butter, stone fruits and nuts, violets, grasses and vegetables
50	100	0–50	50	Acidic	Acid anhydrides organic acids, sulfur dioxide	Vinegar, perspiration, rancid oils, resins, body odor, garbage
100	50–100	50–100	0–100	Halide	Quinones, oxides and ozone, halides, nitrogen compounds	Insecticides, weed killers, musty and moldy odors, husks, medicinal odors, earth, peat
50	50	100	100	Sulfury	Selenium compounds, arsenicals, mercaptans, sulfides	Skunks, bears, foxes, rotting fish and meat, cabbage, onion, sewage
100	50	50	100	Unsaturated	Acetylene derivatives, butadiene, isoprene	Paint thinners, varnish, kerosene, turpentine, essential oils, cucumber
100	50	0–50	100	Basic	Vinyl monomers, amines, alkaloids, ammonia	Fecal odors, manure, fish and shellfish, stale flowers such as lilac, lily, jasmine, and honeysuckle

* Reference 19.
** The degree of odor characteristic perceived is designated as follows: 100 indicates a high level of perception, 50 a medium level of perception, and 0 a low level of perception.

Table 18-6. Odor Threshold Concentrations for Various Chemicals.*

CHEMICAL	NUMBER OF PANELISTS	NUMBER OF OBSERVATIONS	THRESHOLD ODOR LEVEL, mg/l**	
			AVERAGE	RANGE
Acetic acid	9	9	24.3	5.07–81.2
Acetone	12	17	40.9	1.29–330
Acetophenone	17	154	0.17	0.0039–2.02
Acrylonitrile	16	104	18.6	0.0031–50.4
Allyl chloride†	10	10	14,700	3,660–29,300
n-Amyl acetate	18	139	0.08	0.0017–0.86
Aniline	8	8	70.1	2.0–128
Benzene††	13	18	31.3	0.84–53.6
n-Butanol	32	167	2.5	0.012–25.3
p-Chlorophenol	16	24	1.24	0.02–20.4
o-Cresol	13	21	0.65	0.016–4.1
m-Cresol	29	147	0.68	0.016–4.0
Dichloroisopropylether	8	8	0.32	0.017–1.1
2,4-Dichlorophenol	10	94	0.21	0.02–1.35
Dimethylamine	12	29	23.2	0.01–42.5
Ethylacrylate	9	9	0.0067	0.0018–0.0141
Formaldehyde	10	11	49.9	0.8–102
2-Mercaptoethanol	9	9	0.64	0.07–1.1
Mesitylene††	13	19	0.027	0.00024–0.062
Methylamine	10	10	3.33	0.65–5.23
Methyl ethyl pyridine	16	20	0.05	0.0017–0.225
Methyl vinyl pyridine	8	8	0.04	0.015–0.12
B-Naphthol†	14	20	1.29	0.01–11.4
Octyl alcohol†	10	10	0.13	0.0087–0.56
Phenol	12	20	5.9	0.016–16.7
Pyridine	13	130	0.82	0.007–7.7
Quinoline	11	17	0.17	0.016–4.3
Styrene†	16	23	0.73	0.02–2.6
Thiophenol††	10	10	13.5	2.05–32.8
Trimethylamine	10	10	1.7	0.04–5.17
Xylene†	16	21	2.21	0.26–4.13
n Butyl mercaptan	8	94	0.006	0.001–0.06

* Reference 20.
** Threshold values based upon pure substances.
† Threshold of a saturated aqueous solution; solubility data not available.
†† Dilutions started with saturated aqueous solution at room temperature; solubility data obtained from literature for correction back to pure substances.

chlorophenols have a lower odor threshold concentration than phenol has. Chlorination of a treated water can cause formation of chlorophenolic compounds that will cause objectionable tastes and odors in distribution systems.

Ettinger and Ruchhoft reported that the medicinal taste of chlorophenols develops slowly, and that this slow reaction allowed formation of chlorophenols in distribution systems when the ratios of Cl_2:phenol were in the 0.5:1 to 3:1 range.[25] They also reported that phenols were completely oxidized when Cl_2:phenol ratios exceeded 3:1 and sufficient time was provided. Burtschell et al. reported that 2,6-dichlorophenol was always present when a chlorophenolic taste was identified.[26]

Agricultural Runoff

Animal wastes and fertilizers contribute nutrients necessary for the growth of algae and other odor-causing organisms in watercourses. Therefore, they are indirect producers of taste and odor. Insecticides, herbicides, and fungicides can produce objectionable odors in water.

Synthetic organic pesticides have been reported to produce especially objectionable taste and odors. Some chlorinated hydrocarbons, such as endrin and toxaphene, can be detected at concentrations of a few micrograms per liter.[27,28]

Hydrogen Sulfide

This compound, which is produced by the decomposition of natural organic material, industrial wastes, and municipal wastes, is characterized by its rotten egg odor. Anaerobic conditions in deep reservoirs can allow some organisms to convert sulfates to hydrogen sulfide, which may also be found in well water. Sulfates may be reduced to sulfides in the presence of organic matter under anaerobic conditions, with the resultant metallic sulfide being converted to hydrogen sulfide by action of carbonic acid. Hydrogen sulfide is very soluble in water (4,000 mg/l), where it can be detected at concentrations as low as 0.05 mg/l.[29] Sulfate-reducing bacteria (*Desulfovibro desulfurecans*) are anaerobes that convert sulfates and other sulfur compounds to hydrogen sulfide. They can be especially troublesome in distribution systems where they may proliferate in dead-ends.

Inorganic Minerals

The taste of water is affected by mineralization. As the concentration of dissolved solids increases, water becomes less palatable. A laboratory study reported the taste reaction of 15 trained subjects to eight concentrations of $NaCl$, $NaHCO_3$, Na_2SO_4, $CaSO_4$, $CaCl_2$, Na_2CO_3, $MgSO_4$, and $MgCl_2$, ranging from 125 to 2,000 mg/l.[30] The rating method used was the average taste intensity on a 13-point scale where 0 = none and 12 = extremely intense. The results are shown in Fig. 18–1. Taste intensity increases with an increase in dissolved minerals. At concentrations above 750 mg/l, chloride and carbonate tastes are more intense.[31] The predominant tastes for specific minerals were reported as follows:

- CaCl₂—bitter
- MgCl₂—slightly bitter and sweet
- NaHCO₃—very sweet
- Na₂SO₄—salty
- Na₂CO₃—bittersweet
- NaCl—salty

No distinguishing taste was reported for $CaSO_4$ or $MgSO_4$, which had the lowest intensities in Fig. 18–1.

Consumer evaluation rather than taste panel evaluation is the best definition of palatability. Results of a consumer survey of 29 water systems in California were reported by Bruvold et al.[32] The systems were selected to eliminate substances, other than dissolved minerals, that produce taste and odor. The dissolved minerals studied were calcium, magnesium, sodium, bicarbonate–carbonate, chloride, nitrate, and sulfate. No system surveyed had any record

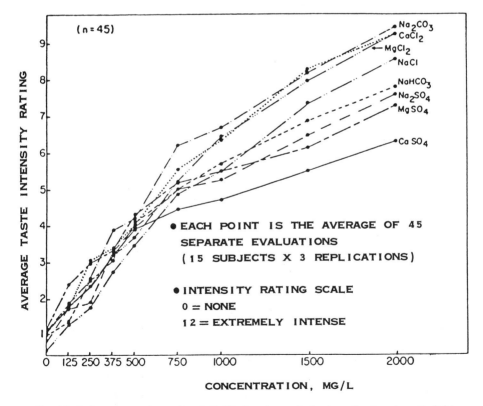

Fig. 18–1. Rated taste intensity of distilled water and of mineralized waters at eight concentrations.[30]

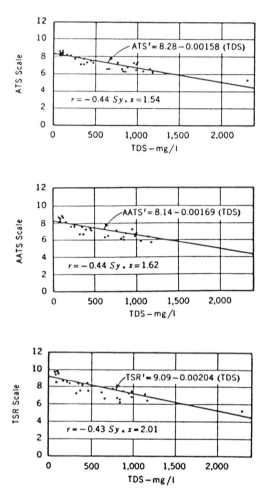

Fig. 18–2. Linear regression of ATS, AATS, and TSR scales. vs. TDS values.[32] *Note:* See Table 18–7 for explanation of ATS, AATS, and TSR scales.

of taste problems other than those attributable to common minerals. Approximately 60 individuals were selected as respondents from each community, and an interview schedule consisting of a series of interview questions, attitude scales, and a taste scale rating procedure was administered to participating respondents in their homes. To complete the taste scale rating, each respondent tasted a sample of his or her tap water under the direction of the interviewer. As shown in Fig. 18–2, the consumer assessment showed that taste quality decreased linearly as mineral content increased. Scale values are described in Table 18–7.

Table 18–7. Sample Items from the
Attitude Taste Scale (ATS), the Attitude
Adjective Taste Scale (AATS), and the
Taste Scale Rating (TSR) Procedure.[*,**]

ITEM	SCALE VALUE
ATS	
Perfect	10.57
Good	7.67
Neither good nor bad	6.00
A little bad	4.33
Bad	2.16
AATS	
Delicious	10.57
Fine	8.04
Average	6.09
Inferior	3.54
Awful	1.94
TSR	
Excellent taste	10.67
Good taste	8.45
Neutral taste	6.00
Bad taste	2.95
Horrible taste	1.16

[*] Reference 32.
[**] For use with Fig. 18–2.

The extent to which a person reacts to dissolved minerals in water can be affected by temperature. Taste intensity was found by one investigator to be greatest at body temperature and room temperature, and significantly reduced by chilling or heating.[33] It was reported to be uncertain whether chilling makes water more acceptable because of perceived coolness alone, or whether perception of the undesirable taste is diminished by cold temperature.

In the study, trained subjects scored the taste intensity of eight mineral solutes (750 and 1,000 mg/l) and six selected natural drinking waters. Taste intensity ratings of mineralized solutions at 750 mg/l and natural drinking waters are shown in Figs. 18–3 and 18–4, respectively, as a function of solution temperature. Generally, carbonates and chlorides showed the greatest taste intensity.

REMOVAL OF TASTE AND ODOR

Many processes and treatments are available and used to remove taste and odor, including:

- Chlorination
- Chlorine dioxide treatment
- Ozonation
- Potassium permanganate treatment
- Microorganism control: copper sulfate and powdered activated carbon
- Activated carbon treatment: powdered or granular
- Aeration

Chlorine, chlorine dioxide, ozone, and potassium permanganate oxidize taste- and odor-causing materials; copper sulfate kills algae that can cause taste and odor; activated carbon adsorbs organic substances; and aeration strips some offending materials and oxygenates the water.

Chlorination

Chlorination has been practiced in the waterworks industry for over 70 years. Although originally applied for disinfection purposes, chlorination is now

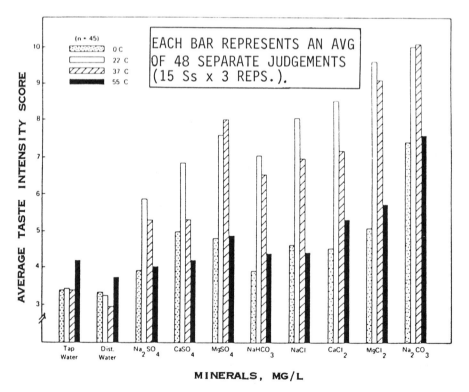

Fig. 18–3. Taste intensity* of tap water, distilled water, and solutions of eight minerals at 750 mg/l. Tested at 32, 72, 99, and 131°F (0, 22, 37, and 55°C).[33] (*13-point scale: 0 = none, and 12 = extremely intense.)

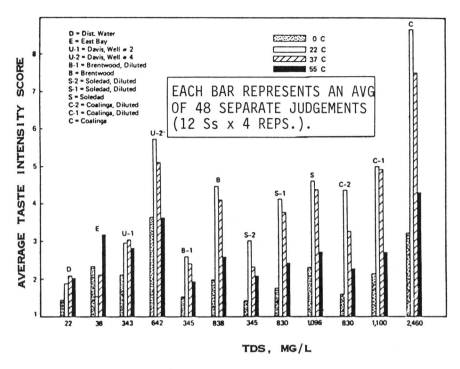

Fig. 18–4. Taste intensity* of distilled water, six natural waters, and mixtures of the two. Tested at 32, 72, 99 and 131°F (0, 22, 37, and 55°C).[33] (* 13-point scale: 0 = none, and 12 = extremely intense.)

widely used for control and removal of taste and odor. Prior to 1925, marginal doses of 0.05 to 0.2 mg/l chlorine were used. Most waterworks officials thought that lower odor levels could be achieved with low chlorine dosages. Unfortunately, the public frequently objected to tastes and odors that were traceable to these chlorination practices.

Chlorination practices are divided into four major groups:[34]

- Combined residual treatment (chlorine–ammonia)
- Free residual chlorination (breakpoint)
- Superchlorination and dechlorination
- Chlorine–chlorine dioxide

Combined Residual Treatment. Combined residual treatment is the least effective method of achieving taste and odor control. It does not oxidize materials that exist in untreated water, and is used primarily as a preventive measure rather than to destroy taste- and odor-causing substances. Ammonia is often added before chlorine, thus preventing chlorine from combining with organic material and producing taste- and odor-causing substances. White and Cox suggest ammonia:chlorine doses of 1:3 to prevent reactions between chlorine and phenols and other taste-producing compounds.[8,35] Cox suggests residual chloramine concentrations of 0.5 to 1.5 mg/l to prevent organism growth in distribution mains.

Controversy exists over how to apply chlorine to a distribution system to control biofouling; some believe that free residual chlorine is the better method, others that a combined residual (mostly monochloramine) is more effective. Proponents of combined residual chlorine argue that:[8]

- Combined residuals, although less potent that free residuals, persist longer and eventually penetrate farther into a system.
- When free residuals are delivered to a badly contaminated system, the residual may eventually be "turned back" toward the hump of the break-point curve and become combined available residual. As this happens, tastes are produced by the inevitable formation of nitrogen trichloride and dichloramines.
- The formation of combined residual chlorine will occur in consumer service pipes long after distribution pipes have been cleaned.

Breakpoint Chlorination. This process is the application of chlorine to water to destroy ammonia or nitrogeneous organic material that may be present so that a free available chlorine residual results.[34] Most taste- and odor-causing substances can be oxidized by chlorine. The breakpoint chlorine demand of potable water varies and can range from 1 to 15 mg/l. Theoretically, an ammonia nitrogen to chlorine ratio of 1:10 by weight will attain breakpoint. Where organic material is present, chlorine requirements will be higher. Breakpoint chlorine dosage is often considered to be the amount of chlorine required to satisfy the immediate chlorine demand plus ten times the ammonia concentration.

Superchlorination and Dechlorination. Superchlorination is the application of chlorine in much greater quantities than would be required to achieve a free chlorine residual. Superchlorination increases taste- and odor-destroying reaction rates.[34] The resulting high chlorine levels will render the water unpalatable, and dechlorination will be required before delivery to the distribution system. Superchlorination is especially useful when detention times are limited and rapid reactions are required. The longer the reaction period for superchlorinated water, the lower the chlorine dose required to achieve taste and odor destruction. Although disinfection may require only short detention times, the desired minimal reaction time for taste and odor control is 2 hours.[35]

Chlorine–Chlorine Dioxide Treatment. In this process, chlorine dioxide is applied to waters previously treated with chlorine. Although it is a powerful oxidant, chlorine dioxide is expensive, and is often used in conjunction with chlorine to reduce costs. Chlorine dioxide is discussed in the following subsection, as well as in Chapter 11 and by White.[8]

Chlorine Dioxide

Chlorination frequently introduces new tastes and odors to water even when breakpoint and superchlorination have been utilized, and there are concerns about the long-term health effects of chlorinated organic compounds that can be formed. Chlorine dioxide has been used to overcome these concerns. It has an oxidative capacity 2.5 times that of chlorine, but is unstable as a gas and must be produced on site.

The most popular method of generating ClO_2 in waterworks practice is by the chlorination of sodium chlorite:

$$2NaClO_2 + Cl_2 \rightarrow 2ClO_2 + 2NaCl \qquad (18-1)$$

One pound (0.45 kg) of chlorine dioxide is produced by the reaction of 1.3416 pounds (0.61 kg) pure sodium chlorite with 0.5 pound (0.23 kg) chlorine. Because sodium chlorite is not 100 percent pure, 1.68 pound (0.76 kg) chlorite will be required in the above reaction if the sodium chlorite is 80 percent pure.

Another common method of chlorine dioxide generation is based on acidification of sodium chlorite:

$$5NaClO_2 + 4HCl \rightarrow 4ClO_2 + 5NaCl + 2H_2O \qquad (18-2)$$

White has described other sources of chlorine dioxide as (1) the hypochlorite–chlorite–sulfuric acid process, and (2) stabilized aqueous ClO_2 solutions that can be delivered to water treatment plants.[8]

Chlorine dioxide has a greenish-yellow color with an odor similar to that of chlorine, but it is more irritating and toxic than chlorine. It is very soluble in water (five times more soluble than chlorine). Chlorine dioxide is explosive, but dilution with inert gases reduces the explosion hazard. It is harmless as an aqueous solution and does not react chemically with water.

Chlorine dioxide is particularly effective in controlling odors associated with phenols and chlorophenols. These compounds are oxidized to odorless and tasteless end products after the addition of chlorine dioxide. Some investigators have suggested that tasteless trichlorophenol is formed; while others concluded that chlorine dioxide ultimately ruptured the benzine ring, forming end products.[36,37]

Chlorine dioxide offers several advantages for treating odors and tastes from phenolic compounds:[8]

- ClO_2, unlike chlorine, will not react with ammonia or nitrogenous compounds. It is always available, as $HOCl$ would be if the latter did not react with ammonia or other nitrogenous compounds to form chloramine.
- Because ClO_2 has 2½ times the oxidative capacity of $HOCl$, it destroys taste-producing phenolic compounds at a faster rate than free available chlorine does.
- In every case studied, it has been reported that ClO_2 destroys any chlorophenol taste caused by prechlorination.
- Unlike $HOCl$, ClO_2 efficiency is not impaired by high pH.

Frequently, chlorine is applied as a pretreatment for disinfection and algae control, and chlorine dioxide is applied after filtration for taste and odor control. As a result, heavy doses of inexpensive chlorine are used on substances that exhibit high chlorine demand. The more expensive chlorine dioxide is used on the filtered water where the dosage is lower. Formation of trihalomethanes may be a consideration in such applications.

It may be economical to use chlorine dioxide as a pretreatment for taste and odor control only where industrial wastes are present in the water source.

The application rate of chlorine dioxide is determined on a case-by-case basis. For general applications, a range of 2 to 3 lb/MG (0.24 to 0.36 kg/Ml) sodium chlorite have been satisfactory.[35] This will yield chlorine dioxide concentrations of approximately 0.2 to 0.3 mg/l. Dosages of 0.3 to 10 lb/MG (0.036 to 1.20 kg/Ml) sodium chlorite have been used for control of phenolic taste and odors.[38] Compounds causing marshy odors and tastes can be oxidized at the rate of approximately 1 mg/l chlorine dioxide for every 5 TON units. Fifty percent more chlorine dioxide may be required if musty tastes and odors are involved.[16] Chlorine dioxide can also be used successfully to treat certain microorganisms such as decaying vegetation, algae, and Actinomycetes (see Table 18–9).

Ozonation

Ozone is a powerful oxidizing chemical. It is unstable and therefore must be produced at the point of application by passing clean dry air between electrically charged surfaces. Typically, about 1 percent of the oxygen is converted to ozone. Values as high as 2.4 percent by weight (1.4 percent by volume) can be achieved.[39] The development of dielectric, ozone-resistant materials has simplified the design and operation of ozone generators, and new designs have increased efficiency.

Desirable properties of ozone and chlorine are complementary in many ways. Ozone produces fast-acting germicidal and viricidal potency, frequently with beneficial results regarding taste, odor, and color. Chlorine provides sustained disinfection action that continues to be beneficial in the distribution system. A combination of ozonation and chlorination provides both disinfection and taste and odor control. At Amsterdam, on the Rhine River, ozone is used initially for phenol oxidation, viral destruction, and general improvement in physical quality. Post-chlorination is then used to assure hygienic quality throughout the distribution system. It is likely that this approach may find increasing use in the United States as improvements in ozone generation are made.

O'Donovan reported the successful application of ozone where taste and odor are caused by algae.[40] Ozone dosages of 1 to 4 mg/l are typically used.[1] For odor oxidation, ozonation of water at the Belmont Plant in Philadelphia resulted in reduction of threshold odors of over 50 percent with an ozone dose of 1.3 mg/l.[41] Powell et al. reported that ozone reduced taste and odors in Iowa River waters when the raw water threshold odor values are high, but was of little use when the values were moderate to low.[42]

The use of ozone does have disadvantages, as summarized by Morris.[43] Because it must be produced electrically as it is needed and cannot be stored, it is difficult to adjust treatment to variations in load or to changes in raw water quality with regard to ozone demand. As a result, ozone historically has been found most useful for supplies with low or constant demand, such as groundwater sources. Moreover, although ozone is a highly potent oxidant, it is quite selective and by no means universal in its action. Some otherwise easily oxidized substances, such as ethanol, do not react readily with ozone.

Where heavily polluted river waters have been ozonated, large organic molecules can be ruptured into fragments more easily metabolized by microorganisms. This fragmentation, coupled with the inability of ozone to maintain an active concentration in the distribution system, can lead to increased slime growths and deterioration of water quality during distribution.

Ozonation rapidly oxidizes iron and manganese into insoluble compounds that can cause color problems. O'Donovan reports that 0.05 mg/l of manganese or 0.1 mg/l of iron can cause difficulties.[40] White indicates that ozone treatment of water containing manganese changes the peaty brown color to a pinkish color, due to traces of permanganate.[8] Permanganate is then changed to colloidal manganese dioxide, which imparts a brown color.

Ozone is more expensive than chlorine in the United States. For each pound (kilogram) of ozone produced and applied, 10 kWh of electricity is consumed at large plants and 12 to 15 kWh at smaller plants. Thus, energy costs alone represent a cost of 50 to 75 cents/lb ($1.10 to $1.65/kg) ozone at most plants.

Potassium Permanganate

Potassium permanganate ($KMnO_4$) is an effective oxidizing agent for destroying taste- and odor-causing substances. In addition to taste and odor removal, it also oxidizes soluble iron and manganese to insoluble oxides that can be removed by coagulation, flocculation, sedimentation, and/or filtration.[44] In water treatment, the manganese in permanganate is reduced from the +7 oxidation state to the +4 state, which is insoluble manganese dioxide. When permanganate is added to water, a pink or purple color results. If reducible material and alkaline conditions are present, this color will change to yellow-brown, which indicates that permanganate has been reduced to the manganic state. A flocculant, insoluble manganese dioxide is formed.

The disappearance of the pink or purple color can be used to indicate the necessary contact time of potassium permanganate with material in the water, and also to regulate feed rates. Normal practice is to allow this coloration to extend approximately two-thirds of the way across a sedimentation basin. Filtration is nearly always required to remove insoluble manganese oxide hydrates. Sedimentation efficiency is important because large amounts of these insoluble manganese compounds can clog filters and produce short filter runs. If overdosed, permanganate may pass through the treatment plant and enter distribution systems where it will form manganese dioxide, with resulting consumer complaints of stained clothing and fixtures. In the pH ranges used in water treatment, the general permanganate reactions are:[45]

$$KMn^{+7}O_4 + H_2O$$
$$+ \begin{bmatrix} \text{Organic compounds} \rightarrow \text{Carbon dioxide} + \text{Water} \\ \text{Soluble iron} \rightarrow \text{Ferric precipitate} \\ \text{Soluble manganese} \rightarrow \text{Manganese dioxide} \\ \text{Hydrogen sulfide} \rightarrow \text{Soluble sulfate} \end{bmatrix} + Mn^{+4}O_2 \quad (18\text{–}3)$$

Potassium permanganate can be added at a number of points in the treatment process, but is normally added in pretreatment at the rapid mix stage. It is usually added prior to chlorination to prevent formation of chloro-organic derivatives (such as chlorophenol) that may produce taste or odor. In this way, permanganate will oxidize these substances prior to contact with chlorine.

When used with powdered activated carbon, permanganate treatment is performed first. If carbon and permanganate are added simultaneously, permanganate consumption can increase.

Potassium permanganate has been shown in some cases to reduce or eliminate the need for treatment by activated carbon. Shull reported that a sizable saving resulted when permanganate and powdered carbon were used in place of carbon alone.[48] Sanks reports a water plant using Lake Michigan water achieved a substantial reduction in chemical costs and an improvement in treated water threshold odor when potassium permanganate replaced powdered activated carbon.[45] Dosages were 0.3 to 0.5 mg/l potassium permanganate. When carbon was fed with permanganate, chemical costs were reported to be less than for use of carbon alone to control odors. For other water supplies, Sanks indicates this combination treatment has been found to be more effective and economical than treatment with either chemical alone.[45] Permanganate has also been used as an algicide in reservoirs.

Potassium permanganate dosages range from 0.5 to 15 mg/l. One manufacturer recommends dosages ranging from 0.5 to 2.5 mg/l to control most taste- and odor-causing chemicals.[46] Visual control is easily effected by observing the characteristic pink color of potassium permanganate in solution.[47]

Microorganism Control: Copper Sulfate and Powdered Activated Carbon

Copper Sulfate. Copper sulfate is the most generally effective and easily applied algicide. Ideally, an algicide should be inexpensive, kill the nuisance microorganisms, be nontoxic to fish and other large organisms, and not degrade water quality. More than any other compound, copper sulfate most nearly meets these requirements. Copper sulfate is most often applied to open reservoirs. Some methods of application include:[1]

- Burlap bags or stainless steel screens containing the chemical that are dragged through the water.
- Scattering of crystals over the water surface.
- Power boats equipped with solution, tank, pump, and distributor pipe.

Distribution is important and is frequently controlled by boat speed and travel pattern. Hopkins and Bean have indicated that parallel paths or triangular patterns are the best travel patterns.[1] Wind, wave diffusion, and gravity currents promote mixing. These systems are illustrated in Fig. 18–5.

Copper sulfate can be toxic to fish if applied improperly or too heavily. Toxic levels of the chemical for selected fish are presented in Table 18–8.[49] When applied to deep layers of a lake or reservoir, copper sulfate will not be toxic because it is precipitated by natural alkalinity and then diffused into the lower levels at nontoxic concentrations. Whipple indicated that proper application depends on seven factors:[50]

- Kind of algae to be destroyed
- Amount of organic matter present
- Hardness of water
- Carbonic acid content
- Water temperature
- Quantity of water
- Species of fish

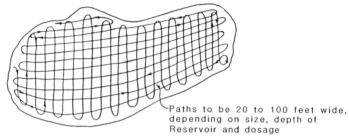

Paths to be 20 to 100 feet wide,
depending on size, depth of
Reservoir and dosage

ADAPTABLE PARTICULARLY TO LARGE RESERVOIRS

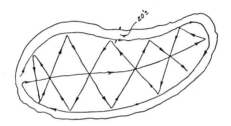

ADAPTABLE PARTICULARLY TO SMALL RESERVOIRS

Fig. 18–5. A suggested system for applying copper sulfate to reservoirs.[1] (Courtesy of Frank E. Hale)

The dosages of copper sulfate, chlorine, and chlorine dioxide required for treatment of different microorganisms are presented in Table 18–9. Often a single application is not sufficient. Application of copper sulfate may kill one type of microorganism (such as blue-green algae), only to be followed by the rapid rise of another type. Supplementary treatment with chlorine

Table 18–8. Killing Dosage of Copper Sulfate for Fish.*

FISH	COPPER SULFATE mg/l	COPPER mg/l	COPPER SULFATE lb/MILLION GALLON	kg/ml
Trout	0.14	0.04	1.2	0.14
Carp	0.33	0.08	2.8	0.34
Suckers	0.33	0.08	2.8	0.34
Catfish	0.40	0.10	3.5	0.42
Pickerel	0.40	0.10	3.5	0.42
Goldfish	0.50	0.12	4.2	0.50
Perch	0.67	0.17	5.5	0.66
Sunfish	1.35	0.34	11.1	1.33
Black bass	2.00	0.50	16.6	1.99

* Reference 49.

Table 18–9. Chemicals Required for Treatment of Different Genera of Microorganisms (mg/l).*

ORGANISMS	ODOR	COPPER SULFATE	CHLORINE	CHLORINE DIOXIDE**
Diatomaceae				
Asterionella	Aromatic, geranium, fishy	0.12–0.20	0.5–1.0	
Cyclotella	Faintly aromatic		1.0	0.4
Diatoma	Faintly aromatic			
Fragilaria	Musty†	0.25		
Melosira	Musty†	0.20	2.0	0.8
Meridion	Aromatic			
Navicula		0.07		0.2
Nitzschia		0.50		
Pleurosigma	Fishy†			
Synedra	Earthy	0.36–0.50	1.0	0.7
Stephanodiscus	Fishy†	0.33		
Tabellaria	Aromatic, geranium, fishy	0.12–0.50	0.5–1.0	
Chlorophyceae				
Actinastrum	Grassy, musty†			
Chlorella	Musty†			
Cladophora	Septic†	0.50		1.2
Closterium	Grassy†	0.17		1.4
Coelastrum		0.05–0.33	1.0	
Conferva		0.25		
Cosmarium	Grassy†			
Desmidium		2.00		
Dictyosphaerium	Grassy, nasturtium, fishy		0.50–1.0	1.2
Draparnaldia		0.33		
Eudorina	Faintly fishy	2.00–10.0		1.6
Enteromorpha		0.50		
Gleocystis	Offensive	0.50		
Hydrodictyon	Very offensive	0.10		
Microspora		0.40		
Palmella		2.00		2.0
Pandorina	Faintly fishy	2.00–10.0		3.8
Pediastrum	Grassy†			
Protococcus			1.0	
Raphidium		1.00		
Scenedesmus	Grassy†	1.00		1.6
Spirogyra	Grassy†	0.12	0.7–1.5	0.3
Staurastrum	Grassy	1.50		0.8
Tetrastrum			1.0	
Tribonema	Fishy			
Ulothrix	Grassy†	0.20		0.3
Volvox	Fishy	0.25	0.3–1.0	
Zygnema		0.50		
Cyanophyceae				
Anabaena	Moldy, grassy, vile	0.12–0.48	0.5–1.0	1.2
Aphanizomenon	Moldy, grassy, vile	0.12–0.50	0.5–1.0	
Clathrocystis	Sweet, grassy, vile	0.12–0.25	0.5–1.0	

Table 18–9 (Continued)*

ORGANISMS	ODOR	COPPER SULFATE	CHLORINE	CHLORINE DIOXIDE**
Coelosphaerium	Sweet, grassy	0.20–0.33	0.5–1.0	0.6
Cylindrospermum	Grassy	0.12		
Gleocapsa (red)		0.24		0.4
Gleotrichia	Grassy†			
Microcystis		0.20		0.5
Nostoc	Septic			
Oscillaria	Musty–spicy†	0.20–0.50	1.1	0.5
Rivularia	Moldy, grassy			
Protozoa				
Bursaria	Irish moss, salt marsh, fishy			
Ceratium	Fishy, vile (rusty-brown color)	0.24–0.33	0.3–1.0	0.5
Chlamydomonas	Fishy, septic†	0.36–1.0		1.6
Chrysosphaerella	Fishy†			
Cryptomonas	Candied violets	0.50		
Dinobryon	Aromatic, violets, fishy	0.18	0.3–1.0	1.5
Endomaeba histolytica (cysts)			25.0–100.0	
Eudorina	Fishy†			
Euglena	Fishy†	0.50	1.0	0.8
Glenodinium	Fishy	0.50		
Gonium	Fishy†			
Mallomonas	Aromatic, violets, fishy	0.50		
Peridinium	Fishy, like clamshells	0.50–2.00		
Synura	Cucumber, muskmelon, fishy (bitter taste)			0.4
Uroglena	Fishy, oily (cod-liver oil taste)	0.05–0.20	0.3–1.0	
Rotifer				
Stentor		0.24		
Crustacea				
Cyclops				
Daphnia		2.00		
Schizomycetes				
Beggiatoa	Very offensive, decayed	5.00		
Cladothrix		0.20		
Crenothrix	Very offensive, decayed, medicinal with chlorine	0.33–0.50	0.5	
Leptothrix				
Sphaerotilis natans	Very offensive, decayed	0.40		
Spirophyllum			0.25	
Thiothrix			0.5–1.0	
Fungus				
Achlya			0.6	
Leptomitus		0.40		
Saphrolegnia		0.18		

Table 18-9 (Continued)*

ORGANISMS	ODOR	COPPER SULFATE	CHLORINE	CHLORINE DIOXIDE**
Miscellaneous				
Chara	Spoiled garlic†	0.10–0.50		
Nitella flexilis	Objectionable	0.10–0.18		
Potamogeton		0.30–0.80		
Nais				1.0
Blood worm,				
chironomus				15.0–50.0
Blood worm, gnats				3.0

* Reference 51.
** Chlorine dioxide data of J. K. G. Silvey, *Texas Manual for Water Works Operators,* Lancaster Press, 1959.
† Odors for abundant conditions, from USPHS Publ. No. 657, 1959.[9]

or activated carbon may be needed. To prevent reoccurring growths, copper sulfate applications may then be required on a regular basis. Hopkins and Bean indicate that treatment may be necessary every 2 weeks in the temperate zone.[1] They also indicate that climate, organisms, nutrients, and temperature affect application frequency. The National Secondary Drinking Water Regulations MCL for copper is 1.0 mg/l. It is recommended that copper sulfate applications not cause this MCL to be exceeded.

Activated Carbon.   Many troublesome microorganisms in the waterworks industry are chlorophyllous and thus require sunlight. They can be controlled by applying powdered activated carbon to create "blackout" conditions that prevent light penetration into water, thereby preventing growth of algae. Typical application rates are 10 to 80 lb/MG (1.20 to 9.59 kg/Ml) of water to be treated. The application of carbon also may remove taste and odors associated with algae. Application is typically by methods similar to those used for copper sulfate. This technique generally is used only in shallow areas. Treatment is only temporary because the carbon gradually settles. Repeated applications usually are not made because of the relatively high cost of activated carbon.

Other Measures.   Algae can also be controlled with organic algicides. Unfortunately, these algicides can be toxic to fish, have as yet unknown effects on humans, and can impart taste and odor even at very low concentrations. Their use requires strict supervision. Application is usually made at remote locations where serious problems arise in only localized areas. A summary of studies on other compounds used as algicides can be found in the *Manual of Water Utility Operations.*[16] The manual stresses the desirability of control of microorganisms by biological means. Unfortunately, it has yet not been possible to manipulate biological systems to achieve effective control of species that grow in water supplies.

Activated Carbon

Activated carbon adsorption is generally considered to be the most effective method of removing taste and odor from water supplies. Both powdered and granular activated carbons are used for taste and odor removal, the powdered form having been the more widely used. (A detailed discussion of activated carbon characteristics is presented in Chapter 12.)

Powdered Activated Carbon. Powdered activated carbon is a finely ground insoluble black powder that has been ground to a size of 50 mesh or smaller. Application is made by dry chemical feeders or by slurry. Capacity, proportioning, constant delivery, wetting, high-rate delivery of slurry, and dust collection are important considerations regarding carbon feeding equipment.[52]

Activated carbon can be added at any point in the treatment process prior to filtration. Locations include raw water intake, rapid mix basins, flocculators, sedimentation tanks, and immediately prior to entrance to filters. Frequently, powdered activated carbon is added at more than one point in the treatment process. Powdered carbon added after filters is not removed and will pass into the distribution system. The most suitable location(s) for carbon addition can be determined by adding equal doses to samples of water from each potential application point and determining residual odors. The following are important considerations in choosing application points:[5]

- Activated carbon functions best at low pH values.
- Carbon must remain in suspension and circulation to provide adequate contact time.
- Care must be exercised to prevent coagulants and other chemicals from sealing or coating the active surface of the carbon.

Regardless of the point of application, sufficient mixing must be provided, together with adequate contact time. Good mixing ensures that the carbon is evenly dispersed in as large a volume of water as possible. Adequate contact time provides the opportunity for the adsorbate to be adsorbed onto the carbon surface.

A raw water intake provides an excellent application point because sufficient contact time is typically available before water reaches the plant. Carbon added to rapid mix basins is dispersed effectively and then has contact time in flocculation and sedimentation basins. Carbon added to the rapid mix offers the advantage of providing a nucleus for floc formation, which can be of value in low-turbidity waters. However, carbon's adsorptive capabilities are reduced once it is incorporated into floc. Softening chemicals that raise the water pH will reduce the effectiveness of carbon.[5] Contact time may also be adequate when carbon is applied at sedimentation tank outlets prior to filtration. However, operational care is required to prevent the carbon from penetrating and escaping filters. Required carbon contact time varies from plant to plant. A contact time of 15 minutes to 1 hour is suggested by AWWA, and contact times exceeding 1 hour are recommended only if mixing is poor.[5] Hyndshaw recommends a contact period of 10 to 15 minutes.[53]

Many treatment plants utilize both activated carbon and chlorine for taste and odor control. Addition of both chemicals simultaneously can reduce the effectiveness of each.[7] Chlorine reacts with activated carbon as indicated:

$$C + 2Cl_2 + 2H_2O \rightarrow 4HCl + CO_2 \qquad (18\text{--}4)$$

When powdered activated carbon and chlorine are used simultaneously, carbon consumes approximately 1 pound (0.45 kg) of chlorine for every 20 pounds (9.1 kg) of carbon added.[8] Carbon can be added before or after chlorine in the treatment process. A survey conducted by Sigworth indicated that 37 percent of reporting plants applied carbon before chlorine, and 28 percent applied chlorine prior to carbon.[7] Hyndshaw indicated that where pH, disinfection, and other factors adversely affected taste and odor control, carbon should be applied first to prevent formation of offending tastes and odors.[53] However, he found that when disinfecting chemicals do not produce offensive substances, carbon addition afterward will prove more economical. Industrial wastes are frequently treated by chlorinating first, followed by activated carbon to adsorb chlorinated organics.[1] Despite recommendations in the literature that carbon and chlorine be added separately, Sigworth reported that 35 percent of plants responding to a survey indicated that both products were added simultaneously.[7] He speculated that the reason was limitations imposed by existing plant designs.

Dosages vary considerably according to point of application, presence or absence of industrial wastewaters, algae, and other odor- and taste-causing constituents. Dosages as low as 0.6 mg/l have successfully removed some taste and odors. The following dosage ranges have been reported:[5]

- Routine continuous application: 2 to 8 mg/l
- Intermittently severe problems: 5 to 20 mg/l
- Emergency spills (chemical spills): 20 to 100 mg/l

A survey by Sigworth found average dosages of 2.1 mg/l.[7] He found that a dosage of 3 to 4 mg/l was representative when odors were present, and that the mean of maximum dosages was 30 mg/l.

Granular Activated Carbon. Granular activated carbon (GAC) is also used for taste and odor control. Taste and odor control has been achieved in some cases by adding a shallow [1 foot (0.31 m) or less] layer of GAC on top of an existing sand filter, or by substituting a properly sized and graded bed of GAC for a rapid sand filter. Unlike powdered activated carbon, which is applied to the waste stream and removed with plant sludges, GAC remains as part of the filter and must be regenerated when its adsorptive capacity is reached. (Detailed information regarding design and operation of GAC treatment is presented in Chapter 12.)

Aeration

Aeration can be used to remove volatile taste- and odor-causing substances from water. It is more effective in removing hydrogen sulfide than most other taste- and odor-producing substances. Aeration is of questionable value when used solely for taste and odor control, and is usually used when oxidation of iron and manganese or removal of carbon dioxide is also required.

One measure of the volatility of a substance is partial pressure. Substances with high partial pressure are volatile and are easily removed by aeration. For example, carbon dioxide has a high vapor pressure, even at low temperatures, and is therefore readily removed. On the other hand, phenol, a notorious taste-producing substance, has a relatively low vapor pressure; so it is not volatile and is not effectively removed. Unfortunately, many odor-producing compounds are not volatile enough to be readily removed by aeration.

Hydrogen sulfide, one of the few taste- and odor-producing compunds to be effectively removed by aeration, hydrolyzes as follows:

$$H_2S \rightleftharpoons H^+ + HS^- \tag{18-5}$$

Hydrosulfide (HS^-) disassociates as follows:

$$HS^- \rightleftharpoons H^+ + S^{-2} \tag{18-6}$$

The existence of H_2S is pH-dependent, as indicated in Fig. 18–6. The pH must be below approximately 7.5 to ensure that a high percentage of the sulfur is in the H_2S form and to encourage stripping by air. Unfortunately, when water containing both carbon dioxide and hydrogen sulfide is aerated, the nonsoluble carbon dioxide is easily removed, thus increasing the pH. Consequently, the HS^- and S^{-2} forms tend to dominate, decreasing the effectiveness of H_2S removal. Removal of hydrogen sulfide by aeration can be enhanced by adding carbon dioxide to lower the pH, so that H_2S predominates. Carbon dioxide is removed by aeration later in the flow scheme. Some investigators have found that hydrogen sulfide is nearly impossible to remove com-

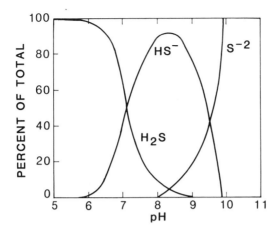

Fig. 18-6. Effect of pH on H_2S–S^{-2} equilibrium.

pletely by aeration, and recommend further oxidation with chlorine for complete removal.[54]

Many types of aerators are employed in waterworks practice. Spray aerators
have fixed nozzles that force water into the air to form fine droplets. Nozzle
designs vary, but typical pressures range from 1 to 10 psi (6.9 to 69.8 kPa).
These aerators are seldom housed, and ventilation is not a problem. This
type of aerator suffers from icing problems in cold weather. Discharge head
requirements are somewhat higher than for other types of aeration equipment.

Cascade aerators force water to flow over a series of obstructions to produce
turbulence. Generally, water is allowed to fall over a number of steps for a
predetermined height. Concrete step structures are commonly used. Water
is spread out in thin sheets and then drops over weirs to the next level.
Head requirements are 3 to 10 feet (0.91 to 3.05 m), and a surface area of
40 to 50 sq ft/mgd (0.98 to 1.22 m²/Ml/d) is needed. Cascade aerators are
not as efficient as tray aerators.

Tray aerators allow water to fall through a series of perforated, slat or
wire mesh bottoms. Water is distributed evenly over the top and collected
in a basin at the bottom. Coarse media can be placed in the trays to increase
gas transfer effectiveness. The number of trays varies from three to nine.
The area of each tray is approximately 25 to 50 sq ft/mgd (0.61 to 1.22
m²/Ml/d). Water is applied at 14 to 28 gpm/sq ft (34.2 to 68.3 m/h). Ventilation is an important design consideration.

Diffused aerators consist of tanks into which perforated pipes or diffusers
are placed to allow air to bubble through the water to be treated. Typical
detention time ranges from 10 to 30 minutes. Air is typically applied at a
rate from 0.01 to 0.15 cu ft/gal (74.8 to 112.2 m³/Ml) of water treated,

and power requirements vary from 0.5 to 2.0 kW/mgd (0.13 to 0.52 KW/Ml/d) of capacity.[5] This type of aerator has the following advantages:

- No housing is required.
- The plant hydraulic grade line is unaffected (no pumping is required).
- There are no problems with cold weather operation.

As previously indicated, aeration is of questionable value when used solely for taste and odor control. Normally, organic compounds are not sufficiently volatile to be removed by aeration, and other techniques are used to remove them. However, exceptions have been reported in the literature. During warm months, a 44 percent reduction in threshold odor number was achieved at Appleton, Wisconsin by using spray aeration.[55] Aeration was not so effective during colder months. At Nitro, West Virginia, water with threshold numbers of 5,000 to 6,000 was effectively reduced using high pressure [55 psi (37.9 kPa)] spray aeration.[56] High nozzle abrasion rates and poor cold weather operation were reported.

REFERENCES

1. Hopkins, E. S., and Bean, E. L., *Water Purification Control,* Robert E. Krieger Publishing Company, Melbourne, Florida, 1975.
2. Fair, G. M., and Geyer, J. C., *Elements of Water Supply and Waste-Water Disposal,* John Wiley and Sons, New York, 1958.
3. American Society for Testing and Materials, *Manual on Water,* 3rd ed., American Society for Testing and Materials, Philadelphia, 1969.
4. American Public Health Association, *Standard Methods,* 14th ed., American Public Health Association, Washington, D.C., 1976.
5. American Water Works Association, *Water Quality and Treatment,* 3rd ed., McGraw-Hill Book Co., New York, 1971.
6. Geldard, F. A., *The Human Senses,* John Wiley and Sons, New York, 1953.
7. Sigworth, F. A., "Control of Odor and Taste in Water Supplies," *J.AWWA* 49:1507, December, 1957.
8. White, G. C., *Handbook of Chlorination,* Van Nostrand Reinhold, New York, 1972.
9. Palmer, C. M., "Algae in Water Supplies," Public Health Service Publication No. 657, 1959.
10. Birge, E. A., and Juday, C., "The Inland Lakes of Wisconsin: The Plankton, I: Its Quantity and Chemical Composition," *Wisconsin Geol. and Natural Hist. Bul.,* 64, 1922.
11. Silvey, J. K. G., Russell, J. C., Redden, D. R., and McCormick, W. C., "Actinomycetes and the Common Tastes and Odors," *J.AWWA* 42:1018, November, 1950.
12. Romano, A. H., and Safferman, R. S., "Studies on Actinomycetes and Their Odors," *J.AWWA* 55:169, 1963.
13. Silvey, J. K. G., "Proceedings of the 5th Sanitary Engineering Conference," University of Illinois, Engineering Experimental Station, Circular No. 81, 1963.
14. Faust, S. D., and Osman, M. A., *Chemistry of Water Treatment,* Butterworths Publishers, Boston, 1983.
15. Silvey, J. K. G., and Roach, A. W., "Actinomycetes in the Oklahoma City Water Supply," *J.AWWA* 45:409, April, 1953.

16. Texas Water Utilities Association, *Manual of Water Utility Operations,* 7th ed., Austin, Texas, 1979.

17. Hartung, H. O., "Taste and Odor," *J.AWWA* 52:1363, November, 1960.

18. Dague, R. R., "Fundamentals of Odor Control," *J.WPCF* 44:5, April, 1972.

19. American Society for Testing and Materials, *1980 Annual Book of ASTM Standards,* Part 31, American Society for Testing and Materials, Philadelphia, 1980.

20. Baker, R. A., "Threshold Odors of Organic Chemicals," *J.AWWA* 55:913, July, 1963.

21. Rosen, A. A., et al., "Odor Thresholds of Mixed Organic Chemicals," *J.WPCF* 34:7, January, 1962.

22. Baker, R. A., "Odor Effects of Aqueous Mixtures of Organic Chemicals," *J.WPCF* 35:728, June, 1963.

23. Hoak, R. D., "Tastes and Odors in Drinking Water," *Water and Sewage Works* 104:243, June, 1957.

24. Hoak, R. D., "Evaulating Taste and Odor Control Problems," *J.AWWA* 52:517, April, 1960.

25. Ettinger, M. B., and Ruchhoft, C. C., "Stepwise Chlorination on Taste and Odor Producing Intensity of Some Phenolic Compounds," *J.AWWA* 43:561, 1951.

26. Burtschell, R. H., et al., "Chlorine Derivatives of Phenol Causing Taste and Odor," *J.AWWA* 51:205, February, 1959.

27. Gaufin, A. R., et al., "Bioassays Determine Pesticide Toxicity," *Water and Sewage Works* 108:355, September, 1961.

28. Middleton, F. M., et al., "Tentative Method for Carbon Chloroform Extract in Water," *J.AWWA* 54:223, February, 1962.

29. McKee, J. E., and Wolf, H. W., *Water Quality Criteria,* 2nd ed., California State Water Quality Control Board, Sacramento, 1963.

30. Pangborn, R. M., Trabue, I. M., and Baldwin, R. E., "Sensory Examination of Mineralized Chlorinated Waters," *J.AWWA* 62:572, September, 1970.

31. Bruvold, W. H., and Pangborn, R. M., "Rated Acceptability of Mineral Taste in Water," *J. Appl. Psychol.* 50:22, January, 1966.

32. Bruvold, W. H., Ongerth, H. J., and Dillehay, R. C., "Consumer Assessment of Mineral Taste in Domestic Water," *J.AWWA* 61:575, November, 1969.

33. Pangborn, R. M., and Bertolero, L. L., "Influence of Temperature on Taste Intensity and Degree of Liking of Drinking Water," *J.AWWA* 64:511, August, 1972.

34. Ryckman, D. W., and Grigoropoulos, S. G., "Use of Chlorine and Its Derivatives in Taste and Odor Removal," *J.AWWA* 51:1268, October, 1959.

35. Cox, C. R., *Laboratory Control of Water Purification,* Case-Shepperd-Mann Publishing Corporation, New York, 1946.

36. Adams, B. A., "Substances Producing Taste in Chlorinated Water," *Water and Sewage Works* 33:109, 1931.

37. Faber, H. A., "A Theory of Taste and Odor Reduction by Chlorine Dioxide," *J.AWWA* 39:691, July, 1947.

38. Granstrom, M. L., and Lee, G. F., "Generation and Use of Chlorine Dioxide in Water Treatment," *J.AWWA* 50:1453, November, 1958.

39. Walker, R., *Water Supply, Treatment and Distribution,* Prentice-Hall, Englewood Cliffs, N.J., 1978.

40. O'Donovan, D. C., "Treatment with Ozone," *J.AWWA* 57:1167, September, 1965.

41. Bean, E. L., "Ozone Production and Costs," *Advances in Chemistry* Series No. 21, American Chemical Society, 1959.

42. Powell, M. P., et al., "Action of Ozone on Tastes and Odors and Coliform Organics," *J.AWWA* 44:1144, December, 1952.

43. Morris, J. C., "Chlorination and Disinfection—State of the Art," *J.AWWA* 63:769, December, 1971.

44. Welch, W. A., "Potassium Permanganate in Water Treatment," *J.AWWA* 55:735, June, 1963.
45. Sanks, R. L., *Water Treatment Plant Design,* Ann Arbor Science, Ann Arbor, Mich., 1978.
46. Carus Chemical Company, "The CAIROX Method for Water Treatment," Form M-1050, La Salle, Ill., 1972.
47. Reidies, A. H., "Potassium Permanganate Treatment of Water," *Water Works Engr.* 116:260, 1963.
48. Shull, K. E., "Operating Experiences at Philadelphia Suburban Water Treatment Plants," *J.AWWA* 54:1332, October, 1962.
49. Moore, G. T., and Kellerman, K. F., Bureau of Plant Industry, Bulletin No. 76, U.S. Department of Agriculture, 1905.
50. Whipple, G. C., *The Microscopy of Drinking Water,* John Wiley and Sons, New York, 1948.
51. Phelps Dodge Refining Corp., *Copper Sulfate in Control of Microscopic Organisms,* New York, 1957.
52. Sigworth, E. A., "The Production of Palatable Water: Part I," *Taste and Odor Control Journal* 27:7, October, 1961.
53. Hyndshaw, A. Y., "Treatment Application Points for Activated Carbon," *J.AWWA* 54:91, 1954.
54. Powell, S. T., and Von Lossberg, L. G., "The Removal of Hydrogen Sulfide from Well Waters," *J.AWWA* 40:1277, December, 1948.
55. Gallaher, W. V. "Control of Algae at Appleton, Wisconsin," *J.AWWA* 32:1165, July, 1940.
56. Haynes, L., and Grant, W., "Reduction of Chemical Odors at Nitro, West Virginia," *J.AWWA* 37:1013, October, 1945.

Chapter 19
Iron and Manganese Removal

INTRODUCTION

Water containing iron and manganese is objectionable to consumers because precipitation of these metals causes the water to turn yellow-brown or black. This rusty-colored water causes yellow-brown stains in plumbing and disagreeable tastes in drinking water and beverages, and stains laundered clothing. Water containing less than 0.1 mg/l iron and 0.05 mg/l manganese is not objectionable to the average customer, although certain industries may require manganese levels as low as 0.01 mg/l. An AWWA task group suggested limits of 0.05 mg/l for iron and 0.01 mg/l for manganese for an "ideal" quality water for public use.[1] The National Secondary Drinking Water Regulations maximum contaminant levels for iron and manganese are 0.3 mg/l and 0.05 mg/l, respectively.[2] Although regulations and suggested limits allow small concentrations of manganese, many waterworks people believe that public water supplies should be free of manganese.

Another problem associated with the presence of iron and manganese is the growth of microorganisms in distribution systems. Slime thicknesses of several centimeters have been observed in distribution pipes. These accumulations, which consist of hydrous iron, manganese oxides, and bacteria, reduce pipeline capacity, require higher chlorine dosages, and deplete dissolved oxygen levels. Sloughing or resuspension of this material by high flows causes high turbidities and complaints of red or black water.[3] Taste and odor complaints can be expected when any of these products decomposes prior to use. Discussions of bacteria associated with the presence of iron and manganese can be found in *Standard Methods* and *Water Quality and Treatment.*[3,4]

Two types of iron and manganese are found in water supplies, and methods for their removal are quite different. Inorganic iron and manganese generally are associated with clearwell water which becomes turbid when oxidized. Organic iron is present in both groundwater and surface waters. The formation of organic complexes and chelates may increase the solubility of iron and manganese. At pH values encountered in natural waters, it is possible that these substances are insoluble but highly dispersed. The natural color of water is frequently due to such highly stabilized colloidal dispersions of iron (II). Manganese (IV) does not readily form complexes with organic or inorganic ligands in water.

OCCURRENCE

Iron and manganese are natural constituents of the earth's crust, with iron one of the most abundant elements. The lithosphere contains approximately 5 percent iron and 0.1 percent manganese; hence iron is found more frequently and in greater concentrations than manganese.[5] Iron occurs in the silicate minerals of igneous rocks, occurring primarily as insoluble ferric oxide and also as slightly soluble ferrous carbonate (siderate). Manganese is found in metamorphic and sedimentary rocks, with only small amounts in igneous rocks.[3] Manganese is present as manganese dioxide and manganese carbonate (rhodochrosite).

Iron and Manganese in Groundwaters

Groundwaters generally require treatment for iron and manganese removal much more often than surface waters; only rarely does free-flowing surface water require such treatment. Iron and manganese are dissolved in groundwater supplies by the action of carbon dioxide on carbonate-bearing minerals (see below). Carbon dioxide is generated by biological decomposition of organic matter leached from soils; the presence of anaerobic conditions and agents such as organic substances and hydrogen sulfide leads to the reduction of iron and manganese oxides to the ferrous Fe(II) and manganous Mn(II) states. Ferrous carbonate and manganous carbonate are dissolved by carbon dioxide as follows:

$$FeCO_3 + CO_2 + H_2O \rightarrow Fe(HCO_3)_2 \qquad (19\text{--}1)$$

$$MnCO_3 + CO_2 + H_2O \rightarrow Mn(HCO_3)_2 \qquad (19\text{--}2)$$

Ferrous bicarbonate is the form of iron most commonly found in water supplies. Iron in natural waters may also be found as ferric hydroxide, ferrous sulfate, or colloidal or organic iron.[6] In waters of low pH, zero dissolved oxygen, and saturated carbon dioxide, the solubility of iron can be 150 mg/l or more expressed as Fe. The concentrations of iron and manganese found in groundwaters generally do not exceed 10 mg/l and 2 mg/l, respectively.

Iron and Manganese in Surface Waters

As previously discussed, free-flowing surface water rarely requires treatment for iron and manganese removal. Exceptions are streams where industrial wastes or acid mine drainage are present. Acid mine drainage may hold considerable amounts of iron and manganese in solution, even in the presence of oxygen.

Free-flowing streams that do not contain iron or manganese can yield troublesome concentrations when impounded. Vegetation and other debris in the hypolimnion of lakes or impoundments decompose to produce conditions of high carbon dioxide and no dissolved oxygen. Iron and manganese present in the soils, vegetation, and sediments are dissolved and distributed throughout the impoundment, especially during the fall overturn. When reduced iron and manganese ions rise to the surface during this period, they are oxidized and precipitated. Thus waters near the surface of a reservoir are likely to contain the lowest concentrations of iron and manganese. Carefully designed raw water intake structures allow selection of water at multiple locations, including those above deep water that may contain iron and manganese and below surface water that may contain troublesome algae. Manganese concentrations in the hypolimnion of reservoirs have been reported to range from 2.0 mg/l to 20 mg/l or more.[7]

Iron and manganese can be particularly troublesome in newly formed impoundments, and problems can continue for many years, especially during summer stagnation periods. On the other hand, some reservoirs may not develop problems initially, and may require as long as 10 years before significant concentrations are developed.[8]

Hopkins and McCall reported that manganese is extracted from bottom muck by dissolved carbon dioxide.[9] Further studies by the same investigators indicated that manganese on the bottom of two reservoirs in Maryland was leached from the original organic material left in the unstripped reservoirs and also from the muck deposited from runoff.[10] Although it would be reasonable to expect a reduction of manganese concentrations with time due to siltation of the originally exposed material on the bottom, this reduction does not often occur. Objectionable concentrations of manganese have persisted indefinitely in many impoundments, especially during certain periods of the year. A survey of 22 reservoirs in the eastern United States concluded that the age of a reservoir apparently has little bearing on manganese content of stored water during the July to November period.[11]

REMOVAL OF IRON AND MANGANESE

There are many methods used for the removal of iron and manganese in public water supplies. The primary method involves oxidation, precipitation, and filtration. Other methods include ion exchange, stabilization, and lime softening. Specifically, chemicals and methods to control iron and manganese include:

- Aeration, precipitation, filtration
- Chlorination or chlorine dioxide oxidation, precipitation, filtration

- Potassium permanganate oxidation, precipitation, filtration
- Ion exchange (zeolite) softening
- Manganese–zeolite filtration
- Stabilization or sequestering
- Lime softening

Aeration, Precipitation, Filtration

The most common method of removing iron and manganese from water supplies involves aeration, precipitation, and filtration. Simple aeration converts ferrous bicarbonate to ferrous hydroxide as follows:

$$Fe(HCO_3)_2 \text{ (aeration)} \rightarrow Fe(OH)_2 + 2CO_2 \qquad (19\text{--}3)$$

Further aeration yields:

$$4Fe(OH)_2 + O_2 + 2H_2O \text{ (aeration)} \rightarrow 4Fe(OH)_3 \qquad (19\text{--}4)$$

The rate of oxidation of Fe(II) by oxygen is slow under conditions of low pH, as indicated by Fig. 19–1.[12] Aeration of water low in dissolved oxygen and high in carbon dioxide will tend to raise the pH slightly because carbon dioxide is easily removed by aeration. Reaction rates are fairly slow at pH less than 7, and a pH of 7.5 to 8.0 may be required to complete the reaction within 15 minutes.[13] In some cases, 1 hour or more of contact time may be needed to complete the reactions. Organic iron is not removed by aeration. Frequently, another strong oxidizing agent such as chlorine or potassium permanganate is required to oxidize and then remove this type of iron.

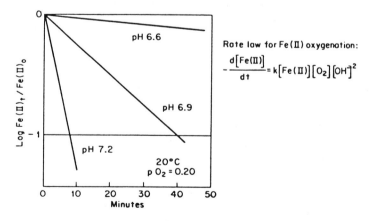

Rate law for Fe(II) oxygenation:

$$-\frac{d[Fe(II)]}{dt} = k[Fe(II)][O_2][OH^-]^2$$

Fig. 19–1. Oxygenation of Fe(II).[12]

The rate of oxygenation of Mn(II) is slow at pH values less than 9.5, as shown in Fig. 19–2.[14] Theoretically, 1 gram of oxygen is required to oxidize 7 grams Fe(II), and 1 gram of oxygen is required to oxidize 3.5 grams of Mn(II).

Self-precipitation of ferric and manganic hydroxides can be slow; therefore, flocculation rates frequently are accelerated by contact and by catalysts. Contact aerators and contact filters are used to accelerate oxidation. Water is trickled over coke, crushed stone, or other contact materials. Deposits of hydrated oxides of iron and manganese accumulate on the surfaces and catalyze further oxidation to ferric and manganic oxides. Limestone and pyrolusite (MnO_2) have been used for this purpose.[15] Contact beds are usually arranged in a series of trays 12 to 16 inches (0.305 to 0.406 m) deep with perforated bottoms to enhance aeration. Loading rates range from 15 to 20 gpm/sq ft (36.6 to 48.8 m/h).[16]

Sufficient detention time must be provided for oxidation and flocculation of iron and manganese compounds. Kinetics of each water supply vary, and field testing is recommended. Filtration is always required for final removal of iron and manganese compounds. Sedimentation basins are rarely provided unless required for other reasons. Detention tanks suffice for reaction/flocculation basins.

Chlorination or Chlorine Dioxide Oxidation, Precipitation, Filtration

Both chlorine and chlorine dioxide are powerful oxidizing agents that are used to oxidize iron and manganese. Chlorine is generally used because of its lower cost, although chlorine dioxide may be preferable if trihalomethane

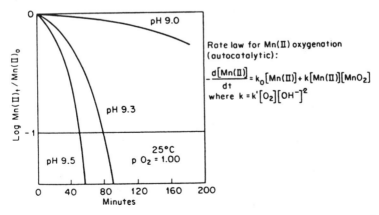

Fig. 19–2. Removal of Mn(II) by oxygenation.[14] Reprinted by permission of John Wiley & Sons, Inc.

formation is a problem. Free or combined chlorine reacts to oxidize ferrous iron as follows:

$$2Fe(HCO_3)_2 + Cl_2 + Ca(HCO_3)_2 \rightarrow 2Fe(OH)_3 + CaCl_2 + 6CO_2 \quad (19\text{--}5)$$

Reaction/flocculation time (similar to aeration) is required to effect removal of both iron and manganese. The above reaction takes place over a wide pH range of 4 to 10, with an optimum of 7.0. At pH 7.0, the reaction requires a maximum of 1 hour.[8] One milligram of iron requires 0.62 mg of chlorine for oxidation. Chelated iron is easily oxidized by chlorine. White reports that where iron is present in the complex organic form, free residual chlorine is more effective than combined chlorine in breaking up the iron complex so that oxidation can proceed.[8] Frequently, preaeration is practiced to strip CO_2 and raise the pH so that oxidation by chlorine is more effective.

Manganese is similarly oxidized by chlorination. The optimum pH for the oxidation of manganese is 7.0 to 8.0.[8] At pH 8.0 the reaction may take 2 to 3 hours, whereas at pH 6.0 as much as 12 hours may be required. At high pH values, oxidation is completed within minutes.[7,17] A dosage of 0.77 mg chlorine is required for each 1 mg of manganese.[17,18]

Chlorine dioxide reacts much more rapidly with manganous compounds than does chlorine, and is therefore successful for manganese removal. The high cost of chlorine dioxide treatment limits its use to applications where manganese concentrations are less than 1 mg/l. The reaction is:

$$2ClO_2 + MnSO_4 + 4NaOH \rightarrow MnO_2 + 2NaClO_2 + Na_2SO_4 + 2H_2O \quad (19\text{--}6)$$

Chlorine dioxide is not normally used for iron removal. Chlorine is preferable unless the water contains ammonia or other nitrogenous compounds that interfere with the formation of free available chlorine.

Potassium Permanganate Oxidation, Precipitation, Filtration

The reaction of potassium permanganate in oxidizing ferrous iron is enhanced by the formation of MnO_2, which acts as a catalyst. The use of potassium permanganate for removal of both iron and manganese is attractive because reactions are rapid and complete. Generally, only a small quantity of chemical is required. The reaction with iron is:

$$3Fe(HCO_3)_2 + KMnO_4 + 7H_2O \rightarrow MnO_2$$
$$+ 3Fe(OH)_3 + KHCO_3 + 5H_2CO_3 \quad (19\text{--}7)$$

Theoretically, 1 gram of $KMnO_4$ oxidizes 1.06 grams of iron. However, because of the catalytic effect of precipitated manganese dioxide, less $KMnO_4$

is generally required. Walker indicates that approximately 0.6 mg $KMnO_4$ per 1.0 gram of ferrous iron normally will suffice.[19]

Potassium permanganate addition is a popular method of removing manganese. Potassium permanganate will oxidize manganous ions to manganese dioxide rapidly and over a wide pH range. The reaction time can be as short as 5 minutes, and the permanganate required for oxidation decreases with increasing pH. Oxidation is accomplished as follows:

$$3Mn\,(HCO_3)_2 + 2KMnO_4 + 2H_2O \rightarrow 5MnO_2 + 2KHCO_3 + 4H_2CO_3 \quad (19\text{--}8)$$

Theoretically, 1 mg $KMnO_4$ oxidizes 0.52 mg divalent manganese to insoluble tetravalent manganese. In practice, however, less permanganate is required unless manganese is organically bound, in which case the theoretical amount is required. Potassium permanganate dosages are compared to other oxidant dosages in Table 19–1. The oxidation reaction of permanganate is rapid. Adams reported completed reactions within 5 minutes.[20] Reaction rates and required dosages of permanganate reported by Adams for a synthetic water containing 1 mg/1 soluble manganese and 0.5 mg/1 total residual chlorine are presented in Table 19–2. As indicated, less permanganate is required at higher pH values. Generally, an alkaline pH of 7.5 to 8.0 is required to effect coagulation, flocculation, and filtration of colloidal MnO_2.

Prechlorination frequently reduces required permanganate dosages for iron and manganese. This is advantageous because it reduces the overall cost of permanganate, which is a more expensive oxidant than chlorine. Adams (Table 19–3) shows the amount of permanganate required to oxidize 1.0 mg/1 manganese at various levels of prechlorination of one type of water at pH 8.0 at Wilkinsburg, Pennsylvania.[20]

Ion Exchange (Zeolite Softening)

Sodium cation zeolite exchange and hydrogen cation exchange units can remove ferrous iron (Fe(II)) and manganous manganese (Mn(II)). This treat-

Table 19–1. Comparison of Oxidants for Oxidation of Iron and Manganese.

	THEORETICAL WEIGHT TO OXIDIZE 1 mg OF:	
OXIDANT	IRON (II)	MANGANESE (II)
Oxygen, O_2	0.14 mg	0.29 mg
Chlorine, Cl_2	0.62	0.77
Chlorine dioxide, ClO_2	1.21	2.45
Potassium permanganate, $KMnO_4$	0.94	1.92

Table 19–2. Dosages and Speed of Oxidation by Permanganate for Synthetic Water Containing 1.0 mg/l Soluble Manganese and 0.5 mg/l Total Residual Chlorine.[*]

pH	KMnO₄ REQUIRED, mg/l	REACTION TIME, MINUTES
5.0	1.8	<5
6.0	1.6	<5
7.2	1.4	<5
8.0	1.4	<5
9.0	1.0	<5

[*] Reference 20.

ment removes hardness as well as iron and manganese. It is important that this process be accomplished in the absence of oxygen to prevent oxidation of any ferrous hydroxide to ferric hydroxide, which can foul the zeolite, stop ion exchange reactions, and clog filters. If the iron or manganese has been preoxidized, ion exchange cannot be used. Frequently, pressure filters with a presand unit to retain any precipitated hydroxides are installed.[16] Generally, this ion exchange process is restricted to individual water supplies containing low concentration of iron and manganese (0.5 mg/l or less) because precipitates reduce the effectiveness of exchange media.[3]

Table 19–3. Amounts of Permanganate Needed to Oxidize 1.0 mg/l Manganese at Various Prechlorination Levels at Wilkinsburg, PA. at pH 8.0.[*,**]

CHLORINE RESIDUAL BY TEST, mg/l		KMnO₄, mg/l
FREE	COMBINED	
0.00	0.00	2.2
0.00	0.50	1.2
0.40	1.70	0.9
4.0	0.90	0.6
12.0		

[*] The prechlorination–permanganate combinations are set up on the basis of a fast reaction; that is, the chlorine contact is approximately 10 minutes, and the mixing time with permanganate is 10 minutes.
[**] Reference 20.

Manganese–Zeolite Filtration

Removal of iron and manganese also is practiced without softening by passing water through manganese-impregnated greensand (zeolite). Manganese greensand is a product manufactured by treating New Jersey glauconite with manganous sulfate and potassium permanganate to provide an active supply of iron and manganese oxides on sand grains. When the oxidizing power of the bed is exhausted, it is regenerated with permanganate, reused, and returned to service. These beds are quite effective in high-carbonate, iron-bearing waters, but are exhausted too quickly if other reducing substances such as organic matter, nitrogenous matter, or hydrogen sulfide are present.[21] Greensand has the ability to oxidize and filter. When the amount of precipitate is large, a layer of crushed anthracite coal is placed over the exchange medium to prolong filter runs.

Because the regeneration step described above is time-consuming, continuous regeneration can be performed. Continuous regeneration has been described by Willey and Jennings.[21] In this procedure, a solution of potassium permanganate is fed continuously into the raw water line ahead of the filter to reduce the amount of soluble iron and manganese applied to the filter. Willey and Jennings reported that the dose of permanganate did have critical limits; undertreatment caused manganese leakage, and overtreatment caused pink effluent from the filter bed. Between these two extremes, however, they described a broad range of dosage within which undertreated and overtreated waters would have had either additional oxygen imparted to them or excess oxidant removed by the filter bed, with a resulting iron- and manganese-free effluent.

Greensand capacity for either iron or manganese removal is 0.09 lb/cu ft (1.46 kg/m^3) of mineral, requiring 0.18 lb/cu ft (2.92 kg/m^3) potassium permanganate for regeneration. Application rates are usually 3 gpm/sq ft (7.32 m/h), and the process is generally limited to water containing approximately 1 mg/l iron or manganese.[13]

Stabilization or Sequestering

An alternative to removing iron and manganese is to hold them in solution by stabilization, sequestering, or dispersion. Iron and manganese must be in the ionic state for this process to be effective. Generally it is appropriate to well waters containing sufficient carbon dioxide to ensure that iron and manganese are present as bicarbonates. The sequestering agent is pumped into the well to contact iron and manganese prior to oxidation.[19] Sequestering agents include sodium hexametaphosphate, trisodiumphosphate, and sodium silicates.

Table 19–4. Relationship of pH Value to the Maximum Precipitation of Manganese and Iron.*

pH	MANGANESE RESIDUAL mg/l	MANGANESE PRECIPITATED mg/l	MANGANESE AMOUNT, MOLE	MANGANESE PERCENTAGE	IRON RESIDUAL mg/l	IRON PRECIPITATED mg/l	IRON AMOUNT, MOLE	IRON PERCENTAGE
7.0	5.49	0.00	0.000000	0.0	5.59	0.00	0.000000	0.0
8.0	5.49	0.00	0.000000	0.0	1.53	4.06	0.000073	72.6
8.4					0.95	4.64	0.000083	83.0
8.8	5.49	0.00	0.000000	0.0	0.47	5.12	0.000093	91.6
8.9	3.00	2.49	0.000045	45.4				
9.0	1.90	3.59	0.000065	65.4				
9.2	1.00	4.49	0.000081	81.7				
9.3	0.28	5.21	0.000094	94.9				
9.4	0.11	5.38	0.000098	98.0	0.06	5.53	0.000099	98.9
9.6	0.03	5.46	0.000099	99.4	0.00	5.59	0.000100	100.0
9.8	0.00	5.49	0.000100	100.0				

* Reference 23.

Sequestering is applicable only when iron and manganese concentrations are less than 2.5 mg/l. Howell describes dosages of 2 mg hexametaphosphate for every 1 mg iron.[22] AWWA suggests dosages of 5 mg sodium hexametaphosphate for every 1 mg Fe plus Mn.[3] Polyphosphate dosages are limited to less than 10 mg/l because phosphorus can stimulate biological growths in distribution systems. Chlorine residuals must be sufficient to control bacterial slime growths when polyphosphates are applied to maintain colloidal dispersions of iron and manganese and to prevent calcium carbonate deposition. Sequestering agents are effective in cold water systems; however, when water is heated or boiled, polyphosphate loses its dispersing properties. Thus, the use of sequestering agents is rare in community water systems.

Lime Softening

Chemical precipitation of iron and manganese is effected at pH values approximating their isoelectric point of 9.4. Manganese and iron are effectively removed by lime addition. In the absence of oxygen, iron precipitates as ferrous hydroxide and manganese as manganous hydroxide. The solubility of $Mn^{+2}(OH)_2$ as calculated from its solubility product, $(Mn^{+2})(OH)^2$, of 7.1×10^{-15} is 3.9 mg/l as Mn at a pH of 9.0. The solubility of ferrous hydroxide $Fe(OH)_2$ is 0.92 mg/l at pH 9.0. The solubility of both $Mn^{+2}(OH)_2$ and $Fe(OH)_2$ decreases 100-fold for each unit rise in pH. Significant precipitation of ferrous hydroxide and manganous hydroxide generally proceeds only above pH values of 9.5 and 10.0, respectively.[15] The relationship of pH value to maximum precipitation of manganese and iron is presented in Table 19–4.[23] As indicated, manganese and iron residuals were 0.03 mg/l or less at a pH value of 9.6. Precipitation of iron and manganese with lime is usually not cost-effective unless lime treatment is also required for hardness reduction.

REFERENCES

1. Bean, E. L., "Progress Report on Water Quality Criteria," *J.AWWA* 54:1313, November, 1962.
2. U.S., Executive Department, National Secondary Drinking Water Regulations, 40 CFR 143, *Federal Register,* p. 42198, July 19, 1979.
3. American Water Works Association, *Water Quality and Treatment,* 3rd ed., McGraw-Hill Book Co., New York, 1971.
4. APHA, AWWA, WPCF, *Standard Methods for the Examination of Water and Wastewater,* 14th ed., American Public Health Association, Washington, D.C., 1976.
5. Robinson, L. R., and Dixon, R. I., "Iron and Manganese Precipitation in Low Alkalinity Groundwaters," *Water and Sewer Works* 115:514, November, 1968.
6. Nordell, E., *Water Treatment for Industrial and Other Uses,* Van Nostrand Reinhold, New York, 1951.
7. Griffen, A. E., "Significance and Removal of Manganese in Water Supplies," *J.AWWA* 52:1326, October, 1960.

8. White, G. C., *Handbook of Chlorination,* Van Nostrand Reinhold, New York, 1972.
9. Hopkins, E. S., and McCall, G. B., "Seasonal Manganese in a Public Water Supply," *Ind. Eng. Chem.* 24:106, January, 1932.
10. Hopkins, E.S., and McCall, G. B., "Manganese in Deep Reservoirs," *Ind. Eng. Chem.* 33:1491, December, 1941.
11. Flentje, M. E., "How Is Your Manganese?," *Water Works Engineering* 113:288, April, 1960.
12. Stumm, W., and Lee, G. F., "Oxygenation of Ferrous Iron," *Ind. Eng. Chem.* 53:143, 1961, American Chemical Society.
13. Conneley, E. J., "Removal of Iron and Manganese," *J.AWWA* 50:697, May, 1958.
14. Morgan, J. J., "Chemical Equilibria and Kinetic Properties of Manganese in Natural Waters," in S. D. Faust and J. V. Hunter, *Proceedings 4th Rudolfs Conference, Principles and Applications of Water Chemistry,* John Wiley & Sons, New York, © 1967.
15. Fair, G. M., and Geyer, J. C., *Elements of Water Supply and Wastewater Disposal,* John Wiley & Sons, New York, © 1958.
16. Hopkins, E. S., and Bean, E. L., *Water Purification Control,* 4th ed., Robert E. Krieger Publishing Co., Huntington, N.Y., 1975.
17. Griffin, A. E., and Baker, R. J., "The Breakpoint Process for the Free Residual Chlorination," *J.NEWWA* 73:250, September, 1959.
18. Edwards, S. E., and McCall, G. B., "Manganese Removal by Breakpoint Chlorination," *Water and Sewer Works* 93:303, August, 1946.
19. Walker, R., *Water Supply Treatment and Distribution,* Prentice-Hall, Englewood Cliffs, N.J., 1978.
20. Adams, R. B., "Manganese Removal by Oxidation with Potassium Permanganate," *J.AWWA* 52:219, February, 1960.
21. Willey, B. F., and Jennings, H., "Iron Removal with Potassium Permanganate," *J.AWWA* 55:729, June, 1963.
22. Howell, D. H., "Sodium Metaphosphate Glass in Water Treatment," *Proceedings American Society of Civil Engineers* 83:1464, December, 1957.
23. Hopkins, E. S., and Whitman, E. R., "Study of the Floc Produced by Chlorinated Copperas," *Ind. Eng. Chem.* 22:79, January, 1930, American Chemical Society.

Chapter 20
Fluoridation and Defluoridation

INTRODUCTION

The element fluorine is found in every water supply used for drinking purposes.[1] Fluorine is required for the formation of bones and teeth, and fluoride ions are essential to the normal growth and development of humans.

This chapter reviews the history of research on fluorides in drinking water, followed by a description of the chemicals used for fluoridation and the techniques used for fluoride addition to drinking water. Also, methods are described for the removal of fluorides in situations where their concentration is too great to allow potable use of the water supply.

FLUORIDE RESEARCH

During the past century, U.S. immigration officials noticed that people arriving from certain parts of Europe were severely afflicted by a disfigurement of the teeth known as mottled enamel or dental fluorosis. This led dental authorities to believe that the disfigurement was due to a local factor endemic to the immigrants' native land. Soon after, reports began to appear of mottled enamel among people native to the United States. These reports came largely from cities in the Great Plains and Rocky Mountain states.

Substantial evidence that fluorides were the cause of mottled enamel was obtained by H. V. Churchill in 1930.[2] The people of Bauxite, Arkansas reported a high incidence of mottled enamel. Churchill, by spectrographic analysis, found appreciable amounts of flouride ion in the Bauxite water supply. In collaboration with F. S. McKay, a dentist, he studied waters from five areas where mottling was endemic and 40 areas where it was not a problem. From these studies, it was concluded that excessive flouride levels in the drinking water caused the mottled enamel.[3] Further proof was reported by Smith et al., who found that mottled enamel could be produced in white rats by adding to their diets either small amounts of fluoride salts or the concentrated residues from waters known to cause mottled teeth in humans.[4]

Finally, Gottlieb in 1934 reported on the relationship between fluoride concentration and mottling.[5] She found that in Kansas communities reporting mottled enamel, the concentrations of fluoride in the municipal drinking water supplies were in excess of 2 mg/l.

In 1938, Dean presented data demonstrating that dental caries were less prevalent when mottled enamel occurred.[6] This led to extensive correlation studies on dental caries versus fluoride levels in drinking waters throughout the United States. The results obtained, as summarized by Dean, are presented in Fig. 20–1.[7] From this information, and the dental fluorosis data, a dental caries–fluoride relationship evolved:

1. When the fluoride level exceeds about 1.5 mg/l, any further increase does not significantly decrease the incidence of decayed, missing, or filled teeth, but does increase the occurrence and severity of mottling.
2. At a fluoride level of approximately 1.0 mg/l, the optimum effect occurs, that is, maximum reduction in caries with no aesthetically significant mottling.
3. At fluoride levels below 1.0 mg/l, some benefit occurs, but dental caries reduction is not great, and it gradually decreases as the fluoride levels decrease until, as zero fluoride is approached, no observable improvement occurs.

From the relationship between dental caries and fluoride concentration, the allowable and recommended fluoride concentrations shown in Table 20–1 were determined. The recommended fluoride concentrations decrease with

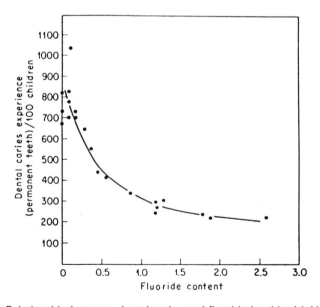

Fig. 20–1. Relationship between dental caries and fluoride level in drinking water.[7]

Table 20–1. Recommended and Approved (Health) Limits for Fluoride in Drinking Water.*

ANNUAL AVERAGE OF MAXIMUM DAILY AIR TEMPERATURES, °F (°C) BASED ON TEMPERATURE DATA OBTAINED FOR A MINIMUM OF 5 YEARS	RECOMMENDED LIMITS, mg/l			APPROVED, mg/l
	LOWER	OPTIMUM	UPPER	LIMIT
50.0 to 53.7 (10 to 12.1)	0.9	1.2	1.7	1.8
53.8 to 58.3 (12.1 to 14.6)	0.8	1.1	1.5	1.7
58.4 to 63.8 (14.6 to 17.7)	0.8	1.0	1.3	1.5
63.9 to 70.6 (17.7 to 21.4)	0.7	0.9	1.2	1.4
70.7 to 79.2 (21.5 to 26.2)	0.7	0.8	1.0	1.2
79.3 to 90.5 (26.3 to 32.5)	0.6	0.7	0.8	1.1

* Reference 8.

increasing temperature because more water is consumed during warm weather periods than at cooler temperatures.

It was noted earlier that all water supplies contain fluoride. Therefore, it can be said that all water supplies are fluoridated. However, because waters with fluoride concentrations of less than 0.7 mg/l do not have appreciable dental significance, they are generally referred to as "naturally fluoridated." For drinking water with fluoride concentrations below 0.7 mg/l, "controlled fluoridation" is used to increase the concentrations.

The effectiveness of controlled fluoridation has been tested in 10-year studies in such cities as Newburgh, New York and Grand Rapids, Michigan. The investigations demonstrated the safety of controlled fluoridation and its effectiveness in controlling dental caries. The program for fluoridation of public water supplies deficient in natural fluorides has been sponsored by many organizations interested in public health, such as the American Dental and Medical Associations. The process of fluoridation is now practiced in approximately 5,000 communities, serving over 80 million persons.[1]

FLUORIDATION

Fluorine is the most chemically active element known. Like chlorine, it is never found in a free state, but always occurs in combination with other elements, in fluoride compounds. In water solution, the compounds dissociate into ions, the form in which fluorine is assimilated by humans. Theoretically, any compound that gives fluoride ions in water solution can be used for increasing the fluoride content of a water supply. However, there are several practical considerations involved in selecting the fluoride source. First, it must be sufficiently soluble to be used in routine water practice. Second, the compound must not have any undesirable characteristics. Third, the material used should be readily available and relatively inexpensive.

The three most commonly used fluoride compounds in water treatment are sodium fluoride (NaF), sodium silicofluoride (Na_2SiF_6), and fluosilicic acid (H_2SiF_6). Other compounds used for fluoridation include calcium fluoride (CaF_2), ammonia silicofluoride ((NH_4)$_2SiF_6$), and hydrofluoric acid (H_2F_2). The most commonly used compounds will be described in detail.

Sodium Fluoride

Sodium fluoride (NaF) was the first fluoride compound used for fluoridation. Although it is one of the most expensive fluoridation compounds for the amount of available F, it is still the most widely used.

NaF is a white, odorless material available either as a powder or in crystalline form. Table 20–2 lists the characteristics of NaF and the other two popular fluoridation compounds. The maximum solubility of NaF is 4.0 percent, resulting in a fluoride concentration of 18,000 mg/l. Its solubility is practically constant over the temperature range generally encountered in water treatment. Solution pH varies with the type and amount of impurities, but solutions prepared from the usual grades of NaF exhibit near-neutral pH.

Powdered NaF is produced in densities ranging from 65 to 90 pounds per cubic foot (1,054 to 1,458 kg/m³). Crystalline NaF is produced in various size ranges, usually designated as coarse, fine, and extra-fine. The crystalline form is preferred when manual handling is involved because the absence of fine powder results in a minimum of dust.

Table 20–2. Characteristics of Fluoride Compounds.

ITEM	SODIUM FLUORIDE, NaF	SODIUM SILICOFLUORIDE, Na_2SiF_6	FLUOSILICIC ACID H_2SiF_6
Form	Powder or crystal	Powder or very fine crystal	Liquid
Molecular weight	42.00	188.05	144.08
Commercial purity, %	90–98	98–99	22–30
Fluoride ion, % (100% pure material)	45.25	60.7	79.2
Lb (kg) required per MG (Ml)	18.8	14.0	35.2
for 1.0 mg/l F at	(2.25)	(1.68)	(4.22)
indicated purity	(98%)	(98.5%)	(30%)
pH of saturated solution	7.6	3.5	1.2 (1% solution)
Solubility, g per 100 g water, at 25°C	4.05	0.762	Infinite

Sodium Silicofluoride

Sodium silicofluoride (Na_2SiF_6) is the cheapest of the compounds currently in use. Its cost makes it very popular for use in fluoridation.

Na_2SiF_6 is a white, odorless crystalline powder. Its solubility varies from 0.44 percent at 32°F (0°C) to 2.45 percent at 212°F (100°C). Saturated solutions exhibit an acid pH, usually between 3.0 and 4.0. The density of Na_2SiF_6 ranges from 55 to 72 lb/cu ft (892 to 1,167 kg/m³). Experience has shown that for best feeding results with mechanical feeders, the Na_2SiF_6 should have a low moisture content plus a relatively narrow size distribution.

Fluosilicic Acid

This fluoridation compound is a 20 to 35 percent solution of H_2SiF_6 in water. When pure it is a colorless, corrosive liquid with a pungent odor and can cause skin irritation. Upon vaporizing, the acid decomposes to form hydrofluoric acid and silicon tetrafluoride. All solutions of fluosilicic acid are characterized by a low pH. A concentration sufficient to produce 1 mg/l of fluoride ion can cause a significant pH depression in poorly buffered waters. For example, in water at pH 6.5 and containing 30 mg/l of total dissolved solids (TDS), the addition of H_2SiF_6 to produce 1 mg/l of fluoride ion caused the pH to drop to 6.2.[1]

Fluosilicic acid is a solution containing a high proportion of water. Consequently, large quantities can be expensive to ship. Economics generally restricts fluoridation by fluosilicic acid to the smaller waterworks.

Other Fluoride Compounds

Calcium fluoride (CaF_2), ammonium silicofluoride (($NH_4)_2SiF_6$), and hydrofluoric acid (H_2F_2) have been used for water fluoridation. Each has particular properties that make the compound desirable in a specific application; however, none has widespread use.

Fluorspar (CaF_2) is the cheapest of the compounds used for fluoridation, but it has the disadvantage of being the least soluble. It has been successfully fed by first dissolving it in an alum solution, and then utilizing the resultant liquid to supply both the alum needed for coagulation and the fluoride for fluoridation. Some attempts have been made to feed fluorspar directly in the form of ultra-fine powder, on the premise that the powder would eventually dissolve or would remain in suspension until consumed.

Ammonia silicofluoride (($NH_4)_2SiF_6$) has the advantage of supplying the ammonium ion necessary for the production of chloramines when this form of disinfectant is preferred to chlorine. Otherwise, it has found little use in fluoridation.

Hydrofluoric acid (H_2F_2) has been used in a few specially designed installations. It is low in cost, but it presents a safety and corrosion hazard and is generally not used for fluoridation.

FLUORIDATION SYSTEMS

No one type of fluoridation system is applicable to all water treatment plants. Selection is based on size and type of water facility, chemical availability, cost, and operating personnel available. For small utilities, usually some type of solution feed is selected, and batches are manually prepared. A simple system consists of a solution tank and a solution metering pump with appropriate piping from the tank to the water main for application (Fig. 20–2). If fluosilicic acid is used, it is either used at full strength (Fig. 20–3) or diluted with water in the feed tank (Fig. 20–4). When sodium fluoride is used, the feed solution may be prepared to a desired strength or as a saturated solution. Because NaF has a maximum solubility of 4.0 percent, a saturated solution is prepared by passing water through a bed containing an excess of NaF (Fig. 20–5). While NaF is quite soluble, calcium and magnesium fluorides form precipitates that can scale and clog feeders and lines. Consequently, dissolution water should be softened whenever the hardness exceeds 75 mg/l.

Large waterworks usually use either gravimetric dry feeders to apply sodium silicofluoride (Fig. 8–13) or solution feeders to apply fluosilicic acid. Often their systems incorporate automatic control systems to regulate flow

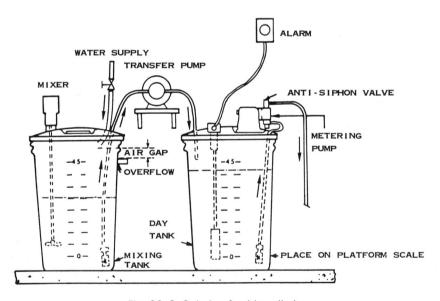

Fig. 20–2. Solution feed installation.

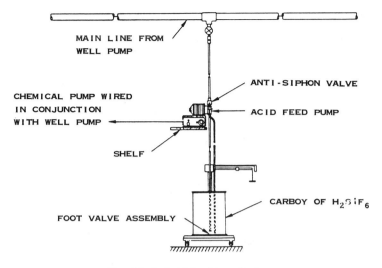

Fig. 20-3. Acid feed installation.

and adjust feed rates. Fluoride feed must be paced to water flow to maintain a consistent fluoride ion concentration.

Fluoride must be injected into all of the water entering the distribution system. If there is more than one supply point, separate fluoride feeding installations are required for each water facility. In a well system, application

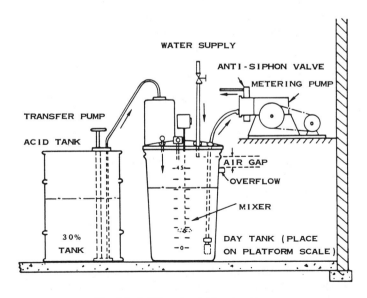

Fig. 20-4. Diluted acid feed system.

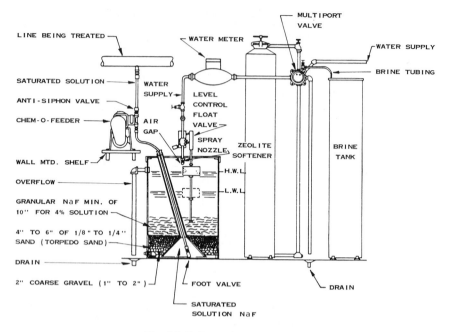

Fig. 20–5. Downflow saturator.

can be in the discharge line of each pump or in a common line leading to a storage reservoir. Fluoride can be applied in a treatment plant in a channel or line from the filters, or directly into the clearwell. Whenever possible, it should be added after filtration to avoid possible losses due to reactions with other chemicals. Of particular concern are coagulation with alum and lime–soda softening. Fluoride injection points should be as far away as possible from the addition of chemicals that contain calcium because of the insolubility of CaF.

Surveillance of water fluoridation involves testing both the raw and treated water for fluoride ion concentration. Records of the weight of chemical applied and the volume of water treated should be kept to confirm that the correct amount of fluoride is being added. The fluoride concentration in the treated water should be that recommended by drinking water standards (Table 20–1).

DEFLUORIDATION

As soon as excessive amounts of fluorides in drinking water supplies had been established as the cause of dental fluorosis, methods for their removal were studied. Comparatively little research has been performed on theoretical design models for the removal of fluorides. Consequently, the design of defluoridation processes is largely based on past experience and empirical models.

Fluoride removal is complicated by the presence of other ions in the water that compete with fluoride for removal. The design of defluoridation systems requires laboratory and pilot-scale work prior to the design of full-scale treatment systems.

Two methods of defluoridation have found practical application. One involves passage of water through defluoridation media such as bone meal, bone char, ion-exchange resins, or activated alumina. The second involves the addition of chemicals such as lime or alumina prior to rapid mixing, flocculation, and sedimentation in a waterworks, for the removal of fluoride only or the concurrent removal of fluoride and other ions (e.g., calcium and magnesium removal for water softening).

The following discussion presents past experience with media filter and chemical addition defluoridation systems.

Defluoridation Media

The uptake of fluoride onto the surface of bone was first reported by Smith and Smith in 1937.[9] They suggested that fluoride was removed by ion exchange in which the carbonate radical of the apatite comprising bone [i.e., $Ca(PO_4)_6 \cdot CaCO_3$] was replaced by fluoride to form an insoluble fluorapatite, according to:

$$Ca(PO_4)_6 \cdot CaCO_3 + 2F^- = Ca(PO_4)_6 \cdot CaF_2 \downarrow + CO_3^{-2} \qquad (20\text{--}1)$$

Similarly, bone char or tricalcium phosphate $(Ca_3(PO_4)_2)$, produced by carbonizing bone at temperatures of 2012 to 2912°F (1100 to 1600°C), has been used for defluoridation. When exhausted, the column is regenerated by application of a 1.0 percent solution of caustic soda, which converts the fluorapatite to hydroxyapatite $(Ca(PO_4)_2 \cdot Ca(OH)_2)$. The fluoride is removed as soluble sodium fluoride. The caustic is followed by a rinse, and then an acid wash to lower the pH.[10] In the regenerated form, the hydroxyl radical becomes the exchange anion in the defluoridation reaction.[11] Bone char has been used successfully for full-scale defluoridation.[12]

Paired cationic and anionic exchange resin beds have also been used for defluoridation. In this process, illustrated by Fig. 20–6, water first passes through a cationic resin (R^+) bed, which exchanges sodium with hydrogen to form the equivalent acid:

$$2NaF + H_2R^+ = H_2F_2 + Na_2R^+ \qquad (20\text{--}2)$$

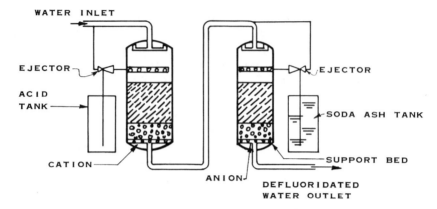

Fig. 20–6. Paired cationic and anionic exchange resin beds for fluoride removal.

The hydrogen fluoride is then removed during passage through the anionic bed (R^-):

$$2R^- + H_2F_2 = 2R^-HF \qquad (20\text{–}3)$$

Periodically, the resins are regenerated with acid and alkaline solutions. Synthetic ion exchange resins for defluoridation are available from several manufacturers.

The most widely used defluoridation method involves beds of granular activated alumina (Al_2O_3). Some researchers have concluded that defluoridation by activated alumina is the result of ion exchange.[13] However, Wu and Nitya recently showed that the process is one of adsorption and follows the Langmuir isotherm.[14]

An example of fluoride removal by activated alumina is the plant at Gila Bend, Arizona, which has been in operation since 1978. This plant has a treatment capacity of 0.625 mgd (2.37 Ml/d), and has successfully reduced raw water fluoride levels of 3.0 mg/l to as low as 0.1 mg/l. The plant consists of two 10-foot (3.05-m)-diameter, rubber-lined vertical pressure vessels, each containing 5 feet (1.5 m), or 380 cubic feet (10.8 m³), of Alcoa F-1 (−28, +48 mesh) activated alumina. Operation consists of adjusting the raw water pH to 5.5 with sulfuric acid prior to treatment, and after treatment, pH adjustment to 7.4 with 50 percent sodium hydroxide. Careful raw water pH control is necessary because of sensitivity of the fluoride removal reaction to pH. Experienced exchange capacities have ranged between 2,000 and 4,000 grains/cu ft (70.7 and 141.4 kilograins/m³), depending upon raw water pH and fluoride concentration. The plant was designed for operation between 5 and 7.5 gpm/sq ft (12.2 and 18.3 m/h). Treatment runs for each exchanger are between 3.5 and 5 MG (13.3 and 18.9 Ml), and treated water is blended in a storage tank to produce a water with 1.0 mg/l of fluoride for distribution.

Regeneration consists of three steps: backwash, regeneration, and neutralization. Backwashing with raw water is performed at a rate of 8 to 9 gpm/sq ft (19.5 to 22 m/h), which expands the bed by 50 percent. The purpose of backwashing is to remove suspended solids that have been filtered from the water and to unpack the bed, thus preventing tendencies toward channeling. Backwashing normally takes 10 minutes.

Regeneration is initiated while the bed is still expanded, by employing a 1 percent sodium hydroxide solution at a rate of 2.5 gpm/sq ft (6.1 m/h), followed by an upflow rinse at a rate of 5 gpm/sq ft (12.2 m/h) for 30 minutes. Next, the unit is drained to the top of the treatment bed, and downflow regeneration with a 1 percent sodium hydroxide solution is performed at the same flow rate as the upflow regeneration for 35 minutes.

The bed pH will be 12 to 13 as a result of the NaOH treatment. Wu and Nitya demonstrated that fluoride removal by activated alumina is strongly pH-dependent, with little removal occurring at pH values above 11 and optimum removal occurring at pH 5 (Fig. 20–7).[14]

Prior to neutralization, the unit is drained to the top of the bed, and raw water is adjusted to pH 2.5 and is fed downflow at the normal treatment rate of 5 to 7.5 gpm/sq ft (12.2 to 18.3 m/h). As the fluoride level in this water drops below the required level, the water is diverted to storage, and is neutralized by water being treated by the second unit.

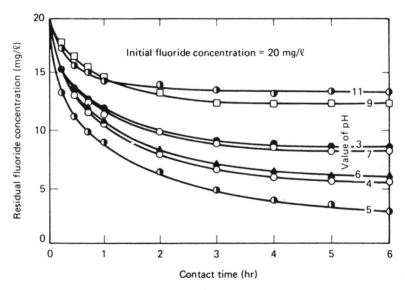

Fig. 20–7. Effect of pH and time of contact on fluoride removal of activated alumina adsorption.[14]

Wastewater is approximately 4 percent of plant throughput and is disposed to a lined evaporation lagoon. Operating costs between start-up in May 1978 and 1981 averaged 11.84¢/1,000 gallons (3.12¢/kl), allocated as follows:

	TREATMENT COST	
ITEM	¢/1,000 GALLONS	¢/kl
Treatment chemicals	5.14	1.36
Operating labor	5.09	1.34
Electricity	0.70	0.18
Media replacement	0.73	0.19
Miscellaneous	0.18	0.05
	11.84	3.12

Chemical requirements are about 15.8 gallons of 66°B′ sulfuric acid/MG treated (15.8 m³/Mm³) and 24.6 gallons of 50 percent sodium hydroxide/MG treated (24.6 m³/Mm³).

Chemical Addition

Lime and alum have been used successfully for fluoride removal. The defluoridation systems generally consists of lime or alum addition to a rapid mix chamber, followed by flocculation and sedimentation.

Alum was one of the first chemicals investigated for use in removing fluoride from drinking water supplies.[10] When added to water, alum reacts with the alkalinity in the water to produce insoluble aluminum hydroxide, according to the following equation:

$$Al_2(SO_4)_3 \cdot 14.3H_2O + 3Ca(HCO_3)_2$$
$$= 2Al(OH)_3 + 3CaSO_4 + 14.3H_2O + 6CO_2 \quad (20\text{--}4)$$

Rabosky and Miller suggest that fluoride is removed by adsorption onto the $Al(OH)_3$ particles.[15] Figure 20–8 presents data from Scott et al. and Culp and Stoltenberg on fluoride removal at various alum dosages.[16,17] The latter reported on the lowering of the fluoride concentration of a soft, highly mineralized water from 3.6 to 1.0 mg/l by the addition of 315 mg/l alum followed by 30 minutes of flocculation.

Finally, fluorides have been observed to be removed during lime softening of drinking water. Fluoride precipitation occurs according to the following reaction:

$$Ca(OH)_2 + 2HF = CaF_2\!\downarrow + 2H_2O \quad (20\text{--}5)$$

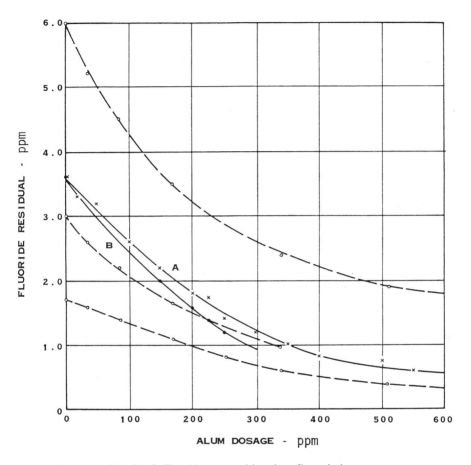

Fig. 20-8. Fluoride removal by alum flocculation.

Theoretical solubility calculations suggest that an effluent fluoride concentration of approximately 7.8 mg/l can be achieved by lime precipitation.[10] However, much lower fluoride concentrations have been observed. Scott et al. reported the defluoridation during lime softening could have been described by the following equation:[16]

$$F_r = 0.07 \cdot F_i \cdot (Mg)^{0.5} \qquad (20\text{--}6)$$

where:

F_r = residual fluoride
F_i = initial fluoride
Mg = magnesium concentration removed

Based on this formula, reduction of fluoride from 3.3 mg/l to 1.0 mg/l requires that 100 mg/l of magnesium be removed. Consequently, lime addition for defluoridation is only appropriate in high-magnesium waters.

REFERENCES

1. Bellack, E., *Fluoridation Engineering Manual,* Office of Water Programs, Water Supply Programs Division, EPA, 1972.
2. Churchill, H. V., "Occurrence of Fluoride in Some Waters of the United States," *Ind. Eng. Chem.* 23:996–998, 1931.
3. McKay, F. S., "The Present Status of the Investigation of the Cause and the Geographical Distribution of Mottled Enamel, Including a Complete Bibliography on Mottled Enamel," *J. Dental Res.* 10:561–568, 1930.
4. Smith, M. C., Lantz, E. M., and Smith, H. V., "The Cause of Mottled Enamel, and Defect of Human Teeth," *University of Arizona Tech. Bulletin* No. 32, 253–282, 1931.
5. Gottlieb, S., "Fluorides in Kansas Waters and Their Relation to Mottled Enamel," *Trans. Kansas Academy of Science* 37:129–131, 1934.
6. Dean, H. T., *U.S. Public Health Reports* 53:1443, 1938.
7. Dean, H. T., *J. AWWA* 35:1161, 1943.
8. "Drinking Water Standards and Guidelines," EPA, Water Division, 1974.
9. Smith, H. R., and Smith, L. C., "Bone Contact Removes Fluoride," *Water Works Engineering* 90:5, 600, 1937.
10. Benefield, L. D., Judkins, J. F., Jr., and Weand, B. L., *Process Chemistry for Water and Wastewater Treatment,* Prentice-Hall, Englewood Cliffs, N.J., 1982.
11. Maier, F. J., "Defluoridation of Municipal Water Supplies," *J. AWWA* 45:8, 879, 1953.
12. Harmon, J. A., and Kalichman, S. G., "Defluoridation of Drinking Water in Southern California," *J. AWWA* 57:2, 245, 1965.
13. Savinelli, E. A., and Black, A. P., "Defluoridation of Water with Activated Alumina," *J. AWWA* 50:1, 33, 1958.
14. Wu, Y. C., and Nitya, A., "Water Defluoridation with Activated Alumina," *J. Environ. Eng. Div. ASCE* 105:EE2, 357, 1979.
15. Rabosky, J. G., and Miller, J. P., "Fluoride Removal by Lime Precipitation and Alum and Polyelectrolyte Coagulation," *Proceedings,* 29th Purdue Industrial Waste Conference, pp. 669–676, 1974.
16. Scott, R. D., et al., "Fluoride in Ohio Water Supplies—Its Effect, Occurrence, and Reduction," *J. AWWA* Vol. 29, 1937.
17. Culp, R. L., and Stoltenberg, H. A., "Fluoride Reduction at La Crosse, Kan.," *J. AWWA* 50:423–431, 1958.

Chapter 21
Lime–Soda Softening

SOFTENING

Hardness

Hardness in water is due principally to calcium and magnesium ions, although iron, manganese, and strontium ions also cause it. Generally, the latter three ions are present in low concentrations, and are not significant in hardness reactions. For practical purposes, calcium and magnesium may be considered to contribute all hardness to water.

Use of the term hardness originates from the characteristics of the reaction products of soap and the hardness-creating ions. A hard, insoluble, gritty curd is produced that prevents the formation of a lather until sufficient soap is added to react with all of the hardness ions.

There are two types of hardness: carbonate or temporary hardness, and noncarbonate or permanent hardness.

Carbonate Hardness

Carbonate hardness includes hardness due to the presence of calcium bicarbonate, $Ca(HCO_3)_2$, and magnesium bicarbonate, $Mg(HCO_3)_2$. At elevated temperatures, carbon dioxide is driven off, and insoluble calcium and magnesium carbonates are formed. Because of this phenomenon, carbonate hardness is sometimes referred to as temporary hardness. Carbonate hardness also includes calcium carbonate, $CaCO_3$, and magnesium carbonate, $MgCO_3$, which are not easily precipitated at concentrations less than 20 to 35 mg/l (milligrams per liter).

Noncarbonate Hardness and Salinity

Noncarbonate hardness can be contributed by calcium sulfate, $CaSO_4$, calcium chloride, $CaCl_2$, magnesium sulfate, $MgSO_4$, and magnesium chloride, $MgCl_2$. These salts also contribute to the salinity of water, together with other neutral salts. The compounds that comprise noncarbonate hardness are not precipitated when the water is heated, but form a very hard scale in boilers when the water is evaporated. Therefore, the term incrustants is sometimes applied to the compounds contributing noncarbonate hardness. Noncarbonate hardness compounds are neutral salts that do not affect the alkalinity of the water.

Alkalinity

Alkalinity in natural water is due to the presence of carbonates and bicarbonates. Thus, carbonate hardness contributes to alkalinity. When the alkalinity (expressed as $CaCO_3$) is equivalent to total hardness (expressed as $CaCO_3$), all hardness is carbonate hardness. When alkalinity is less than total hardness, the difference is noncarbonate hardness.

OBJECTIONS TO HARD WATER

Hard water increases consumption of soap, and may have an adverse effect on clothing and other articles being cleansed. Hardness can also shorten the life and decrease the efficiency of pipes and fixtures and heating and cooling systems, and water with a high hardness concentration is unsuitable for use in many enterprises, such as laundries, canneries, power plants, ice plants, railroads, and industries using boilers. There are definite economic penalties associated with the use of hard water.

It has been well demonstrated that lime softening of all the water for a community is not only more economical, but also provides a better quality water for most purposes than can be obtained from individual domestic or industrial softeners, except where a water of zero hardness is required in a particular industrial application. For general evaluation of relative hardness, the hardness ratings are given in Table 21–1.

The perception of satisfactory hardness varies in different locations throughout the country, as influenced by the hardness of locally available sources of water supply. For example, people living in New England where natural waters are soft, might consider a hardness of 100 mg/l excessive, whereas residents of the Midwest or Southwest with naturally hard water sources might consider such a hardness to be satisfactory.

The cost of water softening is generally not considered justified when the hardness of the source of the water supply is less than 150 mg/l. Public

Table 21–1. Hardness Ratings.

RATING	HARDNESS, mg/l AS $CaCO_3$
Soft	less than 50
Moderately hard	50 to 150
Hard	150 to 300
Very hard	more than 300

water supplies usually are not softened below 30 to 50 mg/l because softer waters are corrosive unless treated with an alkali. Magnesium hardness concentrations greater than 40 mg/l are undesirable because of the potential for magnesium hydroxide scale formation in domestic hot water heaters.

To gain the aesthetic and financial benefits of soft water, communities with hard water supplies generally have used either the lime–soda softening process or the cation exchange process. Economic factors generally limit cation exchange to small water systems.

The design of softening plants is largely determined by the quantity of water to be softened and its physical and chemical characteristics. Other considerations include ease of sludge disposal, available space, final hardness desired, and comparative costs.

THE BASIC LIME–SODA ASH PROCESS

The main function of a lime–soda softening plant is to remove hardness constituents by first converting them to insoluble precipitates and then separating the precipitates from the water by settling and filtration. Hydrated lime (in small plants) or slaked quicklime (in large plants) is added to the water to react with carbon dioxide and calcium bicarbonates to form insoluble calcium carbonate, with magnesium bicarbonates to form insoluble calcium carbonate and magnesium hydroxide, and with any other magnesium compounds to form insoluble magnesium hydroxide plus soluble calcium sulfate and calcium chloride. If necessary, soda ash is then added to further reduce hardness by precipitating remaining magnesium hardness, noncarbonate calcium following lime addition hardness, and nonsettled calcium carbonate. The reactions involved are as follows:

1. $$CO_2 + Ca(OH)_2 \rightarrow \underline{CaCO_3} + H_2O \qquad (21\text{--}1)$$
2. $$Ca(HCO_3)_2 + Ca(OH)_2 = \underline{2CaCO_3} + 2H_2O \qquad (21\text{--}2)$$
3. $$Mg(HCO_3)_2 + Ca(OH)_2 = MgCO_3 + \underline{CaCO_3} + 2H_2O \qquad (21\text{--}3)$$
4. $$MgCO_3 + Ca(OH)_2 = \underline{Mg(OH)_2} + \underline{CaCO_3} \qquad (21\text{--}4)$$
5. $$MgSO_4 + Ca(OH)_2 = CaSO_4 + \underline{Mg(OH)_2} \qquad (21\text{--}5)$$

Equation (21–1) is the first reaction that will occur, and high carbon dioxide concentrations are customarily reduced by pretreatment prior to softening.

The compounds underscored are insoluble, and generally can be removed during subsequent settling. Magnesium carbonate, produced in equation (21–3), is soluble; so more lime must be added to precipitate the magnesium as magnesium hydroxide, as shown in equation (21–4). The addition of 1 pound (0.454 kg) of lime to hard water can lead to the precipitation of up to 3½ pounds (1.59 kg) of mineral solids.

The permanent or noncarbonate hardness due to incrustants is removed by adding both lime and soda ash to the water, resulting in the formation of insoluble calcium carbonate, which is precipitated, and neutral sodium salts. The reactions are as follows:

$$1. \qquad CaSO_4 + Na_2CO_3 = \underline{CaCO_3} + Na_2SO_4 \qquad (21\text{--}6)$$
$$2. \qquad CaCl_2 + Na_2CO_3 = \underline{CaCO_3} + 2NaCl \qquad (21\text{--}7)$$
$$3. \qquad MgSO_4 + Ca(OH)_2 = \underline{Mg(OH)_2} + CaSO_4 \qquad (21\text{--}8)$$
$$4. \qquad MgCl_2 + Ca(OH)_2 = \underline{Mg(OH)_2} + CaCl_2 \qquad (21\text{--}9)$$

In ionic form, the softening and recarbonation reactions are:

$$Ca(OH)_2 \rightarrow Ca^{+2} + 2OH^- \qquad (21\text{--}10)$$
$$CO_2 + OH^- \rightarrow HCO_3^- \qquad (21\text{--}11)$$
$$HCO_3^- + OH^- \rightarrow CO_3^{-2} + H_2O \qquad (21\text{--}12)$$
$$Ca^{+2} + CO_3^{-2} \rightarrow CaCO_3 \qquad (21\text{--}13)$$
$$Mg^{+2} + 2OH^- \rightarrow \underline{Mg(OH)_2} \qquad (21\text{--}14)$$

In calculating the requisite amounts of chemical to be added, the assumption is made that the reactions go to completion. In the presence of magnesium, the amount of lime applied must be sufficient not only to convert the free carbon dioxide and bicarbonates to carbonate, but also to produce magnesium hydroxide by adding the required excess of hydroxide.

Following chemical coagulation, these precipitates are flocculated, settled, and removed continuously from the settling basin. The water is then filtered through single-, dual-, or mixed-media filters. For surface waters the softening process is usually carried out in separate rapid mix, flocculation, and settling basins, as is common in water clarification plant design. With well waters of constant chemical quality that are to be treated at a constant flow rate, solids contact basins that combine mixing, flocculation, and settling with internal sludge recirculation in a single basin may be considered as an alternative.

PRETREATMENT AND OTHER PROCESS VARIATIONS

Prior to softening, some preliminary pretreatment may be advisable if:

1. Raw water turbidities exceed 3,000 NTU at times.
2. Raw water has a high concentration of free carbon dioxide (more than 10 mg/l).
3. The raw water is high in organic colloids of a type that impedes crystallization of calcium carbonate.

4. Raw water quality is highly variable over short periods of time.
5. Recalcining of sludge is to be practiced.

Otherwise, the clarification and softening process trains can be combined. Basically, the applicable design standards for mixing, flocculation, and sedimentation are the same for the lime–soda process as for conventional clarification. When the softening and clarification processes are combined, the clarification criteria should govern.

There are a number of process variations in the basic lime–soda softening process. These modifications are aimed at:

- Improving process efficiency
- Improving the stability of the treated water
- Accommodating widely varying composition of hard waters

These variations include:

- Single or two-stage recarbonation after conventional lime–soda treatment
- Sludge recirculation
- Excess lime treatment with split treatment or recarbonation
- Post-treatment with polyphosphates
- Coagulation with alum, activated silica, or polymers
- The use of three-stage treatment
- The substitution of cation exchangers for soda ash to remove noncarbonate hardness
- The use of caustic soda instead of soda ash

The reactions of caustic soda with the carbonate and noncarbonate hardness are:

$$CO_2 + 2NaOH = Na_2CO_3 + H_2O \qquad (21\text{--}15)$$
$$Ca(HCO_3)_2 + 2NaOH = \underline{CaCO_3} + Na_2CO_3 + 2H_2O \qquad (21\text{--}16)$$
$$Mg(HCO_3)_2 + 4NaOH = \underline{Mg(OH)_2} + 2Na_2CO_3 + 2H_2O \qquad (21\text{--}17)$$
$$MgSO_4 + 2NaOH = \underline{Mg(OH)_2} + Na_2SO_4 \qquad (21\text{--}18)$$

The caustic soda removes free CO_2 and carbonate hardness, producing insoluble calcium carbonate and soluble sodium carbonate. Caustic soda also reacts with noncarbonate hardness, producing insoluble magnesium hydroxide. The above equations show that caustic soda removes both carbonate and noncarbonate hardness. Therefore, it not only takes the place of soda ash, but it satisfies part of the lime requirement as well.

The advisability of using caustic soda instead of soda ash in the lime–

soda ash process depends on comparative delivered costs of the three chemicals and the chemical composition of the raw water.

PROCESS DESIGN CONSIDERATIONS

Flow Patterns

Lime–soda softening plants may be divided into four major categories with respect to process flow patterns. As shown in Fig. 21–1, they are:

- Single-stage softening
- Split treatment
- Two-stage softening
- Three-stage softening with recovery and reuse of lime by recalcination

Figure 21–1 also shows a number of pretreatment options and variations as previously mentioned. The designer must select from these available processes and treatment elements the particular combination that best suits local raw water conditions in order to develop the optimum softening plant.

Although it is not shown in Fig. 21–1, it is possible for well waters of constant temperature and physical and chemical quality, which are to be treated at a constant plant throughput rate, to be softened in solids contact basins. In this type of basin, the elements of rapid mixing, flocculation, settling, and contact with previously precipitated calcium carbonate are combined and carried out within a single structure. The potential cost advantages of softening in solids contact basins may be lost if there are variations in water temperature, flow, or quality that require separate individual control.

Lime Recalcining and Reuse

Lime recalcining has been utilized to some extent in waterworks practice. Long used in many industrial applications, it has been employed in advanced wastewater treatment plants that use lime treatment. The recalcining process is simple, and consists of heating the calcium sludge to about 1850°F (1009°C). This drives off water and carbon dioxide, leaving only the calcium oxide or quicklime plus some inert material. Some lime sludge must be wasted periodically to avoid a buildup of inerts. The amount to be wasted is an economic balance between solids-handling capacity and desired characteristics of the reclaimed lime. In treating waters high in calcium hardness, an excess of recalcined lime may be produced. In small water treatment plants (5 to 10 mgd range, or 18.93 to 37.85 Ml/d), the cost of reclaimed lime may be only slightly less than the cost of new lime, but recalcining still may be warranted because it also reduces the cost of disposing of the lime sludge.

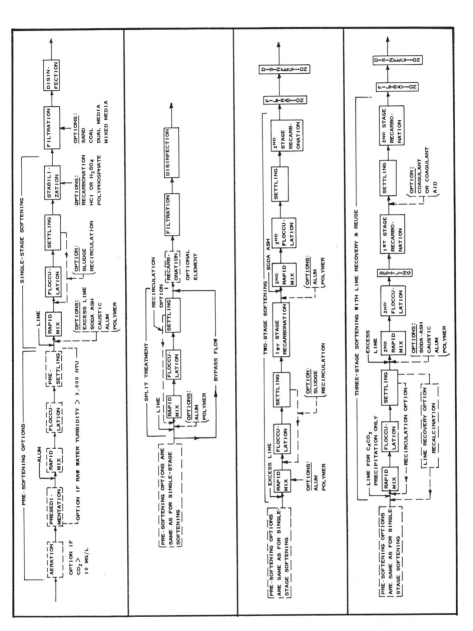

Fig. 21-1. Softening flow patterns.

In large plants (over 10 mgd or 37.85 Ml/d), recalcining may be justified by the lower cost of the reclaimed lime relative to that for new lime.

Recalcining can be accomplished by use of a multiple hearth furnace, a fluidized bed, rotary kilns, or flash drying and calcination systems. The use of rotary kilns is limited to very large systems.

To prepare lime sludge for recalcination, some preliminary treatment is required, usually in the form of gravity thickening and mechanical dewatering. One common method of dewatering is centrifugation to produce a cake containing 50 to 55 percent solids.

Magnesium Carbonate Coagulation and Recovery

This is a relatively new coagulant recovery technique based on a combination of water softening and conventional coagulation techniques that can be applied to all types of waters. Magnesium carbonate is used as the coagulant, with lime added to precipitate magnesium hydroxide as the active coagulant. The resulting sludge is composed of $CaCO_3$, $Mg(OH)_2$ and the turbidity removed from the raw water. The sludge is carbonated by injecting CO_2 gas, which selectively dissolves the $Mg(OH)_2$. The carbonated sludge is filtered, with the magnesium being recovered as soluble magnesium bicarbonate in the filtrate, which is then recycled to the point of addition of chemicals to the raw water and reprecipitated as $Mg(OH)_2$, as a new treatment cycle begins. The resulting solids requiring disposal are the filter cake, composed of $CaCO_3$, and the raw water turbidity. In large plants, it is possible to reduce the waste solids even more by slurrying the filter cake and separating the raw water turbidity from the $CaCO_3$ in a flotation cell. The purified $CaCO_3$ can then be dewatered and recalcined as high-quality quicklime. The stack gases from the recalcining operation then provide CO_2 gas to carbonate the sludge for magnesium recovery and to recarbonate the water in the treatment plant. When this latter lime recovery step is practiced, the waste solids are reduced to only those that constituted the raw water turbidity.

Many waters contain enough magnesium that 100 percent magnesium recovery is achievable. With soft surface waters, magnesium recovery may be limited to about 80 percent.

Black et al. described a technique by which 20 major American cities softening hard, turbid waters could produce about 150,000 tons (136,050 metric tons) of $MgCO_3 \cdot 3H_2O$ for use as a coagulant by other plants in the process described above.[1] By doing so, the problem of disposal of sludges from the softening plants would be resolved. The lime softening of water with high magnesium concentrations results in objectionably large quantities of magnesium hydroxide appearing in the calcium carbonate sludges. Prior to 1957, it was considered impossible to recalcine the sludge produced by softening high-magnesium waters; but it was then demonstrated that it was possible to selectively dissolve the $Mg(OH)_2$ from the $CaCO_3$ by carbonation with CO_2 gas. A plant employing this concept has been in use in Dayton, Ohio since 1957. It is possible to recover the magnesium from the carbonated sludge, as shown in the following equations:

Sludge carbonation:
$$Mg(OH)_2 + 2CO_2 \rightarrow Mg(HCO_3)_2 \qquad (21\text{–}19)$$
Product recovery:
$$Mg(HCO_3)_2 + 2H_2O \xrightarrow[\text{air}]{35\text{–}45°C} \underline{MgCO_3 \cdot 3H_2O} + CO_2 \qquad (21\text{–}20)$$

The magnesium bicarbonate solution clarified by either settling or filtration passes to a heat exchange unit where it is warmed to 35 to 45°C, after which it is aerated by compressed air in a mechanically mixed basin. The $MgCO_3 \cdot 3H_2O$ precipitates rapidly (about 90 minutes), and the resulting product, $MgCO_3 \cdot 3H_2O$, is vacuum-filtered, dried, and bagged for shipment, hopefully to other plants where it may be used as a coagulant, as described in more detail below. The calcium carbonate sludge remaining after the sludge carbonation step is dewatered and recalcined to produce a high-quality quicklime.

The use of magnesium carbonate as a coagulant offers the potential for eliminating the problem of alum sludge disposal by replacing the alum sludges with a readily recoverable sludge. The chemical equations describing the coagulation process resulting from the recycling of magnesium bicarbonate (as described above in this section) are as follows:

$$Mg(HCO_3)_2 + Ca(OH)_2 \rightarrow MgCO_3 + CaCO_3 + 2H_2O \qquad (21\text{–}21)$$
$$MgCO_3 + Ca(OH)_2 \rightarrow \underline{Mg(OH)_2} + CaCO_3 \qquad (21\text{–}22)$$

The magnesium carbonate hydrolyzed with lime is as effective as alum for the removal of organic color and turbidity in many waters. The flocs formed are large and settle rapidly. In treating soft surface waters, this approach produces soft waters with sufficient calcium alkalinity to reduce corrosion. High-carbonate-hardness waters are softened by the process.

The lime doses required for the removal of free CO_2 and carbonate hardness are stoichiometric, but the amount to be added for the precipitation of magnesium depends on a number of factors. This may be illustrated by the following example offered by Black et al.[1] A water containing both organic color and turbidity has an alkalinity of 40 mg/l and contains 30 mg/l magnesium as $CaCO_3$. Jar tests show that the minimum effective amount of $Mg(OH)_2$ to be precipitated will be provided by a dosage of 60 mg/l $MgCO_3 \cdot 3H_2O$. The most economical treatment of this water requires that the entire dosage of 60 mg/l $MgCO_3 \cdot 3H_2O$ be added, and that very little of the 30 mg/l of magnesium present in the water be used. There are four reasons for this:

1. The dosage of lime required for the precipitation of the required amount of $Mg(OH)_2$ from 90 mg/l of $MgCO_3 \cdot 3H_2O$ is significantly less than that required to precipitate all of the $Mg(OH)_2$ from the 30 mg of $MgCO_3$ present plus 30 mg/l added, and the pH is significantly lower.
2. The amount of CO_2 required to redissolve the $Mg(OH)_2$ from the sludge will be the same in either case, but the amount of CO_2 required to recarbonate the high-pH water of the second procedure will be much greater than the amount required for the lower-pH water of the first procedure.
3. After a few cycles a pH value is identified at which complete recovery and recycling of coagulant is achieved, after which no new coagulant is added.
4. The unused 30 mg/l of magnesium remaining in the water represents reserve coagulant that can be used if necessary with increased lime addition.

Although this example, using a water containing a significant amount of magnesium, is not typical of surface waters as a class, there are a number of cities—including Chicago, Detroit, Cleveland, Philadelphia, Pittsburgh, Indianapolis, Washington, D.C., and many smaller-sized cities—treating such waters. They should be able to substantially reduce their chemical treatment cost and at the same time eliminate or greatly reduce the extent of their sludge problems by adopting this technology.

Data on jar tests of magnesium carbonate coagulation on several natural waters have been presented. For example, magnesium carbonate doses of 50 mg/l in conjunction with 125 mg/l of $Ca(OH)_2$ provided essentially the same degree of color and turbidity removal from one water as did a dose of 20 mg/l alum for a soft surface water. The magnesium carbonate technique often provides enough alkalinity in the finished water that it can be effectively stabilized by pH adjustment, whereas alum treatment may produce a water of such low alkalinity that it cannot be completely stabilized. The reduction or elimination of corrosion problems may be a significant point in many cases. Table 21–2 summarizes the required magnesium doses observed for 17 natural waters. A linear regression analysis made on the data shown in Table 21–2 yielded the following equation for estimating the required magnesium dosage:

$$\begin{aligned}
\text{Minimum magnesium dosage (mg/l of } MgCO_3 \cdot 3H_2O) = & \\
(8.33 + 0.33 \times \text{turbidity}) + (0.46 \times \text{organic color}) & \quad (21\text{–}23) \\
- (0.03 \times \text{total alkalinity}) + (0.14 \times \text{total hardness}) &
\end{aligned}$$

Table 21–2. Required Magnesium Dose as Related to Physical
and Chemical Characteristics for 17 Natural Waters.

Mg REQUIRED* (mg/l AS MgCO$_3$ · 3H$_2$O)	TURBIDITY (NTU)	COLOR (PtCo)	ALKALINITY (mg/l)	HARDNESS (mg/l)
40	165	50	13	16
30	105	30	17	17
10	4	26	4	5
15	10	12	74	83
22	13	4	54	84
10	14	10	17	17
25	106	11	51	110
24	50	15	41	71
30	24	30	27	43
35	7.5	27	10	12
35	6	5	92	127
25	2.5	0	80	100
24	41	14	34	69
32	104	38	11	13
21	2	4	12	40
20	7.5	8	71	86
22	15	24	48	71

* Precipitated as Mg(OH)$_2$.

The economics of applying the magnesium carbonate technology have been estimated for 17 major cities and compared to the cost of coagulation for present treatment techniques using alum (and lime in some cases). If lime recovery is practiced so the CO_2 is obtained at no cost, 13 of the 17 cities show a reduction in coagulant costs. If lime is purchased at $20/ton ($22.05/metric ton) and CO_2 is obtained at no cost from a source in or near the plant (a 50-hp or 37.3-kW natural gas engine provides sufficient CO_2 for a 1-mgd or 3.785-Ml/d plant), then 10 of the 17 cities still show a reduction in cost. However, if lime and CO_2 each must be purchased at a cost of $20/ton ($22.05/metric ton), then only 2 of the 17 show a reduction in coagulant cost. These comparisons reflect only the cost of coagulants and do not reflect any credits related to the cost of sludge disposal. Figure 21–2 illustrates a potential flow sheet for the process. The sludge is carbonated in a basin with a detention time of about 1 hour to completely solubilize the magnesium, with the separation from the solids being achieved by a vacuum filter. If the lime is not to be recovered, the resulting filter cake can be handled easily and disposed of in landfill or as an agricultural aid as a pH stabilizer for soil. If lime recovery is practiced, then the processes shown within the dashed lines are added. As discussed earlier, the calcium carbonate can be separated from the raw water turbidity in a flotation cell. With this approach, only the flotation unit skimmings require disposal, and

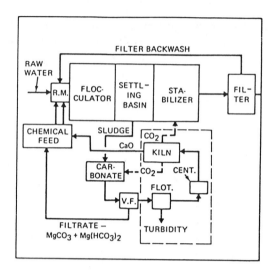

Fig. 21–2. Flow diagram of magnesium carbonate coagulation system.

they are readily handled as they are primarily the raw water solids. The purified calcium carbonate sludge is dewatered and calcined, producing CaO and CO_2 needed in the process.

UNIT PROCESS DESIGN CRITERIA

Mixing, Flocculation, and Sedimentation

In lime–soda softening the applicable design standards are the same as for conventional clarification except that the minimum time for settling is 2 hours. These criteria are outlined below, and are described in more detail in Chapters 7 and 9.

Mixing.

Rapid Mixing. Rapid mixing provides the rapid dispersal of chemicals throughout the water to be treated, by violent agitation.

Mechanical mixers, hydraulic jump, and Parshall flume are recommended methods. Baffled mixing chambers are not recommended and are acceptable only when provisions are made for proper mixing under anticipated variations in flow.

Detention. The detention period should be not less than 30 seconds in mechanical mixing chambers. In-line mixers need less detention time.

Flocculation (Slow Mixing) Basins, Primary or Secondary.

Basin Design.   Inlet and outlet design should prevent short-circuiting and destruction of floc. Series compartments are recommended to prevent short-circuiting and to provide decreasing mixing energy with time. A drain should be provided.

Detention.   Minimum flow-through velocity should be not less than 0.5 or greater than 1.5 ft/sec (0.15 to 0.46 m/s), with a recommended time for floc formation of at least 30 minutes.

Mean Velocity Gradient (G).   *G* values of 20 to 70 seconds are recommended. Flocculation in three basins in series is suggested with *G* values of 50 to 100, 5 to 50, and 1 to 10 in basins 1, 2, and 3, respectively.

Equipment.   Agitators should be driven by variable-speed drives with the peripheral speed of paddles ranging from 0.5 to 2.0 ft/sec (0.15 to 0.60 m/s). External, nonsubmerged drive equipment is preferred. Cross-flow, axial-flow, propeller, or turbine-type equipment is acceptable.

Piping.   Flocculation and sedimentation basins should be as close together as possible. The velocity of flocculated water through pipes or conduits to settling basins should be not less than 0.5 or greater than 1.5 ft/sec (0.15 to 0.46 m/s). Allowances should be made to minimize turbulence at bends and changes in direction.

Other Designs.   Baffling may be used to provide for flocculation in small plants only after consultation with the state. The design should be such that the velocities and flows noted above will be maintained.

Sedimentation.   The requirements for effective clarification are dependent upon a number of factors related to basin design and the nature of the raw water (turbidity, color and colloidal matter, taste- and odor-causing compounds, or other material to be removed). Also to be considered is the character of floc formed. (Chapter 9 discusses some of these topics in detail.)

Inlet Devices.   Inlets should be designed to distribute the water equally and at uniform velocities. Open ports, submerged ports, and similar entrance arrangements are recommended. A baffle should be constructed across the basin close to the inlet end and should project several feet below the water surface to dissipate inlet velocities and provide uniform flows across the basin.

Outlet Devices.   Outlet devices should be designed to maintain velocities suitable for settling in the basin and to minimize short-circuiting. The use of submerged orifices is recommended in order to make the volume above the orifices available for storage of any excess water pumped to the plant beyond that passed through the filters.

Overflow Rate. The recommended maximum surface overflow rate is 600 gpd/sq ft (1.02 m/h), and the rate of flow over the outlet weir should not exceed 20,000 gpd/ft (248 m³/d/m) of weir length. Where submerged ports are used as an alternate for overflow weirs, they should be no lower than 3 feet (0.91 m) below the flow line.

Drainage. Basins should be provided with a means for dewatering. Basin bottoms should slope toward the drain at a slope of not less than 1 in 12 where mechanical sludge collection equipment is not used. Minimum diameter of the drain line is 8 inches (20.32 cm).

Covers. Covers are acceptable at certain plant locations. However, where basins have mechanical equipment, a superstructure should be provided in place of a cover. Where covers are used, manholes should be provided as well as drop light connections so that observation of the floc can be made at several points within the basin.

Velocity. The velocity through settling basins should not exceed 0.5 ft/min (0.15 m/s). The basins should be designed to minimize short-circuiting. Baffles should be provided as necessary.

Overflow. An overflow weir (or pipe) should be installed that will establish the maximum water level desired on top of the filters. It should discharge with a free fall at a location where the discharge will be seen.

Inlet to Outlet Distance. The minimum horizontal distance from basin inlet to outlet should be 10 feet (3.05 m) except as approved by the state.

Safety. Permanent ladders or handholds should be provided for safety on the inside walls of basins above the water level. Guard rails should be included. Flushing lines or hydrants should not include interconnection of the potable water with nonpotable water.

Sludge Collection. Mechanical sludge collection equipment should be provided.

Sludge Disposal. Facilities are often required by the state agencies involved for disposal of sludge. Provision should be made for the operator to observe or sample sludge being withdrawn from the unit.

Solids Contact Units

Units are acceptable for combined softening and clarification where water characteristics are not variable, and flow rates are uniform. Before such units are considered as clarifiers without softening, specific approval of the reviewing agencies should be obtained. Clarifiers should be designed for the maximum uniform rate and should be adjustable to changes in flow that are less than the design rate and for changes in water characteristics. A minimum of two units is recommended.

Installation of Equipment. Supervision by a representative of the manufacturer should be provided with regard to all mechanical equipment at the time of installation and initial operation. There should be adequate piping with suitable sampling taps so located as to permit the collection of samples of water from critical portions of the units.

Chemical Feed. Chemicals should be applied at such points and by such means as to ensure satisfactory mixing of the chemicals with the water.

Mixing. Mixing devices employed should be so constructed as to:

- Provide good mixing of the raw water with previously formed sludge particles.
- Prevent deposition of solids in the mixing zone.

Flocculation. Flocculation equipment used in lime–soda softening should have the following features:

- The system should be adjustable for varying energy input.
- The system should provide for coagulation to occur in a separate chamber or baffled zone within the unit.
- The system should provide a flocculation and mixing period of not less than 30 minutes.

Sludge Concentrators and Recirculation. The equipment should provide either internal or external recirculation and concentrators in order to obtain a concentrated sludge with a minimum of wastewater.

Sludge Removal. Sludge removal design should provide that:

- Sludge pipes are not less than 3 inches (7.62 cm) in diameter and are arranged to facilitate cleaning.
- The entrance to sludge withdrawal piping prevents clogging.

- Valves are located outside the tank for accessibility.
- The operator may observe or sample sludge being withdrawn from the unit.

Cross-Connections. Cross-connections must be avoided, and the following should be considered during design:

- Blow-off outlets and drains should terminate and discharge at locations satisfactory to the state.
- Cross-connection control should be included for the potable water lines used to backflush sludge lines.

Upflow Rates. Unless supporting data are available to justify other rates, clarifier upflow rates should not exceed:

- 1.75 gpm/sq ft (4.27 m/h) of area at the slurry separation line, for units used for softeners.
- 1.0 gpm/sq ft (2.44 m/h) of area at the sludge separation line for units used for clarifiers.

Settling Tube Clarifiers

Shallow depth sedimentation devices or tube clarifier systems of the essential horizontal or steeply inclined types may be used under appropriate conditions. (These devices are described in detail in Chapter 9.)

The loading rates for settling tube installations in horizontal flow basins are listed in Table 21-3, and for solids contact units in Table 21-4.

Location of Tube Settling Modules. The tube settling modules must be properly placed to afford maximum benefit from their use. Tube settler installations should be located:

- In a zone of stable hydraulic conditions.
- In areas nearest effluent launders in basins not completely covered by tubes.
- With the top of tubes 2 to 4 feet (0.61 to 1.22 m) below the water surface.

Tube Support System. The tube support system should be able to carry the weight of the tube modules when the basin is drained plus the weight of a worker standing on the tube modules. Alternatively, it should support not less than 7.5 lb/sq ft (kg/m²). The minimum bearing surface width of a support member should be 1 inch (2.54 cm), and support members should be located about 6 inches (15.24 cm) from the module end.

Table 21-3. Recommended Maximum Overflow Rates for Settling Tube Installations in Horizontal Flow Basins.

WATER TEMPERATURE, DEGREES C	RAW WATER TURBIDITY RANGE, NTU	BASED ON TOTAL CLARIFIER SURFACE AREA		BASED ON TOTAL AREA COVERED BY TUBES	
		GPM/SQ FT	M/H	GPM/SQ FT	M/H
Frequently 4 or	0–100	2	4.88	3	7.32
lower	100–1,000	2	4.88	2.5	6.10
Usually above 4	0–100	3	7.32	3.5	8.54
	100–1,000	2	4.88	3	7.32

Tube Cleaning. Provisions should be made to drop the water level occasionally for tube cleaning, or to provide a water or air jet cleaning system for the top surface of the tube modules. Permanent air scour or water jet systems have been installed in many installations, with success. However, at least two clarifiers should be provided to allow removing one of them from service for cleaning.

Laboratory Control

Chemical precipitation plants for water softening should provide an electric pH meter; equipment for determining alkalinity, hardness, temperature, and residual disinfectant; and a continuously recording turbidimeter on each filter effluent line.

It is also highly desirable to furnish equipment for coagulation control. Pilot filters are preferred, but jar test apparatus, zeta potential meters, streaming current detectors, and colloid titration apparatus may be useful in some situations.

Table 21-4. Recommended Maximum Overflow Rates for Settling Tube Installations in Solids Contact Units.

WATER TEMPERATURE, DEGREES C	BASED ON TOTAL CLARIFIER SURFACE AREA		BASED ON TOTAL AREA COVERED BY TUBES	
	GPM/SQ FT	M/H	GPM/SQ FT	M/H
Frequently 4 or lower	2	4.88	3	7.32
Usually above 4	2	4.88	4	9.76
	2.5	6.10	2.5	6.10

RECARBONATION

Purpose

Recarbonation is a unit water treatment process that has been in use for many years in numerous municipal and industrial lime–soda softening plants throughout the world.[2-11] The term recarbonation refers to the addition of carbon dioxide to a lime-treated water. When carbon dioxide is added to high-pH, lime-treated water, the pH is lowered, and the hydroxides are reconverted to carbonates and bicarbonates. Thus, the term recarbonation is descriptive of the result of adding carbon dioxide to water.

The basic purpose of recarbonation is the downward adjustment of the pH of the water. Properly done, this places the water in calcium carbonate equilibrium, and avoids problems of deposition of calcium scale that would occur without the reduction in pH accomplished by recarbonation. In waterworks practice, the carbon dioxide is added to the water ahead of the filters in order to avoid coating the grains of filter media with calcium carbonate, which would eventually increase the grain size to the point that filter efficiency would be reduced. In waterworks, it is also important to lower the pH of the lime-treated water to the point of calcium carbonate stability to avoid deposition of calcium carbonate in pipelines.

There are several reasons for adjusting the pH of water during treatment. Coagulation, flocculation, and clarification can be accomplished in a rather wide range of pH values, but particular coagulants or coagulant aids ordinarily yield optimum results within a rather narrow pH range. Lime usually gives good results in waters at pH values above 9.6 or 9.8. Alum is ordinarily good at pH values below 7.3 or above 8.6. The effective pH range for coagulants usually can be broadened by use of the correct polymer, activated silica, or other coagulant aids. Generally, filtration through granular media is best in the pH range 6.5 to 7.5, but the filterability of water depends upon many other factors, including the physical and chemical characteristics of the water, the coagulant used, and the chemicals employed as filter aids, if any.

Activated granular carbon adsorption of organics from water usually takes place in the pH range 5 to 9, with better adsorption at values below 7. Often, if the pH is above 9, desorption will occur; that is, organics previously adsorbed on the carbon at low pH will be released to the high-pH water. For example, a column of granular activated carbon that has been operating at pH 7 for several days with good removal of color from the applied water may have the capacity for further color removal at pH 7. However, if water with a pH of 10 is passed through the column, the color of the effluent water from the column will undoubtedly be higher than that of the influent because of desorption.

Massive lime treatment of waters for softening can raise the pH to a value of about 10.4. Primary recarbonation is used to reduce the pH from 10.4 to 9.3, which is near the pH of minimum solubility for calcium carbonate. Primary recarbonation to pH 9.3 results in the formation of a floc that is principally calcium carbonate. If sufficient reaction time, usually about 15 minutes in cold water, is allowed for the primary recarbonation reaction to go to completion, the calcium carbonate floc does not redissolve with subsequent further lowering of pH in secondary recarbonation. However, there is a tendency for the magnesium salts to do so. If lime is to be reclaimed by recalcining and reuse, this settled primary recarbonation floc is a rich source of calcium oxide, and may represent as much as one-third of the total recoverable lime.

The second stage of recarbonation, to pH 7, is beneficial in several ways:

- It prepares the water for filtration.
- It lowers the pH to a value that increases the efficiency of carbon adsorption of organics.
- It is an excellent range for effective disinfection by chlorine.
- It stabilizes the water with respect to scale formation in pipelines.

If the pH were not reduced to less than about 8.8 before application to the filters and carbon beds, extensive deposition of calcium carbonate would occur on the surface of the grains. This could reduce filter efficiency, and could also drastically reduce the adsorptive capacity of granular activated carbon for organics. It would produce rapid ash buildup in the carbon pores upon regeneration of the carbon, and would lead to early replacement of the carbon.

SINGLE-STAGE VS. TWO-STAGE RECARBONATION

It is possible to reduce the pH of a treated water from 11 to 7 or to any other desired value in one stage of recarbonation. Single-stage recarbonation eliminates the need for the intermediate settling basin used with two-stage systems. However, by applying sufficient carbon dioxide in one step for the total pH reduction, little, if any, calcium is precipitated, with the bulk of calcium remaining in solution, thus increasing the calcium hardness of the finished water. In addition, this causes the loss of a large quantity of calcium carbonate, which could otherwise be settled out, recalcined to lime, and reused. If lime is to be reclaimed or if calcium reduction in the effluent is desired, then two-stage recarbonation is required. Otherwise, single-stage recarbonation may be used with some savings in initial cost, and some reduction in the amount of lime sludge to be handled.

SOURCES OF CARBON DIOXIDE

In some water treatment plants, carbon dioxide for recarbonation may be obtained from stack gas from a nearby incinerator or from in-plant recalcination furnaces. Other sources include the use of commercial liquid carbon dioxide; or the burning of natural gas, propane, butane, kerosene, fuel oil, or coke.

The stack gas from an incineration furnace that is fired with natural gas will contain about 16 percent carbon dioxide on a wet basis or about 10 percent on a dry basis. About 10 percent CO_2 on a dry basis is usually used for design purposes. The burning of 1,000 cubic feet (28.3 m³) of natural gas produces about 115 pounds (52.2 kg) of CO_2.

The stack gas from a lime recalcining furnace contains not only the CO_2 produced by combustion of the fuel, but also the CO_2 driven off of the calcium carbonate sludge in the recalcining process. For design purposes, a value of 16 percent CO_2 in lime furnace stack gas is a conservative figure.

Kerosene and No. 2 fuel oil will yield about 20 pounds (9.08 kg) of CO_2 per pound of fuel. Coke will produce approximately 3 pounds (1.36 kg) of CO_2 per pound of coke burned. Commercial liquid CO_2 is 99.5 percent CO_2.

QUANTITIES OF CARBON DIOXIDE REQUIRED

In recarbonation, one molecule of CO_2 is required to convert one molecule of calcium hydroxide (caustic alkalinity) to calcium carbonate. In addition, it takes one molecule of CO_2 to convert one molecule of calcium carbonate to calcium bicarbonate. It follows then, that two molecules of CO_2 are required to convert one molecule of calcium hydroxide to calcium bicarbonate. These reactions are represented by the following equations:

$$Ca(OH)_2 + CO_2 \rightarrow CaCO_3 + H_2O \qquad (21\text{--}24)$$
$$CaCO_3 + CO_2 + H_2O \rightarrow Ca(HCO_3)_2 \qquad (21\text{--}25)$$

Because all forms of alkalinity are expressed in terms of calcium carbonate (molecular weight = 100), the calculations are as follows:

1. Calcium hydroxide to calcium carbonate:

(molecular weight of $CO_2 = 44$, and 1 mg/l = 8.33 lb/MG)

$$CO_2 \text{ (in lb/MG)} = \frac{44}{100} \times 8.33 \times \text{(Hydroxide Alk. in mg/l as } CaCO_3)$$

$$= 3.7^* \times \text{(Hydroxide Alk. in mg/l as } CaCO_3)$$

2. Calcium carbonate to calcium bicarbonate:

$$CO_2 \text{ (in lb/MG)} = \frac{44}{100} \times 8.33 \times (\text{Carbonate Alk. in mg/l as } CaCO_3)$$

$$= 3.7* \times (\text{Carbonate Alk. in mg/l as } CaCO_3)$$

3. Then, for calcium hydroxide to calcium bicarbonate:

$$CO_2 \text{ (lb/MG)} = 7.4** \times (\text{Hydroxide Alk. in mg/l as } CaCO_3)$$

SAMPLE CALCULATIONS FOR THE AMOUNT OF CO₂ REQUIRED

Assume that 400 mg/l of calcium oxide (CaO) has been added to a sample of water and that the stirred and decanted liquor has the following characteristics:

pH = 11.7
Alkalinities (in mg/l as $CaCO_3$): Hydroxide (OH^-) = 380
Carbonate (CO_3^{-2}) = 120
Bicarbonate (HCO_3^-) = 0

After first recarbonation in the laboratory using bottled carbon dioxide, analysis of the lime-treated water shows the following:

pH = 9.3
Alkalinities (in mg/l as $CaCO_3$): OH^- = 0
CO_3^{-2} = 180
HCO_3^- = 380

Then after secondary recarbonation in the laboratory, the water has the following analysis:

pH = 8.3
Alkalinities (in mg/l as $CaCO_3$): OH^- = 0
CO_3^{-2} = 0
HCO_3^- = 750

---

* 3.7 factor becomes 0.44 for kg/Ml.
** 7.4 factor becomes 0.88 for kg/Ml.

To change all caustic alkalinity and all carbonate alkalinity to bicarbonates, the amount of CO_2 required is as calculated below:

$$
\begin{aligned}
7.4 \times 380 &= 2{,}812 \text{ lb/MG } (334 \text{ kg/Ml}) \text{ of } CO_2 \\
3.7 \times 120 &= \underline{444 \text{ lb/MG } (53 \text{ kg/Ml})} \\
\text{Total} &= 3{,}256 \text{ lb/MG } (387 \text{ kg/Ml}) \text{ of } CO_2
\end{aligned}
\tag{21-26}
$$

For a 1-mgd (3.785-Ml/d) flow, 3,256 pounds (1,478 kg) of CO_2 per day are required. If it is assumed that the CO_2 content of the stack gas to be used for recarbonation is 10 percent, then $(100/10) \times 3{,}256 = 32{,}560$ pounds (14,780 kg) of stack gas must be compressed in order to supply the necessary CO_2 to recarbonate 1 mgd (3.785 Ml/d) of water.

At standard conditions of 14.7 psia (101.28 kPa) and 60°F (15.5°C), assume that the density of the stack gas is the same as for CO_2, or 0.116 lb/cu ft (1.88 kg/m³). Then:

$$
\begin{aligned}
\frac{32{,}560 \text{ lb}}{0.116 \text{ lb/cu ft}} &= 280{,}700 \text{ cu ft/day} \\
&= 195 \text{ cu ft/min } (5.52 \text{ m}^3/\text{min})
\end{aligned}
\tag{21-27}
$$

If it is assumed that the stack gas is cooled in scrubbing to 110°F (43.3°C), then the temperature correction is:

$$
\begin{aligned}
\frac{110 + 460}{60 + 460} \times 195 &= 1.1 \times 195 \\
&= 214 \text{ cu ft/min/mgd } (1 \text{ m}^3/\text{min/Ml/d})
\end{aligned}
\tag{21-28}
$$

With the same gas temperature, for a plant at 6,300 feet above sea level (11.6 psia), the altitude correction is:

$$
\begin{aligned}
\frac{14.7}{11.6} \times 215 &= 1.26 \times 215 \\
&= 270 \text{ cu ft/min/mgd } (1.26 \text{ m}^3/\text{min/Ml/d})
\end{aligned}
\tag{21-29}
$$

Because part of the CO_2 is not absorbed in the water but escapes at the water surface, it is customary to add about 20 percent to the theoretical requirements. If this is done, then at sea level 260 cu ft/min/mgd (1.21 m³/min/Ml) of blower or compressor capacity is required, and at 6,300 ft (1,920 m) of altitude 325 cu ft/min/mgd (1.52 m³/min/Ml/d) of plant capacity is needed.

It must be emphasized that these calculations are based on the following assumptions:

1. The water to be recarbonated has a pH $= 11.7$ with OH$^-$ Alk. $= 380$ mg/l as $CaCO_3$ and with a CO_3^{-2} alkalinity of 120 mg/l as $CaCO_3$, all of which is to be converted to bicarbonates.
2. There is 10 percent CO_2 in the gas.
3. The flue gas temperature is 110°F (43.3°C).
4. An excess of 20 percent is added to the theoretical values to allow for absorption losses.

Conditions at each installation undoubtedly will differ from the assumptions used here, and calculations must be based on actual values rather than those given.

In water softening plants that add lime to pH $= 10.6$, the amount of CO_2 required will be much less than in the example above.

NONPRESSURIZED CARBON DIOXIDE GENERATORS

If stack gas from a furnace operating at atmospheric pressure is to be used as a source of CO_2, the gas should be passed through a wet scrubber. Wet scrubbers provide contact between the gas and a flow of scrubbing water. Particulate matter is removed from the gas, and the gas is cooled. Wet scrubbers may be one of three general types: impingement, Venturi, or surface area. Water jet impingement-type scrubbers are efficient in removing potential air pollutants from the exhaust gas, and provide some protection of the CO_2 compression equipment against plugging or scaling by particulates. The scrubbers cool the hot stack gas down to about 110°F (43.3°C).

When stack gas alone is used as the source of CO_2, the stack gas supply must exceed the maximum demands for CO_2. With this situation, control of the amount of CO_2 applied to the water is simple. Air may be admitted through a valve into the suction line leading to the compressor as required to reduce the amount of CO_2 to that desired. Alternatively, part of the compressed gas may be bled off to the atmosphere through a valve in the compressor discharge line. As another method of control, compressed gas may be recirculated from the compressor discharge line back to the suction line through a bypass line and control valve. However, this method has the serious disadvantage of warming the gas by compression, and excessive recirculation can lead to compressor damage by overheating or increased corrosion at the elevated temperatures.

COMPRESSOR SELECTION

Even with thorough scrubbing, stack gas from incineration furnaces or lime recalcining furnaces will contain sufficient particulate matter to cause plugging and seizure problems in some types of blowers and compressors, particularly those with limited clearance between moving metal parts. This problem is less severe with stack gas from atmospheric furnaces, which burn fuel primarily for production of CO_2.

Water-sealed compressors similar to wet vacuum pumps are a good selection for handling dirty, corrosive gases. This type of compressor consists of a squirrel-cage-type rotor that revolves in a circular casing containing water. The rotor shaft is located off center toward the bottom of the casing. As the rotor revolves, centrifugal force pushes the water out against the sides of the casing, leaving a doughnut-like hole of air in the center. Because the center drum of the rotor is located eccentrically, there is a large air space above it and very little below. Thus, as air is driven from the large space into the smaller one, it is compressed. This is a simple, reliable piece of equipment with only one moving part. It has increased capacity when handling hot, saturated vapors because the vapors are condensed by the cool liquid compressant. The water-sealed compressor is a relatively quiet-running unit, free from pulsations and vibrations.

If the CO_2 distribution grids are submerged a minimum of 8 feet in water, as they usually are, the CO_2 compressor must deliver against a differential pressure across the machine of about 6 to 8 psi (41.3 to 55.12 kPa). The exact rating must be determined by calculation, taking into account not only the depth of submergence of the distribution piping but also orifice losses and pipe friction losses. This is discussed in more detail later because it is common to all types of CO_2 systems. The compressors may be of cast iron construction, or may be supplied with a bronze rotor and cones at considerable extra cost.

The following accessories are commonly required with water-sealed compressor units: water separator with gauge glass and bronze float valve; discharge check valve; expansion joints for inlet and outlet piping; water seal supply line with adjusting cock and orifice union, water line strainer, inlet water spray nozzles, and sealing water line solenoid valve. In addition, the discharge line is usually fitted with an automatic pressure relief valve and a bleed-off valve, both of which would discharge to the free atmosphere. It is not good practice to install shutoff or isolation valves on either the compressor suction or discharge lines because of the possibility of serious damage to the compressor or pipelines in the event that the compressor is operated in error with one or both of the valves closed.

In selecting CO_2 compressor units to meet total capacity requirements, it is a good idea, except in very small installations, to provide at least three compressor units. By properly sizing them, it is then possible to satisfy two needs: to secure a range in output and to provide standby service. If it is assumed that the total CO_2 capacity required is 1,500 cu ft/min (42.45 m³/min), then units with individual capacities of 500 cu ft/min (14.15 m³/min), 1,000 cu ft/min (28.3 m³/min), and 1,500 cu ft/min (42.45 m³/min) would

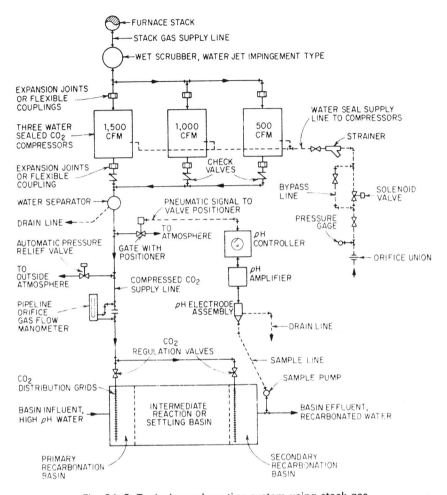

FURNACE STACK
STACK GAS SUPPLY LINE
WET SCRUBBER, WATER JET IMPINGEMENT TYPE
EXPANSION JOINTS OR FLEXIBLE COUPLINGS
WATER SEAL SUPPLY LINE TO COMPRESSORS
THREE WATER SEALED CO_2 COMPRESSORS
1,500 CFM
1,000 CFM
500 CFM
STRAINER
EXPANSION JOINTS OR FLEXIBLE COUPLING
CHECK VALVES
WATER SEPARATOR
DRAIN LINE
PNEUMATIC SIGNAL TO VALVE POSITIONER
BYPASS LINE
SOLENOID VALVE
PRESSURE GAGE
AUTOMATIC PRESSURE RELIEF VALVE
GATE WITH POSITIONER
TO ATMOSPHERE
pH CONTROLLER
TO OUTSIDE ATMOSPHERE
COMPRESSED CO_2 SUPPLY LINE
pH AMPLIFIER
ORIFICE UNION
PIPELINE ORIFICE GAS FLOW MANOMETER
pH ELECTRODE ASSEMBLY
DRAIN LINE
CO_2 REGULATION VALVES
SAMPLE LINE
CO_2 DISTRIBUTION GRIDS
SAMPLE PUMP
BASIN INFLUENT, HIGH pH WATER
INTERMEDIATE REACTION OR SETTLING BASIN
BASIN EFFLUENT, RECARBONATED WATER
PRIMARY RECARBONATION BASIN
SECONDARY RECARBONATION BASIN

Fig. 21–3. Typical recarbonation system using stack gas.

represent a good choice. This combination gives a range of 500 to 1,500 cu ft/min (14.15 to 42.45 m³/min) to match plant needs, and it supplies standby with the largest unit out of service by using the two smaller units together.

A typical recarbonation system using stack gas at atmospheric pressure is illustrated by Fig. 21–3. As indicated in this figure, automatic pH control of the recarbonated effluent can be provided by continuously monitoring an effluent sample for pH. Changes in pH will operate the controller, which in turn positions a bleed-off valve in the CO_2 compressor discharge line to limit the amount of CO_2 to that necessary to maintain the desired pH.

PRESSURE GENERATORS AND UNDERWATER BURNERS

Generators designed specifically for the production of carbon dioxide for recarbonation are usually either pressure generators or submerged underwater burners. Most early installations were of the atmospheric type, in which the fuel is burned at atmospheric pressure and the off-gas is scrubbed and compressed. These systems are expensive to maintain because of the corrosive effects of the hot, moist combustion gases, and atmospheric generators have largely been replaced by pressure generators and underwater burners, except where waste stack gas is available from another source. Both types of pressure CO_2 generation equipment are now in commercial manufacture.

Pressure or forced draft generators produce CO_2 by burning natural gas, fuel oil, or other fuels in a pressure chamber. The fuel and excess air are first compressed and injected, and then burned at a pressure that is sufficiently high to allow discharge directly into the water to be recarbonated. The compressors handle only dry gas or dry air at ambient temperatures, and thus the corrosion problems involved in handling the hot, moist stack gases are avoided. One difficulty with this type of pressure generator is its limited capacity range, which may be 3 to 1, or at best 5 to 1. This low turndown ratio may necessitate the installation of two or more units in order to secure the required flexibility and process control. A wide range of sizes is commercially available in pressure CO_2 generators. This commercial equipment is

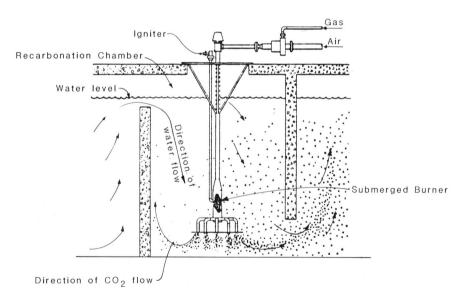

Fig. 21-4. Typical recarbonation system using submerged combustion burner.

well designed and reliable, and includes all auxiliaries and safety controls.

Submerged combustion of natural gas is another method of CO_2 generation. A unit is shown schematically in Fig. 21–4, and a picture of a submerged gas burner is shown in Fig. 21–5. Air and natural gas are compressed and then burned under water at the point of application; that is, in the recarbonation basin. Automatic underwater electric ignition equipment is used to start combustion. Submerged combustion is a simple, efficient means of CO_2 generation that provides good control of recarbonation, and requires a minimum of maintenance. The turndown ratio of this type of burner is only about 2 to 1, so that it is necessary to provide enough burner assemblies to obtain the desired range of control in the amount of CO_2 applied.

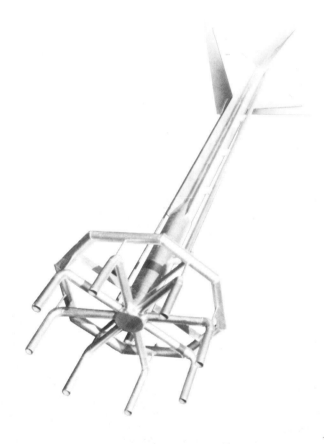

Fig. 21–5. Submerged combustion burner. (Courtesy of TOMCO$_2$ Systems Equipment Company.)

LIQUID CARBON DIOXIDE

Commercial liquid CO_2 has found increasing use for recarbonation in water softening plants in the last few years, primarily because of its steadily decreasing cost. However, the price of liquid CO_2 depends greatly on the distance from the source of supply, and the first factor to be investigated is the cost of liquid CO_2 delivered to the plant under consideration. Even in favorable locations, high cost is still the principal disadvantage of using liquid CO_2. Its advantages include flexibility, ease of control, high purity and efficiency, and the smaller piping required because of its high CO_2 content, 99.5 percent, relative to the 6 to 18 percent obtained from other sources.

Liquid CO_2 may be delivered to customers in insulated tank trucks ranging from 10 to 20 tons (9.1 to 18.2 metric tons) in capacity. Rail car shipments of 30 to 40 tons (27.3 to 36.4 metric tons) are available to large-volume users. Some manufacturers will lease tank cars so that they may be used for storage at the site, thus eliminating the need and expense of bulk storage tanks and auxiliaries at the plant. For small plants, liquid CO_2 is also available in 20- to 50-lb (9.1- to 22.7-kg) cylinders.

Bulk storage tanks may be purchased or leased. Capacities range from 1 to 100 tons (0.91 to 90.7 metric tons), although the common sizes are 4 to 48 tons (3.6 to 43.5 metric tons). A picture of a typical liquid CO_2 storage tank is shown in Fig. 21–6, and Fig. 21–7 shows the valves and compressor in the cabinet. Storage tanks must be insulated and equipped with Freon

Fig. 21–6. A 14-ton carbon dioxide receiver. (Courtesy of AirCo Industrial Gases.)

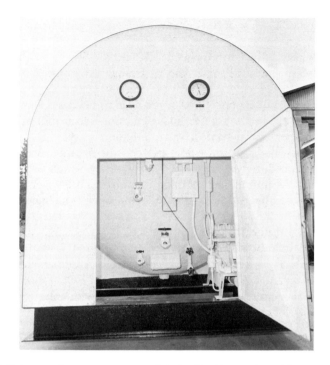

Fig. 21–7. Valves and compressor associated with CO_2 receiver. (Courtesy of AirCo Industrial Gases.)

refrigeration and electric or steam vaporization equipment. The working pressure for storage tanks is 350 psi (2,411.5 kPa), and the ASME Code for Unfired Pressure Vessels requires hydrostatic testing to 525 psi (3,617 kPa), or 1.5 times the working pressure. The tanks may be insulated with pressed cork or polyurethane foam. The cooling and vaporizing systems are designed to maintain the liquid CO_2 at about $0°F$ ($-17.8°C$) and 300 psi (2,067 kPa). If temperature and pressure rise, the cooling system comes on; and with falling pressure, the vaporizer comes into service. In the event either of these systems fails, or in the event of fire or other accident, the storage tanks are fitted with high and low pressure alarms, two safety pop valves, a manual bleeder relief valve, and a bursting disk.

Either liquid or gas feed systems may be used to apply the liquid CO_2 to the water to be treated. As CO_2 is withdrawn from the storage tank, the pressure is reduced. This pressure reduction cools the CO_2, with the danger of dry ice formation if the expansion is too rapid. Consequently, it is common practice to reduce the pressure in two stages, from the 300 psi (2,067 kPa) tank pressure to the 20 psi (138 kPa) pressure ordinarily required for feeding the CO_2. Vapor heaters may also be used just ahead of the pressure reducing valves.

For CO_2 gas feed, an orifice plate in the feed line with a simple manometer may be used to measure flow, and a manual valve installed downstream may be used to regulate or control the amount of CO_2 applied. Automatic control to a manual set point could be provided by using a differential pressure transmitter on the feed line orifice, and connecting it to an indicating controller that would operate the control valve. Optionally, the CO_2 feed could be made fully automatic by providing pH control. In this case, an electrode would be installed to measure the pH of the recarbonated water. This signal would be amplified and sent through a controller that would throttle the control valve on the feed line at low pH and open it at high, as set on the controller.

For solution feed of CO_2, equipment similar (except for materials of construction) to solution feed chlorinators may be used. Chlorinator capacity is reduced about 25 percent when feeding CO_2. Approximately 60 gallons of water is required to dissolve 1 pound (0.454 kg) of CO_2 at room temperature and atmospheric pressure. Absorption efficiency with solution feed of CO_2 approaches 100 percent.

CARBON DIOXIDE PIPING AND DIFFUSION SYSTEMS

Because in recarbonation systems the gas temperature is usually in the 70 to 100°F (21.1 to 37.7°C) range, and pressure is about 6 to 8 psi (41.3 to 55.1 kPa), and because CO_2 pipe runs are usually less than 100 feet (30.48 m), it is convenient to use Table 21–5 to estimate the pipe size required. The pipe sizes obtained from Table 21–5 are, of course, an approximation. For greater accuracy, or for long lines, the following modification of the Darcy-Weisbach formula may be used for pressure loss in air piping:

$$\Delta p = \frac{f}{38,000} \frac{LTQ^2}{pD^5} \qquad (21\text{--}30)$$

where:

Δp = pressure drop, psi (kPa)

$f = \dfrac{0.048\ D^{0.027}}{Q^{0.148}}$ (Note: usual values for f = 0.016–0.049)

L = pipe length, feet (m)
T = absolute temperature of the gas, °R (°F + 460)
Q = gas flow, cu ft/min (m³/min)
p = absolute pressure of the gas, psi (kPa) (or line pressure in psi + 14.7)
D = pipe diameter, inches (cm)

Table 21–5. Approximate Gas-Carrying Capacity.

PIPE DIAMETER		CAPACITY	
INCHES	cm	CU FT/MIN	m³/MIN
1	2.54	45	1.27
2	5.08	250	7.1
3	7.02	685	19.4
4	10.16	1,410	39.9
6	15.24	3,870	109.5

The pressure loss in elbows and tees can be approximated by use of the following formula:

$$L = \frac{7.6\,D}{1 + \dfrac{3.6}{D}} \tag{21-31}$$

where:

L = equivalent length of straight pipe, feet (m)
D = pipe diameter, inches (cm)

The loss in globe valves is about:

$$L = \frac{11.4\,D}{1 + \dfrac{3.6}{D}} \tag{21-32}$$

where L and D are as defined above.

Carbon dioxide absorption systems often consist of a grid of perforated pipe submerged in the water. The recommended minimum depth of submergence is 8 feet (2.44 m). With lesser depths of submergence some undissolved CO_2 will escape at the water surface. Properly designed absorption systems will put into solution 85 to 100 percent of the applied CO_2. PVC pipe is suitable for the perforated CO_2 grid pipes. Current practice is to use $\frac{3}{16}$-inch (0.48-cm)-diameter orifices drilled in the bottom of the pipe at an angle of 30 degrees to the right of the vertical centerline, then 30 degrees to the left, alternating at a spacing of about 3 inches (7.62 cm) along the centerline of the pipe. Another arrangement is to point the orifices straight up at the top of the pipe and to direct a jet of water from a header down at the CO_2 orifice, in order to form fine bubbles of the gas, which dissolve more readily in the water than larger bubbles. Because PVC does not corrode under acidic conditions, the openings are not subject to plugging as they are in metal pipes.

If $\frac{3}{16}$-inch (0.48-cm) orifices are used, each opening is often rated at 1.1 to 1.65 cu ft/min (0.03 to 0.05 m³/min), which corresponds to headlosses through the orifice of 3 and 8 inches (7.6 to 20.3 cm) of water column, respectively. This is sufficient loss through the orifice to ensure good distribution of the CO_2 to each opening. Carbon dioxide laterals must be laid with the same depth of submergence on each orifice. If the size of the pipe changes, then eccentric reducers should be used to keep the bottom of the pipe level (assuming that the holes are in the bottom of the pipe). Horizontal spacing between CO_2 diffusion laterals should be at least 1.5 feet (0.46 m) in order to get good absorption. To convey cool, dry CO_2, plain steel or cast iron pipe may be used; but for hot, moist CO_2 gas, the use of stainless steel or other acid-resistant metal is suggested. Special pipe is also required to convey liquid CO_2 in water; a 1.5-inch (3.8-cm) cotton fabric hose with openings of controlled size or porosity has been used successfully. Basin hydraulics must take into account raised water levels caused by CO_2 injection.

CARBON DIOXIDE REACTION BASINS
OR INTERMEDIATE SETTLING BASINS

Contrary to many early reports in the literature, the recarbonation reaction is not instantaneous. Although about 90 percent of the applied CO_2 does dissolve in its very short upward journey from the distribution grid through 8 feet (2.44 m) of water to the water surface, the time for complete reaction between the dissolved CO_2 and hydroxide and carbonate ions may be as great as 15 minutes in cold water. In the primary phase of two-stage recarbonation, if the reaction is allowed to go to completion at a pH near 9.3, the calcium carbonate formed is not redissolved in the second phase of recarbonation to a low pH—say to pH 7.0. Magnesium salts do tend to redissolve under these conditions. In the recarbonation of water, a floc is formed following first-stage recarbonation. This is a rich source of calcium carbonate from which lime (CaO) can be reclaimed and reused by recalcining at temperatures of about 1,850°F (1,009°C). In this case, then, it is desirable to allow not only for reaction time (15 minutes) but for enough time to provide some separation of the calcium carbonate by settling. This will require a settling basin with at least 30-minute detention at maximum flow rate, and a basin surface overflow rate of not more than 2,400 gal/sq ft/day (4.1 m/h) at peak flow rates. This intermediate settling basin should be fitted with continuous mechanical sludge removal equipment. Figure 21–8 shows a two-stage recarbonation system with intermediate reaction and settling.

Single-stage recarbonation systems should be followed by 15 minutes of

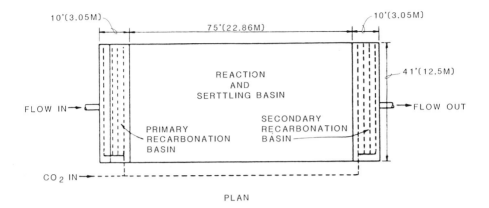

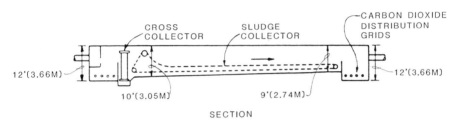

Fig. 21–8. Plan and section showing two-stage recarbonation basin with intermediate settling.

detention for completion of the chemical reactions, but no provisions for settling or sludge collection are required. The light, cloudy floc that may be formed at times with single-stage recarbonation is removed quite readily by mixed-media filtration with little effect on filter effluent turbidity, headloss, or length of filter run.

Recarbonated lime-treated water should not be applied directly to beds of granular activated carbon without filtration. Even at low pH (say 7.0) there is still sufficient deposition of calcium carbonate to cause serious problems in the carbon treatment, which can easily be avoided by prior filtration.

OPERATION AND CONTROL OF RECARBONATION

The operation and control of recarbonation systems is easy and simple. Automated control systems ordinarily use a single point of pH measurement following the last stage of recarbonation as the basis of control. In two-stage recarbonation systems, the split of total CO_2 flow between the two stages of treatment is fairly constant once it is established for a given flow and the particular set of pH values desired, and control based on the final pH alone is satisfactory without readjustment of the valves supplying the first- and second-stage CO_2 supply headers.

An indirect but more sensitive control of recarbonation is provided by alkalinity measurements. Continuous reliable automatic monitoring and control equipment is available for either the pH or the alkalinity method, but the alkalinity measuring equipment is considerably more expensive than the pH equipment. Manual control is also quite satisfactory, based either on grab sampling and analysis or on observation of continuous automatic monitoring of pH or alkalinity of the recarbonated water. The CO_2 demands do not vary rapidly or widely, and manual control of dosage is better than might be expected.

SAFETY

Under certain conditions carbon dioxide can be dangerous, and there are safety precautions that must be observed. Prolonged exposure to concentrations of 5 percent or more CO_2 in air may cause unconsciousness and death. The maximum allowable daily exposure for a period of 8 hours is 0.5 percent CO_2 in air. Carbon dioxide is 1.5 times as dense as air, and therefore will tend to accumulate in low, confined areas. Filter-type gas masks are not useful in atmospheres containing excess CO_2, and self-contained breathing apparatus and hose masks must be used. Contact of the skin with liquid CO_2 can cause frostbite. Recarbonation basins must be located out of doors, and enclosed structures must not be built above them, because of the danger of excessive amounts of CO_2 accumulating within the structures. Before repairmen enter recarbonation basins, the CO_2 supply should be turned off and the space thoroughly ventilated. In the use of liquid CO_2 there are many other safety considerations too numerous and detailed to be covered completely here. Complete, published information can be obtained from liquid CO_2 suppliers or from the Compressed Gas Association, Inc., 500 Fifth Avenue, New York, New York.

REFERENCES

1. Black, A. P., Shuey, B. S., and Fleming, P. J., "Recovery of Calcium and Magnesium Values," *J.AWWA* 63:616, October, 1971.
2. Anonymous, "Submarine Burners Make CO_2 for Softening Recarbonation," *Water Works Engineering,* p. 182, 1963.
3. Bulletin No. 7-W-83, *Carball CO_2 for Recarbonation,* Walker Process Company, May, 1966.
4. Compressed Gas Association, *Handbook of Compressed Gases,* Van Nostrand Reinhold, New York, 1962.
5. Fair, M. F., and Geyer, J. C., *Water Supply and Waste Disposal,* John Wiley and Sons, New York, 1954.

6. Haney, P. D., and Hamann, C. L., "Recarbonation and Liquid Carbon Dioxide," *J.AWWA*, p. 512, 1969.
7. Hoover, C. P., *Water Supply Treatment*, 8th ed., Bulletin 211, National Lime Association, Washington, D.C.
8. Pamphlet G-6, *Carbon Dioxide*, 2nd ed., Compressed Gas Association, New York, 1962.
9. Pamphlet G-6, IT, *Tentative Standard for Low Pressure Carbon Dioxide Systems at Consumer Sites*, Compressed Gas Association, New York, 1966.
10. Ross, R. D., *Industrial Waste Disposal*, Van Nostrand Reinhold, New York, 1968.
11. Scott, L. H., "Development of Submerged Combustion for Recarbonation," *J.AWWA*, p. 93, 1940.

Chapter 22
Small Water Treatment Systems

INTRODUCTION

Small water treatment systems have a unique set of requirements that distinguish them from larger facilities. These particular requirements are related to the problem of treating water to the same quality as in larger community water systems, while costs must be distributed over a much smaller customer base. As drinking water quality standards become more stringent, costs will increase, further compounding these problems. An economical alternative to the high costs of conventional treatment for small systems is package plants, which are available within the size range 0.01 to 2 mgd (0.038 to 7.6 Ml/d).

Presently, it is estimated that over 650 package plants, ranging in capacity from 5 gpm (0.3 l/s) to 2 mgd (7.6 Ml/d), are in service in the United States. These represent only a fraction of the potential applications of package plants. Assuming an annual average per capita water consumption of 150 gpd (570 l/d), water systems serving a population of 10,000 persons would require a 1.5-mgd (5.7-Ml/d)-capacity treatment plant, which is within the capacity range stated above. In 1983, there were over 35,000 small community water systems that served populations of 10,000 or less. For many of these communities, treatment is required to meet the Safe Drinking Water Act standards.

Package plants can be used to treat water supplies for communities as well as recreational areas, state parks, construction camps, ski resorts, remote military installations, and other locations where potable water is not available from a municipal supply. Several state agencies have mounted package plants on trailers for emergency water treatment. Their compact size, low cost, minimal installation requirements, and ability to operate virtually unattended make them an attractive option in locations where revenues are not sufficient to pay for a full-time operator.

The package plant is designed as a factory-assembled, skid-mounted unit generally incorporating a single tank or, at the most, several tanks. A complete treatment process typically consists of chemical coagulation, flocculation, settling, and filtration.

Package plants are most widely used to treat surface supplies for removal of turbidity, color, and coliform organisms prior to disinfection. A limited market for package plants is for treatment of well water for removal of iron and manganese. Additionally, package plants can remove many inorganic chemicals for which maximum contaminant levels (MCL's) have been established. Inorganic contaminants such as arsenic, cadmium, chromium, lead, inorganic and organic mercury, selenium, silver, and fluoride can be partially or totally removed under proper treatment conditions by chemical coagulation followed by filtration,[1] as further described in Chapter 11.

Package plants are also effective in removing contaminants related to the aesthetics of drinking water. Although organic color may be treated adequately by simple chlorination, it can be coagulated with alum or iron salts and removed in package plants by filtration, which produces a water of more appealing quality. Organics contributing to taste and odor in drinking water supplies can also be removed, by adsorption on powdered activated carbon followed by filtration to remove the residual carbon. In this regard, package plants have application to treatment of a wide range of water supplies and can greatly improve the safety and overall acceptance of the finished water to the consumer. (A more detailed discussion of taste and odor removal is included in Chapter 18, and of iron and manganese removal in Chapter 19.)

Design Considerations Unique to Small Systems

A major problem confronting small water distribution systems is the high unit cost of providing good-quality water to a limited number of customers. A small system cannot benefit from economies of scale because of the small number of connections served. Certain costs, such as maintaining a chlorinator, a supply pump, and so on, are fixed regardless of the size of the system; and if the connections are few, enough revenue often cannot be collected to pay for the needed operation and maintenance. Many small systems, for example, cannot afford full-time operators and even have difficulty paying for part-time employees. To further compound the problem, the Safe Drinking Water Act requires these small systems to meet certain quality standards, which in many cases may require use of even more labor and materials, further increasing the cost of treatment.

In order to provide adequate treatment facilities for these small communities at an affordable cost, they must be designed with economy of materials and construction techniques in mind. Traditional labor- and material-intensive cast-in-place concrete structures are not cost-competitive with factory-built and -assembled steel tankage.

Another problem inherent in small systems is the general lack of skilled personnel for operating the plant. For this reason, treatment facilities serving small community systems should be simple in design and simple to operate. Several package plant manufacturers have recognized this requirement and have designed equipment that after adequate training, generally provided by the equipment supplier, can be operated by an individual with no prior water treatment plant operating experience.

Another design consideration for small systems is the need to automate the treatment equipment as much as is practical, to minimize the amount of operator attention required for day-to-day operation of the treatment plant. Also, plant automation reduces the amount of skill and judgment the operator must use to run the equipment. For example, water filtration equipment should be designed to backwash automatically in the absence of the operator and return to service on a preprogrammed basis. Also, sludge wasting from the clarifier should be automated to eliminate another function requiring attention and judgment.

The complexity of the equipment used in a small system should be a major consideration during design. Totally automated equipment does not necessarily need to be complex. The development of solid-state control components has greatly improved the reliability of automatic control systems. Equipment should be used for which readily attainable spare parts are available. To the extent that it is practical, a sufficient inventory of spare parts should be obtained with the construction bid. The availability of spare parts greatly simplifies repair and maintenance work during a plant emergency. This is especially important to a small system, which, unlike a larger utility, may not have the ability to obtain spare parts quickly.

PACKAGE PLANTS

The package plant is factory-built, and its quality and cost can be controlled closely to produce an efficient and economical treatment facility. Package treatment plants have been offered for many years by a number of manufacturers.[2] Treatment concepts in package plants, by and large, have duplicated those used in larger conventionally designed water treatment plants. Where packaged treatment plants have been properly applied and operated, performance has been satisfactory.

The term package plant, defined earlier, refers to a predesigned, factory-assembled, skid-mounted unit. Many manufacturers offer additional treatment components, such as pressure filters, clarifiers, and so forth, that may be skid-mounted and then assembled at the site; however, these installations are not regarded as true package water treatment plants. These individual treatment components are described in other chapters of this book.

Package water treatment plants are available from several manufacturers in a wide range of capacity, incorporating a complete treatment process (coagulation, flocculation, settling, and filtration). Design criteria used for these package plants vary widely. Some manufacturers adhere closely to accepted conventional design practices such as 20- to 30-minute flocculation detention time, a 2-hour sedimentation detention time, and rapid sand filters rated at 2 gpm/sq ft (4.88 m/h). Other manufacturers have utilized new technology including tube settlers and high-rate dual- and mixed-media filters to reduce the size of a plant and hence extend the capacity range of single factory-assembled units. Often, state regulatory agencies dictate the design criteria that must be met by package plant manufacturers and exclude units using new technology that does not meet arbitrary standards adopted by these agencies.

In the mid-1960s, coupled with the development of short-detention-time tube settlers, a package plant was introduced using the high-efficiency, short-detention-time clarification devices.[3] The use of tube settlers effectively reduces the settling detention time by a factor of 10 to 1 over conventional settling basins. This significant reduction in the volume of the settling basin, when used in conjunction with mixed-media filters rated at 5 gpm/sq ft (12.2 m/h), led to the development of a compact package plant with significantly greater capacity per unit volume than other equipment available at the time. The sizable reduction in package plant tankage greatly increased the capacity attainable in a single truck-transportable unit. For example, using tube settlers and mixed-media filters, a single factory-assembled truck-transportable package unit can be built with a capacity of 1 mgd (3.79 Ml/d). Following conventional design criteria, a plant of the same physical dimensions would produce less than 0.25 mgd (0.95 Ml/d).

A flow diagram for a package plant incorporating these newer techniques is shown in Fig. 22–1. Plants of this type are manufactured in two versions: Type A, (Microfloc Products "Water Boy") with a treatment capacity range of 10 to 100 gpm (0.63 to 6.3 l/s), and Type B (Microfloc Products "Aquarius") (Fig. 22–2), which generally consists of dual units with a capacity of 200 to 1,400 gpm (12.6 to 88.4 l/s). As illustrated in the flow diagrams, raw water is either pumped or flows by gravity to the treatment plant. The influent flow is adjusted to the desired rate; the control system is designed to start and stop the treatment plant according to a clearwell level that reflects system demand. The coagulant and disinfectant chemicals are added at the influent control valve. A polyelectrolyte coagulant aid is applied as the water enters the flash mix chamber. After the treatment chemicals are added and mixed, the water is introduced into a mechanical flocculator designed to form a quick-settling floc. Flocculation detention time can vary from 10 minutes in small units to 20 minutes in larger units. The flocculated water is then distributed through a bank of tube settlers, which consist of many 1-inch (25.4-mm)-deep, 39-inch (0.99-m)-long split-hexagonal-shaped passageways that provide an overflow rate, related to available settling surface area, of less than 150 gpd/sq ft (366 m/h). This overflow rate, together with a settling depth of only 1 inch (25.4 mm), results in effective removal of flocculated turbidity with a detention time of less than 15 minutes.

698

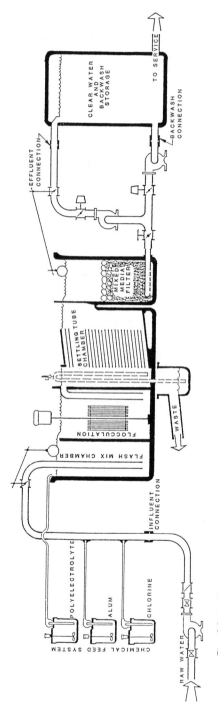

Fig. 22-1. Flow diagram of a Type A package plant. (Courtesy of Microfloc Products Group, Johnson Division, UOP)

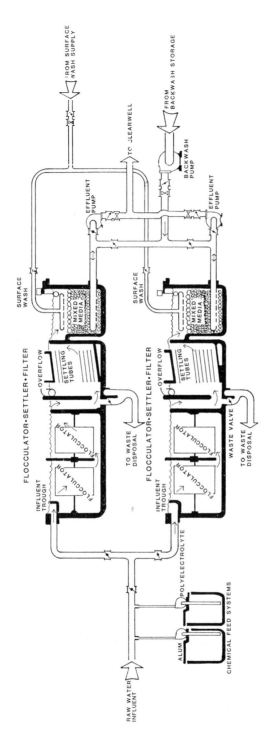

Fig. 22-2. Flow diagram of a Type B package plant. (Courtesy of Microfloc Products Group, Johnson Division, UOP)

699

After passing through the tube settlers, the clarified water flows to a gravity mixed-media filter. Type A and B package plants utilize a 30-inch (0.76-m)-deep mixed-media filter bed consisting of 18 inches (0.46 m) of 1.0 to 1.2 mm effective size anthracite coal, 9 inches (0.23 m) of 0.45 to 0.55 mm effective size silica sand, and 3 inches (76.2 mm) of 0.25 to 0.35 mm effective size garnet sand, designed so that there is a uniform gradation from coarse to fine in the direction of filtration. In all units, the design filtration rate is 5 gpm/sq ft (12.2 m/h).

The filters are designed to operate at a constant flow rate. Rate control is accomplished with a low-head filter effluent transfer pump discharging through a float-operated valve. With this means of filter rate control, once the plant raw water flow is established, there is no change in the filter rate throughout the entire filtration cycle, provided that the inflow to the plant remains the same. If there is a slight change in flow, increases or decreases in the filtration rate are accomplished slowly, minimizing filter surging and the chance for turbidity breakthrough.

The package plant filter is designed to backwash automatically once a preset filter headloss is reached, or the operator may override the automatic controls and backwash the filter manually. During backwash, the material accumulated in the tube settlers is automatically drained from the unit. Combining backwashing with draining of the tube settlers for sludge removal eliminates the need for an operator to judge how often or how much sludge should be wasted from the clarifier. This particular feature simplifies operation and reduces the required skill level of the operator.

Prior to the end of filter backwash, the drain valve on the tube settler basin is closed, allowing the remaining backwash water to refill the settling tube compartment. Upon completion of the backwash cycle, the treatment plant is returned automatically to service. Operational requirements are only to replenish chemical feeds, establish proper dosages, conduct routine water quality tests, and carry out routine daily maintenance activities. Technical data describing a Type A package plant are presented in Fig. 22–3, and a typical installation is shown in Fig. 22–4.

Figure 22–5 is a flow diagram of a Type C (Water Tech, Inc. "MT") package treatment plant of slightly different design. Again, low-detention-time, high-efficiency tube settlers are used to reduce the detention time and hence the volume of the clarification portion of the plant. A dual-media filter rated at 4.76 gpm/sq ft (11.6 m/h) also contributes to the compactness of this plant. Figure 22–6 lists the design criteria for the Type C plant, and Fig. 22–7 is a photograph of the equipment. These plants are designed to operate automatically, and the same amount of operator attention and skill is required for this equipment as for the Type A plant.

Another package plant is available in both conventional and high-rate modes. The Type D plant has capacities of up to 700 gpm (44.2 l/s). This

General Data
Nominal flow: 100 gpm; maximum flow: 115 gpm.
Inlet pressure required at shutoff: 10 psig.
Effluent pressure available at rated flow: 10 ft of water (from base of
   unit
Overall shipping dimensions:
   length—16 ft 3 in.; width—7 ft 2 in.; height—6 ft 10 in.
Weights:
   shipping weight—11 500 lb; operating weight—40 000 lb

Equipment Included With Basic Unit
Influent rate set valve: 2½ in.
Rapid-mix chamber: 39 cu ft volume.
Flocculator: 120 cu ft volume; drive: ¼ hp, 12 rpm.
Tube settler: 150 cu ft volume; overflow rate: 300 gpd/sq ft
Filter area: 20 sq ft; media type: MF-162; underdrain: header and
   lateral
Pumps:
   effluent—3 hp, 100 gpm; backwash—7½ hp, 360 gpm (flooded
      suction required)
Controls:
Plant flow rate is set manually by an adjustable influent float valve.
Another float valve on the effluent pump discharge maintains con-
stant filter level to match incoming flow. A low filter level protection
switch is provided for effluent pump protection. The control provides
automatic backwash initiated by preset filter head loss or by manual
pushbutton. Return to service is automatic. Load center and motor
starters for pumps also provided and mounted on unit.
Finish:
   interior—6 mils dry film thickness vinyl; color—aqua
   exterior—4.5 mils dry film thickness vinyl; color—aqua
Chemical Feed:
   alum—100 gal tank, 60 gpd pump, max. feed 100 mg/L
   hypochlorite—50 gal tank, 60 gpd pump, max. feed 4 mg/L
   polyelectrolyte—50 gal tank, 60 gpd pump, max. feed 1 mg/L, 1/3
      hp mixer
   soda ash (optional)—100 gal tank, 120 gpd pump, max. feed 40
      mg/L
Installation Data
Designed for indoor installation.
Base pad design: 600 lb per sq ft
Electrical supply (to load center on unit): 120/240 V, 3 phase, 70-A
   circuit.
Connections:
   inlet—3 in.
   effluent—2½ in.
   backwash—4 in.
   waste line for backwash flow—8 in.
   Minimum backwash water storage required: 3500 gal
   Overhead clearance required: 4 ft 0 in.

Fig. 22–3. Technical data sheet for Type A package plant. (Courtesy of Microfloc
Products Group, Johnson Division, UOP)

Fig. 22–4. Typical Type A package plant installation. (Courtesy of Microfloc Products Group, Johnson Division, UOP)

package plant utilizes the hydraulic energy of incoming water to mix chemicals and provide flocculation, as shown in Figure 22–8. Flocculation is accomplished in a 15-minute detention time, three-cell flocculator. This unit includes a high rate, tube settler-equipped clarifier, with an overflow rate of about 2.2 gpm/sq ft (5.37 m/h). Also, the unit includes an open gravity filter rated at up to 5 gpm/sq ft (12.2 m/h). The filter is supplied with either dual media or tri media for improved performance. Effluent from the filter can be by gravity flow or pumped. Sludge is withdrawn automatically from the clarifier through a solenoid valve. Plant flow is controlled by an influent throttling valve that operates off the water level in the treatment plant. The plant is designed to operate unattended, requiring only replacement of treatment chemicals and adjustment of chemical dosage as raw water conditions change.

In early 1980, a package plant manufacturer introduced a new concept in package water treatment plant design, utilizing an upflow filter of low-density plastic bead media, termed an adsorption clarifier, followed by a mixed-media filter for final polishing (Type E plant). The adsorption clarifier

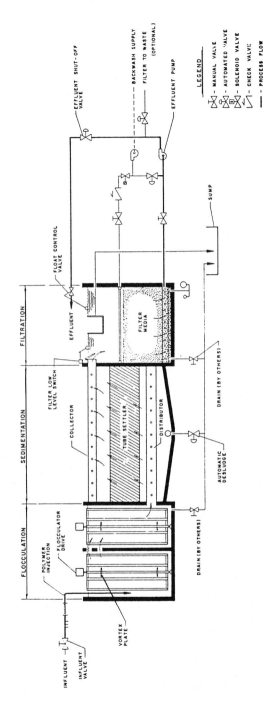

Fig. 22–5. Flow diagram of Type C package plant. (Courtesy of Water Tech, Inc., Vancouver, Washington)

703

Net output	100 gpm
Flocculator detention time	15.2 min.
Sedimentation chamber detention time	25 min.
Tube settlers (60-deg)	44.3 sq ft
Hydraulic loading	2.26 gpm/sq ft
Dual-media sand filter	
Area	21 sq ft
Filtering rate	4.76 gpm/sq ft
Surface wash rate (fixed jet type)	2 gpm/sq ft
Backwash rate	15 gpm/sq ft

Filter media	Depth-in.
Support bed base: no. 1 gravel (1½ by ¾ in)	8
no. 2 gravel (¾ by ½ in)	3
no. 3 gravel (½ by ¼ in)	3
no. 4 gravel (2.0 mm E.S.)	3
Sand bed: no. F5 sand (2.0 mm E.S.)	3
no. F9 sand (1.0 mm E.S.*)	3
no. F16 sand (0.45 mm)	9
Coal: no. C10 anthracite (1.0–1.2 mm E.S.*)	18
TOTAL	50

* F16 filter sand to have a uniformity coefficient of 1.5.

Fig. 22–6. Type C package plant design criteria. (Courtesy of Water Tech, Inc.)

replaces the flocculation and settling processes and results in an extremely compact unit. The manufacturer claims that it further reduces chemical coagulant dosage requirements over other systems. Figure 22–9 is a flow diagram of the Trident Water System manufactured by Microfloc Products, illustrating the various operating cycles. During operation, chemically coagulated water is introduced into the botton of the adsorption clarifier compartment where it passes upward through a bed of buoyant adsorption media. The adsorption clarifier combines the processes of coagulation, flocculation, and settling into one unit process.

In passing through the adsorption media, the chemically coagulated water is subjected to: (1) mixing, (2) contact flocculation, and (3) clarification. At operating flow rates, the mixing intensity, defined by the mean temporal velocity gradient value G, ranges from 150 to 300 sec^{-1}. Flocculation is accomplished by turbulence as water passes through the adsorption media, and is enhanced by contact with flocculated solids attached to the media. Estimates of the mixing parameter Gt in the adsorption clarifier range from about 1×10^4 to 3×10^4, depending upon flux rate and the rate of head-loss development. These values compare well with recommended optimum

Fig. 22–7. Type C package plant. (Courtesy of Water Tech, Inc.)

Gt values for mechanical flocculation systems, which are discussed in Chapter 7.

Turbidity removal in the adsorption clarifiers is accomplished by adsorption of the coagulated, flocculated solids on the surfaces of the adsorption media and on previously attached solids. The adsorption clarifier provides excellent pretreatment, which frequently is better than the performance achievable with complete flocculation and settling processes. Turbidity removal in this stage ranges up to 95 percent.

The material used for the adsorption media was selected as a result of experimentation with various materials, all with a specific gravity of less than 1 and of various sizes and shapes. The buoyant media are retained in the adsorption clarifier by a screen over the compartment.

Cleaning of the adsorption clarifier is accomplished by flushing. This flush cycle is initiated by a timer, but the equipment also includes a pressure switch that monitors headloss across the adsorption media and can automatically initiate a flushing cycle if required. Figure 22–9 illustrates the operation

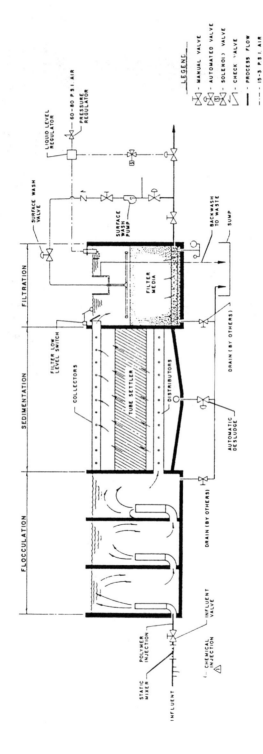

Fig. 22–8. Type D package plant. (Courtesy of Water Tech, Inc.)

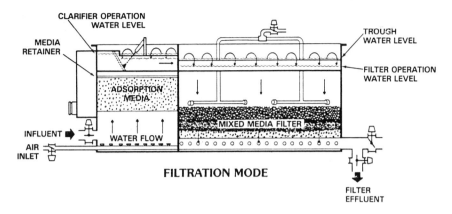

FILTRATION MODE

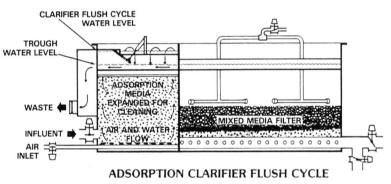

ADSORPTION CLARIFIER FLUSH CYCLE

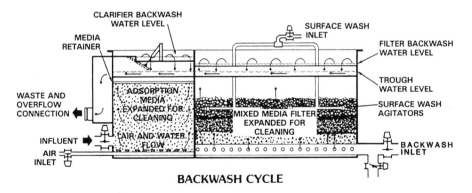

BACKWASH CYCLE

Fig. 22–9. Operating cycles of Type E package plant. (Courtesy of Microfloc Products Group, Johnson Division, UOP)

of the unit during a flushing cycle. When a cycle is initiated, the plant effluent valve closes, causing the water level to rise in the plant as the influent flow continues. When the water level reaches a predetermined level, a switch causes the influent valve to close. Air in the valve opening is distributed through perforated laterals beneath the adsorption media. This causes an immediate expansion in the adsorption media, and a vigorous scrubbing action takes place. Dislodged solids are then hydraulically flushed out of the top of the adsorption clarifier to waste. Influent water is used to flush the adsorption clarifier. Flushing frequency may vary, depending upon influent water quality. Typically, the controls are set up to initiate a flushing cycle every 4 to 8 hours. Unlike conventional filters, complete cleaning of the adsorption clarifier is not required, as the majority of solids are removed by the violent agitation provided during the first minutes of the flush cycle. Also, more efficient performance of the adsorption clarifier occurs if some residual solids are left on the media.

The mixed-media filter is backwashed similarly to a conventional filter. Although the filter may not necessarily be backwashed each time the adsorption clarifier is flushed, the equipment is designed to ensure that a backwash cycle is always preceded by a flushing cycle. The backwash cycle is illustrated in Fig. 22–9.

Extensive pilot evaluations of the Type E package plant design have been conducted. Table 22–1 provides a summary of test results obtained at three sites. More than 20 installations of the Type E package plants are presently in service. The Type E package plant is available in four models with capacities ranging from 350 to 4,200 gpm (1.91 to 22.89 Ml/d). Figure 22–10 is a typical installation of a Type E package plant. A technical data sheet for four Type E models is presented in Fig. 22–11.

The space-saving potential of the equipment is evident because the rise rate through the adsorption clarifier is generally twice the filtration rate [10 versus 5 gpm/sq ft (24.4 versus 12.2 m/h)]. An adsorption clarifier therefore requires about one-half of the space of a typical high-rate filter.

APPLICATION CRITERIA AND REQUIREMENTS

Before selecting a package plant for a particular application, a potential user must be certain that it can produce the required quality and quantity of water from the proposed raw water supply. Package plants characteristically have limitations (especially those employing high-rate unit processes) related to the quality limitations of the raw water supply. These limitations must be recognized when one is considering a package plant. For example, such factors as low raw water temperature, high or flashy turbidity, excessive color, or atypical coagulant dosages (higher than expected based upon normal turbidity levels) may influence the selection and rating of a particular package

Table 22–1. Summary of Results of Type E Plants (Microfloc Products Group).

LOCATION	RATES, gpm/sq ft (m/h)		WATER TEMP. °F (°C)	COLOR UNITS	TURBIDITY, NTU		
	ADSORPTION CLARIFIER	MIXED-MEDIA			INFLUENT	ADSORPTION CLARIFIER EFFLUENT	MIXED-MEDIA EFFLUENT
Corvallis, Oregon	5 (12.2)	9 (22)	68 (20)	0	100	8	0.27
	10 (24.4)	10 (24.4)	68 (20)	0	46	10	0.52
	8 (19.5)	7 (17.1)	57 (14)	0	103	21	0.30
Rainier, Oregon	10 (24.4)	5 (12.2)	41 (5)	30	9.3	1.4	0.24
	15 (36.6)	5 (12.2)	41 (5)	30	8.2	1.6	0.22
	20 (48.8)	5 (12.2)	41 (5)	30	8.1	1.7	0.20
Newport, Oregon	10 (24.4)	5 (12.2)	45 (7)	8	19	4.3	0.13
	15 (36.6)	7.5 (18.3)	45 (7)	8	15	3.7	0.11
	20 (48.8)	10 (24.4)	45 (7)	9	9	3.8	0.23

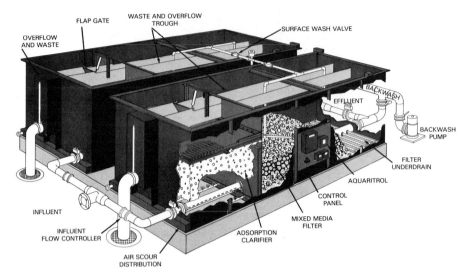

Fig. 22–10. Pictorial view, Type E package plant. (Courtesy of Microfloc Products Group, Johnson Division, UOP)

plant. The manufacturer's nameplate capacity of a package plant may have to be downrated or a larger unit selected to handle difficult treatment conditions. Water supplies of consistently high turbidity (greater than 200 NTU) may require presedimentation prior to treatment in a package plant. A misapplication can occur if the user is not informed of particular equipment limitations for a given treatment requirement.

It is recommended that all records of raw water quality be reviewed to determine the full range of treatment conditions to be expected before a particular-capacity package plant is selected. Especially valuable are laboratory analyses of representative raw water supplies to provide information critical to a proper application. Under certain conditions, on-site pilot tests may be justified and warranted to verify the suitability of a package plant. This is especially important because many of the new package plant designs employ high-rate, short-detention-time unit processes that require close control in order to perform effectively. Advance information on the quality of the proposed raw water supply and its treatment characteristics helps to ensure a successful installation.

OPERATIONAL CONSIDERATIONS

Equally important to the success of package plant applications is the quality of plant operation. Regardless of the size of the facility, if operating personnel do not possess an adequate understanding of the process and equipment

TRIDENT™
WATER SYSTEMS

TECHNICAL DATA SHEET

TECHNICAL DATA	MODEL	TR-105A	TR-210A	TR-420A	TR-840A
TYPICAL DESIGN FLOW	GPM	350-525	700-1,050	1,400-2,100	2,800-4,200
DIMENSIONS (each tank)	LENGTH	10 ft. 1 in.	14 ft. 6 in.	27 ft. 10 in.	39 ft. 10 in.
	WIDTH	6 ft. 11 in.	8 ft. 11 in	8 ft. 11 in.	11 ft. 11 in.
	HEIGHT	8 ft. 5 in.	8 ft. 5 in.	8 ft. 5 in.	10 ft. 1 in.
WEIGHTS (per tank)	OPERATING	44,000 lbs.	85,000 lbs.	165,000 lbs.	353,000 lbs.
TANK CONNECTIONS	INFLUENT	4 in.	6 in.	10 in.	14 in.
	EFFLUENT	8 in.	12 in.	12 in.	20 in.
	WASTE/OVERFLOW	8 in.	10 in.	14 in.	20 in.
	AIR	2 in.	2½ in.	3 in.	4 in.
PIPE SIZES	INFLUENT (Combined)	6 in.	8 in.	12 in.	16 in.
	EFFLUENT (Combined)	6 in.	8 in.	10 in.	14 in.
	BACKWASH	6 in.	8 in.	10 in.	16 in.
	SURFACE WASH	2 in.	2 in.	2½ in.	3 in.
VALVE SIZES	INFLUENT	4 in.	6 in.	8 in.	12 in.
	EFFLUENT	4 in.	6 in.	8 in.	12 in.
	BACKWASH	6 in.	8 in.	10 in.	16 in.
	BACKWASH RATE SET	6 in.	8 in.	10 in.	16 in.
	SURFACE WASH	2 in.	2 in.	2½ in.	3 in.
	SURFACE WASH RATE SET	2 in.	2 in.	2½ in.	3 in.
	AIR SUPPLY	2 in.	2½ in.	3 in.	4 in.
ADSORPTION CLARIFIERS	TOTAL AREA	35 sq.ft.	70 sq.ft	140 sq.ft	280 sq.ft.
MIXED MEDIA FILTERS	TOTAL AREA	70 sq.ft	140 sq.ft.	280 sq.ft.	560 sq.ft
	BACKWASH RATE (NOM.)	525 gpm	1,050 gpm	2,100 gpm	4,200 gpm
TOTAL VOLUME PER WASH CYCLE	ADSORPTION CLAR. (NOM.)	900 gal.	1,800 gal.	3,500 gal.	7,000 gal.
	MIXED MEDIA (NOM.)	4,200 gal	8,700 gal.	16,800 gal.	33,600 gal.

NOTES:

1. Nominal Mixed Media backwash rate shown is 15 gpm/ft² as required at 60 °F. Required rate varies linearly with temperature from 10 gpm/ft² at 32 °F. to 18 gpm/ft² at 75 °F.
2. All valves supplied with system are automatic except for manual backwash and surface wash rate set valves.
3. The adsorption clarifier is normally washed (using influent water) one or more times between Mixed Media backwashes, as well as during a Mixed Media backwash (which uses treated water). The waste holding system should be sized to handle a total of two complete wash volumes from each compartment.
4. Chemical feed systems are included which provide 24 hr. storage of alum and polymer solutions with capability to feed 50 mg/L alum (12% solution), and 1.0 mg/L polymer (0.25% solution).
5. Tanks are furnished with finish potable water grade protective coating on interior; universal primer on exterior, and bare bottom for installation on coal tar or asphaltic type base mastic compound.
6. The influent pumping system should provide a range of 25-35 ft. head at plant inlet.
7. Compressed air is required at 2 scfm (6 scfm peak) at 60 psi for control system.

Fig. 22–11. Technical data sheet for Type E package plant. (Courtesy of Microfloc Products Group, Johnson Division, UOP)

they are responsible for operating, production of safe and palatable finished water may be a hit-or-miss proposition. Some manufacturers have incorporated automatic controls such as effluent turbidity monitors that shut down the plant when the turbidity of the filtered water exceeds a preset limit. This fail-safe device (assuming that it is not bypassed by the operator) ensures that if the plant produces any water at all, it meets a given turbidity standard. An effluent turbidity control device should be mandatory on all package plants, especially if the raw water source is contaminated.

Package plants from several manufacturers have an accessory that automates the chemical feed system to maintain a specified finished water turbidity. This is advantageous where plants do not have full-time operators, and raw water conditions change frequently. The accessory is a microprocessor-based system that receives plant influent and effluent turbidities as input, and adjusts chemical dosages to optimum levels. Filter effluent turbidity is compared to the plant set point turbidity (for example, 0.1 NTU), and alum dosages are adjusted based upon the deviation observed. If the effluent turbidity is 5 percent higher than the set point turbidity, the alum dosage is proportionately increased. Likewise, if the turbidity is less than the set point, the alum dosage is reduced, thus reducing chemical consumption and related operating costs. The unit can be paced to flow and pH levels, and will adjust chemical dosages of coagulant and pH control chemicals accordingly. Figure 22–12 is a process schematic of the "Aquaritrol" automatic chemical dosage control accessory suitable for use with package plants.

No matter what control systems are used, the burden of producing a safe palatable water supply rests with the operator. In this regard, there is no substitute for a well-trained operator who has the necessary skills and dedication to operate the equipment properly.

Most package plant manufacturers' equipment manuals include at least brief sections on operating principles, methods for establishing proper chemical dosages, instructions for operating the equipment, and troubleshooting guides. An individual who studies these basic instructions and receives a comprehensive start-up and operator training session from the manufacturer's start-up technicians should be able to operate the equipment satisfactorily. These services are vital to the successful performance of a package water treatment plant and should be a requirement of the package plant manufacturer. The engineer designing a package plant facility should specify that start-up and training services be provided by the manufacturer, and also should consider requiring the manufacturer to visit the plant at 6-month and 1-year intervals after start-up to adjust the equipment, review operations, and retrain operating personnel. Further, this program should be ongoing, and funds should be budgeted every year for at least one revisit by the package plant manufacturer.

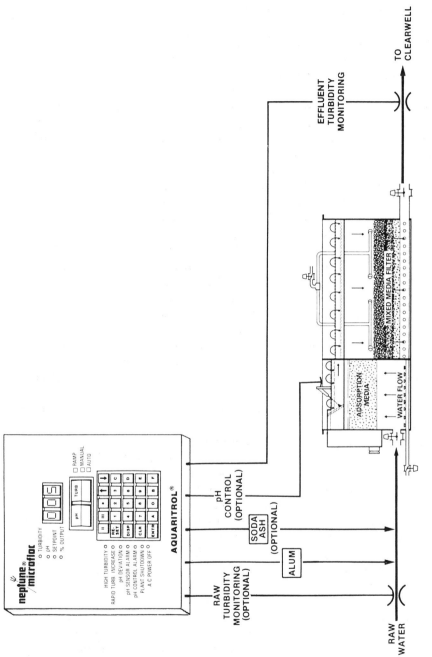

Fig. 22-12. Proprietary automatic coagulant dosage control system. (Courtesy of Microfloc Products Group, Johnson Division, UOP)

713

PACKAGE PLANT PERFORMANCE

The widespread acceptance of package plants as an economical solution to the water treatment needs of small systems resulted in construction of a significant number of plants during the 1970s. The quality of water produced by these plants was of concern, and led to an on-site investigation at six selected facilities.[4] The six selected plants were in year-round operation, used surface water sources, and served small populations. Plants were monitored to assess performance and ability to supply water meeting the interim primary drinking water regulations.

At each facility, grab samples of the raw water, treated water, and water from the distribution system were collected intermittently over a 2-year period and analyzed. Data on effluent turbidity, total coliforms, and chlorine residuals were recorded on all visits. Table 22-2 is a description of the six water treatment facilities surveyed during the study. Only three of the plants (C, T, and W) consistently met the 1 NTU effluent standard. Three of these plants obtained their raw water from a relatively high quality source. The other three plants (P, V, and R) met the turbidity standard on fewer than one-half of the visits. Table 22-3 is a listing of the treatment process characteristics for the six plants surveyed. Table 22-4 compares raw to filtered effluent turbidities measured during the visitations.

According to the authors of the survey, the performance difficulties of plants P, V, and R were related to the short detention time inherent in the design of the treatment units, the lack of skilled operators with sufficient time to devote to operating the treatment facilities, and (in the cases of plants V and R) the wide-ranging variability and quality of the raw water source. It is reported that the raw water turbidity at the site of plant V often exceeded 100 NTU. Later, improvement in operational techniques and methods resulted in substantial improvement in effluent quality. After the adjustments were made, plant filters were capable of producing a filtered water with turbidities less than 1 NTU even when influent turbidities increased from 17 to 100 NTU within a 2-hour period.

To illustrate the importance of raw water quality knowledge prior to the application of a package plant, Plant M was subjected to raw water turbidities ranging from 3.5 to greater than 500 NTU for short durations. It was specifically stated by the manufacturer's representative and supported by the product literature that a water with turbidity this high could not be treated satisfactorily by the plant selected at Plant Site M.

Even though the raw water supply for Plant Site P was a reservoir where raw water turbidities seldom exceeded 17 NTU, filtered effluent exceeded the 1 NTU standard on six of the nine samples taken during the visitation. At this site, however, equipment was not the source of the problem in obtain-

Table 22–2. Water Treatment Facilities Surveyed in Field Study.*

SITE	MANUFACTURER, MODEL, YEAR	DESIGN FLOW RATE gpm (l/s)	POPULATION SERVED/NO. OF METERS	AVERAGE VOLUME PER DAY gpd (Ml/d)	GROUP SERVED	TYPE OF DISTRIBUTION PIPE USED	SOURCE
W	Neptune Microfloc AQ-40** 1973	200 (12.6)	1,500/552	110,000 (0.42)	City	PVC	Surface impoundment
T	Neptune Microfloc AQ-40 1973	200 (12.6)	1,000/360	78,000 (0.30)	City	PVC, cast iron, asbestos cement	Surface impoundment
V	Neptune Microfloc AQ-40 1976	200 (12.6)	—/423	72,000 (0.27)	PSD†	PVC	River
M	Neptune Microfloc AQ-112 1972	560 (35.3)	—/1,680	330,000 (1.25)	PSD†	PVC	River
P	Neptune Microfloc Water Boy‡ 1972	100 (6.31)	—/411	82,000 (0.31)	PSD†	PVC	Surface impoundment
C	Permutit Permujet 1971	200 (12.6)	State park	57,000 (0.22)	State park	Asbestos cement	River

* Reference 5.
** Type B package plant.
† PSD—Public Service District.
‡ Type A package plant.

715

Table 22-3. Treatment Process Characteristics.*,**

SITE	PRE-CHEMICAL ADDITION	RAPID MIX		FLOCCULATION		SEDIMENTATION		FILTRATION		NOTES
		TYPE	d.t.† (SEC.)	TYPE	d.t.† (MIN.)	TYPE	LOADING (gpm/sq ft)	MEDIA	RATE (gpm/sq ft)	
W	Cl, alum, soda ash	In pipe	3	Paddle	12.8	Tubes	100	Mixed anthracite, 18 in. silica sand, 9 in. garnet sand, 3 in.	5	Polymer added before tubes
T	Cl, alum, soda ash, poly	In pipe	3	Paddle	12.8	Tubes	100	Mixed same as above	5	Post soda ash
V	Cl, alum, soda ash, poly	In pipe	3	Paddle	14	Tubes	100	Mixed same as above	5	Post sodium hexameta-phosphate
M	Cl, alum, soda ash, poly	Chamber	30	Paddle	10	Tubes	100	Mixed same as above	5	
P	Cl, alum, soda ash, carbon (summer)	Chamber not used	—	Paddle	10	Tubes	150	Mixed same as above	5	
C	Cl, alum, soda ash, poly	In pipe	—	Upflow solids contact 2 hr d.t. rise rate—1 gpm/sq ft				Silica sand, 24 in.	2	Soda ash added before filtration

* Reference 5.
** Conversion factors may be found in Appendix A.
† d.t. = detention time.

Table 22-4. Plant Turbidity Values (NTU).*,**

PLANT C		PLANT W		PLANT T		PLANT V		PLANT M		PLANT P	
RAW	CLEARWELL EFFLUENT	RAW	CLEARWELL EFFLUENT	RAW	CLEARWELL EFFLUENT	RAW	CLEARWELL EFFLUENT	RAW	CLEARWELL EFFLUENT	RAW	CLEARWELL EFFLUENT
8.5	0.3	—	0.9	10.0	1.9	4.0	1.8	—	0.2	12.0	0.8
6.2	0.2	5.0	0.3	8.0	0.2	12.0	2.8	39.0	3.8	4.4	2.4
1.2	0.3	4.2	0.4	6.0	0.4	—†	9.6	40.0	2.6	—	7.0
1.6	0.1	19.0	0.8	3.2	1.1	35.0	1.5	27.0	2.4	3.5	1.5
2.2	0.1	9.2	2.0	3.2	0.2	42.0†	2.0	6.0	1.2	2.0	0.1
4.0	0.1	11.5	0.3	3.2	0.2	10.0†	2.4	3.8	0.1	1.2	0.5
12.6	0.7	12.0	0.2	5.8	0.2	90.0†	8.5	73.0	11.0	15.6	9.7
5.2	0.2	11.0	0.3	10.4	3.2	28.0†	5.4	3.6†	0.1	3.1	2.2
2.2	0.2	29.7	0.9	3.4	0.7	19.0†	0.3	3.8	0.3	17.2	1.9
		12.8	0.2			47.0†	1.2	6.0	0.5		
						13.0†	0.8	70.0†	16.0		
						8.0†	0.3	25.0†	3.4		
						6.0†	0.3	>100.0†	55.0		
						>100.0†	0.5	>100.0†	31.0		
						60.0†	0.5	8.5†	2.2		
						24.0	1.2	4.3	0.4		
						13.0	0.3	4.0†	1.0		
						2.7	1.2	9.6†	1.9		
						1.2	1.0	19.1	1.1		
						3.3	0.5	64.0	6.9		
								8.2	1.0		

* Reference 5.
** Conversion factors may be found in Appendix A.
† Averaged values for day.

717

Table 22–5. Water Sources of the Surveyed Plants.*

TYPE OF SOURCE WATER	NUMBER OF PLANTS USING THIS TYPE OF SOURCE WATER
Groundwater	3
Impounded spring water	3
Free flowing surface water	7
Impounded surface water	18

* Reference 6.

ing a satisfactory filtered water quality. The poor record was traced to lack of adequate operator attention.

One of the major conclusions of this survey was that package water treatment plants with competent operators can consistently remove turbidity and bacteria from surface waters of a fairly uniform quality. Package plants applied where raw water turbidities are variable require a high degree of operational skill and nearly constant attention by the operators. Further, it was pointed out that regardless of the quality of the raw water source, all package plants require a minimum level of maintenance and operational skill if they are to produce satisfactory water quality.

The University of Cincinnati and the U.S. EPA initiated a field study in which they evaluated the performance of 19 municipal and 17 recreational package plants located in the states of Kentucky, West Virginia, and Tennessee.[6] These plants were all the product of a single manufacturer and employed flocculation, high-rate tube settlers, and mixed-media filters. Table 22–5 lists water sources of the treatment plants surveyed. This survey revealed that of 31 plants for which turbidity measurements were made, as indicated in Table 22–6, eight did not meet the Federal Maximum Contaminant Level (MCL) of 1 NTU. The principal reason for noncompliance with the turbidity

Table 22–6. Plants Meeting Turbidity Maximum Contaminant Level.*

TURBIDITY OF SOURCE NTU	NUMBER OF PLANTS	FINISHED WATER	
		<1 NTU	>1 NTU
<5	15	11	4**
6–15	8	8	0
16–50	6	2	4
51–100	0	0	0
>100	2	2	0

* Reference 6.
** One plant did not add coagulation chemicals.

standard appeared to be related to inadequate chemical coagulant dosage. The failure of plant operators to react to greater coagulant dosages brought about by higher raw water turbidities appeared to be the reason for excessively high filtered effluent turbidities.

The investigators concluded from their evaluation of these facilities that when package plants are properly operated, they are capable of meeting finished water turbidity standards. They further concluded that the combination of proper disinfection with clarification can produce a coliform-free water. A major shortcoming of those package plants that did not meet water quality standards was a lack of proper operator attention.

REFERENCES

1. Sorb, Thomas J., "Treatment Technology to Meet the Interim Primary Drinking Water Regulations for Inorganics," *J.AWWA* 70:2:105, February, 1978.
2. Hansen, Sigurd P., "Package Plants: One Solution to Small Community Water Supply Needs," *J.AWWA* 71:6:315, June, 1979.
3. Hansen, Sigurd P., and Culp, Gordon L., "Applying Shallow Depth Sedimentation Theory," *J.AWWA* 59:9:1134, September, 1967.
4. Morand, James M., and Young, Matthew J., "Performance Characteristics of Package Water Treatment Plants," Project Summary, Municipal Environmental Research Laboratory, Cincinnati, Ohio, EPA Project 600/52–82–101, March, 1983.
5. Hemphill, Brian W., "Trident Water Systems—An Innovative Approach to Water Treatment," Corporate Technical Article, unpublished, Neptune Microfloc, March, 1983.
6. Clark, Robert M., and Morand, James M., "Package Plants: A Cost-Effective Solution to Small Water System Treatment Needs," *J.AWWA* 73:1:24, January, 1981.

Chapter 23
Sludge Handling and Disposal

INTRODUCTION

During water treatment, sludges are produced by a number of processes that remove undesirable constituents from raw water. Among the undesirable constituents are sand and silt, organics in solution, suspended material, ions that cause hardness, bacteria and other organisms, and other substances that detract from product water quality. The most commonly used treatment processes that remove these materials and subsequently produce a sludge include chemical coagulation, lime–soda softening, removal of iron and manganese, taste and odor control, filtration, and many special conditioning processes.

The sludges may be discharged nearly continuously, as from clarifiers, or infrequently, from plain (no sludge collection equipment) settling basins. The sludges may contain clay, silt, sand, carbon, chemical precipitates, and organic substances. The composition of sludge from a treatment plant treating surface water is likely to vary daily, seasonally, and annually, as raw water quality changes occur. Sludges from one plant may be significantly different from sludges at a nearby plant that uses the same raw water source, because of differences in treatment technique and chemical types and dosages.

CHARACTERIZATION OF SLUDGES

Types of Sludge

Coagulant Sludges. Chemical coagulation and subsequent flocculation are widely used water treatment processes for removing clay, silt, dissolved or colloidal organic material, microscopic organisms, and colloidal metallic hydroxides. Aluminum sulfate (alum) is the most widely used coagulant, although iron salts—ferric chloride, ferrous sulfate, and ferric sulfate—also are used as coagulants. Coagulation sludges consist mainly of the hydrous oxide of the coagulant and materials removed from the raw water.

The coagulant sludge characteristics vary with increasing or decreasing proportions of material coagulated from the water. High-turbidity waters from rivers and streams will usually result in sludges that are relatively concentrated and fairly easy to dewater. Low-turbidity water from lakes or reservoirs

720

will produce fewer solids, but will often present a difficult sludge processing problem. In general, iron sludges will have a higher solids content than alum sludges, while the addition of polymer or lime increases the solids concentration of each.

Softening Sludges. The softening process removes a portion of the calcium and magnesium compounds from raw water to reduce hardness to a predetermined value. Use of lime and/or soda ash in the softening process results in sludges that are mainly calcium carbonate (80 to 95 percent by weight of solids); other components include magnesium hydroxide, silt, and minor amounts of unreacted lime and organic matter. Softening sludges normally are easy to concentrate and dewater.

Filter Backwash Water. The filtration process removes suspended matter such as silts, hydrous oxides, clays, colloids, algae, bacteria, and viruses by passing the water through a porous filter medium. The materials removed by filtration are periodically cleansed from the filters by backwashing. Filter backwash waters are difficult to handle. The solids concentration of filter backwash water may vary from 10 to 200 mg/l, and polyelectrolyte may be required to enhance thickening. Filter backwash water is important from the standpoint of both solids disposal and the value of the water recovered.

Presedimentation Sludge. Many river supplies carrying large quantities of suspended solids are presettled, resulting in sludges containing silts, sands, and, if a coagulant is used to aid presedimentation, inorganic precipitates.

Quantities of Sludge

The quantities of water treatment plant sludges are measured by pounds of solids on a dry weight basis and the volume of the liquid sludge resulting from plain sedimentation.

Coagulant Sludges.

Alum Sludges. Alum sludge is voluminous because of its poor compactibility. Alum forms a gelatinous sludge that will concentrate from 0.5 to 2.0 percent (5,000 to 20,000 mg/l) in sedimentation basins. Filter alum ($Al_2(SO_4)_3 \cdot 14H_2O$) when added to water forms aluminum hydroxide ($Al(OH)_3$). For every pound (kilogram) of alum added, 0.26 pound (0.26 kg) of aluminum hydroxide is formed. Some aluminum hydroxide will escape the sedimentation basin and be captured by the filters.

The suspended matter in the raw water supply is usually reported in turbidity units (NTU). There is no absolute correlation between turbidity units and dry weight total suspended solids (TSS); however, based on observed values where both parameters have been measured, the ratio of TSS in mg/l to NTU varies from 0.5 to 2.5, with a typical range of from 1.0 to 2.0. To estimate the solids residue from alum coagulation, two examples are shown below:

EXAMPLE 1 (Impounded water supply)

Raw water turbidity	10 NTU		
Filter alum dose	30 mg/l		
Aluminum hydroxide sludge	30 mg/l (8.34) (0.26)	=	65 lb/MG

$$\text{or } 30 \text{ mg/l} \left(1{,}000\,\frac{1}{m^3}\right) \left(\frac{1\text{ kg}}{1{,}000\text{ mg}}\right) (0.26) = 7.8 \text{ kg/Ml}$$

Raw water solids	10 NTU (1.5*) (8.34)	= 125 lb/MG
	or 10 NTU (1.5*)	= $\underline{\;\;15 \text{ kg/Ml}}$
		TOTAL = 190 lb/MG
		or 22.8 kg/Ml

EXAMPLE 2 (River supply)

Raw water turbidity	150 NTU		
Filter alum dose	60 mg/l		
Aluminum hydroxide sludge	60 mg/l (8.34) (0.26) =	130 lb/MG	
	or 60 mg/l (0.26)	=	16 Ml
Raw water solids	150 NTU (1.5*) (8.34) =	1,876 lb/MG	
	or 150 NTU (1.5*)	=	$\underline{\;\;225 \text{ kg/Ml}}$
		TOTAL =	2,006 lb/MG
		or	241 kg/Ml

* 1.5 is assumed ratio of TSS/NTU.

Generally, sludges resulting from the treatment of raw waters having high turbidities will thicken to higher concentrations than will sludges treating low-turbidity waters. For Examples 1 and 2, the sludge volume at 1.0 percent concentration will represent 2,300 gal/MG (2,300 m³/Mm³) and 24,000 gal/MG (24,000 m³/Mm³) treated, respectively, or 0.2 and 2.4 percent of the treated water flow.

Determination of coagulant dosage and sludge quantities should be derived from field experience at existing facilities using the same raw water source or from a series of jar tests on the proposed raw water source. Suspended solids should be measured to determine sludge quantities. Jar tests and sludge quantity analysis should be made seasonally to account for raw water quality variations. Reported alum dosages and resultant quantities of sludge solids for eight water treatment plants are shown in Table 23–1.

Iron Sludges. Iron salt coagulants include ferric sulfate ($Fe_2(SO_4)_3$), ferrous sulfate ($FeSO_4 \cdot 7H_2O$), and ferric chloride ($FeCl_3$). The precipitate formed

Table 23–1. Coagulation Plant Sludge Quantities.*

PLANTS	RAW WATER SOURCE	AVERAGE RAW WATER TURBIDITY, NTU	CHEMICAL DOSAGES, mg/l		SLUDGE QUANTITIES	
			ALUM	OTHER	lb/MG	kg/Mm³
Erie County, NY						
Sturgeon Point	Lake Erie	2–25	15	—	100–175	12–21
Monroe County, NY						
Shorement Plant	Lake Ontario	1–10	18	7**	116	14
Rochester, NY	Lake Ontario	1–10	25	17†	210	25
Monroe County, NY						
Eastman Kodak Co.	Lake Ontario	1–10	24	10**	143	17
Denver, CO	South Platte River					
Foothills Plant	(Strontia Springs Dam)	6.6	10	—	83	10
Birmingham, AL						
Shades Mountain	Cahaba River	19	NA	—	215	26
Washington, DC						
Patuxent River	Patuxent River	6.9	NA	—	100	12
Potomac River	Potomac River	120	NA	—	630	76

NA—Not available.
* Reference 1, 2, and 3.
** Clay.
† Clay, carbon, starch.

is ferric hydroxide ($Fe(OH)_3$). When one pound (0.45 kg) of ferric sulfate is added to water, 0.54 pound (0.25 kg) of ferric hydroxide is formed. Like alum sludge, ferric hydroxide is hydrophilic and thickens poorly. The amount of sludge formed should be determined from experience or from jar tests conducted on the proposed water supply. To estimate the solids residue from ferrous sulfate coagulation, an example follows.

EXAMPLE 3 (Impounded water supply)
Raw water turbidity	10 NTU
Ferric sulfate dose	15 mg/l
Ferric hydroxide sludge	15 mg/l (8.34) (0.54) = 68 lb/MG
	or 15 mg/l (0.54) = 8.1 kg/Mm³
Raw water solids	10 NTU (8.34) (1.5*) = 125 lb/MG
	or 10 NTU (1.5) = 15 kg/Mm³
	TOTAL = 193 lb/MG
	or 23.1 kg/Mm³

* 1.5 assumed ratio of TSS/NTU.

At a 1.0 percent concentration, the volume of sludge for the above example is 2,600 gallons/MG (2,600 m³/Mm³), or 0.26 percent of the treated water flow.

Softening Sludges. Chemicals used for lime softening include quicklime (CaO), hydrated lime ($Ca(OH)_2$), soda ash (Na_2CO_3), and sodium hydroxide (NaOH). The residues from softening may vary from a nearly pure chemical compound to a mixture of several compounds, depending on raw water characteristics. Lime softening sludges are composed mostly of calcium carbonate ($CaCO_3$) and magnesium hydroxide ($Mg(OH)_2$), as controlled by the hardness removal reactions. (These reactions are discussed in detail in Chapter 21.)

In addition to calcium carbonate and magnesium hydroxide, the sludge may include residues resulting from aluminum or iron coagulation of colloidal particles and unreacted lime. Theoretically, the amount of sludge solids is based on the type and quantity of hardness removed and the treatment chemicals applied, as shown in Table 23–2.

If only the sludge resulting from carbonate hardness removal with lime is considered, it is possible to estimate the dry weight of sludge solids produced as: [5]

$$S = 8.336(Q)(2.0 \text{ Ca} + 2.6 \text{ Mg}) \qquad (23\text{--}1)$$

where:

S = sludge produced, lb/day (kg/d)
Q = plant flow, mgd (m^3/s)
Ca = calcium hardness removed as $CaCO_3$, mg/l
Mg = magnesium hardness removed as $CaCO_3$, mg/l
8.336 = constant for use with English units (86.4 is the constant for use with the metric units shown)

For surface water supplies, use of a coagulant in addition to softening may significantly increase sludge quantities. The formula used to estimate quantities should be modified to include the coagulant and raw water suspended solids: [5]

$$S = 8.143(Q)(2.0 \text{ Ca} + 2.6 \text{ Mg} + 0.44 \text{ Al} + 1.9 \text{ Fe} + \text{SS} + A) \quad (23\text{--}2)$$

where:

S = sludge produced, lb/day (kg/d)
Q = plant flow, mgd (m^3/s)
Al = alum dose as 17.1 percent Al_2O_3, mg/l
Fe = iron dose as Fe, mg/l
SS = raw water suspended solids, mg/l
A = additional chemicals such as polymer, clay, or activated carbon, mg/l
8.143 = constant for use with English units (84.4 is the constant for use with the metric units shown)

Table 23-2. Theoretical Solids Production—lb Dry Solids/lb Hardness Removed as $CaCO_3$.*

TREATMENT CHEMICALS	CARBONATE HARDNESS		NONCARBONATE HARDNESS	
	CALCIUM	MAGNESIUM	CALCIUM	MAGNESIUM
Lime and soda ash	2.0	2.6	1.0	1.6
Sodium hydroxide	1.0	0.6	1.0	0.6

* Reference 4.

Although the above formulas are useful for estimating the quantities of sludge solids, actual data from water treatment plants can also be useful in projecting sludge quantities. Reported softening sludge quantities for numerous water treatment plants in the United States are shown in Table 23-3.

Table 23-3. Softening Plant Sludge Quantities.*,**

	FLOW, mgd	CHEMICAL DOSAGES, lb/MG		RATIO, lb SOLIDS/lb LIME DOSE	REPORTED SLUDGE, lb/day
		LIME	SODA ASH		
Austin, TX	45	750	—	3.8	91,200
Corpus Christi, TX	56.3	428	—	2.6	46,600
Dallas, TX	36.5	342	—	2.2	80,000
Des Moines, IA	30.2	1,830	197	2.8	560,000
El Paso, TX	19.0	825	145	3.4	—
Fort Wayne, IN	27.0	1,746	268	2.0	135,000
Grand Rapids, MI	6.0	1,350	—	2.0	18,000
Kansas City, MO	97.9	1,410	—	1.8	—
Louisville, KY	109.8	348	—	4.5	116,000
Minneapolis, MN	73.5	1,014	284	2.0	138,000
New Orleans, LA	120.3	637	143	1.6	304,000
Oklahoma City, OK	11.9	1,045	—	2.5	183,000
Oklahoma City, OK	21.0	336	—	1.4	174,000
Oklahoma City, OK	14.7	906	—	2.1	272,000
Omaha, NB	46.8	705	63	2.0	70,600
Toledo, OH	80.4	602	24	2.0	168,000
Topeka, KS	15.7	1,500	250	1.8	—
Wichita, KS	34.2	900	—	1.5	12,300
Pontiac, MI	10.1†	2,200	—	2.5	—
Miami, FL	180.0†	1,800	—	2.2	—
Lansing, MI	20.0†	2,200	—	2.3	—
Dayton, OH	96.0†	2,140	—	2.5	—
St. Paul, MN	120.0†	990	—	2.4	—

* References 6 and 7.
** Conversion factors may be found in Appendix A.
† Plant capacity.

A survey of softening plants in the United States by the AWWA Sludge Disposal Committee analyzed information from 84 plants. The suspended solids (SS) concentration withdrawn from the sedimentation basins varied widely, as shown in Table 23-4.[4]

The volume of sludge produced averaged 1.87 percent of the average plant flow with a standard deviation of 2.1 percent. Ninety percent of the plants were between 0.4 percent and 1.5 percent, with an average of 1.2 percent.[4] Another reference describes softening sludge volumes as ranging from 0.3 to 5 percent of the volume of raw water treated.[1]

Filter Backwash Water   Filter backwash water represents a large volume of liquid with a relatively low solids content. Filter backwash water typically represents 1 to 5 percent of the total water processed or 10,000 to 50,000 gal/MG (10,000 to 50,000 m³/Mm³) treated. Backwash rates may be 2 to 20 times the filtration rate.[1]

The solids content of filter backwash water can vary widely from plant to plant, depending on the raw water quality, efficiency of preliminary treatment units, and duration of filter run and backwash cycle. Table 23-5 provides data on filter backwash solids production and concentration from several water treatment plants.[1]

In some raw waters low coagulant dosages are required, and the flocculated water is applied directly to the filter (direct filtration), whereas in other cases sedimentation precedes the filters. In direct filtration applications, the solids loading is a function of the coagulant dosage and the raw water turbidity. Where sedimentation precedes the filter, typical suspended solids concentrations escaping the sedimentation basin range from 2 to 10 mg/l with typical turbidities of 2 to 6 NTU. Filter backwash water will contain 15 to 85 pounds of solids/MG (1.8 to 10.2 kg/Ml) processed. If direct filtration of coagulated water is practiced, the solids content will be higher.

Filter backwash solids characteristically are difficult to separate from the

Table 23-4. Softening Plant Sludge Concentrations.*

SUSPENDED SOLIDS CONCENTRATION, PERCENT	PERCENTAGE OF PLANTS
<5, avg. 2.4	52
5 to 10	24
11 to 15	11
16 to 25	6
>25	7

* Reference 4.

Table 23–5. Filter Backwash Solids Data.*

PLANT	SOLIDS PRODUCTION		SOLIDS CONCENTRATION
	lb/MG	kg/Mm³	mg/l
Birmingham, AL			
Shades Mountain	5	0.6	15
Putnam	2	0.24	7
Western	0.5	0.06	3
H. Y. Carson	0.5	0.06	7
Monroe County, NY			
Rochester	20	2.4	160
Monroe County Water Authority	24	2.9	120
Eastman Kodak Co.	22	2.6	100
Erie County, NY			
Sturgeon Point	35	4.2	170

* Reference 1.

liquid. Wash water recovery ponds sized to hold backwash water for 24 hours or more may recover up to 80 percent of the solids with the aid of polymers or other coagulant aids. The recovered water is reprocessed through the treatment plant or discharged to a surface water.

Presedimentation Sludge. The quantity of sludge removed during presedimentation is a function of the quantity and type of solid material present in the raw water supply. Quantities should be expected to vary widely between different plants and different times of year. Pilot level testing is recommended to determine quantities of presedimentation sludge.

Characteristics of Sludges

The significant characteristics of water treatment plant solids are those that affect handling and disposal. The general goal is to reduce the bulk of the sludge and produce a material that is suitable for disposal or recovery processes.

Coagulant Sludges

Alum Sludges. In the absence of significant organic pollution of the raw water, coagulant sludges are essentially biologically inert and retain a near-neutral pH. The sludge is generally thixotropic (the plastic nature of the sludge changes with agitation) and gelatinous; however, coagulant sludges from plants obtaining raw water from river supplies with a fairly high silt content are not as gelatinous as sludge from plants obtaining raw water from lakes or quiescent reservoirs.

Coagulant sludges such as alum sludge may be characterized at varying solids contents, as shown in Table 23–6.

Various raw water quality parameters and sludge characteristics are shown in Table 23–7 for five water treatment plants in the United States. The Moline,

Table 23–6. Alum Sludge Characteristics.

SOLIDS CONTENT	SLUDGE CHARACTER
0 to 5%	Liquid
8 to 12%	Spongy, semisolid
18 to 25%	Soft clay
40 to 50%	Stiff clay

Table 23–7. Raw Water Quality Parameters and Alum Sludge Characteristics.*

	TEST LOCATIONS				
	INDIANAPOLIS, IN	CONCORD, CA	TAMPA, FL	MOLINE, IL	WASHINGTON, DC
Raw Water Parameters					
Turbidity, NTU	45	42	0.63	71	18
Color, Pt-Co	30	7	100	26	4.2
Average alum dose, mg/l	24	41	100	43	19.8
Sludge Characteristics					
Initial solids concentration, %	1.7	1.7	1.6	1.7	12.1
Dissolvable inorganic solids, %	26	36	61	7	9
Nondissolvable inorganic solids, %	52	18	6	79	49
Dissolvable organic solids, %	12	18	25	2	26
Nondissolvable organic solids, %	9	28	8	13	16
Total aluminum, mg/l	2,400	2,400	3,500	295	3,750

* Reference 8.

Illinois water also required a lime dose of 141 mg/l. The Washington, D.C. water was low in color, but had a high suspended solids level and turbidities ranging as high as 160 NTU. In considering recovery of alum, the aluminum content and dissolved inorganic and organic solids are of importance.[8]

Iron Sludges. For a laboratory study on the recovery and reuse of iron coagulants reported by Pigeon et al., concentrated samples of water treatment plant iron sludges were obtained from the St. Louis County Water Company Central Plant No. 3 and the Kingsport (Tennessee) Water Treatment Plant.[9] In both facilities the sludges had been retained in the sedimentation facilities for about 3 months before they were collected for study. Both sludges were diluted with deionized water to total solids concentrations of 2 percent for all studies. Characteristics of the two iron coagulant sludges are shown in Table 23-8. Of particular interest is the difference in the iron concentration between the sludges. This is attributed to the difference in suspended solids levels of the water influent to the iron coagulation basins at the two plants, and the differences in iron doses used to coagulate the solids.

Softening Sludges. Softening sludges are generally white, have no odor, and have a low biochemical oxygen demand (BOD_5) and chemical oxygen demand (COD). The chemical constituents of the sludge vary with the composition of the raw water and the chemicals added. The results of chemical analyses of dry solids from eight water softening plants are presented in Table 23-9. It is evident that for these waters, the precipitates are about 85 to 95 percent calcium carbonate plus 0.4 to 7 percent magnesium oxide. These results are not entirely typical because many water softening plants produce residues with a higher proportion of magnesium oxide.[4]

Softening sludges should be analyzed periodically for excess lime, and the calcium to magnesium (Ca:Mg) ratio should be calculated. Excess lime is

Table 23-8. Iron Coagulant Sludge Characteristics.*

SLUDGE CHARACTERISTIC	ST. LOUIS COUNTY	KINGSPORT
Total solids	20,000 mg/l (2.0%)	20,000 mg/l (2.0%)
Total volatile solids	1,000 mg/l (5% of total solids)	2,800 mg/l (14% of total solids)
Total suspended solids	19,800 mg/l (1.98%)	19,900 mg/l (1.99%)
Volatile suspended solids	1,000 mg/l (5.1% of total suspended solids)	2,800 mg/l (14.1% of total suspended solids)
Iron content	930 mg/l as Fe (4.65% of total solids)	4,120 mg/l as Fe (20.6% of total solids)
pH	8.5	7.4

* Reference 9.

Table 23-9. Lime Softening Sludge Characteristics for Eight Plants.

	BOULDER CITY, NV*	MIAMI, FL*	WRIGHT AERO CORP., CINCINNATI, OH	ST. PAUL, MN*	LANSING, MI*	WICHITA, KS**	VANDENBERG, CA†	COLUMBUS, OH§
				PERCENT BY WEIGHT				
Silica, iron, aluminum oxides	2.6	1.5	4.4	2.0	—	0.6–2.0	7	3.3–3.6
Calcium carbonate	87.2	93.0	88.1	85.0	80–90	89–98	85	80–85
Calcium hydroxide	—	—	—	—	—	—	1	—
Magnesium oxide	7.0	1.8	2.2	6.2	4–6	0.4–3.5	7	5.2–8.6

* Reference 10.
** Reference 11.
† Reference 12.
§ Reference 13.

an indicator of incomplete reaction in the softening process. If CaO or $Ca(OH)_2$ is present in the solid phase, it is an indication of poor slaking or dissolving which results in an increase in chemical costs. If the lime does not dissolve prior to incorporation into the sludge, it might remain as $Ca(OH)_2$ if present at a concentration greater than 1,300 mg/l, thus causing poor dewaterability and ultimately an increase in sludge quantities. Corrective action should be undertaken to eliminate these conditions.[4]

The Ca:Mg ratio of a sludge is an indicator of its ability to thicken and dewater. Generally a sludge with a Ca:Mg ratio less than 2 will be difficult to dewater, whereas a sludge with a Ca:Mg ratio greater than 5 will dewater relatively easily. A plot of Ca:Mg molar ratio versus the settled solids and the filter cake solids concentration is shown in Fig. 23–1.[14] High-magnesium softening sludges can be considered to be nearly equivalent to mixed coagulant–softening residues because of similar poor dewaterability. Studies at Johnson County, Kansas showed that carbon dioxide could be used to dissolve the magnesium hydroxide present in the sludge, thereby reducing, by a factor of 3, the thickening area required to produce an underflow solids content of 15 percent.[15]

The dewatering properties of sludges may be characterized by the rate of dewatering (specific resistance) and the extent of dewatering (filter cake solids

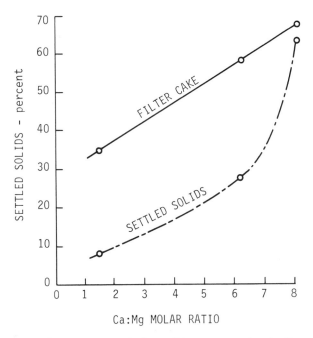

Fig. 23–1. Effect of Ca:Mg ratio on sludge solids concentration for lime sludges.[14]

or settled solids concentration). Values for specific resistance of a solids slurry can be expected to vary from 1.49×10^{11} ft/1b (1×10^{11} m/kg) for a readily dewaterable residue to 745×10^{11} ft/lb (500×10^{11} m/kg) for poorly dewatering coagulant sludges. Table 23–10 shows specific resistance values for several lime and lime–coagulant sludges.[16] Higher magnesium concentrations adversely affect specific resistance, as shown in Fig. 23–2.[14]

Of particular importance in the physical characterization of water treatment residues is the solids concentration necessary to produce a sludge that can be easily handled, transported, and disposed of. As a sludge dewaters, it becomes an increasingly viscous fluid and eventually forms a solid cake. When the sludge is a fluid, it continually deforms upon subjection to a shear stress. The point at which the slurry becomes a solid is not distinct, and the extent to which a sludge must be dewatered depends on the method of handling. If a sludge is dewatered by vacuum filtration and handled by a conveyor belt, then a lower shear stress may be sufficient to permit handling than if the sludge is drained on a drying bed and is removed from the bed by a loader.[17]

The solids concentration of a dewatered sludge is a poor indicator of its handleability. While an alum sludge may be sufficiently dewatered for handling at 30 to 40 percent solids, a lime sludge dewatered in a lagoon to 50 percent solids may not be handleable with earth-moving equipment. The character of lime sludges at varying solids contents is generalized as shown in Table 23–11.

The solids concentration necessary for a sludge to become handleable is shown in Fig. 23–3.[14]

Recent studies have shown that the size of the calcium carbonate particle also significantly affects sludge thickening and dewatering.[18,19] The mean

Table 23–10. Lime Sludge Specific Resistances.*

TYPICAL SLUDGE COMPOSITION	SPECIFIC RESISTANCE	
	10^{11} FT/LB	10^{11} M/KG
Lime	0.30	0.20
Lime	1.42	0.95
Lime and iron	3.20	2.15
Lime and iron	9.12	6.12
Lime and iron	10.12	6.79
Lime and iron	10.43	7.00
Lime and alum	8.70	5.84
High-magnesium softening sludge	8.12	5.45
High-magnesium softening sludge	37.40	25.10

* Reference 16.

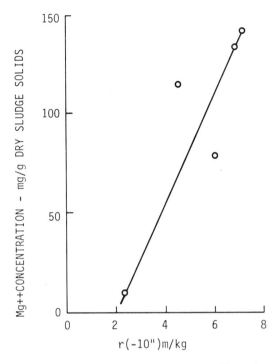

Fig. 23-2. Effect of magnesium concentration on the specific resistance of softening sludges.[14]

diameter of precipitated calcium carbonate has been found to range from < 1 μm to > 50 μm, depending on the degree of solids recycled and other physical–chemical parameters that affect reaction kinetics. In general, the larger the size of the calcium carbonate particle, the easier the sludge will be to thicken and dewater.

Many utilities report that lime sludge cakes in the 50 to 65 percent moisture content range are sticky and difficult to discharge cleanly from dump trucks. Lime sludges are also biologically inert, but have a high pH due to unspent lime and high alkalinity.

Table 23-11. Lime Sludge Characteristics.

SOLIDS CONTENT	SLUDGE CHARACTER
0 to 10%	Liquid
25 to 35%	Viscous liquid
40 to 50%	Semisolid, toothpaste consistency
60 to 70%	Crumbly cake

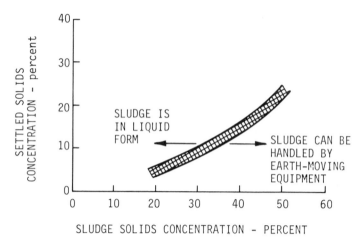

Fig. 23–3. Solids concentration at which a sludge becomes handleable.[14]

MINIMIZING SLUDGE PRODUCTION

A 1975 study of the water supply industry found that few of the 128 plants visited made an effort to minimize sludge volumes, and thus they increased water waste and placed an added hydraulic burden on the subsequent sludge handling facilities.[20] Sludge production can be minimized either by the removal of water to reduce the volume or by the reduction of the amount of solids present in the sludge, or by a combination of the two. Reducing the amount of solids has the greatest impact on sludge-disposal operations because intermediate dewatering processes normally remove water and reduce sludge volumes. In addition, if a useful by-product can be obtained from the residual material, some of the sludge handling costs can be recovered.

One method of reducing the amount of solids is to reduce the amount of chemicals used for coagulation and/or softening. The quantity of chemical coagulant used can be reduced in some plants by substituting organic coagulants (polymers) for inorganic coagulants, either partially or entirely. However, Bishop has cautioned that polymers are not effective in removing color and can create problems in alum recovery processes.[21]

The use of polymers as a possible replacement for alum was tested in laboratory-scale jar tests at the Orange Water and Sewer Authority in Orange County, North Carolina.[22] The raw water source is a protected reservoir with raw water turbidities between 5 to 50 NTU, which contains significant concentrations of organic matter of natural origin (TOC from 6 to 10 mg/l). Three coagulant chemicals were selected for study: alum $(Al_2(SO_4)_3 \cdot 18H_2O)$, a cationic polymer with a low molecular weight and a high charge density (polymer A), and a cationic polymer with a high molecular weight and a low charge density (polymer B). Based upon the laboratory testing, the necessary alum dosage was reduced from 60 to 30 mg/l when 0.05 mg/l of polymer B was also used. These tests indicated that the alum–polymer B combination resulted in improved removals of turbidity and TOC, reduced sludge volume, and lower chemical costs, compared to the use of alum alone. Best results were obtained by adding alum first, followed by the polymer.[22]

One method of reducing solids production in a softening plant is to replace soda ash and some or all of the lime with sodium hydroxide. Based on the theoretical solids production data shown above in equations (23–1) and (23–2), solids loads could be reduced by up to 50 percent. Substitution of sodium hydroxide has not been widely practiced because it generally is more expensive than the lime and soda ash it replaces.[23] However, the higher chemical costs for sodium hydroxide are at least partially offset by lower solids production and sludge disposal costs.

Split treatment is another method used to reduce softening sludge quantities when high magnesium hardness removals are required. This method normally is justified because it eliminates lime treatment of the bypassed water and minimizes recarbonation requirements.[24,25] It also minimizes sludge production because the calcium carbonate solids created by recarbonation of excess lime are eliminated.

Operation at a pH less than 10 to 10.5 will selectively remove calcium hardness, leaving magnesium in solution; waste volumes can be reduced, and the dewatering characteristics of the softening residues can be improved. However, this method results in incomplete softening that may not be acceptable in some cases.

Optimization of chemical feed systems and mixing also can reduce solids loads by maximizing the efficiency of chemical utilization and by minimizing the amount of unreacted lime in the waste solids. Improved mixing in feeders, flash mixers, and flocculation zones (through proper baffling) reduces excess lime usage. Facilities with well-mixed solids contact clarifiers use only 2 to 3 percent excess lime.[26]

Sludge recirculation from the clarifier back to the rapid mixer improves the efficiency of calcium carbonate precipitation and reduces excess lime usage.[26,27] Operating data from a plant with sludge recirculation confirmed the improvement in performance and the reduction in lime and soda ash usage by 11 and 5 percent, respectively.[12] A full-sized sludge recirculation system was installed with a sludge recirculation capacity of 0.77 mgd for a 3.1-mgd plant, or a recycle rate of 0.25.

A laboratory study of the recycle of calcium carbonate sludge to serve as seed crystals for the further precipitation of hardness from solution showed sludge dewaterability, as quantified by specific resistance, improved significantly.[19] The effects of reactor mixing and seed recycle rate also were investigated.

The recycle of filter backwash and clarified water (supernatant, filtrate, or centrate) from the dewatering process will reduce solids loads because this water already has been softened. These process wastewaters represent 3 to 5 percent of the total plant flow; hence, their recycle would reduce solids loads by a similar amount.[4]

Sludge volumes can be minimized prior to subsequent treatment and dewatering by controlling sludge withdrawls from the settling basins to increase the solids content. By increasing the solids content 2 to 5 percent, the sludge volume would be reduced by 60 percent.[4] Similarly, alum sludge volumes may also be reduced by controlling sludge withdrawals, although the increase in solids content may be only from 0.5 to 0.75 percent since alum sludges are much more dilute than calcium carbonate sludges.

Direct filtration has been used in some plants where the raw water supply is of high quality. This process generally allows lower chemical feed rates than conventional flocculation, settling, and then filtration, and therefore produces less sludge.

Another method of minimizing waste solids production is to reevaluate finished water quality needs, such as reducing the amount of softening. For example, if a plant is removing 150 mg/l of hardness, it could reduce its waste solids load by 16 percent by only removing 125 mg/l. Not only would the sludge quantity be reduced, but chemical usage costs would decline by a similar amount.

Other, broader possibilities for reducing the quantity of sludge produced include selecting alternative water supplies that result in the least sludge production, and reducing the finished water demand through use of water conservation techniques.

THICKENING AND DEWATERING ALTERNATIVES

Often, sludge disposal can be to either a sanitary sewer or a waterway, eliminating any requirement for on-site thickening and dewatering. These highly desirable disposal techniques are discussed later in this chapter. For plants where these disposal options are not available, thickening and/or dewatering may be necessary.

Gravity Thickening

Thickening, which begins with concentrating the sludge in the bottom of the clarifier, is an effective and inexpensive method and generally the first phase of reducing sludge volume and improving sludge dewatering characteristics. Thickening, however, is most effectively accomplished as a separate operation. Thickening will improve the consistency of feed material for subsequent dewatering units, and in many cases will reduce the size of dewatering equipment.

Coagulant Sludge Thickening Results. Coagulant sludge, which is usually withdrawn from clarifiers at less than 1 percent solids, can be thickened in mechanical thickeners to about 2 percent solids. Aluminum and iron hydroxides may be conditioned with the aid of polymers; but the polymers will have a minimal effect on the ultimate degree of compression. Polymers will affect particle size and zone settling velocity and will likely improve capture efficiency. Experienced settling curves are shown in Fig. 23–4.[28] Typical design parameters reported for alum sludge thickening are 100 to 200 gal/sq ft/day (4.07 to 8.14 m³/m²/d) when sludges are conditioned with polymers. Alum sludges that are mixed with clay or lime have exhibited thickened concentrations of 3 to 6 percent and 9 percent, respectively, at higher overflow rates than sludges without clay or lime addition.[2,29]

Unlike coagulant sludge, which thickens poorly, softening sludge can often be concentrated to greater than 5 percent solids in the clarifier. An increase in solids concentration from 1 to 5 percent reduces the sludge volume by about 80 percent, providing a cost savings in sludge discharge piping and appurtenances as well as the cost for supernatant recycling.

Lime Sludge Thickening Results. Lime sludge thickening is used to provide a concentrated sludge and a consistent product for feed material to dewatering units. Solids loadings of 60 to 200 lb of solids/sq ft of thickener surface area/day (293 to 976 kg/m²/d) are commonly practiced. Reported

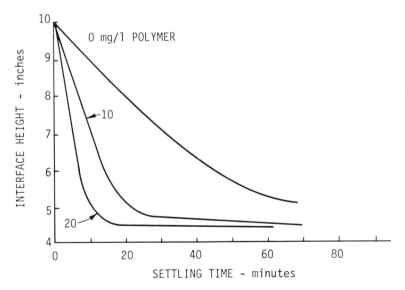

Fig. 23–4. Effects of polymer dosage on the rate of thickening of metal hydroxide suspensions.[28]

thickening results for lime softening sludges are presented in Table 23–12.

Thickening requirements increase as the softening sludge composition varies to include magnesium hydroxide, turbidity, or coagulants. Bench-scale thickening tests that are both simple and inexpensive to perform provide a good estimate of sludge thickening characteristics and design requirements. Care must be taken to ensure that existing sludge handling equipment is adequate for the higher solids levels.

Provisions to recycle the underflow to the thickener feed are sometimes provided to prevent the solids from becoming too thick, with subsequent handling problems. In the design of a thickener, storage requirements, particularly when a dewatering device is used, must be considered, along with the thickener area required to produce a desired solids underflow.

Nonmechanical Dewatering

Lagoons, sand drying beds, and freezing are the most feasible nonmechanical dewatering methods available. Descriptions of these alternatives and typical operational data from actual installations and pilot-scale studies are presented below.

Lagoons. The most common method currently used for handling of water treatment plant sludges is lagooning. The operating costs of this technique are low, but the land requirements are high. Because of the space requirement, lagooning may be most attractive for small, isolated plants.

Lagoons are generally built by enclosure of a land area with dikes or berms, or by excavation, with no attempt to maximize drainage with underdrains or by a sand layer. However, it is usually desirable for lagoons to have good drainage. This is best accomplished by constructing them with the bottom of the lagoon at natural ground level; an exception occurs where the existing ground surface is a tight clay soil with poor drainage initially. Preferably, sludge lagoons should not be built in excavated pits, and never

Table 23–12. Gravity Thickening Performance of Lime Softening Sludges.*

LOCATION	SOLIDS INPUT (%)	SOLIDS OUTPUT (%)
Boca Raton, FL	1 to 4	28 to 32
Dayton, OH	2 to 4	15 to 25
Lansing, MI	12 to 16	20 to 25
Ann Arbor, MI	9	20
Miami, FL	30	40
Cincinnati, OH	5	15

* Reference 10. (Courtesy of Scranton Gillette Communications, Inc.)

with their bottoms below groundwater level. Sludge should have drainage into the subsoil, and its surface should be open to evaporation.[30]

Lagoons typically are earthen basins with 4- to 12- foot (1.2- to 2.7-m) sidewater depths, covering from 0.5 to 15 acres (0.2 to 6.0 ha), which are equipped with inlet control devices and overflow structures. The best design practice is to place the inlet and outlet structures as far apart as possible. Sludge is added until the lagoon is filled with solids; and then it is removed from service until the solids have dried to the point at which they can be removed for landfill disposal, if in fact the sludge can dewater to this point. Lagooning is not a disposal technique; rather it is a technique for thickening, dewatering, and temporary storage.

The sludge removed from coagulation basins can readily be disposed of in lagoons, preferably on the treatment plant site, although, if necessary, the sludge can be pumped for relatively long distances to the lagoon site. The distance between filter plants and lagoons, while adding to the cost of a pipeline, is a minor cost item in total water filtration plant costs. At York, Pennsylvania the sludge is pumped 2 miles to a lagoon, and at Appleton, Wisconsin the sludge lagoons are 2.1 miles (4.03 km) from the softening plant. The Louisville (Kentucky) Water Company, which clarifies and softens Ohio River water, has three lagoons located 7 miles from the major treatment plant. No difficulty has been experienced at any of these locations due to distance.[30]

Alum Sludge Lagoon Operation and Design. Alum sludges have proved difficult to dewater in lagoons to a concentration at which they can be land-filled. Neubauer reported a detailed study of a lagoon receiving alum sludges from a 32-mgd (121-Ml/d) plant in Rochester, New York.[31] The 400 × 300-foot (122 × 91-m)-wide lagoon had been in operation for 3 years. Core samples indicated that the solids concentration increased from about 1.7 percent at the sludge interface to a maximum of 14 percent at the lagoon bottom. The average solids concentration of the sludge was 4.3 percent, with a majority of the sludge having less than 10 percent solids concentration. It was concluded that the lagoon did not produce a sludge suitable for landfill disposal without further dewatering.

Other plants reported removing thickened alum sludge by dragline or clamshell and dumping the sludge in thin layers on the lagoon banks to air-dry; dumping the thickened sludge on land disposal areas or on roadsides; or transporting the thickened sludge to a specially prepared drying bed.[32]

At the Somerville, New Jersey plant, which has a capacity of 170 mgd (643.5 Ml/d), lagooning of alum sludges has provided a satisfactory solution to the sludge handling problem.[32] A low-turbidity raw water (1 to 30 NTU) receives an alum dose of 40 mg/l prior to sedimentation and rapid sand filtration. The sludge is removed twice a year from the sedimentation basin and pumped to a lagoon 400 × 1,200 feet (122 × 396 m) in area [total land used = 12 acres (4.8 ha)] formed by a previously existing roadway fill on three sides and a dike on the fourth. Supernatant is decanted from the lagoons by overflow pipes. The total available sludge storage depth of 6 feet (1.8 m) had filled to a depth of 3.5 feet (1.07 m) after 10 years of operation. The sludge was dry enough to walk on, and was estimated to have a solids content of 30 percent. It was estimated that the lagoon would have a total useful life of 20 years with slight additions to the dike.

In general, alum sludges do not consolidate under water, but they do dry readily when exposed to air and when drainage through the soil occurs. When lagoons are built above ground, the berms should be from 10 to 15 feet (3 to 5 m) high and far enough from property lines so that, if necessary, their top elevation can be raised. This can be done by removing dried sludge from the interior of the lagoons for use as embankment material. Lagoon berms for larger plants or those with softening should be about 12 feet (4 m) wide at the top to facilitate the use of construction equipment for lagoon cleaning.[30]

Two or more lagoons should be provided for alternating use, to allow between 6 months and 1 year for decanting, evaporation, and drainage. Such a drying period reduces the sludge to the consistency of the soil at the site.[30] At Green Bay, Wisconsin the filter plant operated with a single, 2-acre (0.81-ha) lagoon for its first 9 years, during which it filtered approximately 30,000 MG (110,000 Ml) of Lake Michigan water for an average of about 9 mgd (34.1 Ml/d). At that time 2,000 cu yd (1,530 m³) of dried sludge was removed and the lagoon deepened.[30]

The State of Kansas Bureau of Water Supply has sludge storage lagoon design criteria requiring Kansas water treatment plants to have as a minimum the following storage lagoon capacity:[33]

- Two cells to be provided.
- Clarification sludge: Each cell to be designed for storage of 16 cu ft of sludge/MG (0.119 m³/Ml) of raw water treated during an 18-month period.
- Softening sludge: Each cell to be designed for storage of 85 cu ft of sludge/MG (0.64 m³/Ml) treated during an 18-month period for the first 100 mg/1 of total hardness removed; each cell designed for storage of 82 cu ft/MG (0.62 m³/Ml) treated for each additional 100 mg/1 of total hardness removed.
- Iron and manganese removal: Each cell to be designed for storage of 1.0 cu ft (0.028 m³) of sludge per 1.0 mg/1 of Fe and Mn removed during an 18-month period. If $KMnO_4$ is used, add to either Fe or Mn (or both) present in raw water 0.35 mg/1 Mn for each 1.0 mg/1 $KMnO_4$ added.

- Each cell to have a 2-foot (0.61-m) depth of supernatant above the sludge storage design level.
- Each cell embankment to have a 1.5-foot (0.46-m) freeboard above the clear water zone.
- Facilities for return of water from the clear zone to be sized between 5 and 10 percent of the plant design flow rate.
- Cells to be designed and constructed so that they can be mechanically cleaned. Consideration must be given to the movement of heavy equipment such as trucks, draglines, etc. over the top of the embankments.

Lime Sludge Lagoon Operation and Design. Lime sludges are more easily dewatered in lagoons than are alum sludges. The town of Wauseon, Ohio has successfully used a lagoon system to dewater lime sludge since 1968.[4] The town operates two lagoons; one provides storage for current sludge discharges while the other is in the drying phase. The sludge in the second lagoon is allowed to freeze in the winter and dry through the summer. In August or September a front-end loader and dump truck remove the dewatered sludge at 40 percent solids concentration. The cleaned lagoon is then placed in operation to accept sludge while the other lagoon begins its drying phase. The sludge in the lagoons typically accumulates to a maximum depth of 30 inches (0.76 m).

Lime sludges form lagoons were used to fill in lowland areas around the Miami, Florida water treatment plant, and were later covered with 1 to 2 inches (25.4 to 50.8 mm) of soil and sown with grass.[32] However, lagooned lime sludges are generally considered to be a poor fill material, and final disposal of the dried material may still present a problem.[32]

Sand Drying Beds. Sand drying beds, similar to those used for drying of sewage sludges, have also been used for drying water plant sludges. Drying beds generally consist of a shallow structure with a 6- to 9-inch (0.15- to 0.22-m) layer of sand over a 12-inch (0.305-m)-deep gravel underdrain system. Sand sizes of about 0.5 mm are typically used with a uniformity coefficient of less than 5. Excessively coarse sands result in too great a loss of solids in the drying bed filtrate. The gravel underdrain system used is typically ⅛ to ¼ inch (3.2 to 6.4 mm) graded gravel overlying drain tiles.

In dry climates, shaped, shallow earthen basins are used that rely solely upon evaporation to separate solids from the water. These basins are more similar to lagoons than sand drying beds, with the exception that the depth of sludge application is similar to that used for sand drying beds. Sludge is applied in 1- to 3-foot (0.3- to 0.9-m) layers and allowed to dewater. With either drying bed type, sludge storage facilities may be necessary for periods when climatic conditions prevent effective dewatering.

The rate at which sludges placed on sand drying beds will dewater depends upon the air temperature and humidity, wind currents, and the viscosity and specific resistance of the sludge.

The sludge dewatering process occurs by two mechanisms: (1) gravity drainage through the sludge cake and sand-filter and (2) air drying from the surface of the sludge cake by evaporation. Usually both processes must be functioning for the sludge to reach a condition in which it may be removed from the drying bed for transport to a point of ultimate disposal. The design of drying beds should consider sludge characteristics affecting gravity drainage and drying rates and the extent to which sludge may penetrate into and through the sand bed during the initial drainage phase. Such penetration requires frequent sand replacement and produces unacceptable direct filtrate discharge.[1]

Novak and Montgomery have shown that the degree of sludge penetration into and through a sand bed depends on sludge compressibility.[34] Organic polymer conditioning increases compressibility and reduces penetration.

Gravity drainage rates for water treatment sludges vary considerably with the nature of the sludge, the extent of conditioning, and the applied depth. Generally, softening sludges drain rapidly, iron-based coagulant sludges show intermediate drainage properties, and unconditioned alum sludges show relatively poor drainage characteristics. The specific resistance of the sludge correlates well with the gravity drainage rate; thus, physical or chemical conditioning significantly improves drainage characteristics of poorly draining sludges.[1]

Sand Drying Bed Operating Results.   Effective organic polymer conditioning substantially decreases the time required for the gravity drainage phase of dewatering. For example, King et al. found that for well-conditioned alum sludge the time can be decreased by 50 to 75 percent.[35] Similar results have been reported by Novak and Langford, who also discussed the relationship between drainage rate and applied depth.[36] For sludges with a low specific resistance, drainage can be satisfactory at applied depths of 2 to 3 feet (0.6 to 0.9 m). For poorly draining sludges, applied depths of 1 foot (0.3 m) or less are required unless conditioning agents are used.

Air drying is normally necessary for a drained sludge on a sand drying bed to reach a state in which it can be removed. Although sludge drying rates vary through the depth of sludge, with the top layers drying most rapidly, Novak and Langford found that sufficient moisture is normally lost in air drying to render the entire cake handleable.[36] However, they also note that some sludges, especially those that have not been conditioned, may form a dry surface crust that prevents further evaporation. Novak and Langford believe that the completeness of drying throughout the sludge cake is dependent on the completeness of drainage, particularly for well-conditioned sludges. This concept is supported by the data of King et al.[35]

Neubauer reports that with a 5 mph (8.1 km/h) wind, temperatures of 69 to 81°F (20.6 to 27.2°C), and humidities of 72 to 93 percent, solids concentrations of 20 percent were achieved from an alum sludge in 70 to 100 hours with 97 percent capture of solids and a solids loading of 0.8 lb/sq ft (3.9 kg/m²).[31] Use of effective sand sizes of 0.38, 0.50, and 0.66 mm made little difference in total drying time.

The sizing of drying beds should be based on the effective number of uses per year that may be made of each bed and the depth of sludge that can be applied to the bed.

$$A = \frac{V}{N \times D \times 7.48} \qquad (23-3)$$

where:

A = drying bed area, sq ft (m²)

7.48 = constant for use with English units (1,000 is the constant for use with the metric units shown)

N = number of times that beds may be used each year

D = depth of sludge to be applied, ft (m)

V = annual volume of sludge for disposal, gal (1)

The number of times that the beds may be used is dependent upon the drying time and the time required to remove the solids and prepare the bed for the next application. The bed is usually considered dewatered when the sludge can be removed by earth-moving equipment (such as a front-end loader) and does not retain large quantities of sand. Alum sludges generally attain solids concentrations of 15 to 30 percent, and lime softening sludges attain 50 to 70 percent solids content. Alum sludges require from 3 to 4 days to drain, but drainage may be accelerated by the use of polymers to 1.5 to 3 days.[36] These times are optimum and do not reflect realistic field conditions. Both field tests and a detailed study of the climatic variations are required. The number of bed uses will range from 10 to 20 times per year, depending upon the climate. The usage rate may be increased if polymers are used.

The drying time required will increase with greater sludge depths. Alum sludge at Kirksville, Missouri required 20 hours per percent solids concentration for an 8-inch (0.2-m) application and 60 hours per percent solids concentration for a 16-inch (0.41-m) application.[36] In order to obtain a dewatered cake on the bed with a thickness suitable for removal with a front-end loader, at least 16 to 24 inches (0.41 to 0.61 m) of sludge should be applied. For example, with a 1 mgd (3.785 Ml/d) average treated water quantity, 2,000 lb of sludge/MG (239.6 kg/Ml) treated, and 20 bed uses per year, a 2 percent concentration sludge applied at a 16-inch (0.41 m) depth will require:

$$A = \frac{(2,000 \text{ lb/MG}) (1 \text{ MG/day}) (365 \text{ day/yr})}{(0.02) (8.34 \text{ lb/gal}) \dfrac{20 \text{ uses}}{\text{yr}} \dfrac{16 \text{ in.}}{12 \text{ in./ft}} (7.48 \text{ gal/cu ft})} \quad (23\text{-}4)$$

$$A = \frac{4,376,000}{(20) (1.33) (7.48)} = 22,000 \text{ sq ft } (2,044 \text{ m}^2)$$

Sand drying beds have low construction costs if land is readily available. Operating costs can be moderate to high depending upon the difficulties encountered in sludge removal operations. A sludge that is adequately conditioned and does not penetrate deeply into the sand layer may be fairly easily removed with a front-end loader.

Freezing. In climates where freezing temperatures occur frequently, the freezing and thawing of lagooned alum sludges may result in a marked improvement in the dewatering of the sludges. Freezing of waste alum soilds causes the water in the gelatinous material to crystallize, and upon thawing, the water does not return to the sludge, but leaves a granular solid of coffee-ground consistency. Artificial freezing has been applied, but the electrical energy costs are prohibitive [$85/ton ($94/metric ton) @ $0.05/kWh].[37] The sludge essentially must be totally frozen for the freezing technique to be effective, and thus requires shallow depths.

A 1974 report prepared for the Denver Board of Water Commissioners described a testing program for the proposed Foothills water treatment plant.[3] Pilot ponds were used to show the effect of freezing on 6- to 14-inch (0.15- to 0.36-m) depths of alum sludge. The initial sludge placed in the pilot ponds was taken from an existing sludge lagoon at the Moffat water treatment plant, and had initial solids concentrations ranging from 5.2 to 11.7 percent total solids. The final solids concentrations of samples taken from the pilot ponds after 2 to 5 months during the winter ranged from 10 to 68 percent solids. Eight of the twelve ponds showed final solids concentrations of 26 to 68 percent solids; of these eight ponds, six had final solids concentrations of greater than 44 percent solids.

Important design and operating considerations for shallow freezing and drying ponds were described in the 1974 report based on pilot testing:[3]

- Provide means for uniform application of solids.
- Provide means of complete decanting of separated water.
- Maximize porosity of pond bottoms to increase drying.
- Design to minimize capture of drifting snow, since snow acts as insulation and markedly reduces freezing of solids.

• Provide means of snow removal and/or mixing of solids into any snow layer.

The results of this testing program were promising enough that a system to allow wintertime freezing of alum sludge was designed into the 125-mgd (473-Ml/d) Foothills plant. The alum sludge handling system consists of six 200 x 400-foot (61 x 122-m) ponds with decant capability, a gravel drainage system, and spray application of sludge in thin layers. A large basin is provided that will allow storage of alum sludge during periods when it is too cold for spray application of sludge.[38]

Tests conducted in New York State indicated that a 0.3 percent solids sludge that was placed in a lagoon in January with a depth of 30 inches (0.76 m) and subjected to natural freezing, dewatered to 35 percent solids by the next August by decanting the liquid.[39] Allowing the sludge remaining after decanting to stand for 1 week in 80°F (26.6°C) weather then increased the solids content to about 50 percent, suitable for handling and disposal in a landfill.

The Onondaga County Water District, New York, facility applies lagooned thickened alum sludge to freeze–thaw beds at a depth of from 3 to 9 inches (76 to 229 mm). The applied sludge concentration is 8.0 percent. During one season, sludge was applied at 2.5 lb/sq ft (12.2 kg/m²), and after freezing, thawing, and decanting, the sludge had concentrated to 25 percent. Final disposal of the sludge is to landfill.[40]

Mechanical Dewatering

In recent years various mechanical dewatering systems have been tested on all types of water treatment plant sludge. Centrifugation, belt press filtration, vacuum filtration, and pressure filtration have been the most widely accepted methods.

Centrifugation. Centrifugation is basically a settling process compressed into a shallow depth. It utilizes centrifugal force created by rotating a liquid at high speeds to increase the settling rate of solids. Among the different types of applicable commercial centrifuges are the scroll-discharge, the solid-bowl decanter, the plow-discharge, and the basket-bowl.

Solid-Bowl Centrifuge. The most commonly used centrifuge for dewatering of water treatment sludges is the continuously discharging solid-bowl decanter centrifuge. The principles upon which this machine is based are illustrated in Fig. 23–5. The sludge is introduced into the rotating bowl through a stationary feed tube at the center of rotation. The solids are thrown against the wall of the bowl, with the lighter liquid forming a concentric layer inside

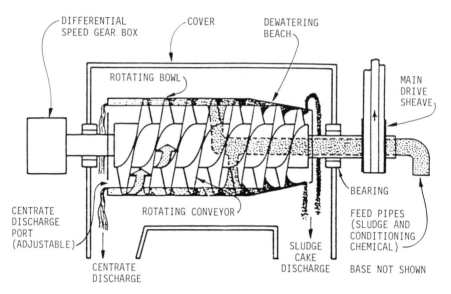

Fig. 23-5. Continuous countercurrent solid-bowl centrifuge.

the solids layer in the bowl. Inside the bowl is a helical screw conveyor or "scroll" that rotates in the same direction as the bowl but at a slightly different speed. This conveyer moves the solids deposited against the bowl toward the small-diameter end of the bowl; there they are "plowed" up the dewatering "beach" and out of the liquid layer, being discharged from the bowl through suitably located ports.

The typical construction of a horizontal, solid-bowl machine is shown in Fig. 23–6. Ports in the bowl head act as overflow weirs for discharge of clarified effluent. The location of these ports with respect to the axis is adjustable, and determines the level of slurry or "pool depth" retained in the bowl. Usually the pool depth is set so that the liquid in the bowl submerges all but a portion of the conical drainage deck. A solid-bowl centrifuge must carry out the dual functions of clarifying the incoming sludge and conveying the solids out of the bowl. Increasing the centrifugal force and lowering the liquid depth in the bowl, for example, theoretically will improve clarification, but in many instances may act to the detriment of the machine by hindering the conveying of solids.

Most solid-bowl machines employ the countercurrent flow of liquid and solids described above and illustrated in Fig. 23–5, and are appropriately referred to as "countercurrent" centrifuges. A second variation of the solid-bowl centrifuge is the concurrent model shown in Fig. 23–7. In this unit, liquid sludge is introduced at the opposite end of the bowl from the dewatering beach, and sludge solids and liquid flow in the same direction. General con-

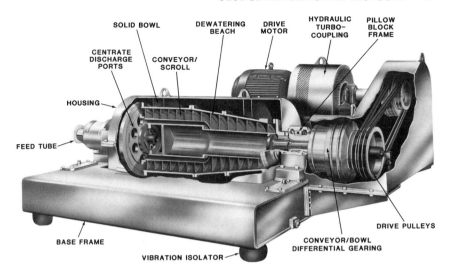

Fig. 23–6. Cross section of a horizontal, solid-bowl centrifuge. (Courtesy of Humboldt Wedag.)

struction is similar to the countercurrent design except that the centrate does not flow in a different direction from that of sludge solids. Instead, the centrate is withdrawn by a skimming device or return tube located near the junction of the bowl and the beach. Clarified centrate then flows into channels inside the scroll hub and returns to the feed end of the machine, where it is discharged over adjustable weir plates through discharge ports built into the bowl head.

A relatively new development in solid-bowl decanter centrifuges is the use of a backdrive to control the speed differential between the scroll and

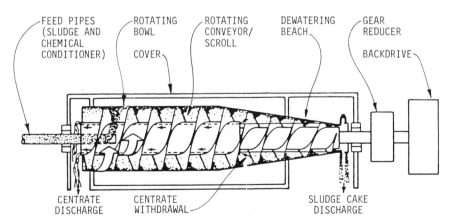

Fig. 23–7. Continuous concurrent solid-bowl centrifuge.

the bowl. Figures 23–8 and 23–9 show operating systems with their respective backdrive units. The objective of the backdrive is to control the differential to give the optimum solids residence time in the centrifuge and thereby produce the optimum cake solids content. A backdrive of some type if considered essential in dewatering alum sludges because of the fine particles present. The backdrive function can be accomplished with a hydraulic pump system, an eddy current brake, a dc variable speed motor, or a Reeves-type variable speed motor. The two most common backdrive systems are the hydraulic backdrive and the eddy current brake, which are shown in Figs. 23–9 and 23–8, respectively. A typical bowl assembly using a hydraulic backdrive is shown in Fig. 23–10.

Cake solids content increases of 4 percent or more relative to machines without a backdrive are achievable, although it must be recognized that the effective capacity of the machine is decreased by utilizing a backdrive to produce a higher-solids-content cake. A backdrive unit generally will not reduce the quantity of polymer required, but it will increase overall stability of centrifuge performance when the feed solids characteristics vary.

The eddy current brake backdrive is commonly provided by one high-G centrifuge manufacturer. The eddy current brake is attached to the pinion shaft of the gearbox and consists of a stationary field coil and a brake rotor

Fig. 23-8. Solid-bowl centrifuges in operation at Lancaster, Pennsylvania. (Courtesy of Sharples-Stokes Division of Pennwalt Corp.)

Fig. 23–9. Solid-bowl centrifuge with hydraulic backdrive system. (Courtesy of Humboldt Wedag.)

on the shaft. When a dc voltage is applied to the stationary field coil, magnetic flux lines are created in the brake rotor. The amount of flux in the rotor is a function of the speed differential between the rotor and the field coil as well as the dc current applied to the field coil. This flux produces eddy currents that create a resistance to turning, or a braking action. Thus, varying the dc voltage applied to the stationary field coil results in a change in the speed differential between the bowl and the scroll.

An automatically controlled variable speed hydraulic backdrive is commonly provided by several low-G centrifuge manufacturers to control the speed differential between the scroll and the bowl. The differential is controlled to maintain a constant torque on the scroll shaft, with the resulting production of a high-solids-content sludge cake. A hydraulic pump and a hydraulic backdrive motor are the two principal components of the hydraulic backdrive unit. The hydraulic backdrive is a noise-producing operation, whereas the eddy current brake is silent.

Most centrifuge installations have the centrifuge mounted a few feet above the floor, and use a belt conveyor to move dewatered cake away. Other methods of installing a solid-bowl centrifuge are to put the centrifuge on the second floor of a two-story building and drop the dewatered cake into either trucks or a storage hopper on the first level; to mount the centrifuge about a foot off the floor and drop the cake into a screw conveyor built into the floor; or to let the centrifuge cake drop into an open-throated progressive cavity-type pump for transfer of the cake to a truck, incinerator, or storage.

Centrifuge performance is measured by the percent solids of the sludge

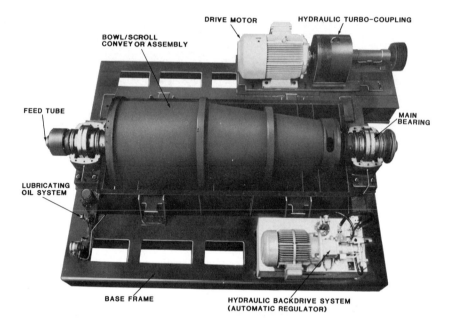

Fig. 23-10. Top view of centrifuge bowl and backdrive unit. (Courtesy of Humboldt Wedag.)

cake, the percent solids capture, the overall quality of the centrate, the solids loading rate, and the polymer requirement. The performance of a particular centrifuge unit will vary with the sludge feed rate and the characteristics of the feed sludge, including percent solids, and sludge temperature.

Centrifuge performance is also affected by polymer type, the dosage utilized, and its point of introduction. Centrifuge performance on a particular sludge will also vary with bowl and conveyor design, bowl speed, differential speed, and pool volume. Bowl and conveyor design are not variable after installation. Although pool depth is variable on solid-bowl units, up to several hours of labor may be required to change the pool depth. Increasing the pool depth will normally result in a wetter sludge cake but better solids recovery, but this is not necessarily true on newer machines equipped with an automatic backdrive.

Bowl speed is not normally varied on most centrifuge models once a centrifuge is installed. An increase in bowl speed normally results in a drier sludge cake and better solids recovery, although in some cases it may result in shearing of the sludge floc and a reduction in solids capture. With the addition of polymer internally into the bowl of the centrifuge, a capability available from several manufacturers, no shearing occurs because both the polymer and the solids are up to bowl speed when the formation of the floc occurs.

Conveyor differential speed normally can be varied, but it may require some disassembly of the machine. On centrifuges equipped with an automatic back-drive, the differential speed can be easily varied. Increasing the differential between the bowl speed and the scroll speed normally results in a wetter sludge cake, poorer solids recovery, and higher machine throughput. On the other hand, reducing the differential speed produces a drier cake, increases solids capture, and decreases machine throughput. Operating at too low a differential speed can cause the pile of solids formed in front of the scroll conveyor blade to increase in overall height such that it infringes on the clarified liquid area. This may result in the skimming of some fine solids from the top of the cake pile to the centrate, lowering solids capture. Too low a differential speed, unless it is adequately controlled, can also result in plugging the centrifuge, if solids are removed more slowly than they are fed to the machine.

The feed rate to the centrifuge is always a critical factor. The best performance data have been obtained at about 75 to 85 percent of the total solids or hydraulic capacity of the centrifuge. At slightly below maximum capacity, the lowest polymer consumption is observed, and the driest cake is obtained.[1]

An advantage of the solid-bowl centrifuge for larger plants is the availability of equipment with the largest sludge throughput capability for single units of any type of mechanical dewatering equipment. The larger centrifuges are capable of handling 300 to 700 gpm (19 to 44 l/s) per unit, depending on the sludge's characteristics. The centrifuge also has the ability to handle higher than design loadings, such as temporary increases in hydraulic loading or solids concentration, and the percent solids recovery can usually be maintained with the addition of more polymer; while the cake solids concentration will drop slightly, the centrifuge will stay on-line.

There has been controversy over the benefits of low-G and high-G centrifuges. The G-force is proportional to the square of the bowl speed; therefore, high-G centrifuges create a greater centrifugal force by using higher bowl speeds. Low-G decanter centrifuge manufacturers claim that their machines consume less energy, have a lower noise level, and require less maintenance than comparable high-G machines. On the other hand, high-G decanter centrifuge manufacturers claim that their machines require less polymer and achieve a higher throughput because of the higher G-forces utilized. Whether or not low-G centrifuges have a lower total annual cost than high-G centrifuges can only be determined after side-by-side tests are conducted with a particular sludge, and the design parameters are known for each machine.

Solid-Bowl Centrifuge Performance Data. The solid-bowl centrifuges that have been applied to alum sludges show widely varying results. Feed solids concentrations of alum sludges dewatered by centrifuges generally range from 2 to 6 percent solids; however, 0.4 to 1.0 percent alum sludges have been successfully dewatered. Well-controlled feed concentration usually produces polymer savings and good performance, while variation in feed solids concentration produces increased polymer consumption, variation in cake dryness, and solids recovery rate fluctuation.[1]

It was reported that at Lancaster, Pennsylvania, using 1 to 2 lb/ton of polymer/ton of solids (0.5 to 1.0 kg/metric ton), 30 percent cake solids are achieved with a 98 percent solids recovery.[2] However, the feed solids concentration to the centrifuge is 5 to 8 percent, which significantly aids dewatering performance. Raw water is obtained from the Susquehanna River, and alum and hydrated lime are both added to the water during clarification.[41] Figure 23–8 illustrates the solid-bowl centrifuges in operation at Lancaster.

Overall raw water characteristics affect the dewatering property of coagulant sludge. For alum sludge generated from processing raw water with a turbidity of 4 to 8 NTU, cake dryness will generally reach 15 to 16 percent, which is considered good performance for a centrifuge.[1] An example of this is the Sobrante filter plant located in El Sobrante, California, approximately 15 miles (24 km) north of Oakland. The plant is operated by the East Bay Municipal Utility District. The raw water source is reservoir, and the raw water quality is generally very good, with turbidities normally in the range of 2 to 10 NTU. Alum sludge is dewatered by solid-bowl centrifuges, resulting in a 16 to 18 percent dry solids cake. The dry polymer feed rate is typically 3 to 6 lb/ton (1.5 to 3 kg/metric ton).[42]

The Erie County, New York, Sturgeon Point plant reported a 24 to 28 percent cake solids content with about 98 percent recovery using 3.0 lb/ton 1.5 kg/metric ton) of polymer. Alum sludges containing high raw water turbidity, clay additives, or lime may be expected to produce higher cake solids concentrations with lower polymer requirements than pure alum sludges.

Lime softening sludge dewaters with relative ease because of its calcium carbonate content. Softening sludge typically contains 80 to 85 percent calcium carbonate, 5 to 10 percent magnesium hydroxide, and 3 to 5 percent iron or aluminum hydroxide.[10] The key parameters for evaluating centrifuge performance in dewatering softening sludges are final concentration of cake solids and the percent of solids captured in the sludge cake. High values are desirable for each, especially in relation to centrate quality. Large amounts of suspended solids in the centrate may lead to the buildup of solids within the treatment plant and possible upset of certain plant operations.

It was reported that a thickened lime sludge could be dewatered on a solid-bowl centrifuge to a cake solids concentration of 55 percent with 78 to 93 percent solids capture.[43] An improvement in solids capture efficiency (90 to 100 percent solids capture) was produced by increasing the sludge residence time; however, a slightly wetter cake of 45 percent suspended solids resulted.

Table 23–13. Solid-Bowl Centrifuge
Performance Data.*

PARAMETER	TYPICAL RANGE
Feed, percent solids	10 to 25
Cake solids concentration, percent solids	55 to 70
Centrate, percent solids	1.0 to 1.5
Solids recovery, percent	91 to 96
Force, gravities	3,500 to 4,000
Scroll differential speed, rpm	20 to 28

* Reference 45. (Courtesy of Scranton Gillette Communications, Inc.)

A summary of operational data from four plants using centrifugation showed that the cake solids concentrations were in the range of 55 to 65 percent suspended solids by weight, with a centrate suspended solids concentration of 500 to 10,000 mg/l.[44] These values are comparable to the summary of typical operational data from the centrifugation of lime softening sludges presented in Table 23–13.[45] Another study showed a consistent cake solids concentration of 50 to 60 percent suspended solids with a total suspended solids recovery greater than 90 percent.[13] The apparent relationship between cake solids concentrations and centrate quality and how each was affected by centrifuge operation can be explained by a plot of characteristic centrifuge results (Fig. 23–11).[46,47] At very low residence times the only solids captured in the cake are those of high density such as calcium carbonate. Solids of

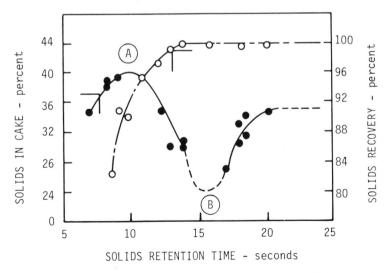

Fig. 23–11. Effect of solids retention time on centrifuge efficiency.[46,47]

a lighter, more flocculent nature pass into the centrate, yielding a high centrate solids concentration and a correspondingly low solids capture. As the solids residence time in the centrifuge increases, a greater percentage of the lower-density solids is incorporated into the cake, often yielding a decreased cake solids concentration. It is only after increased time of centrifugation that these solids are sufficiently compressed to yield a further increase in cake solids concentration.

Water softening applications of centrifuges may have one of two goals:

- Dewatering with the objective of maximum capture of suspended solids.
- Dewatering and classification to produce as pure a calcium carbonate cake as possible.

When lime is added to waters with a high magnesium content, large quantities of magnesium hydroxide are precipitated, rendering the sludge more difficult to thicken and dewater. If lime recalcining is practiced, the resulting magnesium oxide will appear as recycled inert material; so there will be an ever increasing amount of solids in the system. Another approach involves carbonation of the lime sludges with recalciner flue gas rich in carbon dioxide to convert the magnesium hydroxide to dissolved magnesium carbonate. An alternate approach is to use a centrifuge to classify the sludge into its calcium carbonate and non-calcium carbonate components. This classification can be accomplished because of the difference in the specific gravities of calcium carbonate and the impurities. With proper operation, most of the lighter impurities can be classified in the centrate. Control of the pool depth permits this classification to be adjusted as desired while the machine is in operation. Ideally, the cake produced would be 100 percent calcium carbonate, with 100 percent of the impurities rejected in the centrate. Recovery levels with good classification will be 75 to 85 percent of the suspended solids, and the cake will be drier because gelatinous material, much of it in the form of magnesium hydroxide, is eliminated from the cake. The ability of the centrifuge to provide magnesium classification is a major advantage of centrifuges over vacuum filters when recalcining is practiced.

Basket Centrifuge. The imperforate basket centrifuge is a semicontinuous feeding and solids discharging unit that rotates about a vertical axis. Schematic diagrams of a basket centrifuge in the sludge feed and sludge plowing cycles are shown in Fig. 23–12. Sludge is fed into the bottom of the basket, and sludge solids form a cake on the bowl walls as the unit rotates. The liquid (centrate) is displaced over a baffle or weir at the top of the unit. Sludge feed is continued either for a preset time or until the suspended solids in the centrate reach a preset concentration.

After sludge feeding is stopped, the centrifuge begins to decelerate, and

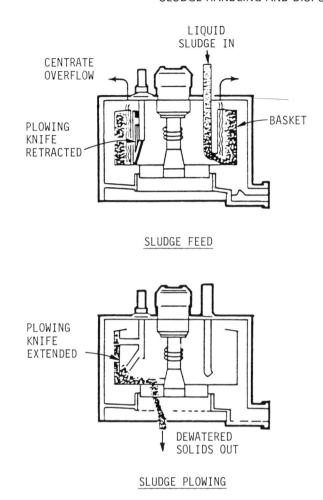

Fig. 23-12. Basket centrifuge in sludge feed and sludge plowing cycles.

a special skimmer nozzle moves into position to skim the relatively soft and low-solids-concentration sludge on the inner periphery of the sludge mass. After the skimming operation, the centrifuge slows further to about 70 rpm, and a plowing knife moves into position to cut the sludge away from the walls; the sludge cake then drops through the open bottom of the basket. After plowing terminates, the centrifuge begins to accelerate, and feed sludge is again introduced. At no time does the centrifuge actually stop rotating.

Performance of a basket centrifuge is measured by the cake solids content, the solids capture, the required polymer dosage, and the average feed rate or solids throughput. Cake solids concentration is the average solids content, since the solids content is at its maximum at the bowl wall and decreases toward the center. The polymer required for a basket centrifuge is generally lower than that required by other mechanical dewatering equipment. The average feed rate includes the period of time during a cycle when sludge is not being pumped to the basket (acceleration, deceleration, discharge). Therefore, dividing total gallons pumped per cycle by total cycle time gives the average feed rate. Solids throughput can be determined using the average feed rate, the percent feed solids, and the solids capture.

Basket Centrifuge Performance Data.   The basket centrifuge has not proved as reliable a method for dewatering alum sludge as the solid-bowl centrifuge. Performance obtained during pilot tests has ranged from poor to acceptable. In one report, sludge generated from treating low-turbidity water could not be dewatered by the basket centrifuge. However, with turbidities of 30 to 50 NTU, a 15 percent dry cake could be produced by using 4 to 5 lb of polymer/ton (2. to 2.5 kg of polymer/metric ton) of solids. The average cycle time was approximately 20 minutes.[48]

Belt Press Filtration.   Belt filter presses employ single or double moving belts to continuously dewater sludges through one or more stages of dewatering. All belt press filtration processes include three basic operational stages: chemical conditioning of the feed sludge, gravity drainage to a nonfluid consistency, and shear and compression dewatering of the drained sludge.

Figure 23–13 depicts a simple belt press and shows the location of the three stages. Although present-day presses are usually more complex than this, they follow the same principle indicated in Fig. 23–13. The dewatering process is made effective by the use of two endless belts of synthetic fiber. The belts pass around a system of rollers at constant speed and perform the function of conveying, draining, and compressing. Many belt presses also use an initial belt for gravity drainage, in addition to the two belts in the pressure zone.

Good chemical conditioning is very important for successful and consistent performance of the belt filter press. A flocculent (usually an organic polymer) is added to the sludge prior to its being fed to the belt press. Free water drains from the conditioned sludge in the gravity drainage stage of the press.

The sludge then enters a two-belt contact zone, where a second upper belt is gently set on the forming sludge cake. The belts with the captured cake between them pass through rollers of decreasing diameter. This stage subjects the sludge to continuously increasing pressures and shear forces. Pressure can vary widely by design, with the sludge in most presses moving from a low pressure section to a medium pressure section. Some presses include a high pressure section that provides additional dewatering. Progressively, more and more water is expelled throughout the roller section to

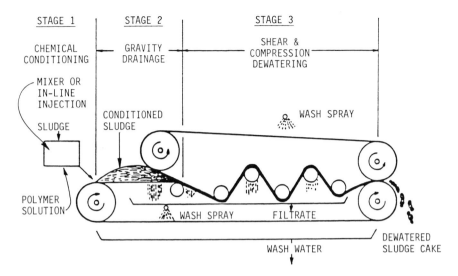

Fig. 23-13. The three basic stages of a belt filter press.

the end where the cake is discharged. A scraper blade is often employed for each belt at the discharge point to remove the cake from the belts. Two spray-wash belt cleaning stations are generally provided to keep the belts clean.

Belt press performance is measured by the percent solids of the sludge cake, the percent solids capture, the solids and hydraulic loading rates, and the required polymer dosage. Several machine variables, including belt speed, belt tension, and belt type, influence belt press performance.

Belt speed is an important operational parameter that affects cake solids, polymer dosage, solids recovery, and hydraulic capacity. Low belt speeds result in drier sludge cakes. At a given belt speed, increased polymer dosages result in higher cake solids. With an adequate polymer dose, solids recoveries are improved by lowering belt speeds. Hydraulic capacity increases at higher belt speeds, but the solids capture drops. Depending on desired performance, the belt speed setting can be used to produce a variety of different results.

Belt tension has an effect on cake solids, maximum solids loading, and solids capture. In general, a higher belt tension produces a drier cake but causes a lower solids capture, at a fixed flow rate and polymer dose. A drawback of using higher tension is increased belt wear. For sludges with a large quantity of alum sludge, the belt tension must be reduced to contain the sludge between the belts. The maximum tension that will not cause sludge losses from the sides of the belts should be used. The high pressure zones on belt presses may cause problems with some alum sludges and may be unusable to require the lowest pressure setting possible.

Belt type is an important factor in determining overall performance. Most belts are woven of polyester filaments, and they are available with weaves of varying coarseness and strength. A belt with one of the coarser and stronger weaves may require high polymer dosages to obtain adequate solids capture.

Failure of the chemical conditioning process to adjust to changing sludge characteristics can cause operational problems. If sludge is underconditioned, improper drainage occurs in the gravity drainage section, and either extrusion of inadequately drained solids from the compression section or uncontrolled overflow of sludge from the drainage section may occur. Most manufacturers' belt presses can be equipped with sensing devices that may be set to automatically shut off the sludge feed flow in case of underconditioning. Both underconditioned and overconditioned sludges can blind the filter media. In addition, overconditioned sludge drains so rapidly that solids cannot be distributed across the belt. Vanes and distribution weirs included in the gravity drainage section help alleviate the problem of distribution of overconditioned sludge across the belt, and inclusion of a sludge blending tank before the belt press can also reduce this problem.

Scraper units and filtrate trays are sites where solids build up. A belt press installation should be designed for daily washdown by hosing; therefore, drainage and safe walking areas around the press are important.

The flow rate required for belt washing water is usually 50 to 100 percent of the flow rate of sludge to the machine, and the pressure is typically 100 psi (690 kPa) or more. Some belt presses recirculate washwater from the filtrate collection system, but usually potable water is used. The combined flow of washwater and filtrate contains between 500 and 2,000 mg/l of suspended solids and is typically returned to the plant raw water inlet.

Belt presses have numerous moving parts, including up to 25 to 30 rollers and 50 to 75 bearings. Spare parts should be kept available to prevent prolonged unit downtime. Belts, bearings, and rollers can deteriorate quickly if maintenance is inadequate. However, most parts are small and easily accessible, so that even small facilities should have little difficulty in maintaining replacement parts.

Belt Filter Press Performance Data.   At the water treatment plant at Somerset, Kentucky, belt filter presses are reported to produce an average cake solids content of 34 percent form a feed solids content of 3.1 to 5.3 percent.[49] Raw water is obtained from a lake and has average turbidities ranging from 16 to 31 NTU. Coagulation is accomplished with 14 to 15 mg/l of alum, 1.0 mg/l of cationic polymer, and 6 to 6.5 mg/l of lime. Polymer dosage is 3.2 lb/ton (1.6 g/kg) of solids at a cost of $7.96/ton ($8.84/metric ton).

A demonstration test of a belt filter press was conducted at the Western Pennsylvania Water Company plant in New Castle, Pennsylvania.[50] The raw river water received coagulation treatment with alum, polymer, and lime (if required to raise pH). The feed solids content averaged 3.8 percent, and the sludge cake produced averaged 32 percent solids. Polymer requirements were 2.1 lb/ton (1.1 g/kg) of dry solids. Solids recovery averaged about 90 percent.

A demonstration test conducted on alum sludge at the San Jacinto Water Purification Plant in Houston, Texas illustrates the difficulty in dewatering some alum sludges.[51] With polymer dosages of 7.3 to 10.1 lb/ton (3.7 to 5.1 g/kg) of solids, cakes solids contents of only 13.5 to 15.6 percent were obtained from a feed solids concentration of 2.5 percent. With the addition of 12.5 percent diatomaceous earth, cake solids increased to 19.4 percent solids. With the addition of 23 percent lime for conditioning, cake solids contents of 23 percent were obtained.

Dewatering of a combined alum/lime sludge at the Gastonia, North Carolina water treatment plant is conducted on a belt filter press.[52] The belt press dewaters approximately 60 to 80 gpm (230 to 300 l/min) of sludge from a feed solids concentration of 2 percent to a cake solids concentration of 25 percent. Typical polymer requirements are 3 to 8 lb/ton (1.5 to 4 kg/g) of dry solids. The unit has been in operation since 1982.

The belt press installation at Gastonia is shown in Fig. 23–14, and the dewatered alum/lime sludge cake coming off the unit is shown in Fig. 23–15.

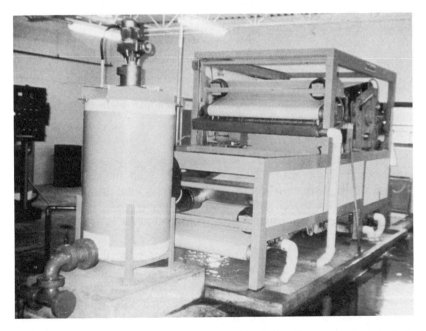

Fig. 23–14. Belt press installation at Gastonia, North Carolina water treatment plant. (Courtesy of Parkson Corp.)

Fig. 23–15. Alum/lime sludge discharging from belt filter press at Gastonia, North Carolina. (Courtesy of Parkson Corp.)

Typical performance data of belt filter presses on lime softening sludge at three water plants have been summarized.[53] Feed sludge concentrations are about 20 to 25 percent solids, and cake solids concentrations are 60 to 70 percent. The solids recoveries are 90 to 95 percent, and polymer requirements are typically 2 to 3 lb/ton (1 to 1.5 kg/g) of dry solids.

Vacuum Filtration. A vacuum filter consists basically of a horizontal cylindrical drum that rotates partially submerged in a vat of sludge. The filter drum is divided into multiple compartments or sections by partitions (seal strips), and each compartment is connected to a rotary valve by a pipe. Bridge blocks in the valve divide the drum compartments into three zones: the cake formation zone, the cake drying zone, and the cake discharge zone. The filter drum is submerged to about 25 percent of its depth (variable) in a vat of conditioned sludge, and this submerged zone is the cake formation zone. Vacuum applied to the submerged drum section causes filtrate to pass through the drum covering (filter medium) and sludge cake to be retained on the covering. As the drum rotates, each section is successively carried through the cake forming zone to the vacuum drying zone (see Fig. 23–16). This zone begins when the filter drum emerges from the sludge vat. The cake drying zone represents from 40 to 60 percent of the drum surface and ends at the point where the internal vacuum is shut off. At this point, the sludge cake and drum section enter the cake discharge zone, where sludge cake is removed from the filter medium.

Two types of rotary drum vacuum filters are used in water treatment: traveling medium and precoated medium filters. The traveling belt filter allows continuous removal of the drum covering from the drum by continuous washing on both sides by a high pressure spray, without diluting the sludge in the vat. This permits filter operation at maximum loading rate with a constantly clean filter medium.

Figure 23–17 shows a schematic cross section of a fiber cloth, traveling-belt-type rotary vacuum filter. The medium on this type of unit leaves the drum surface at the end of the drying zone and passes over a small-diameter discharge roll to facilitate cake discharge. Washing of the medium occurs after discharge and before it returns to the drum for another cycle. This type of filter normally has a small-diameter curved bar between the point where the belt leaves the drum and the discharge roll. This bar aids in main-

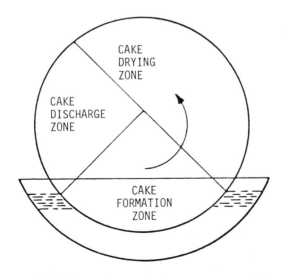

Fig. 23–16. Operating zones of a rotary vacuum filter.

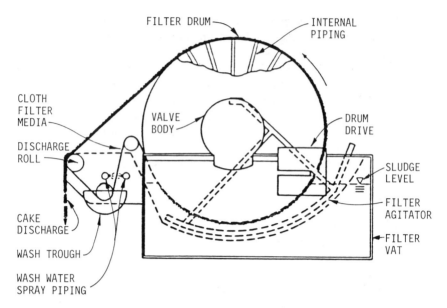

Fig. 23–17. Cross-sectional view of a fiber cloth, traveling-belt-type rotary vacuum filter.

taining belt dimensional stability. In practice, it is frequently used to ensure adequate cake discharge. Remedial measures, such as addition of scraper blades or use of excess chemical conditioner, are sometimes required to obtain cake release from the cloth medium.

To prevent localized thickening and settling of sludge in the filter vat, an agitator keeps the solids in suspension. The agitator operates with a pendulum motion between the drum and the vat. The filter drum and the agitator are of variable speed to maintain optimum operation.

Although vacuum filters are widely used in dewatering sewage sludges, they have had limited application in the water treatment field. Alum sludges have proved difficult to dewater by vacuum filtration, as their gelatinous nature almost precludes the use of vacuum filtration without precoating of the filter with diatomaceous earth. The precoated filter is used mainly for dewatering coagulation sludges. The filter drum is coated with the filter medium, which is slowly shaved off. Minimal design precoat cake shaving is approximately 0.005 inch (0.127 mm), which corresponds to the maximum penetration of sludge solids into the precoated bed. The drum rotates slowly, with rotation speed ranging from 5 to 12 rpm depending on the permeability of the deposited cake and the grade of precoat material. An average precoat layer of 2 to 3 inches (50.8 to 76.2 mm) is applied, and is shaved in very small increments. Approximately 50 to 60 minutes is needed to precoat a filter.[1]

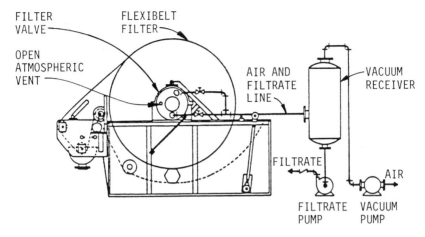

Fig. 23-18. Schematic drawing of typical vacuum filter installation. (Courtesy of Komline-Sanderson Engineering Corp.)

A vacuum filter installation includes a sludge feed pump, chemical feeders, sludge conditioning tanks, vacuum filter, vacuum pump and receiving tank, filtrate pump, and filter cake conveyor and hopper. Figure 23–18 is a schematic diagram of a typical vacuum filter installation. It is sometimes necessary to precondition sludges or slurries with coagulating chemicals such as ferric chloride, lime, or polyelectrolytes. This may be accomplished in a rotating conditioning tank (see Fig. 23–19). Two chemical application points are pro-

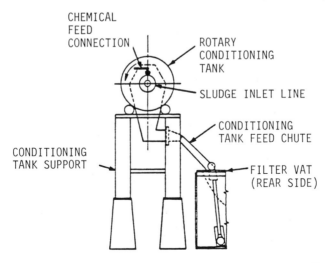

Fig. 23-19. Schematic drawing of a sludge conditioning system. (Courtesy of Komline-Sanderson Engineering Corp.)

vided, and the rotating drum (variable speed) is equipped with nonclog mixing blades and an annular baffle, which produces isolated mixing chambers.

Vacuum filters are available in a wide range of sizes with diameters of 3 to 12 feet (0.91 to 3.66 m) and lengths of 1 to 20 feet (0.3 to 6.1 m). Table 23–14 provides a summary of the filtering area associated with a variety of vacuum filter sizes.

Vacuum Filter Performance Data. Because the dewatering characteristics of alum sludge vary from day to day, and because alum sludges are moderately compressible, the applicability of the best filter medium has some limitations. It is difficult to dewater coagulation alum sludge generated from raw water with turbidities between 4 and 10 NTU. The specific resistance of this sludge can be as high as 2.9 to 11.7×10^{17} ft/lb (2 to 8×10^{13} cm/g); also, because the sludge is compressible, the vacuum applied during filtration must be set and controlled to avoid blinding of the filter medium.[1]

The selection of the filter medium is a critical element in dewatering alum sludge. A polypropylene monofilament belt medium, rated at an air flow of 300 cfm/sq ft (91.5 m³/min/m²) at 15 inches (0.38 m) of mercury vacuum, results in improved filtrate quality and a solids loading rate of 1.4 lb/sq ft/ hr (0.28 kg/m²/hr).[1]

Successful operation of traveling belt vacuum filters requires the use of filtration aids before dewatering. The most conventional conditioning aids are polymers, lime, and a combination of the two. Results of vacuum filtration using a traveling belt filter medium are shown in Table 23–15

Precoat filters have proved to be more effective than traveling belt vacuum filters in dewatering alum sludge, even under the most severe conditions. Typical performance data obtained by precoat rotary vacuum filters are given in Table 23–16.

At most water treatment plants, a combination sludge that is a mixture of coagulation alum sludge and residue from filter backwash water is produced. The dewatering characteristics of the mixture are similar to those of conventional coagulation alum sludge, although the designer must give special attention to fluctuations in solids concentration.

Vacuum filtration has been more successful for lime sludges than for alum sludges. The most common filter medium is a multifilament polypropylene belt medium. Vacuum filter installations at Boca Raton, Florida and Minot, North Dakota both are reported to dewater thickened lime sludges of 30 percent solids to a cake of 45 to 65 percent solids, adequate for use as a landfill.[32] Cloth belt lives of 6 to 9 months were reported. At the Boca Raton plant, 120 to 130 mg/l lime is added to soften a water with 210 to 250 mg/l hardness to provide a 58 mg/l total hardness effluent. The raw water contains only 6 mg/l magnesium. Average plant flow was 7.6 mgd (28.8 Ml/d) during the study period reported.[32] Sludge is removed continuously

Table 23–14. Filtering Area of Vacuum Filters.*

FILTERING AREA OF DRUM FILTERS IN SQUARE FEET

DIAMETER (IN FEET)	FACE (IN FEET)																			
	1	2	3	4	5	6	7	8	9	10	11	12	13	14	15	16	17	18	19	20
3	9.4	18.8	28.2	37.7	47.1	56.5														
4½		—	42.4	56.5	70.6	84.8	98.9	103	127											
6				—	94.2	113	132	151	170	188	207	226								
8						—	176	201	226	251	277	302	327	352	377	402				
10								—	283	314	345	377	409	440	471	502	534	565	596	628
12										377	415	452	490	528	565	603	641	679	716	754

* Reference 54.

765

Table 23–15. Performance and Operating Data Obtained by the Traveling Belt Vacuum Filter on Alum Sludge Dewatering.*

PARAMETER	DATA RANGE
Feed concentration, percent	2 to 6
Flow rate: l/m²/hr	0.7 to 1.4
gal/sq ft/hr	2 to 4
Dry solids yield: kg/m²/hr	0.15 to 0.25
lb/sq ft/hr	0.75 to 1.25
Cake concentration with polymer, percent	15 to 17
Cake concentration with lime, percent	30 to 40
Filtrate solids, mg/l	100 to 200
Polymer dosage: kg/tonne dry solids	3 to 6
lb/ton dry solids	6 to 12
Lime dosage, percent	30 to 60
Drum speed, rpm	0.2 to 0.5
Operating vacuum: mm Hg	254 to 381
in. Hg	10 to 15

* Reference 55.

Table 23–16. Typical Precoat Performance Data on Alum Sludge Dewatering.*

PARAMETER	DATA RANGE
Feed concentration, percent	2 to 6
Flow rate: l/m²/hr	0.7 to 2.1
gal/sq ft/hr	2 to 6
Dry solids yield: kg/m²/hr	0.2 to 0.3
lb/sq ft/hr	1.0 to 1.5
Cake concentration, percent	30 to 35
Filtrate suspended solids, mg/l	10 to 20
Solids recovery, percent	99+
Precoat recovery, percent	30 to 35
Precoat rate: kg/m²/hr	0.02 to 0.04
lb/sq ft/hr	0.1 to 0.2
Precoat thickness: mm	38.1 to 63.5
in.	1.5 to 2.5
Drum speed, rpm	0.2 to 0.3
Operating vacuum: mm Hg	127 to 508
in. Hg	5 to 20

* Reference 55.

from upflow clarifiers at 1 to 4 percent solids content and pumped to a 25-foot-diameter thickener. The sludge has a low magnesium hydroxide content, which facilitates dewatering. Thickened sludge (28 to 32 percent solids) is pumped to a vacuum filter that has been in use since 1964 to 1965. The cake discharged consists of 65 to 70 percent solids and is directly used as a road stabilizer. Solids loadings of 60 lb of dry solids/sq ft/hr (293 kg/hr/m²) are reported as the average operating condition. A small lagoon serves as an emergency sludge storage basin to be used when the vacuum filter is removed from service for repair. The solids production is about 1 ton of dry solids/MG.

The Minot plant processes 10 mgd (37.9 Ml/d) during summer months with raw water that has a high hardness (about 115 mg/l). About 3 tons of dry solids/MG (0.72 metric ton/Ml) are produced. The sludge passes through a thickener prior to vacuum filtration, which produces a dewatered cake with 44 percent solids at loadings of 30 lb/sq ft/hr (147 kg/m²/hr). The sludge cake is placed in a landfill on the plant site. Experiences with hauling to a town dump indicated that rubber seals in the dump truck box were necessary to minimize sludge leakage during transport.

Typical operational data from the literature on vacuum filter dewatering of lime softening sludges are presented in Table 23–17. Figure 23–20 illustrates a vacuum filter in operation on lime sludges.

Lime-dominant softening sludges dewater so well by vacuum filtration that the application of sludge conditioning chemicals often is not required. Only when extreme problems of filtrate solids occur may conditioning chemicals be warranted. Often under such conditions, operators have used precoat vacuum filtration systems instead to minimize solids concentrations in the filtrate liquor.

Table 23–17. Vacuum Filter Performance Data*

PARAMETER	TYPICAL RANGE
Feed solids, percent	5 to 30
Cake concentration, percent	40 to 70
Cake yield: kg/m²/hr	0.8 to 4.0
lb/sq ft/hr	4 to 20
Filtrate solids, mg/l	950 to 1,500
Solids recovery, percent	95 to 99
Filter speed, rpm	0.2 to 0.5
Operating vacuum: mm Hg	381 to 635
in. Hg	15 to 25

* Reference 56.

Fig. 23–20. Vacuum filter operating on lime sludge. (Courtesy of Komline-Sanderson Engineering Corp.)

A study of the effect of various sludge floc properties on vacuum filter dewatering showed floc size distribution to be a key parameter in determining the dewatering rate.[10] Coagulation sludges were found to be dominated by flocs of extremely small size (6 to 8 μm). In comparison, the more easily dewatered softening sludge had average floc sizes of 40 to 50 μm.

Pressure Filtration. The application of pressure filtration systems to water treatment plant sludges is rather recent. The filter press is an applicable dewatering technique for both metal hydroxide coagulant and lime softening sludges; however, its application has been primarily for the metal hydroxide coagulant sludges. The filter press offers advantages for difficult sludges because it is a batch operation where the sludge can be kept under pressure on the filter for extended periods of time until it dewaters to the desired consistency. In addition, filtrate liquors produced are usually low in suspended solids content. Since lime softening sludges dewater readily to a dry cake with other devices, and since centrifuges have the ability to classify the magnesium hydroxide from the calcium carbonate, the filter press has not found significant application for softening sludges.

A filter press consists of a number of plates or trays that are held rigidly in a frame to ensure alignment and are pressed together either electromechanically or hydraulically, between a fixed and a moving end. Figure 23–21 illustrates a typical filter press.

The two types of filter presses that are commonly available to dewater sludges from water treatment plants are the fixed volume recessed plate filter press and the variable volume recessed plate filter press, also referred to as the diaphragm filter press. The recessed plate filter press is also called the chamber filter press.

In the fixed volume recessed plate filter press, liquid sludge is pumped by high pressure pumps into a volume between two filter plates. On the

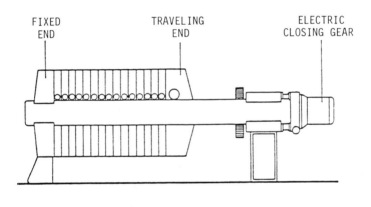

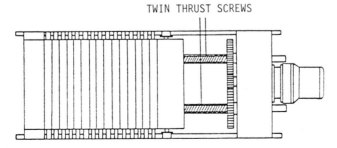

Fig. 23–21. Typical filter press. (Courtesy of Nichols Engineering.)

face of each individual plate a filter cloth is mounted. As a result of the high pressure that the sludge is under, a substantial portion of the water in the feed sludge passes through the filter cloth and drains from the press. Sludge solids and the remaining water eventually fill the void volume between the filter cloths, so that continued pumping of solids to the press is no longer productive. At this point, pumping is stopped, and the press is opened to release the dewatered sludge cake prior to initiation of a new "pressing cycle."

In a variable volume recessed plate or diaphragm filter press, sludge is pumped into the press at a low pressure until the volume of the press has been filled with a loosely compacted cake; then sludge pumping is stopped, and the diaphragm is inflated for a preset time. For the diaphragm press, although most of the water removal occurs when sludge is being pumped into the press, a significant quantity of water is also removed after the diaphragm is inflated.

In the fixed volume recessed plate press, a cloth filter medium is used on both sides of the filtering volume. As shown in Fig. 23–22, sludge is pumped

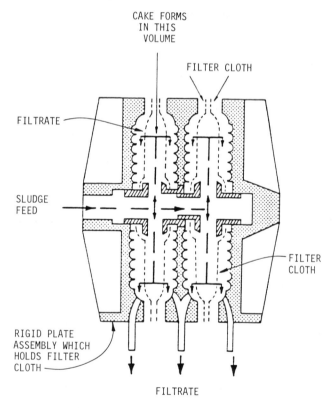

Fig. 23–22. Cross section of a fixed volume recessed plate filter press assembly.

into the volume between the cloth sides, and water is expelled through the medium. Sludge pumping is at relatively high pressures, up to 225 psi (1,550 kPa), and the driving force for movement of water through the cloth is this high pressure. Low pressure recessed plate presses are also available, operating at about 100 psi (690 kPa). When little or no additional filtrate is being produced, the pumping is stopped, the press is opened, and the sludge cake falls from the press.

The diaphragm press is a relatively new innovation. An inflatable diaphragm is used to further compress the sludge solids after low pressure [100 psi (690 kPa)] sludge pumping into the press is ineffective in promoting further dewatering. The diaphragm is expanded by pumping either air or water into it at pressures up to between 215 psi (1,480 kPa) and 285 psi (1,965 kPa), depending upon the manufacturer. After a preset time has elapsed, the diaphragm is deflated, and the press opens allowing the cake to drop out the bottom. The filter cloth is washed periodically, by permanent spray nozzles. Figure 23–23 shows the basic configuration of one cell of a diaphragm press and the four separate stages of operation.

The diaphragm press has several advantages over the fixed volume recessed plate press. First, a drier cake with a relatively uniform moisture content is produced. This uniformity generally does not occur in the fixed volume press, because low-solids-content feed sludge, which produces the filtering pressure, is being added continually; thus the inner part of the cake in each cell is generally of low solids content. The second key advantage of the diaphragm press is an overall shorter cycle time and therefore a higher production throughput. The primary reason for this shorter cycle is that the diaphragm creates a more effective and uniform pressure on the sludge cake than occurs when liquid sludge is pumped into the chamber. Two other advantages of the diaphragm press are the lower operation and maintenance requirements for the sludge feed pumps, and the ability to dewater a marginally conditioned sludge to a high solids content. Generally, a fixed volume recessed plate press cannot dewater a marginally conditioned sludge to a satisfactory cake concentration. Another advantage of the diaphragm press is that it does not require a precoat, while a precoat is frequently necessary with a fixed volume press.

The principal disadvantage of the diaphragm press is its higher initial cost, which can be two to three times the cost of a fixed volume recessed plate press with the same daily throughput. Another disadvantage is that although the diaphragm press has a lower cycle time, the capacity of the largest diaphragm filter press is generally less than that of the largest fixed volume recessed plate filter press.

Control of filter presses may be manual, semiautomatic, or fully automatic. Labor requirements for operation will vary dramatically, depending on the degree of instrumentation utilized for control. In spite of automation, operator

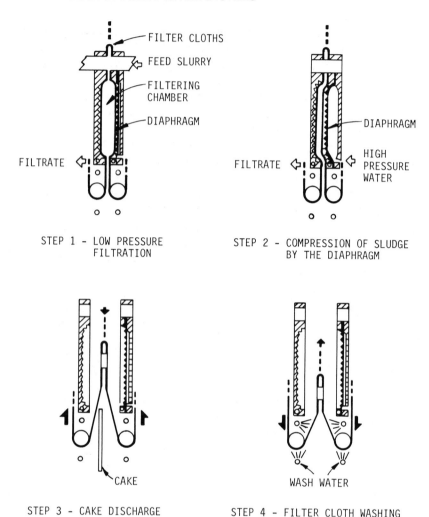

Fig. 23-23. Operational cycle for a Lasta diaphragm filter press. (Courtesy of Ingersoll-Rand.)

attention is often needed during the dump cycle to ensure complete separation of the solids from the media of the filter press. Process yields can typically be increased 10 to 30 percent by carefully controlling the optimum cycle times with a microcontroller. This is important because the capital costs for filter presses are very high.

Filter presses are normally installed well above floor level so that the cakes can drop out into trailers positioned underneath the presses. Alternatively, conveyers can be installed under the presses to transport the cakes to the storage area.

In order for pressure filtration to be economical, alum sludges must be conditioned to lower resistance to filtration. Lime is an effective conditioner and is in use in several full-scale plants now operating; it is added to the sludge as a slurry at a 7.5 percent solution. However, during pilot filtration runs, it has been found that mechanical conditioning agents, such as fly ash, can be equally effective.[1] The choice of conditioning agent is strictly an economic one that should be investigated for each application.

Unlike wastewater sludges, where conditioning agents are added on a percentage basis to dry suspended solids, lime is added to alum sludge until the pH of the slurry is raised to about 11, a pH at which the alum sludge is satisfactorily conditioned. In addition, there must be sufficient residence time of the lime-treated sludge to allow complete reaction of the lime with the sludge. A minimum time requirement is 30 minutes. Insufficient residence time may produce premature plating-out of lime on the filter media and the interior of pipelines.[1]

Two-stage conditioning systems have been demonstrated to be somewhat more economical than one-stage lime addition. In these systems only a portion of the lime is added to the incoming untreated sludge, followed by a small polymer addition. When the mixture is then allowed to age, an appreciable supernatant is formed, which is decanted and returned to the head of the plant. After sufficient residence time, the remainder of the lime is added, and the sludge is then ready for filtration. The total lime addition is less than that used in a single-phase system.[1] Each case should be tested to determine the effectiveness of the type of conditioning system.

The capacity of a pressure filter is determined by the number of filter plates, the size of filter plates, and the cake thickness provided for in the filter plate chamber. Filter cake thickness is critical in the design of a pressure filter; cake thicknesses have been standardized to 0.98, 1.18, and 1.57 inches (25, 30, and 40 mm) thickness.[1] Filtration tests determine the most economical cake thickness for any given application.

The basic requirements for a filter press system include (see Fig. 23–24):

- Storage and mixing tanks for chemical reagents.
- A storage and conditioning tank, to provide a consistent feed to the filter presses for the duration of the filtration cycle. In this tank chemical reagents are introduced to improve the filterability characteristics of the sludge. Means for agitation are provided to prevent segregation of particles and also to prevent size degradation and breakdown in the flocculated feed.
- Feed pumps.

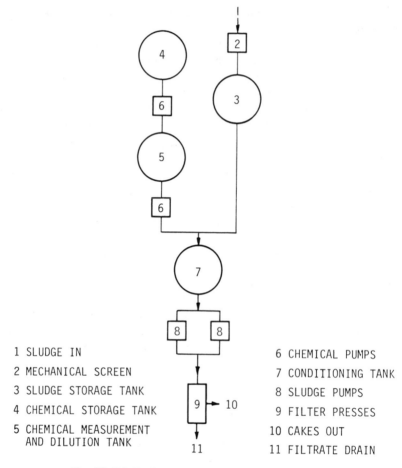

Fig. 23–24. Typical filter press system components.

1 SLUDGE IN

2 MECHANICAL SCREEN

3 SLUDGE STORAGE TANK

4 CHEMICAL STORAGE TANK

5 CHEMICAL MEASUREMENT
  AND DILUTION TANK

6 CHEMICAL PUMPS

7 CONDITIONING TANK

8 SLUDGE PUMPS

9 FILTER PRESSES

10 CAKES OUT

11 FILTRATE DRAIN

- Filter presses with ancillaries, including filter media and filtrate collection trays or launders. The selection of the correct filter media is an important factor in filtration.
- Interconnecting pipework, valves, and air vessels for the feed system and filtrate disposal. Air vessels serve to even out pressure variations caused by the action of the feed pumps.
- Filter cloth washing machines and drying racks.
- Means for collecting and delivering filter cake to a disposal point in a form that is acceptable for any further processing required.

Present-day filter media are almost exclusively of the monofilament type; that is, woven from a single filament rather than a yarn that is twisted together of many fibers. The monofilament media do not blind from swelling of the yarn, which was one of the problems with multifilament media in the past. Media are rated by particle retention and permeability.[1]

Mechanical filter-feed pumps must satisfy pressure filter-feed requirements for the relatively high flow volume at the beginning of the filtration cycle and a low flow volume at the end of the cycle. The discharge pressure of a feed pump is relatively low when the flow from the pumps is high, and terminates at relatively high pressure at the end of the filtration cycle.

Pneumatic sludge feeding has been used successfully, with the raw sludge fed into pressure vessels and lime slurry added to adjust the pH to above 11. After sufficient residence time, compressed air is applied to the tank and the sludge pressed into the filter. Depending upon the sludge concentration, one or more pneumatic tanks are necessary to complete a cycle. Although this system is alleged to be more maintenance-free than mechanical systems, the capital cost of pneumatic systems is appreciably higher.[1]

Any filter medium eventually blinds and must be washed periodically. If sludge conditions cause calcium carbonate to precipitate on the filter cloth, an acid wash system—either an acid soaking system or an acid recirculating system—is used. The latter is preferred because it not only reacts with the calcium carbonate deposits, but also offers a surface scrubbing action to wash out loose material. In a soaking system, the filter is filled with acid and allowed to stand overnight. One drawback of this system is the formation of gas pockets in the upper portion of the filter chambers that prevent the acid solution from reaching the upper portion of the filter plates.[1]

The major purpose of a precoat system is to reduce the frequency with which a filter must be washed by water or acid wash. Thus, it is strictly a convenience for the operator. A precoat system consists of a pump that circulates clear water through the filter at a high rate at the beginning of each filtration cycle. A small amount of filter aid, such as diatomaceous earth or fly ash added before filtration, deposits in a uniformly thin layer over the entire filter cloth surface. Without interruption of the flow, the sludge feed is then started, and sludge will deposit against the interface of the precoat material. The sludge never directly contacts the filter medium; so time intervals between washings are lengthened.

Fly ash is a less costly precoat material than diatomaceous earth. Although fly ash precoating requires about 10 lb/100 sq ft (0.49 kg/m²) of filter area as compared to diatomaceous earth precoating at 6 lb/100 sq ft (0.29 kg/ m²), the costs are approximately $20 per ton and $300 per ton ($22 and $330/metric ton), respectively.[57] In addition, diatomaceous earth is available only in large, minimum-quantity bulk shipments, and requires significantly larger transfer and storage facilities than fly ash.

Disposal of filtrate produced during pressure filtration is a problem because of the chemical characteristics of the material. The conditioned sludge has a pH of about 11.5, which causes a significant fraction of insoluble aluminum hydroxide to be converted to soluble aluminate. In addition, precoat material can contribute potentially significant concentrations of trace metals to the filtrate.

The design of several pressure filtration dewatering systems in New York reflects the resistance of the New York State Department of Health, Bureau of Public Water Supply, to returning the filtrate to the head of the water treatment process. The concern of this regulatory agency is based on uncertainty about the effects of significant increases in pH and trace metals.[1]

Other possible filtrate disposal methods include direct discharge to a waterway, discharge to a sanitary sewer system, and treatment of the filtrate prior to disposal. Direct discharge to a waterway requires compliance with effluent standards, while discharge to a sanitary sewer requires compliance with local sewer use ordinances. A filtrate disposal method acceptable to the regulatory agencies includes neutralization of the filtrate, sedimentation, and returning the filtrate to the head of the sludge treatment process or discharging it to a waterway.

Filter press performance is measured by the solids content in the feed sludge, required chemical conditioning dosages, cake solids content, total cycle time, solids capture, and yield, in lb/sq ft/hr (kg/sq m/hr). These performance parameters are all interrelated; for example, as the feed solids content increases, the required chemical dosages and total cycle time usually decrease, while the filter yield, or throughput, usually increases. As the chemical conditioning dosage is increased up to the optimum level, the cake solids content, solids capture, and yield all increase, while the cycle time decreases.

Filter press testing at several Monroe County, New York water treatment plants and at the Erie County Water Authority's Sturgeon Point plant was conducted on alum sludges.[2] Filter cake concentrations of 40 to 50 percent solids were obtained in laboratory experiments and in a trailer-mounted pilot plant. Filtrate quality was suitable for its inclusion as raw water at the treatment plants. Lime requirements amounted to approximately 25 percent of the waste solids on a dry weight basis, and the precoat was approximately 2 percent of the waste solids. The pressure filtration cycle time ranged from 90 to 120 minutes.

Operating data from the alum sludge treatment system in Atlanta, Georgia indicate that filter presses are capable of producing a lime-conditioned filter cake with an average of 46 percent solids. The filtrate contains less than 10 mg/l of suspended solids.[29]

An AWWA committee on sludge disposal reported in 1978 that there were five pressure filtration systems at U.S. water treatment plants by 1976, and that at least four more were in various stages of construction.[1] They reported that alum sludge is usually gravity-thickened to about 2 to 6 percent solids (by weight) and then dewatered mechanically to 40 to 50 percent solids.

Lime Sludge Pelletization.   Sludge pelletization occurs during the suspended-bed cold-softening water treatment process, used primarily in the southeastern United States. The softening process seems to work best on high-calcium, warm-temperature groundwater.

The detention time in a suspended-bed softening reactor is approximately 8 to 10 minutes. The conical reactor vessel is constructed with sides approximately 80 degrees from horizontal (see Fig. 23–25), and the vessel is charged initially with a 0.20 to 0.25 mm effective size silica catalyst. The high-velocity, upward spiral flow of the raw water suspends the granular catalyst, which is essential for the efficient removal of hardness. Upward velocity is limited to about 3 ft/min (0.015 m/s) at the top of the cone to prevent carryover of catalyst particles.[4]

Lime is injected into the reactor while the raw water flow is gradually increased from a low initial rate to design capacity. The lime reacts with calcium bicarbonate and carbon dioxide to form calcium carbonate, which precipitates on the suspended particles.

Claims have been made that the size of the calcium carbonate–coated particles can reach 1.6 mm diameter; however, operating experience has shown that maximum sizes are in the 0.7 to 1.0 mm range.[4]

Theoretically, reactors should be capable of continuous operation. This requires a fine balance between the blowdown of sludge pellets and the addition of new, granular catalyst to maintain a constant-volume bed. In practice,

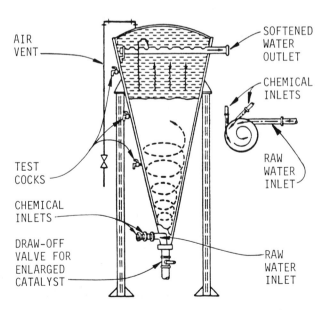

Fig. 23–25. Schematic diagram of a reactor for lime sludge pelletization. (Courtesy of Permutit Co.)

the balance is difficult to accomplish; so the reactors are generally operated in a batch mode. At Hollywood, Florida, 16 hours is required after reactor startup until stable operating conditions are reached.[4]

Treated water turbidity is used as a measure of treatment effectiveness. High turbidity, which can interfere with subsequent treatment processes, is the signal to terminate a reactor run. It occurs first at the beginning of a run while the process reaches equilibrium; when high turbidities are experienced again after about 40 days of operation, the reactor run is terminated. At the end of the run, the contents of the reactor, water, and sludge pellets are discharged into a storage and drainage facility. After drainage, the pellets can be treated as solids.

The pelletized sludge is approximately 60 percent solids by weight as it leaves the reactor. The entrained water can be readily drained, with the resulting product being 90 percent solids by weight. If the weight of the entrained catalyst is accounted for, each 1.0 cu ft (0.028 m³) of drained sludge contains approximately 105 pounds (48 kg) of calcium carbonate.[4]

A comparison of the volumes of conventional and pelletized sludge shows that the volume of pelletized sludge is 10 to 20 times less than the volume of undewatered conventional sludge. For a flow of 1 mgd (3.8 Ml/d) over a 40-day period, and a 100 mg/l reduction in hardness, 33 tons (29.9 metric tons) of calcium carbonate would be produced. The volume of a conventional sludge at 3 percent solids would be approximately 35,500 cu ft (1,000 m³), weighing approximately 1,140 tons (1,034 metric tons) before dewatering. The pelletized sludge would occupy 635 cu ft (18 m³) and weigh, before dewatering, less than 56 tons (51 metric tons). The pelletized sludge dewaters readily; so the drained weight would be reduced to about 36 tons (32.7 metric tons).[4]

The limitations on this approach are: magnesium content should be less than 85 mg/l as $CaCO_3$: turbidity should be less than 10; and, in cold climates, the reactors must be enclosed in heated structures. Excessive magnesium forms magnesium hydroxide, which does not plate out on the nuclei and will quickly clog downstream filters. Also, upflow rates of about 10 gpm/sq ft (24.4 m/h) are too high to permit removal of suspended solids, which will also pass on to downstream filters. However, the resulting economies in sludge handling may be great enough for the designer to consider adding the reactor ahead of a conventional clarifier.

The pelletized sludge particles can be disposed of in a landfill, but they may cause transportation problems due to their small, round size. Should a spill occur on a roadway, the water utility would be responsible for any accidents resulting from it.

Japanese Pellet Flocculation Process. The pellet flocculation process has been used successfully to treat alum sludge in Japan.[21] It involves multistage gravity thickening of the sludge, chemical treatment using sodium silicate and a polymer, and a dewatering process using a large horizontal rotating drum called a dehydrum.

This process is reported to be capable of producing a sludge with 25 to 30 percent solids without the need for mechanical vacuum or pressing equipment. It is also reported that the process appears to be most appropriate for treating large volumes of sludge.[58]

The Japanese have used the process in industrial applications for dewatering other types of solids and have only recently applied it to dewatering alum sludge. The sludge is sufficiently thickened to allow conveyor handling and truck transport to landfill disposal.[21]

RECOVERY OF COAGULANTS

The recycling and recovery of coagulants has long seemed a promising means of resolving the waste disposal problem of water treatment plant solids. Recalcination of spent lime is a proven technology at many locations, as discussed later in this section. The recovery and reuse of coagulants has been more elusive. Recent technology in alum recovery shows great promise, and research efforts are being made to recover and reuse spent iron coagulants.

Alum Recovery

Recovery By Acidification. Aluminum recovery from sludges produced in potable water coagulation plants has been studied by many researchers since the early 1950s.[8] The traditional scheme for alum recovery consists of thickening sludge from settling basins and filter backwashing, reducing the pH of the sludge by acid addition, and separation of the dissolved aluminum (in the form of aluminum sulfate) by decanting it from the residual solids. The recovery of alum by acidification with sulfuric acid is shown in the following equation:[54]

$$2Al(OH)_3 + 3H_2SO_4 \rightleftarrows Al_2(SO_4)_3 + 6H_2O \qquad (23-5)$$

As indicated by the above equation, about 1.9 g of sulfuric acid is required for each gram of sludge treated. This assumes, generally, that aluminum recovery in excess of 80 percent can be expected at a pH at or below 2.5.[59] Figure 23–26 presents a potential layout for an alum recovery system with direct acidification of alum sludge. Following acidification, the waste solids are separated from the recovered alum in a gravity settling step. The solids requiring ultimate disposal are significantly reduced by alum recovery, and the remaining solids can be more easily dewatered for ultimate disposal.

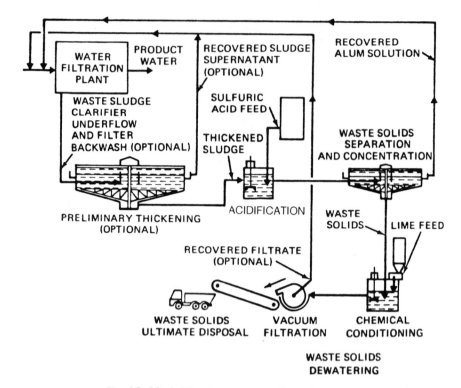

Fig. 23–26. Acidic alum recovery flow diagram.[54]

The acidic alum recovery process presents a potentially serious problem because it is vulnerable to the concentration of impurities in the recovered alum.[59] Such concentration in a recovered alum used for water treatment might cause a degradation of plant filtered water. The potential impurities include those capable of being converted to soluble form in the acidification process, such as iron, manganese, chromium, and other metals; a wide variety of organic materials; and impurities from the sulfuric acid.

For some time, the acidification of alum sludge was considered beneficial prior to pressure filtration at many water treatment plants in Japan. Fifteen Japanese water treatment plants, five in the Tokyo area, use an acid alum recovery process ahead of pressure filtration. All of these plants were built between 1965 and 1972. Concern over the possible recycling and concentrating of heavy metals has halted this procedure at sludge treatment facilities built since 1972.[59]

A 14-week study using a specially designed alum recovery pilot facility was performed at the Erie County, New York, Water Authority's Sturgeon Point water treatment plant in 1972.[55,60] This plant takes raw water from

Lake Erie and has a maximum capacity of 90 mgd (340 Ml/d). All waste alum sludge produced by the pilot plant was captured, concentrated, and reacted with sulfuric acid to produce recovered alum, which was used as the pilot plant primary coagulant. Small amounts of purchased liquid alum were used for makeup as needed. Thus the pilot plant was a closed system with respect to alum. The only system exhaust was through disposal of the residue from the alum recovery reaction. Periodic analyses of bacteriological, physical, and chemical characteristics were made on the raw water, the pilot plant filtered water, and the Sturgeon Point treatment plant filtered water. In addition to the usual determinations for water treatment plant control, tests were run for ten heavy metals, total microscopic count, and nine other chemical constituents.

This study reached the following conclusions:

1. There was no indication in the pilot plant filtered water of buildup of any of the constituents identified as indicators of contamination through 18 cycles of alum recovery and reuse as a primary coagulant.
2. There was a general indication that the use of recovered alum as a primary coagulant may increase filtered water turbidity by about O.1 NTU. This increase may have resulted from less efficient physical treatment capacity in the pilot plant as compared with that of the Sturgeon Point plant. It also is suspected that low aluminum concentration in the recovered alum may have affected coagulation.
3. There was a general indication that aluminum concentration was higher in the pilot plant filtered water than in the treatment plant filtered water.

The lack of contaminant buildup appeared to result from the relatively good raw water quality and the ability of the system to discharge organic and inorganic contaminants by way of the alum recovery process residue. The recovered alum was suitable for use as a primary raw water coagulant.

In the study, alum recovery combined with vacuum filtration of the residue was considered a feasible alternative, but it was not the most cost-effective process for use at the Sturgeon Point plant.[61]

Recovery By Liquid–Liquid Extraction. An alternative method of alum recovery has been investigated at Michigan State University with the financial and technological support of the AWWA Research Foundation.[8,59,62,63] Basically, the method would use organic solvents for the extraction, by liquid ion exchange, of high-purity concentrated alum from sludge. The technology for solvent extraction comes from the chemical process industries, where it has been used for over 20 years.[62]

In general, solvent extraction (more appropriately called liquid–liquid extraction) is the separation of the constituents of a liquid solution by contact with an immiscible liquid. The operation is dependent on the differential solubilities of the individual species in the two liquid phases. For water, wastewater, and industrial waste treatment applications, the solvent from which the extraction is made is water, in which the other solvent must be both insoluble and immiscible.

A special type of liquid–liquid extraction is termed liquid ion exchange because of its similarity to resin ion exchange. In liquid ion exchange a small quantity of an organic-soluble chemical called the extractant is dissolved in a second organic liquid called the diluent. The mixture is often referred to as the organic phase or the solvent. The diluent may be a material such as kerosene or a similar hydrocarbon. During the extraction operation, the extractant reacts chemically with the desired metal in the aqueous phase, forming a metal–extractant complex that is soluble in the diluent.[62]

The basic process for alum recovery by liquid ion exchange is shown in diagram form in Fig. 23–27.[59] Thickened aluminum hydroxide sludge is acidified to a pH of about 2 by concentrated sulfuric acid. Gravity settling in the sedimentation tanks is rapid for the acidified sludge. The liquid overflow from the sedimentation tank is dilute alum from the reaction of sulfuric acid and aluminum hydroxide in the sludge. It may contain contaminants not removed by gravity settling.

Liquid ion exchange takes place in the extraction circuit. For alum recovery,

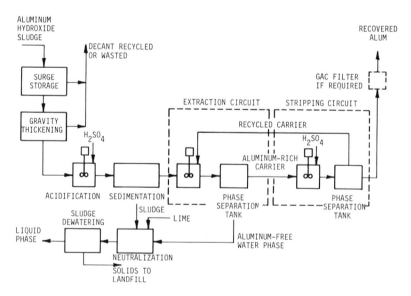

Fig. 23–27. Schematic flow diagram of basic process for alum recovery by liquid ion exchange.[59]

liquid ion exchange is used to separate and concentrate the aluminum from the acidulated sludge. A special hydrocarbon is used as the diluent. The solvent is insoluble and immiscible with the aqueous sludge solution and the ion exchange occurs at the interface of the two liquids. The system is operated by agitating the solvent and water mixture to form a dispersion of small water droplets in the organic acid. The aluminum exchanges for the positive hydrogen ion, the hydrogen ion enters the aqueous phase, and the positively charged aluminum ion moves to the carrier–water interface. The dispersed phases flow to a separation tank where rapid coalescence occurs, resulting in an aluminum-rich carrier phase and an aluminum-free aqueous phase.

The second half of the liquid ion exchange operation is the stripping circuit, where the aluminum is recovered from the carrier phase in a concentrated form, and the carrier is recycled. In general, the extraction circuit performs the major purification because of the selectivity of the solvent used, whereas most of the concentration is done in the stripping circuit.

The stripping circuit operates similarly to the extraction circuit, except that a stripping agent is chosen that draws the aluminum from the carrier phase. Sulfuric acid is the stripping agent selected. Aluminum in the carrier phase exchanges for protons in the acid phase, producing aluminum sulfate and regenerated solvent for recylcing to the extraction circuit. The stripping circuit is operated at an organic-to-acid volume ratio that concentrates the aluminum to the desired level.[59]

Based upon testing of the liquid ion exchange process for alum recovery on continuous-flow laboratory equipment, the following conclusions were reached:[59]

- The process will recover more than 90 percent of the aluminum from influent aluminum hydroxide sludge.
- The recovered alum is of the same quality and concentration as commercial liquid alum, or higher.
- The extractant is highly selective for aluminum over potential heavy metal contaminants (copper, cadmium, manganese, zinc, iron, and chromium).
- It may be feasible to operate a system at zero cost or with a net operating cost credit at plants using a high alum feed rate; that is, the operation and maintenance costs and the amortization of the capital cost may be offset by the value of the recovered alum. At plants using lower feed rates, the value of the alum recovered will help offset the annual operating costs.

A modified, more cost-effective process than that described above and shown in Fig. 23–27 was also tested on the laboratory scale, and is shown in Fig. 23–28.[59] The most significant modification is elimination of the initial acidification and settling step ahead of the extraction circuit. The residual solids are contained in the bleed stream from the separation tank of the extraction circuit. The modified process has a much lower solids flow (5 vs. 20 percent of influent), reducing the size of the neutralization and residue dewatering systems. It also conserves potential aluminum loss from the sludge for a higher net recovery. With the modified process, lime demand is reduced, and gypsum is not produced in neutralization of the aluminum-free water phase. (In the basic process, gypsum is produced, negating the reduced amount of residue solids produced in the aluminum recovery process.)

A demonstration plant for the testing of the modified liquid ion exchange alum recovery process was built at Tampa, Florida.[64] The City of Tampa's Hillsborough River water treatment plant is a 65 mgd (246 Ml/d) conventional coagulation plant that treats a highly colored raw water. At the time of this project, Tampa utilized alum coagulation with sodium silicate addition as a settling aid (polymer has now replaced the sodium silicate). The sludge from the sedimentation basins had about a 0.3 percent suspended solids concentration, which can be gravity-thickened to a suspended solids concentration of 1 to 1.5 percent. Between 60 and 90 percent of the suspended solids are acid-dissolvable. (See Reference 8 for details on the characteristics of this

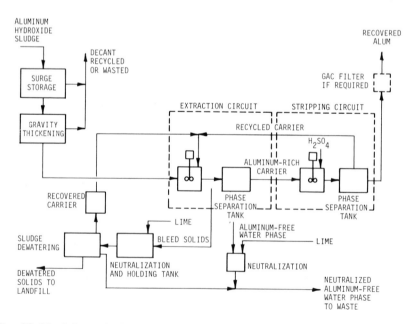

Fig. 23–28. Schematic diagram of modified process for alum recovery by liquid ion exchange.[59]

sludge.) The Tampa sludge was chosen for the demonstration plant for two reasons. First, the 22 tons/day (20 metric tons/d) of alum used at the plant made coagulant recovery an attractive treatment method. Second, because of the large amount of organic matter present in the sludge, it was thought that treating this sludge would be a demanding test of the liquid ion exchange process.

The demonstration plant was designed for a 10 gpm (0.63 l/s) sludge flow, having been scaled up from a laboratory pilot plant operating at 0.0026 gpm (10 ml/min). The major components of the system are sludge collection and feed, extraction, stripping, and residual sludge treatment.

Two equipment setups were tested: one providing a rapid mix of sludge and solvent followed by settling, and the other providing a slow mix. Both were able to recover more than 90 percent of the aluminum in the sludge. The recovered alum had the same characteristics as commercial alum. The sludge reduction accompanying aluminum recovery was as predicted by direct sludge acidification. The slow mix equipment setup, which made use of an 18-inch (45-cm)-diameter contactor for aluminum extraction, performed better than the rapid mix system. The tradeoff is in capital investment; the contactor requires a longer detention time and has a higher construction cost than the rapid mix system.

One conclusion made from the demonstration testing was that any size plant can break even on operating costs, independently of the alum dose. The economic key is to obtain an aluminum concentration greater than 1,200 mg/l in the sludge. Most sludges tested at Michigan State University could be thickened to 2,000 mg/l Al^{3+} either in the sedimentation basin or with external thickening.[64]

These costs do not include the cost of treating residual sludge, which should be reduced in dry weight and exhibit improved drainability and settling characteristics. The Tampa residual sludge showed a 50 percent reduction of dry weight solids and readily settled to a 10 percent solids concentration.[64] An overall economic analysis should compare the cost of treating the raw sludge to the cost of treating the residual sludge, plus or minus any costs of recovering the alum.

A 1 gpm (0.06 l/s) capacity alum recovery pilot plant was operated by the Niagara County, New York, Water District in 1980.[64] The pilot plant used a contactor and two countercurrent strippers. Data indicate an 80 to 90 percent aluminum recovery and a 35 percent dry weight reduction in sludge. The residual sludge readily thickened to a 10 to 15 percent solids concentration. Dewatering studies by a centrifuge manufacturer showed that the residual sludge could be dewatered to a 30 to 40 percent cake in a scroll centrifuge. Solvent losses after centrifugation of the floating solids were 0.9 gal/1,000 gal for a 2 percent suspended solids feed sludge to the recovery process. The most significant result for this water district has been that as a result of the improved thickening of the residual solids, the sludge volume was reduced by a factor of 50, following sedimentation of the residual sludge, compared to the process feed material. If this alum recovery process were operated, the currently estimated 3- to 5-year storage life of existing sludge lagoons could be greatly extended.

Iron Coagulant Recovery

The recovery of iron coagulants involves acidification of ferric hydroxide and a recovery technique very similar to that described for the acidic alum recovery process. The pH of the iron sludge is lowered by acid addition to a range where the solubility of ferric iron is significantly increased, and the iron is released back into solution. The principle of this method is shown in the simplified relationship:

$$2Fe(OH)_3 + 3H_2SO_4 \rightarrow Fe_2(SO_4)_3 + 6H_2O \qquad (23-6)$$

The pH must be reduced to 1.5 to 2.0 to attain 60 to 70 percent recoveries of iron.[9]

On a laboratory scale this method has been improved by adding a reducing agent to the sludge before acid addition to convert the form of the iron from the precipitated ferric ion (Fe^{+3}) to the more soluble ferrous ion (Fe^{+2}). Of four reducing agents tried, sodium sulfide (Na_2S) was the most effective. Recoveries of as much as 60 percent of the sludge iron were achieved at a pH of 3.0 when the reducing agent was used. The recovery processes also had a marked effect on the settling characteristics of the residual sludges. For both residual sludge samples the solids settled to about 20 percent of the original volume in a period of less than 30 minutes, and the sludge solids concentrations increased from the original 2 percent total solids before recovery to 7 to 9 percent total solids in the settled sludge. In addition, the weight of sludge solids requiring ultimate disposal was reduced.

The presence of a sulfide residual in the recovered iron solution adversely affected the coagulation performance of the recovered iron. The problem was remedied by adding potassium permanganate to the recovered solution before reuse. If this were done in a plant application, sulfate buildup might be a problem with repeated iron recycling. Another alternative for removing the sulfide would be to strip it as H_2S by aerating the recovered iron solution at the low recovery pH. This was proved effective in the laboratory although substantial foam developed.[9] If such aeration were practiced on a larger scale, it would be necessary to control the H_2S release in the off-gas. While these problems can be resolved, they do complicate the process.

Recalcination of Lime Softening Sludge

Lime recovery by recalcination has been widely used for years. Recalcination has the potential to recover substantially more lime than used in the original lime–soda water softening process, while at the same time producing carbon dioxide for use in recarbonation and greatly decreasing the volume of sludge requiring ultimate disposal. However, a review of recent literature has failed to show a distinct trend toward more widespread adoption of this sludge management alternative.[4]

Quicklime (CaO) can be produced from lime softening residues by recalcination following dewatering and drying. The basic step in the recalcination process is the burning of softening sludges at a temperature of 1,850°F (1,010°C). The reactions during recalcination are:

$$CaCO_3 \rightarrow CaO + CO_2 \qquad (23\text{--}7)$$
$$Mg(OH)_2 \rightarrow MgO + H_2O \qquad (23\text{--}8)$$

Since the lime–soda process produces approximately two parts of lime for every part of calcium carbonate applied, it is theoretically possible to recover up to twice as much lime as originally used. In practice, the yield is somewhat reduced by side reactions involving impurities and inefficiencies associated with preliminary dewatering, as well as capture of the recalcined lime.[4]

Among the factors that lead to decreased lime recovery, the effects of magnesium, silica, and typical surface water suspended solids must be considered. Of these materials, the magnesium content of the raw water is the most important. It is appropriate to recarbonate softening sludges to redissolve magnesium selectively. At Lansing, Michigan, recarbonation to a pH of approximately 9.0 lowered the magnesium content of the sludge cake from 3.5 to 1.8 percent as magnesium oxide, and greatly improved recalcination performance.[10]

The magnesium oxide recovered from recalcination will not slake but will pass through the softening process as the oxide. It has been noted that one part of silica will combine with six parts of calcium oxide to form an inert complex that will not slake.[13] A significant loss of lime can occur if high concentrations of silica are evident. In addition, when turbid surface water was softened by the lime—soda process, the clarification efficiency associated with alum treatment of the influent water significantly influenced the purity of the recovered lime. It was observed that a coagulation stage effluent turbidity of about 10 NTU or less was needed prior to softening to yield recovered lime with a minimum purity of about 86 percent available calcium oxide. This limit must be determined by the amount of calcium and impurities (Mg, Fe, SiO_2) removed in the softening sludge.[4]

Calcium carbonate is a high-density solid, whereas the hydroxides of metals such as magnesium, iron, and aluminum are more light and flocculent by nature. These three metal hydroxides are undesirable contaminants in a lime recalcination process.

Lime recalcining plants generally consist of the following components:[4,7]

- Sludge thickening from an initial 3 to 10 percent solids to 18 to 30 percent.
- Recarbonation using stack gases, 15 to 27 percent CO_2, to redissolve magnesium hydroxide selectively.
- Dewatering, usually by either centrifuges or vacuum filters, to 45 to 65 percent dry solids by weight.
- Flash dryers and cyclone separators using hot off-gases from the recalciner.
- Recalcining furnace.

Available furnace types include the rotary kiln, the flash calciner, the fluidized-bed calciner, and the multiple-hearth calciner. Features of some of the existing recalcining plants are summarized in Table 23–18.[7]

Recalcining yields a calcium oxide product of approximately 90 to 93 percent purity at a fuel rate of about 8.5 to 12 bil Btu/ton (8,967 to 12,660 bil J/907 kg).[10] High energy use apparently has contributed to the current lack of enthusiasm for the recalcination process, but it should be noted that calcination of limestone to produce lime for softening also is energy-intensive.

Magnesium Carbonate Coagulation and Recovery

A coagulant recovery technique for lime softening plants has also been developed.[54,65] It is based on a combination of water softening and conventional coagulation procedures that can be applied to all types of waters. Magnesium carbonate is used as the coagulant, with lime added to precipitate magnesium hydroxide as the active coagulant. The resulting sludge is composed of $CaCO_3$, $Mg(OH)_2$, and the turbidity removed from the raw water. The sludge is then recarbonated by injecting CO_2 gas, which selectively dissolves the $Mg(OH)_2$. The recarbonated sludge is filtered, with the magnesium being recovered as soluble magnesium bicarbonate in the filtrate. The magnesium bicarbonate coagulant is then recycled to the point of chemical addition in the raw water, where it is reprecipitated as $Mg(OH)_2$, and a new cycle is initiated. The filter cake produced in the separation step contains $CaCO_3$ and the turbidity removed from the raw water.

In large plants the process may be extended with a further reduction in waste solids. The filter cake ($CaCO_3$ plus turbidity particles) is slurried and processed in a flotation unit to separate the turbidity particles from the $CaCO_3$.

Table 23-18. Recalcining Plants.*

LOCATION	APPROXIMATE CONSTRUCTION DATE	RATED CAPACITY PRODUCT OUTPUT, TONS/DAY	TYPE OF DEWATERING EQUIPMENT[1]	TYPE OF RECALCINER[2]	HEAT REQUIREMENT, MIL BTU/TON OUTPUT	CO_2 RECOVERED	WATER PLANT RATED CAPACITY, mgd	LIME FED, LB/MG	SLUDGE/ LIME RATIO
Marshalltown, IA	—	4.2	C	FC	9.66	—	3.5	—	—
Pontiac, MI	—	—	C	FC	—	—	10	2,200	2.50
Miami, FL	1948	80	C	RK	8.5	Yes	180	1,800	2.20
Lansing, MI	1954	30	C	FB	8.0	—	20	2,200	2.27
Salina, KS	1957	—	C	FC	—	—	—	—	—
Dayton, OH	1960	150	C	RK	9.0	Yes	96	2,140	2.47
San Diego, CA	1961	25	C	RK	9.0	—	—	—	—
S. D. Warren Company[3]	1963	70	VF	FB	7.2	—	—	—	—
Merida, Yucatan	1965	40	C	RK	—	Yes	24	—	—
Ann Arbor, MI	1968	24	C	FB	7.0	—	—	—	—
Lake Tahoe, NV[4]	1968	10.8	C	MH	7.0	Yes	7.5	—	—
St. Paul, MN	1969	50	C	FB	8.2	Yes	120[5]	990	2.40

* Reference 7. Reproduced by permission of McGraw-Hill Book Co.
[1] C = centrifuge; VF = vacuum filter.
[2] RK = rotary kiln; FB = fluidized bed; FC = flash calcination; MH = multiple hearth.
[3] Paper mill owned by Scott Paper Company.
[4] Tertiary lime treatment at wastewater treatment plant. Lime treatment system to be discontinued in 1986.
[5] Lime recalcining plant designed to accommodate future water treatment plant capacity of 170 mgd.

The purified $CaCO_3$ can then be dewatered and recalcined to high-quality quicklime for reuse in the process. The stack gases from the recalcining operation provide the CO_2 gas for recarbonating the sludge in the magnesium recovery process and for recarbonating the water in the water treatment process. In large plants utilizing this lime and CO_2 recovery process, the only waste solids to be disposed of are those that were introduced to the plant as raw water turbidity.

Many waters contain enough magnesium that 100 percent magnesium recovery is achievable. With soft waters, magnesium recovery may be limited to about 80 percent. (This process is discussed in detail in Chapter 21.)

ULTIMATE SOLIDS DISPOSAL

Surveys of Disposal Methods

Water treatment plant sludges historically have been discharged either directly or indirectly into a surface water. In 1953, 92 percent of 1,600 coagulation and softening plants surveyed disposed of their sludges in streams or lakes.[66] A 1969 survey of 80 primarily large plants showed that the disposal of water treatment plant wastes in surface waters had decreased to 39 percent for softening plants and 49 percent for coagulation plants.[66] Filter backwash waters were discharged to streams and lakes by 83 percent of the plants responding in 1953 and 49 percent in 1969.

Results from a 1979 survey of 75 alum coagulation plants and a 1981 survey of 100 softening plants are shown in Table 23–19.[4,8] The percentage

Table 23–19. Methods for Disposal of Water Treatment Plant Waste.

	PERCENT OF PLANTS USING INDICATED DISPOSAL METHOD	
	SOFTENING* SLUDGE	COAGULATION** SLUDGE
Sludge lagoon	34†	43
Sanitary sewer	8	27
River or lake	13	20
Recalcination	5	—
Direct land application	5	—
Other	—	10

* Reference 4.
** Reference 8.
† Fifty-six percent of plants surveyed had sludge lagoons, 60 percent of which were considered "permanent lagoons"; thus 34 percent of plants used sludge lagoons for disposal.

of softening plants discharging sludge to rivers or lakes had decreased to 13 percent, while 20 percent of alum coagulation plants still practiced this method of sludge disposal. Substantial numbers of water treatment plants continue to discharge sludge to surface waters; however, it is obvious that this practice is steadily being restricted with the increased emphasis by regulatory agencies on controlling wastewater discharges.

From the information in Table 23–19, it can also be seen that mechanical dewatering is not commonly practiced at existing water treatment plants. The lime recalcination plants (5 percent of the softening plants surveyed) have mechanical dewatering, and the "other" category for coagulation plants (10 percent of plants surveyed) may include mechanical dewatering and landfill disposal.

Disposal Options

There are eight basic sludge disposal options that can be used by water treatment plants:

1. Discharge to waterway.
2. Discharge to sanitary sewers.
3. Codisposal with sewage sludge at a wastewater treatment plant.
4. Lagooning with and without natural freezing, requiring ultimate disposal of the residue.
5. Mechanical dewatering with landfilling of residue.
6. Coagulant recovery.
7. Land application, especially of softening sludge.
8. Use for building or fill materials.

Discharge to Waterway. The oldest disposal method is to discharge sludges to the nearest available waterway: stream, pond, lake, or ocean. This method, although still widely used, is being discontinued under the pressure of state regulatory agencies and federal laws. Although most of the material in water treatment plant sludges originates from the raw water, chemical coagulant solids are usually present as well. These sludges are potential pollutants because they may produce undesirable sludge deposits in the receiving water that are aesthetically objectionable or may form deposits harmful to bottom organisms. Generally, the settleability of the material removed from the water is increased in the water treatment plant so that when it is returned to the same watercourse, it is more likely to form a sludge deposit than it originally was.

A survey of softening plants conducted in 1981 found that only 13 percent of the responding plants still used direct discharge to a river as their disposal method, and that the management of half of those plants intended to implement a sludge treatment method in the near future.[4]

Discharge to Sanitary Sewers. The practice of disposal of water treatment plant solids to sanitary sewers increased substantially between 1953 and 1968; a 1953 survey showed that 0.3 percent of the plants surveyed discharged solids to sanitary sewers, while a 1968 survey showed that 8.3 percent of plants did so.[67,68] A 1979 survey of alum users found that 27 percent of these plants discharged sludge to the sanitary sewer,[8] but a 1981 survey of softening plants found that only 8 percent of these plants discharged softening sludges to sanitary sewers.[4] Therefore, this practice appears to have increased for coagulation plants, but not for softening plants, perhaps because of the greater solids quantity in softening plants, which would be a larger load on the wastewater treatment plant.

This technique of sludge disposal transfers the solids handling problem from the water treatment plant to the waste treatment plant. However, inclusion of the necessary capabilities in the solids handling facilities of the waste treatment facility may result in an overall cost savings by consolidating the equipment and reducing the number of personnel required for total solids handling. Many wastewater utilities are concerned that the water treatment plant solids will adversely affect their treatment process. However, these same chemicals are used extensively in waste treatment to remove phosphorus, and no adverse effects result. A number of factors must be evaluated if this approach receives serious consideration for a given application.

A major consideration is the ability of the sewage collection system and wastewater treatment plant to accept the increased hydraulic and solids load imposed by the addition of the water treatment plant wastes. The direct discharge of filter backwash flows into the sewer system, for example, could cause a hydraulic overload of the collection system, or a hydraulic surge large enough to cause the wastewater treatment plant clarifier performance to deteriorate. Surge storage at the water plant, with gradual release, may be needed if the volume of water plant waste is large in proportion to the sewage flows. Release during low-sewage-flow periods (midnight to 6:00 A.M.) may be desirable. Another aspect to consider is that the sewer receiving the water plant sludges must be of adequate capacity and should provide velocities adequate to prevent deposition of the sludge in the sewer. Studies at Detroit report that a velocity of 2.5 ft/sec (0.76 m/s) is adequate to prevent settling of the sludges in the sewer.[32]

The bulk of the solids from the water plant sludges will be removed in the primary clarifier. Obviously, the solids handling system at the wastewater treatment plant must be capable of handling the additional solids load. It will probably be a rare but fortunate circumstance for an existing waste treatment plant solids handling system to be able to handle the unplanned addition of water plant sludges if the water plant and wastewater plant are of comparable size. For example, disposal of large amounts of gelatinous hydroxide floc in an anaerobic digestion–sand drying bed system may cause difficulties in obtaining proper solids dewatering. A careful evaluation of the dewatering characteristics of the combined water and sewage sludges should be made before this approach is adopted. The large quantities of inert, dense materials found in softening sludge may lead to filling of poorly mixed anaerobic digesters at a wastewater treatment plant.

In 1949, waste lime sludge was disposed of via sanitary sewers to the Daytona Beach, Florida wastewater treatment plant.[69] The addition of 200 to 300 mg/l of waste lime sludge provided 45 percent BOD removal and 75 percent suspended solids removal in the primary clarifier, which are somewhat higher efficiencies than typical. The resulting sludge was dewatered on a vacuum filter. Dallas, Texas, in an attempt to dispose of waste lime sludge and to use it effectively for wastewater treatment, employed a similar process at its wastewater treatment facility.[70]

Culp and Wilson studied the effect of adding alum sludge to an activated sludge wastewater treatment facility and reported no significant benefit or detriment to the treatment process or the anaerobic digester.[71] There was an increase in sludge handling quantities in proportion to the increased water treatment solids.

Codisposal. Codisposal of lime sludges could also include a situation where lime sludge is of some value in disposing of another waste. For example, it would be very attractive to use a lime waste to treat another waste. Using high-pH liquid or semisolid waste for neutralizing an acidic waste is very desirable, but the major opportunity for use is more likely to be in the industrial field. At one water treatment plant in Europe, lime softening is practiced in conjunction with hydrogen-ion exchange softening; the lime-softening waste is used to neutralize the acidic rinse water from regeneration of the resin. Such imaginative design is rarely possible, but waste-lime sludge probably could be used in codisposal schemes to alter the characteristics of other wastes. Lime sludge could be used for many reasons:[4]

- Elevation of pH.
- Bulking agent.
- Neutralization of acid wastes to bring them within National Pollutant Discharge Elimination System (NPDES) permit limits.
- Assistance in pretreatment of industrial wastes.
- Incineration to produce high alkaline ash.

Lagoons. A detailed description of lagoon dewatering appears earlier in this chapter. Mechanical dewatering is expensive, especially for small plants with water treatment capacities less than 50 mgd (189 Ml/d). The most viable alternative for small plants may be lagooning, which is also an option for larger plants where large tracts of inexpensive land can be obtained.

However, in many instances storage of dilute or concentrated water treatment plant solids in lagoons is considered the ultimate disposal. In effect, this is a postponement of the inevitable ultimate disposal requirement.

Mechanical Dewatering/Landfill. Detailed descriptions of various mechanical dewatering processes were given earlier in this chapter. Sanitary landfills are also used for disposal of solids dewatered in lagoons or on drying beds. Disposal of water treatment plant solids to landfills requires concentration of the solids to a semisolid or cake form. The problems that occur in landfills are related to the semisolid nature of the sludge discharged. The landfilling operation must be controlled, with adequate provision against pollution from runoff contamination or from the leachate.

The disposal of coagulation sludge in a sanitary landfill has become a point of increasing concern among water utilities and state regulatory agencies because landfills are anaerobic systems operating in the acid fermentation phase to produce leachate in the acidic pH range (between 5.5 and 7.0).[1] The leachate is somewhat buffered and may redissolve some of the heavy metals contained in the sludge. In 1975, a lysimeter study for the Monroe County Water Authority in New York State evaluated the characteristics of leachate resulting from landfill disposal of dewatered alum sludge.[1] The aluminum concentration in the leachate was low enough that even minimal dilution with groundwater would reduce it to acceptable ranges.

The data obtained from the lysimeter study indicate that landfilling of dewatered alum sludge is feasible in a special landfill where only dewatered sludge is deposited, or at a sanitary landfill site in combination with other waste.[1]

Coagulant Recovery. Even with aluminum or iron coagulant recovery there is a remaining solids residue to dispose of. Dry solids remaining may amount to 50 to 65 percent of the original solids. Lime recalcination also results in some chemical sludge to dispose of, although the quantity of remaining solids may be up to 20 times less than the original quantity of lime sludge.

Land Application. Land application of softening sludge, the mixing of a waste material into the natural environment, is not a new method of disposal. Farmers were allowed to remove dewatered softening sludge from a plant in Ohio approximately 30 to 40 years ago.[72] Today, land application is the potential use of a resource discarded at great expense, and it should be an economical and beneficial solution to waste disposal problems for many softening plants.

The solids content of softening sludge discharged from clarifiers is 1 to 5 percent. If land application of softening sludge is employed, this sludge should be thickened and applied for soil conditioning as a liquid at 8 to 10 percent solids or as a solid after dewatering to about 40 percent solids.[20] If the solids content of the sludge is between these values, and if conventional farming equipment is used, handling problems will be encountered.

In farming regions, the application of nitrogen fertilizers causes a reduction in soil pH. If optimum pH conditions do not exist, crop yields will be reduced. Therefore, farmers must apply sufficient quantities of calcium carbonate as a means of counteracting the fertilizer applications. For each 100 pounds (45.4 kg) of ammonia fertilizer, 3 to 4 pounds (1.4 to 1.8 kg) of limestone must be applied.[73]

Several state agencies have evaluated the neutralizing power of softening sludges versus commercial limestone. In 1969, the Ohio Department of Health reported that the total neutralizing power (TNP) of lime sludge is greater than that of marketed liming materials.[23] To bring the soil pH into the desirable range, 3 tons/acre (0.67 kg/m^2) lime, or about 10 tons (9.07 metric tons) of lime sludge at a 30 percent solids concentration, are required. Subsequent lime applications are required to maintain the desired pH.[4]

In Illinois, a calcium carbonate equivalent test (CEE) performed on several softening sludges indicated that the softening sludges were superior to agricultural limestones available locally.[27] Because softening sludges contain a high quantity of calcium carbonates and offer a high degree of neutralization, this resource should be used when it is practical for soil conditioning. The addition of softening sludge also increases the porosity of tight soils, making them more workable for agricultural purposes.[73]

Scambilis has found that both alum and lime sludge increase the cohesiveness of soils, but, more important, both sludges also widen the moisture content over which soil remains cohesive.[74] Alum sludge increases soil cohesiveness at high moisture contents, while lime sludge increases soil cohesiveness at lower moisture contents. The effects of 10 percent additions of alum and lime sludge on soil cohesiveness are shown in Fig. 23–29.

Use of lime and alum sludge on agricultural land has had limited success because farmers are unfamiliar with its use as a source of lime and with the logistics of its transportation and applications to their lands. In addition, the availability of dewatered sludge needs to be scheduled to coincide with farmers' demands. Because of the neutralizing power of dewatered lime sludge, farmers may be willing to pay for it, depending on demands and costs. At a minimum, agricultural use of dewatered sludge would eliminate the need to pay for its disposal.

A demonstration program was recently initiated by the Mahoning Valley Sanitary District, which operates a 60-mgd (227-Ml/d) softening plant in Ohio. Softening sludge is being spread on 25 acres (10 ha) of farmland at a concentration of 6 to 7 percent solids.[4] The slurry is being applied at a rate corresponding to 3.5 to 4 tons (3.2 to 3.6 metric tons) of dry softening

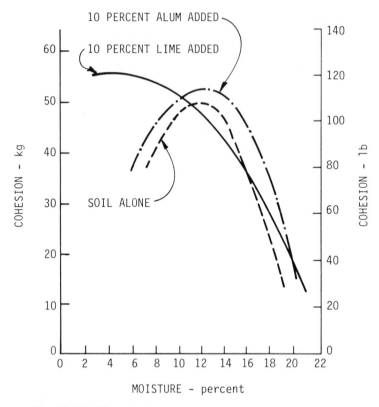

Fig. 23–29. Effect of alum and lime sludges on soil cohesion.[74]

solids per 1.0 acre (0.4 ha) of farmland. The calcium carbonate content and magnesium hydroxide content of this material are about 93 percent and 4 percent, respectively. The trace metals in the slurry do not appear to have any adverse effects on the crops.[72]

In Illinois, an 28.8-acre (11.5-ha) farmsite was selected for spreading a 50 percent solids softening sludge with a modified fertilizer spreader.[73] Although handling and spreading problems were encountered and the soil pH was increased to above the optimum, the farmer was pleased with the crop yield. Later experiments showed that the handling and spreading problems could be resolved by mixing the thixotropic sludge to a fluid consistency without adding water. Wastewater sludge disposal equipment also can be utilized for spreading dry softening sludges.[75] The water utility ultimately signed an agreement where an equipment manufacturer assumed responsibility for marketing, handling, and spreading the softening sludge on farmland. When costs to the farmers increased, demand for this material decreased.

Softening sludges also aid in the reclamation of unusable lands by neutralizing acid soils. Disposal of softening sludge on strip-mine land will minimize and possibly eliminate the discharge of acid compounds. The Ohio Environmental Protection Agency allows the disposal of dewatered softening sludge in abandoned strip-mines as long as runoff quality meets stream standards.[23] More specifically, it must be demonstrated that the disposal of this material will not adversely affect water quality, water flow, or vegetation; cause hazards to public health and safety; or cause instability to the backfill area. The quantity of softening sludge that can be applied to strip-mine land generally is 10 tons (9,070 kg) of dry solids per 1.0 acre (0.4 ha).

The costs involved with hauling and spreading this material over strip-mine land are substantial. However, if higher-solids-content sludge could be supplied, hauling costs could be reduced.

Use for Building or Fill Material. Alum sludge has been suggested for use as a plasticizer in the ceramics industry, as a constituent of refractory bricks, and as a road-stabilizing agent.[66] In Atlanta, dewatered alum sludge from the Hemphill alum sludge treatment facility is transported to a residential building site where it is used as fill.[1] Sludge cake is spread and compacted by a bulldozer to depths as great as 6 feet (1.8 m). No problems have been reported with driving loaded trucks over the compacted sludge cake.

REFERENCES

1. AWWA Committee on Sludge Disposal, "Water Treatment Plant Sludges—An Update of the State of the Art," *J.AWWA,* pp. 498–503 and 543–554, September and October, 1978.
2. Gruninger, Robert M., "Disposal of Waste Alum Sludge from Water Treatment Plants," *J.WPCF* 47:3:543–556, 1975.
3. Chapman, Robert L., *Solids Disposal Study Summary Report,* Denver Board of Water Commissioners Foothill Treatment Plant, June, 1974.
4. AWWA Committee on Sludge Disposal, "Lime Softening Sludge Treatment and Disposal," *J.AWWA,* pp. 600–608, November, 1981.
5. Cornwell, D. A., and Westerhoff, G. P., "Management of Water Plant Sludge," *Sludge Management,* Ann Arbor Science Publishers, Ann Arbor, Mich., 1981.
6. AWWA Water Treatment Plant Waste Committee Report, unpublished data, December, 1972.
7. AWWA, *Water Quality and Treatment,* 3rd ed., McGraw-Hill Book Company, New York, 1971.
8. Cornwell, David A., and Susan, James A., "Characteristics of Acid-Treated Alum Sludges," *J.AWWA,* pp. 604–608, October, 1979.
9. Pigeon, Paul E., et al., "Recovery and Reuse of Iron Coagulants in Water Treatment," *J.AWWA,* pp. 397–403, July, 1978.
10. Singer, P. C., "Softener Sludge Disposal—What's Best," *Water and Waste Engineering,* p. 25, December, 1974.
11. Black and Veatch, *Report on Water Treatment Plant Waste Disposal,* Wichita, Kans., December, 1969.
12. Laurence, Charles, "Live Soda Sludge Recirculation Experiments at Vandenberg Air Force Base," *J.AWWA* 55:2:177, 1963.

13. Burris, Michael A., et al., "Softening and Coagulation Sludge—Disposal Studies for a Surface Water Supply," *J.A WWA*, 68(s):247, May, 1976.
14. Calkins, D. C., and Novak, J. T., "Characterization of Chemical Sludges," *J.A WWA* 65:6:423, 1973.
15. *Recovery of Lime and Magnesium in Potable Water Treatment*, USEPA Technol. Series, EPA-600/2–76–285, December, 1976.
16. O'Connor, J. T., and Novak, J. T., "Management of Water Treatment Plant Residues," AWWA Water Treatment Waste Disposal Seminar *Proceedings*, AWWA Annual Conference, 1978.
17. Novak, J. T., and Calkins, D. C., "Sludge Dewatering and Its Physical Properties," *J.A WWA* 67:1:42, 1975.
18. Shimek, Jacobs, Finklea & CH2M Hill, Pilot Plant Studies at the Elm Fork Treatment Plant, unpublished report, Dallas Water Utilities, Dallas, Texas, April, 1980.
19. Judkins, J. F., Jr., and Wynne, R. H., Jr., "Crystal-Seed Conditioning of Lime-Softening Sludge," *J.A WWA* 64:5:306, 1972.
20. Draft Document Guidelines for the Water Supply Industry, Southern Research Institute, March, 1975.
21. Bishop, Stephen L., "Alternate Processes for Treatment of Water Plant Wastes," *J.A WWA*, pp. 503–506, September, 1978.
22. James, Carol Ruth, and O'Melia, Charles R., "Considering Sludge Production in the Selection of Coagulants," *J.A WWA*, pp. 148–151, March, 1972.
23. Ohio Department of Health, *Supplement to Report on Waste Sludge and Filter Washwater Disposal from Water Softening Plants*, September, 1969.
24. Black, A. P., "Split-Treatment Water Softening at Dayton," *J.A WWA* 58:1:97, 1966.
25. Shuey, B. S., "Economics of Split-Treatment Water Softening," *J.A WWA* 58:1:107, 1966.
26. Singley, J. E., and Brodeur, T. P., "Control of Precipitative Softening," AWWA Water Quality and Technology Conference, 1980.
27. Malcolm Pirnie, Inc., Evaluation of Water Treatment Facilities, unpublished Report for the Mahoning Valley Sanitary District, Youngstown, Ohio, May, 1980.
28. Knocke, W. R., "Thickening and Conditioning of Chemical Sludges," *Proceedings* of ASCE Environmental Engineering Conference, San Francisco, Calif., July, 1979.
29. Foster, W. S., "Get the Water Out of Alum Sludge," *American City and County*, September, 1975.
30. Howson, Louis R., Chapter 10, "Sludge Disposal," in *Water Treatment Plant Design*, Robert L. Sanks, editor, 1978, pp. 183–193, Ann Arbor Science Publishers, Ann Arbor, Mich.
31. Neubauer, W. K., "Waste Alum Sludge Treatment," *J.A WWA* 60:819, July, 1968.
32. AWWA Research Foundation Report, "Disposal of Wastes from Water Treatment Plants," *J.A WWA* 61:541, October, 1969.
33. Kansas Department of Health and Environment, Design Criteria—Sludge Production and Storage Requirements, Division of Environment, Bureau of Water Supply, Topeka, Kans., 1983.
34. Novak, J. T., and Montgomery, G. E., "Chemical Sludge Dewatering on Sand Beds," *J.ASCE—Environmental Engineering Division* 101:1, February, 1975.
35. King, P. H., et al., "Treatment of Waste Sludges from Water Purification Plants," Bulletin 52, Virginia Water Resources Research Center, VPI, Blacksburg, Va., 1972.
36. Novak, J. T., and Langford, Mark, "The Use of Polymers for Improving Chemical Sludge Dewatering on Sand Beds," *J.A WWA* 69:106, February, 1977.
37. Wilhelm, J. H., and Silverblatt, C. E., "Freeze Treatment of Alum Sludge," *J.A WWA* 68:312, June, 1976.
38. Parsons, Jack, Project Manager, Denver Board of Water Commissioners, Denver, Colo., personal communication, June, 1983.

39. Fulton, G. P., "Disposal of Wastewater From Water Filtration Plants," *J.AWWA* 61:322, July, 1969.
40. "Water Plant Waste Treatment," *American City and County,* March, 1979.
41. Freedman, Michael B., "Centrifuge Ends Drying Bed Chores for Pennsylvania Water Plant," *American City and County,* November, 1977.
42. Nielson, Hubert L., "Alum Sludge Disposal—Problems and Success," *J.AWWA* 69:335–341, June, 1977.
43. Albertson, O. E., and Guidi, E. E., Jr., "Centrifugation of Waste Sludges," *Water Pollution Control Federation* 41:4, 607, 1969.
44. AWWA Research Foundation Report, "Disposal of Wastes from Water Treatment Plants: Part 2," *J.AWWA* 61:11:619, 1969.
45. Kramer, A. J. and Whitaker, J., "Sludge Handling," *Water and Waste Engineering,* May, 1975.
46. Vesilind, P. A., *Treatment and Disposal of Wastewater Sludges,* Ann Arbor Science Publishers, Ann Arbor, Mich., 1979.
47. Rohen, M., "Characterization of Lime Sludges from Water Reclamation Plants," D.Sc. dissertation, University of Pretoria, South Africa, 1978.
48. Nielsen, H. L., Carns, K. E., and DeBoice, J. N., "Alum Thickening and Disposal. Processing Water-Treatment-Plant Sludge," *AWWA Manual,* 20108, Denver, Colo., 1974.
49. Finney, J. Wiley, Jr., "Dewatering of Alum Sludge at Somerset, Kentucky," presented at Kentucky–Tennessee AWWA annual meeting, Owenshoro, Ky., November 11–14, 1979.
50. Parkson Corporation, "Summary of Test Results on Magnum Press," Western Pennsylvania Water Company, New Castle, Pa., March 2–6, 1981.
51. Tait-Andritz, "Sludge Dewatering Test Report," Houston, Tex., March 23 to April 15, 1976.
52. Michael G. Schill, Business Manager, Parkson Corporation, Fort Lauderdale, Fla., personal communication, March, 1984.
53. Hambor, J. Michael, "Dewatering of Water Treatment Sludges—The Belt Filter Press," presented at the AIChE Joint Meeting—Central and Peninsular Sections, Clearwater Beach, Fla., May 28, 1983.
54. Culp, Gordon L., and Culp, Russell L., *New Concepts in Water Purification,* Van Nostrand Reinhold, New York, 1974.
55. Westerhoff, G. P. and Daly, M. P., "Water-Treatment-Plant Wastes Disposal—Part 2," *J.AWWA* 66:6:378, June, 1974.
56. *Operating and Plant Data,* Ingersoll Rand, Nashua, N.H., 1970–1977.
57. Report on Pressure Filter Filtrate Disposal at the Central Alum Sludge Processing Facilities, Rochester, N.Y., unpublished report, Malcolm Pirnie, Inc., September, 1976.
58. Ide, T., and Yusa, M., "Solid–Liquid Separation by Pelletting Flocculation Process," 2nd Pacific Chemical Engineering Congress, August, 1977.
59. Westerhoff, Garrett P., and Cornwell, David A., "A New Approach to Alum Recovery," *J.AWWA,* pp. 709–714, December, 1978.
60. Westerhoff, Garrett P., "Alum Recycling: An Idea Whose Time Has Come?," *Water and Waste Engineering,* December, 1973.
61. Westerhoff, Garrett P., and Daly, Martin P., "Water Treatment Plant Wastes Disposal—Part 3," *J.AWWA* 66:7:441, 1974.
62. Cornwell, David A., "An Overview of Liquid Ion Exchange With Emphasis on Alum Recovery," *J.AWWA* 71:12:741–744, 1979.
63. Cornwell, David A., and Lemunyon, Roger M., "Feasibility Studies on Liquid Ion Exchange for Alum Recovery From Water Treatment Plant Sludges," *J.AWWA* 72:1:64–68, 1980.
64. Cornwell, David A., et al., "Demonstration Testing of Alum Recovery by Liquid Ion Exchange," *J.AWWA,* pp. 326–332, June 1981.

65. Black, A. P., and Thompson, C. G., "Plant Scale Studies of the Magnesium Carbonate Water Treatment Process," EPA Publication EPA-660/2-75-006, May, 1975.
66. AWWA Committee Report, "Disposal of Water Treatment Plant Wastes," *J.AWWA*, pp. 814–820, December, 1972.
67. Dean, R. B., "Disposal of Wastes from Filter Plants and Coagulation Basins," *J.AWWA* 45:1226, November, 1953.
68. Krasauskas, J. W., "Review of Sludge Disposal Practices," *J.AWWA* 61:225, May, 1969.
69. Williamson, J., Jr., "Something New in Sewage Treatment," *Water and Sewage Works* 96:159, 1949.
70. Benjes, H. H., Jr., "Treatment of Overflow from Sanitary Sewers," presented at the 9th Texas WPCA Conference, Houston, July, 1970.
71. Culp, R. L., and Wilson, W. I., "Is Alum Sludge Advantageous in Wastewater Treatment?," *Water and Wastes Engineering,* p. 16, July, 1979.
72. Reeves, T., "The Mahoning Valley Sanitary District Plans to Spread Slurry on Farmland," *Farm and Dairy,* Salem, Ohio, p. 10, November 13, 1980.
73. Russell, G. A. "Agricultural Application of Lime Softening Residue," Ill. AWWA Section Meeting, March, 1980, and Mo. AWWA Section Meeting, April, 1980.
74. Scambilis, N., "Effect of Chemical Sludge on Soils," unpublished Ph.D. thesis, University of Missouri, Columbia, 1977.
75. Russell, G. A. "From Lagooning to Farmland Application: The Next Step in Lime Sludge Disposal," *J.AWWA* 67:10:585, 1975.

Chapter 24
Upgrading Existing Water Treatment
Plants

INTRODUCTION

Many existing water treatment plants are faced with concurrent problems of increased water demand, increased regulatory requirements, decreased water quality, and increased public concern over finished water quality. In 1984, over 80 percent of the U.S. population was served by approximately 20,000 water treatment plants, many of which were constructed prior to World War II. Unless these aging facilities are upgraded, many cannot be expected to supply ever increasing quantities of treated water meeting more stringent quality standards.

Experience has shown that many existing water treatment plants can be upgraded to provide significantly increased output by application of the newly developed treatment techniques described elsewhere in this book. In many cases these techniques can be applied to improve performance while also increasing capacity. If the existing structures are sound, and if historically accepted design parameters were applied to the original installation, plant output often can be increased by as much as 100 to 200 percent without major plant structural additions, and at minimal cost. The increased capacity often can be achieved at costs substantially less than would be associated with construction of new facilities. This chapter describes full-scale application of techniques described earlier in this book to upgrade existing water treatment plants.

EVALUATING A PLANT'S POTENTIAL FOR UPGRADING

The first step in determining whether the capacity of an existing plant can be increased is a thorough investigation of the physical and mechanical condition of the facilities. Some of this information can be obtained from construction drawings, manufacturers' equipment drawings, plant operating records, and any reports or studies that have been made on the facilities.

A second step is to perform a field survey to obtain any information not available in plant documents. This information is vital to establish whether the plant is hydraulically capable of handling a greater throughput. Existing facilities often have hydraulic restrictions that are extremely difficult and costly to modify. These limitations must be identified, as they may limit the capacity of the renovated facility.

During the field survey, dimensions and other information shown on as-built drawings should be verified. Measurements of all basins should be obtained, and care should be taken to note high water marks in each basin as water levels fluctuate, especially in older plants. The elevations of all basin walls, weirs, launders, pipe intakes and discharges, high and low water marks, channel walls, sluice gate openings, and other important elevations should be obtained by running a level circuit around the plant. This information is necessary for making plant hydraulic computations.

The physical condition of the facilities and treatment units should be noted. Also, the capacity of the existing equipment should be established. This procedure should begin at the intake structure and include all elements in the treatment plant. This inspection should be done with experienced operating personnel who can point out trouble spots in the plant and relate their experiences with process and equipment deficiencies. During the inspection, manufacturers' names and model numbers and other identifying numbers on all mechanical equipment should be noted. With this information, the equipment manufacturers can be contacted for recommendations for extending the service life expectancy of motors, gear boxes, pumps, valves, rate-of-flow controllers, and other mechanical items. This is important because replacement parts may not be available for equipment that is more than 20 to 30 years old. Improvements to the plant may require replacement of aging equipment.

All components of the treatment process should be analyzed and evaluated on the basis of treatment efficiency, maximum and minimum hydraulic capacity, sustainable average capacity, operating deficiencies, and serviceability. In addition, increasing the capacity of the existing facilities may be subject to limitation by allowable design criteria established by regulatory agencies. For example, a regulatory agency may require a minimum settling basin detention time of 4 hours. If the detention time at existing flows is only slightly less than 4 hours and no departure from this standard is permitted, the plant cannot be expanded significantly without construction of a new settling basin. Arbitrary design standards present the greatest barrier to expanding an existing water treatment plant by the use of new technology.

The information obtained from the field survey and review of construction and manufacturers' drawings, plant operating records, and discussions with plant personnel must be considered in developing a recommendation for plant upgrading. All recommendations for increasing plant capacity should be based on life-cycle cost analyses to support the recommendations for plant modifications. Also, the modifications developed during the evaluation of the facilities should be assigned a priority level based upon contribution to plant improvements at minimum costs.

PROCESS MODIFICATIONS

Rapid Mixing

The rapid mixing process usually requires improvement if it is to perform adequately at higher plant throughputs. Effective coagulation requires instantaneous mixing of the coagulant with the incoming water. If the coagulant does not come in contact with all colloidal turbidity particles, the uncoagulated turbidity will pass through the filters.

Improvements to rapid mixing facilities usually involve adjustments to the means and point of application of the coagulated chemical. These improvements can be accomplished at little capital cost, but can reduce chemical dosages and related operating costs substantially and improve finished water quality. Tests performed by the Metropolitan Water District of Southern California (MWD) have shown that filter effluent turbidities could be reduced by as much as 50 percent with the same coagulant dosage by properly diffusing the coagulant into the flowstream.[1] Figure 24–1 illustrates a coagulant diffuser used successfully by MWD at one of its facilities.

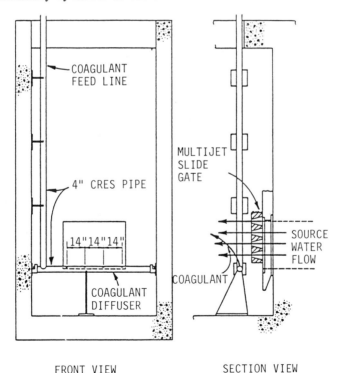

Fig. 24–1. Coagulant diffuser.

At a water treatment plant in Rio de Janeiro, the alum solution application point was moved from 25 feet (7.6 m) upstream of a hydraulic jump in the influent channel to the bottom of the hydraulic jump, which is the zone of maximum turbulence.[2] Approximately 5 to 7 tons (4.5 to 6.4 metric tons) of alum per day was saved by this improvement. It also resulted in better settled and filtered water quality. A reported $100,000 (1981 dollars) in chemical savings and operating costs was realized through reduction in alum use. These results indicate the importance of applying the coagulant solution at the point of maximum turbulence in the rapid mix process.

A 1978 report on upgrading of the 200-mgd (757-Ml/d) Potomac filtration plant, prepared by the Washington Suburban Sanitary Commission (WSSC), recommended that the mechanical mixers be replaced with diffusers to provide more effective dispersion of the coagulant with the incoming raw water.[3] In the evaluation of the rapid mix arrangement at the Potomac plant, it was found that single pipe diffusers placed in large influent lines would not adequately and completely disperse the coagulant in the incoming raw water. Also, there was excessive travel time between the point of coagulant addition and rapid mixers that permitted hydrolysis of the coagulant to occur before the raw water reached the mixer. A third problem was excessive detention time in the rapid mix chamber that prevented rapid dispersion of the coagulant. Rapid dispersion is required in order to prevent completion of the hydrolysis reaction before contact is made with all colloidal matter in the water.

The modification selected to solve these problems is shown in Fig. 24-2. The mechanical mixers were removed and replaced with concentric diffusers located downstream of the existing mix chambers. The diffusers were positioned with the orifices facing upstream into the flow to affect complete and almost instantaneous dispersion across the entire cross section of the pipe.

Other investigators have recognized shortcomings of turbine or impeller mixers in rapid mixing basins and suggested that this inefficiency may be due to back mixing, which generally results in less efficient use of coagulation chemicals. Mixing inefficiencies have resulted in a move toward in-line blenders using a principle known as "in-line jet injection." A jet injection rapid mixing system is illustrated in Fig. 24-3. In this system alum solution is introduced to the incoming raw water through a series of jets positioned radially around a central injection pipe. The jets discharge in a plane perpendicular to the incoming water. The dilution water is provided by a pump taking suction from the raw water line.

In this system the impact of diluting the alum solution with the raw water must be considered. This dilution effect was evaluated at a treatment plant in Florida, where it was found that dilution of the delivered liquid alum solution in ratios of 1:80 had no adverse impact on coagulation efficiency.[4]

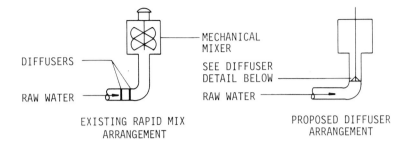

EXISTING RAPID MIX
ARRANGEMENT

PROPOSED DIFFUSER
ARRANGEMENT

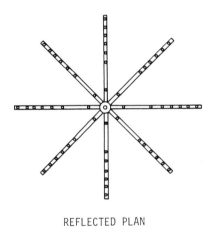

REFLECTED PLAN

CROSS SECTION OF DIFFUSER

Fig. 24-2. Diffuser for coagulant application.

Based upon these experiences, the supply pump for the injection jets should be sized so that a dilution ratio of 1:80 is not exceeded.

Flocculation Basins

Upgrading water treatment plants often requires improvements to the existing flocculation basins. This is especially true if the capacity of the existing facilities is to be expanded greatly. For example, a 100 percent expansion in

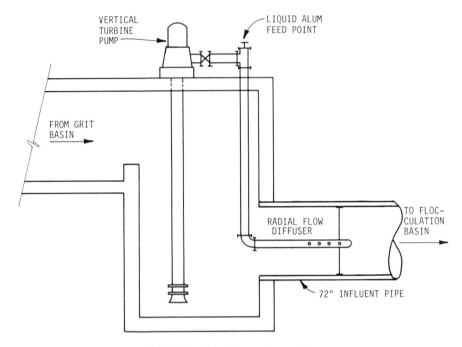

Fig. 24–3. Alum dispersion system.

plant capacity is possible by changing the existing rapid sand media operated at 2 gpm/sq ft (4.88 m/h) to mixed or dual media operated at 4 to 5 gpm/ sq ft (9.76 to 12.2 m/h). This would reduce the flocculation detention time from 30 minutes to 15 minutes or less. At these shortened detention times, it is vitally important that all factors that might contribute to poor flocculation performance be identified and corrected during the modifications.

The Manatee County, Florida, Water Treatment Plant was designed with an initial capacity of 27 mgd (102.2 Ml/d).[4] A laboratory and plant test was carried out to investigate the feasibility of increasing the capacity to 50 mgd (189.3 Ml/d) by process changes. Both laboratory and plant tests were carried out to enable the designers to identify process deficiencies and develop modifications to permit operation at higher plant throughput. During the evaluation, it was established that flocculation was inadequate at the expanded flow.

To obtain state approval for increasing the capacity of the flocculation facilities without major construction, a plant test was necessary. Modifications were made to the existing flocculation basins to minimize short-circuiting at the higher flows. An existing single-compartment flocculator was modified into an eight-compartment flocculator with alternate connecting top and bottom ports. These modifications are illustrated in Fig. 24–4. In addition, a

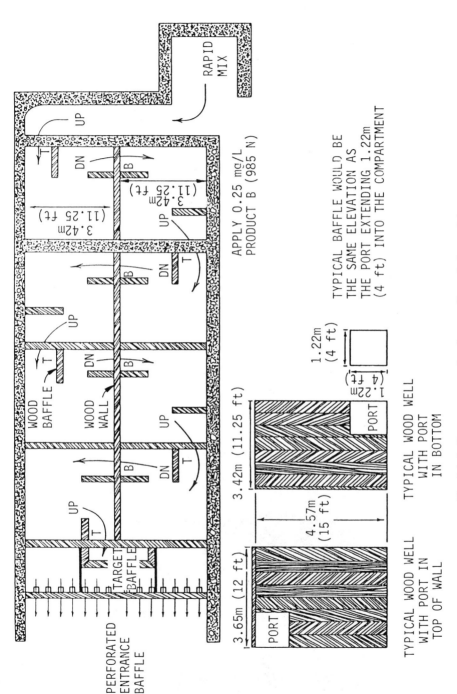

RAPID MIX

UP

APPLY 0.25 mg/L PRODUCT B (985 N)

DN

B

3.42m (11.25 ft)

3.42m (11.25 ft)

UP

DN

T

B

WOOD BAFFLE

T

WOOD WALL

DN

B

UP

B

DN

T

UP

T

TARGET BAFFLE

PERFORATED ENTRANCE BAFFLE

TYPICAL BAFFLE WOULD BE THE SAME ELEVATION AS THE PORT EXTENDING 1.22m (4 ft) INTO THE COMPARTMENT

1.22m (4 ft)

1.22m (4 ft)

PORT

TYPICAL WOOD WELL WITH PORT IN BOTTOM

3.42m (11.25 ft)

4.57m (15 ft)

3.65m (12 ft)

PORT

TYPICAL WOOD WELL WITH PORT IN TOP OF WALL

Fig. 24–4. Compartment plan for one flocculation unit.[4]

807

perforated baffle wall was installed at the flocculation basin outlet to improve inlet conditions to the settling basin. The perforated baffle wall is illustrated in Figs. 24–5a and 24–5b.

During plant tests at a flow of 63 mgd (238.5 Ml/d), the settled water turbidities from the portion of the plant containing the modified flocculator were slightly better than those from the unmodified plant at a flow of 37 mgd (140 Ml/d). On the basis of these experiments, a state agency granted approval of essentially doubling the flow through the modified portion of the plant. Doubling of the capacity was achieved by eliminating two mechanical rapid mixers, dividing the existing single compartment into an eight-compartment flocculator, and installing perforated baffles for better distribution of flocculated water into the settling basins.

These improvements reduced the average alum dosage from 70 to 50 mg/l. The settled water color was reduced from the previous 18 to 23 pcu to 10 pcu or less. The estimated savings in capital costs (1983 dollars) were

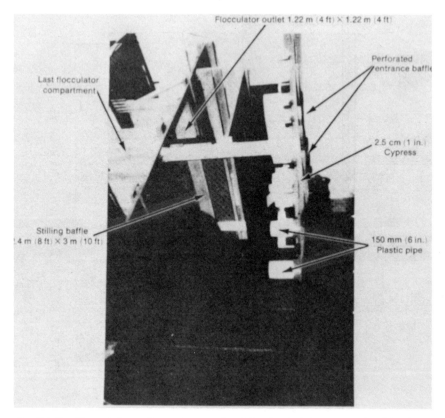

Fig. 24–5a. Flocculator outlet system and settling basin inlet system.[4]—side view

Fig. 24-5b. Flocculator outlet system and settling basin inlet system.[4]—end view

approximately $2,000,000, and the estimated power savings were $10,000 per year.

Improving Inlet and Outlet Conditions. Sufficient time and intensity of flocculation are essential to effective operation. However, a common cause of flocculation and sedimentation basin deficiencies is improper physical arrangement of the basins, allowing unequal flow distribution and short-circuiting of the influent flow. Maximum use of the available flocculation basin detention time must be made by distributing the coagulated raw water evenly across the flocculation basin inlet to prevent short-circuiting. This may require modifying the existing influent flumes, either by adding openings or perhaps by installing a secondary baffle across the inlet to the flocculation basin. These inlet baffles are commonly constructed of wood with openings sized and placed to impart enough headloss so that the flow is distributed uniformly across the inlet end of the flocculation basin. An elaborate perforated flocculation basin inlet baffle was installed in Rio de Janeiro's Guandu treatment plant, as illustrated in Fig. 24-6. It should be noted that the velocities through these openings should not provide a G-value greater than the G-value provided by the flocculation equipment in the first chamber.

Fig. 24–6. A perforated baffle with square-nozzled ports distributes flow from the flocculators at Rio de Janeiro's Guandu treatment plant.[2]

Correcting Short-Circuiting.   Short-circuiting is a major problem with many existing flocculation basins. Elimination of the short-circuiting can be accomplished by installing either around-the-end or over–under baffles within the basin. A serious short-circuiting problem was identified in the Guandu plant in Rio de Janeiro.[2] The problem was solved by installing baffles and changing the flocculator from a single-stage to a six-stage flocculator. A secondary improvement involved installation of a perforated timber baffle with square nozzle ports between the flocculation and settling basin. The headloss imparted by these nozzles resulted in a more even flow distribution to the settling basins.

Similar techniques were applied to the Potomac filtration plant in Washington, D.C. At this facility, existing two-compartment flocculation basins were converted to six-compartment basins using a perforated baffle to distribute flow evenly across the width and depth of the downstream settling basin. It was noted that to accomplish even distribution, the headloss through the ports must be significantly greater than the kinetic energy of the water moving laterally past the orifices. Care must be exercised so that the velocities through the ports are kept low enough to prevent shearing of the floc particles.

Installing High-Energy Flocculators.   Many existing conventional water treatment plants were designed with low-energy, paddle-style, mechanical flocculators. Often, these were designed with either variable or constant speed drives. With these original designs, emphasis was on formation of a large floc, which might or might not settle rapidly. At high-rate treatment plants, at reduced flocculation and downstream settling detention times, it is imperative that the floc formed in the flocculation basins settle at the fastest possible velocity. The existing flocculation equipment must be evaluated with respect to its capability of meeting the new requirements. If the existing equipment is inadequate, consideration should be given to replacing existing flocculators with new high-energy turbine-style units. Where turbine units have been installed and operated side by side with slow-speed, paddle-style units, they have provided a more dense floc with more rapid settling characteristics than were achieved with the old-style, slow-speed units.

The City of San Diego upgraded two of its major water treatment plants by modifications to the existing flocculation facilities.[5] In both plants, the rapid mixer and the first two rows of the old-style paddle flocculators were replaced with new high-energy mixers. Figure 24–7 shows the basin with the new- and the old-style flocculation equipment. The above-described modifications, along with conversion of the existing rapid sand to dual-media filters, effected a 30 percent increase in the treatment capacity of these two treatment plants.

Before recommending replacement of existing low-intensity, reel-style horizontal flocculators with new high-energy units, a careful plant evaluation should be made to determine whether this modification is necessary. The existing equipment should be inspected to determine whether a higher energy gradient could be produced by simple modifications to the existing equipment. These modifications may include the addition of paddles or new mechanical drives to impart greater mixing intensity. The condition as well as the capability of the existing drive equipment will determine whether a modification of this type is physically possible or economically practical.

Coagulation Aids. The performance of many water treatment plants can be improved by the addition of a coagulant aid to overcome some of the deficiencies of a poorly designed flocculation basin. Flocculation is a time-dependent process that can be enhanced and accelerated by the addition of organic coagulants. The resulting floc generally settles more rapidly than unaided alum floc, permitting the treatment plant to perform more efficiently at higher plant throughput. It should be noted, however, that application of coagulant aids is no substitute for adequate upstream mixing of coagulant with incoming raw water.

A coagulant aid dosage is influenced by numerous operational parameters and raw water characteristics. In general, organic polymer coagulant aids are applied at an alum-to-polymer ratio ranging from 100:1 to 50:1. Plant jar tests must be run, however, to establish the proper dosage ratio for optimum results. For a relatively modest investment in capital costs, the applica-

Fig. 24-7. New- and old-style flocculators.[5]

tion of a coagulant aid can provide marked improvement in the flocculation and sedimentation process in an existing plant. The improved sedimentation will reduce turbidity loadings to the filters, extending filter operation cycles, and in some instances can provide a higher-clarity filtered effluent. (Equipment for feeding coagulant aids is discussed and illustrated in Chapter 8.)

Sedimentation Basins

Increasing treatment plant capacity may require modifications and improvements to the existing settling basins. In most cases, a significant increase in plant capacity reduces settling basin detention times and increases clarification rates so that the basins will not perform efficiently at the expanded flows. Problems caused by poor entry and exist conditions as well as inadequate sludge collection and removal may also have to be corrected when the capacity of existing sedimentation basins is increased.

The manner in which flocculated water is delivered to the settling basin influences the efficiency of the clarification process. Many older plants have separate flocculators that require transfer of the flocculated water to the settling basin in open or closed conduits. The velocity and associated turbulence in these channels break up floc and thus prevent good clarification. Wherever possible, these channels should be eliminated, and the flocculated water should be introduced directly into the settling basin through a perforated baffle wall. Such improvements should be considered if observations during plant tests indicate potential major problems in obtaining good inlet flow distribution. Care must be exercised in the design of perforated inlet baffles so that velocities through these openings will not be high enough to break up floc. Proper distribution of the flocculated water at the settling basin inlet is critical to realizing maximum efficiency from the basin.

A well-designed settling basin should have an effluent collection system to uniformly withdraw water from the clarification zone. In rectangular basins, finger launders extending inward from the end wall and located at uniform spacing provide the best assurance against short-circuiting. Weir loading rates should be within the range of 10,000 to 20,000 gpd (124 to 248 m^3/min/d) per foot of weir length. These launders should be designed with adjustable V-notch weir plates, or orifices should be placed in the sides of the launders at uniform centers over the length of the trough.

Circular, radial-flow basins should have a continuous peripheral collection launder with a sufficient length of weir to match the loading criteria for rectangular basins. The sludge collection equipment should be inspected, and it should be determined whether it will adequately handle greater quantities of sludge at the increased plant throughput. Sludge piping and pumps should also be inspected to verify adequate capacity.

Where clarification rates at expanded flows are too high for adequate clarification, tube or plate settlers can be installed to increase the effective surface settling area and thus increase the capacity of the basin. (Guidelines for the application of tube and plate settlers were discussed in Chapter 9.) Tube settlers can be installed in most conventionally designed settling basins to permit a significant increase in capacity without loss of clarification efficiency. In many cases, settling capacity can be more than doubled by introduction of tube settlers. A number of case histories describing water treatment plants where the capacity was increased significantly by installing tube settlers and other modifications are presented later in this chapter.

Increasing Filtration Capacity

Conventional water treatment plants employing rapid sand filters have historically been designed for a filtration rate of 2 gpm/sq ft (4.88 m/h). Replacing the existing rapid sand with either dual or mixed media can easily result in a doubling of plant capacity while generally yielding a better-quality finished water.

To determine the feasibility of expanding plant capacity by filter modifications, a thorough filter hydraulic study should be completed. Because the plant originally was designed for 2 gpm/sq ft (4.88 m/h) filtration rates, settled water collection and distribution systems may reflect these rates and, in some instances, may not have adequate capacity for expanded throughput. A hydraulic analysis of transfer channels, pipes, opening, weirs, etc. can determine the maximum amount of water that can be delivered to the existing filters. In some cases, minor changes in piping, gates, weir openings, etc. will permit these greater flows. In other plants, entire segments of the settled water distribution piping may have to be modified or replaced in order to deliver the water to the filters. If possible, field tests should be carried out to establish the maximum carrying capacity of influent piping and channels.

Existing filter boxes should be examined to determine whether there are any potential problems in converting the filters to high-rate filtration. In general, most well-designed rapid sand filters are easily converted to high-rate dual- or mixed-media filters. Backwash rates for dual- and mixed-media are the same as those for rapid sand filter beds. In some instances, filter wash troughs may have to be modified in order to provide the required clearance from the surface of the filter media to the lip of the wash trough. If the clearance is inadequate in the existing filter, there is a potential for excessive loss of the lighter anthracite coal media. As a general rule, where rotary surface wash is used, a filter should have 24 to 27 inches (0.61 to 0.69 m) of clearance between the filter surface and the lip of the wash trough and 8 inches (0.20 m) between the underside of the wash trough and the filter surface.

The condition of the existing underdrain can often be established by observing the condition of the filter media. The presence of boils during backwashing or an uneven mounded appearance of the filter could indicate an underdrain failure. Only removal of the filter media and gravel and inspection of the underdrain will establish its physical condition. If the underdrain is found to be damaged or in poor condition, its replacement must be a part of the filter renovation project.

If the existing filters are not equipped with surface wash facilities, they should be installed during the rebuilding project. Dual- and mixed-media filters operating at filter rates of 4 to 6 gpm/sq ft (9.8 to 14.6 m/h), with polyelectrolyte filtration aids used to control floc breakthrough, are more difficult to clean than rapid sand filters. A well-designed surface wash system is needed to scour the upper layers of the filter during or just prior to backwashing. This procedure breaks up surface accumulations and prevents formation of "mud balls." Either a rotary surface sweep with nozzles that penetrate the surface as the arm rotates or a fixed jet surface wash system should be installed. Some advocate a two-arm rotary surface wash system with one arm placed at the surface and the other located at the coal–sand interface in a dual-media filter. The theory of operation of this configuration is that "mud balls" have a tendency to form at the coal–sand interface, and additional agitation is required at this location to assist in adequate cleaning of the filter bed.

Air scour has been used in a limited number of plants to supplement the backwashing procedure and to achieve the same purpose as surface wash equipment. In general, air scour must be employed only with underdrain systems that do not require supporting gravel layers. Specially designed nozzles with retaining screens on which the filter media can be placed directly are typically used with air scour. Introducing air beneath the sand and coal in a dual-media filter thoroughly mixes and scrubs the filter media. Backwashing at a flow of about 15 gpm/sq ft (36.8 m/h) reclassifies the media. Advocates of air-scour systems claim that less backwash water is used with them.

Although most well-designed rapid sand filter plants have adequately sized backwash supply and waste piping, rate-of-flow controllers and other flow-limiting devices in filter effluent piping may not pass the higher flows possible with dual- or mixed-media filters. Replacement of undersized rate-of-flow controllers with new pneumatically or electrically controlled controllers may be required to handle the higher flows. Frequently, filter effluent piping is also undersized and will not handle flows associated with filter rates of 5 to 10 gpm/sq ft (12.2 to 24.4 m/h), and must be replaced. To offset filter gallery space limitations, more compact butterfly valves and fabricated steel piping can be used in place of bulky cast iron gate valves and piping.

A modification of the method of filter rate control can also be used to increase the capacity of an existing rapid sand filter. Variable declining rate filtration can be adapted to an existing plant at minimal expense. Figure 24–8 shows a desirable design for variable declining rate operations. This modification eliminates rate-of-flow controllers which contribute to troublesome operation and in old plants are often totally inoperative. Another advantage is that the filter media are submerged at all times, eliminating negative heads that cause air binding. As illustrated in Fig. 24–9, the rate is initially constant on a clean filter, and declines as the filter becomes dirty. Proponents of the variable declining rate method of filter flow control regard the simplicity of operation and the lack of expensive and troublesome rate-of-flow controllers as the overriding reason why this technique should be seriously considered in any plant that must be upgraded.

Polyelectrolyte Filtration Aids. Conversion to dual- or mixed-media filters that operate at a filter rate of 4 to 10 gpm/sq ft (9.76 to 24.4 m/h) should be accompanied by installation of filtration aid application equipment.

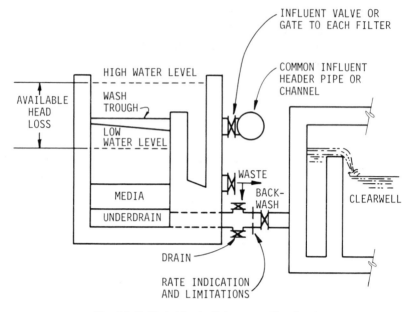

Fig. 24-8. Variable declining rate filtration.[2]

Filtration aids, which are typically nonionic polyelectrolytes, are a useful tool in optimizing the performance of high-rate filters. These materials prevent premature filter turbidity breakthrough by controlling floc penetration into the filter. A dosage sufficient only to retain floc in the bed until the maximum operating headloss is reached represents the optimum condition. This condition is illustrated by the hypothetical headloss/turbidity breakthrough curve in Fig. 24-10. Typically, polyelectrolyte dosages ranging from 0.02 to 0.1 mg/l are appropriate, depending upon applied water characteristics. Polyelectrolytes are especially valuable in preventing premature floc breakthrough where waters are cold.

Turbidity Monitoring. Placement of dual- or mixed-media filters should be accompanied by installation of turbidity-monitoring instrumentation. With this equipment, the performance of each filter can be monitored to assure that it is operating properly. As a minimum, at least one continuously recording nephelometric turbidimeter should be provided for measuring filtered effluent turbidity. Ideally, each filter should have a continuously recording turbidimeter equipped with an alarm that is initiated in the event the filtered effluent turbidity exceeds a preset maximum. Filter effluent turbidity breakthrough near the end of a filter run can be detected quickly, indicating the need for backwashing. It can also be used to optimize the filter aid dosage

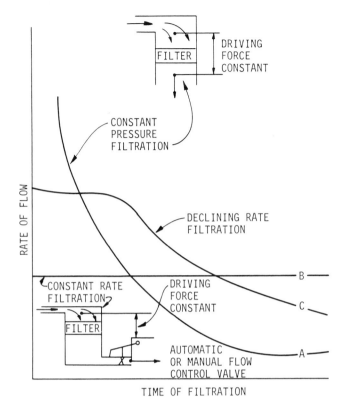

Fig. 24-9. Typical rate of filtration patterns during a filter run.[2]

by noting the trend in filter effluent turbidity as a function of headloss over an entire filter run.

Surface scatter turbidimeters should be considered for monitoring the turbidity of raw and settled water. Where raw water is subject to daily variations, raw water turbidity data are especially useful. Information obtained from these locations can be used to adjust chemical dosages or make other in-process changes to optimize plant performance.

Coagulation Control Pilot Filters. Another potentially valuable addition to an upgraded facility is a proprietary monitoring and coagulation control device that uses pilot filters and turbidimeters to determine whether the proper coagulant dosage is being applied to the raw water, as described in Chapter 7. This unit contains twin high-rate pilot filters, one of which is constantly fed by a sample of coagulated water obtained immediately after rapid mixing while the other filter is backwashing. The filtrate is monitored continuously for turbidity and in some applications for pH.

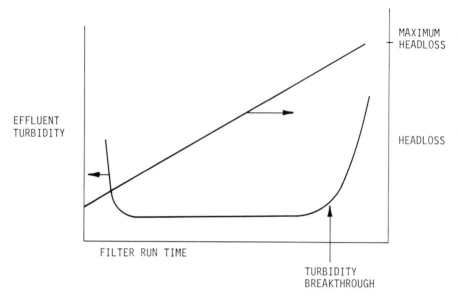

Fig. 24–10. Hypothetical optimum filter headloss/turbidity breakthrough curve.

The pilot filter is overdosed with filter aid to prevent breakthrough. A change in turbidity generally signifies a need to adjust the coagulant dosage to compensate for changing raw water conditions. With this device, the operator has virtually instantaneous control over the coagulation process.

CASE HISTORIES

Sacramento, California

The City of Sacramento has two water treatment plants with a combined treatment capacity of 140 mgd (530 Ml/d). The newer of these two water treatment plants is located on the American River. It was constructed in 1963, and has a design capacity of 60 mgd (227 Ml/d). Increased demand, especially during the summer irrigation season, placed a strain on both facilities, and additional water capacity was needed to meet present as well as future water requirements.

Although both water treatment plants were amenable to expansion, the city selected the newer American River Plant for the first phase of the project. A comprehensive study was carried out by the city's consultant to determine how much additional capacity could be gained from the American River Plant by a staged expansion program.[6] The preliminary review indicated that replacing the existing rapid sand filter media with either more efficient dual- or mixed-media would accomplish a significant expansion in capacity and could be achieved with no major plant structural modifications or additions. The existing filters were designed at 3 gpm/sq ft (7.3 m/h) and consisted of 24 inches (0.61 m) of silica sand over a graded gravel underdrain. At the expanded flow of 105 mgd (398 Ml/d), the filter rate was increased to 5.3 gpm/sq ft (12.9 m/h), which is well within the capabilities of a well-designed dual- or mixed-media filter.

The existing 60-mgd (227-Ml/d) conventional plant, at design flow, provided 25 minutes of flocculation and 115 minutes of settling prior to filtration. At 105 mgd (398 Ml/d), the flocculation and settling times would be reduced to 14 and 65 minutes, respectively. Although the amount of pretreatment would be substantially reduced at the expanded flow, either dual- or mixed-media filters could handle anticipated higher turbidity loadings. The highest turbidities occur during the months of October, November, December, and January. These are months of low water demand. Lowest periods of turbidity occur from April through September, which are months of high water demand. This is a fortunate situation in that periods of highest demand do not occur simultaneously with high raw water turbidity. These interrelationships are illustrated by Fig. 24–11 and Table 24–1. For the most part, raw water turbidity seldom exceeds 5 NTU during the season of peak demand.

Following preliminary assessment of the feasibility of increasing the capacity of the American River Plant by converting the existing rapid sand filters to either dual- or mixed-media filters, extensive pilot filtration tests were

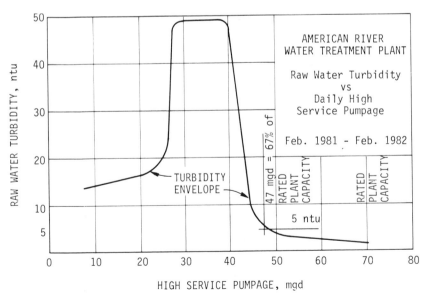

Fig. 24–11. Raw water turbidity as a function of water demand. (Courtesy of Public Works Magazine[6])

Table 24–1. Maximum Daily Raw Water Turbidity by Months.*

MONTH	AMERICAN RIVER MAXIMUM TURBIDITY, NTU		
February, 1981	7.1		Maximum turbidity less than 7.1 NTU for 8 months
March	6.9		
April	3.2	Maximum turbidity less than 3.2 NTU for 6 months	
May	3.1		
June	2.6		
July	1.4		
August	1.8		
September	2.3		
October	28.0		
November	42.3		
December, 1981	15.0		
January, 1982	54.0		

* Reference 6. Courtesy of Public Works Magazine.

carried out by the City of Sacramento. The specific purpose of the tests were to evaluate the performance of various types of filter media and to determine the economics of their use in expanding the capacity of the plant. These tests were done in February through May of 1982.

Four types of filter media were evaluated simultaneously in four side-by-side pilot filters. Each filter was equipped with instrumentation for measuring turbidity and headloss. Figure 24–12 is a flow diagram showing the general layout of the pilot filtration equipment. Feedwater to the pilot filters could be drawn from either of two sources: (1) from the raw river water pipeline (before alum addition), or (2) from the plant filter influent flume following alum addition, mixing, flocculation, and settling. Although the principal objective of the test was to qualify a filter media for replacing the existing rapid-sand materials, the test had a secondary objective, to establish the feasibility of direct filtration of the American River supply for a future plant expansion.

During the 3 months that round-the-clock testing was done, a total of more than 360 filter runs were made. During the tests, influent turbidities varied from 3 to 33 NTU. For the controlled variables: (1) alum dosages ranged from 12 to 21 mg/l; (2) polymer dosages varied from 0 to 0.3 mg/l; and (3) filter rates were set at 5, 6.5, or 10 gpm/sq ft (12.2, 15.9, or 24.4 m/h). Test runs were terminated for one of two reasons, either high headloss or high turbidity. The City of Sacramento established a maximum turbidity goal for drinking water of 0.1 NTU. During a particular pilot test run, an upper turbidity limit of 1 NTU was selected as the point at which a run was terminated.

All of the pilot filtration tests performed on plant settled water clearly

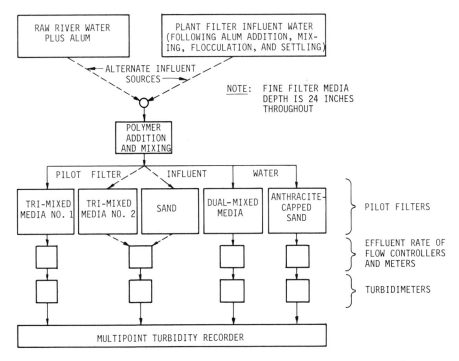

Fig. 24-12. Flow diagram of American River pilot filtration equipment. (Courtesy of Public Works Magazine[6])

demonstrated that the turbidity goal of 0.1 NTU could be obtained at all filter rates examined [5 to 10 gpm/sq ft (12.2 to 24.4 m/h)] by either dual- or mixed-media filters. Mixed-media possessed a greater reliability in achieving the turbidity goal. During filter surging tests, the mixed-media demonstrated a greater resistance to turbidity breakthrough, hence providing greater assurances that potentially harmful bacteria or viruses sheltered by turbidity particles would not pass through the filter. In addition, the mixed-media filter appeared to return more quickly to producing the desired effluent turbidity after the completion of a backwash. The test results also indicated less chemical usage by the mixed-media than the dual-media filter. These results were extrapolated to an annual chemical cost savings of approximately $8,000.

Based upon the results of the pilot filtration study, it was recommended that the treatment plant be expanded by replacing the existing sand filter media with either dual- or mixed-media. It was also recommended that equipment be installed for preparing and feeding a polyelectrolyte filtration aid.

Bids were received in the spring of 1983 for removing the existing sand and replacing it with mixed media. The sand was removed, but the existing underdrain gravel was retained in the eight plant filters, for a total filter bed area of 13,940 square feet (1,295 m²). The entire operation of removing and replacing the sand in the eight plant filters, each 1,743 square feet (162 m²) of bed area with mixed-media, was accomplished over a 3-month period in the summer of 1983. The total cost of the filter conversion was approximately $680,000 (1983 prices). The bid price for furnishing and installing the polymer filtration aid system was approximately $140,000. With these modifications, an additional 45 mgd (170 Ml/d) of treatment capacity was realized at a cost of less than $20,000 per mgd ($5,300/Ml/d).

The treatment plant has been operated successfully at 101 mgd (382 Ml/d), which is the maximum capacity of the existing raw water pumps. Before the treatment plant can be operated at sustained flows beyond 100 mgd (379 Ml/d), additional raw water pumps and high service finished water pumps will be required. During qualifying tests, effluent turbidities averaged less than 0.1 NTU, and filter cycles between backwashes exceeded 48 hours.

Erie County, New York

The application of mixed-media to the plant in Erie County, New York is of special interest because it provides side-by-side comparisons of the efficiency of mixed-media filters and rapid sand filters in a large plant. The results have been reported by Westerhoff and are summarized below.[7]

The Erie County Water Authority serves filtered water to approximately 390,000 persons in all or part of 12 towns and the city of Lackawanna in Erie County, New York. The average water production in 1969 was 44.3 mgd (168 Ml/d). Water is taken from Lake Erie, treated in the Water Authority's two filtration plants (Sturgeon Point and Woodlawn Filter Plants), and pumped into the transmission and distribution systems. Most of the water is treated at the Water Authority's Sturgeon Point Filter Plant, located in the town of Evans, approximately 30 miles (48 km) south of the city of Buffalo. This plant was originally constructed in 1961.

In 1965, the water demands of a burgeoning population in the suburban Buffalo area necessitated expansion. The first stage [to a capacity of 60 mgd (227 Ml/d)] was needed immediately.

The Water Authority decided during the first-stage expansion to install mixed media in six new filters. The four original filters contain fine-to-coarse graded-sand media.

Special data were collected from a total of 48 parallel-filter runs during a 12-month study period. The sand filters were operated at the standard 2 gpm/sq ft (4.88 m/h) filtration rate as a control. The filtration rate on the mixed-media filters was varied from 2 to 10 gpm/sq ft (4.88 to 24.4 m/h).

The water treatment processes at the Sturgeon Point Filter Plant consist of aeration, chemical addition, rapid mixing, flocculation and sedimentation, filtration, and chlorination.

Flocculation and sedimentation are accomplished in five units, each of which consists of a rectangular flocculation tank followed by a rectangular settling tank. Individual flocculation tanks have a liquid capacity of 46,500 cu ft (1,316 m³) [about 348,000 gallons (1.32 Ml)]. At a capacity of 8 mgd (30.3 Ml/d), the detention time is nearly an hour. Two of the tanks are equipped with mechanical paddle-wheel-type flocculators, and three of the tanks have walking-beam-type flocculators.

From the flocculators, flow passes through rectangular settling tanks, each with a surface area of 12,700 sq ft (1,180 m²) and a volume of about 1.4 million gallons (5.3 Ml). At a rated capacity of 8 mgd (30.3 Ml/d), the settling period is slightly longer than 4 hours with a surface loading of 630 gpd/sq ft (1.07 m/h).

The four original rapid sand filters consisted of a single-media bed of hydraulically graded fine-to-coarse sand. When rated at 2 gpm/sq ft (4.88 m/h), their combined capacity was 16 mgd (60.6 Ml/d).

The six new filters contain mixed media and have a total area of 8,400 sq ft (780 m²). The plant capacity, assuming a filtration rate of 2 gpm/sq ft (4.88 m/h) for all ten filters, is 40 mgd (151 Ml/d). At a filtration rate of 3 gpm/sq ft (7.32 m/h), the plant capacity is 60 mgd (227 Ml/d).

During the parallel tests, the filter runs were terminated at a headloss of 8 feet (2.4 m). Typical turbidities applied to the filters were 2 to 4 NTU with little variation as the clarifier overflow rate was varied from 625 to 940 gpd/sq ft (1.06 to 1.59 m/h). Alum feed rates were typically 14 to 16 mg/l. The following conclusions were reached:

- More than 80 percent of the time, the filtered water turbidity from the sand filters was 0.10 NTU or less.
- Nearly 88 percent of the time, the filtered water turbidity from the mixed-media filters was 0.10 NTU or less.
- The mixed-media filters operating at filtration rates of 2 to 10 gpm/sq ft (4.88 to 24.4 m/h) consistently produced a lower filtered water turbidity than the sand filters operating at a filtration rate of 2 gpm/sq ft (4.88 m/h).
- At filtration rates up to 6 gpm/sq ft (14.6 m/h), the mixed-media-filter effluent had a lower total microscopic count than the sand filters operating at a 2 gpm/sq ft (4.88 m/h) rate.
- The mixed-media filters operating at 5 gpm/sq ft (12.2 m/h) had an average filter run length of 29 hours. At 6 gpm/sq ft (14.6 m/h) they had an average filter run of 20 hours.
- The mixed-media filters operating at 5 to 6 gpm/sq ft (12.2 to 14.6 m/h) used a considerably lower proportion of wash water than the sand filters operating at 2 gpm/sq ft (4.88 m/h) (1.8 percent as compared to 2.5 percent on the average).

As a result of these tests, it was concluded that the mixed-media filters at 6 gpm/sq ft (14.6 m/h) produced an end product superior to that from the rapid sand beds at 2 gpm/sq ft (4.88 m/h) with concurrent economic advantage. The finished water goals adopted for filtered water at this plant were established for the mixed-media filters at 6 gpm/sq ft (14.6 m/h) as follows:

Turbidity	Less than 0.10 NTU average
	Less than 0.50 NTU maximum
Total microscopic count	Less than 200 su/ml average
	Less than 300 su/ml maximum
Color	Less than 1 unit
Odor	Not detectable
Aluminum	Less than 0.05 mg/l
Iron	Less than 0.05 mg/l

Corvallis, Oregon

Corvallis has increased the capacity of its municipal water treatment plant by two and one-half times without increasing the size of the plant. Results of this application of advanced concepts have included a one-third saving of construction costs, an improvement in water quality, and a considerable extension of the city's long-range plans for structural expansion of its water treatment facilities.

Using the Willamette River as a source of supply, the original treatment plant was designed and built in 1949 as a conventional rapid sand filtration plant. It was designed in increments of 4 mgd (15.2 Ml/d) capacity, each increment consisting of two 16 × 130-foot (4.9 × 39.6-m) flocculation–sedimentation basins and two rapid sand filters, each with an area of 484 sq ft (45 m²).

The initial plant capacity of 4 mgd (15.1 Ml/d) rose to 8 mgd (30.3 Ml/d) with the addition of a second increment in 1961. Under the original plan, the plant was scheduled for expansion to its ultimate capacity of 16 mgd (60.6 Ml/d) with construction of the third and fourth increments. This, plus 4.5 mgd (17 Ml/d) available from the city's second source on Rock Creek, was expected to take care of municipal water requirements through 1975, when construction of another treatment facility would have been necessary.

With application of new techniques to the existing plant, Corvallis increased its water supply to 25 mgd (95 Ml/d). Capacity of the Willamette River treatment plant had been increased without structural addition from 8 mgd to 21 mgd (30.3 to 80 Ml/d), and there remained space for the third and fourth increments called for under the original plan. Thus, the ultimate capacity potential had been boosted to 42 mgd (159 Ml/d). This expansion of capacity was achieved by application of: (1) mixed-media filtration, (2) shallow-depth sedimentation with tube settlers, and (3) coagulation control techniques.

The turbidity of the raw water drawn from the Willamette River normally ranges from 15 to 30 NTU, with surges up to 1,000 NTU. The water is soft (15 to 30 mg/l hardness) and exhibits periodic taste and odor problems. Typically, the following chemical dosages are used: alum, 20 to 40 mg/l; lime, 10 to 20 mg/l; chlorine, 2 mg/l; polymer (as coagulant aid), 0.1 to 0.2 mg/l; activated carbon, 5 to 10 mg/l (for taste and odor control).

A diaphragm pump feeds alum into the existing flash-mix basin, $10 \times 10 \times 14$ feet ($3.05 \times 3.05 \times 4.27$ m), in liquid form. The outlet of the flash-mix basin was modified to accommodate the increased hydraulic capacity.

Virtually all of the treatment piping had to be enlarged to handle the increased plant flow. From the flash-mix to the sedimentation basin, two welded steel pipelines were increased from 18 to 24 inches (0.46 to 0.61 m); from the sedimentation basin to the filters, one 24-inch (0.61 m) line increased to 36 inches; in the filter gallery, the influent line increased from 16 to 24 inches and the effluent line, from 12 to 16 inches (0.305 to 4.88 m). Butterfly valves have replaced all gate valves.

Settling tube modules were installed over about 60 percent of the 3,500 sq ft (325 m²) of rectangular settling basin area. The tubes were installed on simple "I" beams spanning the width of the basins and are located at the discharge end of the basin. New effluent weirs and launders were also installed to provide good flow distribution through the area covered by the tubes. An example of tube settling modules is shown in Fig. 24–13.

Overflow rate in the sedimentation basin area covered by the tubes is 4.2 gpm/sq ft (10.3 m/h). This compares with the 1.05 gpm/sq ft (2.6 m/h) loading on the basin before they were installed.

Utilization of the existing basins with the tubes reduces the load on filters, especially during periods of high river turbidity, and provides more economical operation by increasing filter runs during normal river conditions. With an extreme turbidity range of 2 to 1,000 NTU, filter runs have been increased from the 40 hours experienced with the old conventional basins at much lower plant throughput rates to 60 to 65 hours. Operating on raw water turbidities of 15 to 30 NTU, the tube effluent has had turbidities of 1 to 2 NTU.

The capacity of the filtration portion of the plant was increased by modifying the filter piping and by replacing the rapid sand filter media with mixed media. Two of the four filters had Leopold underdrains, and the other two had Criscrete underdrains. At the 21 mgd (79.5 Ml/d) rate, the filtration rate is 7.5 gpm/sq ft (18.3 Ml/d).

Fig. 24–13. Clarifiers showing tube settling modules. (Courtesy of Microfloc, Products Group, Johnson Division, VOP)

The pilot filter coagulation control system described in Chapter 7 is used in conjunction with a Hach 1730 CR low-range continuous reading turbidimeter. A pilot effluent turbidity of 0.2 NTU is considered the maximum value associated with proper coagulation. The full-scale plant continuously produces a finished water turbidity of 0.2 NTU or less.

The cost to expand the plant from 8 to 21 mgd (30.3 to 79.5 Ml/d) capacity [including a new 5 MG (18.9 Ml) reservoir, a new high-service pump station, and a crosstown 16-inch (0.41-m) transmission line to the new reservoir] was $430,000 (1969 prices). It was estimated that the same expansion through the addition of new settling basins and filters would have been at least $650,000 (1969 prices). Thus, improved performance and increased capacity were achieved while a substantial savings in cost was realized.

Fairfax County, Virginia

The Fairfax County (Virginia) Water Authority supplies water to about 500,000 people in the suburban Washington, D.C., area. The area served is one of the most rapidly growing areas in the United States, and the time that would have been required for construction of added, conventional treatment facilities was not compatible with meeting the increasing water demands.[9]

The raw water supply is the Occoquan Reservoir, which has an impounded watershed of approximately 570 square miles (1,478 km²). This reservoir of 9.8 billion gallons (37.1 Gl) has a dependable yield of 65 mgd (246 Ml/d). Upstream discharges of secondary sewage effluent from several sewage treatment plants, prior to the construction of an AWT plant in 1978 (described in Chapter 4), contributed to the eutrophication of the reservoir. Taste and odor problems in the reservoir water are severe during summer months.

Water treatment is provided at two interconnected plants. The treatment units consist of 17 circular steel combination coagulating, settling, and filtering units that had a nominal capacity of 46.4 mgd (113 Ml/d) based on a filter rate of 2.0 gpm/sq ft (4.88 m/h). The maximum rated capacity based on 1.5 times the nominal rate was 69.6 mgd (169.8 Ml/d). Principal chemicals used in the treatment process are liquid alum, pre- and post-lime, pre- and post-chlorine, fluoride, activated carbon, and potassium permanganate.

The existing units at the River Station Treatment Plant are circular steel units exactly 86 feet (26.2 m) in diameter. The center portion is an upflow mixing, flocculation, and settling tank. The filter unit extends around the outer circumference. Original design features of the unit furnished by Dorr-Oliver, which was modified to test performance at higher filter rates, are as follows:

Nominal rating = 4.0 mgd (15.1 Ml/d) at 2.0 gpm/sq ft (4.88 m/h)
Filter area = 1,400 sq ft (130 m²)
Settling tank area = 3,848 sq ft (357.5 m²)
Side wall depth (S.W.D.) = 17 feet 3 inches (5.26 m)
Overflow rate = 1,040 gpd/sq ft (1.76 m/h)
Detention time = 2.98 hours at 4.0 mgd (15.1 Ml/d)
Wash rate = 15 gpm/sq ft (36.6 m/h)

In September, 1968, the Authority entered into a contract with Neptune Microfloc, Inc., for furnishing and installing tube settler modules in Unit No. 1 of the River Station Treatment Plant. In addition, the existing media, consisting of sand and anthracite, was to be removed and replaced with an inverse graded bed of ilmenite, sand, and coal. The contract documents required certain performance prior to acceptance by the Authority, as follows:

- Unit No. 1 must produce 10.0 mgd (37.9 Ml/d) continuous flow with effluent characteristics at least equal to the averages for the other four units at 4 mgd (15.1 Ml/d) during the period January 1, 1968 through August 31, 1968.

- Maximum rate of backwashing shall not exceed 15.0 gpm/sq ft (36.6 m/h) of filter surface area.
- Filtered water turbidity shall not exceed 0.2 NTU at 5.0 gpm/sq ft (12.2 m/h). Headloss shall be 8.0 feet (2.4 m) or less.
- Backwash water shall not exceed 3 percent when water applied to the filter has a turbidity not in excess of 10.0 NTU.
- At a rate of 10.0 mgd (37.9 Ml/d) [5.0 gpm/sq ft (12.2 m/h)], the turbidity, odor, and taste of water leaving the tube modules of Unit No. 1 shall be less than or equal to the turbidity, odor, and taste of water leaving Units No. 2, 3, 4, and 5 when they are operating at 4.0 mgd (15.1 Ml/d) [2.0 gpm/sq ft (4.88 m/h)]. The applied turbidity to the filter shall never exceed 10.0 NTU.

In addition, the Virginia Department of Health placed the following requirements on the project:

- The addition of chemicals and coagulation of the raw water would be controlled by zeta potential and a zetameter.
- Continuous indicating and recording turbidimeters would be used to monitor and record raw water, applied water, and filter effluent turbidity.
- Applied water should have an average turbidity of 2.0 NTU and a maximum of 5.0 NTU. Filtered water would have an average turbidity of 0.1 to 0.2 NTU with a maximum of 0.5 NTU.
- The unit would be operated continuously for a full year in parallel with the other four existing units in order to compare performance results.

In order to increase the capacity of the settling portion of the treatment units (Dorr-Hydro-Treators), tube settling modules were installed around the outer periphery of the settling tanks. The filter media in the test unit (Unit No. 1) was replaced with 18 inches (0.46 m) of coal (1.0 mm effective size), 7 inches (0.18 m) of sand (0.5 mm effective size), and 3 inches (76 mm) of ilmenite (0.18 mm effective size). The four standard units contained dual-media filters made up of 24 inches (0.61 m) of coal (effective size 1.0 mm) and 6 inches (0.15 m) of sand (effective size 0.5 mm). Hach CR surface scatter turbidimeters and Hach CR turbidimeters were installed in the pipe gallery below the Hydro-Treators to record raw water, applied, and effluent turbidities. Some difficulties were encountered in the original testing due to the critical nature of the tube settler location in relation to the sludge blanket maintained in the units. Although no additional published data were available at the time of this writing, it was reported that testing subsequent to the preliminary report proved satisfactory, and additional plant clarifier-filter units were modified in a manner similar to that described above.[9]

North Marin County Water District, Novato, California

In 1973, the North Marin County Water District, which provides water service to Novato, California and nearby areas, faced impending summer-season water shortages due to limited delivery capacity and rapidly increasing demands.[10] Typical seasonal demands on the Water District vary, from 60 percent of the average annual demand on the system during winter to 160 percent in the peak month of the summer season.

Because long-range plans to increase the supply from the Russian River could not be implemented for at least 3 years, it was necessary to find some means to upgrade the plant to meet water demands during the summers of 1974, 1975, and 1976. The projected demands were then beyond the capability of existing facilities.

Within the time available, the only practical possibility of accomplishing this was to expand the Stafford Lake Water Treatment Plant, provided that it could be done in time to have the plant back in operation to meet summer demands. Such a compressed time schedule ruled out any major, time-consuming new construction.

Lake Water Quality. The turbidity of the Stafford Lake supply varies from about 2 NTU to about 35 NTU, with a usual range of about 10 to 20 NTU. Coliform MPN values range from 5 to 72,400 per 100 ml. The pH of the raw water is between 7.6 and 8.2. The temperature range is from 47 to 70°F (8.3 to 21.1°C), and dissolved oxygen may be as low as 5.5 or as high as 10.5 mg/l. The lake water color ranges from 25 to 50 units. Iron and manganese each also vary from 0.1 to 1.0 mg/l.

Experience at the Stafford plant had not indicated any operating problems with high raw water turbidities or coliform densities. However, high color and low turbidity in the range of 5 to 10 NTU in raw waters produced problems of clarification. Further, high plankton populations caused taste and odor problems, and other biological forms had seriously shortened filter runs on occasion. The installation of aeration equipment in the lake had improved raw water quality by increasing dissolved oxygen, reducing thermal stratification, and reducing manganese concentrations by preventing the development of anaerobic conditions in the reservoir.

The basic steps in the existing treatment process were as follows:

- Raw water was disinfected with chlorine and coagulated with alum in a downflow hydraulic mixing chamber.
- Flocculation was provided using additions of recirculated sludge and lime.
- Flocculated water was clarified, and sludge was removed and sprayed on land for disposal.

- Activated carbon was added immediately prior to filtration in rapid sand filters.
- Dechlorination to the desired residual was accomplished by the use of sulfur dioxide in the filtered water clearwell.
- Elevated storage of water for backwashing filters was provided.
- Water used in washing filters was stored in a recovery pond and pumped at a controlled rate back to the head of the plant for reprocessing.

Plant Modifications. In March, 1973, the district retained a consultant to prepare a report on the design of plant improvements to increase the capacity from 3.75 to 6.2 mgd (14.2 to 23.5 Ml/d). Previous preliminary studies by the district engineering staff had indicated that it might be possible to accomplish this in a short time period without the need for major plant additions. The consultants' design report confirmed these findings and recommended construction of the necessary improvements. The firm capacity or output of the plant in terms of sustained 30-day production was to be increased to 6.2 mgd (23.5 Ml/d).

The major changes required were modification of the clarifier by adding settling tube modules, replacement of the existing filter sand with new mixed media, and the installation of a coagulant control and turbidity monitoring center. These changes were expected not only to allow increased production but also to improve water quality.

To handle the higher flow rates it was necessary to increase the capacity of certain pumps, pipelines, meters, chemical feeders and valves, and modify controls.

By making an immediate start on the project, it was believed that the additional capacity would be ready within the time available. Consumption records indicated the plant could be out of service from about October 1, 1973, to late May, 1974. The district's board of directors immediately implemented the recommended plan, authorizing design of the facilities, and followed this shortly by awarding a construction contract in the amount of $337,445 (1974 prices). This made possible an increase of 2.45 mgd (9.3 Ml/d) in plant capacity at $137,800 per mgd ($36,300/Ml/d), a reasonable figure, particularly because about 25 percent of the project cost was for modernization and other improvements not directly related to increasing capacity.

Through excellent efforts on the part of the contractor and plant operating personnel, the expanded plant was ready for operation on May 24, 1974, just before Memorial Day weekend. Consumption records over that weekend and the days following show that any later starting date could have resulted in a serious water shortage. In spite of the initial difficult operating conditions and the lack of opportunity for pretesting, high performance standards were maintained from the outset.

In describing the improvements according to the plant flow scheme, a primary consideration was to provide more raw water to the plant. This was done by adding larger pumping units and making piping modifications. The mixing chamber, previously operating hydraulically, was equipped with a vertical mechanical rapid mixer. Provisions were made to add polymer as an aid to flocculation. Facilities were furnished for storing and feeding sodium hydroxide solution, instead of using lime for pH adjustment.

A major change was to increase the capacity of the settling basin. This was done by installing tube settling modules. As a general rule, installing tube settling modules in an existing basin will permit clarifying two to four times more water than can be clarified within the same basin volume with no loss in efficiency. In this particular case, because of the light alum floc present, it was recommended that the flow rate be increased from 4.0 mgd (15.1 Ml/d) to not more than 6.5 mgd (24.6 Ml/d), or 1.6 times the existing capacity. This would increase the surface overflow rate from 1.52 to 2.5 gpm/sq ft (3.7 to 6.1 m/h), decrease the detention time from 1.3 to 0.8 hour, and increase the weir loading from 7 to 11 gpm/ft (125 to 196 m³/min/d). Good operation was obtained at these loadings.

The tube settling modules were installed in the annular outer settling compartment of the circular clarifier. Figure 24–14 illustrates the arrangements of tube modules in the existing basin. The existing sludge collection mechanism still has enough room to operate in the modified basin. The conduit capacity from the settling basin to the filters was increased by adding a second connecting pipe.

The sand filters were completely rebuilt. The filter bottoms were removed and replaced with Leopold vitrified clay blocks. The single-media sand bed was replaced with mixed media supported on a gravel bed with an intermediate layer of coarse garnet. New rotary surface washers were installed. Filter

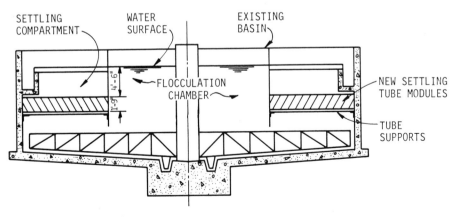

Fig. 24–14. Settling tube installation in existing flocculation–settling basin.

piping and valves were modified as necessary to accommodate the increased flows. Most of the gate valves were replaced with butterfly valves.

Coagulation Improvements. A new coagulant control center with turbidity monitoring was added to the existing facilities. The coagulant control center was installed to provide the operators with a precise method for optimizing the coagulant dosage. Continuous filter effluent turbidity monitoring assures that the California Department of Health Services mandated standard of 0.5 NTU is never exceeded.

Operating Experience. Since 1974, the plant operating experience has been good. The expanded facilities have been operated routinely at their maximum design capacity. The settling tubes have functioned well at surface overflow rates of 2.5 gpm/sq ft (6.1 m/h), and the mixed-media filters have performed very satisfactorily at rates up to 6.5 gpm/sq ft (15.9 m/h). There has been some improvement in the quality of the finished water, and plant reliability has been increased as a result of continuous monitoring and other additions and changes. Finished water turbidities are consistenly less than 0.15 NTU, and the bacteriological quality has been excellent.

A demonstration test was conducted for 2 weeks in June, 1975, to show that the design flow criteria and water quality requirements of the State Department of Health Services would be achieved and maintained. By that time, most of the pumps, motors, and automatic controls had been installed and were operative. During the previous year, the plant had been tested at 4.0 mgd (15.1 Ml /d) as well as 5.0 mgd (18.9 Ml/d) under manual control. The results of the June, 1975 test are shown in Table 24–2.

The Department of Health Services had very strict requirements regarding operational controls and maximum limits for turbidity. Each of the four individual filter effluents and clearwell effluent had to have continuous monitoring and recording of turbidity on the water being processed. If, at any time during operation, the turbidity reached 0.5 NTU on any one of the

Table 24–2. Turbidity Ranges during Demonstration Test.*

	FILTER DESIGNATIONS				
RANGE	1	2	3	4	CLEARWELL
High	0.35	0.36	0.34	0.32	0.27
Low	0.05	0.06	0.07	0.06	0.06
Mean	0.15	0.14	0.14	0.14	0.14

* All values expressed in turbidity units as registered with a Hach 2100 Nephelometer.

five instruments, the plant would automatically shut down, and the operator would have to correct the situation before operation could be resumed. As can be seen from Table 24–2, the turbidity levels remained below the minimum set by the state.

REFERENCES

1. Bowers, A. Eugene, and Beard, James D., II, "New Concepts in Filtration Plant Design and Rehabilitation," *J.AWWA* 73:9:457, 1981.
2. Forbes, Robert E., Nickerson, Gary L., Hudson, Herbert E., Jr., and Wagner, Edmund G., "Upgrading Water Treatment Plants: An Alternative to New Construction," *J.AWWA* 72:5:254, 1980.
3. Upgrading Study, Potomac Water Filtration Plant, Washington Suburban Sanitary Comm., Hyattsville, MD, September, 1978.
4. Brodeur, Timothy P., "Upgrading to Increase Treatment Capacity," *J.AWWA* 73:9:464, 1981.
5. King, Richard W., and Crossley, Eugene I., "Upgrading Water Treatment Plants in San Diego," *J.AWWA* 73:9:476, 1981.
6. Sequeira, James A., Harry, Lee, Hansen, Sigurd P., and Culp, Russell L., "Pilot Filtration Tests at the American River Water Treatment Plant," *Public Works* 114:1:36, 1983.
7. Westerhoff, Garrett P., "Experiences with Higher Filtration Rates," *J.AWWA* 63:6:376, 1971.
8. Collins, F., and Shieh, C. Y., "More Water from the Same Plant," *The American City,* p. 96, October, 1971.
9. Eunpu, Floyd F., "High Rate Filtration in Fairfax County, Virginia," *J.AWWA* 62:6:340, 1970.
10. Culp, R. L., "Increasing Water Treatment Capacity with Minimum Additions," *Public Works,* August, 1976.

Chapter 25
Plant Hydraulics

INTRODUCTION

Hydraulic performance is an important aspect of water treatment plant operation. If the headloss caused by a particular plant component or process unit is not accurately predicted, the plant's and/or unit's operation may be adversely affected or entirely inhibited. Conversely, if the predicted headloss through each plant component or process unit is too conservative or exaggerated, the cost of plant operation may be needlessly increased by higher pumping costs.

Analyzing water treatment plant hydraulics involves a myriad of hydraulic formulas and theoretical and empirical concepts. The designer should be aware of all the principles involved in the formulas so that a proper application of the formulas is made, and the limitations of the formulas are recognized. Because hydraulics is not an exact science, the designer may utilize a particular design philosophy for hydraulics. The design philosophy would include where and when to use safety factors and when to determine the magnitude of the safety factors.

This chapter will discuss the general design considerations necessary to calculate hydraulic headloss properly, and also will review the typical hydraulic applications encountered in water treatment plant design. The theoretical development of hydraulic concepts is discussed in the references included at the end of this chapter.

GENERAL DESIGN CONSIDERATIONS

Capacity

The initial task in determining plant hydraulics is to determine the headloss through each plant component. Once the design capacity of the plant is determined, the maximum and minimum flows anticipated for each unit should be defined. Considerations for determining the maximum flow should be based on how the plant or component is controlled, as the maximum flow is often greater than the capacity of the plant. For instance, if a particular unit is controlled by a pump, the maximum flow to that unit would be the pump's maximum flow, which might not be the same as the plant capacity.

There may also be recycle flows within the plant that would affect the maximum flow to a unit. For instance, a backwash decant recycle system may operate on level control and return to a clarifier periodically at a constant rate. Because this may occur at a time when maximum flow is being experienced by the clarifier, the recycle flow would have to be considered in the hydraulic design of the clarifier. Consideration should be given to such temporary peak conditions when determining the headloss through a particular unit.

When multiple units operate in parallel and one is out of service for maintenance, the remaining units must be capable of hydraulically handling the flow without major disruption to the plant performance. This would apply to all related appurtenances such as feed pipes, valves, and so on.

Units, such as filters, that normally are taken out of service for backwashing should be capable of handling the flows with some units out of service for backwashing, as well as allowing for some units to be out of service for repair or maintenance.

Available Head

Each water plant has unique parameters affecting the available head. These parameters may influence the design of a process, the plant layout, the type of plant control, and the method of water transport. The available head will determine the cost of the headloss. In mountainous or hilly terrain it may be feasible to design a plant without need for pumping the main flow. In such cases the cost of headloss is negligible. In fact, excessive available head may be used to generate power.

In flat areas or in the treatment of well water, additional head will result in increased operating costs for pumping. At a power cost of $0.075/kWh, and a wire-to-water efficiency of 70 percent, it costs $123/year to pump 1 mgd (3.78 Ml/d) 1 foot (0.31 m). At 8 percent interest over a 20-year design life, the present worth cost is approximately $1,200. Therefore a 50-mgd (189.3-Ml/d) plant with 3 feet (0.91 m) of wasted head results in an equivalent additional present worth cost of approximately $180,000.

When an existing water plant is expanded or upgraded, the available head is known accurately. Additional new processes in parallel with the existing processes must be designed to match the available head.

COMPONENT DESIGN CONSIDERATIONS

There are two types of flow encountered in water plant design: pipe flow and open channel flow. Pipe flow occurs when a conduit flows full, and open channel flow occurs when the flow has a free surface subject to atmospheric pressure. Figure 25-1 illustrates the two types of flow and their compo-

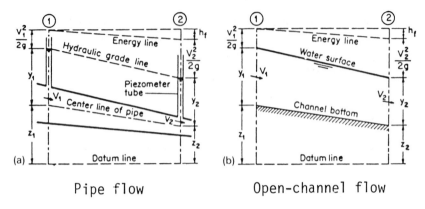

Pipe flow   Open-channel flow

Fig. 25–1. Comparison of (a) pipe flow and (b) open-channel flow.[1] Reproduced by permission of McGraw-Hill Book Co.

nents. The hydraulic grade line for pipe flow is represented by the water level in the piezometer (vertical) tubes shown in Fig. 25–1a. The water levels in the tubes are maintained by pressure exerted by the water in each section of pipe. The hydraulic grade line in an open channel where the flow is parallel with uniform velocity distribution and a slight slope is the water surface.

The energy grade line, or energy line, represents the total energy in the flow of a particular section. When referenced to an arbitrary datum line as illustrated in Fig. 25–1, the energy line is the sum of the pipe centerline height (z) (or channel bottom height), the piezometric height (y), and the velocity head ($V^2/2g$). The loss of energy that results when the liquid flows from one point to another is the headloss (H_L).

PIPES

In calculating the headloss in a pressure pipe system, three general conditions are: the pipe entrance, the piping system, and the pipe outlet. Typically, the headloss for each component of the system is determined in terms of velocity head or equivalent length of pipe:

$$H_L = K_L \cdot \frac{V^2}{2g} \qquad (25–1)$$

where:

H_L = headloss, ft (m)
K_L = headloss coefficient, dimensionless
V = velocity, ft/sec (m/s)
g = acceleration of gravity
= 32.174 ft/sec² (9.82 m/s²)

Table 25–1 lists typically accepted values of losses due to turbulence in terms of velocity head. These losses are in addition to pipe friction losses. The nomograph in Fig. 25–2 illustrates the equivalent pipe length method for some fittings and valve types.

Entrance

The entrance configuration is commonly a pipe in a tank wall, such as a reservoir outlet box, chemical mixing box, flow split box, junction box, and so forth. A typical piping system is shown in Fig. 25–3.

Table 25–1. Special Losses of Head (K_L) in Terms of $V^2/2g$.

LOSSES IN PIPES, PIPE FITTINGS AND VALVES*

APPURTENANCE, ALPHABETICALLY	K_L
BENDS	
90° Elbow	Smooth – Rough
Flanged, regular	0.21 – 0.30
Flanged, long radius	0.14 – 0.23
90° Bend	
Screwed, short radius	.9
Screwed, medium radius	.75
Screwed, long radius elbow	.60

Losses in terms of $V^2/2g$

VELOCITY, FT/SEC	LENGTH OF BEND, IN FEET								
	0	0.5	1	2	6	8	10	20	30
2	1.03	0.31	0.21	0.19	0.18	0.21	0.26	0.45	0.53
5	1.30	0.38	0.26	0.23	0.22	0.26	0.32	0.57	0.67
10	1.54	0.46	0.31	0.28	0.28	0.31	0.38	0.68	0.79

Intersection of two cylinders	
(not rounded, e.g., welded pipe)	1.25 – 1.8
45° Elbow	
Screwed, regular	0.30 – 0.42
Flanged, long radius	0.18 – 0.20
Flanged, regular	0.20 – 0.30
General rule to use ¾ of	
loss for 90° bend of same radius.	
22½° Bend	
Use ½ of loss for 90° bend	
of same radius	
Standard 45° bend (4–18″)	0.20 – 0.30
Obtuse-Angled	
Deflection of pipe ($\theta°$) < 90°	$1.5 \left[\dfrac{\theta°}{90°} \right]^2$

Table 25-1 (*Continued*)

LOSSES IN PIPES, PIPE FITTINGS, AND VALVES*

APPURTENANCE, ALPHABETICALLY	K_L
Return Bend	
Flanged, regular	0.38
Flanged, long radius	0.25
Screwed, regular	2.2
Any Bend (except as above)	
θ = angle of bend	$0.25\sqrt{\dfrac{\theta}{90°}}$
Wye Branches or 45° Laterals	
Use ¾ of the loss for a tee, or	1.0
Tee	
Standard, bifurcating	1.50 – 1.80
Standard, 90° turn	1.80
Standard, run of tee	.60
Reducing, run of tee of ½	.90
Reducing, run of tee of ¼	.75
Use losses for 90° bend with zero radius	
Miter Bends	

	K_s (SMOOTH SURFACE)	K_r (ROUGH SURFACE)
5° deflection angle	0.016 –	0.024
10°	0.034 –	0.044
15°	0.042 –	0.062
22.5°	0.066 –	0.154
30°	0.130 –	0.165
45°	0.236 –	0.320
60°	0.471 –	0.684
90°	1.129 –	1.265

CONDUITS, Closed Pipes or Open Channels

90° Bends (Velocity 2 to 6 ft/sec)	
0.0 R	1.0 – 1.4
0.25 R	0.5 – 0.6
0.50 R	0.3 – 0.4
1'–8' R	0.2 – 0.3
10' R	0.3 – 0.35
15' R	0.4 – 0.5
20' R	0.5 – 0.6
25' R	0.55 – 0.65

CONTRACTION, SUDDEN

$d/D = \frac{1}{4}$	.42
$d/D = \frac{1}{2}$	.33
$d/D = \frac{3}{4}$	.19

ENLARGEMENT, SUDDEN (due to turbulence)

(V_1 = downstream velocity)

Sharp-cornered outlet	$1.0\,(V_2^2/2g - V_1^2/2g)$

Table 25-1 (*Continued*)

LOSSES IN PIPES, PIPE FITTINGS, AND VALVES*

APPURTENANCE, ALPHABETICALLY	K_L
Bell-mouthed outlet	$0.1\,(V_2^2/2g - V_1^2/2g)$
In terms of velocity of small end	
$d/D = \frac{1}{4}$	.92
$d/D = \frac{1}{2}$	.56
$d/D = \frac{3}{4}$	.19
ENTRANCE	
Entrance Losses	
Pipe projecting into tank	$0.83 - 1.0$
(Borda entrance)	
End of pipe flush with tank	0.5
Slightly rounded	0.23
Bell-mouthed	0.04
GATES	
Sluice	
Submerged port in 12″ wall	2.5
Contraction in a conduit	0.5
Width equal to conduit width	
and without top submergence	0.2
Shear	
Wide open (orifice)	1.80
INCREASERS	
(V_1 = velocity of small end)	$0.25\,(V_1^2/2g - V_2^2/2g)$
Bushing or Coupling	$1.4\,(V_2^2/2g - V_1^2/2g)$
OBSTRUCTIONS IN PIPES	
Where A_1/A_0 = ratio of area of pipe to area of opening in obstruction	*Note:* Values of coefficients below are for the corresponding A_1/A_0 ratios listed (in terms of pipe velocities)

A_1/A_0	COEFFS.
1.05	0.10
1.1	0.19
1.2	0.42
1.4	0.96
1.6	1.54
1.8	2.17
2.0	2.70
2.2	3.27
2.5	4.00
3.0	5.06
4.0	6.75
5.0	8.01
6.0	9.4
7.0	10.4
8.0	11.3
9.0	12.5
10.0	13.5

Table 25-1 (Continued)

LOSSES IN PIPES, PIPE FITTINGS, AND VALVES*

APPURTENANCE, ALPHABETICALLY	K_L
Diaphragm of Thin Material	
(With concentric hole in pipe such as projecting washer or gasket)	
Diameter of hole = 0.91 diam. of pipe	0.34
Diameter of hole = 0.8 diam. of pipe	1.88
OPEN CHANNELS	
Sudden Contraction or Inlet Losses	
(In terms of downstream velocity)	
Sharp-cornered entrance	$0.5 \ (V_1^2/2g - V_2^2/2g)$
Round-cornered entrance	$0.25 \ (V_1^2/2g - V_2^2/2g)$
Bell-mouthed entrance	$0.05 \ (V_1^2/2g - V_2^2/2g)$
Turns Around Baffles	3.2
ORIFICE METERS	
Orifice to Pipe Diameter Ratio	
(In terms of pipe velocities)	
1:4 (0.25)	4.8
1:3 (0.33)	2.5
1:2 (0.50)	1.0
2:3 (0.67)	0.4
3:4 (0.75)	0.24
OUTLET	
(V_1 = velocity in pipe)	
Outlet	
From pipe into still water or atmosphere (free discharge)	1.0
From pipe to well	$0.9 \ (V_1^2/2g - V_2^2/2g)$
Bell-mouthed outlet	$0.1 \ (V_1^2/2g - V_2^2/2g)$
REDUCERS	
(Velocity of small end)	
Ordinary	0.25
Bell-mouthed	0.10
Standard	0.04
Bushing or coupling	0.05 – 2.0
VALVES	
Angle	
Wide open	2.1 – 3.1
Butterfly	
Fully open	0.30
$\theta = 10°$	0.46
$\theta = 20°$	1.38
$\theta = 30°$	3.6
$\theta = 40°$	10
$\theta = 50°$	31
$\theta = 60°$	94
$\theta = 70°$	320
$\theta = 80°$	1750

Table 25–1 (*Continued*)

LOSSES IN PIPES, PIPE FITTINGS, AND VALVES*

APPURTENANCE, ALPHABETICALLY	K_L
Check	
Horizontal lift type	8 – 12
Ball type	65 – 70
Swing check	0.6 – 2.3
Swing check (fully open)	2.5
Diaphragm	
Fully open	2.3
¾ open	2.6
½ open	4.3
¼ open	21.0
Foot	1.5
Gate	
Fully open	0.19
¼ closed	1.15
½ closed	5.6
¾ closed	24.0
Globe	
Fully open	10.0
Plug, Screwed	
(¼ turn from closed to fully open)	
Fully open	0.77
99% open	0.86
98% open	0.95
95% open	1.45
90% open	2.86
80% open	9.6
70% open	28.0
Plug, Globe, or Stop, 600 psi	
Fully open	4.0
¾ open	4.6
½ open	6.4
¼ open	780.0
"Y" or Blow-Off	2.9

VENTURI METERS

The loss of head occurs mostly in, and downstream from the throat.

(Losses are in terms of throat velocities.)

Loss between Upstream End and Throat		.03 – .06	
Total Loss through Meter for:	>0.5 DIAM. OF PIPE		0.33 TO 0.5 DIAM. OF PIPE
Total Angle of Divergence = + 5°	1/7		1/10
Total Angle of Divergence = +15°	1/3		1/16

Long Tube
(upstream angle of 10.5° and a downstream
angle of 2.5°)

Table 25-1 (*Continued*)

LOSSES IN PIPES, PIPE FITTINGS, AND VALVES*

APPURTENANCE, ALPHABETICALLY	K_L
Pipe Diameter	
6"	.135
10"	.126
16"	.122
20"	.119
24"	.116
30"	.113
36"	.111
42"	.110
48"	.109
60"	.107
Throat-to-Inlet Diameter Ratio in terms of Inlet (pipe) Velocity	
(1:3) 0.33	1.0 – 1.2
(1:2) 0.50	0.44 – 0.52
(2:3) 0.67	0.25 – 0.30
(3:4) 0.75	0.20 – 0.23
Eccentric or Flat Invert	.129
Concentric	
(Throat-to-Inlet Diameter Ratio)	
$^4/_{12} = .33$	.27
$^6/_{12} = .50$	.18
$^7/_{12} = .58$	.143
$^8/_{12} = .67$	.14
$^9/_{12} = .75$	.135
Eccentric or Flat Invert	.283

* *Note:* These losses may be considered as turbulence losses in excess of friction over the same length of straight pipe or conduit.

The intake may be a sluice gate as shown in Fig. 25–3, a simple sharp-edged entrance, or a streamlined bell-mouthed entrance. The designer must decide which type to use, depending upon upstream head, the importance and cost of headloss, velocities, and downstream condition. Table 25–1 includes the K_L values for the various entrance conditions that could be used with equation (25–1). If headloss is critical and must be kept to a minimum, the radius of the bell-mouthed entrance should be $D/7$ with D = diameter of the pipe downstream.

The entrance may include a sluice gate or slide gate that is used for isolation of downstream units. For example, if the gate is large enough that it does not constrict the opening, the headloss coefficient (K_L) can be estimated to

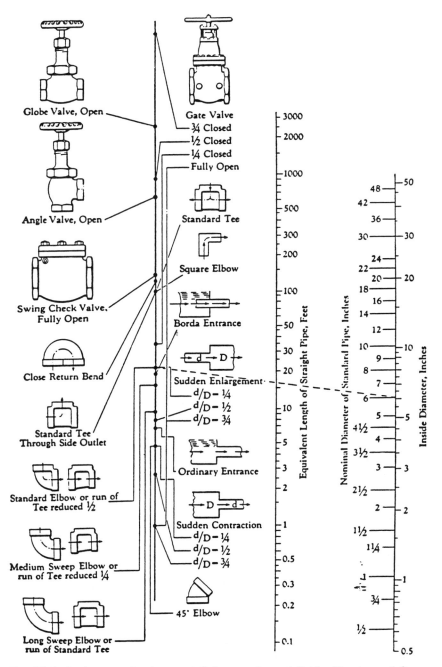

Fig. 25-2. Resistance of valves and fittings to flow of fluids. (Courtesy of Crane Co.)

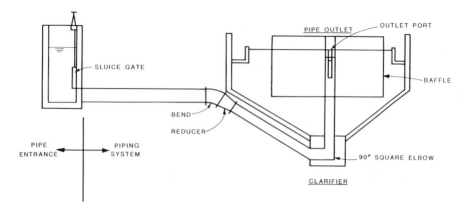

Fig. 25–3. Typical piping system.

be approximately 2.0 times the velocity head. If the sluice gate is partially submerged, the K_L value will be approximately 2.5.

Piping System

The headlosses incurred in the piping system may result from any combination of:

- Pipe friction
- Bends
- Reducer (constricting and enlarging)
- Valves
- Controls
- Branches
- Intersections
- Orifices
- Nozzles
- Venturis
- Manifolds

The resistance to liquid flow in a pipe results in friction headloss or friction pressure loss. The resistance is caused by turbulence occurring along pipe walls from interior pipe roughness and viscous shear stresses within the liquid.

The amount of headloss for a given pipe system depends on several factors:

- Size of pipe
- Pipe interior surface roughness
- Pipe length
- Liquid viscosity
- Loss through appurtenances within the system

The headloss in pipe due to friction can be approximated by several formulas, including Hazen-Williams, Darcy-Weisbach, or Manning's. The Hazen-Williams formula is:

$$V = 1.318 C_{HW} R_h{}^{0.63} S^{0.54}$$
$$V = 0.8949 C_{HW} R_h{}^{0.63} S^{0.54} \text{ (metric)} \tag{25-2}$$

V = velocity, ft/sec (m/s)
C_{HW} = Hazen-Williams roughness coefficient, dimensionless
R_h = hydraulic radius, ft (m)
S = energy slope, ft/ft (m/m)

Table 25–2 lists typical values of C_{HW} for pipes of various materials and pipe condition.
The Darcy-Weisbach formula is:

$$H_{DW} = f \frac{L}{D} \frac{V^2}{2g} \tag{25-3}$$

where:

H_{DW} = headloss, ft (m)
  f = roughness coefficient, which varies with pipe sizes, roughness, velocity, and kinematic viscosity, dimensionless
  D = diameter of pipe, ft (m)
  L = length of pipe, ft (m)
  V = average pipe velocity, ft/sec (m/s)
  g = acceleration of gravity, 32.174 ft/sec² (9.81 m/s²)

The Hazen-Williams formula can also be expressed in a more convenient form as:

Table 25–2. Hazen and Williams Friction Factor C.[*,**]

| | VALUES OF C | | |
TYPE OF PIPE	RANGE— HIGH = BEST, SMOOTH, WELL LAID— LOW = POOR OR CORRODED	AVERAGE VALUE FOR CLEAN, NEW PIPE	COMMONLY USED VALUE FOR DESIGN PURPOSES
Cement—Asbestos	160–140	150	140
Fibre	—	150	140
Bitumastic-enamel-lined iron or steel centrifugally applied	160–130	148	140
Cement-lined iron or steel centrifugally applied	—	150	140
Copper, brass, lead, tin or glass pipe and tubing	150–120	140	130
Wood-stave	145–110	120	110
Welded and seamless steel	150–80	130	100
Interior riveted steel (no projecting rivets)	—	139	100
Wrought-iron, Cast-iron	150–80	130	100
Tar-coated cast-iron	145–50	130	100
Girth-riveted steel (projecting rivets in girth seams only)	—	130	100
Concrete	152–85	120	100
Full-riveted steel (projecting rivets in girth and horizontal seams)	—	115	100
Vitrified, Spiral-riveted steel (flow with lap)	—	110	100
Spiral-riveted steel (flow against lap)	—	100	90
Corrugated steel	—	60	60

[*] Reference 22.
[**] *Note:* The Hazen-Williams friction factor C must not be confused with the Darcy-Weisbach-Colebrook friction factor f; these two friction factors are not related to each other.

$$H_{HW} = 0.002083L \frac{100^{1.85}}{C_{HW}} \frac{q^{1.85}}{D^{4.8655}}$$

$$H_{HW} = .01334L \frac{100^{1.85}}{C_{HW}} \frac{q^{1.85}}{D^{4.8655}} \text{ (metric)}$$

(25–4)

where:

H_{HW} = headloss, ft (m)
L = length of pipe, ft (m)
C_{HW} = Hazen-Williams roughness coefficient, dimensionless
q = flow, gpm (l/s)
D = pipe diameter, ft (m)

The values of f can be obtained from the Moody diagram, Fig. 25–4.

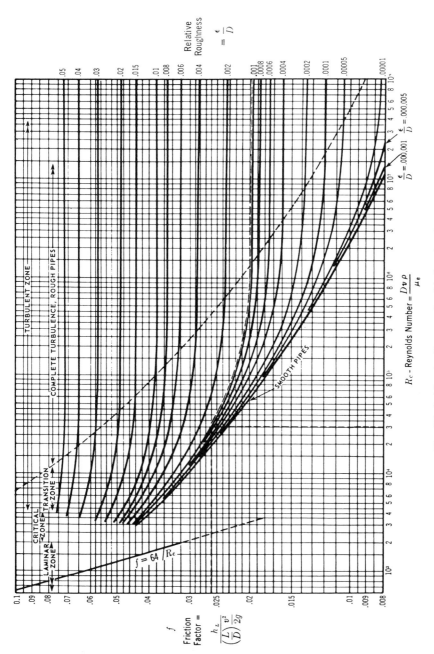

Fig. 25-4. Moody diagram for friction in pipes.

The Manning formula is:

$$V = \frac{1.49}{n} R^{2/3} S^{1/2}$$

$$V = \frac{1.0}{n} R^{2/3} S^{1/2} \text{ (metric)}$$

(25-5)

where:

V = velocity, ft/sec (m/s)
n = Manning roughness coefficient, dimensionless
R = hydraulic radius, ft (m)
  = cross-sectional area of liquid divided by wetted perimeter
S = energy slope, ft/ft (m/m)

Table 25–3 lists typical value of n, for closed conduits flowing partly full for various pipe materials.

Table 25–3. Values of the Manning Roughness Coefficient.*

TYPE OF CHANNEL AND DESCRIPTION	MINIMUM	NORMAL**	MAXIMUM
A. CLOSED CONDUITS FLOWING PARTLY FULL			
A–1. Metal			
a. Brass, smooth	0.009	0.010	0.013
b. Steel			
1. Lockbar and welded	0.010	0.012	0.014
2. Riveted and spiral	0.013	0.016	0.017
c. Cast iron			
1. Coated	0.010	0.013	0.014
2. Uncoated	0.011	0.014	0.016
d. Wrought iron			
1. Black	0.012	0.014	0.015
2. Galvanized	0.013	0.016	0.017
e. Corrugated metal			
1. Subdrain	0.017	0.019	0.021
2. Storm drain	0.021	0.024	0.030
A–2. Nonmetal			
a. Lucite	0.008	0.009	0.010
b. Glass	0.009	0.010	0.013
c. Cement			
1. Neat, surface	0.010	0.011	0.013
2. Mortar	0.011	0.013	0.015
d. Concrete			
1. Culvert, straight and free of debris	0.010	0.011	0.013

Table 25-3 (*Continued*)

TYPE OF CHANNEL AND DESCRIPTION	MINIMUM	NORMAL**	MAXIMUM
2. Culvert with bends, connections, and some debris	0.011	0.013	0.014
3. Finished	0.011	0.012	0.014
4. Sewer with manholes, inlet, etc., straight	0.013	0.015	0.017
5. Unfinished, steel form	0.012	0.013	0.014
6. Unfinished, smooth wood form	0.012	0.014	0.016
7. Unfinished, rough wood form	0.015	0.017	0.020
e. Wood			
1. Stave	0.010	0.012	0.014
2. Laminated, treated	0.015	0.017	0.020
f. Clay			
1. Common drainage tile	0.011	0.013	0.017
2. Vitrified sewer	0.011	0.014	0.017
3. Vitrified sewer with manholes, inlet, etc.	0.013	0.015	0.017
4. Vitrified subdrain with open joint	0.014	0.016	0.018
g. Brickwork			
1. Glazed	0.011	0.013	0.015
2. Lined with cement mortar	0.012	0.015	0.017
h. Sanitary sewers coated with sewage slimes, with bends and connections	0.012	0.013	0.016
i. Paved invert, sewer, smooth bottom	0.016	0.019	0.020
j. Rubble masonry, cemented	0.018	0.025	0.030
B. LINED OR BUILT-UP CHANNELS			
B–1. Metal			
a. Smooth steel surface			
1. Unpainted	0.011	0.012	0.014
2. Painted	0.012	0.013	0.017
b. Corrugated	0.021	0.025	0.030
B–2. Nonmetal			
a. Cement			
1. Neat, surface	0.010	0.011	0.013
2. Mortar	0.011	0.013	0.015
b. Wood			
1. Planed, untreated	0.010	0.012	0.014
2. Planed, creosoted	0.011	0.012	0.015
3. Unplaned	0.011	0.013	0.015
4. Plank with battens	0.012	0.015	0.018
5. Lined with roofing paper	0.010	0.014	0.017
c. Concrete			
1. Trowel finish	0.011	0.013	0.015
2. Float finish	0.013	0.015	0.016
3. Finished, with gravel on bottom	0.015	0.017	0.020
4. Unfinished	0.014	0.017	0.020
5. Gunite, good section	0.016	0.019	0.023

Table 25–3 (Continued)

TYPE OF CHANNEL AND DESCRIPTION	MINIMUM	NORMAL**	MAXIMUM
6. Gunite, wavy section	0.018	0.022	0.025
7. On good excavated rock	0.017	0.020	
8. On irregular excavated rock	0.022	0.027	
d. Concrete bottom float finished with sides of			
1. Dressed stone in mortar	0.015	0.017	0.020
2. Random stone in mortar	0.017	0.020	0.024
3. Cement rubble masonry, plastered	0.016	0.020	0.024
4. Cement rubble masonry	0.020	0.025	0.030
5. Dry rubble or riprap	0.020	0.030	0.035
e. Gravel bottom with sides of			
1. Formed concrete	0.017	0.020	0.025
2. Random stone in mortar	0.020	0.023	0.026
3. Dry rubble or riprap	0.023	0.033	0.036
f. Brick			
1. Glazed	0.011	0.013	0.015
2. In cement mortar	0.012	0.015	0.018
g. Masonry			
1. Cemented rubble	0.017	0.025	0.030
2. Dry rubble	0.023	0.032	0.035
h. Dressed ashlar	0.013	0.015	0.017
i. Asphalt			
1. Smooth	0.013	0.013	
2. Rough	0.016	0.016	
j. Vegetal lining	0.030	—	0.500
C. EXCAVATED OR DREDGED			
a. Earth, straight and uniform			
1. Clean, recently completed	0.016	0.018	0.020
2. Clean, after weathering	0.018	0.022	0.025
3. Gravel, uniform section, clean	0.022	0.025	0.030
4. With short grass, few weeds	0.022	0.027	0.033
b. Earth, winding and sluggish			
1. No vegetation	0.023	0.025	0.030
2. Grass, some weeds	0.025	0.030	0.033
3. Dense weeds or aquatic plants in deep channels	0.030	0.035	0.040
4. Earth bottom and rubble sides	0.028	0.030	0.035
5. Stony bottom and weedy banks	0.025	0.035	0.040
6. Cobble bottom and clean sides	0.030	0.040	0.050
c. Dragline-excavated or dredged			
1. No vegetation	0.025	0.028	0.033
2. Light brush on banks	0.035	0.050	0.060
d. Rock cuts			
1. Smooth and uniform	0.025	0.035	0.040
2. Jagged and irregular	0.035	0.040	0.050
e. Channels not maintained, weeds and brush uncut			

Table 25-3 (Continued)

TYPE OF CHANNEL AND DESCRIPTION	MINIMUM	NORMAL**	MAXIMUM
1. Dense weeds, high as flow depth	0.050	0.080	0.120
2. Clean bottom, brush on sides	0.040	0.050	0.080
3. Same, highest stage of flow	0.045	0.070	0.110
4. Dense brush, high stage	0.080	0.100	0.140
D. NATURAL STREAMS			
D–1. Minor streams (top width at flood stage <100 ft)			
a. Streams on plain			
1. Clean, straight, full stage, no rifts or deep pools	0.025	0.030	0.033
2. Same as above, but more stones and weeds	0.030	0.035	0.040
3. Clean, winding, some pools and shoals	0.033	0.040	0.045
4. Same as above, but some weeds and stones	0.035	0.045	0.050
5. Same as above, lower stages, more ineffective slopes and sections	0.040	0.048	0.055
6. Same as 4, but more stones	0.045	0.050	0.060
7. Sluggish reaches, weedy, deep pools	0.050	0.070	0.080
8. Very weedy reaches, deep pools, or floodways with heavy stand of timber and underbrush	0.075	0.100	0.150
b. Mountain streams, no vegetation in channel, banks usually steep, trees and brush along banks submerged at high stages			
1. Bottom: gravels, cobbles, and few boulders	0.030	0.040	0.050
2. Bottom: cobbles with large boulders	0.040	0.050	0.070
D–2. Flood plains			
a. Pasture, no brush			
1. Short grass	0.025	0.030	0.035
2. High grass	0.030	0.035	0.050
b. Cultivated areas			
1. No crop	0.020	0.030	0.040
2. Mature row crops	0.025	0.035	0.045
3. Mature field crops	0.030	0.040	0.050
c. Brush			
1. Scattered brush, heavy weeds	0.035	0.050	0.070
2. Light brush and trees, in winter	0.035	0.050	0.060
3. Light brush and trees, in summer	0.040	0.060	0.080

Table 25–3 (Continued)

TYPE OF CHANNEL AND DESCRIPTION	MINIMUM	NORMAL**	MAXIMUM
4. Medium to dense brush, in winter	0.045	0.070	0.110
5. Medium to dense brush, in summer	0.070	0.100	0.160
d. Trees			
1. Dense willows, summer, straight	0.110	0.150	0.200
2. Cleared land with tree stumps, no sprouts	0.030	0.040	0.050
3. Same as above, but with heavy growth of sprouts	0.050	0.060	0.080
4. Heavy stand of timber, a few down trees, little undergrowth, flood stage below branches	0.080	0.100	0.120
5. Same as above, but with flood stage reaching branches	0.100	0.120	0.160
D–3. Major streams (top width at flood stage >100 ft). The n value is less than that for minor streams of similar description, because banks offer less effective resistance.			
a. Regular section with no boulders or brush	0.025	—	0.060
b. Irregular and rough section	0.035	—	0.100

* Reference 1. Reproduced by permission of McGraw-Hill Book Co.
** Values typically used in practice.

The Hazen-Williams formula is the most commonly used formula for water plant hydraulic calculations.[6] Tabulated values of headloss per 1,000 feet can be found in Reference 6. The headloss is given for various pipe sizes, C_{HW} values, and flows, which allows simple calculation of the expected friction losses.

The Darcy-Weisbach equation is sometimes used because the headloss is expressed in terms of a constant times the velocity head, as are many of the other system losses.

Although the Manning formula is more typically used with open channel flow, it can also be applied to pipes flowing full.

The headloss incurred by bends, valves, controls, reducers, branches, and intersections can be approximated by equation (25–1).

The values of K_L vary within the published literature, and median values are shown in Table 25–1. The designer should adopt a set of K_L values for each condition and use them in a consistent manner.

Headloss coefficient values for valves and control devices should be obtained from the manufacturer. This is especially important for control devices. The headloss coefficient will vary widely with the type of control and may even vary significantly for the same type of control produced by different manufacturers. The cavitation potential for control devices must also be investigated thoroughly. Reference 21 is an excellent source of information on cavitation parameters for valves and control devices.

Headloss in pipes that are enlarged or expanded can be approximated by:

$$H_L = K_L \frac{V_1{}^2 - V_2{}^2}{2g} \qquad (25\text{--}6)$$

Where V_1 = velocity at the entrance to the enlargement, ft/sec (m/s), and V_2 = velocity at the exit from the enlargement, ft/sec (m/s). In this case, values for K_L depend on the expansion rate, as shown in Fig. 25–5b. (Figure 25–5a includes coefficients for sudden contractions.)

The K_L value for a sudden large increase is approximately 1.0. Sudden enlargements are very effective head reducers. They can also withstand cavitation without significant vibration and damage to the system because the cavitation will take place in the fluid. Figure 25–6 illustrates the fluid action in a sudden enlargement. The diameter ratio of a sudden enlargement controls the severity of the headloss and cavitation potential. The angle of enlargement should be kept below 4 degrees for minimal headloss and cavitation.

Orifices are used in pipelines to create a headloss and/or measure a discharge. They are often used in filter underdrain design to create a headloss during the backwash cycle to achieve an even distribution of backwash water to filter media. The headloss can be approximated using equation (25–1) with K_L values ranging from 0.15 to 100. The K_L values are based on the pipe-to-orifice area ratios, which typically range from 0.9 to 0.15. Figure 25–7 illustrates K_L values based on area ratios. The velocity used in equation (25–6) for an orifice is the mean pipe velocity. If an orifice is used, the cavitation potential should be investigated.

Nozzle headloss can be defined similarly to that for orifices. The headloss coefficient values (K_L) range from 0.15 to 100 for area ratios from 0.8 to 0.1. Figure 25–7 illustrates K_L values based on area ratios for nozzles. Nozzles, like orifices, should be checked for cavitation potential.

Headlosses for Venturi tubes are determined in a similar manner to those for orifices and nozzles. The headlosses in Venturi tubes are typically lower than those of orifices and nozzles for the flow-measuring Venturi tubes. Typical K_L values vary from 0 to 10 for area ratios of 0.8 to 0.1 for recovery angles from 5 to 7 degrees. Figure 25–8 illustrates the range of K_L values based on area ratios and recovery angles.

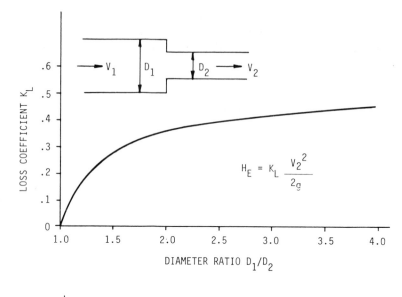

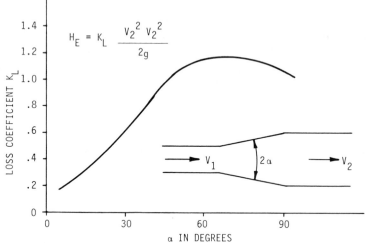

Fig. 25-5. Headloss coefficients for (a) contractions and (b) enlargements.[20] Reprinted with permission from the Hydraulic Institute Engineering Data Book (1979).

Manifolds are typically used in filter underdrains and for filter influent and effluent gallery piping. The manifold may be defined as a series of lateral pipes that connect to a main pipe. There are two types of manifolds: dividing flow and combining flow. Combining flow manifolds act differently from dividing flow manifolds. The headloss in the main pipe of a combining flow manifold is caused by the mixing of flows, and is expressed as a headloss

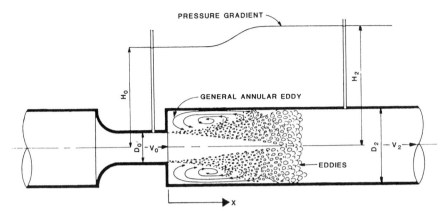

Fig. 25-6. Flow action in sudden enlargement.[20] Reprinted with permission from the Hydraulic Institute Engineering Data Book (1979).

coefficient times the downstream velocity head. Blaisdell and Mansen have verified the following equation for the headloss coefficient in the main pipe caused by a junction:[13]

$$K_j = 2 \frac{Q_B}{Q_M} = \left(1 + 2 \frac{A_M}{A_B} \cos \theta\right) \left(\frac{Q_B}{Q_M}\right)^2 \qquad (25\text{-}7)$$

where:

A_M = area of the downstream main pipe, sq ft (m²)
Q_M = flow of the downstream main pipe, cu ft/sec (m³/s)
A_B = area of the branch or lateral pipe, sq ft (m²)
Q_B = flow of the branch or lateral pipe, cu ft/sec (m³/s)
θ = angle of convergence between the main and branch pipes, degrees

   When manifolds are used for filter underdrain systems, they serve as combining flow manifolds under normal operation, and dividing flow manifolds under backwashing conditions. Filter inlet manifolds and backwash line manifolds are dividing flow types, while filter effluent manifolds are combining flow manifolds. Dividing flow manifolds are often used to distribute flow to a series of successive lateral outlet ports. Dividing flow manifolds commonly discharge to filters or sedimentation basins, or distribute backwash water to filter bottoms.

   Although it has been common practice to attempt to distribute flow evenly in this manner, it is not recommended without the use of a control device. Studies have been performed to attempt to define the mechanics of dividing flow manifolds for the purpose of determining the flow splits.[23,24] It can be

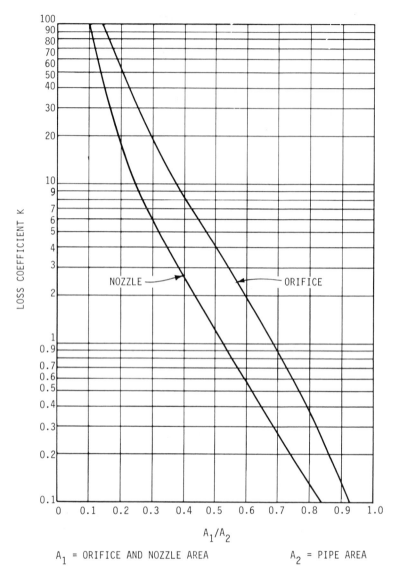

Fig. 25-7. Loss coefficients for orifices and nozzles.[20] Reprinted with permission from the Hydraulic Institute Engineering Data Book (1979).

concluded from these analyses that it is not practicable to divide the flow by manifold construction only. It is recommended that a control device such as an orifice or throttling valve be used to impart sufficient headloss in the laterals to equalize the imbalanced manifold losses. For example, in a five-lateral manifold the flow split between the first and last lateral may be 14

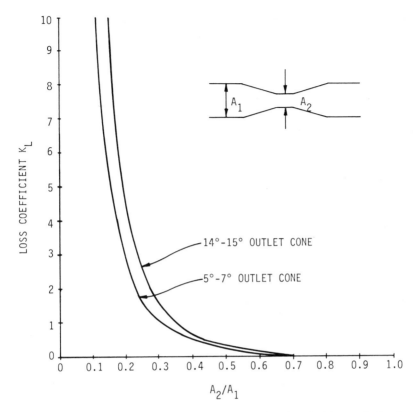

Fig. 25-8. Headloss coefficients for Venturi tubes.[20] Reprinted with permission from the Hydraulic Institute Engineering Data Book (1979).

and 25 percent, respectively. A control device would be necessary to throttle Lateral 5 sufficiently to reduce its flow to 20 percent while allowing Lateral 1 to increase to 20 percent.

The manifold should be analyzed for all extreme flow conditions of its operation. The operating flows may vary considerably and result in significant difference in headloss. The manifold should be sufficiently large to minimize velocities and thereby reduce inaccuracies in the calculation.

Outlets

The outlets to a piping system typically involve a control structure or a basin. Basins, such as flocculation basins, clarifiers, filters, and wet wells, will result in the loss of all velocity head available in the pipe system. The headloss coefficient is simply 1.0. Control structures may include such devices as gates, butterfly valve, fixed cone valves, tube valves, needle valves, or hollow-jet valves. The headloss attributable to a control device at the end of a conduit that discharges to atmosphere can be given as:[20]

$$H_C = \left(\frac{1}{C^2} - 1\right)\frac{V^2}{2g} \tag{25-8}$$

where:

H_C = headloss of control device, ft (m)
C = coefficient of discharge
V = pipe velocity, ft/sec (m/s)
g = acceleration of gravity, 32.174 ft/sec² (9.81 m/s²)

The coefficient of discharge (C) varies with the type of control opening. Equation (25-8) expressed in terms of a headloss coefficient (K_L) would be:

$$H_C = K_L \frac{V^2}{2g} \tag{25-9}$$

Various headloss and discharge coefficients for several types of outlet control devices are given in Table 25-4.

Open Channels

Open channel flow is used to convey water through many of the water plant processes. Open channel flow occurs in basins, conveyance channels, split

Table 25-4. Headloss and Discharge Coefficients.*

TYPE OF GATE OR VALVE	FREE DISCHARGE COEFFICIENT	HEADLOSS COEFFICIENT
Slide gate	0.97 to 0.93	0.07 to 0.16
Jet flow gate	0.80	0.56
Cylinder gate	0.90 to 0.80	0.23 to 0.56
Gate valve	0.95	0.11
Sphere valve	1.00	0
Butterfly valve	0.95 to 0.60	0.11 to 1.78
Tube valve	0.55 to 0.50	2.31 to 3.00
Needle valve	0.60 to 0.45	1.78 to 3.95
Fixed cone valve	0.85	0.38
Hollow-jet valve	0.73 to 0.70	0.88 to 1.04

* Reference 20.
Reprinted with permission from the Hydraulic Institute Engineering Data Book (1979).

boxes, flumes, launders or troughs, and over weirs. Under most circumstances, the flow can be assumed to approach parallel flow with a uniform velocity distribution. The types of flow typically encountered approach steady uniform flow, steady varied flow, or spatially varied (nonuniform) flow. Figure 25–9 illustrates the typical types of flow. Although these types rarely are accurately achieved, they provide reasonable approximations in most circumstances. There may be special circumstances in a plant process or design configuration resulting in other types of flow. In these cases, a more detailed investigation of the open channel hydraulics is warranted than is presented here. Several references listed in this chapter may be used.[1,2,4,9,10,17]

Conveyance Channels

Open flow conveyance channels may be pipes flowing partially full or concrete channels of various shapes. These are termed artificial flow channels as opposed to natural flow channels. Natural channels may also be encountered,

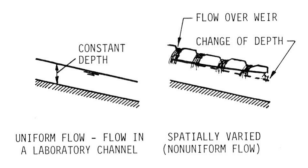

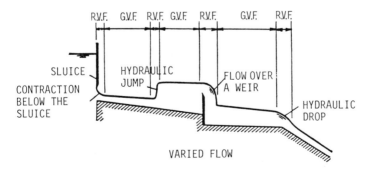

Fig. 25–9. Various types of open-channel flow.[1] G.V.F. = gradually varied flow. R.V.F. = rapidly varied flow. Reproduced by permission of McGraw-Hill Book Co.

particularly in plant influent raw water channels. Most applications for conveyance channels approach uniform flow conditions.

By far, the most widely used formula for computations involving open channel flow is the Manning formula, introduced in 1889 by Robert Manning and later modified to its present form as presented in equation (25–5).

Values of n may be obtained from Table 25–3. Manning's formula may be related to channel discharge by using the continuity equation:

$$Q = VA \qquad (25\text{--}10)$$

where:

$Q =$ discharge, cu ft/sec (m³/s)
$A =$ cross-sectional area, sq ft (m²)
$V =$ mean velocity, ft/sec (m/s)

Thus:

$$Q = \frac{1.49}{n} AR^{2/3} S^{1/2}$$
$$Q = \frac{1.0}{n} AR^{2/3} S^{1/2} \text{ (metric)} \qquad (25\text{--}11)$$

The calculation of headloss in open channels is basically the determination of the water surface elevations. These elevations are controlled by either upstream, downstream, or artificial conditions. Mild or flat slopes result in subcritical flow, and surface profiles are generally controlled by downstream or artificial conditions. Steep slopes result in supercritical flow, and surface profiles are generally controlled by upstream or artificial conditions. Steep slopes and supercritical flows are not typically encountered in water treatment plant conveyance means.

It may be necessary to calculate the water profile in a pipe flowing partially full. Since this is classified as open channel flow, the relation between the hydraulic elements as computed by Manning's equation must be known. Figure 25–10 illustrates the relationships between the hydraulic elements of a circular pipe.

King and Brater's *Handbook of Hydraulics*[3] contains tables (reprinted here as Tables 25–5 and 25–6) that can be used to calculate the depth of flow in a pipe flowing partially full, as the following example shows.

Example: Determine the depth of flow in a 36-inch (9.1-m)-diameter cast iron pipe flowing partially full with a slope of 0.010, an n value of 0.013, and a flow of 20 cu ft/sec (0.566 m³/s).

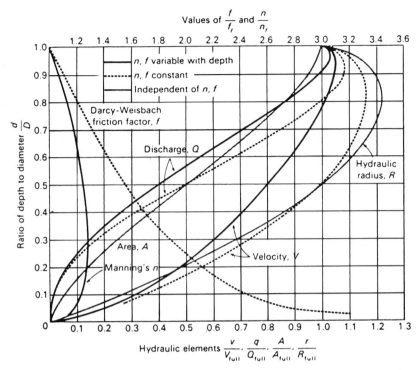

Values of $\dfrac{f}{f_t}$ and $\dfrac{n}{n_t}$

Fig. 25–10. Hydraulic elements for circular pipes.[3] Reproduced by permission of McGraw-Hill Book Co.

Table 25–5. Values of K for Circular Channels in the Equation Shown.[3] Based on Water Depth

$$Q = \frac{K}{n} D^{8/3} S^{1/2}*$$

(D = DEPTH OF WATER: d = DIAMETER OF CHANNEL)

$\dfrac{D}{d}$	0.00	0.01	0.02	0.03	0.04	0.05	0.06	0.07	0.08	0.09
0.0		15.02	10.56	8.57	7.38	6.55	5.95	5.47	5.08	4.76
0.1	4.49	4.25	4.04	3.86	3.69	3.54	3.41	3.28	3.17	3.06
0.2	2.96	2.87	2.79	2.71	2.63	2.56	2.49	2.42	2.36	2.30
0.3	2.25	2.20	2.14	2.09	2.05	2.00	1.96	1.92	1.87	1.84
0.4	1.80	1.76	1.72	1.69	1.66	1.62	1.59	1.56	1.53	1.50
0.5	1.470	1.442	1.415	1.388	1.362	1.336	1.311	1.286	1.262	1.238
0.6	1.215	1.192	1.170	1.148	1.126	1.105	1.084	1.064	1.043	1.023
0.7	1.004	0.984	0.965	0.947	0.928	0.910	0.891	0.874	0.856	0.838
0.8	0.821	0.804	0.787	0.770	0.753	0.736	0.720	0.703	0.687	0.670
0.9	0.654	0.637	0.621	0.604	0.588	0.571	0.553	0.535	0.516	0.496
1.0	0.463									

* where: Q = flow rate, n = friction factor, and S = slope of conduit. Reproduced by permission of McGraw-Hill Book Co.

Table 25–6. Values of K' for Circular Channels in the Equation Shown.[3] Based on Diameter of Channel

$$Q = \frac{K'}{n} d^{8/3}S^{1/2}*$$

$(D = \text{DEPTH OF WATER}; \; d = \text{DIAMETER OF CHANNEL})$

$\dfrac{D}{d}$	0.00	0.01	0.02	0.03	0.04	0.05	0.06	0.07	0.08	0.09
0.0		0.00007	0.00031	0.00074	0.00138	0.00222	0.00328	0.00455	0.00604	0.00775
0.1	0.00967	0.0118	0.0142	0.0167	0.0195	0.0225	0.0257	0.0291	0.0327	0.0366
0.2	0.0406	0.0448	0.0492	0.0537	0.0585	0.0634	0.0686	0.0738	0.0793	0.0849
0.3	0.0907	0.0966	0.1027	0.1089	0.1153	0.1218	0.1284	0.1352	0.1420	0.1490
0.4	0.1561	0.1633	0.1705	0.1779	0.1854	0.1929	0.2005	0.2082	0.2160	0.2238
0.5	0.232	0.239	0.247	0.255	0.263	0.271	0.279	0.287	0.295	0.303
0.6	0.311	0.319	0.327	0.335	0.343	0.350	0.358	0.366	0.373	0.380
0.7	0.388	0.395	0.402	0.409	0.416	0.422	0.429	0.435	0.441	0.447
0.8	0.453	0.458	0.463	0.468	0.473	0.477	0.481	0.485	0.488	0.491
0.9	0.494	0.496	0.497	0.498	0.498	0.498	0.496	0.494	0.489	0.483
1.0	0.463									

* where: Q = flow rate, n = friction factor, and S = slope of conduit. Reproduced by permission of McGraw-Hill Book Co.

Step 1. Calculate K' in the equation:

$$Q = \left(\frac{K'}{n}\right) d^{8/3} S^{1/2}$$

Rearranging and substituting:

$$K' = \frac{nQ}{d^{8/3} S^{1/2}} = \frac{0.013 \, (2.0)}{(3)^{8/3} \, (0.01)^{1/2}} = 0.0139$$

Step 2. Determine the depth of flow:
Find D/d value for $K' = 0.0139$:

$$D/d = 0.12$$

Therefore the depth of flow is:

$$D/d \times d = 0.12 \times 36 = 4.32 \text{ inches or } 0.36 \text{ foot (110 mm)}$$

If desired, the velocity can then be obtained using Fig. 25–10. The required pipe size can also be calculated to match a given flow line if desired.

Weirs

The headloss through a basin, whether it is a flow split basin, flocculation basin, or settling basin, is generally controlled by weirs at the outlet. Weirs are also used for flow splitting and flow measurement. The headloss over a weir depends on the type, length, and flow rate. There are a number of different types of weirs, as illustrated in Fig. 25–11. The headloss over weir

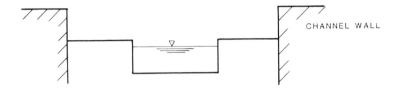

a. Rectangular – contracted

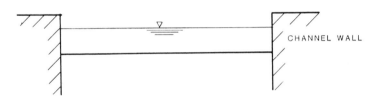

b. Rectangular–suppressed

c. Cipolletti

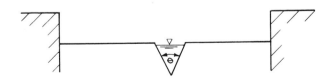

d. V-notch– Θ = 90°, 60°, 45°, 22 1/2°

Fig. 25-11. Weir types.

for given flows and weir shapes is widely published in hydraulic and water resource handbooks. The formulas for the weir types illustrated in Fig. 25–11 are as follows:

- Rectangular—contracted weir:

$$Q = 3.33 (L - 0.2H) H^{3/2}$$
$$Q = 1.84 (L - 0.2H) H^{3/2} \text{ (metric)}$$
(25–12)

- Rectangular—suppressed weir:

$$Q = 3.33 LH^{3/2}$$
$$Q = 1.84 LH^{3/2} \text{ (metric)}$$
(25–13)

- Cipolletti weir:

$$Q = 3.367 LH^{3/2}$$
$$Q = 1.858 LH^{3/2} \text{ (metric)}$$
(25–14)

- V-notch weir:

$$Q = XH^{2.48}$$
(25–15)

where:

Q = discharge, cu ft/sec (m³/s)
L = length of weir, ft (m)
H = head on weir crest, ft (m)
X = 2.49 (1.34) for α = 90°
X = 1.443 (0.778) for α = 60°
X = 1.035 (0.558) for α = 45°
X = 0.497 (0.268) for α = 22½°

Baffles

Some basins contain a number of baffles that are used for mixing. When the flow velocity is not negligible, the headloss through the basin must be considered. Experimental tests have determined that the K_L value to be used in equation (25–1) is approximately 3.3. This was determined for 180-degree bends in flow.

Launder and Troughs

The flow in the clarifier launder and backwash trough is considered spatially varied flow or nonuniform flow. To determine the maximum water level in a launder, the desired solution is the upstream head in the launder at the farthest point from the discharge point, as shown in Fig. 25–12. The general formula is:[1]

$$H_L = \sqrt{(2h_c{}^3/h_0) + \left(h_0 - \frac{1}{3} iL\right)^2} - \frac{2}{3} iL \qquad (25\text{–}16)$$

where:

H_L = headloss, ft (m)
h_c = critical depth, ft (m) = $(Q^2/gb^2)^{1/3}$ for rectangular channels
L = length of channel, ft (m)
h_0 = depth at discharge location, ft (m)
i = launder slope, ft/ft (m/m)
g = acceleration of gravity, 32.174 ft/sec² (9.81 m/s²)
b = channel width, ft (m)
Q = discharge, cu ft/sec (m³/s)

When the channel is level, as most clarifier landers are, $i = 0$, and the formula may be reduced to:

$$H_L = \sqrt{(2h_c{}^3/h_0) + h_0{}^2} \qquad (25\text{–}17)$$

If the discharge is free, that is, if $h_0 = h_c$, then the equation may be further reduced:

$$\begin{aligned} H_L &= \sqrt{3h_c{}^2} \\ &= h_c \sqrt{3} \\ &= 1732\, h_c \end{aligned} \qquad (25\text{–}18)$$

Effluent launders should be sized to provide at least 4 inches (0.10 m) freeboard from the bottom of the V-notches at the upstream end.

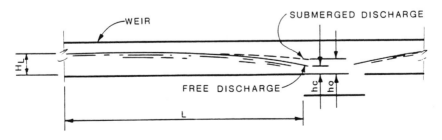

Fig. 25–12. Clarifier launder flow dimensions.

Filter washwater troughs act identically to clarifier launders, and the same criteria apply. A rectangular washwater trough with free discharge can be sized by the following equation:

$$Q = 2.49\ bH_L{}^{3/2}$$
$$Q = 1.37\ bH_L{}^{3/2}\ \text{(metric)}$$

(25–19)

where b = width of launder, ft (m).

Filters

The determination of headloss through filters is made for both filter operation and filter backwashing. The operational headloss consists of two components: headloss through a clean bed and the headloss allowed between filter runs before backwashing. The headloss allowed between backwashes can be chosen on the basis of several factors, including cost of available headloss, filter media design, and filter breakthrough. Typical design headloss values range from 5 to 10 feet (1.524 to 3.048 m).

The headloss through a clean filter bed can be calculated through the use of a number of equations including the Carmen-Kozeny, Fair-Hatch, and Rose equations, which are given below for reference. However, unless a unique filter system is designed and headloss is critical, the benefit of these equations is questionable. Typical clean filter beds and underdrain systems, whether mixed- or dual-media, generally impart 1 to 1½ feet (0.31 to 0.46 m) of headloss. Specific media manufacturers or suppliers should be consulted when headloss values are determined.

The Carmen-Kozeny equation, for use with uniform-size media, is:

$$H_L = \epsilon \left(\frac{l}{\phi d}\right)\left(\frac{1-f}{f^3}\right)\left(\frac{V^2}{2g}\right)$$

(25–20)

where:

H_L = headloss, ft (m)
  l = depth of media, ft (m)
  ϕ = particle shape factor, dimensionless
    = 1.0 for sphere
    = 0.28 for mica flakes
    = 0.82 for rounded sand
    = 0.73 for angular sand
  d = particle diameter, ft (m)
  ϵ = bed porosity, dimensionless
  V = filtration rate of water, ft/sec (m/s)

g = acceleration of gravity, 32.174 ft/sec² (9.81 m/s²)
f = friction factor, dimensionless

$$= 150 \frac{1-f}{Re} + 1.75$$

Re = Reynolds number, dimensionless

$$= \frac{\phi d \overline{Vs} \, \rho}{\mu}$$

$\overline{Vs}$ = average superficial velocity through an empty bed, ft/sec (m/s)
ρ = density, slugs/cu ft (kg/m³)
μ = dynamic viscosity, lb · sec/sq ft (N · s/m²)

The Fair-Hatch equation for nonuniform media is:

where:

$$\frac{H_L}{l} = \frac{6K}{g} \, \nu \, \frac{1 - \epsilon^2}{f^3 \phi} \sum_{i=1}^{2n} \frac{P_i}{d_p{}^2} \qquad (25\text{--}21)$$

H_L = headloss, ft (m)
l = depth of media, ft (m)
K = Kozeny constant, ≈ 5
ν = kinematic viscosity, sq ft/sec (m²/s)
g = acceleration of gravity, 32.174 ft/sec² (9.81 m/s)
V = filtration rate of water, ft/sec (m/s)
ϵ = bed porosity, dimensionless
ϕ = shape factor, dimensionless
P_i = sand fraction within adjacent sieve sizes, by weight
d_p = geometric mean size, ft (m) = $[(d_1)(d_2)]^{1/2}$

The Rose equation for uniform-size media is:

$$H_L = f \left(\frac{l}{\phi d} \right) \left(\frac{1}{f^4} \right) \left(\frac{V^2}{g} \right) \qquad (25\text{--}22)$$

The Rose equation for nonuniform-size media is:

$$H_L = \left(\frac{1.067}{\phi} \right) \left(\frac{l}{g} \right) \left(\frac{V^2}{f^4} \right) \sum_{i=1}^{n} \frac{C_D P_i}{d_p} \qquad (25\text{--}23)$$

where:

H_L = headloss, ft (m)
l = depth of media, ft (m)

ϕ = shape factor, dimensionless
d = particle diameter, ft (m)
f = porosity, dimensionless
g = acceleration of gravity, 32.174 ft/sec² (9.81 m/s)
V = filtration rate of water, ft/sec (m/s)
P_i = sand fraction within adjacent sieve sizes, by weight
d_p = geometric mean size, ft (m) = $[(d_i)(d_2)]^{1/2}$
f = friction factor, dimensionless
  = 1.067 C_D

$$C_D = \frac{24}{Re} + \frac{3}{\sqrt{Re}} + 0.34 \text{ for transitional flow}$$

$$C_D = \frac{24}{Re} \text{ for laminar flow}$$

Re = Reynolds number

The headloss through a filter bed during backwash can be calculated as simply the sum of headlosses of components of the backwash system:

$$H_L = h_b + h_g + h_u + h_f \qquad (25\text{–}24)$$

where:

H_L = total headloss, ft (m)
h_b = headloss through the expanded bed, ft (m)
h_g = headloss through the gravel support bed, ft (m)
h_u = headloss through the underdrain system, ft (m)
h_f = headloss through the remaining backwash system, ft (m)

The headloss through the underdrain and backwash system can be determined from previous discussions in this chapter. As in the clean filter bed, the headloss through the expanded bed and gravel support system is more difficult to calculate. Typical values are 2 to 5 feet (0.61 to 15.24 m), but the bed and gravel support headloss may be calculated by the following formula:

Expanded bed headloss:

$$h_b = l_b (1 - f)(S_s - 1) \qquad (25\text{–}25)$$

where:

h_b = headloss through the expanded bed, ft (m)
l_b = depth of bed (at rest), ft (m)

f = porosity of bed, dimensionless
S_s = specific gravity of the filter media, dimensionless

Gravel support bed headloss:

$$h_g = 200 \, l_g \left(\frac{V\mu}{\rho g \, \phi^2 d_{10}{}^2} \right) \left(\frac{(1-f_g)^2}{f_g{}^3} \right) \qquad (25\text{--}26)$$

where:

h_g = headloss through the gravel support bed, ft (m)
l_g = depth of bed, ft (m)
V = backwash rate, ft/sec (m/s)
μ = dynamic viscosity, slugs/ft-sec (n · s/m²)
ρ = density, slugs/cu ft (kg/m³)
g = acceleration of gravity, 32.174 ft/sec² (9.81 m/s)
ϕ = shape factor of gravel, dimensionless
d_{10} = size of gravel of which 10% by weight is less than, ft (m)
f_g = porosity of gravel bed, dimensionless

Some of the information needed to utilize the above formulas on filter bed headloss may not be available to the designer. In that case, it is recommended that one calculate all of the system headloss that can be determined from the available information. The remaining unknown headloss values can then be estimated based on experience or on the operation of similar installations. Filter manufacturers can also be consulted when headloss determinations are critical.

REFERENCES

1. Chow, Ven Te, *Open Channel Hydraulics*, McGraw-Hill, New York, 1959.
2. Merris, H. M., and Wiggert, J. M., *Applied Hydraulics in Engineering*, The Ronald Press, New York, 1972.
3. King, H. W., and Brater, E. F., *Handbook of Hydraulics*, 6th ed., McGraw-Hill, New York, 1976.
4. Daugherty, R. L., and Franzini, J. B., *Fluid Mechanics with Engineering Application*, 6th ed., McGraw-Hill, New York, 1965.
5. Moody, Lewis F., "Friction Factor for Pipe Flow," *Trans. ASME*, Vol. 66, 1944.
6. Williams, G. S., and Hazen, A., *Hydraulic Tables*, 3rd ed., John Wiley and Sons, New York, 1920.
7. U.S. Dept. of Interior, Bureau of Reclamation, *Water Measurement Manual*, Water Resources Technical Publication, 2nd ed., Denver, 1967.
8. Gill, W. N., Cole, R., Davis, J., Estrin, J., Nunge, R. J., and Littman, H., "Fluid Dynamics," *Industrial and Engineering Chemistry* 62:4:49–79, April, 1970.
9. ASCE Task Force, "Friction Factors in Open Channels," *Journal of the Hydraulics Division, ASCE*, pp. 97–143, March, 1963.

10. Hendersen, F. M., *Open Channel Flow,* Macmillan, New York, 1966, p. 522.
11. Streeter, V. L., "Steady Flow in Pipes and Conduits," in *Engineering Hydraulics,* Hunter Rouse, editor, John Wiley and Sons, New York, 1950, pp. 387–443.
12. Davis, C. V., and Sorenson, K. E., *Handbook of Applied Hydraulics,* 3rd ed., McGraw-Hill Book Co., New York, 1969.
13. Blaisdell, F. W., and Mansen, P. W., "Energy Loss of Pipe Junction," *Journal of the Irrigation and Drainage Division, ASCE,* pp. 59–78, September, 1967.
14. Cross, Hardy, *Analysis of Flow in Networks on Conduits in Conductors,* University of Illinois Bulletin 286, November, 1936.
15. Piggott, R. J. S., "Pressure Losses in Tubing, Pipe, and Fittings," *Trans. Amer. Soc. Mechanical Engrs.* 72:679, July, 1950.
16. Rouse, Hunter, editor, *Engineering Hydraulics.* John Wiley and Sons, New York, 1950.
17. Soliman, M. M., and Tinney, E. Roger, "Flow Around 180° Bends in Open Rectangular Channels," *Journal of Hydraulics Division, ASCE,* pp. 893–908, July, 1968.
18. Rouse, Hunter, "Discharge Characteristics of the Free Overfall," *Civil Engineering* 6:257–260, 1936.
19. Hydraulic Institute, *Standards of the Hydraulic Institute,* 10th ed., New York, 1955.
20. Ball, James W., "Cavitation Design Criteria," Control of Flow in Closed Conduits, *Proceedings* of the Institute held at Colorado State University, August, 1970.
21. Tullis, J. Paul, and Marachner, B. W., "Review of Cavitation Research or Valves," *Journal of the Hydraulics Division, ASCE,* pp. 1–16, June, 1968.
22. Westaway, C. R., and Loomis, A. W., editors, *Cameron Hydraulic Data,* Ingersoll Rand, Woodcliff Lake, N.J., 1981.
23. Hudson, Herbert E., Jr., *Water Clarification Processes Practical Design and Evaluation,* Van Nostrand Reinhold, New York, 1981.
24. McMann, J. S., "Mechanics of Manifold Flow," *Trans. ASCE,* pp. 119–1103, 1954.

Chapter 26
Storage

PURPOSE

Finished-water storage is an essential element of any sizable public water supply system. Storage on the distribution system can serve several functions. It can be used to equalize demands on sources of supply, treatment works, production facilities, pumping stations, transmission lines, and distribution mains. Through the use of storage to equalize demands throughout the day, it is possible to reduce the design capacities and sizes of these other system elements. Further, storage makes it easier to maintain uniform pressures throughout the water service area. Storage also provides a reserve supply that can be drawn on during emergencies such as power outages, fires, and equipment failures.

In brief, distribution storage serves three principal purposes: it meets hourly variations in daily water system demands, it stores water for firefighting, and it provides a reserve supply for other emergency use. Storage requirements for these three functions are considered individually and then combined to compute the total amount of storage capacity to be provided. References 1 through 5 provide further descriptions of the benefits and purposes of providing storage.

CAPACITY

General

In all but the very smallest water service areas, the costs for constructing and operating water storage facilities are partially or totally offset by the savings storage enables in building and operating other parts of the water system and in fire insurance premiums. Further discussions of the capacity of storage tanks are included in References 6 through 9.

In extremely small communities, finances may not permit construction of storage reservoirs. Systems without storage must be designed to meet peak hourly (rather than maximum daily) demands, and are usually provided with small hydropneumatic tanks as a limited means of equalizing flow and pressure. Without storage or emergency power, fire protection and other emergency services obviously are greatly reduced.

The biggest variable in sizing storage reservoirs is the daily quantity of water consumed, which is influenced by the population served. Fire and other emergency reserve requirements are less variable. In large cities, the amount of storage required for peaking service is greater than the storage needed for fire reserve. In cities with populations of 10,000 to 20,000, the storage needs for fire reserve are about equal to those for hourly peaking on the day of maximum water use. In small cities, the fire protection storage capacity required exceeds that needed for flow equalization. The amount of other emergency reserve storage needed depends upon the danger of interruption of water production and the time required to make repairs.

Equalizing or Operating Storage

Common practice in the United States is to provide enough equalizing storage to allow the supply, treatment, and other water production facilities to operate at a uniform rate at all hours during the day of maximum annual use. Because water use is greater during daylight hours than at night, water is withdrawn from storage during the peak demand hours of the day and is replenished during the minimum demand hours of the night. Reference 10 provides further details on equalizing and operating storage.

Figure 26–1 illustrates the hourly variations in daily water use that might occur in a small residential community on the day of maximum water use for the year. The maximum hourly rate of use is 175 percent of the average

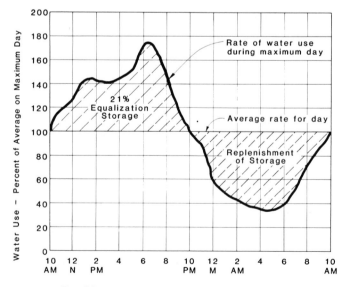

Fig. 26–1. Hourly variation in daily water use.

rate of use for the day, and the minimum hourly rate of use is 35 percent of the daily average. In this instance, water is being withdrawn from storage between the hours of 10:00 A.M. and 10:00 P.M., and the reservoir is being refilled from 10:00 P.M. to 10:00 A.M. The amount of storage required to supply hourly demands greater than the average hourly demand for the day is represented by the cross-hatched area under the demand curve and above the line representing average hourly rate of use on the maximum day. In this example, storage amounting to 21 percent of the water volume used on the maximum day is needed for flow equalization. In other words, if the maximum daily demand is 10 MG (37.9 Ml), then the storage needed for equalization is 0.21 × 10 MG = 2.1 MG (7.95 Ml). This example is based on 24 hours per day operation of water production facilities. If water production were restricted to only one or two shifts, then more storage would be required. Hourly rates of flow follow different patterns in various cities. Design data should be derived from the records of the community under study. However, the 21 percent figure obtained in the example is typical for small residential areas in many parts of the country.

Fire Reserve

Storage requirements for firefighting should be calculated according to the latest *Guide for Determination of Required Fire Flow,* published by the Insurance Services Office, 160 Water Street, New York, New York 10038.

The minimum municipal fire flow requirement is 500 gpm (1.31 Ml/d) for 1 hour, and the maximum for a single fire is 12,000 gpm (31.5 Ml/d) for 10 hours. Typical fire reserve storage requirements are given in Table 26-1.

These storage capacities are given for illustrative purposes only. Actual capacity needs should be calculated for the specific local prevailing conditions.

As previously mentioned, the fire reserve storage capacity needs must be added to the equalizing storage requirement and the emergency reserve in order to obtain the total storage required. Further information concerning reserve and equalizing storage may be found in Reference 11.

Table 26-1. Storage Requirements for Fire Protection.

TYPE OF DEVELOPMENT	STORAGE VOLUME	
	GALLONS	ML
Residential, low density, 2 hours at 500 gpm	60,000	0.227
Residential, built-up, 2 hours at 1,000 gpm	120,000	0.454
Commercial, 4 hours at 2,000 gpm	480,000	1.817
Commercial, 4 hours at 4,000 gpm	960,000	3.634

Emergency Reserve

The storage capacity needed for emergency reserve depends upon the danger of interruption of the water production facilities and the length of time needed to make repairs. Interruptions of supply may occur as a result of well, intake, pipeline, treatment plant, pump, or power failures.

Interruptions may also be due to intentional shutdowns for inspection, maintenance, or short-term water quality degradation in surface supplies. The probable length of such outages should be estimated on the basis of the best available local information, and the necessary storage added to the needs for flow equalization and fire reserve.

Hydropneumatic Tanks

In some small service areas, it may not be financially feasible to provide storage in the water system. In such cases, hydropneumatic tanks often are installed as a limited means for equalizing flow and pressure and reducing cycling of pressure pumps.[12]

For the supply of water from a pressure tank system to be adequate, the storage and mains must be of the proper size to meet peak hourly or momentary demands. In small cities (population under 5,000), peak rates of flow may be as much as ten times greater than the average annual rate of use. Actual meter readings, where available, should be used in preference to an estimate. Pump capacities must be adequate to deliver 125 percent of the peak hourly or momentary maximum flows. Hydropneumatic tanks typically are sized to have a capacity equal to 30 times the pump capacity (in gallons per minute or liters per second). Another rule of thumb is to provide 10 gallons (37.9 l) of storage per person served. In general, only about 10 to 20 percent of the total volume of a pressure tank is available storage.

A simple and direct method for determining the recommended pump size and pressure tank capacity was given by J. A. Salvato, Jr.[13] It is based on the formula:

$$V = \frac{V_m}{\left[1 - \dfrac{P_1}{P_2}\right]} \tag{26-1}$$

where:

V = pressure tank volume, gallons (m³)
V_m = 15 minutes storage at the peak hourly demand rate, gallons (m³)

P_1 = minimum absolute operating pressure, psi (kPa)
  = gauge pressure plus 14.7 (101.3 kPa)
P_2 = maximum absolute pressure, psi (kPa)
  = gauge pressure plus 14.7 (101.3 kPa)

NUMBER AND LOCATION OF RESERVOIRS

The capital cost to construct a single reservoir to store a given volume of water is, of course, less than the cost to build several smaller reservoirs to store the same volume. However, the use of several reservoirs in different parts of the system may make it possible to use smaller pipelines, and may also reduce pumping heads. If this is the case, it may be advantageous to provide several smaller scattered units rather than a single unit of the same capacity at a central location. Additional details are included in References 14–16.

Four factors to consider in selecting the location of storage reservoirs are: (1) ground elevation, (2) effects of location on pressure variations, (3) location with respect to the center of consumption, and (4) location with respect to the main pumping station. The latter two factors may influence the size of an area to be provided with two-directional flow at times of heavy demands for water. It is evident that flow from opposite directions will increase the capacity of the distribution system, and that two-directional flow in the main transmission lines will increase the flow of water that they can deliver. Obtaining two-directional flow in transmission mains may require locating the main pumping station and the storage facilities on opposite sides of the center of consumption, but this arrangement can result in considerable differences in system pressures when pumps are on or off. Hence, many reservoirs are located close to the center of consumption. This minimizes pressure variations, which is desirable, but it limits the area receiving two-way feed, which is undesirable.

The elevation of the reservoir site is an important factor. If the topography is such that the storage reservoir can be located on a hill of the proper height, considerable savings in construction cost can be realized because of the reduced height or the type of structure that can be utilized.

Water systems that receive part or all of their supply from a water treatment plant usually provide part of the storage at the treatment plant in the form of a clearwell. The capacity of the clearwell must be sufficient to store filter backwash water for the plant and equalizing storage for the plant's high service pumping station. In most cases, a clearwell capacity equal to 10 percent of the maximum design capacity of the treatment plant is adequate, but capacity needs should be calculated for specific local conditions when background data for this purpose are available. The balance of the water system's storage needs would be provided on the distribution system.

TYPES OF STORAGE

Three types of storage, classified with respect to their elevation, are: (1) underground, (2) ground-level, and (3) elevated. Selection of the best type of storage for a particular situation depends upon topography, hydraulic grade lines, economics, freezing conditions, aviation hazards, and sabotage potential. The bottom of ground-level reservoirs and standpipes should be placed at the normal ground surface and above the 100-year flood level. References 17 and 18 include additional discussions on the types of storage.

Where the bottom is to be below normal ground surface, it should be placed above the groundwater table. Sewers, drains, standing water, and similar sources of contamination should be kept at least 50 feet (15.3 m) from the reservoir. Mechanical-joint water pipe, pressure-tested in place to 50 psi (344 kPa) without leakage, may be used for gravity sewers at lesser separations. The top of a ground-level reservoir should not be less than 2 feet (0.61 m) above the normal ground surface and any possible flood level.

High storage requirements may favor ground-level or underground tanks. The existence of a hill in the right location and at the right elevation may provide elevated storage at the cost of ground-level reservoirs. In level terrain, consideration might be given to ground-level or underground tanks with provisions for repumping.

Underground tanks may be used when it is necessary to conceal the tanks. This may be done because of sabotage or vandalism potential, for aesthetic reasons, or to facilitate multiple uses of the site, such as tennis courts, parking, or other uses.

DESIGN CONSIDERATIONS

General

The materials and designs used for finished water storage structures should provide stability and durability as well as protect the quality of the stored water. Steel structures should conform to AWWA D100 79, *AWWA Standard for Welded Steel Elevated Tanks, Standpipes, Reservoirs, and Elevated Tanks for Water Storage*. Concrete and other materials of construction are acceptable when properly designed. Further details are included in References 19–27.

Protection

To protect the stored water against chance contamination and quality deterioration, all potable water reservoirs should be covered. This protection is described in References 28 and 29.

Stored, treated water in the distribution system may be contaminated by substances that fall into uncovered finished water storage tanks or reservoirs, by windblown material entering vents, and by ground or surface water seepage. Open storage is subject to pollution from birds, animals, windblown contaminants, human activities such as bathing, fishing, and deliberate contamination, and many other sources. The best way to prevent deterioration of the quality of water stored in tanks and reservoirs is to provide watertight storage facilities that are covered by roofs. In the absence of such cover, disinfection of all water fed to the system from storage is essential, and tends to offset, but does not prevent, the potential effects of introduced contamination.

Covers and Roofs. All finished water storage structures should have suitable watertight roofs or covers that exclude birds, animals, insects, and excessive dust.

Protection from Trespassers. Fencing, locks on access manholes, and other necessary precautions should be provided to deter trespassing, vandalism, and sabotage.

Drains. No drain on a water storage structure should have a direct connection to a sewer or storm drain.

Overflow. The overflow pipe of a water storage structure should be brought down near the ground surface and discharged over a drainage inlet structure or a splash plate. Overflows should not be connected directly to a sewer or storm drain, and should adhere to the following guidelines:

- When an internal overflow pipe is used, it should be located in the access tube.
- The overflow of a ground-level structure should be high enough above the normal or graded ground surface to prevent the entrance of surface water.
- Overflow capacity should equal the maximum inlet flow.

Access. Finished water storage structures should be designed with reasonably convenient access to the interior for cleaning, maintenance, and sampling. Manholes on scuttles above the waterline should:

- Be framed at least 4 inches (0.1 m), and preferably 6 inches (0.15 m), above the surface of the roof at the opening. On ground-level structures, manholes should be elevated 24 to 36 inches (0.61 to 0.91 m) above the top or covering sod.
- Be fitted with a solid watertight cover that overlaps the framed opening and extends down around the frame at least 2 inches (50.8 mm).

- Be hinged at one side.
- Have a locking device.

Vents. Finished water storage structures should be vented by special vent structures. Open construction between the sidewall and the roof is not recommended. These vents should:

- Prevent the entrance of surface water.
- Exclude birds and animals.
- Exclude insects, rain, and dust by a vent cap or other means, to the extent that this function can be made compatible with effective venting. For elevated tanks and standpipes, 8-mesh noncorrodible screen may be used.
- On ground-level structures, terminate in an inverted "U" construction, the opening of which is 24 to 36 inches (0.61 to 0.91 m) above the roof or sod and is covered with 24-mesh noncorrodible screen cloth.
- Contain screens that swing on hinges or collapse when clogged.

Roof and Sidewalls. The roof and sidewalls of all structures should be watertight with no openings except properly constructed vents, manholes, overflows, risers, drains, pump mountings, control ports, or piping for inflow and outflow. The following additional factors should be considered:

- Any pipes running through the roof or sidewall of a finished water storage structure should be welded or properly gasketed in metal tanks, or connected to standard wall castings that are poured in place during the forming of a concrete structure; these wall castings should have flanges embedded in the concrete.
- Openings in a storage structure roof or top, designed to accommodate control apparatus or pump columns, should be curbed and sleeved with proper additional shielding to prevent the access of surface or slop water to the structure.
- Valves and controls should be located outside the storage structure so that valve stems and similar projections will not pass through the roof or top of the reservoir and will not be subject to contamination by surface water.

Drainage for Roof or Cover. The roof or cover of the storage structure should be well drained, but downspout pipes should not enter or pass through the reservoir; parapets or similar constructions that would tend to hold water and snow on the roof are not recommended.

Freezing. All finished water storage structures and their appurtenances, especially the riser pipes, overflows, and vents, should be designed to prevent freezing, which will interfere with proper functioning. Consideration should be given to heating and/or insulation of exposed pipes and valves. This subject is discussed in more detail in References 31 and 32.

Internal Catwalk. Every catwalk over finished water in a storage structure should have a solid floor with raised edges so designed that shoe scrapings and dirt will not fall into the water.

Grading. The area surrounding a ground-level structure should be graded in a manner that will prevent surface water from standing within 50 feet (15.2 m) of the structure.

Painting and/or Cathodic Protection. The AWWA standards for painting exclude the use of paints that might add toxic materials to the stored water through leaching or other action. Cathodic protection devices must be regularly inspected and maintained for satisfactory performance. Painting and maintenance of storage tanks are critical elements in ensuring public safety, and are discussed in References 33–39.

Proper protection should be given to metal surfaces by paints or other protective coatings, by cathodic protective devices, or by both. Paint systems consistent with AWWA Standard D102–78, or otherwise acceptable to regulatory agencies, should be used. Additionally:

- Cathodic protection should be designed, installed, and maintained by competent technical personnel.
- Proper painting and lighting should be provided for aircraft visibility.

Silt Protection. Reservoirs should be protected against entry of silt through discharge pipes.

Fire Protection. Elevated steel tanks should be protected against damage or structural failure due to fire in the vicinity of the tanks.

Materials. Steel and concrete are permitted by most states. Other suitable materials may be approved at the discretion of the regulatory agencies. Further details are summarized in References 40–47. Steel water tanks should conform to AWWA Standard D100–79. Examples of steel water tanks are shown in Fig. 26–2 and 26–3.

Current design and construction practices for prestressed concrete storage tanks are summarized in References 43, 44, and 46. These designs and construction practices should be followed whenever possible.

Fig. 26–2. Photo of water tank.

Safety

The safety of employees should be considered in the design of the storage structure. As a minimum, such matters should conform to pertinent laws and regulations of the area where the reservoir is constructed. Further information is included in Reference 30, and as follows:

- Ladders, ladder guards, balcony railings, and safe location of entrance hatches should be provided where applicable.
- Elevated tanks with riser pipes over 8 inches (0.2 m) in diameter should have protective bars over the riser openings inside the tank.

Disinfection

New reservoirs and reservoirs that have been emptied for service or repair should be disinfected. This will ensure that any contamination introduced by workmen or materials during the course of construction or maintenance, that has not been physically removed by cleaning the floor, walls, and pipelines of the reservoir, is rendered harmless to water users. Proper disinfection is critical to the safety of public health, and is described in detail in References 48–52.

Detailed procedures, equivalent to those outlined in the current American

Fig. 26–3. Photo of water tank.

Water Works Association Standard D102 for painting and repairing steel tanks, standpipes, reservoirs, and elevated tanks, should be used for tank disinfection. Also, the following provisions should be made:

• Two or more successive sets of samples, taken at 24-hour intervals, should indicate bacteriologically satisfactory water before the facility is released for use.
• Smooth end taps in riser pipes and connecting mains where disinfectant is added should be provided.

Plant Storage

Water storage may be provided at the treatment plant for in-plant uses (the principal in-plant is for filter backwashing) or to reduce the total amount of distribution system storage required for flow equalization from the treatment plant.

Washwater Tanks. Washwater tanks should be sized, in conjunction with available pump units and finished water storage, to give the backwash water required.

The capacity in storage required for this in-plant use is the quantity of water required to backwash a filter for 15 minutes at the design backwash flow rate. However, consideration should be given to the possible need to wash more than one filter at a time, or several filters in succession.

Clearwell. Clearwell storage should be sized, in conjunction with distribution system storage, to relieve the filters from having to meet all fluctuations in water use or to meet peak demands, including the use of filter backwash water and the loss of capacity due to filter outages. When finished water storage is used to provide proper contact time for disinfection, special attention should be given to size and baffling.

Adjacent Compartments. Finished water should not be stored or conveyed in a compartment adjacent to unsafe water when the two compartments are separated by a single wall. The possibility of leakage of unsafe water into potable water storage through cracks or leaks in the division wall cannot be risked.

Basins and Wet Wells. Receiving basins and pump wet wells for finished water should be designed as finished water storage structures.

Pressure Tanks

Hydropneumatic (pressure) tanks are acceptable in some small water systems. When used, they should meet ASME Code requirements for unfired pressure vessels. Also, they should comply with the requirements of state and local laws and regulations for such construction.

The requirements here have been derived empirically for the most part, based on operating experience over many years. These requirements are:

- The tank should be located above the normal ground surface. In areas where freezing occurs, the tank should be completely housed, or earth-mounded with one end projecting into an operating house.
- The tank should have bypass piping or a duplicate unit to permit operation of the system while the tank or the tank accessories are being repaired or painted.

- Each tank should have an access manhole, a drain, and control equipment consisting of pressure gauge, water sight glass, automatic or manual air blow-off, mechanical means for adding air, and pressure-operated start–stop controls for the pumps.
- The total capacity of the wells and pumps in a hydropneumatic system should be at least six to ten times the average daily consumption rate of the community. A minimum of two wells or pumps should be provided.
- The gross volume of the hydropneumatic tank, in gallons, is often sized to have a capacity equal to 10 times the capacity of the largest pump. For example, a 250-gpm (15.8-l/s) pump should have a minimum 2,500- to 2,750-gallon (9.46- to 10.41-kl) pressure tank.

Distribution Storage

The requirements for distribution storage reservoirs are based on the need to maintain adequate quantities and pressures of water in the distribution system in order to protect the sanitary quality of the water. As a minimum, distribution system storage capacity should be sufficient to take care of hourly variations in water demands and pressures during the day plus the required reserve for fire protection and minor contingencies. For the required fire reserve see the above-mentioned *Guide for Determination of Required Fire Flow,* of the Insurance Services Office (formerly National Board of Fire Underwriters). Storage in the distribution system should be capable of replenishment each night during low-demand periods. More detail is included in References 53–55.

Pressure Variation.   The maximum desirable variation between high and low levels in storage structures that float on a distribution system is 30 feet (9.1 m).

Drainage.   Storage structures that float on the distribution system should be designed for draining, cleaning, and maintenance without loss of pressure in the distribution system. The drains should discharge to the ground surface with no direct connection to a sewer or storm drain.

Level Controls.   Adequate controls should be provided to maintain levels in distribution system storage structures, as follows:

- Telemetering equipment should be used for storage structures when pressure-type controls are employed in control stations and any appreciable headloss occurs in the distribution system between the source and the storage structure.

- Altitude valves or equivalent controls may be desirable for a second and subsequent structure on the system. The high-water level should be set 3 feet (0.91 m) below overflow.

Miscellaneous Design Considerations.   Circulation of water through reservoirs can assist in the control of tastes and odors, maintenance of residual disinfectant, and control of slime growths. These and other factors are discussed in References 56 and 57. Reservoirs should be cleaned periodically to prevent accumulation of sediments and the creation of chlorine-demanding residues. Consider also that:

- Water circulation through reservoirs should be provided by the use of baffles, or by placing inlets and outlets on opposite sides of the reservoir with outlets near the bottom.
- All pipes except overflow should have valves.
- The minimum reservoir water depth should be 12 feet (3.66 m).
- More than one reservoir should be provided, to give storage during outages; or, if there is a single reservoir, it should be divided into compartments so at least one section is available for use at all times.
- Reservoirs should be drained and cleaned every 2 years.

COVERING EXISTING RESERVOIRS

Importance

Public health protection is one of the principal benefits that can be provided by public water supply systems. Three basic elements in assuring the safety of drinking water supplies are: (1) the covering of all finished water storage reservoirs, (2) the filtration of all surface water supplies, and (3) provisions for continuous disinfection of all public water supplies. Water purveyors in the United States are justifiably proud of the safety record of their supplies. Many have gone to great lengths and expense to satisfy vague, sophisticated concerns about water quality. At the same time, a sizable number of water systems still operate under the hazards of open storage reservoirs or unfiltered surface supplies, or without disinfection. Operation of these systems poses a constant potential threat to the health of water drinkers. These deficiencies are basic, obvious, and serious, but can be corrected easily at reasonable costs. Many people are surprised to learn that these important, simple problems were not corrected long ago. These issues are discussed further in References 58–63.

Reservoir covers protect the stored water against quality deterioration by airborne or windblown contamination, birds and animals, human activities such as bathing and fishing, and deliberate contamination or sabotage. In addition to the public health protection, covers provide other benefits. They prevent development of tastes and odors by limiting or preventing algal growths. By excluding sunlight they reduce the amount of chlorine lost during storage and may eliminate the need for rechlorination of water leaving the reservoir. The frequency of reservoir cleaning is reduced, and water waste by draining is avoided.

TYPES OF COVERS SUITABLE FOR INSTALLATION ON EXISTING OPEN RESERVOIRS

There are numerous ways to provide satisfactory covers on existing tanks and reservoirs. Covers may be fixed, or they may float on the water surface. They can be made from reinforced concrete, steel, aluminum, wood, Hypalon, or other materials. They can be flat, conical, or dome-shaped. The type of cover that is best suited for a particular installation can be selected on the basis of a number of factors, including: the location, size, shape, and materials used in construction of the existing reservoir; footing and other support conditions; snow, wind, and seismic loads; aesthetics; the length of time, if any, that the reservoir can be removed from service for renovation; and capital cost, maintenance cost, and other economic factors such as durability and service life. Concrete and synthetic covers are described in more detail in References 64–66.

In the past, the capital cost of covering existing open reservoirs has been a deterrent to such improvement in many places. However, over the past decade the advent and acceptance of floating covers has substantially reduced the costs for many installations, and the number of tanks covered has increased greatly. Floating covers not only are lower in first cost than rigid covers, but require little maintenance, are totally resistant to seismic loads, can be installed rapidly, eliminate the need to piece the floor, eliminate evaporation and loss of chlorine, reduce algal growths and taste and odor problems, and can be installed without taking the reservoir out of service. A floating cover can be installed on a reservoir of any size or shape. Figure 26–4 shows a section through an earthen reservoir fitted with a floating cover.

Another type of cover now seeing increased use is the all-aluminum geodesic dome, which is limited to covering circular reservoirs. The dome consists of a skeleton of aluminum trusses and a skin of aluminum panels. The largest dome installed to date has a span of 230 feet (70.1 m). The advantages of the aluminum geodesic dome are long life (50 to 100 years); no maintenance; no painting; light weight; a bottom tension ring that transmits only vertical forces to the reservoir walls; easy reservoir access; fast construction; no interior columns or supports; and the fact that it can be installed while the reservoir is in service. Disadvantages include high initial cost, and span and shape

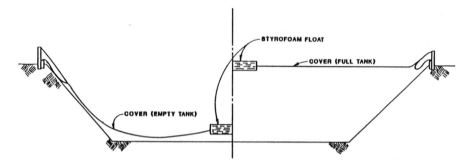

Fig. 26–4. Floating cover—section through reservoir.

limitations. Figure 26–5 shows several views of aluminum geodesic dome covers.

Another recent development in covering water storage reservoirs is the air-supported roof. This type of cover is also used for swimming pools, tennis courts, greenhouses, sludge drying beds, and similar enclosures (see Fig. 26–6). The covers can be circular, square, or rectangular.

Air-supported covers are inflated and kept in place by blowers that operate continuously to maintain about 2 pounds per square inch (psi) (13.8 kPa)

a. Aerial view of domes covering circular tanks.

Fig. 26–5. Aluminum geodesic dome covers. (Courtesy of Temcor.)

b. View of dome from the inside

c. Geodesic dome showing structural members.

Fig. 26-5 (Continued)

Fig. 26–6. Fabric air-supported cover over rectangular tank. (Courtesy of Thermo-Flex, Inc., Salina, Kansas)

more pressure inside the cover than outside. Ordinarily, two blowers and a standby generator are employed for this purpose. One blower operates continuously to supply air lost through leaks. The other comes on intermittently at times when one blower cannot maintain pressure. The standby generator powers the blowers at times of power outages. Air-supported reservoir covers are not designed to support snow load; instead, the crown of the cover must be kept warm by hot air circulation to melt the snow upon contact with the cover. Portions of the cover below the crown that have a slope greater than 30 degrees with the horizontal do not require heat to be kept snow-free. A double layer of fabric is provided in the crown area, and hot air is circulated between the layers. A flexible duct carries the hot air from the heater to the crown of the roof.

Air-supported covers can be fitted with a revolving door for personnel entry and an $8 \times 8 \times 15$-foot ($2.4 \times 2.4 \times 4.6$-m) air lock door for equipment and supply access. There is a choice of several types of fabric that can be used in air-supported enclosures. Three that represent the range of materials available are: polyester-reinforced vinyl, Hypalon, and Teflon-coated fiberglass, which reportedly have service lives of 5 to 10, 10 to 20, and 25 or more years, respectively. Costs ascend in the order of increasing durability.

For existing reservoirs (new reservoirs are now required to be covered in most states), the alternatives to installing adequate covers are: continued more frequent draining and cleaning of the reservoir; and higher dosages of chlorine. The disadvantages to open reservoirs are: higher chemical costs;

greater vulnerability to deliberate contamination; and overall reduced health protection.

REFERENCES

1. Fair, G. M., and Geyer, J. C., *Water Supply and Wastewater Disposal,* John Wiley and Sons, New York, 1959.
2. Hartman, B. J., "Types and Uses of Water Storage," *J.AWWA,* p. 395, March, 1959.
3. Pluntze, James C., "Water Utilities—A Victim of Success," *J.AWWA* 63:9:551, September, 1971.
4. Davis, C. V., *Handbook of Applied Hydraulics,* McGraw-Hill Book Co., New York, 1952.
5. Stearns, D. E., "Advantages of Additional Storage Facilities for Improving Distribution Systems," *J.AWWA,* p. 661, May, 1958.
6. Brock, D. A., "Determination of Optimum Storage in Distribution System Design," *J.AWWA,* p. 1027, August, 1963.
7. Freese, S. W., "Peak Demand Storage," *J.AWWA,* p. 263, March, 1957.
8. EPA, *Manual for Evaluating Public Drinking Water Supplies,* Superintendant of Documents, Washington, D.C., 1971.
9. Culp, Russell L., *EPA Technical Guidelines for Public Water Systems,* Chapter VI, "Storage," NTIS, PB-255–217, U.S. Department of Commerce, Springfield, Va., June, 1976, p. 195.
10. Farmer, Ed, "Storage and Distribution for Peak Loads," *J.AWWA,* p. 135, February, 1962.
11. Neumayer, C. A., Jr., "Determining Reserve and Equalizing Storages," *J.AWWA,* p. 1499, November, 1957.
12. Salvato, J. A., Jr., *Environmental Sanitation,* John Wiley and Sons, New York, 1958.
13. Salvato, J. A., Jr., "The Design of Pressure Tanks for Small Water Systems," *J.AWWA,* p. 532, June, 1949.
14. Haskew, G. M., Jr., "Storage Tank Location," *J.AWWA,* p. 296, March, 1963.
15. Hardenbergh, W. A., *Water Supply and Purification,* International Textbook Co., Scranton, Pa., 1945.
16. Smith, M. C., "Advantages of Ground Level Distribution Storage," *J.AWWA,* p. 1559, December, 1959.
17. Fee, J. R., "Planning Distribution Storage in Nonhilly Areas," *J.AWWA,* p. 714, June, 1960.
18. Foland, D. L., "Ground-Level Versus Elevated Storage," *J.AWWA,* p. 524, October, 1981.
19. AWWA, *Water Treatment Plant Design,* AWWA, Denver, Colo., 1969.
20. Babbitt, H. E., and Doland, J. J., *Water Supply Engineering,* McGraw-Hill Book Co., New York, 1939.
21. AWWA Committee Report, "Potable Water Storage Reservoirs," *J.AWWA,* p. 45, October, 1953.
22. Patterson, A. B., "Design of Underground Storage Reservoirs," *J.AWWA,* p. 1499, November, 1957.
23. Great Lakes–Upper Mississippi River Board of Sanitary Engineers, *Recommended Standards for Water Works,* New York State Health Education Services, Albany, N.Y., 1972.
24. McDermott, James H. "Federal Drinking Water Standards—Past, Present and Future," *Journal Environmental Engineering Division ASCE* 99:EE4:469, August, 1973.
25. Wade, J. A., "Design Guidelines for Distribution Systems as Developed and Used by an Investor-Owned Utility," *J.AWWA,* p. 346, June, 1974.
26. Jorgenson, I. B., and Close, S. R., "Arvada 10-mil gal Treated-Water Storage Reservoir," *J.AWWA,* p. 181, April, 1976.
27. AWWA Committee Report, "Multiple Use of Covered Distribution Reservoirs," *J.AWWA,* p. 262, May, 1978.

28. Emigh, F. A., "Protection of Open Reservoirs against Birds," *J.AWWA,* p. 1353, November, 1962.

29. USPHS, *Manual of Recommended Water Sanitation Practice,* USPHS Publication No. 196, Government Printing Office, Washington, D. C., 1946.

30. Texas Water Works and Sewerage Short School, *Manual for Water Works Operators,* Texas State Department of Health, Austin, 1951.

31. Larsen, L. A., "Cold Weather Operation of Elevated Tanks," *J.AWWA,* p. 17, January, 1976.

32. Wormald, L. W., "Water-Storage-Tank Failure Due to Freezing and Pressurization," *J.AWWA,* p. 173, March, 1972.

33. Harper, W. B., "Inspecting, Painting, and Maintaining Steel Water Tanks," *J.AWWA,* p. 584, November, 1982.

34. Brotsky, Bob, "Interior Maintenance of Elevated Storage Tanks," *J.AWWA,* p. 506, September, 1977.

35. AWWA, *AWWA Standard for Painting and Repainting Steel Tanks, Standpipes, Reservoirs, and Elevated Tanks for Water Storage,* AWWA D102–64, AWWA, Denver, Colo., 1964.

36. AWWA, *AWWA Standard for Inspecting and Repairing Steel Water Tanks, Standpipes, Reservoirs, and Elevated Tanks, for Water Storage,* AWWA D101–53, AWWA, Denver, Colo., 1953.

37. Shelton, F. G., "Water-Tank Maintenance and Repair Practices," *J.AWWA,* p. 472, August, 1974.

38. Jackson, J. E., "Better Performance from Paints for Steel Water Tanks," *J.AWWA,* p. 577, September, 1970.

39. Elkins, H. B., "Lead Content of Water from Tanks Painted with Red Lead," *J.AWWA,* p. 570, May, 1959.

40. AWWA, *AWWA Standard for Welded Steel Elevated Tanks, Standpipes, and Reservoirs for Water Storage,* AWWA D100–73, AWWA, Denver, Colo., 1973.

41. Crowley, F. X., "Maintenance Problems and Solutions for Prestressed Concrete Tanks," *J.AWWA,* p. 579, November, 1976.

42. Hertzberg, L. B., and Westerback, A. E., "Maintenance Problems with Wire-Wound Prestressed Concrete Tanks," *J.AWWA,* p. 652, November, 1976.

43. Woodside, R. D., "Prestressed Concrete Storage Tanks: Design and Construction," *J.AWWA,* p. 389, May, 1972.

44. Harem, F. E., "Maintenance and Repair of Concrete Reservoirs," *J.AWWA,* p. 1160, September, 1964.

45. Dykmans, M. J., "Prestressed Concrete Tanks—Design Factors and Economy," *J.AWWA,* p. 591, May, 1965.

46. Walters, C. G., "Prestressed Concrete Tanks—Maintenance and Repair," *J.AWWA,* p. 596, May, 1965.

47. Jackson, J. O., "Stronger Steels for Tanks," *J.AWWA,* p. 1390, November, 1965.

48. Jones, F. E., and Greenberg, A. E., "Coliform Bacteria in Redwood Storage Tanks," *J.AWWA,* p. 1489, November, 1964.

49. APHA Committee Report, "Bacterial Aftergrowths in Water Distribution Systems," *American Journal Public Health,* p. 43, 1930.

50. Lippy, E. C., and Erb, J., "Gastrointestinal Illness at Sewickley, PA," *J.AWWA,* p. 606, November, 1976.

51. AWWA Joint Discussion, "Maintaining Water Quality in Distribution Systems," *J.AWWA,* p. 54, January, 1962.

52. AWWA Committee Report, "Deterioration of Water Quality in Large Open Distribution Reservoirs," *J.AWWA,* p. 313, June, 1983.

53. McPherson, M. B., and Wood, G., "Operating Options for Pumped Equalizing Storage," *J.AWWA,* p. 869, July, 1965.

54. AWWA, *Water Distribution Training Course,* Manual No. M8, AWWA, Denver, Colo., 1962.
55. Carpenter, C. H., "Constructing and Maintaining Distribution Storage Structures," *J.AWWA,* p. 580, November, 1982.
56. Harem, F. E., et al., "Reservoir Linings," *J.AWWA,* p. 238, May, 1976.
57. McCaugan, F. A., "Pneumatic Application of Cement—Mortar Tank Coatings," *J.AWWA,* p. 1145, September, 1965.
58. Chin, A. G., "Covering Open Distribution Reservoirs," *J.AWWA,* p. 763, December, 1971.
59. Dellaire, Gene, "The Floating Cover: Best Way to Cover a Finished-Water Reservoir?," *ASCE Civil Engineering,* p. 75, June, 1975.
60. Elwell, F. H., "Flexible Reservoir Covers: A Case Study," *J.AWWA,* p. 210, April, 1979.
61. Pluntze, J. C., "Health Aspects of Uncovered Reservoirs," *J.AWWA,* p. 432, August, 1975.
62. Pluntze, J. C., "Survey of State Practices and Attitudes for Covering Water Distribution Reservoirs," unpublished, June, 1974.
63. Orgerth, Henry J., "Quality Control in Distribution Systems," in *Water Quality and Treatment,* 3rd ed., McGraw-Hill Book Co., New York, 1971.
64. Crowley, F. X., "Spherical Concrete Dome Covers for Circular Tanks and Reservoirs," *J.AWWA,* p. 568, November, 1976.
65. Prazer, S. J., "Reservoir Renovation: Butyl-Rubber Lining and Floating Cover," *J.AWWA,* p. 151, February, 1973.
66. Brown, J. H., "Flexible Membrane: An Economical Reservoir Liner and Cover," *J.AWWA,* p. 328, June, 1979.

Chapter 27
Distribution Systems

INTRODUCTION

A water distribution system conveys water from the treatment or production facilities to the user. The distribution system should supply water, without impairing its quality, in adequate quantities and at sufficient pressures to meet system requirements.

The facilities that make up the distribution system may include finished water storage, pumping, large-scale transmission and distribution piping, supply mains, and appurtenant valves. This chapter will discuss these facilities, the basis for their sizing and design, considerations for construction and maintenance, and techniques to evaluate the impacts of water demands on existing systems. Storage facilities are discussed in Chapter 26, and pumps in Chapter 28.

The water distribution system must be capable of supplying water needed for basic domestic purposes and commercial and industrial uses, and where possible, the flows necessary for fire protection. Fire protection is a desirable feature for any water system; however, many of the smaller systems serving 50 to 100 homes may find it economically infeasible to provide the facilities required for such protection. The safety and palatability of potable water should not be degraded as it flows through the distribution system.

SYSTEM CAPACITY

Distribution system design or analysis requires the determination of system flows resulting from consumption or system loading. Water use or water consumption typically is determined on the basis of average day demand, maximum day demand, and maximum hour demand. The determination of these data may be a straightforward procedure, or it may be difficult, depending upon the existence and type of records available. Most larger and some small utilities will have metering records for both the treatment facilities and individual consumers. Data from these meters may be used to determine average daily demands.

A comparison of annual records for water produced at the treatment or production facility and water metered at the consumer's meter may show a variance, which usually results in less consumption recorded than production recorded. This variance is often referred to as unaccounted-for-water. This unaccounted-for-water results from metering inaccuracies and water system losses due to leakage and unmetered water use, such as flushing water mains. Systems that are well-maintained, have a customer meter testing and replacement program, and keep accurate records may experience an unaccounted-for-water loss of 2 to 5 percent of total annual metered water use. Systems with limited maintenance or a large percentage of older pipe and home meters may have unaccounted-for-water losses in excess of 10 percent of total annual metered water use.[1] This unaccounted-for-water must be considered in the sizing of system facilities.

Records that enable the computation of maximum daily demand or maximum hour demand are far less prevalent in the industry than annual consumption records, and must be based on water production and changes in system storage.

Historical data can be used to project demands into new areas where new system facilities are to be constructed. Water system demands can be based on a per capita, per acre, or per unit basis.

In most water systems there are a variety of customer classes and system needs. Water use or demand will vary, depending on customer class. Customer classes may be broken down into residential and commercial/industrial or may be more finely divided into single-family residential, multifamily residential, public and parks, schools, heavy industrial, light industrial, commercial, and so forth. The finer the breakdown and the more accurate the data, the better the determination of demand requirements.

Once the various system demands are determined, it is then possible to begin design and/or analysis of the water distribution system.

SYSTEM DESIGN

Nonhydraulic Design Considerations

A number of factors unrelated to system hydraulics must be considered in water distribution system design. These nonhydraulic design considerations are summarized in the following paragraphs.

The ability to isolate parts of the system is important, especially during emergency operation conditions (e.g., main breaks). All water distribution systems should be provided with sufficient isolation and drain valves to permit necessary repairs without undue interruption of service over any appreciable area.

Valves should be placed in numbers and locations that give the best possible control of the system consistent with cost limitations. Valves in smaller mains are typically more numerous than valves in larger mains.

Valves should be located where they will be readily accessible in the event of a main failure. Branch mains, connecting to larger primary or secondary feeders that cross under arterial highways or streets, are usually valved close to the larger main before the crossing.

In transmission lines, valve spacing is usually determined by operating requirements, and thus is a matter of individual design. In feeder mains, valves are usually spaced so that each feeder loop can be effectively isolated. Usually this spacing will not exceed 3,000 feet (914 m) in 16-inch (0.41-m) feeders or 4,000 feet (1.2 km) in 20-inch (0.51-m) feeders.

Service main valves will be spaced so that adequate shutdown capability is provided without putting large numbers of customers out of service. In residential areas or districts, valves in 6- and 8-inch (0.15- and 0.20-m) mains should be spaced no more than 1,000 feet (305 m) apart, and in 10- and 12-inch (0.25- and 0.31-m) mains no more than 2,000 feet' (610 m) apart. Normally, each smaller main will be valved in intersections involving 10- and 12-inch (0.25- and 0.31-m) mains connecting to smaller mains. Service mains crossing creeks, railroads, and expressways will be valved on each side of the crossing. Each fire hydrant branch should be equipped with a control valve.

Valve sizes are normally the same as the water main in which they are installed, except that in mains 30 inches (0.76 m) or larger, line valves are often one size smaller than the main size.

Hydraulic Design Considerations

The design of water distribution systems may rely entirely on detailed calculations of system hydraulics, or may depend in part on minimum design standards defined by the owner of the distribution facilities. Design based on minimum standards typically will meet or exceed hydraulic requirements under all the ordinary system needs. Minimum design standards generally define materials and construction methods, minimum pipeline sizes, maximum pipeline headloss or velocities, and various criteria for ensuring adequate system flows and pressures without detailed hydraulic analysis of system extensions.

An example of such a set of design standards can be found in the Engineering Standards of the Board of Water Commissioners, Denver, Colorado.[2] These standards ensure that all mains are sized large enough to provide for domestic, irrigation, and fire protection flows to the area to be served. The Denver Water Department standards require that all mains be of standard sizes, which are defined as 6, 8, and 12 inches (0.15, 0.20, and 0.31 m). Mains larger than 12 inches (0.31 m) are considered a part of the transmission system. The sizing of the distribution mains follows a standardized grid (shown in Fig. 27–1) that requires a 12-inch (0.31-m) main every half mile (0.81

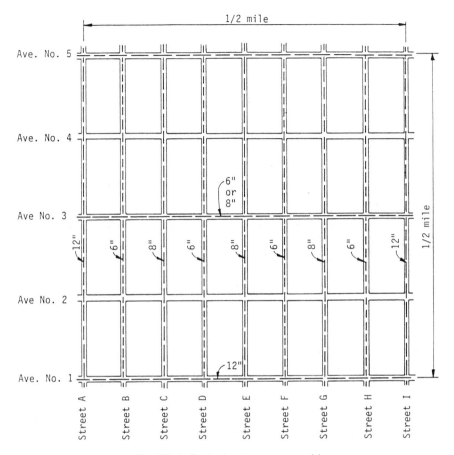

Fig. 27–1. Typical water system grid.

km) with alternating 6- and 8-inch (0.15- and 0.20-m) mains in the streets within quarter sections. The standards also require 6- or 8-inch (0.15- or 0.20-m) mains running perpendicular to the street mains at the 1/16 section line. The Denver Water Department's standards impose an upper limit on acceptable maximum headlosses for 6-, 8-, and 12-inch (0.15-, 0.20-, and 0.31-m) mains as 2 feet of headloss per 1,000 feet (0.61 m/304.80 m) of main under maximum hour flows. In typical residential areas, the department requires 8-inch (0.20-m)-diameter mains and allows 6-inch (0.15-m)-diameter piping only to complete looping or grid piping where 6-inch (0.15-m) pipe will be no longer than 600 feet (183 m) in length. The purpose of these standards is to enable the design and construction of water system extensions and improvements without detailed hydraulic engineering studies. These standards also ensure the best available fire insurance rating for the water system.

The Denver Water Department does conduct detailed hydraulic engineering studies on all pipelines 16 inches (0.41 m) in diameter and larger. These hydraulic analyses help to ensure that distribution system growth and expansion does not negatively impact the transmission system. In special cases, the department does a computerized hydraulic analysis of main extensions for new development areas.

The above minimum design standards exemplify those often used in large water systems that have evolved over a long period of time. In addition to pipe sizes, headloss, and velocity constraints, other system criteria also must be considered, as follows.

System Pressures. It is important to establish criteria for minimum and maximum system pressures occurring during the maximum hour demand. In some cases, it is also desirable to establish maximum pressure fluctuations within specific areas of the water distribution system. Typically, minimum acceptable water system pressures are 35 to 40 psi (1,190 to 1,360 kPa), and maximum pressures are 100 to 120 psi (3,400 to 4,080 kPa). A minimum system pressure of 35 to 40 psi (1,190 to 1,360 kPa) ensures adequate flows to the individual consumers and allows for reasonable operation of home-type irrigation/sprinkler systems. A more desirable system pressure may be in the range of 50 psi (1,700 kPa). This level of system pressure provides adequate flows and working pressures for most typical residential and commercial uses. Maximum pressure limitations are desirable to minimize the additional cost of providing piping materials with adequate strength to cope with the high pressure. In addition, high pressures can injure existing residential and commercial plumbing systems. Where main pressures exceed 100 to 110 psi (3,400 to 3,740 kPa), individual pressure-reducing valves should be installed on each service.

Minimizing the pressure fluctuations in the distribution system provides reliability for the consumer. Irrigation or in-house water-based appliances operate more consistently when pressure fluctuations do not exceed certain limits. Wide system pressure variations make proper design and selection of these appliances difficult and can also create operational difficulties.

Water system design may be based entirely on a hydraulic analysis that selects pipe sizes to ensure adequate pressures within all areas of the water distribution system under a set of system demand or loading conditions. For some water systems, a detailed hydraulic analysis may be done for every development area or main extension. In other systems, design may be based on a combination of minimum design standards and hydraulic analysis. Hydraulic analyses may be as complex as a computer-based water distribution system simulation, or as simple as a hydraulic evaluation of a single pipe utilizing standard hydraulic tables.

WATER SYSTEM SIMULATION AND HYDRAULIC ANALYSIS

Water system simulation and hydraulic analysis are based on fluid mechanics. The analysis and the simulations rely on basic equations used to determine friction headloss. Equations that have been applied in hydraulic analysis and simulation include the Darcy-Weisbach, Hazen-Williams, Manning, and Darcy-Weisbach/Colebrook-White formula pipe flow equations. (The fundamentals of fluid mechanics are described in several basic texts[3-5] and previously in Chapter 25.)

The design of a single pipe or main may be accomplished using any of the above equations or by the use of standard hydraulic tables or nomographs based on the above equations.[6] However, except for large conduits or transmission mains, the designer or engineer does not often evaluate a single water system pipeline to determine design or operating characteristics. More frequently, the designer evaluates a detailed pipe network consisting of a number of pipes branching to form complex, looped systems. The analysis and design of complex piping networks can be tedious, especially if the network consists of a large number of pipes and system appurtenances. Therefore, in any water system simulation or hydraulic analysis, it is usually necessary to reduce the complexity of the system enough for it to be analyzed, yet to maintain enough detail to provide a meaningful result. This approach to "skeletonizing" the water distribution system will be elaborated on later in this section.

The oldest, and probably the most widely utilized, method for pipe network analysis is the Hardy-Cross method. This method was developed before the use of computers, and uses hand computations to solve pipe network problems.

A number of other "office study" methods are available, as described in early water engineering books. The earliest "electric" methods for hydraulic analyses were described by Howland, Camp, and Hazen, and a commercially available electric analyzer was developed by McIlroy.[7-10]

The Hardy-Cross method is based on trial and error, with corrections applied to an initial set of assumed flows or heads until the network is balanced hydraulically. This type of iterative calculation is amenable to computerization; however, the Hardy-Cross method is cumbersome for the computer because the mathematics hinder the speed at which solutions can be obtained. The method is still used for analyzing small water systems and sections of large water systems and for larger systems where sufficient computer storage and computer time are available to handle the mathematics associated with the approach.

Subsequent computer-based hydraulic analysis techniques employ pipe friction formulas with less cumbersome mathematical approaches to reduce both memory requirements and computational time. Like the Hardy-Cross method, the pipe friction equations are nonlinear and require solution by trial and error. These trial-and-error solutions may require a significant amount of computer memory and computational time; hence, various mathematical methods have been applied to accelerate solution of the equations. Jeppson provides a detailed discussion of the Newton-Raphson method, one of the most widely used, as it applies to hydraulic simulation.[11] This method requires an initial guess of the solution. If the initial guess or the initialization provided in the Newton-Raphson equation is not reasonably accurate, convergence may be difficult. However, when the initial values are sufficiently close to the solution, the convergence may be quadratic. The method obtains the solution to a system of nonlinear equations by iteratively solving a system of linear equations.

The number of computer programs currently available to perform hydraulic analysis and water system simulation is significant and expanding. These programs range in capability and complexity from those that operate on hand-held calculators for solving problems dealing with up to four-loop systems, to commercially available programs written for small personal computers capable of analyzing networks in the range of 750 pipes (Table 27–1 lists some of the available programs for small computers), to programs that operate on high-speed, high-volume storage computers capable of analyzing distribution networks containing thousands of pipes and other water system appurtenances, including pumps, pressure-reducing valves, and system storage reservoirs.[12] Available programs have output capabilities ranging from flows and direction of flow in pipelines and pressures at given points within the distribution system, to programs capable of summarizing all the desired system characteristics and providing a graphical plot of the distribution system, including plotting of isopressure contours. Available programs are for sale or for lease on time-sharing systems, or may be proprietary and offered for use only by the program owner.

Many of the currently available computer programs provide steady state, or static, simulation of the water system under given conditions of supply

Table 27–1. Examples of Hydraulic Analysis Programs for Small Computers.

PROGRAM NAME	COMPANY
Pipe Network/Water Distribution System Analysis	Microcomp, Solana Beach, CA
Dynamic Response Module for Pipe Network Analysis	Microcomp, Solana Beach, CA
Hydraulics	Disco-Tech, Santa Rosa, CA
Flow Network Analysis	Kelix Software Systems, Baton Rouge, LA
NETWK- (Pipe Network Analysis Program)	Civilsoft, Anaheim, CA
Pipe Network Analysis	Doran Engineering, Pleasantville, NJ
Pipe Network Analysis	Computapipe, Huntington Beach, CA

and demand. Under this steady state analysis, the solution is achieved when the flow rate in each pipe is determined. The output is based on fixed system characteristics, including single-condition values for pressure-reducing valves and single head increases for pumping and booster stations. Alternatively, some computer programs offer extended-period, or dynamic, solutions based on time period simulation of the water system under varying supply and loading conditions and varying system characteristics resulting from changing pressure-reducing-valve set points and variation in pump curve characteristics. A dynamic solution model may simulate a 24- to 48-hour period.

It is important to plan and operate the distribution system so as to ensure an adequate level of service at all points in the system under varying conditions of loading. Criteria for determining the level of service include:

- The maintenance of flows and pressures (heads) at various points in the system within limits. It may be acceptable to vary these limits with time.
- The management of storage to balance the supply and the distribution.

In system design and operation, it is beneficial to know the impact of changes in demand on the level of service. This knowledge aids in the development of design and control strategies to maintain the level of service. There is also a need to evaluate the adequacy of storage or of proposed network additions with respect to an increase in total demand. Both of these needs can be met by simulating the behavior of the system over a period of 24 to 48 hours under changing demand patterns. A static solution will not be so informative in terms of changing conditions because a static solution provides only a one-time evaluation or a "snapshot" of the system characteristics.

Another hydraulic simulation technique is the application of linear theory.[13] This method is easily applied if all external flows to the system are known. The linear theory method has several advantages over the Newton-Raphson or the Hardy-Cross methods because it does not require an initialization, and it converges in relatively few iterations. Linear theory transforms the nonlinear loop equations into linear equations by approximating the head in each pipe. A system of linear equations is developed that can be solved by linear algebra.

An excellent discussion of hydraulic analysis of pipe networks is presented by Jeppson.[11] In this reference, the reader may find detailed discussion of several methods, the mathematics and examples of which are beyond the scope of this chapter.

Skeletonizing the Water Distribution System

As previously discussed, water systems may be large networks consisting of several hundred pipes and different sources of supply. To analyze such systems, it is important to achieve a representation of the network consistent with computational capabilities. This representation must then be detailed enough so that the system can be accurately analyzed to meet the desired objectives of the study.

The above goals are accomplished by first skeletonizing the system, which reduces the number of pipes in consideration before the system is defined to the computer. This skeletonizing may involve the total dismissal of an unimportant or minor pipe, the replacement of a series of pipes of varying diameter with one equivalent pipe, or replacing a system of parallel pipes with an equivalent pipe. An equivalent pipe is one in which the loss of head for a specified flow is the same as the loss in head of the pipes that it replaces.

The decision on how various pipes are treated is based largely on the experience of the individual conducting the analysis. If the water system being evaluated is small, and if the computer and program have adequate capabilities, it may be desirable to include all of the system piping. If the system is very large, it may be desirable that pipes under a specific size, 12 inches (0.31 m) for example, not be considered, unless they are necessary to complete a loop.

When skeletonizing the system, one must keep in mind that the backbone of the water distribution system is the transmission mains. Large water systems may contain transmission mains that are 16 inches (0.41 m) and larger. Connections to transmission mains are typically held to a minimum; so it is typical to utilize all of the existing transmission piping with no skeletonization. Connections from the transmission mains to the distribution system can then be considered the demand, or load points on the system. Water distribution piping is interconnecting pipes, which are the principal components of the distribution system. When water system analysis is conducted on the distribution system, significant skeletonizing of the system is done.

For medium-size systems, one must consider the effect of ignoring an existing pipe. If problems become apparent in areas of the system where pipes have been ignored, then the decision to ignore those pipes must be reevaluated. It is important to note that when no well-defined feeder system is apparent, significant error may result from skeletonizing.

Figure 27–2 shows an existing complete water distribution system, and Fig. 27–3 shows the skeletonized system that was defined to the computer. This is just one example of skeletonizing a water distribution system. In this example, the computer capabilities were sufficient to include more of the system than might have been included if such capabilities had not been available. The same accuracy of the output data might have been obtained if additional skeletonizing had been accomplished.

Another factor affecting the amount of system skeletonizing is the intent of the simulation analysis. When the purpose is to examine the impacts of fire flow in large residential areas, it is necessary to include a significant

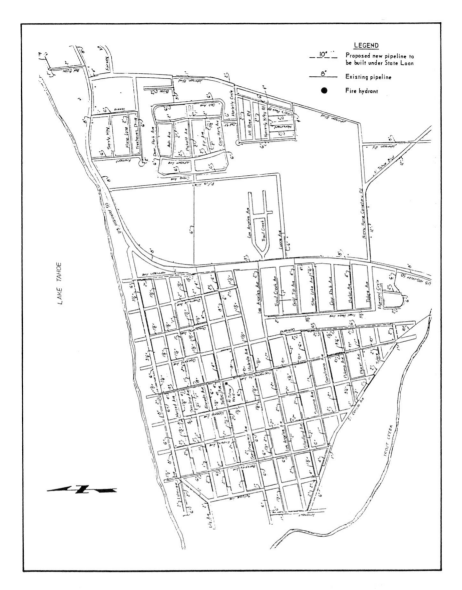

Fig. 27-2. Water distribution piping, complete system.

portion of smaller-diameter piping. However, if the intent is to evaluate the impact of a large development on the transmission system, the majority of the small-diameter piping can be ignored.

In the early analysis techniques, minor losses were included in the analysis. Water flowing in a straight pipe under pressure at a constant velocity experi-

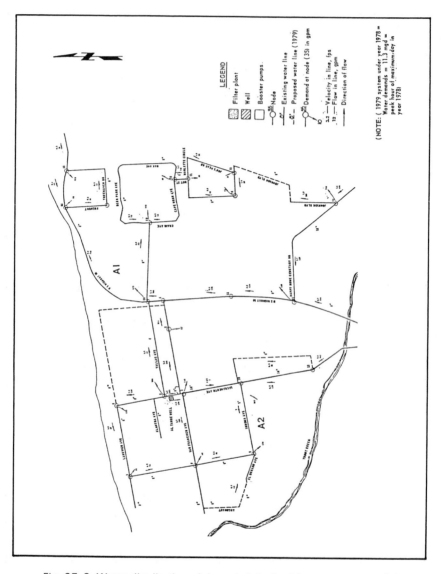

Fig. 27-3. Water distribution piping, skeletonized for computer model.

ences a constant friction loss, and a change in the velocity or direction of the flow creates additional losses. Such losses are classified as minor, as they are a small percentage of the total loss in relatively long pipelines—a pipeline with a length in excess of 500 diameters usually being classified as long.[14] Minor losses become important when small-diameter pipes and high velocities are considered in the total piping and appurtenances relating to a pumping station. Minor losses are produced by flow through bends, fittings, valves, and openings such as fire hydrants, and in and out of reservoirs.

In large-scale network simulations, the minor losses may become insignificant and can be ignored. Deciding when to include minor losses is a matter of judgment; when minor losses are included, they can easily be introduced as equivalent lengths of pipe, although some programs provide for inclusion of these losses separately. Where C values (coefficient of pipe roughness) are determined from field measurements, they invariably include a component due to the various minor losses encountered.

Pipe Friction Factor

One of the input requirements for most water system simulation models is the pipe friction factor appropriate to the pipe flow equations upon which the model is based.

The interior of a pipeline changes with time, and these changes affect the pipe friction factor. Typically, over time, the friction head losses in a section of pipe will increase with the pipe's age, because of various physical and chemical characteristics of the water that change the finish or roughness of the inside of the pipe. The inside diameter of the pipe also may be reduced. Actions that affect the line capacity include sedimentation, scaling, organic growth, tuberculation, and corrosion. The effect of these actions may be partially reversed by pipeline cleaning.

When performing water system simulation, it is important to consider the age and condition of the pipeline to determine the appropriate friction factor. The hydraulic capacity of pipelines may be determined by conducting flow tests on representative sections of the pipelines. Procedures for such tests are based on a measured flow through a straight section of pipeline (as long as possible) with the pressure drop recorded by gauges installed at both ends of the section.

Once the input data are developed and the water system is skeletonized, the system is described to the computer in a format compatible with the simulation model. A particularly effective water systems simulation model was prepared for the Office of Water Resources Research of the U.S. Department of the Interior.[15] This model, known as WATSIM, is widely available. The WATSIM program is capable of both static and dynamic solutions and, without modification, can handle networks with up to 1,500 nodes and 2,500 lines. The original program was written in Fortran IV, and it can be modified to operate on a number of systems. The model employs a modified Newton-Raphson procedure. The program calculates pressures at given points within the distribution system, quantity, and the direction of flow in each pipe.

The water distribution system is defined to the computer as a set of "nodes" and "lines." Nodes are connected in pairs by network elements (lines) such as pumps, pipes, one-way valves, pressure-reducing valves, and so forth. The nodes are specified as having either a fixed head or a fixed flow. An outflow is considered to be a loading or demand on the system; an inflow is considered to be a supply (well, treatment plant, or reservoir) to the system. Nodes may have flow values of 0—thus no inflow or outflow. These nodes are used solely for the purpose of connecting different distribution system facilities. Figure 27–4 shows the line and node representation for a small system. This schematic representation is then translated onto a computer coding form for input into the computer. Figure 27–5 presents the typical node input summary from a WATSIM printout, and Fig. 27–6 presents the typical line input summary.

The desired output data from WATSIM may be selected from several alternatives. Figure 27–7 shows the summary output for a static solution. The figure shows data giving the flow rate and direction in each line; the losses through each line section; pressures at each node; the pipe characteristics including pipe size, length, and friction coefficient for the individual sections; and the demand or supply at the node. WATSIM will employ either the Manning equation or the Hazen-Williams equation for pipe friction, as selected by the user.

A water distribution model that operates on a small computer has been used by the authors for several projects. The model, called NETWORK, was developed by Computapipe Computer Design Systems, which has been developing programs for the waterworks and sanitary industries since 1976. The program is the sixth version of the University of Kentucky's program, originally developed in the 1960's for mainframe computers. The program is available for personal computers and requires a minimum memory capability of 485 kb.

NETWORK is an interactive program that can be used to analyze steady-state flows and pressures in a piping network (static solution) or to simulate the operation of the network over an extended period of time. The static solution usually is quite adequate to determine a distribution system's ability to meet all critical demand conditions.

The basis of the program is a direct solution of basic pipe system hydraulic equations using a linearization scheme and sparse matrix methods to handle the nonlinear terms in the energy equations. The program is capable of simulating networks of up to 750 nodes and lines. The NETWORK model handles pipe sizes 2 inches and larger and can simulate any portion of a network for which boundary conditions can be described. The model is capable of incorporating storage tanks, pumps, pressure reduction valves, pressure sustaining valves and multiple system water sources. The program is being modified to simplify the input requirements, and improve the internal diagnostics for incorrect input data or instructions. An example of the input and output

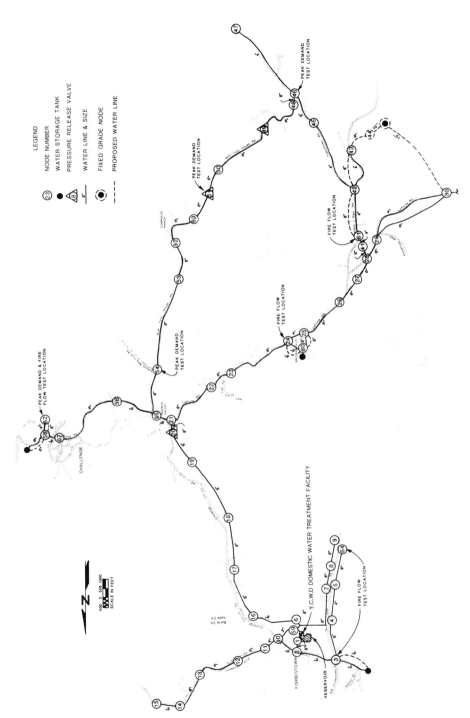

Fig. 27-4. Water piping system showing lines and nodes for computer model.

TYPE	ID	ZONE	FLOW	ELEVATION	HEAD	LO LIMIT	HI LIMIT
NQ	94		-0.0200	3670.	3800.	0.	0.
NQ	95		-0.0500	3650.	3800.	0.	0.
NQ	96		0.0000	3630.	3800.	0.	0.
NQ	97		-0.0200	3650.	3800.	0.	0.
NQ	M97		-0.0200	3640.	3800.	0.	0.
NQ	98		-0.0200	3645.	3800.	0.	0.
NQ	99		-0.0200	3640.	3800.	0.	0.
NQ	100		-0.0100	3630.	3800.	0.	0.
NQ	101		-0.0400	3610.	3800.	0.	0.
NQ	102		-0.0700	3640.	3800.	0.	0.
NQ	103		-0.2000	3635.	3800.	0.	0.
NQ	104		-0.1000	3595.	3800.	0.	0.
NQ	105		-0.0100	3590.	3800.	0.	0.
NH	205		0.0000	3590.	3755.	3750.	3760.
NQ	106		0.0000	3725.	3800.	0.	0.
NQ	206		0.0000	3725.	4100.	0.	0.
NQ	107		-0.0100	3750.	4100.	0.	0.
NQ	108		0.0000	3810.	4100.	0.	0.
NQ	109		0.0000	3850.	4100.	0.	0.
NQ	110		0.0000	3890.	4100.	0.	0.
NQ	111		-0.0100	3920.	4100.	0.	0.
NQ	112		-0.0100	4000.	4100.	0.	0.
NQ	113		0.0000	3960.	4100.	0.	0.
NH	114		0.0000	4000.	4190.	4160.	4194.
NQ	115		0.0000	3960.	4100.	0.	0.
NH	215		0.0000	3400.	3540.	3500.	3600.
NQ	116		-0.0200	3940.	4100.	0.	0.

Fig. 27-5. Typical node input summary, WATSIM program.[27]

files for an analysis are shown in Figures 27–8 and 27–9, respectively, and refer to the piping system in Figure 27–4. Only a partial printout is included here.

Computer programs simulate the conditions within a water distribution system based on data provided by the user. For best results, it is important that the input data be as accurate as possible, and that all assumptions regarding input data be carefully made. It is also necessary that the model being used describe the behavior of the existing system. This can be checked by "calibrating" the model, which requires running the model under a set of known conditions. Typically, a number of recording pressure gauges will be in operation on an existing water system while demands are carefully monitored. It is then possible to compare the computer output results with those obtained in the field. Adjustments can then be made in the input assumptions to make the model simulate the real system.[16]

When applying the model output, one must not assume that the output data provided by the computer represent the real conditions without reflecting on the effect of assumptions in the input data. In municipal water supply, loads at individual nodes are usually not precisely known. The flows into the system are usually metered so that the total load is known, but the problem lies in the proper distribution of individual demands to nodes. Statistical methods based on customer meter records can help. The Denver Water Department has produced a number of computer programs that develop the load or demands on the system for input into the model.[17] The demands are based on carefully collected data.

In many water systems, there may also be considerable uncertainty about pipe roughness characteristics and actual pipe diameters. In the development of WATSIM, the program was compared to an existing water system in San Jose, California. In comparison to the existing system, the program predicted pressures at specific points in the system within an accuracy of 5 to 10 percent. Our experience indicates that the pressure prediction accuracy in a carefully calibrated program may be 1 to 3 percent at specific points in the system. A predictive accuracy of system pressures in the 10 percent range should not cause problems in the interpretation of the model output results or ultimate decisions based on these results. However, we strongly recommend that calibration of the model be performed.

Computer- or non-computer-based water distribution systems simulation can serve many purposes in the design and operation of water distribution systems. Linear programming techniques have been used to develop a system for optimizing gravity-fed water distribution systems.[18] Simulation modeling also can be used to determine the impact of existing demand conditions on an existing water system. When a problem has been detected in the field or is predicted by the model output, alternative solutions may be investigated to obtain the best and most economical solution to the problem. Simulation

TYPE	ORG.ID	ORG.ZONE	DES.ID	DES.ZONE	PAR.	LENGTH (FEET)
L	M97		102			350.000000
L	91		99			800.000000
L	99		98			200.000000
L	99		100			200.000000
L	100		101		2	350.000000
L	84		100			1600.000000
L	85		101			1600.000000
L	101		61			550.000000
L	98		103			950.000000
L	102		103		2	500.000000
L	103		104			550.000000
L	104		105			375.000000
R	105		205			50.000000
R	66		66A			50.000000
R	58		58A			50.000000
R	60		60A			50.000000
L	205		71			3400.000000
L	66A		71			1400.000000
L	71		70			1100.000000
L	58A		62			1400.000000
L	62		67			950.000000
L	67		70			700.000000
L	67		68			520.000000
L	70		68			1200.000000
L	62		63			1250.000000
L	68		63			750.000000
L	60A		63			1050.000000
PP	106		206			-611.990000
L	206		107			1200.000000
L	107		108			3550.000000
L	108		109			1800.000000
L	109		110			1000.000000
L	110		111			3500.000000
L	111		112			1250.000000
L	111		113			2000.000000
L	112		113			750.000000
C	113		114			2750.000000
L	115		114			2750.000000
PP	215		115			-578.820000
L	110		116			1800.000000
L		8		9		.18

Fig. 27-6. Typical line input summary, WATSIM program.[27]

modeling can be invaluable in system operations, as alternative operation strategies can be evaluated. As previously indicated, sophisticated models are capable of incorporating pumping stations, valves, and appurtenances along with the system piping, and these capabilities allow evaluation of operational alternatives as well as construction alternatives.

The water system simulation model may, by varying the demands on the system, determine effects at critical periods. These may include maximum hour demand, fire demands consistent with maximum day demands, and storage replenishment conditions. This information can provide valuable assistance in the cost-effective design of distribution systems. When major development is planned within an existing water system, water system simulation modeling not only can show the impact of that development on the existing system, but can assist in the sizing of the lines within the development itself.

DIAMETER (INCHES)	HAZEN-WILLIAMS COEFFICIENT	PRV SETPOINT OR PUMP HEAD INCREASE
8.000000	100.000000	0.000000
4.000000	100.000000	0.000000
6.000000	100.000000	0.000000
6.000000	100.000000	0.000000
6.000000	100.000000	0.000000
4.000000	100.000000	0.000000
6.000000	100.000000	0.000000
6.000000	100.000000	0.000000
8.000000	100.000000	0.000000
8.000000	100.000000	0.000000
8.000000	100.000000	0.000000
8.000000	100.000000	0.000000
8.000000	100.000000	165.000000
6.000000	100.000000	160.000000
6.000000	100.000000	115.000000
6.000000	100.000000	95.000000
8.000000	100.000000	0.000000
6.000000	100.000000	0.000000
8.000000	100.000000	0.000000
6.000000	100.000000	0.000000
6.000000	100.000000	0.000000
8.000000	100.000000	0.000000
8.000000	100.000000	0.000000
6.000000	100.000000	0.000000
6.000000	100.000000	0.000000
8.000000	100.000000	0.000000
8.000000	100.000000	0.000000
164.930000	263.330000	0.000000
10.000000	100.000000	0.000000
10.000000	100.000000	0.000000
8.000000	100.000000	0.000000
8.000000	100.000000	0.000000
8.000000	100.000000	0.000000
8.000000	100.000000	0.000000
8.000000	100.000000	0.000000
8.000000	100.000000	0.000000
8.000000	100.000000	0.000000
8.000000	100.000000	0.000000
97.240000	1070.970000	0.000000
8.000000	100.000000	0.000000
.177365	1.39765	

Fig. 27-6. (Continued)

Water Hammer

There are simulation programs for evaluating various special problems in pipelines and water distribution systems. One such problem is water hammer (surge or transient flow), which may be caused by rapid changes in flow within closed conduit systems. These rapid changes in flow cause pressure waves that travel both upstream and downstream from the point of origin. Water hammer problems can be complex and require specialized analysis for solution. The condition of the water hammer is usually analyzed by the equations of motion and continuity.

Computer methods have been applied to solve a number of water-hammer-causing flow situations, including: pump failures and power outages, extremely long water supply lines that may contain multiple pumping points, and the

TYPE	NODE	ZONE	PAR	FLOW MGAL/DAY	HEAD-FT (DROP)	PSI AT ELEV (DEV-HEAD-FT)	ELEV-FT (LENGTH)	DIAM-IN	LO HWC	HI RES
NQ	102			-0.07	3818.136	77.186	3640.00			
L	M97		2	-0.18	-0.228		350.0	8.00	100.0	
L	103			0.17	0.273		500.0	8.00	100.0	
L	98			-0.05	-0.128		470.0	6.00	100.0	
NQ	103			-0.20	3817.863	79.235	3635.00			
L	102		2	-0.17	-0.273		500.0	8.00	100.0	
L	104			0.11	0.141		550.0	8.00	100.0	
L	98			-0.14	-0.401		950.0	8.00	100.0	
NQ	104			-0.19	3817.723	96.506	3595.00			
L	103			-0.11	-0.141		550.0	8.00	100.0	
L	105			0.01	0.001		375.0	8.00	100.0	
NQ	105			-0.01	3817.722	98.672	3590.00			
L	104			-0.01	-0.001		375.0	8.00	100.0	
R	205			0.00	62.722	165.0	50.0	8.00	100.0	
NQ	106			0.12	3832.413	46.542	3725.00			
PP	206			0.17	273.734	-256.	107.0	263.		
L	5			-0.05	-0.041		650.0	8.00	100.0	
NQ	107			-0.01	4106.121	154.307	3750.00		40.	140. LIMIT
L	108			0.04	0.050		3550.0	10.00	100.0	
L	206			-0.05	-0.025		1200.0	10.00	100.0	
NQ	108			-0.00	4106.072	128.288	3810.00			
L	107			-0.04	-0.050		3550.0	10.00	100.0	
L	109			0.04	0.074		1800.0	8.00	100.0	
NQ	109			-0.00	4105.998	110.924	3850.00			
L	108			-0.04	-0.074		1800.0	8.00	100.0	
L	110			0.04	0.040		1000.0	8.00	100.0	

Fig. 27-7. Summary output for a static solution, WATSIM program.[27]

FLOWRATE IS EXPRESSED IN GPM AND PRESSURE IN PSIG

A SUMMARY OF THE ORIGINAL DATA FOLLOWS

THERE IS A PRESSURE REGULATOR AT JUNCTION 20 FOR LINE 22 WHICH MAINTAINS A GRADE OF 2551.00

THERE IS A PRESSURE REGULATOR AT JUNCTION 31 FOR LINE 37 WHICH MAINTAINS A GRADE OF 1794.00

THERE IS A PRESSURE REGULATOR AT JUNCTION 32 FOR LINE 38 WHICH MAINTAINS A GRADE OF 1490.00

THERE IS A PRESSURE REGULATOR AT JUNCTION 51 FOR LINE 58 WHICH MAINTAINS A GRADE OF 2441.00

THERE IS A PRESSURE REGULATOR AT JUNCTION 49 FOR LINE 56 WHICH MAINTAINS A GRADE OF 2157.00

PIPE NUMBER	NODE NUMBERS		LENGTH (FEET)	DIAMETER (INCHES)	ROUGHNESS	MINOR LOSS K	FIXED GRADE
1	1	2	750.0	8.0	140.0	11.50	
2	2	3	2200.0	6.0	140.0	5.20	
3	0	3	700.0	6.0	140.0	7.50	2948.00

THERE IS A CHECK VALVE IN LINE NUMBER 3

4	3	4	1750.0	6.0	140.0	1.60	
5	4	5	1800.0	4.0	140.0	.80	
6	59	6	400.0	8.0	140.0	.60	
7	6	7	3700.0	6.0	140.0	1.60	
8	7	8	1200.0	6.0	140.0	1.60	
9	8	9	1500.0	6.0	150.0	1.80	
10	9	23	14500.0	8.0	150.0	4.50	

LINE 10 IS CLOSED

11	10	2	1200.0	6.0	150.0	5.50	
12	10	11	775.0	4.0	150.0	1.20	
13	11	12	1625.0	6.0	150.0	1.80	
14	12	13	2710.0	6.0	150.0	3.60	
15	13	14	1750.0	6.0	150.0	2.40	
16	14	15	1313.0	6.0	140.0	1.80	
17	16	6	2925.0	8.0	140.0	1.00	
18	16	17	3250.0	8.0	140.0	1.80	
19	17	18	2750.0	8.0	110.0	1.20	
20	18	19	4250.0	8.0	110.0	1.20	
21	19	20	1700.0	8.0	120.0	1.20	
22	20	21	120.0	8.0	120.0	.00	
23	21	22	3100.0	6.0	120.0	1.80	
24	22	23	1300.0	6.0	120.0	1.20	
25	23	24	4400.0	6.0	120.0	.60	
26	24	60	780.0	6.0	120.0	.60	
27	24	25	1000.0	4.0	120.0	.00	

LINE 27 IS CLOSED

28	60	25	1080.0	6.0	140.0	2.60	
29	25	26	2600.0	6.0	140.0	2.00	
30	26	27	1500.0	6.0	140.0	.80	
31	27	28	1250.0	6.0	140.0	.20	

JUNCTION NUMBER	DEMAND	ELEVATION	CONNECTING PIPES			
1	.00	2750.00	1	73		
2	.00	2760.00	1	2	11	75
3	.00	2812.00	2	3	4	83
4	20.00	2750.00	4	5	84	
5	40.00	2725.00	5	80	84	
6	80.00	2735.00	6	7	17	75
7	54.00	2725.00	7	8		
8	20.00	2680.00	8	9		
9	10.00	2525.00	9	10		
10	20.00	2780.00	11	12		
11	20.00	2785.00	12	13		
12	30.00	2775.00	13	14		
13	20.00	2772.00	14	15		
14	30.00	2772.00	15	16		
15	20.00	2720.00	16			
16	14.00	2690.00	17	18		
17	14.00	2680.00	18	19		
18	14.00	2705.00	19	20		
19	14.00	2450.00	20	21		
20	.00	2300.00	21	22	68	
21	.00	2285.00	22	23	63	68
22	10.00	2240.00	23	24		
23	20.00	2325.00	10	24	25	
24	12.00	2290.00	25	26	27	85
25	48.00	2235.00	27	28	29	86
26	20.00	2200.00	29	30		
27	20.00	2175.00	30	31		
28	.00	2120.00	31	32	47	
29	40.00	2110.00	32	33	34	
30	36.00	2072.00	33	34	35	36
31	.00	2076.00	36	37	69	

A SUCCESSFUL GEOMETRIC VERIFICATION HAS BEEN COMPLETED

Fig. 27-8. Typical input summary, NETWORK program.

THE FOLLOWING RESULTS ARE OBTAINED AFTER 5 TRIALS WITH A RELATIVE ACCURACY = .00457

PEAK MONTH FLOW

PIPE NO.	NODE NUMBERS		FLOWRATE	HEAD LOSS	PUMP HEAD	MINOR LOSS	VELOCITY	HL/1000
1	1	2	209.68	.66	.00	.32	1.34	.88
2	2	3	69.68	1.02	.00	.05	.79	.47
3	0	3	2.20	.00	.00	.00	.02	.00
4	3	4	60.00	.62	.00	.01	.68	.35
5	4	5	5.57	.06	.00	.00	.14	.03
6	59	6	557.00	2.16	.00	.12	3.55	5.39
7	6	7	84.00	2.44	.00	.02	.95	.66
8	7	8	30.00	.12	.00	.00	.34	.10
9	8	9	10.00	.02	.00	.00	.11	.01
LINE 10 IS CLOSED								
11	10	2	-140.00	-1.79	.00	-.22	-1.59	-1.49
12	10	11	120.00	6.26	.00	.17	3.06	8.08
13	11	12	100.00	1.30	.00	.04	1.13	.80
14	12	13	70.00	1.12	.00	.04	.79	.41
15	13	14	50.00	.39	.00	.01	.57	.22
16	14	15	20.00	.06	.00	.00	.23	.05
17	16	6	-393.00	-8.26	.00	-.10	-2.51	-2.82
18	16	17	379.00	8.58	.00	.16	2.42	2.64
19	17	18	365.00	10.59	.00	.10	2.33	3.85
20	18	19	351.00	15.22	.00	.09	2.24	3.58
21	19	20	337.00	4.80	.00	.09	2.15	2.83
22	0	21	337.00	.34	.00	.00	2.15	2.83
23	21	22	337.00	35.57	.00	.41	3.82	11.47
24	22	23	327.00	14.11	.00	.26	3.71	10.85
25	23	24	307.00	42.48	.00	.11	3.48	9.65
26	24	60	93.79	.84	.00	.01	1.06	1.07
LINE 27 IS CLOSED								
28	60	25	92.47	.85	.00	.04	1.05	.79
29	25	26	247.00	12.61	.00	.24	2.80	4.85
30	26	27	227.00	6.22	.00	.08	2.58	4.15
31	27	28	207.00	4.37	.00	.02	2.35	3.50

JUNCTION NUMBER	DEMAND	GRADE LINE	ELEVATION	PRESSURE
1	.00	2950.09	2750.00	86.70
2	.00	2949.08	2760.00	81.93
3	.00	2948.00	2812.00	58.93
4	20.00	2947.37	2750.00	85.53
5	40.00	2947.32	2725.00	96.34
6	80.00	2851.33	2735.00	50.41
7	54.00	2848.87	2725.00	53.68
8	20.00	2848.75	2680.00	73.13
9	10.00	2848.73	2525.00	140.28
10	20.00	2947.07	2780.00	72.40
11	20.00	2940.64	2785.00	67.44
12	30.00	2939.30	2775.00	71.20
13	20.00	2938.15	2772.00	72.00
14	30.00	2937.75	2772.00	71.82
15	20.00	2937.69	2720.00	94.33
16	14.00	2842.97	2690.00	66.29
17	14.00	2834.22	2680.00	66.83
18	14.00	2823.54	2705.00	51.37
19	14.00	2808.22	2450.00	155.23
20	.00	2803.33	2300.00	218.11
21	.00	2550.66	2285.00	115.12
22	10.00	2514.69	2240.00	119.03
23	20.00	2500.32	2325.00	75.97
24	12.00	2457.73	2290.00	72.68
25	48.00	2455.99	2235.00	95.76
26	20.00	2443.14	2200.00	105.36
27	20.00	2436.83	2175.00	113.46
28	.00	2432.44	2120.00	135.39
29	40.00	2431.55	2110.00	139.34
30	36.00	2431.46	2072.00	155.76
31	.00	2431.46	2076.00	154.03

THE NET SYSTEM DEMAND = 1203.00
SUMMARY OF INFLOWS(+) AND OUTFLOWS(-) FROM FIXED GRADE NODES

PIPE NUMBER	FLOWRATE
3	2.20
43	100.00
51	41.00
67	96.01
73	209.68
74	557.00
83	-11.88
87	208.99

THE NET FLOW INTO THE SYSTEM FROM FIXED GRADE NODES = 1214.88
THE NET FLOW OUT OF THE SYSTEM INTO FIXED GRADE NODES =   -11.88

Fig. 27-9. Typical output summary, NETWORK program.

control of high-head pipelines by pressure-reducing/throttling valves. Where detailed analysis of water hammer problems may be required, computer solutions are probably the most reasonable approach. One such program, LIQT (liquid transient) has the capabilities of analyzing simple single line problems or complex network problems and can evaluate the impact of water hammer and possible control techniques including valves, surge tanks, and bypasses.[19] This particular program is proprietary, and is available for use from the owner and through several large time-sharing networks. Another surge analysis program for IBM-compatible personal computers is the Waterhammer and Transients program, by Computapipe, Huntington Beach, California.

Summary

In choosing a water system simulation model, it is necessary to evaluate the available input data, the size and complexity of the system to be analyzed, the desired uses of the model output, and the number of alternatives that are to be evaluated. One may find a simple Hardy-Cross evaluation of a small water system to be more efficient than pursuing a detailed and complex computer evaluation of the system. However, if a number of alternatives are to be evaluated, or if the simulation model will serve as a major design tool for future system expansion, the use of a highly efficient computerized model will be the best choice.

Computers and computer-based simulation models provide the ability to analyze complex systems quickly. This is an important benefit in view of the costs of operation of water systems, especially energy costs. Simulation can also reduce construction costs by allowing a determination of the most cost-effective of several possible solutions.

The use of computer-based simulation models has increased over the past few years, and the available programs have become more and more advanced. Some of these programs interface several peripheral programs to allow for a more exact analysis and easier data base handling.[20]

PIPELINE LOCATION, PROTECTION, AND MATERIALS OF CONSTRUCTION

Water distribution system networks are expensive; and they are expected to be in service for long periods of time, without significant cost for maintenance, repair, or replacement. For these reasons, not only are the proper sizing and design of systems important, but so are their construction and maintenance.

In general, all mains should be installed in dedicated public streets or in other public access ways with a minimum of 30 to 50 feet (9.1 to 15.3 m) of right-of-way. This will ensure adequate access for normal and routine maintenance as well as emergency repair procedures. Unless it is absolutely unavoidable, mains should never be located on private property, under structures, or under or in lakes.

In evaluating potential pipeline routes, it is desirable to select direct routes. Data on topography, soil, and geology should be considered. Rough or difficult terrain should be avoided, as should areas that may be susceptible to land or mud slide, 100-year flood, or other hazards that could cause breakage or outage. Tunneling should be considered only when there are no feasible alternate routes and when it is economically justified. All line installations should give consideration to future construction and the need for repair and maintenance.

When water mains are installed, consideration must be given to providing adequate separation from sanitary sewers. In parallel installations, water mains should be at least 10 feet (3.1 m) horizontally from any sanitary sewer, storm sewer, or sewer manhole. The distance should be measured edge-to-edge. When conditions prevent a horizontal separation of 10 feet (3.1 m), a water main may be laid closer to a storm or sanitary sewer if the bottom of the water main is at least 18 inches above the top of the sewer. If this vertical separation cannot be obtained, the sewer should be constructed of materials and with joints that are equivalent to water main standards of construction, and should be pressure-tested to assure watertightness prior to backfilling.

In perpendicular crossings when water mains cross house sewers, storm sewers, or sanitary sewers, a separation of at least 18 inches (0.46 m) between the bottom of the water main and the top of the sewer should be provided. When local conditions prevent such a vertical separation, sewers passing over or under water mains should be constructed of the materials equivalent to water main standards of construction, and should be pressure-tested to ensure watertightness prior to backfilling. Water mains passing under sewers should, in addition, be protected by a vertical separation of at least 18 inches (0.46 m) between the bottom of the sewer and the top of the water main with adequate structural support for the sewers. This will help prevent excessive deflection of joints and settling of the sewer on the water mains.

A full-length water pipe should be centered at the point of crossing so that the joints will be equidistant and as far as possible from the sewer. Figure 27-10 shows a typical crossing of this type.

No water pipe should pass through or come into contact with any part of a sewer or sewer manhole.

The top of all water pipes should be at least 6 inches (0.15 m) below the maximum recorded depth of frost penetration in the area of installation. The minimum depth of water mains should be 5 feet (0.91 m) from the ground surface to the top of the pipe. Pipes always should have adequate cover for external design loads. In designing buried pipelines, simplified tables have been developed based on formulas established by Marston and Spangler,

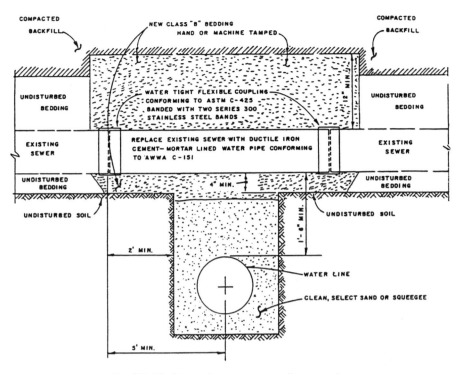

Fig. 27–10. Sewer line over water line crossing.

and Young and Smith. Buried pipeline design procedures may be found in several books, handbooks, and standards.[21-26] Pipe classes should be capable of handling the exterior loading resulting from the backfill and surface loading, and should be capable of handling interior working pressures of a minimum of 150 psi (1033.5 kPa) and water hammer surge pressures of 100 to 120 psi (689 to 827 kPa). The typical minimum acceptable pipe class is usually Class 22 for cast iron, Class 50 for ductile iron [except at 4- and 20-inch (0.1- and 0.51-m) diameters, where Class 51 is typical], Class 200 for asbestos cement pipe, and Class 200 for PVC, where 4-, 6-, and 8-inch (0.1-, 0.15-, and 0.20-m) diameters are generally acceptable and 12-inch (0.31-m) diameters may be acceptable under certain conditions.[2] Installation of mains through hazardous areas or at depths greater than 10 feet (3.05 m) in roadways may also require pressure classes in excess of the above minimum.

Whenever the installation of metallic pipe is contemplated, a soil resistivity survey of the area should be performed. The survey data and calculations should be evaluated together with the history of existing pipes in the area to determine if a nonmetallic pipe should be used. These data will also indicate what level of protection should be provided for metallic fittings and appurtenances. Typically, where resistivities are less than 2,500 ohms-cm, all metallic pipe should be polyethylene-wrapped. If resistivities are less than 1,000 ohms-cm, nonmetallic pipe should be used. The corrosive effects of finished water on the interior of the pipe must also be given consideration.

All piping, joints, and fittings should conform to applicable AWWA specifications and should be rated at least the pressure rating of the straight pipe involved. Table 27–2 presents a list of applicable AWWA standards. All

Table 27–2. Summary of Material Standards

STANDARD	NUMBER
American National Standard for Cement-Mortar Lining for Ductile-Iron and Gray-Iron Pipe and Fittings for Water	C104/A21.4–80
American National Standard for Polyethylene Encasement for Gray and Ductile Cast-Iron Piping for Water and Other Liquids	C105/A21.5–82[R77]
American National Standard for Gray-Iron and Ductile-Iron Fittings, 3 in. through 48 in., for Water and Other Liquids	C110/A21.10–82
American National Standard for Rubber-Gasket Joints for Ductile-Iron and Gray-Iron Pressure Pipe and Fittings	C111/A21.11–80
American National Standard for Flanged Ductile-Iron and Gray-Iron Pipe with Threaded Flanges	C115/A21.15–83
American National Standard for the Thickness Design of Ductile-Iron Pipe	C150/A21.50–81
American National Standard for Ductile-Iron Pipe Centrifugally Cast in Metal Molds or Sand-Lined Molds for Water or Other Liquids	C151/A21.51–81
AWWA Standard for Steel Water Pipe 6 in. and Larger	C200–80
AWWA Standard for Coal Tar Protective Coatings and Linings for Steel Water Pipelines—Enamel and Tape—Hot Applied	C203–78
AWWA Standard for Chlorinated Rubber-Alkyd Paint System for the Exterior of Aboveground Steel Water Piping	C204–75
AWWA Standard for Cement-Mortar Protective Lining and Coating for Steel Water Pipe—4 in. and Larger—Shop Applied	C205–80
AWWA Standard for Field Welding of Steel Water Pipe	C206–82
AWWA Standard for Steel Pipe Flanges for Waterworks Service—Sizes 4 in. through 144 in.	C207–78
AWWA Standard for Dimensions for Fabricated Steel Water Pipe Fittings	C208–83
AWWA Standard for Cold Applied Tape Coatings for Special Sections, Connections, and Fittings for Steel Water Pipelines	C209–76
AWWA Standard for Coal Tar Epoxy Coating System for Interior and Exterior Steel Water Pipe	C210–78
AWWA Standard for Fusion-Bonded Epoxy Coating for the Interior and Exterior of Steel Water Pipelines	C213–79
AWWA Standard for Tape Coating Systems for the Exterior of Steel Water Pipelines	C214–83
AWWA Standard for Reinforced Concrete Pressure Pipe, Steel Cylinder Type, for Water and Other Liquids	C300–82
AWWA Standard for Prestressed Concrete Pressure Pipe, Steel Cylinder Type, for Water and Other Liquids	C301–79
AWWA Standard for Reinforced Concrete Pressure Pipe, Non-Cylinder Type, for Water and Other Liquids	C302–74

Table 27-2 (*continued*)

STANDARD	NUMBER
AWWA Standard for Reinforced Concrete Pressure Pipe, Steel Cylinder Type, Pretensioned, for Water and Other Liquids	C303–78
AWWA Standard for Asbestos-Cement Distribution Pipe, 4 in. through 16 in., for Water and Other Liquids	C400–80
AWWA Standard Practice for the Selection of Asbestos-Cement Distribution Pipe, 4 in. through 16 in., for Water and Other Liquids	C401–83
AWWA Standard for Asbestos-Cement Transmission Pipe, 18 in. through 42 in., for Water and Other Liquids	C402–77
AWWA Standard Practice for the Selection of Asbestos-Cement Transmission and Feeder Main Pipe, Sizes 18 in. through 42 in.	C403–78
AWWA Standard for Gate Valves—3 in. through 48 in. NPS—for Water and Sewage Systems	C500–80
AWWA Standard for Sluice Gates	C501–80
AWWA Standard for Dry-Barrel Fire Hydrants	C502–80
AWWA Standard for Wet-Barrel Fire Hydrants	C503–82
AWWA Standard for Rubber-Seated Butterfly Valves	C504–80
AWWA Standard for Backflow Prevention Devices—Reduced Pressure Principle and Double Check Valve Types	C506–78
AWWA Standard for Ball Valves, Shaft (or Trunnion) Mounted—6 in. through 48 in. for Water Pressures up to 300 psi	C507–73
AWWA Standard for Swing-Check Valves for Ordinary Waterworks Service	C508–82
AWWA Standard for Resilient-Seated Gate Valves, 3 in. through 12 in. NPS—for Water Systems	C509–80
AWWA Standard for Protective Interior Coatings for Valves and Hydrants	C550–81
AWWA Standard for Installation of Gray and Ductile Cast-Iron Water Mains and Appurtenances	C600–82
AWWA Standard for Disinfecting Water Mains	C601–81
AWWA Standard for Cement-Mortar Lining of Water Pipelines—4 in. and Larger—in Place	C602–76
AWWA Standard for Installation of Asbestos-Cement Pressure Pipe	C603–78
AWWA Standard for Grooved and Shouldered Type Joints	C606–81
AWWA Standard for Polyvinyl Chloride (PVC) Pressure Pipe, 4 in. through 12 in., for Water	C900–81
AWWA Standard for Polyethylene (PE) Pressure Pipe, Tubing, and Fittings, ½ in. through 3 in., for Water	C901–78
AWWA Standard for Polybutylene (PB) Pressure Pipe, Tubing, and Fittings, ½ in. through 3 in., for Water	C902–78
AWWA Standard for Glass-Fiber-Reinforced Thermosetting-Resin Pressure Pipe	C950–81
AWWA Standard for Welded Steel Tanks for Water Storage	D100–79
AWWA Standard for Inspecting and Repairing Steel Water Tanks, Standpipes, Reservoirs, and Elevated Tanks, for Water Storage	D101–53[R79]
AWWA Standard for Painting Steel Water Storage Tanks	D102–78
AWWA Standard for Factory Coated Bolted Steel Tanks for Water Storage	D103–80
AWWA Standard for Disinfection of Water Storage Facilities	D105–80

valves should conform to AWWA specifications and, as previously indicated, are typically the same size as the main on mains smaller than 30 inches (0.76 m), and are one size smaller than the main on mains 30 inches (0.76 m) and larger.

When distribution system piping is installed, proper trenching, bedding, and backfilling are required to properly protect the piping. A typical example of a standard trenching and bedding requirement is shown in Fig. 27-11. This is a modification of Standard Class B bedding. Figure 27-12 shows Standard Class A through D bedding for pipeline installation. Preparation of the trench bottom to give an even bedding for the barrel of the pipe and proper alignment of the pipes is of primary importance. The trench width will depend to some extent on the ground conditions and depth of laying, but should be kept to practical minimums.

As previously indicated, the depth at which the pipeline is to be laid will normally have been determined by frost depths, backfill, and surface loadings. Backfilling should take place as soon as possible after the joints have been made, and whenever possible the joints should be left uncovered until completion of pressure testing.

Properly conducted pressure testing is important for future system operation. Excess leakage is an economic waste and should be avoided.

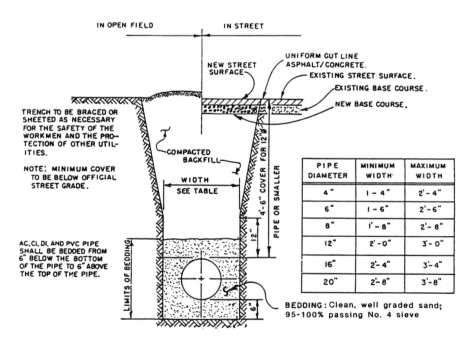

Fig. 27-11. Typical trenching and bedding requirements—trench section.

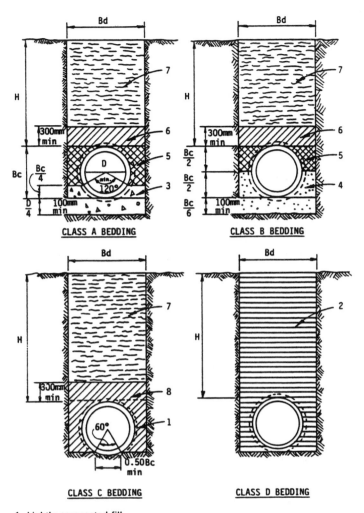

1. Lightly compacted fill.
2. Loose fill.
3. 20.5 N/mm² at 28 days concrete well packed under pipe.
4. Selected granular material well tamped under and alongside the pipe.
5. Selected material well tamped by hand in 75 mm to 150 mm layers.
6. Selected material lightly tamped by hand in 150 mm layers.
7. Normal fill.
8. Lightly compacted by hand.
Bd = Width of trench at crown of pipe
H = Depth of cover over crown of pipe
Bc = Outside diameter of pipe

Fig. 27–12. Bedding for pipes in trenches.

Watermain Appurtenances

All tees, bends, plugs, and hydrants should be properly anchored and protected from movement by tying or thrust blocking. When concrete anchor blocking is provided, the size and bearing area are dependent on the type and condition of the surrounding soil. The concrete thrust blocks should always be placed against undisturbed earth. Typical thrust blocking requirements are shown in Fig. 27–13. Figure 27–14 presents a nomograph for determining thrust block size and bearing area. When undisturbed earth cannot be obtained, alternative methods of restraint should be investigated. Straps, tie bars, interlocking devices, and other types of restrained joints may be used for various types of pipes. The number of joints to be restrained will depend on soil resistance and friction calculations.

New mains and repaired main sections should be adequately disinfected before being placed in or returned to service. The AWWA "Standard for Disinfecting Water Mains" should be followed. Before new lines and appurtenances are placed in service, the absence of pollution should be demonstrated by bacteriological sampling and examination.

Pressure-reducing valves are provided for maintaining downstream pressure at a uniform pressure less than the upstream main pressure. These valves are usually sized so that the velocity through the valve at maximum hour demand does not exceed 15 ft/sec (4.57 m/s). If a wide range of flows is anticipated, more than one valve may be required. When pressure differentials across the valve are greater than 45 psi (310 kPa), or when downstream pressure will be low relative to the differential pressure, special valving materials should be considered.

System Maintenance

After the water transmission and distribution system has been designed and installed, it is necessary to maintain the system to continue optimal performance. Proper maintenance will include repair of breaks and damaged components, and the routine operation and maintenance of system pumps, valves, and appurtenances. Maintenance of optimal transmission and distribution capacities may require a scheduled pipe-cleaning program. Pipe cleaning may be accomplished by electromechanical methods, pigging, or high pressure water jetting. Careful selection of the cleaning method and equipment is an essential step if pipes are to be cleaned efficiently. Many water departments have been successful in maintaining a Hazen-Williams friction coefficient in the 120 to 140 range by planned cleaning of pipelines. Other aspects of good distribution system care include leakage testing and control as well as routine evaluation of system performance.

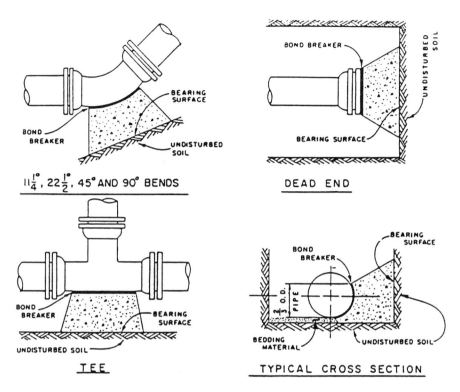

$11\frac{1^{\circ}}{4}$, $22\frac{1^{\circ}}{2}$, 45° AND 90° BENDS

DEAD END

TEE

TYPICAL CROSS SECTION

MINIMUM BEARING SURFACE AREA
(IN SQUARE FEET)

SIZE OF PIPE	B E N D S				TEE OR DEAD END
	11¼°	22½°	45°	90°	
4"	1.00	1.00	1.00	N A	1.5C
6"	1.00	1.25	2.25	N A	3.00
8"	1.00	2.00	4.00	N A	5.25
12"	2.00	4.25	8.25	N A	11.00
16"	3.50	6.50	12.50	23.00	16.50
20"	5.00	10.00	19.50	35.50	25.00

NOTES

1. ON 16" & 20" TRANSMISSION MAINS ALL BENDS SHALL BE BOTH RODDED & KICK-BLOCKED
2. BEARING SURFACE AREAS SHOWN IN CHART ARE MINIMUM.
3. BASED ON 150 P.S.I. INTERNAL PIPE PRESSURE PLUS WATER HAMMER. 4", 6" & 8" WATER HAMMER = 120 PSI.
   12" WATER HAMMER = 110 PSI.
   16" & 20" WATER HAMMER = 70 PSI.
4. SOIL BEARING CAPACITY = 3,000 LB./SQ. FT.
5. ALL 90° BENDS SHALL BE RODDED.
6. N.A. = NOT APPLICABLE.

Fig. 27–13. Typical thrust block details.

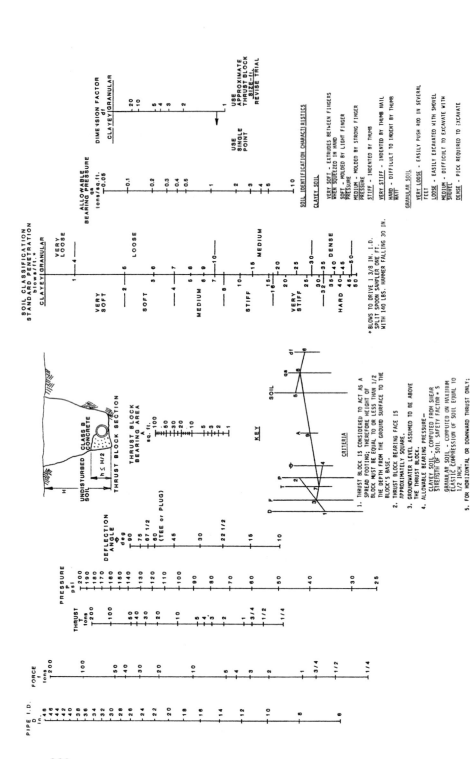

Fig. 27-14. Thrust block nomograph.[14]

REFERENCES

1. Sullivan, John P., Jr., "Maintaining Aging Systems—Boston's Approach," *J.AWWA* 74:555, November, 1982.
2. *Engineering Standards,* 3rd ed., The Board of Water Commissioners, Denver, Colo. February 1, 1980.
3. Olson, Reuben M., *Essentials of Engineering Fluid Mechanics,* 2nd ed., International Textbook Company, Scranton, Pa., 1966.
4. Vennard, John K., *Elementary Fluid Mechanics,* 3rd ed., John Wiley & Sons, New York, 1959.
5. Daugherty, Robert L., and Franzini, Joseph B., *Fluid Mechanics with Engineering Applications,* 6th ed., McGraw-Hill Book Co., New York, 1965.
6. Williams, Gardner S., and Hazen, Allen M., *Hydraulic Tables,* 3rd ed. rev., John Wiley & Sons, New York, 1960.
7. Fair, Gordon M., and Geyer, John C., *Elements of Water Supply and Waste-Water Disposal,* John Wiley & Sons, New York, 1958.
8. Howland, W. E., "Expansion of the Freeman Method for the Solution of Pipe Flow Problems," *J. New England Water Works Assoc.* 48:408, 1934.
9. Camp, T. R., and Hazen, H. L., "Hydraulic Analysis of Water Distribution Systems by Means of an Electric Network Analyzer," *J. New England Water Works Assoc.* 48:383, 1934.
10. McIlroy, M. S., "Direct-Reading Electric Analyzer for Pipeline Networks," *J.AWWA* 42:347, 1950.
11. Jeppson, Roland W., *Analysis of Flow in Pipe Networks,* Ann Arbor Science Publishers, Ann Arbor, Mich., 1983.
12. "Water Distribution System Analysis—Hardy Cross Method," a Hewlett Packard HP-97 Program by Bernard L. Golding.
13. Wood, Don J., and Charles, Carl O. A., "Hydraulic Network Analysis Using Linear Theory," *J. Hydraulics Div. ASCE* 98(HY7):1157–1170, July, 1972.
14. Report of the Task Committee on Engineering Practice in the Design of Pipelines, Committee on Pipeline Planning—Pipeline Division, *Pipeline Design for Water and Wastewater,* American Society of Civil Engineers, New York, N.Y., 1975.
15. Systems Control, Inc., *WATSIM—Extended Period Simulation of Water Distribution Networks,* Office of Water Resources Research, U.S. Department of the Interior, OWRR Contract No. 14–31–001–9027, Project No. C-4164.
16. Walski, Thomas M., "Why Calibrate Water Distribution System Models?," *WATER/Engineering & Management,* p. 27, October, 1983.
17. Cesario, Lee A., "Computer Modeling Programs: Tools for Model Operations," *J.AWWA* 72:508, September, 1980.
18. Bhave, Pramod R., "Optimization of Gravity-Fed Water Distribution Systems: Application," *J. Environmental Engineering* 109:383, April, 1983.
19. "LIQT Service Users Guide," Stoner Associates, Inc., Carlisle, Pa.
20. Cesario, Lee A., "Computers in the Distribution System: Today and Tomorrow," *Proceedings AWWA Distribution System Symposium,* Milwaukee, Wis., March 7–10, 1982.
21. *Handbook of Valves, Piping and Pipelines,* Gulf Publishing Company, Houston, Tex., 1982.
22. *The Uni-Bell Plastic Pipe Association Handbook of PVC Pipe—Design and Construction,* Uni-Bell Plastic Pipe Association, Dallas, Tex., 1982.
23. *Handbook—Ductile Iron Pipe, Cast Iron Pipe,* 5th ed., Cast Iron Pipe Research Association, Oak Brook, Ill., 1978.
24. *American National Standard for the Thickness Design of Ductile-Iron Pipe,* ANSI/AWWA C150/A21.50–81, American Water Works Association, Denver, Colo.

25. *AWWA Standard Practice for the Selection of Asbestos-Cement Distribution Pipe 4 in. through 16 in. (100 mm through 400 mm), for Water and Other Liquids,* ANSI/AWWA C401–83, American Water Works Association, Denver, Colo.
26. *Concrete Pipe Handbook,* American Concrete Pipe Association, Vienna, Va., 1980.
27. Culp/Wesner/Culp, and Ott Water Engineers, Water Distribution System Planning–Emergency Water Supply, prepared for City of Broomfield, Colorado and United States Department of Energy, December, 1980.

Chapter 28
Pumping Systems

INTRODUCTION

This chapter addresses a range of applications for systems that are commonly used in the waterworks industry to pump water and sludges; chemical feed pumping applications were described in Chapter 8. Application of pumping systems requires an understanding of the pumping unit as well as an understanding of how the pumping unit functions within the system. It is necessary to determine how the inlet conditions will affect the performance of the pump, as well as to understand conditions that cause water hammer and vibration in pumps.

Figure 28–1 shows pump classifications. Each class finds application in various industries; however, the centrifugal pump is the workhorse of the waterworks industry. Positive displacement pumps are used for low-volume applications where precise delivery is required, such as chemical feed applications, and where high-viscosity fluid pumping is required. Certain classes of rotary positive displacement pumps are invaluable in fluid power transmission.

In a centrifugal pump, the liquid from the suction side is drawn into the pump by atmospheric pressure or artificial pressure. The rotating impeller within the centrifugal pump discharges the liquid at its periphery at a higher velocity. This velocity is converted into pressure energy by means of a volute or diffusion vanes. Figure 28–2 shows the two types of casings used in centrifugal pumps.

SPECIFIC SPEED

Impellers in centrifugal pumps are classified as radial-flow, axial-flow, or mixed-flow, as shown in Fig. 28–3. Also in Fig. 28–3 the relationship between the impeller type and "specific speed" is shown. Specific speed is the speed (in rpm) at which a given impeller would operate if reduced proportionately in size so as to deliver a capacity of 1 gpm (0.063 l/s) against a total head of 1 foot (0.31 m).[1]

Specific speed is defined as:

$$N_s = \frac{\text{rpm }\sqrt{O}}{H^{3/4}} \qquad (28-1)$$

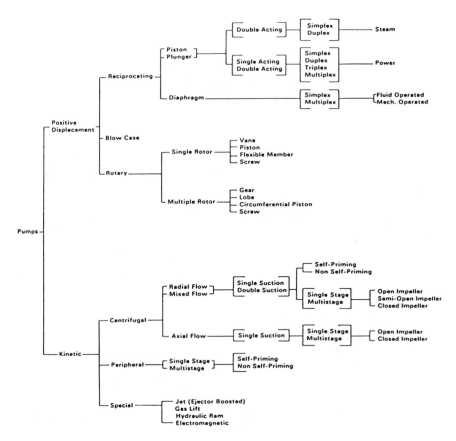

Fig. 28–1. Classification of pumps. (Courtesy of Hydraulics Institute.)

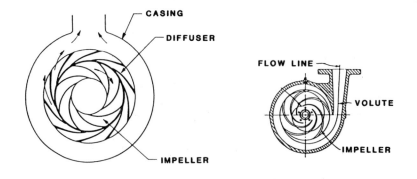

TYPICAL DIFFUSER-TYPE PUMP

SECTION THROUGH IMPELLER
AND VOLUTE

Fig. 28–2. Centrifugal pump casings—volute and diffusion vane.

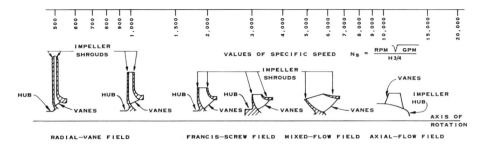

Fig. 28–3. Pump description, impellers and specific speed.

where:

N_s = specific speed
Q = flow, gpm (l/s)
H = pumping pressure, ft (m)
rpm = revolutions per minute

Specific speed is calculated at the point of peak efficiency for the maximum impeller diameter. It is an index number that assists the user in assessing the maximum efficiency, as shown in Fig. 28–4. Specific speed has also found

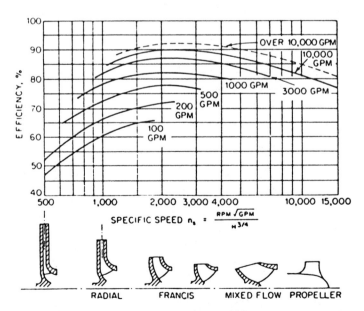

Fig. 28–4. Specific speed vs. efficiency. (Courtesy of Worthington Pump Division, Dresser Industries, Inc.)

use in relating pump operating conditions to potential for cavitation. The Hydraulic Institute Standards set upper limits for specific speed of pumps to minimize the potential for cavitation, as shown in Fig. 28–5.[2]

More than one engineering firm has found the Hydraulic Institute Standards for specific speed to be too liberal, and has reverted to more conservative standards developed over 40 years ago.[6,7] A comparison of these values is shown in Fig. 28–6.

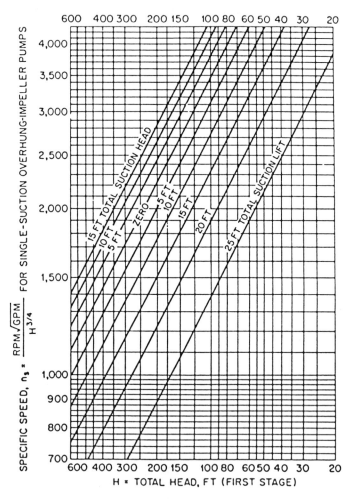

Fig. 28–5. Upper limits of specific speeds for single suction overhung impeller pumps handling clear water at 85° (185°C) at sea level.

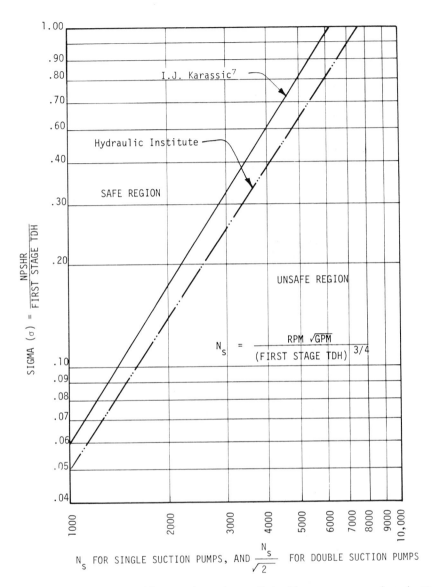

Fig. 28-6. Maximum specific speed vs. sigma. *Note:* These curves apply only at the pump's best efficiency point.

AFFINITY LAWS

The characteristics of centrifugal pumps are shown in Fig. 28–7. This representation shows geometrically similar pumps at a constant speed. Only the pump impeller diameter varies. Most manufacturers furnish information on their pumps in a similar manner. Another common way to present information on pumps is shown in Fig. 28–8, where the physical characteristics are the same, but the impeller speed is varied.

In both presentations, the mathematical relationships between the family of curves are known as the affinity laws, and are expressed as follows:

For Constant-Diameter Impeller	For Constant Impeller Speed	
$Q_1/Q_2 = N_1/N_2$	$Q_1/Q_2 = D_1/D_2$	
$H_1/H_2 = (N_1/N_2)^2$	$H_1/H_2 = (D_1/D_2)^2$	(28–2)
$Bhp_1/Bhp_2 = (N_1/N_2)^3$	$Bhp_1/Bhp_2 = (D_1/D_2)^3$	
$E_1 = E_2$	$E_1 = E_2$	

where:

Q_1, Q_2 = capacity at condition 1 and condition 2
H_1, H_2 = head at condition 1 and condition 2
D_1, D_2 = impeller diameter at condition 1 and condition 2

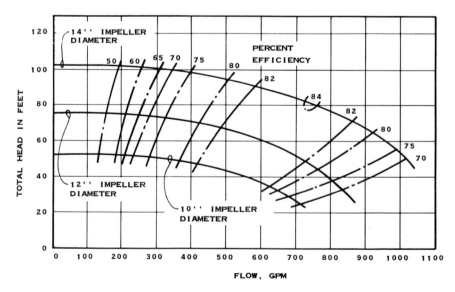

Fig. 28–7. Relating pump operation to impeller diameter.

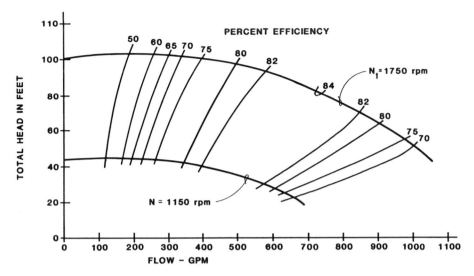

Fig. 28–8. Relating pump operation at different speeds (using the affinity laws).

N_1, N_2 = impeller rpm at condition 1 and condition 2
E_1, E_2 = pump efficiency at condition 1 and condition 2
Bhp_1/Bhp_2 = brake horsepower at condition 1 and condition 2

An example of the application of the affinity laws showing a reduced speed is given in Table 28–1 and in Fig. 28–8.

NET POSITIVE SUCTION HEAD

Each liquid has its own vapor pressure, which varies with temperature. For example, water has a vapor pressure of 0.6 foot (0.18 m) at 60°F (15.6°C).

Table 28–1. Example of Application of Affinity Laws.

				$N_1/N_2 = 0.66$ $(N_1/N_2)^2 = 0.43$ $(N_1/N_2)^3 = 0.28$			
	$N_1 = 1{,}750$ rpm $N_2 = 1{,}150$ rpm						
Q_1	H_1	Bhp_1	E_1	Q_2	H_2	Bhp_2	E_2
0	100	7.5	—	0	43	2.1	—
200	102	9.9	52	130	44	2.8	52
400	100	13.5	75	260	43	3.8	75
600	92	17.0	82	400	40	4.8	82
800	75	18.3	83	530	32	5.1	83
1,000	52	18.8	70	660	22	5.3	70

If the pressure falls below the water vapor pressure, the liquid will boil. In pump and piping systems, it is typical for the lowest pressure to occur at the pump impeller inlet. Should the pressure at this point be below the vapor pressure of the liquid, the water will boil, and cavities will form. The cavities will collapse rapidly as they move into higher pressure regions in the pump. The collapse of these cavities, termed cavitation, is accompanied by noise and vibration. Often the collapse of the cavities occurs against the impeller or pump casing and causes erosion of the pump itself.

In order to design a pumping system that assures that the water will flow to the pump and operate without cavitation, it is necessary to assure that there will be a positive suction pressure at the pump inlet. Also, as mentioned above, the specific speed of the pump must be limited, to avoid cavitation. The net positive suction head (NPSH) is defined in the following equation:

$$\text{NPSH} = \frac{144}{W} (P_a - P_{vp}) + hs$$

$$\text{NPSH} = \frac{102}{W} (P_a - P_{vp}) + hs \text{ (metric)}$$

(28–3)

where:

NPSH = available net positive suction head, ft (m)
P_a = atmospheric pressure, psia (kPa)
P_{vp} = liquid vapor pressure, psia (kPa)
W = specific weight of liquid, lb/cu ft (kg/m³)
hs = static head on pump suction, ft (m)

The "available" net positive suction head is the minimum suction condition required to prevent cavitation. The "required" NPSH must be determined by a test by the pump manufacturer. The available NPSH at the installation must be at least equal to or greater than the required NPSH for a proper installation. Figure 28–9 illustrates calculation of the NPSH.

Centrifugal pumps produce variable flows as total pumping pressure changes, as shown in Fig. 28–8. Therefore, with centrifugal pumps it is inappropriate to indicate a specific pump capacity unless the pumping head is also described. The pump curve shown in Fig. 28–8 also illustrates that the efficiency of pumping is variable for different head or pressure conditions. This particular pump curve indicates that the horsepower requirement increases with increasing capacity. Other centrifugal pumps may require greater horsepower at lower pumping capacities because of the characteristics of the pump.

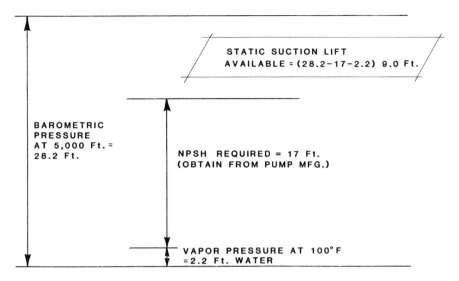

Fig. 28-9. Example of NPSH calculation.

PUMPING SYSTEM CURVE

In order to define the conditions to which the pump will be applied, the following information is required:

- Suction reservoir water level and level variation
- Suction piping friction losses
- Discharge reservoir water level and level variation
- Discharge piping friction losses
- Desired water delivery rate

Table 28-2 illustrates the calculations required for the simple pumping system shown in Fig. 28-10. The calculations are reduced to a diagram, called a system curve, that relates the variation in flow and system conditions to a discharge pressure. More detail on pumping systems can be found in Reference 3.

The prudent designer will make calculations based on both the initial and the future anticipated friction losses within the piping system, and will consider the extreme minimum and maximum water level conditions at both the suction and discharge reservoirs. Knowledge of these variations is important to ensure that the pump will be stable under all operating conditions.

The system curve (Fig. 28-11) is constructed from the calculations. The capacity requirement for a specific operating condition is determined in order to select a pumping unit. Pump curves are provided by pump manufacturers.

Table 28-2. Pumping System Calculations.

c = 100

FLOW RATE, GPM	SUCTION PIPING		DISCHARGE PIPING			TDH, FT	
	FITTINGS	PIPING	FITTINGS	PIPING	TOTAL	MIN	MAX
0	0	0	0.	0	0	30	60
200	0.01	0.05	0.10	1.5	1.7	31.7	61.7
400	0.06	0.18	0.44	5.6	6.3	36.3	66.3
600	0.13	0.39	1.01	11.4	12.9	42.9	72.9
800	0.25	0.66	1.78	19.5	22.2	52.2	82.2
1,000	0.39	1.00	2.80	29.6	33.8	63.8	93.8

c = 140

FLOW RATE, GPM	SUCTION PIPING		DISCHARGE PIPING			TDH, FT	
	FITTINGS	PIPING	FITTINGS	PIPING	TOTAL	MIN	MAX
0	0	0	0	0	0	30	60
200	0.01	0.03	0.10	0.8	0.9	30.9	60.9
400	0.06	0.10	0.44	2.9	3.5	33.5	63.5
600	0.13	0.21	1.01	6.1	7.5	37.5	67.5
800	0.25	0.35	1.78	10.4	12.8	42.8	72.8
1,000	0.39	0.54	2.80	15.9	19.6	49.6	79.6

System Description (see Fig. 28–10)

Suction Piping		K
	1—10" elbow	0.3
	1—10" gate valve	0.3
	1—10" entrance	0.5
	100' 10" pipe	—
	$\Sigma K =$	1.1

Discharge Piping		K
	1—8" check valve	2.5
	1—8" gate valve	0.3
	2—8" elbows	0.6
	1—8" exit	1.0
	1,000' 8" pipe	—
	$\Sigma K =$	4.4

Fitting Losses $H = Kv^2/2g$; Piping Losses $H/L = (Q/0.43\ Cd^{2.63})^{1.85}$

where:

H = headloss, ft (m)
K = fitting loss factor, dimensionless
V = velocity, ft/sec (m/s)
L = pipe length, ft (m)
Q = flow, cu ft/sec (m³/s)
C = Hazen-Williams friction factor
d = pipe diameter, ft (m)

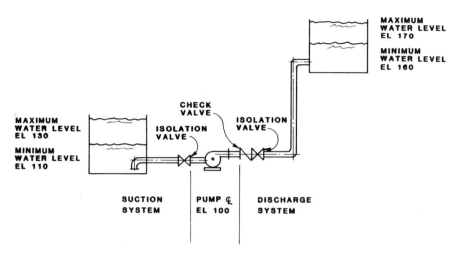

Fig. 28-10. Simple pump system.

By reviewing several pump curves, the engineer selects the pump or alternative pumps that most efficiently meet the operating conditions and that are stable throughout the conditions that will be experienced. One such selection is shown in Fig. 28-11. It is apparent that the desired flow rate will be achieved at only one operating condition. The flow rate at other operating conditions should be acceptable.

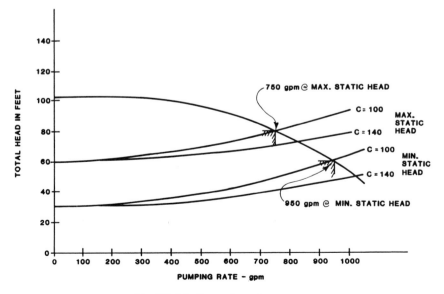

Fig. 28-11. Example system curve.

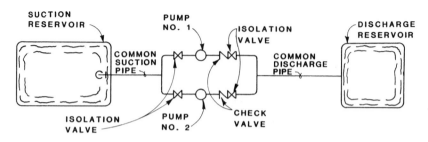

Fig. 28-12. Parallel pump system.

When pumps are placed in parallel, which is a typical application, the calculations for the system curve are similar; however, the application of the pump curves to the system curve is modified. An example system is shown in Fig. 28-12. In this example, the system curve will be similar to the one presented in Fig. 28-10. The pumping curve provided by the manufacturer is modified to reflect the connecting piping between the common suction and the common discharge piping associated with each individual pump. These calculations are shown in Table 28-3. In order to graphically show the pumping rate of both pumps operating in parallel, the pumping flow

Table 28-3. Parallel Pumping System.*

PUMP CURVE INFORMATION		SUCTION		DISCHARGE		TOTAL	REVISED
Q, GPM	HEAD, FT	$v^2/2g$	HEADLOSS, FT	$v^2/2g$	HEADLOSS, FT	HEADLOSS, FT	PUMP HEAD, FT
0	100	0	0	0	0	0	100
200	102	0.01	0.01	0.03	0.09	0.1	102
400	100	0.04	0.04	0.11	0.38	0.4	100
600	92	0.09	0.10	0.24	0.85	1.0	91
800	75	0.16	0.18	0.42	1.51	1.7	73
1,000	52	0.25	0.28	0.66	2.37	2.6	49

		SUCTION	DISCHARGE
Individual Pump Piping Losses	1—10" tee	0.5	—
	1—10" ell	0.3	—
	1—10" gate valve	0.3	—
	1—8" check valve	—	2.5
	1—8" gate valve		0.3
	1—8" ell		0.3
	1—8" tee		0.5
	$\Sigma =$	1.1	3.6

* Modification of pump curve to reflect individual pump piping losses.

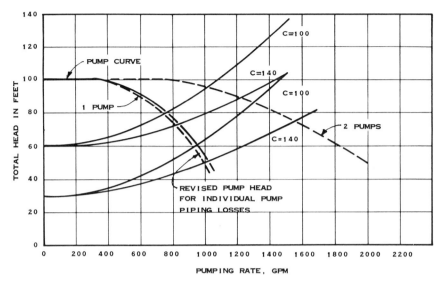

Fig. 28-13. Parallel operations of pumps.

rates for each pump at an equal head are added. Figure 28–13 shows the pumping rate for equal-size, parallel pumps.

In some cases it is necessary to operate pumping units in series, as shown schematically in Fig. 28–14. In this case, the pumping heads at equal flow rates are added. Figure 28–15 graphically shows example pump and system curves for such a system.

POWER

The useful power required to pump a liquid is calculated as follows:

$$\frac{Q \times 8.34 \times H \times sg}{33,000} = \frac{Q \times H \times sg}{3,960} = \text{useful power} \qquad (28\text{-}4)$$

where:

Q = pump flow rate, gpm (l/d)
H = pumping head, ft (m)
sq = specific gravity of liquid being pumped

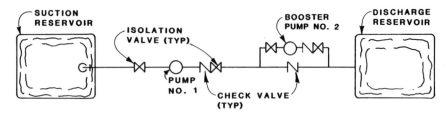

Fig. 28-14. Series pump system.

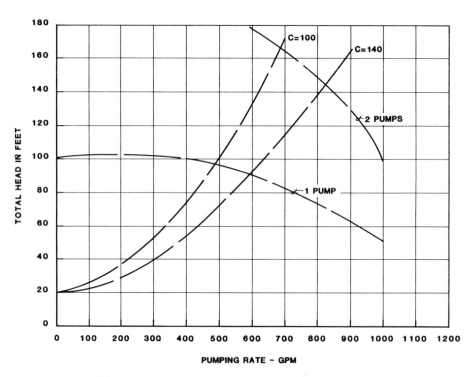

Fig. 28-15. Pumping system curves for series pump operation.

The *brake* horsepower required to drive the pump is the useful power divided by the pump efficiency:

$$Bhp = \frac{Q \times H \times sg}{3,960 \times E_p} \qquad (28\text{--}5)$$

where:

Bhp = brake horsepower
E_p = pump efficiency expressed as a decimal fraction

The electrical use on the electrical power supply to the water is:

$$\frac{Bhp \times 0.746}{E_m} = \text{electrical use} \tag{28-6}$$

where E_m = motor efficiency expressed as a decimal fraction.

PRESSURE SURGE

Pressure surges or water hammer occurs when the fluid flow momentum is changed. The magnitude of the forces resulting from pressure surges can be extremely large, and may result in pipeline failures in extreme cases. A detailed discussion of this subject is presented in Reference 4.

A common type of pressure surge happens when pumping equipment is stopped. The sudden cessation of flow results in the water attempting to continue to move without the driving force of the pump. A negative pressure results that moves along the pipeline causing collapsing pressures in the pipe. When it reaches the discharge reservoir, the entire length of pipe is subject to collapsing pressures, and the fluid velocity is zero. Because the pressure in the pipe is below that in the reservoir, water flows into the line, setting up a positive wave front traveling back toward the pump. Typically the pump discharge is provided with a check valve to prevent backflow through the pump. The positive wave reaches the valve, and the kinetic energy is converted to static pressure, which is the pressure surge.

For propagation of the full magnitude of the water hammer pressure wave to occur, the valve must be closed instantly. Because it is virtually impossible to close a valve instantly, the pressure wave created may be less than the full wave pressure that would be created by instantaneous valve closure. However, if the valve is completely closed before the first positive pressure wave returns to it, the pressure in the pipe is continuously increasing, and the pressure wave created will be the same as for instantaneous valve closure. Therefore, when the valve closure time is equal to or less than $2L/a$ [where L is pipe length, feet (m), and a is pressure wave velocity, ft/sec (m/s)], the wave pressure will be the same as for instantaneous valve closure. When the valve closure time is longer than $2L/a$, the pressure wave will be less than instantaneous valve closure.

The velocity of the water hammer wave and the increase in pressure due to water hammer may be obtained from the following equations, where the valve closure or stoppage of flow due to a pump failure occurs in less than $2L/a$:

$$a = \frac{4,660}{\sqrt{1 + K_e D/Et}} \tag{28-7}$$

$$h = \frac{a\, V_0}{g} \qquad\qquad (28\text{--}8)$$

where:

a = water hammer wave velocity, ft/sec (m/s) (acoustic velocity)
K_e = bulk modulus of elasticity of water, 294,000 psi (kPa)
D = pipe diameter, in. (cm)
E = modulus of elasticity of pipe wall material, psi (kPa)
b = thickness of pipe wall, in. (cm)
h = head due to water hammer (in excess of static head), ft (m)
g = acceleration of gravity, ft/sec^2(m/s^2)
t = pipe wall thickness, in. (cm)
L = length of pipe, ft (m)

The modulus of elasticity for various pipes is:

- Cast iron pipe, about 13.5×10^6 psi
- For most steel pipe conduits, about 30×10^6 psi
- For concrete pipe or transite, about 3×10^6 psi

Pipes of composite materials, such as concrete-lined steel cylinder pipe, will have wave velocities more nearly those of the inner material.

From the previous illustration of wave propagation for instantaneous valve closure, it is apparent that if the valve were not completely closed by the end of the cycle $2L/a$, the intensity of the pressure wave reflected from the valve would be suppressed. The maximum intensity of the pressure wave for a valve closure time of longer than $2L/a$ may be obtained by Kulmann's nomograph charts.[5]

Typical solutions to reducing pressure surge magnitude include:

- *Surge tanks:* The top of the surge tank must be above the hydraulic grade line. The surge tank accumulates water in response to the pressure surge and supplies water to moderate the negative pressure.
- *Air chamber:* The lower portion of the air chamber contains water, and the upper portion contains compressed air. Like the surge tank, the air chamber provides water and a place to accumulate water during a surge.
- *Slow-closing check valve:* By slowly closing the pump check valve over a period of more than one cycle $2L/a$, the water hammer pressure is moderated.
- *Surge suppressor:* A surge suppressor consists of a pilot-operated valve that opens quickly upon pump outage through the use of a solenoid

valve. The valve releases water from the pipeline and is closed later at a slower rate by using the pipeline pressure acting on a diaphragm valve.

SUCTION CONDITIONS

The suction conditions for pumps are often critical. Water should be introduced to the pump in a uniform manner without induced swirl. Improperly designed intakes may result in unbalanced load on pump impellers, poor pump efficiency, or vortexing and air introduction into the pump. The Hydraulic Institute Standards provide guidelines for arrangement of sumps for larger pumping units.[2] Some of these are shown in Fig. 28–16.

Pump manufacturers often have recommended inlet dimensions that should be considered in layouts for pumping facilities. The Hydraulic Institute Standards also provide dimensions and recommended arrangements for pumps.[2]

PUMP LAYOUT

In making pump layouts, there are several considerations that the designer should keep in mind:

- At least a 3-foot clearance between obstacles should be provided to permit wheeled hand truck movement on the pumping station floor.
- Suction piping should be arranged to avoid high points where air or gas may collect. Where reducers are required in horizontal piping, they should be eccentric.
- Discharge piping also should avoid high points. Dissolved gas will have less tendency to come out of solution on the discharge side than on the suction side; but if air pockets are allowed to accumulate, they will restrict flow. Provide air relief valves where high points occur. Pipe the air relief to a drain.
- Provide piping supports to keep all weight off of the pump. Support all discharge piping rigidly to prevent movement.
- Provide suction couplings on the pump to permit removal and replacement of the pump. The coupling is best located on the suction side because the suction typically does not experience pressure variations that are common to the discharge.
- Locate the discharge isolation valve downstream from the check valve. The check valve, which operates frequently, usually experiences more trouble and requires more maintenance than the discharge isolation valve. The operators should be able to isolate the check valve from the rest of the system.
- When pumping sludges, locate the check valve in the horizontal. When idle, solids will settle against the closed check valve. Many times, the

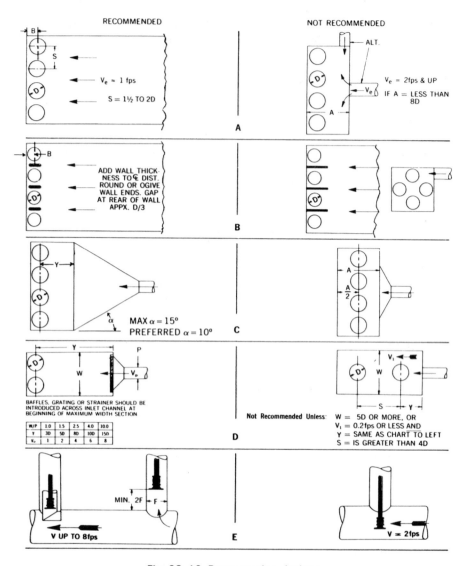

Fig. 28–16. Pump suction designs.

pump pressure will force the solids away from the check valve and allow it to operate; however, there may be times when solids will lodge on the downstream side of the check valve and prevent it from opening.

- Provide embedded lifting eyes or hoist rails above pumps and a passageway to allow their removal.
- On small service areas where diversification of water demand is small

and a pumped system may experience periods of zero flow, provide a small relief bypass to prevent overheating the water in the pump.
- Arrange piping to avoid situations where pipes are less than 8 feet above the floor or obstruct passageways to valve actuators.

REFERENCES

1. *Hydraulic Handbook,* Colt Industries, Fairbanks Morse & Co., Pump Division, Kansas City, KA, 1954.
2. *Hydraulic Institute Standards,* Hydraulic Institute 1230 Keith Bldg., Cleveland, Ohio. 13th edition, 1975.
3. Karassik, Igor, J., et al., editors, *Pump Handbook,* McGraw-Hill, New York, 1976.
4. Parmakian, J., *Water Hammer Analysis,* Dover Publications, New York, 1963.
5. Kulmann, C. Albert, *Nomograph Charts,* McGraw-Hill, New York, 1951.
6. Benjes, H. H., Sr., *Black & Veatch Design Procedures,* Black and Veatch, Kansas City, MO, February 18, 1983.
7. Karassic, I. J., et al., "Cavitation Characteristics of Centrifugal Pumps Described by Similarity Considerations," *Transactions of the ASME,* Vol. 62, pp. 155–156, February, 1940.

Chapter 29
Energy Conservation for
Water Treatment Facilities

INTRODUCTION

During the past decade, dramatic increases in energy prices have made the public acutely aware of the significance of energy costs. "Energy," once a term used by relatively few scientists and engineers, is not only a well-known term to waterworks managers, but is also a household topic of discussion. In the past, energy was rarely considered in the design and operation of waterworks facilities. Now, energy plays an important role in decisions that pertain to design and operation of water treatment plants, pumping stations, and distribution systems. This chapter examines energy fundamentals, historical energy rates, and how these factors relate to other capital and operating costs for waterworks facilities. A generalized approach to an energy audit is then presented, followed by a discussion of energy conservation equipment and techniques.

Electrical Energy Fundamentals

Electrical systems are either the direct current (dc) or alternating current (ac) type. In direct current systems, the voltage remains constant, and current always flows in the same direction. In alternating current systems, voltage and current follow sine wave patterns, reversing direction regularly as shown in Fig. 29–1, which represents instantaneous power with coinciding voltage and current waveforms.

Apparent power delivered to an alternating current circuit is calculated by the vector dot product of resistive power and reactive power. This relationship is illustrated in the power triangle shown in Fig. 29–2. Apparent power is expressed as kilovolt amperes (kVA); this is the demand placed on the electrical utility's system by a customer. The resistive load actually performs work, and is known as *active power,* which is the power actually delivered to the customer and is the value measured by a customer's power meter as kilowatts (kW). Reactive load does not perform work but is necessary to provide energy for changes in magnetic flux. The reactive demand on a circuit is the algebraic sum of capacitive and inductive demands. When these demands are equal, the sum is zero, and the reactive demand is zero. Under this

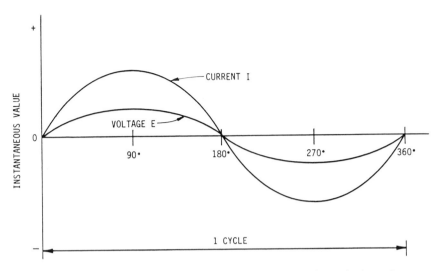

Fig. 29–1. Instantaneous power in an ac circuit with coinciding voltage and current waveforms.

condition the current and voltage functions coincide, and a wave pattern, as shown in Fig. 29–1, results, with all power delivered to a circuit available as active power. When circuit inductance is greater than capacitance, current will lag voltage, as shown in Fig. 29–3. When capacitance is greater than inductance, current will lead voltage. The amount by which the current lags or leads the voltage is expressed as an angle, with one full cycle being 360 degrees. Power factor is the cosine of this angle and is the ratio of the active

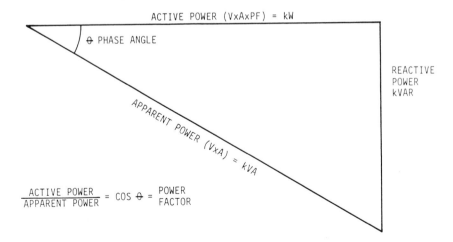

Fig. 29–2. Power triangle.

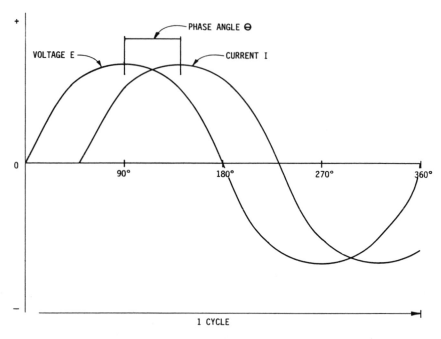

Fig. 29-3. Instantaneous power in an ac circuit with current lagging voltage.

power to the apparent power. Partially loaded induction motors create the largest reactive demand on most systems, and current tends to lag voltage on most.[1]

Useful electrical relationships are shown in Table 29-1.[2]

Energy Costs

The average retail cost of energy has increased dramatically since 1972. Figure 29-4 shows the average price of energy delivered to the industrial sector from 1970 to 1981.[3] The figure shows the increasing price trend for coal, natural liquified petroleum gas (LPG), and electricity. The upward trend after the 1973 Arab oil embargo is clearly evident. Figure 29-5 shows the average cost of electricity in ¢/kWh delivered to the industrial sector from 1972 to 1983. The cost had increased from approximately 1.16¢/kWh (1972) to approximately 4.97¢/kWh (1983). However, when inflation is taken into account, the price had increased from 1.16¢/kWh in 1972 to approximately 2.30¢/kWh in 1983, as shown in Fig. 29-5. Although the latter increase is smaller, the inflation-adjusted cost of electricity nearly doubled from 1972 to 1983. Energy prices shown in Figures 29-4 and 29-5 are national averages. Prices vary by type of account, region, and utility. The nationwide electricity

Table 29–1. Useful Electrical Conversion Formulas.*

TO FIND	DIRECT CURRENT	ALTERNATING CURRENT	
		SINGLE-PHASE	THREE-PHASE
Amperes when horsepower known	$\dfrac{hp \times 746}{V \times e}$	$\dfrac{hp \times 746}{V \times e \times PF}$	$\dfrac{hp \times 746}{V \times \sqrt{3} \times e \times PF}$
Amperes when kilowatts known	$\dfrac{kW \times 1{,}000}{V}$	$\dfrac{kW \times 1{,}000}{V \times PF}$	$\dfrac{kW \times 1{,}000}{V \times \sqrt{3} \times PF}$
Amperes when kVA known		$\dfrac{kVA \times 1{,}000}{V}$	$\dfrac{kVA \times 1{,}000}{V \times \sqrt{3}}$
Kilowatts	$\dfrac{A \times V}{1{,}000}$	$\dfrac{A \times V \times PF}{1{,}000}$	$\dfrac{A \times V \times \sqrt{3} \times PF}{1{,}000}$
kVA		$\dfrac{A \times V}{1{,}000}$	$\dfrac{A \times V \times \sqrt{3}}{1{,}000}$
Power factor		$\dfrac{kW \times 1{,}000}{A \times V}$ or $\dfrac{kW}{kVA}$	$\dfrac{kW \times 1{,}000}{A \times V \times \sqrt{3}}$ or $\dfrac{kW}{kVA}$
Horsepower (Output)	$\dfrac{A \times V \times e}{746}$	$\dfrac{A \times V \times e \times PF}{746}$	$\dfrac{A \times V \times \sqrt{3} \times e \times PF}{746}$

* Reference 2.
Note: Power factor and efficiency when used in above formulas should be expressed as decimals.

V = voltage
kW = kilowatts
e = efficiency
PF = power factor
A = amperes
kVA = kilovolt amperes

rate variation in March, 1984 is shown in Table 29–2; rates ranged from a low of 2.0¢/kWh in Tacoma, Washington, to a high of 11.1¢/kWh in San Diego, California, and New York, New York.[4] The prices represent averaged cost in ¢/kWh calculated from data reported for a manufacturer using a hypothetical 200,000 kWh per month at a maximum demand of 500 kW.

ENERGY CONSUMPTION IN WATER TREATMENT PLANTS

Energy is a major cost in the waterworks industry. Figure 29–6 shows typical budget allocations for capital, energy, labor, and chemicals and materials for water treatment facilities ranging in size from 5 to 200 mgd (18.9 to 757 Ml/d). The distribution shown in Fig. 29–7 represents facilities without sludge handling that are operating at 95 percent of capacity.[5] Assumed life

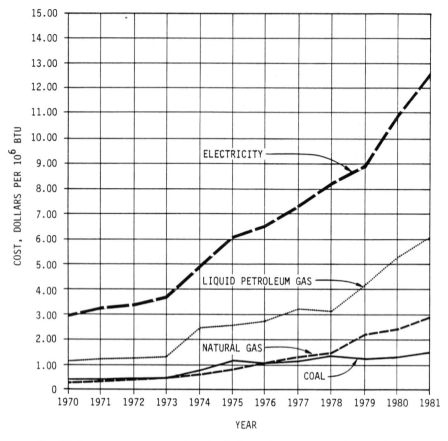

Fig. 29–4. Average energy prices for industrial customers, 1970–1981.[3]

of equipment is 20 years, interest rate 10 percent, energy 6¢/kWh, and labor $15.00/hr. In this case, energy represents 24.5 percent of budget requirements for a 5-mgd plant, and increases to 40.5 percent of budget requirements for 100- and 200-mgd (378.5- and 757-Ml/d) plants.

Distribution of energy by process in water treatment plants of various sizes is presented in Table 29–3.[5] Design criteria for the processes listed in Table 29–3 are detailed in Table 29–4.

As shown in Table 29–3, finished water pumping is by far the largest energy consuming process. Typical unit energy consumption is 1,700 to 1,900 kWh/MG (449 to 502 kWh/Ml). A graphical representation of the relative distribution of energy at a 10-mgd (37.9-Ml/d) facility is shown in Fig. 29–7.

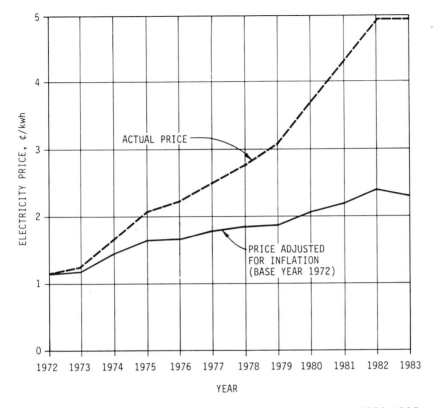

Fig. 29-5. Average price of electricity sold to industrial customers, 1972–1983.[3]

ENERGY AUDITS FOR WATERWORKS FACILITIES

An energy audit of a water treatment plant or system is a survey of major areas of energy usage followed by a cost-effectiveness analysis of methods that can be used to reduce energy usage. This section discusses a general approach to conducting audits. Specific energy conservation measures are discussed later in the chapter.

Successful energy audits include:

- Examination of existing plant (or system) operations and energy data.
- Examination of utility billing schedules.
- Detailed field investigation of all facilities.
- Preparation of energy consumption inventory.
- Operating overview and evaluation of potential operational changes to reduce energy use.

- Report of findings, conclusions, and recommendations.
- Diligent monitoring and follow-up.

Operations and Energy Data

The audit team should be familiar with the daily operation of the facility prior to conducting the investigation. This is easily accomplished by operations personnel. However, if the audit is performed by staff members not directly involved in operations or by outside consultants, this data gathering and operations overview step assumes special importance. Energy summaries,

Table 29-2. Industrial Electricity Prices By City, March, 1984.*

STATE AND CITY	PRICE ¢/kwh	UTILITY
ALABAMA:		
Birmingham	6.2	Alabama Power Co.
Huntsville	4.7	Public agency
ARIZONA:		
Phoenix	4.7	Public agency
ARKANSAS:		
Little Rock	5.2	Arkansas Power & Light
CALIFORNIA:		
Long Beach	7.8	Southern Calif. Edison
Los Angeles	5.2	Municipality
Sacramento	2.6	Public agency
San Diego	11.1	San Diego Gas & Elec. Co.
San Francisco	7.1	Pacific Gas & Elec.
COLORADO:		
Colorado Springs	3.1	Municipality
Denver	4.4	Public Service Co. of Colo.
CONNECTICUT:		
Bridgeport	9.5	United Illuminating Co.
Groton	7.1	Municipality
Hartford	7.8	Conn. Light & Power Co.
Wallingford	6.6	Municipality
Waterbury	7.8	Conn. Light & Power Co.
DISTRICT OF COLUMBIA:		
Washington	5.5	Potomac Elec. Power Co.
FLORIDA:		
Jacksonville	6.4	Public agency
Lakeland	6.3	Municipality
Miami	6.2	Fla. Power & Light Co.
Orlando	5.6	Municipality
GEORGIA:		
Atlanta	5.3	Georgia Power Co.
IDAHO:		
Boise	3.8	Idaho Power Co.

Table 29-2. Continued.

STATE AND CITY	PRICE ¢/kWh	UTILITY
ILLINOIS:		
Chicago	7.1	Commonwealth Edison Co.
Springfield	5.3	Municipality
INDIANA:		
Anderson	5.0	Public agency
Fort Wayne	4.8	Indiana–Mich. Electric
Indianapolis	4.1	Indianapolis P. & Light
Richmond	5.3	Public agency
KANSAS:		
Kansas City	5.6	Municipality
Wichita	6.5	Kansas Gas & Elec. Co.
KENTUCKY:		
Louisville	4.3	Louisville Gas & Elec. Co.
LOUISIANA:		
New Orleans	5.1	La. Power & Light Co.
MAINE:		
Portland	5.5	Central Maine Power Co.
MARYLAND:		
Baltimore	5.6	Baltimore Gas & Elec. Co.
MASSACHUSETTS:		
Boston	8.3	Boston Edison Co.
Chicopee	6.4	Municipality
Reading	9.0	Municipality
Taunton	7.9	Municipality
Worcester	7.3	Massachusetts Elec. Co.
MICHIGAN:		
Detroit	6.4	Detroit Edison Co.
Flint	5.6	Consumers Power Co.
Lansing	4.3	Municipality
MINNESOTA:		
Minneapolis	4.1	Northern States Power Co.
Rochester	4.7	Municipality
MISSISSIPPI:		
Jackson	5.5	Miss. Power & Light Co.
MISSOURI:		
Independence	4.4	Municipality
Kansas City	5.7	Kans. City P. & Light Co.
St. Louis	4.2	Union Electric Co.
NEBRASKA:		
Omaha	4.1	Municipality
NEW JERSEY:		
Newark	7.1	Pub. Service Electric
NEW MEXICO:		
Albuquerque	7.1	Pub. Serv. of N.M.
NEW YORK:		
Buffalo	5.8	Niagara Mohawk P. Co.

Table 29-2. Continued.

STATE AND CITY	PRICE ¢/kWh	UTILITY
New York	11.1	Consolidated Edison Co.
Rochester	5.3	Rochester Gas & Elec.
NORTH CAROLINA:		
Charlotte	4.6	Duke Power Co.
OHIO:		
Canton	5.3	Ohio Power Co.
Cincinnati	5.2	Cincinnati Gas & Light
Cleveland	6.7	Cleveland Electric Co.
Cleveland	5.4	Municipality
Youngstown	6.5	Ohio Edison Co.
OKLAHOMA:		
Oklahoma City	3.9	Okla. Gas & Elec. Co.
Vinita	3.6	Public agency
OREGON:		
Portland	4.7	Pacific Power & Light Co.
PENNSYLVANIA:		
Philadelphia	6.8	Philadelphia Elec. Co.
Pittsburgh	6.2	Duquesne Light Co.
Scranton	4.8	Pa. Power & Light Co.
SOUTH CAROLINA:		
Conway	3.8	Public agency
TENNESSEE:		
Chattanooga	4.7	Municipality
Memphis	4.8	Municipality
TEXAS:		
Austin	8.1	Municipality
Dallas	5.2	Dallas Power & Light Co.
Fort Worth	5.0	Texas Elec. Service Co.
Houston	6.5	Houston Lighting & P. Co.
San Antonio	5.5	Municipality
UTAH:		
Salt Lake City	5.1	Utah Power & Light Co.
VERMONT:		
Burlington	6.2	Municipality
VIRGINIA:		
Arlington	5.3	Va. Electric & Power Co.
WASHINGTON:		
Seattle	2.4	Municipality
Tacoma	2.0	Municipality
WEST VIRGINIA:		
Charleston	4.6	Appalachian Power Co.
WISCONSIN:		
Kaukauna	4.1	Public agency
Manitowoc	4.4	Public agency
Milwaukee	5.3	Wisconsin Elec. Power Co.

* Reference 4.
Note: Average prices based on monthly bill of 200,000 kWh at a maximum rate of 500 kWh.

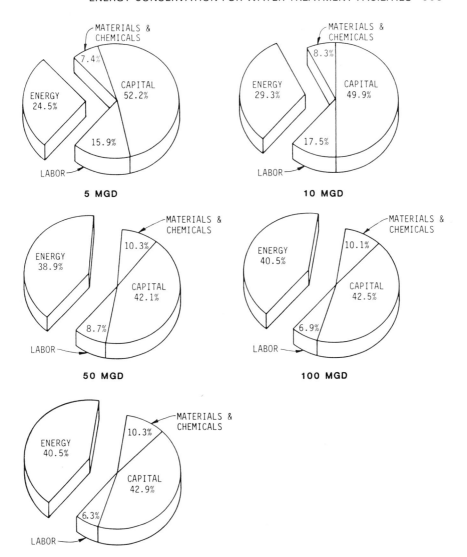

Fig. 29–6. Typical distribution of water treatment plant operating budget, 1984.[5] (*Note:* Based on constructing new plants in 1984 with interest rate at 10%, labor at $15.00/hr, electricity at 6¢/kWh. Energy costs will represent larger percentages in existing plants because capital portion is lower. Data are for treatment plants; distribution system costs are not included.)

equipment inventories, electrical construction drawings, design data, and operating data are items to be gathered in this early stage of the audit. The overall philosophy of facility operation is then discussed to determine exactly why a plant or system is operated in a certain manner. Specific data to be acquired in this stage are:

- Yearly hours of operation of each major piece of equipment.
- Manufacturers' performance curves for all major pumps.
- Routine maintenance schedules for changing such items as impeller wear rings and packing.
- Previous equipment test reports.

Utility Bill Schedules and Energy Pricing

It is important to understand the electric utility rate schedules to maximize the effectiveness of the energy audit. Charges are normally based on kilowatt-hours consumed (energy) and demand. Many utilities include adjustments for power factor, and large users (over 500 kW demand) are often charged by time of day. These charges are discussed in detail later in this chapter.

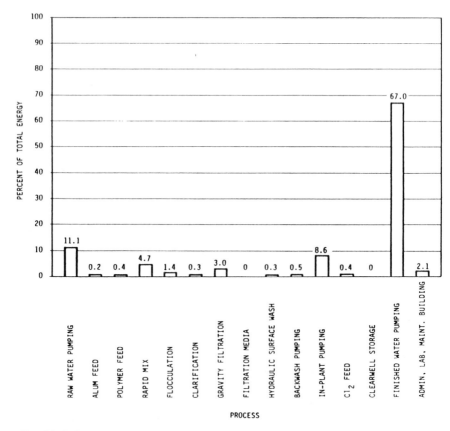

Fig. 29-7. Relative distribution of energy by process at a 10-mgd water treatment plant.[5] (Hypothetical plant operating at 95% capacity.)

Table 29–3. Distribution of Energy By Process in Water Treatment Plants, 5 to 200 mgd.[*,**]

	ENERGY, kWh/YR AT LISTED PLANT CAPACITY				
PROCESS	5 mgd	10 mgd	50 mgd	100 mgd	200 mgd
Raw water pumping	356,250	712,500	3,562,500	7,125,000	13,500,000
Alum feed	10,250	13,250	33,375	49,875	62,750
Polymer feed	23,250	23,375	23,875	24,375	25,375
Rapid mix	149,875	299,750	1,498,125	2,995,000	5,990,000
Flocculation	44,500	89,000	444,625	889,000	1,775,000
Clarification	16,500	21,375	93,125	186,250	372,000
Gravity filtration	103,750	192,500	825,375	1,487,000	2,795,000
Filtration media	0	0	0	0	0
Hydraulic surface wash	9,500	18,750	91,125	181,000	373,625
Backwash pumping	15,750	31,500	151,625	315,250	630,250
In-plant pumping	283,875	551,000	2,677,250	5,322,000	10,104,125
Cl_2 feed	16,250	25,750	94,625	119,750	204,875
Clearwell storage	0	0	0	0	0
Finished water pumping	2,137,500	4,275,000	21,375,000	42,750,000	81,000,000
Admin., lab., maint. bldg.	83,000	131,625	394,875	509,250	642,250
TOTAL	3,250,250	6,385,375	31,265,500	61,953,750	117,475,250
Unit consumption, kWh/MG	1,875	1,841	1,803	1,787	1,693

* Reference 5.
** Hypothetical plants operating at 95% capacity.

Field Investigation and Preparation of Energy Consumption Inventory

After operational data and utility schedules are consolidated and reviewed, a field investigation is conducted to document actual operating conditions. The greatest energy consumers in waterworks facilities are pumps. Detailed operation can be recorded and calculated on a pump data sheet similar to that shown in Fig. 29–8. Date, nameplate information, flow, head, and power input information are measured and entered. Based on this information, overall wire-to-water efficiency can be calculated by the following equation:

$$e = \frac{H \cdot Q}{hp \cdot 3,960} \qquad (29\text{–}1)$$

where:

e = wire-to-water efficiency
  = pump efficiency $\times$ drive efficiency $\times$ motor efficiency

H = differential head across pump, ft (m)
Q = flow rate, gpm (l/s)
hp = horsepower

Manufacturers' performance curves can be used to estimate pump flows when field flow measurements are not possible. However, it should be recognized that pump wear can have a significant effect upon pump performance.

It is important to understand that measured motor current is the current associated with apparent power (kVA) rather than current associated with real power (kW), which is doing the work. To illustrate, consider a 15-hp 460-volt motor with a nameplate, full load current of 20 amperes. If a field measurement of 10 amperes is made, it might be assumed that the motor is operating at 50 percent of rated load. However, the motor actually may be operating at approximately 30 to 35 percent of rated load because of a lower power factor. Most utilities charge energy users for real power (kW). Therefore, a kilowatt meter (which measures real power in kW) is the preferable field energy measurement device.

Although the calculated wire-to-water efficiency is an important consideration for energy analysis, volume per kWh provides the most useful energy conservation information. Where multiple pumps are available in a given situation, the unit(s) with the highest flow per kWh would be operated to minimize energy consumption.

Table 29–4. Treatment Plant Design Criteria.

PROCESS	REMARK
Raw water pumping	TDH = 50 ft
Alum feed	Liquid stock
Polymer feed	
Rapid mix	$G = 900$
Flocculation	Horizontal Paddle, $G = 80$
Clarifier	Rectangular $FeCl_3$ and alum sludge
Gravity filtration structure	HVAC
Filtration media (mixed media)	Mixed media, no energy
Hydraulic surface wash	Rotary arms with pumped source
Backwash pumping	
In-plant pumping	Pump to clear storage, TDH = 35 ft
Cl_2 feed	Gas, 1-ton cylinders
Clearwell storage	No energy
Finished water pumping	TDH = 300 ft
Administration, laboratory, maintenance building	

CWC
CULP•WESNER•CULP
CONSULTING ENGINEERS

Date _____

Time _____

Location _____

Pump Manufacturer _____

Pump Model _____

Pump Speed, rpm _____

Pump Type, Number _____

Motor Manufacturer _____

Motor Volts _____

Motor Frame _____

TEST DATA

Static Water Level Below Pump, C_L (Pump Off) _____ ft

Pumping Water Level Below Pump, C_L (Pump On) _____ ft

Pump Discharge Head _____ ft

Total Pump Pressure (Lift) _____ ft

Flow Rate _____ gpm

kW Input to Motor _____ kw

hp Input to Motor _____ hp

kWh/1,000 Gallons Pumped _____ kWh/1,000 gal

Wire-to-Water Efficiency _____ %

Remarks _____

_____

_____

_____

Fig. 29–8. Pump test data sheet.

Electrical energy data, including nameplate, field data, and yearly hours of operation for all equipment, can be recorded on the electrical equipment inventory sheet shown in Fig. 29–9. The calculation of yearly energy consumption is useful for preparation of energy consumption data for each area of a facility. Energy reduction can then be emphasized in areas with the highest energy use.

Data Analysis, Operations Overview, Presentation of Findings

When data acquisition and the energy inventory have been completed, representatives from management, engineering, and operations can confer to deter-

PROCESS AREA, EQUIPMENT	MOTOR NAMEPLATE DATA					FIELD MEASUREMENTS					
	hp	VOLTS	FULL LOAD AMPS	SPEED	PHASES	AMPS	VOLTS	kW	POWER FACTOR	HOURS OPERATED PER YEAR	kW/YR

Fig. 29–9. Electrical equipment inventory sheet.

mine a list of proposed energy conservation measures. Input from the operational staff during this phase is vital to determine the effect of proposed changes on operation of the facility. Analysis of conservation ideas can be done in many ways, including simple payback period, present worth, and cost/benefit. One method of summarizing screened energy conservation suggestions would include the following information in tabular form:

- Facility, equipment
- Recommended change
- Effects on operation
- Cost for recommended change
- Yearly savings
- Simple payback
- Utility rebate (if applicable)
- Simple payback with utility company rebate
- Comments

Monitoring Energy Consumption and Follow-up

This is perhaps the most important and most frequently overlooked step in an energy audit. Without strong commitment by management, the overall energy efficiency of a facility can deteriorate after initial conservation measures. Energy conservation must be given continued attention to ensure operation at the lowest energy cost. Management should encourage energy conservation by staff members and give recognition for *all* suggestions, whether or not they are implemented. To ensure a successful energy conservation program, field energy data should be taken at least every 6 months. Computerized energy and load management can provide this information on a continuous basis for early identification of problem areas.

ENERGY CONSERVATION EQUIPMENT AND TECHNIQUES

Electric Motors

The vast majority of energy in the waterworks industry is consumed by electric motors. Therefore, it is essential that electric motors be sized, applied, and operated to ensure maximum efficiency. Two approaches are used to improve efficiency in existing and proposed systems:

- Use of more efficient motors
- Operation of motors at or near the nameplate rating

Premium Efficiency Motors. Premium efficiency motors have gained widespread use in recent years. They offer motor efficiency gains over standard efficiency motors ranging from 1.0 percent at 200 hp (150 kW) to 9.0 percent at 1 hp (0.75 kW) at 1800 rpm. Table 29–5 summarizes typical standard and premium efficiencies (at 100 percent load) for motors of 1 through 200 hp (0.75 through 150 kW).[6,7] Figure 29–10 illustrates full load efficiency of standard and premium efficiency 1800-rpm motors for one manufacturer. Efficiency differences are greater than those shown in Table 29–5 when motors are operated at less than rated load.

Premium efficiency motors generally cost more than comparable standard efficiency motors. The extra cost of premium efficiency motors can usually be recovered within a few years or less if they are operated more than 50 percent of the time. In fact, where power costs are above 5¢/kWh and motors are operated more than 50 percent of the time, premium efficiency motors offer exceptional value. Savings realized from the difference in operating costs between standard and premium efficiency motors can be calculated as follows:

$$S = 0.746 \times hp \times C \times N \left(\left[\frac{100}{E_s} \right] - \left[\frac{100}{E_p} \right] \right) \qquad (29\text{-}2)$$

where:

S = savings, \$/yr
hp = nameplate horsepower
C = energy cost, \$/kWh
N = running time, hr/yr
E_s = standard motor efficiency, percent
E_p = premium motor efficiency, percent

In addition to lower energy consumption, premium efficiency motors offer the advantages of lower operating temperature, longer insulation life, higher power factor, and reduced full load amp rating.

To increase efficiency, emphasis is given to reducing the five major types of motor losses:

- Magnetic losses
- Windage and friction losses
- Stator losses

Table 29-5. Premium Efficiency vs. Standard Efficiency Motor Comparison.*

| | NOMINAL EFFICIENCY | | | | | |
| | 1200 RPM | | 1800 RPM | | 3600 RPM | |
HP	PREMIUM	STANDARD	PREMIUM	STANDARD	PREMIUM	STANDARD
1	78.5	73.0	84.0	75.0	81.5	74.5
2	86.5	78.5	84.0	79.5	84.0	79.0
3	88.5	76.0	88.5	80.5	82.5	80.0
5	88.5	79.0	88.5	84.0	88.5	83.0
7½	88.5	83.0	90.2	82.0	86.5	84.5
10	90.2	82.0	90.2	82.0	87.5	83.0
15	90.2	83.0	91.7	84.0	90.2	86.0
20	91.0	86.0	91.7	85.0	90.2	87.0
25	91.7	86.0	93.0	87.0	91.0	87.0
30	91.7	87.0	93.0	87.0	91.0	87.0
40	93.0	89.0	93.0	87.0	91.7	88.0
50	93.0	90.0	94.1	88.0	91.7	88.0
60	93.0	90.0	94.1	90.5	92.4	89.0
75	94.1	91.0	94.1	90.5	93.6	90.5
100	94.1	91.5	94.5	92.5	93.6	91.5
125	94.1	91.5	95.0	92.0	93.6	91.5
150	94.5	91.5	95.0	93.0	94.1	92.0
200	94.5	93.0	95.0	94.0	93.0	91.5

* References 6 and 7.

- Rotor losses
- Stray load losses

To accomplish this, premium efficiency motors may include some or all of the following design features:[8]

- High-grade electrical steel to reduce magnetic losses.
- Longer core to lower flux density.
- Increased cooling capacity to reduce magnetic and load losses.
- Thinner laminations to reduce eddy currents and thus reduce magnetic losses.
- Improved fan design to reduce windage losses.
- More copper in windings to improve cooling to reduce stator losses.
- Increased conductor cross section to lower stator load losses.
- Larger rotor bars and rings to reduce resistance and thereby lower rotor load losses.

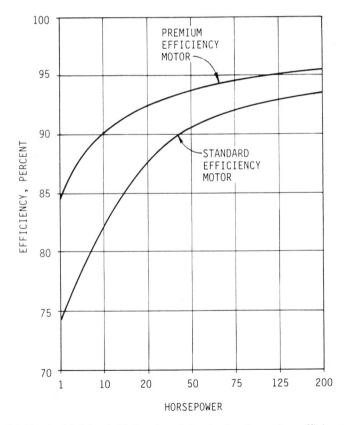

Fig. 29–10. Nominal full load efficiencies of standard and premium efficiency motors at 1800 rpm.[6]

It is extremely difficult to accurately compare efficiencies between motor manufacturers because of:[9]

- Lack of a single test standard that is used throughout the industry.
- Manufacturing variations.
- Test equipment.
- Ambiguity of published data.

The Institute of Electrical and Electronic Engineers (IEEE) 112A is the most commonly used test method in the United States. The preferred procedure is method B, in which the motor is operated at full load and power is measured directly. The greatest advantage of this method is that the results are accurate, and efficiencies will not be overstated. Consider Table 29–6, which shows 7.5-hp and 20-hp motors rated using three international methods and one U.S. test method.[10] As indicated, IEEE values range from 2.0 to 4.7 efficiency points lower then the international standards. A user making a motor selection on the basis of the international rating methods could be greatly overestimating potential savings.

Manufacturing variations also contribute to efficiency variations, even with identical designs. A 2 to 3 percent variation is not uncommon. Types and accuracy of test equipment can also greatly affect efficiency determinations.

There is considerable confusion about quoted performance data. Two typical methods of reporting efficiency are guaranteed and nominal. NEMA has adopted standard MG1–112.53b, an efficiency standard based on probability. The minimum or guaranteed efficiency is the value that all motors meet or exceed. The nominal efficiency is the value that is exceeded by 50 percent of the motors produced. The new standard indicates a minimum and nominal efficiency to be expected from a motor design and population of motors as shown in Table 29–7. The index letter assigned should appear in manufacturers' data and on the motor nameplate.[10]

Table 29–6. Comparison of Efficiencies for the Same Motor Determined by Different Test Standards.*

	FULL LOAD EFFICIENCY	
TEST STANDARDS	7.5 HP	20 HP
International (IEC 34–2)	82.3%	89.4%
British (BS-269)	82.3%	89.4%
Japanese (JEC-37)	85%	90.4%
U.S. (IEEE-112 Method B)	80.3%	86.9%

* Reference 10. (Reprinted by special permission from Fairmont Press, Inc., Atlanta, Georgia.)

Table 29–7. NEMA Efficiency Marking
Standard.*

INDEX LETTER	NOMINAL EFFICIENCY	MINIMUM EFFICIENCY
A	—	>95.0
B	95.0	94.1
C	94.1	93.0
D	93.0	91.7
E	91.7	90.2
F	90.2	88.5
G	88.5	86.5
H	86.5	84.0
K	84.0	81.5
L	81.5	78.5
M	78.5	75.5
N	75.5	77.0
P	72.0	68.0
R	68.0	64.0
S	64.0	59.5
T	59.5	55.0
U	55.0	50.5
V	50.5	46.0
W	—	<46.0

* Reference 10. (Reprinted by special permission from Fairmont Press, Inc., Atlanta, Georgia.)
Note: Index letters should appear in manufacturers' data and on motor nameplate.

Motor Sizing. Frequently, motors are oversized for their actual load. As indicated in Fig. 29–11, the efficiency of induction motors remains fairly constant over a broad range of mechanical loadings, but drops rapidly below about 50 percent of rated load.[11] The smaller motors are particularly susceptible to efficiency deterioration at lower loading.

There are many good reasons for oversizing motors, chief among them longer service life under adverse operating conditions. However, these considerations should be weighed carefully against efficiency, which may be lower if motors are grossly oversized.

Submersible Pumps and Motors. The efficiencies of motors used for submersible well pumps vary considerably by manufacturer, but have generally improved in recent years. When pumps are removed periodically for maintenance, operations personnel can consider replacing older motors with new, more efficient units. The efficiencies of existing units can be compared to efficiencies available from a pump manufacturer, as shown in Table 29–8.[12] Efficiencies comparable to or better than those shown in Table 29–8 may be available from certain manufacturers. Important energy considerations

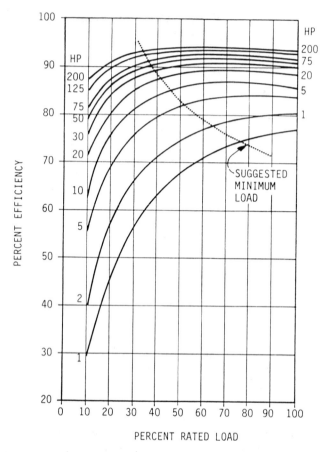

Fig. 29-11. Typical efficiency vs. load curves for 1800 rpm, three-phase squirrel cage induction motors.[11]

in choosing a new or replacement submersible motor include price, efficiency, and efficiency rating test method.

Optimum Conveyance Systems

The design of piping systems has long been guided by intuition and rule-of-thumb. Obviously, larger pipe sizes result in lower pumping energy; however, excessively large pipes are not economical. Therefore, several mathematical models have been developed for optimizing pipeline sizes. Patton and Horsley have developed a simplified approach to pipeline sizing based on the following:[13]

$$V_{opt} = 0.140 \ C^{0.65} D^{0.0595} (PE/KT)^{0.35} \qquad (29\text{--}3)$$

Table 29–8. Full Load Efficiencies of Submersible Motors Available from One Manufacturer in 1984.*

MOTOR SIZE, HP	MOTOR DIAMETER, INCHES				
	6	8	10	12	14
5	81				
7.5	83				
10	84				
15	84				
20	85				
25	86				
30	85				
35	85	87			
40	85	87			
45		87.5			
50		88			
55		88.5			
60		88.5			
65		88.5			
75		88.5	89		
100		88.5	89		
125			89		
150			89	89	
200				89	
250					89
300					89
350					89
400					90
500					90

* Reference 12.
Note: 3600-rpm motors, 460-volt.

where:

V_{opt} = optimum pipeline velocity, ft/sec
C = Hazen-Williams coefficient, dimensionless
D = pipeline diameter, ft
P = unit cost of pipe, \$/linear ft/in. diameter
E = average overall pumping unit efficiency, decimal
K = average price of electricity, \$/kWh
T = design life, years

From this it can be shown that:

$$D_{opt} = 2.92 \ Q^{0.486} C^{-0.316} (KT/PE)^{0.170} \qquad (29\text{–}4)$$

where: Q = average flow, cu ft/sec.

Therefore, for a flow of 5 mgd (7.74 cu ft/sec or 18.9 Ml/d), at a Hazen-Williams coefficient of 100, a pipeline cost of $2/linear ft/in. diameter, an overall pumping system efficiency of 75 percent, electricity at $0.045/kWh, and a design life of 20 years, the optimum pipe size is 20 inches for a velocity of 3.5 ft/sec (1.1 m/s).

Pumping Considerations

Pump Selection and Design. Oversized pumps represent a major source of inefficiencies in waterworks systems. Pumping stations are often designed for maximum flow at ultimate plant capacity. The best pump selection usually maximizes efficiency at average operating conditions rather than at maximum conditions. Consider two examples: one situation with a variable speed, and one situation with a constant speed pump. The first example is a variable speed, two-pump station (one pump standby) that operates at an average flow of 7,000 gpm (442 l/s) at 150 feet (45.7 m) TDH, and a maximum flow of 10,600 gpm (669 l/s) at 190 feet (57.9 m) TDH.[14]

As can be seen from Fig. 29–12, a pump selected for maximum efficiency at the maximum flow will not be the most efficient in this case. Referring to the system curve, the pump will operate most of the time in the region around the average flow point, only rarely reaching the maximum point. A 5 percent difference in efficiency at the average operating point therefore represents about 150,000 kWh/yr (1.6 × 10⁹ Btu/yr), which is valued at $10,500/yr at 7¢/kWh.

The second example is a commonly encountered situation in pump design. A constant flow rate of 3,500 gpm (221 l/s) is desired, and the designer selects a design point and pump based on "worst case" conditions: a Hazen-Williams C value of 100 (compared to C = 120 at project midlife or C = 140 at start-up), maximum possible static head (compared to normal operating conditions), and fitting and valve headloss based on equivalent length (compared to the velocity head method). The worst case system curve and an actual system curve are shown in Fig. 29–13. As indicated, use of the more conservative, worst case design system curve will yield a pump with a best efficiency point at or near the design flow of 3,500 gpm (221 l/s). However, when the pump is placed in operation, it is found that the actual system curve forces the pump to operate far to the right on the performance curve. The efficiency is poor, and the pump is operating in an unstable range that may subject it to damage due to cavitation and vibration. To achieve the desired flow rate of 3,500 gpm (221 l/s), a discharge valve must be throttled to reduce flow. Excess head is unnecessarily "burned up," and an inefficient pumping system results (as measured by volume per unit of energy). The

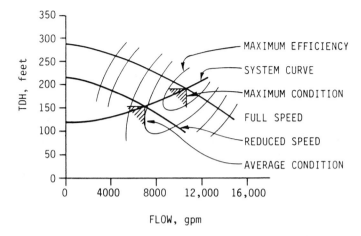

PUMP SELECTION BASED ON ACHIEVING
MAXIMUM EFFICIENCY AT MAXIMUM CONDITIONS

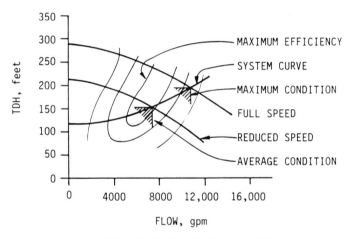

PUMP SELECTION BASED ON ACHIEVING
MAXIMUM EFFICIENCY AT AVERAGE CONDITIONS

Fig. 29–12. Comparison of pump selection approaches with variable speed pumps for variable flow conditions.[14] (Courtesy of Scranton Gillette Communications, Inc.)

required input to the pump at 3,500 gpm is 199 hp (149 kW). The designer would have made a better choice by selecting a pump based on most probable conditions within the first 5 to 10 years after start-up of the facility.

Figure 29–13 also shows how a smaller 20.5-inch (0.52-m) impeller can provide excellent efficiency at a lower operating head for initial operation. Using this impeller, the required power input is only 169 hp (127 kW) at

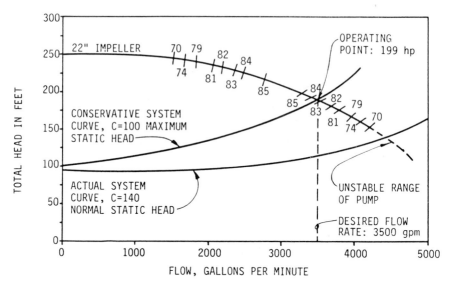

PUMP SELECTION BASED ON WORST CASE CONDITIONS (22" IMPELLER)

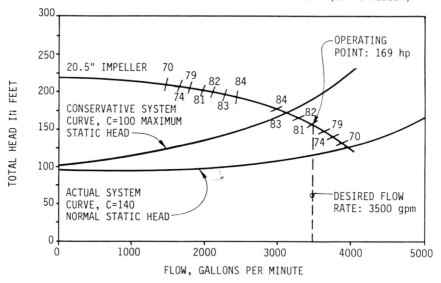

PUMP SELECTION BASED ON ACTUAL CONDITIONS EXPECTED
FOR FIRST 10 YEARS OF SERVICE (20.5" IMPELLER)

Fig. 29-13. Comparison of pump selection approaches for constant speed pumps with constant flow conditions.

3,500 gpm (221 l/s). An inexpensive change to a 22-inch (0.56-m) impeller at a future date will ensure a relatively efficient pumping station even after the distribution system ages and headloss increases. In this case, a motor changeout is avoided by sizing the motor for future conditions.

Design engineers are obligated to ensure that pumping stations will deliver desired flow in worst case situations. However, all too frequently actual operating system curves are much different. Thoughtful pump design calculations show "best case," "worst case," and "most probable" system curves plotted together with the performance curve of the prospective pump. When this is done, an efficient and stable pump can easily be selected.

Pump Testing.   Routine testing is an essential requirement to ensure efficient pump operation. The pump testing procedure and a data sheet (Fig. 29–8) were described previously in this chapter. Pump test data can be compared to manufacturers' performance sheets to identify methods of reducing energy consumption.

Maintenance: Wear Rings, Impellers, and Packing.   Proper maintenance is required to ensure efficient pump operation. Wear on impellers, casings, and wear rings can result in considerable internal recirculation of flow that lowers efficiency. Installation or replacement of wear rings or adjustment of the impeller may enable a pump to regain its original efficiency. Improper adjustment to worn or damaged packing can result in binding of the pump shaft and loss of efficiency. Rather than continually tightening packing on a problem pump to stop leakage, operators can consider replacing the packing on a routine basis or changing to a mechanical seal.

Drive Selection

Traditional approaches to pumping system design, which evolved when low-cost energy was abundant, are based on component analysis and selection, and thus do not provide adequate consideration of overall system operation. As discussed previously, proper pump selection represents an important method of energy conservation in municipal water systems. Since the 1950s, variable speed pumping has become popular to match pumping to variable system conditions. In the past, little attention was focused on the efficiency of variable speed controllers. Increases in energy rates and a corresponding reduction in the cost and complexity of "energy-efficient" drive systems has led to, a dramatic increase in the use of these systems.

The constant speed drive with simple on–off control is the most efficient overall drive unit, provided it operates at or near the most efficient point on the pump curve. Unfortunately, this system is not suitable for all pumping applications, and some method must be used to control the output from a pump, fan, or other equipment.

The throttling valve is perhaps the simplest and lowest-cost flow control device. It relies on "burning up" head to achieve the desired system operating conditions. The relatively low efficiency of this type of control relegates its use to systems where simplicity rather than efficiency is preferable.

Speed variation is another efficient method to control equipment output. The following is a listing of some drives currently available for variable speed control:

- Variable voltage
- Hydraulic clutch or coupling
- Eddy current
- Wound rotor motor with resistance or reactance secondary control
- Pulse width modulated variable frequency drives (voltage source)
- Current source variable frequency drives
- Wound rotor motor with secondary power recovery (slip recovery)
- Silicon controlled rectifier (SCR) dc
- Load commutated inverter (LCI) with synchronous motor

This section will discuss the basic components, efficiency, advantages, and disadvantages of four of the energy-efficient drive systems that are currently available. The efficiencies of other lower-efficiency drive systems are presented, but the units are not discussed in detail. The high-efficiency drives discussed here are:

- Wound rotor motor with secondary power recovery
- Variable frequency
- SCR dc
- LCI with synchronous motors

Efficiencies of variable voltage, hydraulic coupling, eddy current, wound rotor with secondary resistance or reactance, and pulse width modulated inverters are presented in Fig. 29–14.[15] These systems have lower efficiencies than the systems described below.

Wound Rotor Motor with Slip Power Recovery. This system has three components: a wound rotor ac motor, a motor starter, and a slip energy recovery section, which is the controller. A schematic of this system is shown in Fig. 29–15.[16] The wound rotor motor is similar to a squirrel cage induction motor except that in place of the bare rotor bars used in the squirrel cage motor, the rotor is wound, and the insulated conductors are brought out to slip rings. Carbon brushes at these rings allow the rotor to be connected

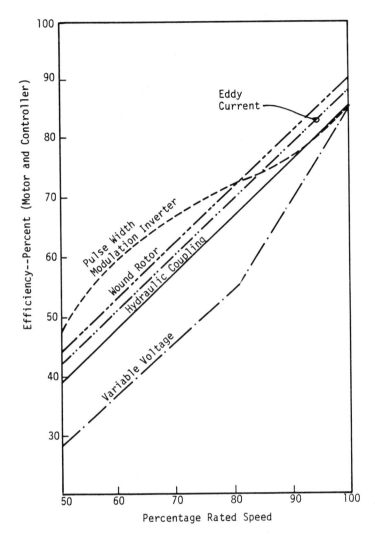

Fig. 29-14. Comparative efficiency of several variable-speed drives at 50 to 100 percent of rated speed at 100 hp.[15]

to external impedance such as resistors or reactors. In the past, control of wound rotor motors was accomplished by stepping resistance or reactance in and out of the circuit. This scheme was limited by the number of speed steps that could be practically achieved, and its efficiency was very low because of secondary circuit losses as heat.

Energy loss can be reduced by recovering slip power and returning it to the power line in a stepless manner. This is accomplished by recovering

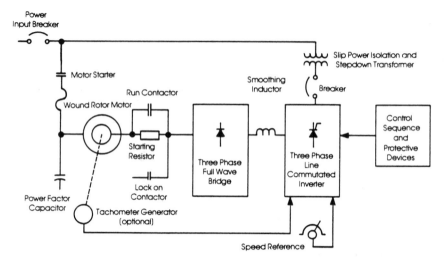

Fig. 29–15. System schematic of wound rotor motor with slip power recovery.[16]

the power previously dumped to resistor banks and rectifying the ac power to dc. Power is then filtered, inverted back to ac, and returned to the primary power lines. The efficiency achieved by this system is presented in Fig. 29–16, which shows line to shaft efficiency vs. speed for four energy-efficient variable speed drive systems.[16-18]

This system has the advantages of relatively high efficiency, and the ability to operate at full speed across-the-line in the event of controller failure. Wound rotor motors are reliable and serviceable at most motor shops. The primary disadvantage is the slightly increased maintenance required for the brushes. The system is considered cost-competitive with other high-efficiency drive systems for the 200-hp (150-kW) sizes and larger.

Variable Frequency. In the variable frequency drive (VFD) system, the incoming ac power is converted to dc. The dc power is then inverted to adjustable frequency and supplied to a conventional squirrel cage induction ac motor. The motor rotates at a speed proportional to the applied frequency. Drives are available from 1 to 1,500 hp (0.75 to 1,125 kW). The variable frequency drive is competitive on a first-cost basis, especially in sizes below 200 hp (150 kW). Although some reliability problems were encountered with earlier generations of VFD units, they are now considered to be a reliable and efficient drive system.[17] Efficiency of the VFD is shown in Fig. 29–16.[17] This system has the advantages of excellent reliability, competitive prices, ability to operate the motor across-the-line in the event of controller failure, use of conventional ac motors, and ease of retrofitting to existing constant speed pumping systems.

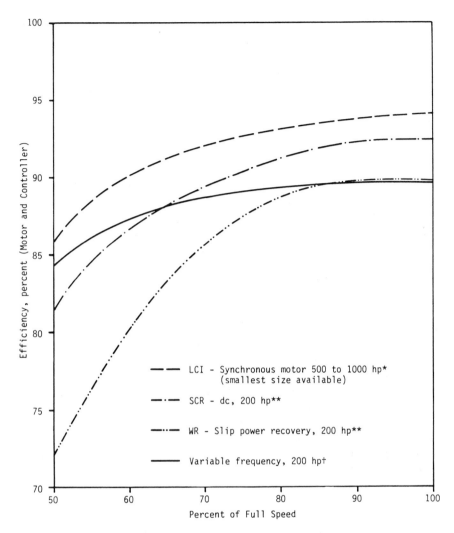

Fig. 29–16. Efficiency vs. speed for four energy-efficient drive systems. (* Reference 18. ** Reference 16. † Reference 17.)

SCR dc.   The SCR dc drive system has not been widely used in waterworks facilities, primarily owing to designers' and operators' unfamiliarity with dc motors and concern about brush maintenance. This is unfortunate because dc drive systems offer efficient speed control, especially at speeds above 70 percent of full speed.

The efficiency of a 200-hp (150-kW) dc system is shown in Fig. 29–16,[16] and a schematic diagram of a dc adjustable speed drive system is shown in

Fig. 29–17.[16] Utility ac power is converted to dc with a silicon controlled rectifier SCR bridge. The SCR dc drives are available from 1 to 2,250 hp (0.75 to 1,687 kW). They have the advantages of low capital cost, high efficiency, reliability, a wide range of sizes, and ability to operate higher than base speed in emergency situations. This system has the disadvantages of a reduced power factor at decreased loads and the inability to run across-the-line if the controller fails. Maintenance is higher for ac motors because of the brushes. The dc motor is inherently noisier than the ac motor.

Load Commutated Inverter (LCI) with Synchronous ac Motor.

This relatively new design is drawing increasing interest for large installations. It uses a synchronous ac motor and a naturally commutated power circuit based on dc technology. The LCI drive controls speed by efficiently regulating motor frequency. Ratings of LCI drives range from 500 to 20,000 hp, and their energy efficiency is unequaled. The efficiency of a 500-hp (375-kW) LCI drive with a synchronous motor is shown in Fig. 29–16. In addition to exceptional efficiency, LCI drives provide very precise speed control.[18]

Power Factor, Demand, and Time-of-Use Charges

Electric rates vary among utilities. Rate schedules applicable to water treatment facilities are composed of some combination of demand, reactive demand (power factor), standby, energy usage, and time-of-day charges.

Demand charges are based on the maximum average demand in kilowatts for a 15-, 30-, or 60-minute interval during a billing period. The maximum demand and charge can carry over for 1-, 3-, 6-, or 12-month (or longer) periods. The basis of the charge is the electric utility's responsibility for providing power generation and transmission capacity to meet peak demands when needed. The charge also reflects the relative inefficiency of utility peaking units. Demand charges can be reduced by making diligent efforts to ensure

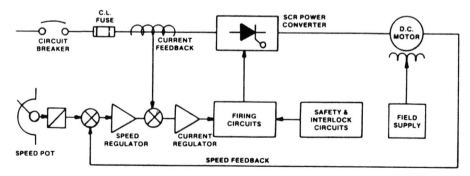

Fig. 29–17. Schematic diagram of SCR dc drive system.[25]

that high-energy devices are not operated simultaneously. For example, additional system storage can reduce peak pumping demands. The water utility can also encourage regulations that mandate that pool filling and lawn watering be rescheduled to low-demand periods. This has been done in many West Coast communities. Additional demand savings are possible by reducing the system head curves of the existing distribution system, by cleaning and lining or paralleling and looping to allow fewer pumps at lower heads to meet water demand.[19]

Power factor charges, which are used by some electric utilities to compensate for the increased cost of supplying energy to customers with certain electrical load characteristics, vary considerably among utilities. For example, induction-type motors commonly used to drive pumps can exhibit unfavorable power factors, especially when the motors are lightly loaded. Two common methods of charging for power factors are a direct charge for reactive demand if the power factor drops to a predetermined value, typically 85 to 95 percent, or a charge for each percentage point that the power factor drops below a stated value. In some cases, customers are also offered a credit if the power factor is above the stated value.

Capacitors have been added to a system to increase the power factor. Their installation requires detailed analysis to determine cost-effectiveness and the best location. Power factor correction frequently is not cost-effective because power factor charges generally are small. Capacitors can be installed on the primary or secondary side of the transformer, at the motor control center, or at the motors. Installation at the motors is the preferred method from an energy standpoint because it minimizes voltage drop from additional current flow. Power factor correction has the advantage of freeing up transformer and feeder capacity, which are limited by the high currents when the power factor is poor.[19]

Many large electrical customers (usually 500 kW demand and over) are placed on time-of-day charge schedules. Facilities can take advantage of this system by scheduling equipment to run during off-peak hours. Operations such as filter backwashing and sludge dewatering can be performed during this period, and if storage is available, pumping can be increased then. Energy management systems are very effective in automatically controlling equipment to operate during off-peak hours.

Distribution System Considerations

Significant energy savings are possible by careful analysis, management, and maintenance of distribution systems. Perhaps the greatest potential for energy conservation is in the identification and correction of unaccounted-for-water; water loss in leaky mains, hydrants, and line breaks can represent large energy losses. Carefully managed systems can operate with a loss of 10 percent or less.[19] Another source of energy loss in systems is partially or fully closed valves, perhaps left in those positions after emergencies. The additional head produced is reflected in higher energy costs.

As previously mentioned, reducing system friction by cleaning and lining as well as looping and paralleling can greatly reduce dynamic head. A cost analysis on each project will determine energy cost-effectiveness.

Energy Management Systems

For a given facility, the energy management system will be a method of implementing, controlling, and monitoring the operating-type energy conservation practices that are put into practice at that facility. Operating-type energy conservation practices are those measures that require operating decisions to realize the savings, as compared to fixed-type practices such as relamping with high-efficiency lamps.

Many of the operations-oriented energy-saving considerations identified in this chapter can be implemented by using some form of energy management system (EMS). The system can be as simple as a 7-day time clock, a wall switch, or a commercially available computer-based EMS. Practical systems for water treatment facilities will take many forms, depending on the size and complexity of the facility and the desires of the owner. Most measures will fall into one of two categories: (1) reduction of energy and demand use, or (2) shifting of energy and demand from peak to off-peak (time-of-day metering). Examples of each will be given.

Experience over a number of years has shown that in many cases nearly all the potential benefit of energy conservation practices can be realized with simple controls such as wall switches, thermostats, 7-day timers, manually set timers, and similar devices. More sophisticated EMS's such as computer-based systems provide a high degree of convenience, but in many cases are sold on high expectations with later disappointments.

Careful and detailed advanced planning will allow the realization of maximum energy conservation benefit with a system that is carefully tailored to the owner's needs and desires. If the facility already has a computer-based plant control system, it may be possible to incorporate many of the energy conservation measures on that system. Although a computer-based system provides a high degree of convenience and is an excellent data base, most control functions are simple and can be accomplished with simple devices.

Area Lighting. Energy management can be accomplished by turning on lighting only when it is needed. This applies to indoor as well as outdoor lighting. The best indoor measure is to use a wall switch and use lights only when needed. Photocell controls and time clocks can be used to automatically cut back lighting during daylight hours or at other times. Exterior lighting can be controlled by photocells where it is needed continuously at night. In outdoor areas where lighting is only needed when someone is in the area, it can be controlled manually from central locations, by time clocks, or by other suitable means.

HVAC Systems. Most energy management systems are applied for the control of HVAC systems. This can involve automatic shutdown or turndown of heating and air conditioning systems when areas are not occupied. Ventilation in areas that are heated or cooled can sometimes be reduced or stopped when such areas are unoccupied. Simple bypass timers can be used to activate ventilation for a preset time period when an operator enters an area. In other cases, ventilation can be placed on an adjustable on–off cycle timer so that it operates only for a portion of each time period (for instance, a portion of each hour).

Time-of-Day Metering. Major potential savings exist in facilities that are on time-of-day metering rate schedules. The savings are realized by shifting demand and energy use from the defined peak periods to "shoulder" or off-peak times. The realization of demand savings requires consistency because the billing demand is based on the highest demand for the billing period. One mistake during a billing period will erase any potential demand savings for that period. Energy savings are much easier than demand savings because credit will be received for each kilowatt-hour that is shifted from the peak period.

Systems to aid in shifting to off-peak use range from such simple ones as rigid schedules for the operation of certain pumps, such as backwash, to relatively sophisticated computer programs that monitor and predict system needs in order to optimize the use of storage to defer pumping.

Implementation. A majority of energy-conservation savings can be realized by using relatively simple control devices. Many of these measures can be planned and implemented by operating personnel.

On the other end of the scale are commercially available energy management systems, which are generally microprocessor-based. These systems can offer convenience, sophisticated measures, and an extensive data base. Typically they can provide:

- HVAC energy management
- Lighting energy management
- Power monitoring and process control to minimize peak use
- Security
- Fire protection
- Maintenance scheduling
- Reports

These systems should be applied with realistic expectations; they are versatile, but there are limits to their capabilities. They can be very cost-effective if integrated with a plant control system.

Most electrical utilities are anxious to assist with planning and implementation of energy conservation measures. Utilities should be contacted early in the planning in order to gain maximum benefit from their advice.

Energy Considerations in the Design of HVAC Systems

The design of an energy-conscious HVAC system begins with determining the minimum acceptable levels of ventilation, heating, cooling, humidification, and filtration for each building or process area. Generally, the most energy-intensive HVAC systems are those that serve areas with high ventilation rates, strict temperature needs, and extreme filtration requirements.

Ventilation Design Criteria. Ventilation rates have been developed to provide personnel safety, comfort, and protection of process equipment. These rates are based on requirements of the Uniform Mechanical Code (UMC), Uniform Building Code (UBC), Occupational Safety and Health Administration Standards (OSHA), National Fire Protection Agency (NFPA), past experience, and good engineering judgment. Sometimes a state agency will have specific recommendations for ventilation rates in areas where hazardous or corrosive gases may be present. In addition, the American Society of Heating, Refrigeration, and Air Conditioning Engineers (ASHRAE) and local state energy codes may have requirements for nonprocess buildings. Table 29–9 lists typical ventilation rates by areas. Rates from this table may need to be increased or decreased for a specific application.

Temperature Design Criteria. Temperature requirements for water treatment plants have been developed to provide personnel safety, comfort, and protection of process equipment. One approach is to group each building or process area into one of four levels:

- Level 1: Ventilation only for areas where freezing is not a concern and there is no need for comfort heating.
- Level 2: Ventilation and minimal heating to prevent freezing or where some heating is required for worker comfort.
- Level 3: Ventilation and minimal heating and cooling for areas similar to level 2 except that cooling is required because of high internal heat gains or outside air temperatures.
- Level 4: Full heating and cooling for areas continuously occupied, including offices, control rooms, laboratories, computer rooms, and so on.

Table 29-9. Suggested Ventilation Rates.

AREA	VENTILATION RATE AIR CHANGES/HR	COMMENTS
Administrative		
Offices	2 to 4	
Conference	2 to 4	
Stairwells	2 to 4	Positive pressure.
Toilet/locker areas	10	100% outside air.
Laboratories	2 to 4	
Process		
Chlorination rooms	10 to 20	100% outside air, double rate for leak condition, exhaust at low level.
MCC rooms	2 to 4	Positive pressure. Do not use evaporative cooling.
Equipment rooms	2 to 4	Check for high heat gains.
Tunnels	2 to 4	
Ozone rooms	4 to 8	100% outside air, double rate for leak condition.
Chemical rooms	4 to 8	
Boiler rooms	2 to 4	Allow for combustion air.
Battery rooms	2 to 6	100% outside air, in low– out high.
Engine rooms	2 to 4	100% outside air, allow for combustion, high heat gain area.

Typical design temperatures for levels 2 through 4 are shown in Table 29–10.

Laboratories and computer rooms have stricter temperature requirements than these.

Typical grouping of areas in a water treatment plant is shown in Table 29–11.

Table 29-10. Typical Design Temperatures for Buildings and Processes.

LEVEL	HEATING	COOLING
2	55°F (13°C)	—
3	55°F (13°C)	90°F (32°C)
4	70°F (21°C)	78°F (26°C)

Table 29–11. HVAC Level Requirements for Water Treatment Plants.

LEVEL 1	LEVEL 2	LEVEL 3	LEVEL 4
Unoccupied areas w/o freezing concerns	Chlorine facilities Chemical rooms	MCC Equipment rooms Tunnels and galleries Ozone rooms Boiler rooms Engine rooms	Administrative Control rooms Laboratories

Energy Considerations for Ventilation System Design. Ventilation systems consist of once-through and recirculation systems. Areas that require once-through ventilation can be made less energy-intensive by reusing the ventilation air in other areas or integrating the HVAC system as described in the following subsection.

Integrated Ventilation Systems. Integrated ventilation systems can be considered for every process building and for the entire plant, if applicable. An integrated ventilation system endeavors to reuse ventilation air as much as possible to minimize makeup air heating and cooling. Generally, ventilation air is transferred from one room to another until it is finally exhausted; also it is transferred from uncontaminated areas to areas where contamination is more likely. For example, heated and cooled outside air would be supplied to a motor control room, transferred to a pump room, transferred to a chemical room, and then exhausted. With the internal heat gain of most process areas, exhausted ventilation air could be warm enough for heat recovery.

Plants with buildings interconnected by tunnels are particularly well suited for integrated ventilation systems. Heated and cooled ventilation air would be supplied to the tunnel system at one or more locations. Air would then be transferred into individual buildings and distributed as previously described. A plantwide integrated system results in larger but fewer HVAC systems, and usually is cost-effective. An integrated ventilation system, when designed properly, will provide adequate ventilation at minimal energy consumption.

Recirculation Ventilation Systems. Recirculation systems are employed for buildings or areas with no requirements for once-through ventilation. The system utilizes a minimal amount of outside air to mix with return air to meet the heating or cooling load. This control concept is commonly referred to as an economizer cycle. A control schematic of an economizer cycle is shown in Fig. 29–18.

Air Handling Units. The type and selection of air handling units can signifi-

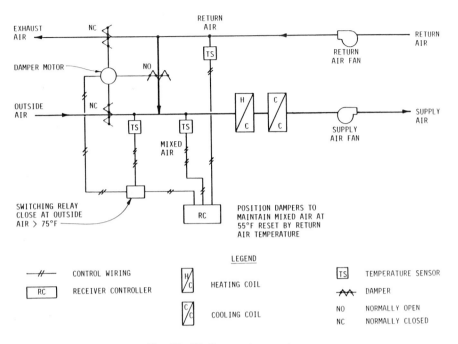

Fig. 29–18. Economizer cycle.

cantly affect system energy efficiency. For a constant flow and static head, required power can be determined from manufacturers' catalog data for any unit. Generally, as units increase in size, efficiency increases. Also, improved efficiency is more significant in the smaller units. An energy-efficient selection is a unit where the next larger unit does not significantly decrease the brake horsepower requirements. Some manufacturers make this selection process easier by publishing efficiency curves and a guide for selecting the optimal unit for a given set of conditions. Generally, a unit operating at peak efficiency will also be operating at the lowest noise and vibration point.

Constant-volume single-zone units are the workhorse air handling units for water treatment plants. They are generally used in all process areas where heated, cooled, or filtered air is required. If equipped with heating and cooling coils, the unit is capable of producing a constant flow of air at a variable temperature. A variation of this is the two-speed unit equipped with inlet guide vanes.

A two-speed unit can save energy by operating at low speed during the heating season, switching to high speed when the cooling load can be met with outside air, and switching back to low speed with mechanical cooling for the larger cooling loads. Theoretically, fan brake horsepower is reduced 87.5 percent when the speed is reduced by 50 percent.

Multizone units are typically used only in administrative areas. These units provide a number of constant-volume air streams at independently adjustable temperatures. This is useful in administration areas because of the variable heating and cooling requirements of adjacent areas (zones) due to constantly changing solar loads, personnel loads, and equipment loads. The standard multizone unit mixes air from a hot and a cold deck to achieve temperature control for each zone. This simultaneous heating and cooling is wasteful of energy, and local energy codes generally do not allow these systems except in small buildings or unless the energy required for heating or cooling is recovered energy. A variation of the standard multizone unit is the triple-deck multizone, which has an unconditioned third-deck air stream that is used for mixing with the hot and cold decks for temperature control. This method of temperature contol does not simultaneously consume heating and cooling energy to meet the temperature requirements of a zone, and so is allowed by energy codes. Although triple-deck multizones are a relatively new concept, they are available from most of the larger air handling unit manufacturers.

Double-duct variable-air-volume units are similar to the multizone units in that they provide individual temperature control for a number of zones. Temperature control is achieved by varying the flow of air to each zone. Hot and cold air ducts are routed to a volume control box that regulates the amount of heated or cooled air into each zone. The system is thermally efficient because there is little or no mixing of heated and cooled air streams. The advantage over triple-deck multizones is a reduction in average fan horse-power requirements. Except for peak loads which require peak flows, this system operates at a reduced flow and horsepower requirement. The disadvantage of this system is the higher cost of the double duct, the volume control boxes, and the controls.

Waste Heat Recovery. Several types of waste heat recovery systems are applicable to water treatment plants. The prime uses for waste heat in a water treatment plant are for space heating and service hot water, both of which require relatively low-temperature heat.

Run-around-loops. Waste heat can be recovered from warm building exhaust air and used to preheat makeup air using run-around-loops. This system consists of a heat recovery water coil in the exhaust air stream, a heating coil in the makeup air, and a pumped water loop connecting the coils. A schematic of a run-around-loop is shown in Fig. 29–19. The three-way control valve at the heating coil is for close temperature control and would normally be used only for a level 4 area. The percentage of recovered heat is proportional to the surface area of the coils, water flow rate, and temperature difference of the makeup and exhaust air.

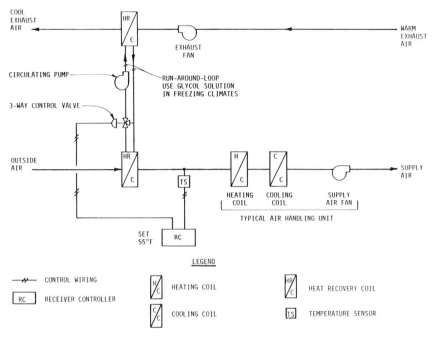

Fig. 29–19. Run-around-loop.

Increasing the coil surface area to increase recovered heat is cost-effective up to a point. Experience has shown that raising the makeup air temperature so that it approximates one-half of the exhaust air and makeup air temperature difference is the most cost-effective design. For example, with 70°F (21°C) exhaust air and 20°F (−7°C) makeup air, the heat recovery system should heat the makeup air to 45°F (7°C) degrees. Most of the large air handling unit manufacturers now have an in-house computer coil selection program that will size coils for run-around-loops. Several solutions should be analyzed to determine the most economical design, and a yearly energy analysis must be performed to determine savings. For this system an average monthly temperature rate is sufficient. (See the ASHRAE system handbook for a discussion on energy analysis methods.[20]) Although they are usually relatively small, the energy costs of the circulation pump and increased fan static should be accounted for in the analysis.

Aeration Air Heat Recovery.   Another source of recoverable heat is compressed air. The compression process increases the air temperature 100 to 200°F (38 to 93°C), depending on the required pressure. Heat recovery is accomplished using a run-around-loop system with the heat recovery coil

984 HANDBOOK OF PUBLIC WATER SYSTEMS

in the aeration air piping. Aeration air velocities are typically high compared to exhaust air in ductwork, because of the relatively high cost of pipe. To bring the pressure drop through the heat recovery coil down to a reasonable level, the pipe must be transitioned to a larger cross-sectional area. Typically, the pipe is transitioned to a square with the coil velocity between 500 and 1,000 ft/min (152 and 305 m/min). Because the aeration air is at a relatively high temperature, the recovered heat can be used for service hot water heating as well as space heating.

Heat Recovery Water Chillers.   Heat recovery water chillers (HRWC) provide hot and chilled water simultaneously. A standard water chiller rejects waste heat to the atmosphere as condensed water. An HRWC operates at a slightly higher condensing temperature and produces a waste heat hot enough to be worth recovering. Normal heat recovery water temperatures are 105 to 120°F (41 to 49°C). This water is useful for space heating and preheating service hot water. Because HRWC operate at higher condensing temperatures, their efficiency or coefficient of performance (COP) is less than that of standard chillers. This difference is usually minor compared to the heating energy saved if they are used in the right application. HRWC are usually cost-effective in buildings when there is a simultaneous heating and cooling demand. Existing administration buildings or other buildings with close temperature requirements may have simultaneous heating and cooling loads.

Central Heating and Cooling Systems.   Large water treatment plants require considerable heating and cooling and can take advantage of high-efficiency central heating and cooling systems. Two or three large chillers and/or boilers will usually have a higher efficiency than many single small units located throughout a facility. Initial cost, operation and maintenance costs, and noise and vibration will also be less. Typically, central plants require less total space. Energy consumption of central systems is lower, not only because of higher equipment efficiencies at rated conditions but also the fact that multiple units in the central system can be utilized more efficiently than the single small units.

Synthetic Lubricants

Synthetic lubricants are petroleum-based, man-made products. They have uniform molecular structures, unlike mineral oils which are made up of thousands of hydrocarbon compounds. Impurities commonly found in mineral oils are absent in these lubricants. The uniform structures of a synthetic lubricant are selected for their stability and excellent lubricating characteristics. Synthetic lubricants have advantages over high-quality mineral oil lubricants:

- Excellent shear and viscosity/temperature properties to provide improved wear protection.
- Stability in high-temperature applications to reduce oxidation and sludge formation.
- Good low-temperature characteristics to improve low-temperature flow properties.
- Long service life.
- Low volatility that reduces consumption.

Synthetic lubricants are more costly than mineral oils and previously were reserved only for problem areas such as extremely high or low temperatures and equipment with marginal lubrication systems. However, because of their superior lubricative properties and low traction coefficients, energy reductions of 2 to 6 percent can be achieved by changing to synthetic lubricants for worm gear, spur gear, and spur gear/chain gearboxes.[21]

Synthetic lubricants have potential for making small but significant energy reductions in gearbox and bearing applications for water treatment plants. After synthetic lubricants are utilized, field electrical measurements can be obtained to determine whether energy reductions occur. The manufacturer of lubricated equipment should be contacted to determine the suitability of a synthetic lubricant before a change is made from a mineral oil lubricant.

Lighting

Lighting can account for 30 to 50 percent of a building's electrical energy consumption. Therefore, lighting energy conservation can be important although it represents only a small fraction of energy at a waterworks facility. Important considerations in lighting energy conservation include:

- Lighting levels
- Controls
- Lamp efficacy and characteristics
- Fixture efficiency
- Ballast efficiency
- Maintenance

Lighting Levels. The practice in many water treatment plants has been to have lights burning continuously regardless of the needs of the lighted areas. Nonwork areas and yards are often highly illuminated. Electricity required for lighting could be reduced by 20 to 30 percent by reducing or eliminating lighting in nonessential areas and by reducing overillumination in work areas. The following suggestions are described in detail in publications of the National Lighting Bureau.[22-24]

Controls. Control modifications can yield substantial cost and energy savings. Manual controls are effective if individuals responsible for using them do so properly. Manual control is provided by wall switches; when applied to control small groups of luminaries, they can permit excellent selectivity, resulting in energy and cost savings. Manually controlled dimmers are available for most types of lighting. Indoors, dimmers applied to a uniform system can generate many of the benefits of nonuniform task-oriented lighting.

Automatic controls are becoming increasingly popular. Used predominantly for outdoor lighting, they can allow week-at-a-time programming and can automatically compensate for changing hours of light and darkness, as well as leap year. Batteries or spring-wound mechanisms maintain accuracy during power interruptions. Time clocks are also used indoors with local switch overrides.

Photocell controls also are used extensively for outdoor purposes. They activate and deactivate lighting based on the amount of ambient light detected. When a photocell/time clock control is used, the time clock keeps lighting off for a certain period; during other periods, the lighting is photocell-controlled. Photocells also are used indoors, particularly for lighting near windows and skylights. Some systems use one photocell to control all lighting in an area; others use one photocell per fixture. Photocell/dimmer controls increase or decrease lighting levels based on the amount of ambient lighting available to maintain a constant illumination level.

Personnel detection controls activate and deactivate lighting based on the presence or absence of people in a space. Ultrasonic controls perform this function through motion detection, passive infrared controls operate by sensing body heat, active infrared controls activate lighting when beams they emit do not return, and acoustic controls are available to detect noise and "human activity." Most personnel detection controls can be integrated with dimmers, and some can be integrated with in-space heating/cooling units or motorized dampers of central multizone systems.

Several types of centralized lighting control systems are available. Many are computer-controlled and allow user programming. They are ideally suited for control of all lighting (and other loads) in and around large buildings or groups of buildings. Other types of centralized systems use existing wiring to conduct signals to actuators mounted inside fluorescent luminaires. Another type sends control signals through wireless radio transmission to receivers mounted in ballast modules.

Almost all the above controls are applicable to new and existing buildings.

Lamp Efficacy and Characteristics. In most applications for a water treatment facility, lamps can be divided into several categories: incandescent, fluorescent, and high intensity discharge (mercury vapor, metal halide, high pressure sodium, and low pressure sodium). Their basic characteristics are shown in Table 29–12. In this discussion of lighting, the term *efficacy* is used to describe lamp light output divided by power input (lumens/watt). The term *efficiency* is used in reference to luminaires and is the ratio of the light emitted by the luminaire divided by the light emitted by the lamp.

Incandescent Lamps. These lamps have the poorest efficacy but are popular because the fixture and lamp are inexpensive. Also no ballast is required to modify the characteristics of the power supply.

Fluorescent Lamps. The fluorescent lamp is the most common light source in water plants. Unlike the incandescent lamp, the fluorescent lamp requires a ballast to strike the electric arc in the tube initially and to maintain the proper voltage and current to the lamp to maintain an arc. Proper ballast selection is important to optimum light output, lamp life, and overall efficiency.

Typical lamp sizes range from 40 watts to 125 watts. The efficacy (lumens per watt) of a lamp increases with lamp length [from 4 to 8 feet (1.2 to 2.4 m)]. The reduced-wattage fluorescent lamps introduced in the last few years use from 10 to 20 percent less wattage than conventional fluorescent lamps.

The cool white and warm white lamps provide very acceptable color and energy efficacy ratings in most locations. New types of fluorescent lamps can produce color that is similar to the incandescent lamps or daylight but at lower efficacy. Several dimming technologies are available for the fluorescent lamp. Some provide full-range dimming; others permit limited dimming only, but require no modification of existing fixtures and ballasts.

Fluorescent lamp life is rated according to the number of operating hours per start; for example, 20,000 hours at 3 hours of operation per start. The greater the number of hours operated per start, the greater the lamp life. Because fluorescent lamp life ratings have increased, the number of starts is less important. If a space is to be unoccupied more than a few minutes, lamps should be turned off.

High Intensity Discharge (HID) Lamps. High intensity discharge or HID is the term commonly used to describe mercury vapor, metal halide, high pressure sodium, and low pressure sodium lamps. Each requires a few minutes (1 to 7) to reach full output. Also, if lamp power is lost or turned off, the arc tube must cool to a given temperature before the arc can be restruck

Table 29–12. Approximate Range of Lamp/Ballast Characteristics.*

TYPE OF LAMP	WATTAGES	AVERAGE LUMENS PER WATT INCLUDING BALLAST LOSSES		% LUMEN MAINTENANCE	AVERAGE RATED LIFE (HOURS)	WARM-UP/ RESTRIKE (MINUTES)
		INITIAL	MEAN			
Low pressure sodium	18–180**	62–150	60–143	95+	10,000–18,000	7.00/0–1
High pressure sodium	35–1,000	51–130	46–118	90–91	10,000–24,000+	3–4/.5–1
Metal halide	175–1,500	69–115	54–92	77–80	10,000–20,000	2–5/10–20
Mercury vapor	50–1,000	30–60	23–46	75–89	16,000–24,000+	5–7/3–6
Fluorescent	40–125	63–100	42–88	66–88	12,000–20,000+	Immediate
Incandescent	60–1,500	15–24	13–23	90–95	750– 2,500	Immediate

* Reference 11.
** Lamp wattage rises as lamp ages.
Note: Approximate range of lamp/ballast combination characteristics, based on data applicable to the most commonly used lamps within each lamp category. Data are provided for general comparative purposes only. Actual performance of a lamp with regard to any of the factors indicated will vary, because of the uniqueness of the specific installation. Energy characteristics of an installation depends on other factors than lamp or lamp/ballast efficacy.

and light produced. Up to 15 minutes may be required for metal halide lamps.

The mercury vapor lamp produces light when the electrical current passes through a small amount of mercury vapor. The lamp consists of two glass envelopes: an inner envelope where the arc is struck, and an outer or protective envelope. The lamp requires a ballast designed for its specific use. Electronic dimming also is available.

Mercury vapor lamps are most commonly used in industrial applications and outdoor lighting because of their low cost and long life (16,000 to 24,000 hours).

The color-rendering qualities of the mercury vapor lamp are not as good as those of incandescent and fluorescent lamps. Because the color rendition and lamp efficacy of phosphor-coated mercury vapor lamps is better than that of their clear (no phosphor-coating) counterparts, the development of phosphor-coated mercury vapor lamps has enabled their application indoors. Mercury vapor lamp sizes range from 50 to 1,000 watts.

The metal halide lamp is similar in construction to the mercury vapor lamp, the major difference being that the metal halide lamp contains various metal halide additives in addition to mercury vapor. The efficacy of metal halide lamps is much higher than that of mercury vapor lamps. Some of the newer metal halide lamps provide color similar to that of incandescent lamps; others emulate daylight. Metal halide lamp sizes range from 175 to 1,500 watts. Ballasts designed specifically for metal halide lamps must be used.

The high pressure sodium (HPS) lamp has the highest lamp efficacy of all lamps commonly used indoors. It produces light when electricity passes through a sodium vapor. The lamp has two envelopes, the inner one being made of a polycrystalline alumina in which the light-producing arc is struck. The outer envelope is protective, and is clear or coated.

The sodium in the lamp is pressurized; so the light produced is not the characteristic bright yellow associated with sodium, but rather a "golden white" light. Although the HPS lamp first found its principal use in outdoor lighting, it now is a readily accepted light source in industrial plants. HPS lamp sizes range from 35 to 1,000 watts. Ballasts designed specifically for high pressure sodium lamps are required.

The low pressure sodium (LPS) lamp has the highest efficacy, providing up to 183 lumens/watt. It is used where color is not important because it has a monochromatic light output: reds, blues, and other colors illuminated by an LPS light source all appear as tones of gray or yellow. Low pressure sodium lamps range in size from 18 to 180 watts. Ballasts designed specifically for LPS are required. These units are best used for outdoor applications; indoor applications are practical only where color is not important.[24]

Lamp Substitution. One type of lamp substitution involves a lamp retrofit, where one lamp is removed from a fixture and another installed to increase efficacy. Some typical substitutions for a water treatment facility are shown in Table 29–13.[22]

Another type of substitution involves replacement of fixtures themselves. Table 29–12 provides information on relative efficacy of various lamps. Lamp efficacy is not the only criterion to be used for selection. Other important factors include color rendition, useful life, light distribution, restrike time, lumen depreciation rate, and disposability.[22]

Fixture Efficiency. Manufacturers of lighting fixtures provide product information regarding fixture efficiency in terms of the coefficient of utilization (CU). This value is the percentage of light emitted by the lamps that is delivered to the work plane. The luminaire direct depreciation (LDD) is the amount of light lost by dirt buildup on the fixtures or lens. This reduction of light with time requires the designer to install more fixtures than would be required if light output did not deteriorate. When luminaires become outdated or damaged, it is best to replace the fixture with a modern, efficient unit with good cleaning capabilities, as well as excellent lumen maintenance characteristics. Simply changing the lens of a fixture can substantially increase luminous efficiency.

Ballast Efficiency. A ballast transforms line voltage and controls lamp current to match the operating characteristics of the lamp. The only lamp type that does not require a ballast is incandescent. Recently, manufacturers have introduced high-efficiency ballasts that reduce energy consumption of fluorescent fixtures by as much as 9 percent. These ballasts have a longer service life than the older design and can provide an overall savings of as much as 27 percent when used with high-efficiency lamps. This energy savings is produced with only a small loss of light. Other ballasts will reduce energy consumption by 20 percent when used with standard lamps, but with a corresponding decrease in light output. They should be used only where reduced lighting is acceptable.[22]

Maintenance. Lighting systems are poorly maintained at most facilities, with consequent waste of both money and energy. Lamp output for most lighting systems decreases with use, as seen in Fig. 29–20. As indicated in the figure, only the low pressure sodium system maintains its original light output. Relamping may be cost-effective before lamps burn out. Consideration should be given to group relamping to reduce maintenance, storage, and energy requirements.

Table 29–13. Energy Savings Lamp Interchangeability Chart.*,**,***

WHERE YOU NOW USE	CHANGE TO	GET THIS MUCH LIGHT†	SAVE THIS MANY WATTS PER LAMP††	KILOWATT HOURS SAVED OVER LIFE OF LAMP	VALUE OF kWh SAVINGS OVER LIFE OF LAMP (AT $0.05/kWh)
F40	F40 reduced-wattage high-efficiency	Same or more	7	140	$ 7.00
F96	F96 reduced-wattage high-efficiency	Same or more	17.5	315	$15.75
F96 HO	F96 HO reduced-wattage high-efficiency	Same or more	17.5	315	$15.75
F96 1500 MA	F96 1500 MA reduced-wattage high-efficiency	Same or more	46	690	$34.50
75 watt	Fluorescent screw-in§	Same or more	53	400	$20.00
100 watt	50R20	Same or more	50	100	$ 5.00
150 watt flood	75ER30	Same or more	75	150	$ 7.50
200 watt	120ER40	Same or more	80	160	$ 8.00
150 watt flood	120ER40	Up to 100% more	30	60	$ 3.00
300 watt flood	120ER40	Same or more	180	360	$18.00
175 watt mercury	Retrofit 150 watt HPS	Up to 120% more	25	300	$15.00
250 watt mercury	Retrofit 215 watt HPS	Up to 130% more	35	420	$21.00
400 watt mercury	Retrofit 325 watt metal halide	Up to 40% more	70	1,050	$52.50
400 watt mercury	Retrofit 400 watt metal halide	Up to 300% more	—	—	$ 0.00
400 watt mercury	Retrofit 360 watt HPS	Up to 300% more	40	650	$32.50
1000 watt mercury	Retrofit 1000 watt metal halide	Up to 440% more	85	850	$42.50
1000 watt mercury	Retrofit 880 watt HPS	Up to 440% more	120	1,440	$72.00

* Reference 22.
** All numbers shown are approximations. Actual results to be derived for any given lamp replacement depend upon factors unique to the installation involved. Consult the manufacturer for details, including any ballast or temperature restrictions that may apply.
† Typically when older standard lamps are group-replaced with reduced wattage, lamps and fixtures are cleaned.
†† Including ballast loss where applicable.
§ Several types of screw-in fluorescent replacements are available for incandescent fixtures. Figures do not include significant lamp replacement savings. (Lamps last five to ten times longer than incandescent.)

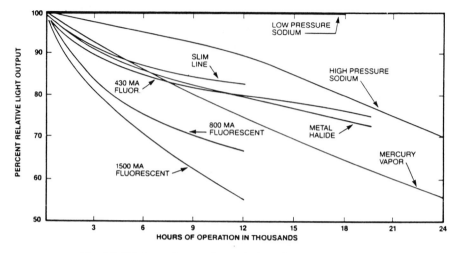

Fig. 29–20. Lamp lumen depreciation of standard lamps.[22]

REFERENCES

1. Fay, J. W., "Applying Capacitors to Improve Operations," *Electrical Consultant*, July/ August, 1982.
2. Water Pollution Control Federation, *Energy Conservation in the Design and Operation of Wastewater Treatment Facilities*, Manual of Practice No. FD-2, Lancaster Press, Lancaster, Pa., 1982.
3. U.S. Dept. of Energy—Energy Information Administration, *Annual Energy Review*, April, 1984.
4. *Energy User News*, April 30, 1984, Fairchild Publications, New York.
5. Based on CWC cost data prepared for EPA Contract No. 68–03–3093: "Estimation of Small Water System Costs."
6. Westinghouse Electric Corporation, catalog data.
7. General Electric Corporation, catalog data.
8. Siemen-Allis, Inc., catalog data.
9. Bonnett, A. H., "Understanding Efficiency in Squirrel Cage Induction Motors," *IEEE Transactions on Industry Applications*, Vol. 1A-16, No. 4, July/August, 1980.
10. Association of Energy Engineers, *Energy Management—A Sourcebook of Current Practices*, Fairmont Press, Inc., Atlanta, Ga., 1981.
11. Knudson, E. C., "Integral Horsepower Motor Loading and Efficiency Considerations in Wastewater Treatment Plants," paper presented at 44th Annual Pacific Northwest Pollution Control Association Meeting, Portland, Oreg., November 2–4, 1977.
12. TRW—PLEUGER Pumps, published catalog information.
13. Patton, J. L., and Horsley, M. B., "Curbing the Distribution Energy Appetite," *J.AWWA*, Vol. 72, No. 6, 1980.
14. Jacobs, A., "Managing Energy at Water Pollution Control Facilities," *Water and Sewage Works*, August, 1980.
15. Doffer, Robert A., and Price, John M., "Pump Systems: What Price Inefficiency?," *J.AWWA* 72:338, June, 1980.
16. WATTMISER ® by Marathon Electronics, Avtek Drives Division, Burlingame, Calif.

17. Personal communication with Arnold Ambrosini, Ambrosini and Chalmers, San Mateo, Calif., June, 1984. (Efficiency data from Autocon Industries, a Control Data Company.)
18. General Electric Company, Drive Systems Operations, Salem, Va.
19. Bradley, D., and Jacobs, A., "Workbook Example, Illustrating Electrical Savings," *Water and Sewage Works,* May, 1980.
20. ASHRAE, *ASHRAE Handbook and Product Directory, 1980 Systems,* American Society of Heating, Refrigerating and Air Conditioning, Inc., Atlanta, Ga., 1980.
21. Mobil Oil Corporation, published technical and catalog data.
22. National Lighting Bureau, *Lighting Energy Audit Workbook,* Washington, D.C.
23. National Lighting Bureau, *Industrial Lighting Handbook,* Washington, D.C.
24. National Lighting Bureau, *Getting the Most From Your Lighting Dollar,* Washington, D.C.
25. PACESETTER by Marathon Electronics, Avtek Drives Division, Burlingame, Calif.

Chapter 30
Cost of Water Treatment Plants

INTRODUCTION

An important factor in selection of a treatment process or treatment technique is a comparison of the cost of various processes that can accomplish the desired degree of water quality enhancement. One of the best methods of accomplishing such comparisons is to use cost data. Such information has been developed by Culp/Wesner/Culp for the U.S. Environmental Protection Agency. The initial work was completed in 1978, and is a four-volume set of documents entitled *Estimating Water Treatment Costs.*[1] These cost data cover the capacity range from 2,500 gpd to 200 mgd (9.46 kl/d to 757 Ml/d). A second project, completed in early 1984, provided more in-depth coverage of treatment costs at plants of less than 1 mgd (3.785 Ml/d) capacity; it was titled *Estimating Small Water System Treatment Costs.*[2] Each report includes a computer program for use in updating and combining costs from several unit processes.

COST DATA DEVELOPMENT

The construction cost data in References 1 and 2 were developed using equipment cost data supplied by manufacturers, cost data from actual plant construction, unit cost take-offs from actual and conceptual designs, and published cost data. The costs were developed and are presented in terms of eight cost categories: excavation and sitework; manufactured equipment; concrete; steel; labor; pipe and valves; electrical and instrumentation; and housing. These categories were selected to facilitate cost development and subsequent cost updating using cost indices specific to each of the cost categories. The cost data in References 1 and 2 are presented in terms of construction costs, and cost additions must be made for the following items to determine capital cost:

- Overall sitework, interface piping, and roads
- General contractors' overhead and profit
- Engineering
- Land
- Legal, fiscal, and administrative costs
- Interest during construction

Operation and maintenance information in References 1 and 2 is presented in terms of energy, maintenance material, and labor. Energy requirements are presented for process energy and building energy. This division is necessary due to variations in building energy in different geographical locations. Energy requirements are presented in energy units as follows:

- Electrical energy—kWh/yr
- Diesel fuel—gal/yr
- Natural gas—cu ft/yr

Maintenance material requirements are presented in dollars/yr, and labor requirements are presented in hrs/yr. Energy and labor requirements can be converted to an annual cost by using the current unit cost for energy and labor.

Accuracy of the cost data in References 1 and 2 is ±15 percent. This is adequate for comparison of costs of several treatment processes and for use in the conceptual or predesign stage of a project.

WATER TREATMENT COST EXAMPLES

Data contained in References 1 and 2 were used to generate costs for a number of different treatment concepts. In developing these costs, the following percentages and unit costs were utilized:

- Engineering—8 percent
- Sitework, interface piping—15 percent
- General contractor overhead and profit—10 percent
- Interest during construction—10 percent interest rate
- Electricity—$0.08/kWh
- Labor—$15.00/hr
- Diesel fuel—$1.25/gal ($0.33/l)
- Natural gas—$0.0055/cu ft ($0.1943/m³)
- Land costs—Not included
- Alum (liquid)—$130/ton ($143.33/metric ton)
- Polymer (dry)—$2.00/lb ($4.41/kg)
- Chlorine (gas)—$250/ton ($275.63/metric ton)
- Sulfuric acid—$65/ton ($71.66/metric ton)
- Sodium hexametaphosphate—$1,400/ton ($1,543.55/metric ton)
- Lime (quicklime)—$70/ton ($77.18/metric ton)
- Granular activated carbon—$1.00/lb ($2.20/kg)
- Powdered activated carbon—$0.50/lb ($1.10/kg)
- Anhydrous ammonia—$200/ton ($220.51/metric ton)
- Sodium chlorite—$2,000/lb ($2,205.07/metric ton)

Subsequent sections of this chapter present capital and operation and maintenance cost data for the following treatment concepts:

- Conventional Alum Coagulation Treatment
- Direct Filtration Treatment
- Reverse Osmosis Treatment
- Lime Softening Treatment
- Granular Activated Carbon Treatment
- Multiple Hearth Carbon Regeneration
- Conventional Package Complete Treatment
- Alum Sludge Disposal
  - Sanitary sewer
  - Lagoons/landfill
  - Sand drying beds/landfill
  - Centrifuge/landfill
- Lime Sludge Disposal
  - Lagoons/landfill
  - Sand drying beds/landfill
  - Centrifuge/landfill
  - Recalcination
- Treatment Modifications for Trihalomethane Reduction
  - Chloramines
  - Chlorine dioxide
  - Seasonal use of powdered activated carbon
  - Increased alum and polymer dosages
- New Reservoir Construction
  - Below-ground-level reinforced concrete storage
  - Ground-level steel storage tanks
  - Wash water storage
  - Covering existing reservoirs
- Package Plant Costs

The capital costs shown in these examples are based on January, 1984 costs. Because water treatment plants seldom operate at 100 percent capacity, the operation and maintenance costs in most of these examples are presented at treatment flow rates equivalent to 20, 55, and 95 percent of full treatment capacity.

It is important to recognize that these examples are for hypothetical situations, and that the design criteria and costs are general in nature, and are not necessarily applicable to all situations. The examples presented may have different costs from actual installations for one or more of the following reasons:

- Actual plant capital costs includes costly site acquisition and/or site improvement costs.
- Actual plant design criteria differ from the examples.
- Actual plant construction includes more standby facilities than in the examples.
- Labor, energy, and chemical unit cost differences exist between the examples and actual plants.

Experience in comparing costs from References 1 and 2 with actual plants indicates a high degree of correlation when there is a common design basis, and when equivalent labor, energy, and chemical unit costs are utilized.

CONVENTIONAL ALUM COAGULATION TREATMENT

Conventional treatment plants are defined as plants employing alum and polymer addition, rapid mix, flocculation, settling, and filtration. Design criteria used in the development of costs were as follows:

Process	Design Criteria
Raw water pumping	100 ft (30.5 m) TDH
Alum feed—liquid	50 mg/l design; 30 mg/l operating
Polymer feed	1 mg/l design; 0.2 mg/l operating
Rapid mix	60 sec detention; $G = 900$
Flocculation	30 min detention; $G = 80$
Clarifiers—rectangular	900 gpd/sq ft (1.525 m/h)
Gravity filtration—mixed media	5 gpm/sq ft (12.2 m/h)
Chlorine feed—gas chlorine	5 mg/l design; 2 mg/l operating
Product water pumping	300 ft (91.44 m) TDH

Below-ground clearwell storage, pumping to clearwell storage, and an administration, laboratory, and maintenance building were also included in the costs. Costs for alum sludge disposal are not included, and are covered subsequently in this chapter.

Figure 30–1 presents capital and operation and maintenance costs for conventional treatment plants. The operation and maintenance costs are shown for operation at 20, 50, and 95 percent of plant capacity.

DIRECT FILTRATION TREATMENT

Direct filtration is a useful treatment technique for water with low turbidity and low color. In the treatment process, lower doses of chemicals are used

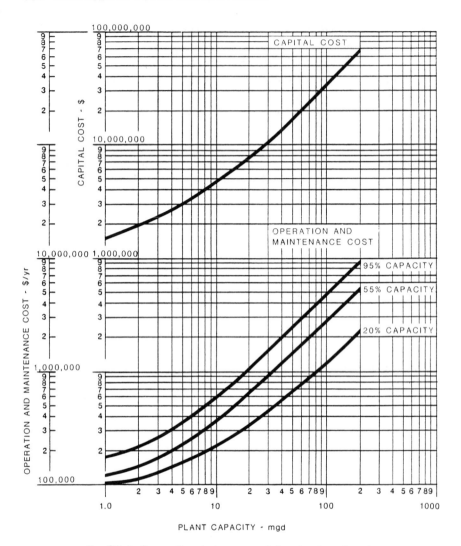

Fig. 30-1. Conventional alum coagulation treatment costs.

than in conventional treatment, and the settling basins are eliminated. All raw water suspended solids are removed by the filter, typically resulting in shorter filter runs than in conventional plants. Overall, the detention time is much less in direct filtration plants than in conventional plants, and the shorter detention time requires greater operator attention to raw and finished water quality.

Design criteria used in the development of costs were as follows:

Process	Design Criteria
Raw water pumping	100 ft (30.5 m) TDH
Alum feed—liquid	35 mg/l design; 20 mg/l operating
Polymer feed	1 mg/l design; 0.2 mg/l operating
Rapid mix	60 sec detention; $G = 600$
Flocculation	20 min detention; $G = 80$
Gravity filtration—mixed media	5 gpm/sq ft (12.2 m/h)
Chlorine feed—gas chlorine	5 mg/l design; 3 mg/l operating
Finished water pumping	300 ft (91.44 m) TDH

Capital and operation and maintenance costs are presented in Fig. 30–2. Operation and maintenance cost data are shown for plants operating at 20, 55, and 95 percent of design capacity. Costs for sludge disposal are not included, and are presented elsewhere in this chapter.

REVERSE OSMOSIS TREATMENT

Reverse osmosis plants include reverse osmosis membrane elements and pressure modules, high pressure pumping, cartridge filters, chemical feed equipment, degasification facilities, clearwell storage, finished water pumping, and an administration, laboratory, and maintenance building. The capital costs are applicable to either spiral-wound or hollow fine-fiber reverse osmosis membranes. Design criteria used in the development of costs were as follows:

Process	Design Criteria
Reverse osmosis	450 psi (3.1 MPa) operating pressure; single pass treatment; 80 to 85 percent water recovery; membrane replacement every 3 years
Sulfuric acid feed	40 mg/l design; 25 mg/l operating
Hexametaphosphate feed	10 mg/l design; 5 mg/l operating
Chlorine feed—gas chlorine	5 mg/l design; 3 mg/l operating

Capital and operation and maintenance costs are presented in Fig. 30–3. Operation and maintenance costs are shown for operation at 20, 55, and 95 percent of plant capacity. The operation and maintenance costs include the cost of membrane replacement every 3 years. Brine disposal costs are not included.

LIME SOFTENING TREATMENT

Lime softening treatment includes lime feed, rapid mix, flocculation, settling, recarbonation, and gravity filtration. Customarily lime softening is used for hardness removal, but it may also be used for removal of several of the inorganic contaminants included in the National Interim Primary Drinking Water Regulations. Design criteria used in the development of the cost data were as follows:

Process	Design Criteria
Raw water pumping	100 ft (30.5 m) TDH
Lime feed—quicklime	300 mg/l design; 150 mg/l operating
Rapid mix	5 min; $G = 600$

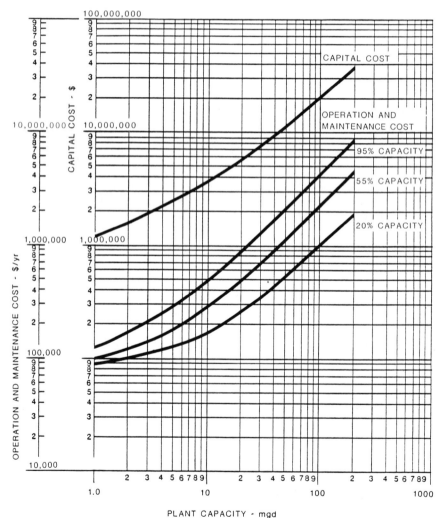

Fig. 30–2. Direct filtration treatment costs.

Process	Design Criteria
Flocculation	60 min; $G = 80$
Clarifiers—circular	1,000 gpd/sq ft (1.70 m/h)
Recarbonation basin	15 min detention; submerged burners
Gravity filtration—mixed media	5 gpm/sq ft (12.2 m/h)
Chlorine feed—gas chlorine	5 mg/l design; 3 mg/l operating
Product water pumping	300 ft (91.44 m) TDH

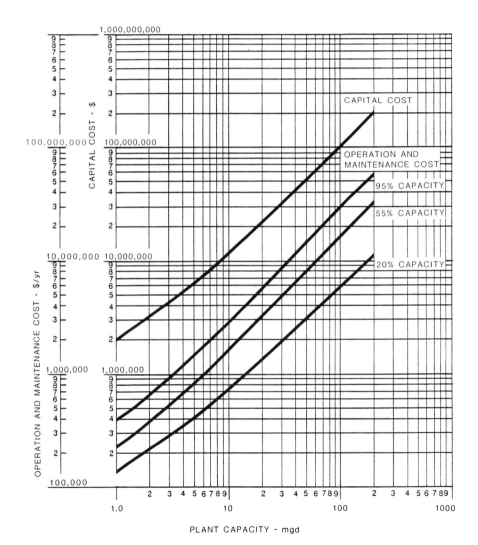

Fig. 30–3. Reverse osmosis treatment costs.

In addition to the above unit processes, below-ground clearwell storage, pumping to clearwell storage, and a building for administration, maintenance, and a laboratory are also included. Sludge handling costs are not included and must be added separately, using costs presented subsequently in this chapter.

Capital and operation and maintenance costs are presented in Fig. 30–4. The figure shows operation and maintenance costs for plants operated at 20, 55, and 95 percent of the rated design capacity.

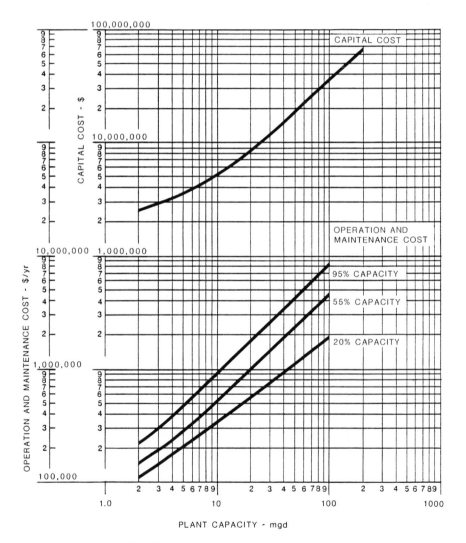

Fig. 30–4. Lime softening treatment costs.

GRANULAR ACTIVATED CARBON TREATMENT

Granular activated carbon treatment systems include initial pumping to the carbon contactors, initial carbon charge, carbon handling and storage facilities, backwash pumping, backwash water storage basin, and an administration, laboratory, and maintenance building. A carbon regeneration furnace is not included although the cost of makeup carbon is included. Generally, open gravity-flow steel contactors result in the lowest construction cost, although pressure steel contactors may also be used. Design criteria used in the development of costs were:

Process	Design Criteria
Initial pumping	Gravity contactors—35 ft (10.67 m) TDH
	Pressure contactors—75 ft (22.86 m) TDH
Carbon contactors	Contact time—15 min empty bed
	Pressure contactors—12 ft (3.66 m) diam, 12 ft (3.66 m) carbon depth
	Gravity contactors—20 and 30 ft (6.1 and 9.1 m) diam, 20 ft (6.1 m) bed depth
Carbon	Density—26 lb/cu ft (417 kg/m³)
Makeup carbon	4 regenerations/yr, 6 percent loss/regeneration

Capital and operation and maintenance costs are presented in Fig. 30–5 for gravity and pressure contactors. The cost of a carbon regeneration furnace is not included, although the cost of carbon handling facilities and makeup carbon is included.

MULTIPLE HEARTH CARBON REGENERATION

The required hearth area of multiple hearth carbon regeneration furnaces depends on carbon regeneration frequency, hearth loading, and anticipated furnace downtime. It is desirable to operate furnaces at or near design capacity.

Figure 30–6 presents capital and operation and maintenance costs for multiple hearth carbon regeneration furnaces. The capital cost includes the furnace, exhaust scrubbing, necessary instrumentation, and housing. Operation and maintenance cost is based on operation at 100 percent capacity.

CONVENTIONAL PACKAGE COMPLETE TREATMENT

Conventional package complete treatment units are factory-fabricated, and provide rapid mix, flocculation, settling, and filtration. Such treatment units are commonly used at flows of less than 1 mgd (3.785 Ml/d). The facilities included in this cost development are raw water pumping, alum and polymer storage and feed equipment, the package treatment plant, gas chlorination, a steel clearwell, and product water pumping. All facilities except the clearwell are housed. Design criteria used in the development of costs were:

Process	Design Criteria
Raw water pumping	50 ft (15.2 m) TDH
Alum feed	50 mg/l design; 20 mg/l operating
Polymer feed	1 mg/l design; 0.2 mg/l operating

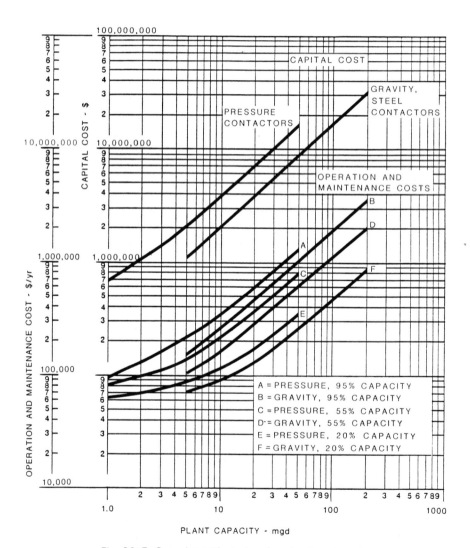

Fig. 30–5. Granular activated carbon treatment costs.

Process	Design Criteria
Package complete treatment	5 gpm/sq ft (12.2 m/h) filtration rate; mixed-media filters; two backwashes/day
Chlorine feed—gas chlorine	3 mg/l design; 1.5 mg/l operating
Product water pumping	160 ft (48.8 m) TDH

Figure 30–7 presents capital cost and operation and maintenance cost, with the latter based on operation at flows equivalent to 20, 55, and 95 percent of the plant design capacity.

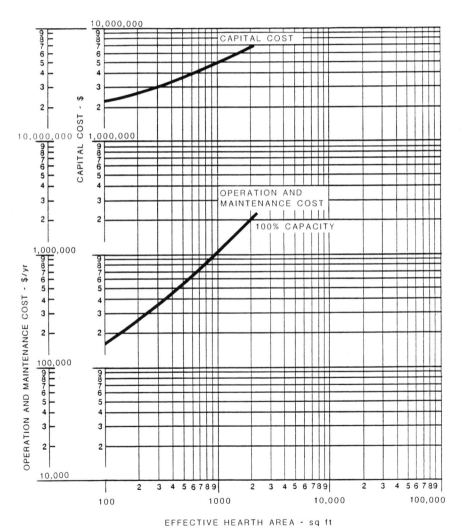

Fig. 30–6. Multiple hearth granular carbon regeneration cost.

ALUM SLUDGE DISPOSAL

Four techniques are particularly applicable for handling and disposal of alum sludges:

- Disposal to sanitary sewer
- Lagooning, air drying, and haul to landfill
- Sand bed drying and haul to landfill
- Centrifuge dewatering and haul to landfill

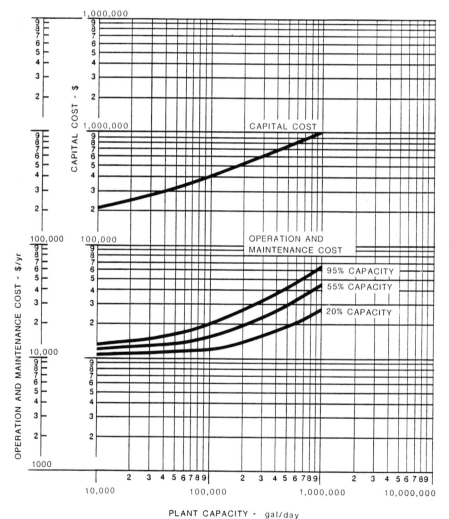

Fig. 30-7. Conventional package complete treatment costs.

These techniques were each discussed in detail in Chapter 23. The basic concepts used to develop costs for the techniques are presented in the following paragraphs.

Disposal to Sanitary Sewer

Alum sludge disposal to sanitary sewers is an effective technique if it is allowed by the agency operating the sewerage system. Figure 30–8 presents

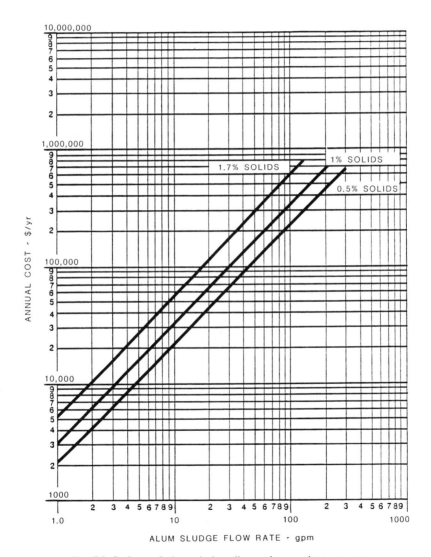

Fig. 30–8. Cost of alum sludge disposal to sanitary sewer.

costs for disposal of sludges with solids concentrations of 0.5, 1.0, and 1.7 percent. These costs are based upon a charge of $678/million gallons ($179/ Ml) discharged, $0.069/lb ($0.15/kg) of suspended solids discharged, and $0.082/lb ($0.181/kg) of BOD_5 discharged. The BOD_5 in alum sludge typically is low, and a concentration of 50 mg/l was assumed.

Lagooning, Air Drying, and Haul to Landfill

Lagooning of alum sludge is a frequently used technique despite the inability of alum sludge to thicken significantly. To produce a solids content high enough for hauling, thickened sludge is removed from the lagoon by clamshell, dragline, or pumping, followed by on-site air drying in a designated area to produce a sludge with approximately 20 percent solids.

The costs presented in Fig. 30–9 are based upon lagoons with 16 cu ft (0.45 m³) of storage per million gallons (3.785 Ml) treated and an 18-month detention time. Following sludge removal by pumping, the sludge is allowed to air-dry to 20 percent solids and is then loaded in trucks using a front-end loader. The one-way haul distance to landfill is 20 miles (32.2 km), and trucks used for hauling have a fuel consumption of 7 mpg (2.98 km/l). The costs do not include land.

Sand Bed Drying, Haul to Landfill

Dewatering to 25 percent solids was assumed on the drying beds. Sludge removal is by front-end loader, with a 20-mile (32.2 km) one-way haul to landfill. Depth of application on the drying beds is 2.5 feet (0.76 m), with eight applications per year. Liquid sludge applied to the drying beds has a solids content of 1.7 percent. Costs are presented in Fig. 30–10.

Centrifuge Dewatering and Haul to Landfill

Centrifuges fed with a 1.7 percent solids content alum sludge were assumed to produce a 15 percent solids sludge. Polymer addition is at a rate of 3 lb/ton (1.5 kg/metric ton) of dry sludge solids. The costs include hauling to a landfill 20 miles (32.2 km) one way, using trucks with a fuel consumption of 7 mpg (2.98 km/l). Costs are presented in Fig. 30–11.

LIME SLUDGE DISPOSAL

Four techniques are particularly applicable for handling and disposal of lime sludges. These techniques are:

- Lagooning and hauling to landfill
- Sand drying beds and hauling to landfill
- Thickening, centrifuging, and haul to landfill
- Recalcination

These techniques were discussed in detail in Chapter 23. The basic concepts used to develop the costs for these treatment techniques are discussed in the following paragraphs.

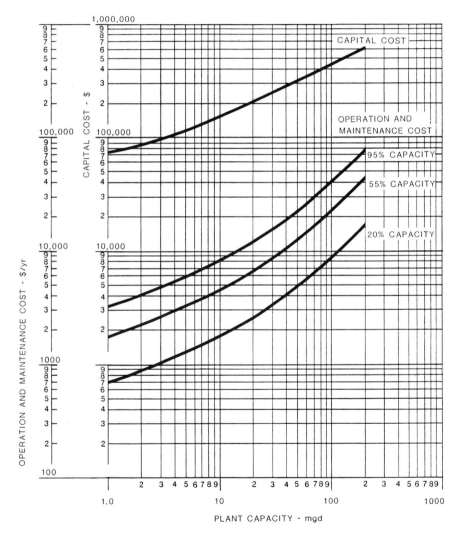

Fig. 30-9. Cost for alum sludge lagooning and hauling to landfill.

Lagooning, Air Drying, Hauling to Landfill

Lagooning is an effective method of concentrating lime sludge. The concept is normally restricted to use at plants less than 10 mgd (37.9 Ml/d) because of land area requirements. Concentration to 40 percent solids was assumed, with subsequent drying in the lagoon and removal by front-end loader and trucking to landfill.

The costs presented in Fig. 30–12 are based on removal of 150 mg/l of hardness, 85 cu ft (2.41 m³) of storage per million gallons (3.785 Ml) of

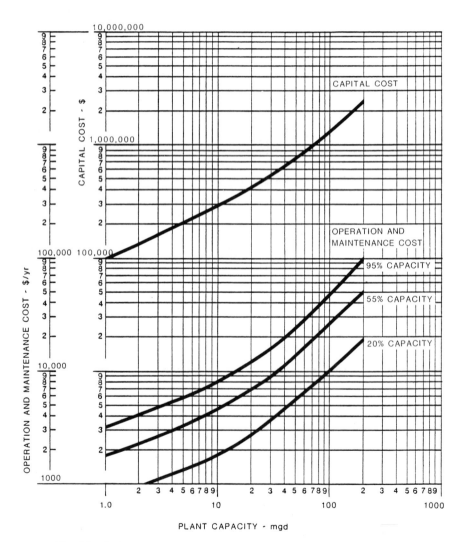

Fig. 30–10. Cost for alum sludge sand bed drying and haul to landfill.

water treated, and an 18-month sludge detention time. Air drying in place is used prior to removal and truck hauling to landfill. A one-way haul distance of 20 miles (32.2 km) and a diesel fuel consumption of 7 mpg (2.98 km/l) are used. The costs do not include land for the lagoons, or a tipping charge at the landfill.

Sand Drying Beds, Hauling to Landfill

Sand drying beds can be used to dewater lime sludges to 50 to 70 percent solids. This concept is applicable at smaller plants where sufficient land is

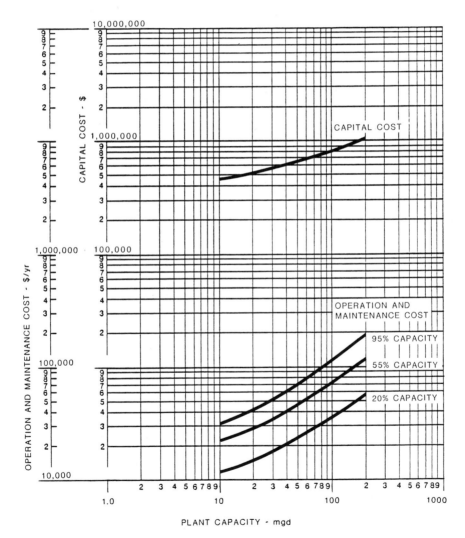

Fig. 30–11. Cost for alum sludge centrifuging and hauling to landfill.

available, and at larger plants where land is available and the concept is economically feasible. To size drying beds, a loading rate of 0.4 lb/sq ft/day (1.95 kg/m²/d) was used, and drying to 60 percent solids was assumed. Dried sludge is removed by front-end loader and hauled to a landfill, a distance of 20 miles (32.2 km) one way. Diesel fuel consumption is based on mileage of 7 mpg (2.98 km/l). Costs do not include land for the sand drying beds or a tipping charge at the landfill. Cost information is presented in Fig. 30–13.

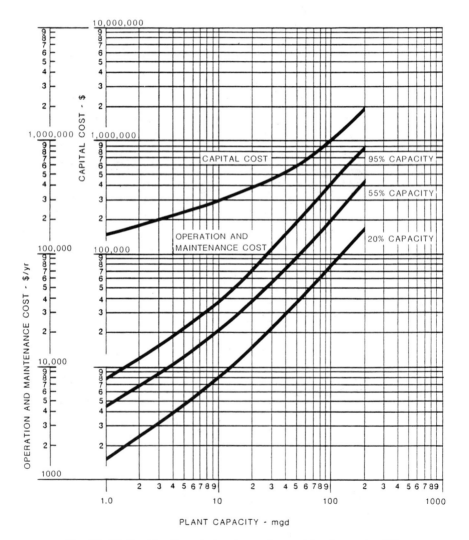

Fig. 30–12. Cost for lime sludge lagooning and hauling to landfill.

Thickening, Centrifuging, Hauling to Landfill

Gravity thickening followed by use of a solid bowl centrifuge can dewater lime sludge to 55 percent solids. This technique is particularly well suited to plants greater than 10 mgd. Development of costs was based on a thickener loading of 100 lb of dry solids/sq ft/day (488.7 kg/m²/d) and a thickener underflow concentration of 15 percent solids. The centrifuge produces a 55 percent solids cake that is hauled 20 miles (32.2 km), one way, to a landfill.

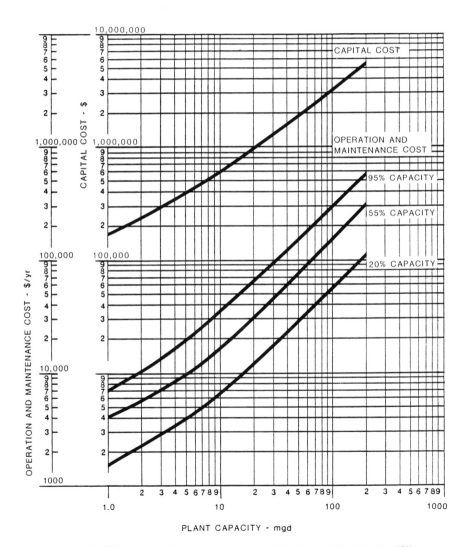

Fig. 30-13. Cost for lime sludge sand bed drying and haul to landfill.

Truck fuel consumption is 7 mpg (2.98 km/l). Landfill tipping charges are not included. Figure 30-14 presents cost information.

Thickening, Centrifuging, Lime Recalcination

At large plants where there is little land available and/or there is no nearby landfill or other disposal location for sludge, recalcination may be an effective solution. Costs for gravity thickening, centrifugation, and lime recalcination

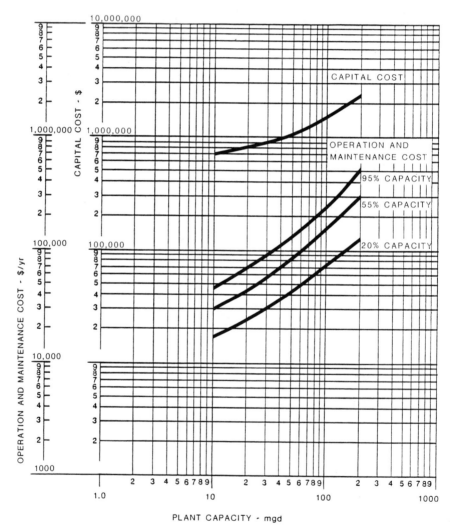

Fig. 30-14. Cost for lime sludge thickening, centrifuging, and hauling to landfill.

are presented in Fig. 30–15. Design criteria and performance for the gravity thickener and solid bowl centrifuge are as presented earlier. The recalcination furnace is sized on the basis of 7 lb of wet solids/sq ft of hearth area/hr (34.2 kg/m²/h), and includes an allowance of 20 percent downtime. It is assumed that recalcined sludge will be reused in the treatment operation. No cost is included for the disposal of excess recalcined sludge produced.

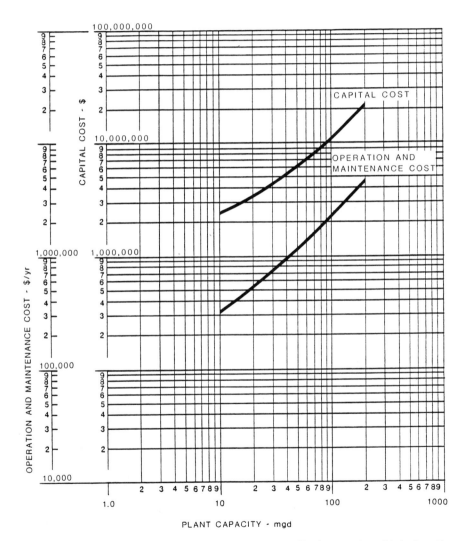

Fig. 30–15. Cost for lime sludge thickening, centrifuging, and multiple hearth recalcination.

TREATMENT MODIFICATIONS FOR TRIHALOMETHANE REDUCTION

A number of treatment modifications for trihalomethane (THM) reduction are discussed in Chapter 14. Among these modifications are use of chlorine dioxide and chloramines as alternate disinfectants, seasonal use of powdered activated carbon, and increased removal of THM precursors by increasing alum and/or polymer dosages. The following sections discuss these treatment modifications.

Use of Alternative Disinfectants

If an alternative to chlorine is being evaluated, one consideration is the current cost of chlorination. The costs of operating chlorination facilities at dosages of 1, 3, and 6 mg/l and operating capacities of 20 and 95 percent are shown in Fig. 30–16. Also shown in Fig. 30–16 are initial capital costs, although they would not be a factor in considering the use of an alternate disinfectant.

Chloramines. If chlorination facilities currently exist, the cost of chloramine formation would be only that required for ammonia feed facilities. Figure 30–17 presents capital and operation and maintenance costs at feed rates of 0.33, 1.0, and 2.0 mg/l and 20 and 95 percent plant utilization capacity.

If chlorination facilities do not exist, both chlorine and ammonia feed facilities would be required. These costs are shown in Fig. 30–18 for production of 0.75, 2.25, and 4.5 mg/l of chloramines at plants operating at 20 and 95 percent capacity.

Chlorine Dioxide. Chlorine dioxide is generated by combination of chlorine and sodium chlorite. Capital and operation and maintenance costs for chlorine dioxide generation at rates of 1 and 3 mg/l and 20 and 95 percent plant utilization are shown in Fig. 30–19. If chlorination facilities already exist, the capital costs can be reduced by approximately 40 percent.

Seasonal Use of Powdered Activated Carbon

Powdered activated carbon (PAC) may be fed seasonally to increase the removal of THM precursor material. Capital and operation and maintenance costs for 5- and 15-mg/l dosages and 20 and 95 percent plant utilization are shown in Fig. 30–20. As the operation and maintenance costs are based upon usage 100 percent of the year, adjustment must be made for the percentage of the year that PAC is actually utilized.

Increased Alum or Polymer Dosages

Increased dosages may be used seasonally (or throughout the year) when the concentration of THM precursor material is high. Generally, there will be no addition to the feed facilities, and the only change in cost will be for increased chemical usage. This cost can be calculated directly, with the incremental increase in the dosage, the plant flow, and the cost of alum and polymer known. No figure is presented for these costs because of the simplicity of the calculation.

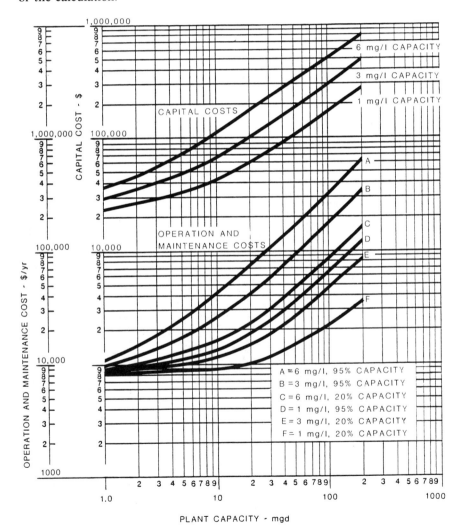

Fig. 30-16. Chlorination cost.

NEW RESERVOIR CONSTRUCTION

Below-Ground-Level Reinforced Concrete Storage

This type of reservoir is used commonly for clearwell storage at water treatment plants. It also is used frequently for distribution storage. Construction involves the highest capital costs, the longest service life, and perhaps the lowest maintenance costs of the various choices available. Table 30–1 gives

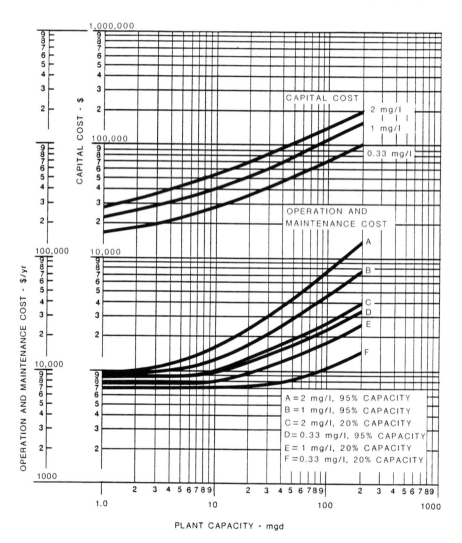

Fig. 30–17. Ammonia feed facilities cost.

the estimated cost for construction of below-ground reinforced concrete storage reservoirs. The figures are for rough preliminary estimates only. Actual construction costs will vary considerably, based on differing local conditions.

Ground-Level Steel Storage Tanks

Table 30–2 lists the estimated construction costs for ground-level steel storage tanks. Again these figures are intended for use only for preliminary planning

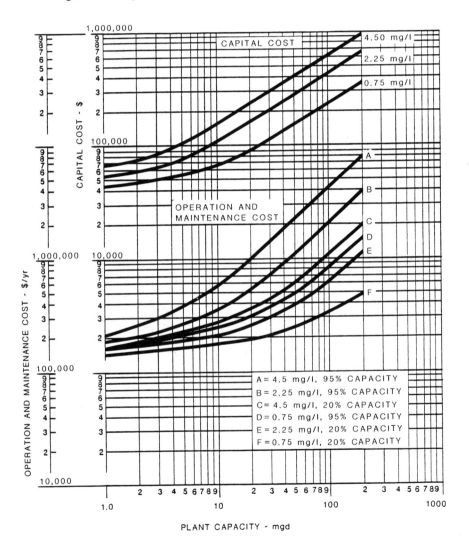

Fig. 30–18. Chloramine production cost.

purposes. They should be refined to take prevailing local conditions into account.

Wash Water Storage for Treatment Plants

This storage may be provided in the water treatment plant underground clearwell for supplying wash water pumps, or it may be provided in a separate wash water tank for a gravity supply. Customarily, a minimum of 15 feet

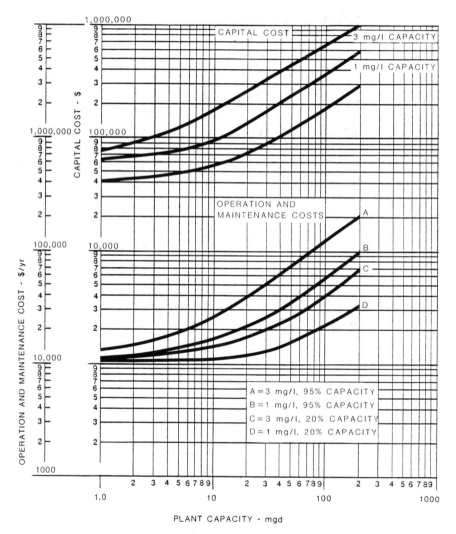

Fig. 30–19. Chlorine dioxide generation and feed cost.

(4.6 m) is provided between the top of the filter wash trough and the bottom of the storage tank, although the actual minimum elevation difference should be determined from calculation of the wash system hydraulics for each particular set of local conditions. Because of the low head difference required, it is often possible to utilize favorable topography for installation of a ground-level wash water storage tank of the type illustrated in Fig. 30–21.

For preliminary planning purposes, the costs given in Table 30–2 may be used for ground-level wash water storage tanks.

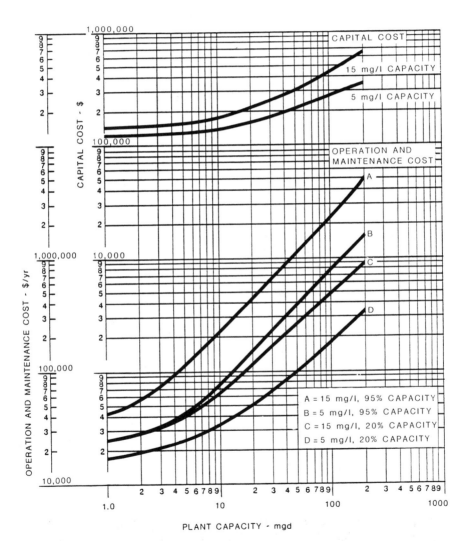

Fig. 30–20. Powdered activated carbon feed cost.

Table 30–1. Estimated Construction Costs at 1983 Prices, Below-Ground Reinforced Concrete Storage Reservoirs.

CAPACITY		ESTIMATED 1983 COSTS		
GALLONS	LITERS	TOTAL	PER GALLON OF STORAGE	PER LITER OF STORAGE
10,000	37,900	$ 66,000	$6.60	$1.74
50,000	189,300	108,000	2.16	0.57
100,000	378,500	184,000	1.84	0.49
500,000	1,892,500	440,000	0.88	0.23
1,000,000	3,785,000	822,000	0.82	0.22
7,500,000	28,387,500	3,240,000	0.43	0.11

Table 30–2. Estimated Construction Costs at 1983 Prices, Ground-Level Steel Storage Reservoirs.

CAPACITY		ESTIMATED 1983 COSTS		
GALLONS	LITERS	TOTAL	PER GALLON OF STORAGE	PER LITER OF STORAGE
10,000	37,900	$ 59,000	$5.90	$1.56
50,000	189,300	100,000	2.00	0.53
100,000	378,500	143,000	1.43	0.38
500,000	1,892,500	269,000	0.54	0.14
1,000,000	3,785,000	464,000	0.46	0.12
7,500,000	28,387,500	2,000,000	0.27	0.07

Covering Existing Reservoirs

The various methods for covering existing open reservoirs were discussed in Chapter 26. Costs for the methods that might be used for this purpose vary even more widely than does the cost of all-new reservoir construction, depending upon local conditions. Some factors influencing capital costs include:

- Type of existing construction.
- Foundation conditions.
- Length of time, if any, that reservoir can be taken out of service.
- Snow, wind, and seismic loads on cover.
- Reservoir shape.
- Other local factors.

To give an approximate idea of costs that may be involved in the different cover systems, Table 30–3 has been prepared. It gives preliminary estimated cost ranges and the estimated service life of each type of cover.

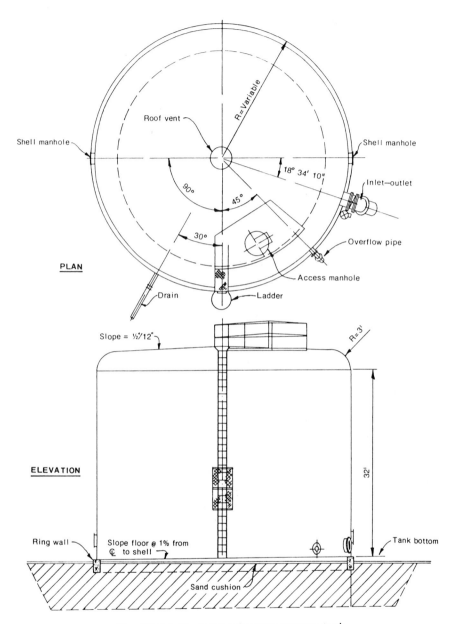

Fig. 30–21. Typical wash water storage tank.

Table 30–3. Estimated Cost Ranges to Cover Existing Reservoirs at 1983 Prices.

| TYPE OF COVER | ESTIMATED COST, 1983 | | ESTIMATED SERVICE LIFE, YEARS |
	$ PER SQ FT	$ PER m³	
Air-supported fabric	$ 1.50 to $ 4.00	$ 16.10 to $ 43.10	5 to 10
Floating, Hypalon	$ 1.75 to $ 5.00	$ 18.90 to $ 53.80	15 to 30
Wood	$ 9.00 to $16.00	$ 96.90 to $172.20	25 to 50
Steel	$10.00 to $15.00	$107.70 to $161.50	30 to 60
Concrete	$15.00 to $20.00	$161.50 to $215.30	50 to 100
Aluminum dome	$10.00 to $25.00	$107.70 to $269.10	50 to 100

PACKAGE PLANT COSTS

Cost estimates for water treatment facilities designed around package plants have been presented by several authors.[3,4] Both of the discussions cited dealt with cost comparisons between package and conventional treatment plants. The comparative cost analysis of a 1 mgd (3.8 Ml/d) plant presented in Table 30–4 indicated that a 50 percent savings would be realized by using a package plant over a conventional plant.[4] The construction cost of the conventional plant was estimated at $1.12 million (1978 cost) versus $488,000 for the package plant. Both studies concluded that package plants have lower construction and operating costs than conventional plants at capacities up to 2 mgd (7.6 Ml/d).

The simplicity of installation is reflected in the economical construction costs of a number of package plant installations. In 1983, a 1 mgd (3.8 Ml/d) Type B package plant was installed in Georgia to treat water from the Chattahoochee River. The installation included a floating raw water intake structure, raw water pumps, package treatment equipment with automatic chemical dosage control, a 2,800-square-foot (260 m²) prefabricated steel building on a slab foundation, a 93,000-gallon (353-kl) steel storage tank,

Table 30–4. Comparative Cost Analysis for 1 mgd (3.8 Ml/d) Plant.

TYPE OF COST	COST ESTIMATE FOR CONVENTIONAL PLANT $1,000	COST ESTIMATE FOR PACKAGE PLANT $1,000
Construction cost	1,120	488
Annual treatment, operation, and maintenance costs	63	40

and a package high service pump station. The installed cost of this facility was less than $450,000.

A 200 gpm (12.6 l/s) package plant was installed in 1981 at a state park in Oklahoma. The total construction cost of this facility was $304,000. The plant included a floating raw water intake structure with two pumps, a 36 × 40-foot (11 × 12.2-m) steel building, a Type B package plant, a 22,000-gallon (83.3-kl) FRP storage tank, and a package high service pump station.

A small community in Tennessee replaced an obsolete and inadequate 25-year-old conventional treatment plant with a 1 mgd (3.8 Ml/d) Type B plant during 1980. The plant treats river water from a reservoir that averages 6 to 8 NTU. The plant is operated 10 to 11 hours per day, and the operator is always in attendance, which is a state requirement. Filter effluent turbidity is continuously monitored and recorded, and never allowed to exceed 0.2 NTU. Filter runs range between 80 and 100 hours. The complete facility excluding the raw water intake structure, pumps, inlet pipeline, finished water, and storage was installed for $499,000.

A 100 gpm (6.3 l/s) package plant was recently placed in service at a central Oregon coast community. The raw water is obtained from a small creek; turbidity and color removal are the major treatment requirements. The plant includes a raw water intake structure, raw water pumps, and a package treatment unit housed in a 45 × 32-foot (13.7 × 9.8-m) concrete block building. The building is designed to accommodate a future 100 gpm (6.3 l/s) treatment unit and has space for laboratory facilities, an office area, and a chemical storage area. Filtered water is stored in a below-grade clearwell, and treated water is distributed to the system by high service pumps taking suction from the clearwell. The total construction cost of this facility was $202,000.

REFERENCES

1. Gumerman, R. C., Culp, R. L., Hansen, S. P., and Lineck, T. S., *Estimating Water Treatment Costs,* Vols. 1–4, EPA-600/2-79-162a, 162b, 162c, and 162d, U.S. Environmental Protection Agency, Cincinnati, Ohio, 1979.
2. Gumerman, R. C., Burris, B. E., and Hansen, S. P., *Estimating Small Water System Treatment Costs,* in publication, U.S. Environmental Protection Agency, Cincinnati, Ohio, 1984.
3. Hansen, Sigurd P., "Package Plants: One Solution to Small Community Water Supply Needs," *J.AWWA* 71:6:315, June, 1979.
4. Clark, Robert M., and Morand, James M., "Package Plants: A Cost-Effective Solution to Small Water System Treatment Needs," *J.AWWA* 73:1:24, January, 1981.

Appendix A
Conversion Factors

TO CONVERT	MULTIPLY BY	TO OBTAIN

A

TO CONVERT	MULTIPLY BY	TO OBTAIN
acres	4.047×10^{-1}	hectares or sq. hectometers
acres	4.35×10^{4}	sq. feet
acres	4.047×10^{3}	sq. meters
acres	1.562×10^{-3}	sq. miles
acres	4.840×10^{3}	sq. yards
acre-feet	4.356×10^{4}	cu. feet
acre-feet	3.259×10^{5}	gallons
atmospheres	7.348×10^{-3}	tons/sq. in.
atmospheres	1.058	tons/sq. ft.
atmospheres	7.6×10^{1}	cms. of mercury (at 0°C)
atmospheres	3.39×10^{1}	ft. of water (at 4°C)
atmoshperes	2.992×10^{1}	in. of mercury (at 0°C)
atmospheres	1.0333	kgs./sq. cm.
atmospheres	1.47×10^{1}	pounds/sq. in.

B

TO CONVERT	MULTIPLY BY	TO OBTAIN
barrels (U.S. dry)	7.056×10^{3}	cu. inches
barrels (U.S. dry)	1.05×10^{2}	quarts (dry)
barrels (U.S. liquid)	3.15×10^{1}	gallons
barrels (oil)	4.2×10^{1}	gallons (oil)
Btu	7.7816×10^{2}	ft.-lbs.
Btu	2.52×10^{2}	gram-calories
Btu	3.927×10^{-4}	horsepower-hours
Btu	1.055×10^{3}	joules
Btu	2.928×10^{-4}	kilowatt-hours
Btu/hr.	2.162×10^{-1}	ft.-lbs./sec.
Btu/hr.	7.0×10^{-2}	gram-cal./sec.
Btu/hr.	3.929×10^{-4}	horsepower
Btu/hr.	2.931×10^{-1}	watts
Btu/min.	1.296×10^{1}	ft.-lbs./sec.
Btu/min.	2.356×10^{-2}	horsepower
Btu/min.	1.757×10^{1}	watts

C

TO CONVERT	MULTIPLY BY	TO OBTAIN
calories, gram (mean)	3.9685×10^{-3}	Btu (mean)
centigrade (degrees)	(°C x 9/5) +32	fahrenheit (degrees)
centigrade (degrees)	°C + 273.18	kelvin (degrees)
centimeters	3.281×10^{-2}	feet
centimeters	3.937×10^{-1}	inches
centimeters	$1. \times 10^{4}$	microns
centimeters	$1. \times 10^{8}$	angstrom units
centimeters of mercury	1.316×10^{-2}	atmospheres
centimeters of mercury	4.461×10^{-1}	feet of water
centimeters/sec.	1.969	feet/min.
centimeters/sec.	3.281×10^{-2}	feet/sec.
centimeters/sec.	3.6×10^{-2}	kilometers/hr.

TO CONVERT	MULTIPLY BY	TO OBTAIN

C Cont.

centipoise	1.0×10^{-2}	gr./cm.-sec.
centipoise	6.72×10^{-4}	pound/ft.-sec.
centipoise	2.4	pound/ft.-hr.
cubic centimeters	3.531×10^{-5}	cu. feet
cubic centimeters	6.102×10^{-2}	cu. inches
cubic centimeters	2.642×10^{-4}	gallons (U.S. liquid)
cubic feet	2.8320×10^{4}	cu. cms.
cubic feet	1.728×10^{3}	cu. inches
cubic feet	7.48052	gallons (U.S. liquid)
cubic feet	2.832×10^{1}	liters
cubic feet/min.	1.247×10^{-1}	gallons/sec.
cubic feet/min.	4.720×10^{-1}	liters/sec.
cubic feet/sec.	6.46317×10^{-1}	million gals./day
cubic feet/sec.	4.48831×10^{2}	gallons/min.
cubic inches	1.639×10^{1}	cu. cms.
cubic inches	4.329×10^{-3}	gallons
cubic inches	1.639×10^{-2}	liters
cubic meters	3.531×10^{1}	cu. feet
cubic meters	1.308	cu. yards
cubic meters	2.642×10^{2}	gallons (U.S. liquid)
cubic yards	7.646×10^{5}	cu. cms.
cubic yards	2.7×10^{1}	cu. feet
cubic yards	2.02×10^{2}	gallons (U.S. liquid)
cubic yards	7.646×10^{2}	liters

F

fathoms	6.0	feet
feet	3.048×10^{-1}	meters
feet	1.894×10^{-4}	miles (statute)
feet of water	2.95×10^{-2}	atmospheres
feet of water	8.826×10^{-1}	in. of mercury
feet of water	4.335×10^{-1}	pounds/sq. in.
feet/min.	5.080×10^{-1}	cms./sec.
feet/min.	3.048×10^{-1}	meters/min.
feet/min.	1.136×10^{-2}	miles/hr.
feet/sec.	3.048×10^{1}	cms./sec.
feet/sec.	1.829×10^{1}	meters/min.
feet/sec.	6.818×10^{-1}	miles/hr.
feet/sec./sec.	3.048×10^{1}	cms./sec./sec.
foot-pounds	1.286×10^{-3}	Btu
foot-pounds	3.241×10^{-1}	gram-calories
foot-pounds	1.356	joules
foot-pounds	3.766×10^{-7}	kilowatt-hrs.
foot-pounds/min.	1.286×10^{-3}	Btu/min.
foot-pounds/min.	3.030×10^{-5}	horsepower

TO CONVERT	MULTIPLY BY	TO OBTAIN

F Cont.

TO CONVERT	MULTIPLY BY	TO OBTAIN
foot-pounds/min.	2.260×10^{-5}	kilowatts
foot-pounds/sec.	1.818×10^{-3}	horsepower
foot-pounds/sec.	1.356×10^{-3}	kilowatts

G

TO CONVERT	MULTIPLY BY	TO OBTAIN
gallons	3.785×10^{3}	cu. cms.
gallons	1.337×10^{-1}	cu. feet
gallons	3.785	liters
gallons (liq.br.imp.)	1.20095	gallons(U.S. liquid)
gallons of water	8.337	pounds of water
gallons/min.	2.228×10^{-3}	cu. feet/sec.
gallons/min.	6.308×10^{-2}	liters/sec.
grains (troy)	6.48×10^{-2}	grams
grains (troy)	2.0833×10^{-3}	ounces (avdp.)
grains/U.S. gallon	1.7118×10^{1}	parts/million
grains/U.S. gallon	1.4286×10^{2}	pounds/millon gallons
grams	9.807×10^{-3}	joules/meter (newtons)
grams	3.215×10^{-2}	ounces (troy)
grams	2.205×10^{-3}	pounds
grams/cm.	5.6×10^{-3}	pounds/in.
grams/cu. cm.	6.243×10^{1}	pounds/cu. ft.
grams/liter	8.345	pounds/1,000 gal.
grams/liter	6.2427×10^{-2}	pounds/cu. ft.
grams/sq. cm.	2.0481	pounds/sq. ft.
gram-calories	3.9683×10^{-3}	Btu
gram-calories	3.086	ft.-lbs.
gram-calories/sec.	1.4286×10^{1}	Btu/hr.
gram-centimeters	9.297×10^{-4}	Btu

H

TO CONVERT	MULTIPLY BY	TO OBTAIN
horsepower	4.244×10^{1}	Btu/min.
horsepower	3.3×10^{4}	ft.-lbs./min.
horsepower	5.50×10^{2}	ft.-lbs./sec.
horsepower	7.457×10^{-1}	kilowatts
horsepower-hours	2.547×10^{3}	Btu
horsepower-hours	1.98×10^{6}	ft.-lbs.
horsepower-hours	7.457×10^{-1}	kilowatt-hrs.

I

TO CONVERT	MULTIPLY BY	TO OBTAIN
inches	2.540	centimeters
inches	1.0×10^{3}	mils
inches of mercury	3.342×10^{-2}	atmospheres
inches of mercury	1.133	feet of water

TO CONVERT	MULTIPLY BY	TO OBTAIN

I Cont.

inches of mercury	4.912×10^{-1}	pounds/sq. in.
in. of water (at 4°C)	2.458×10^{-3}	atmospheres
in. of water (at 4°C)	7.355×10^{-2}	inches of mercury
in. of water (at 4°C)	3.613×10^{-2}	pounds/sq. in.

J

joules	9.486×10^{-4}	Btu
joules	7.736×10^{-1}	ft.-lbs.
joules	2.389×10^{-4}	kg.-calories
joules/cm.	2.248×10^{1}	pounds

K

kilograms	9.807	joules/meter (newtons)
kilograms	2.2046	pounds
kilograms/cu. meter	6.243×10^{-2}	pounds/cu. ft.
kilograms/meter	6.72×10^{-1}	pounds/ft.
kilograms/sq. cm.	9.678×10^{-1}	atmospheres
kilograms/sq. cm.	3.281×10^{1}	feet of water
kilograms/sq. cm.	2.896×10^{1}	inches of mercury
kilometers	3.281×10^{3}	feet
kilometers	6.214×10^{-1}	miles (statute)
kilometers/hr.	2.778×10^{1}	cms./sec.
kilometers/hr.	5.468×10^{1}	feet/min.
kilowatts	5.692×10^{1}	Btu/min.
kilowatts	4.426×10^{4}	ft.-lbs./min.
kilowatts	7.376×10^{2}	ft.-lbs./sec.
kilowatts	1.341	horsepower
kilowatt-hrs.	3.413×10^{3}	Btu
kilowatt-hrs.	2.655×10^{6}	ft.-lbs.
kilowatt-hrs.	1.341	horsepower-hours

L

light year	5.9×10^{12}	miles
liters	3.531×10^{-2}	cu. feet
liters	6.102×10^{1}	cu. inches
liters	2.642×10^{-1}	gallons (U.S. liquid)
liters/min.	5.886×10^{-4}	cu. ft./sec.
liters/min.	4.403×10^{-3}	gals./sec.

M

meters	3.281	feet
meters	6.214×10^{-4}	miles (statute)
meters	1.094	yards
meters/min.	3.281	feet/min.

TO CONVERT	MULTIPLY BY	TO OBTAIN

M Cont.

TO CONVERT	MULTIPLY BY	TO OBTAIN
meters/min.	5.468×10^{-2}	feet/sec.
meters/min.	3.728×10^{-2}	miles/hr.
meters/sec.	2.237	miles/hr.
micromicrons	1.0×10^{-12}	meters
microns	1.0×10^{-6}	meters
miles (statute)	5.280×10^{3}	feet
miles (statute)	1.609	kilometers
miles (statute)	1.760×10^{3}	yards
miles (statute)	1.69×10^{-13}	light years
miles/hr.	4.470×10^{1}	cms./sec.
miles/hr.	8.8×10^{1}	ft./min.
miles/hr.	1.467	ft./sec.
miles/hr.	1.6093	kms./hr.
milligrams	1.5432×10^{-2}	grains
milligrams/liter	1.0	parts/million
millimeters	3.937×10^{-2}	inches
million gals./day	1.54723	cu. ft./sec.
mils	1.0×10^{-3}	inches
mils	2.54×10^{-8}	kilometers

N

TO CONVERT	MULTIPLY BY	TO OBTAIN
newtons	1.0×10^{5}	dynes

O

TO CONVERT	MULTIPLY BY	TO OBTAIN
ounces	4.375×10^{2}	grains
ounces	2.8349×10^{1}	grams
ounces (fluid)	2.957×10^{-2}	liters

P

TO CONVERT	MULTIPLY BY	TO OBTAIN
parts/million	5.84×10^{-2}	grains/U.S. gal.
parts/million	7.016×10^{-2}	grains/imp. gal.
parts/million	8.345	pounds/million gal.
pints (liquid)	4.732×10^{2}	cu. cms.
pints (liquid)	1.671×10^{-2}	cu. feet
pounds	4.448×10^{5}	dynes
pounds	7.0×10^{3}	grains
pounds	4.5359×10^{2}	grams
pounds	4.448	joules/meter (newtons)
pounds	4.536×10^{-1}	kilograms
pounds of water/min.	2.670×10^{-4}	cu. ft./sec.
pound-feet	1.3825×10^{4}	cm.-grams
pounds/cu. ft.	1.602×10^{1}	kgs./cu. meters
pounds/cu. in.	2.768×10^{1}	grams/cu. cms.
pounds/cu. in.	1.728×10^{3}	pounds/cu. ft.

TO CONVERT	MULTIPLY BY	TO OBTAIN

P Cont.

pounds/ft.	1.488	kgs./meter
pounds/sq. ft.	4.725×10^{-4}	atmospheres
pounds/sq. ft.	1.602×10^{-1}	feet of water
pounds/sq. ft.	1.414×10^{-2}	inches of mercury
pounds/sq. ft.	4.882	kgs./sq. meter
pounds/sq. ft.	6.944×10^{-3}	pounds/sq. in.
pounds/sq. in.	6.804×10^{-2}	atmospheres
pounds/sq. in.	2.307	feet of water
pounds/sq. in.	2.036	inches of mercury
pounds/sq. in.	7.031×10^{2}	kgs./sq. meters
pounds/sq. in.	1.44×10^{2}	pounds/sq. ft.
pounds/sq. in.	7.03×10^{-2}	kgs./sq. cms.

R

radians	5.7296×10^{1}	degrees
radians	3.438×10^{3}	minutes
radians/sec.	1.592×10^{-1}	revolution/sec.
radians/sec./sec.	1.592×10^{-1}	revs./sec./sec.
revolutions	6.283	radians

S

seconds (angle)	4.848×10^{-6}	radians
slugs	1.459×10^{1}	kilograms
slugs	3.217×10^{1}	pounds
square centimeters	1.550×10^{-1}	sq. inches
square feet	2.296×10^{-5}	acres
square feet	1.44×10^{2}	sq. inches
square feet	9.29×10^{-2}	sq. meters
square inches	6.452	sq. cms.
square kilometers	2.471×10^{2}	acres
square kilometers	1.076×10^{7}	sq. feet
square kilometers	3.861×10^{-1}	sq. miles
square meters	2.471×10^{-4}	acres
square meters	1.076×10^{1}	sq. feet
square meters	1.196	sq. yards
square yards	1.296×10^{3}	sq. inches
square yards	8.361×10^{-1}	sq. meters

T

temperature (°C)+273	1.0	absolute temperature (°K)
temperature (°F)+460	1.0	absolute temperature (°R)
temperature (°F)-32	5/9	temperature (°C)

1034

TO CONVERT	MULTIPLY BY	TO OBTAIN
	W	
watts	3.4129	Btu/hr.
watts	4.427×10^1	ft.-lbs./min.
watts	1.341×10^{-3}	horsepower
watts (abs.)	1.0	joules/sec.
watt-hours	3.413	Btu
watt-hours	2.656×10^3	ft.-lbs.
watt-hours	1.341×10^{-3}	horsepower-hours
watt-hours	8.605×10^{-1}	kilogram-calories
watt-hours	3.672×10^2	kilogram-meters
	Y	
yards	9.144×10^1	centimeters
yards	9.144×10^{-1}	meters
yards	4.934×10^{-4}	miles (nautical)
yards	5.682×10^{-4}	miles (statute)
yards	9.144×10^2	millimeters

USEFUL PHYSICAL CONSTANTS

Gas Constants (R)

$R = 1.987$ g.-cal./(g.-mole) ($°K$)
$R = 1.987$ Btu/(lb.-mole) ($°R$)
$R = 8.314$ joules/(gm-mole) ($°K$)
$R = 1.546$ (ft.) (lb. force)/(lb.-mole) ($°R$)

ACCELERATION OF GRAVITY (STANDARD)
$g = 32.17$ ft./sec.^{2} = 980.6 cm./sec.^{2}

HEAT OF FUSION OF WATER
79.7 cal./gm = 144 Btu/lb.

HEAT OF VAPORIZATION OF WATER @ 1.0 atm.
540 cal./gm = 970 Btu/lb.

SPECIFIC HEAT OF AIR
$Cp = 0.238$ cal./(gm) ($°C$)

Appendix B
Chemicals Used in Treatment
of Water and Wastewater

This Technical Bulletin contains pertinent data and information of interest to design engineers and supervisors of operations in water and wastewater treatment plants where chemicals are used in treatment processes.

The data presented herein has been compiled from various published sources and the files of B I F, and represent current information on 75 chemicals used in water purification and sewage waste treatment.

Note: Some chemicals can be hazardous-consult with your supervisor for proper safety measures.

SYMBOLS AND ABBREVIATIONS

A number of abbreviations and symbols occur in the accompanying tabulations, as follows:

Activated	Act.	Grains per gal.	gpg
Amount	Amt.	Gravity	Grav.
Approximate	Approx.	Hour	Hr
Aqueous	Aq.	Industrial	Ind.
Available	Avail.	Insoluble	Insol.
Average	Avg.	Less Than Carload	LC/L
Barrel	Bbl.	Maximum	Max.
Carload	C/L	Melting Point	M.P.
Composition	Comp.	Milliliter	ml.
Concentrated	Concd.	Minimum	Min.
Concentration	Conc.	Minute	min.
Cubic Centimeter	cc	Molecular Weight	M.W.
		Packages	Pkgs.
Cubic Foot	cu ft	Parts per million	ppm
Degrees Baume	°Be	Polyvinyl chloride	PVC
Degrees Centigrade	°C	Pound	lb
		Pounds per gallon	lb/gal
Degrees Fahrenheit	°F	Pounds per sq. in.	psi
Dilution	Diln.	Proportioning pump	Proportioneer
Especially	esp.	Saturated	Sat.
Fiberglas Rein. Polyester	FRP	Solution	Soln.
		Specific Gravity	Sp. G.
Gallon	gal	Standard	Std.
Gal. per hr.	gph	Tank Car	T/C
Gal. per min.	gpm	Tank Truck	T/T
Gram	gm	Weight	Wt.

USE OF TABLE

The left hand page of the following table gives information on the various chemicals, their formulas, common names, most common uses, forms in which they can be obtained, sizes of shipping containers and characteristics including appearance and properties, bulk density, commercial strength and solubilities.

The right hand page of the table gives information on recommendations for feeding these chemicals with particular reference to the best form for feeding, the amount of water required for continuous dissolving, types of feeders, accessory equipment required, and suitable materials for handling both the dry chemicals and solutions.

For the most part, the column headings and the entries are self-explanatory. Under "Characteristics", "Weight (lb/cu ft)" is bulk density of the dry chemical, except as noted for liquids and gases. Under the "Feeding Recommendations", "Best Feeding Form" refers to physical form bulk density, mesh, specific gravity or solution as indicated in each case.

The column headed "Chemical to Water Ratio for Continuous Dissolving" refers to dry chemical feeders with dissolving chambers, except where solution feed is used and the data indicates strength or concentration of solution to be used or fed.

SOLUBILITIES

Solubilities are given according to standard Chemical Handbooks in grams per 100 milliliters (gm/100ml) and in general are for the pure chemical compound. Solubilities are generally given at four different temperatures stated in degrees Centigrade. These temperatures and their Fahrenheit equivalents are:

$$0°C . . . . .32°F$$
$$10°C . . . . .50°F$$
$$20°C . . . . .68°F$$
$$30°C . . . . .86°F$$

To convert gm/100 ml to lb/gal, multiply figure (for gm/100 ml) by 0.083. Recommended strengths of solutions for feeding purposes are given in pounds of chemical per gallon of water (lb/gal) and are based on plant practice for the commercial product.

The following table shows the number of pounds of chemical to add to one gallon of water to obtain various per cent solutions:

%Soln.	lb/gal	%Soln.	lb/gal	%Soln	lb/gal
0.1	0.008	2.0	0.170	10.0	0.927
0.2	0.017	3.0	0.258	15.0	1.473
0.5	0.042	5.0	0.440	20.0	2.200
1.0	0.084	6.0	0.533	25.0	2.760
				30.0	3.560

TABLE OF CHEMICAL USES

On the back page of this Technical Bulletin appears a Table of Chemical Uses showing where these 75 chemicals are used in water treatment processes.

BIF a Unit of General Signal. Reproduced with permission.

CHEMICAL	SHIPPING DATA		CHARACTERISTICS			
FORMULA COMMON NAME USE	GRADES OR AVAILABLE FORMS	CONTAINERS AND REQUIREMENTS	APPEARANCE AND PROPERTIES	WEIGHT lb/cu ft. (Bulk Density)	COMMERCIAL STRENGTH	SOLUBILITY IN WATER gm/100 ml
ACTIVATED CARBON (Nuchar, Norit, Darco, Carbodur) Decolorizing, Taste and odor Removal Dosage between 5 to 80 ppm	Powder Granules	Bags - 35 lb (3x21x39 in) Drums - 5 lb & 25 lb Bulk - C/L	Black powder, about 400 mesh Dusty, smoulders if ignited Arches in hoppers; floodable(1) Do not mix with $KMnO_4$, hypochlorite or CaO; pH varies, usually weakly acidic	Powder 8 to 28 Avg. 12	10% C (Bone charcoal) to 90% C (Wood charcoal)	Insoluble - forms a slurry
ACTIVATED SILICA SiO_2 (Silica Soln.) Coagulation Aid Dosage - 1 to 20 ppm Raw H_2O (5-25 ppm-Waste)	Produced at site as needed from sodium silicate and activating agent	—	Clear often opalescent Na_2SiO_3 may be activated by alum, ammonium sulfate, chlorine, carbon dioxide. Sodium bicarbonate, sulfuric acid or Na_2SiF_6	41°Be sod. silicate Approx. 11.6 lb/gal	41°Be sodium silicate is used and diluted to 1% SiO_2 before use as coag. aid	Act. Silica is a colloidal soln. will gel at high conc.
ALUMINUM AMMONIUM SULFATE $Al_2(SO_4)_3 \cdot (NH_4)_2SO_4 \cdot 24H_2O$ (Ammonia Alum) Coagulation, especially with pressure filters	Lump Granular Ground Powder	Fiber Drums Kegs - 100-400 lb Bags - 100 lb Bulk	Colorless crystals Strong astringent taste 1% Soln. -pH 3.5	60 to 70	11% Al_2O_3	3.9 at 0°C 9.5 at 10°C 15.1 at 20°C 20.0 at 30°C
ALUMINUM CHLORIDE $AlCl_3 6H_2O$ Sludge Conditioner	Solution 15 to 35%	Carboys Rubber lined tank trucks	Clear pale yellow liquid Aq. Soln. is acid Always fed in solution.	Sp. Gr. 1.5 to 1.16	32°Be or 35% $AlCl_3$	Completely Miscible
ALUMINUM POTASSIUM SULFATE $Al_2(SO_4)_3 \cdot KaSO_4 \cdot 24H_2O$ (Potassium Alum) Coagulation	Lump Granular Rice Ground Powder	Bags - 100 lb Bbl. - 250 & 350 lb Drums - 100 lb & 350 lb Bulk - C/L	White crystals Low even solubility M.P. - about 200°F 1% Soln. pH 3.5 Above 60-65°C loses $9H_2O$ Crystals are efflorescent.	60 to 70	10 to 11% Al_2O_3	5.7 at 0°C 7.6 at 10°C 11.4 at 20°C 16.6 at 30°C
ALUMINUM SULFATE $Al_2(SO_4)_3 \cdot 14H_2O$ Alum, Filter Alum Coagulation at pH 5.5 to 8.0 Dosage between 0.5 to 9 gpg. Precipitate PO_4(90-110 ppm)	Lump Granular Rice Ground Powder	Bags - 100 lb & 200 lb Bbl. - 325 lb & 400 lb Drums - 25 lb, 100 lb & 250 lb Bulk - C/L	Light tan to gray-green Dusty, astringent Only slightly hygroscopic 1% Soln. - pH 3.4	60 to 75 (Powder is lighter) To calculate hopper capacities, use 60	98% plus or 17% Al_2O_3 (Min.)	72.5 at 0°C 78.0 at 10°C 87.3 at 20°C 101.6 at 30°C
ALUM $Al_2(SO_4)_3 \cdot XH_2O$ Liquid 1 gal 36°Be=5.38 lb of dry Alum - 60°F Coag. at pH 5.5 to 8.0 Sludge Conditioner Precipitate PO_4	Solution 32.2°Be to 37°Be	Manufactured near site; 6000-8000 gal steel T/C; 2000-4000 gal rubber lined steel tank trucks. High freight cost precludes distant shipment	Light green to light brown Soln. F.P. or crystallization point for: 35.97°Be= 4°F 36.95°Be= 27°F 37.7°Be= 60°F 1% Soln. - pH 3.4 Visc. 36°Be at 60°F= 25 cp.	36°Be Sp. G. = 1.33 or 11.1 lb/gal at 60°F	At 60°F 32.2°Be 7.25% Al_2O_3 35.97°Be 8.25% Al_2O_3 37°Be 8.5% Al_2O_3	Completely Miscible

becomes "fluidized" so that it will flow through small openings, like water.

1040

(2) Iron and steel can be used with chemicals in the dry state unless the chemical is deliquescent or very hygroscopic, or in a dampish form and thus corrosive to some degree.
(3) FRP, in every case, refers to the chemically resistant grade (bisphenol A+) of Fiberglass Rein. Polyester.

FEEDING RECOMMENDATIONS

BEST FEEDING FORM	CHEMICAL TO WATER RATIO FOR CONTINUOUS DISSOLVING	TYPE OF FEEDERS FOR	ACCESSORY EQUIPMENT REQUIRED	SUITABLE HANDLING MATERIALS FOR SOLUTIONS (2)
Powder - with bulk density of 12 lb/cu ft; Slurry - 1 lb/gal	According to its bulkiness and wetability, a 10 to 15% solution would be maximum concentration	Volumetric; Helix; Slurry; Rotodip; *Diaphragm Metering Pumps Hydraulic 1700 & Mechanical 1200	Washdown type wetting tank; Vortex mixer; Hopper agitators; Non-flood rotors; Dust collectors; Large storage cap. for liquid feed; Tank agitators; Transfer Pumps; If no surface mixer, need surface jets	316 ss, rubber, bronze, Monel, Hastelloy C, FRP(3) Saran, Hypalon, Titanium
0.6 to 1.0% Soln. to prevent gel formation Dilute for transport by educter or pump	Diluted "SiO2" to process. Max. (100 gal aging tank) Capacity Factor = 143; Alum 16.4 gpm; H_2SO_4 24.9 gpm; $(NH_2)SO_4$ 9.9 gpm; Cl_2 15.5 gpm; $NaHCO_3$ 9.9 gpm	Liquid (continuous) Silica activator or batch systems Diaphragm metering Pump - Series 1700 & 1200 & 5700 Rotodip	Flushing equipment Batch mixing tanks Storage or feeding tank Aging tank	Iron, steel, rubber, stainless steel, Viton, Hypalon (dilute), Tyril
Lump for pressure filters with solution pot. Ground for dry feeders	0.5 lb/gal	Usually fed in pot type feeders for pressure filters Gravimetric or volumetric feeders may be used - see aluminum sulfate feeders	Dissolvers	Lead, rubber, ceramic, PVC-1 Saran, FRP, Uscolite, Carp. 20 ss, Hypalon, Halar, Ti
As shipped. Solution does disassociate.	----	Solution Rotodip Diaphragm Metering Pumps Series 1200 & 1700 5700	Storage tanks Transfer pumps	Carp. 20 ss (10% at room temp.) (pits), rubber, FRP, PVC-1, Vinyl, Saran, Hastelloy B, Hypalon, ceramic, Teflon, Tyril, Uscolite, Ti (25%, Rm) Noryl, Derakane Halar, Polypropylene
Ground	0.5 lb/gal	Usually fed in pot type feeders for pressure filters Gravimetric or volumetric feeders may be used - see aluminum sulfate feeders	----	Lead-lined tanks, rubber, ceramic, FRP, Saran, Hypalon, Epoxy, Carp. 20 ss, Uscolite
Ground, granular or rice. Powder is dusty, arches and is floodable	0.5 lb/gal Dissolver detention time 5 minutes for ground (10 minutes for granules)	Volumetric Helix Solution Diaphragm metering pump Series 1200 & 1700 & 5700	Dissolver Mechanical mixer Scales for volumetric feeders Dust Collectors	Lead, rubber, FRP, PVC-1, 316 ss, Carp. 20 ss, Vinyl, Hypalon, Epoxy, Ni-Resist glass, ceramic, Polyethylene, Tyril, Uscolite, Ti, Noryl Derakane, Halar, Polypropylene
Full strenght under controlled temp. or dilute to avoid crystallization Minimize surface evap. - causes flow problems Keep % dry Album below 50% to avoid crystallization	Dilute to between 3 to 15% - according to application or conditions, mixing, etc.	Solution Rotodip Diaphragm Metering Pump Series 1200 & 1700 & 5700	Tank gauges or scales Transfer pumps Storage tank Temperature control Eductors or dissolvers for dilution	Lead or rubber-lined tanks, Duriron, FRP, Saran, PVC-1, Vinyl, Hypalon, 316 ss, Carp. 20 ss, Tyril, Ti, Derakane, Halar, Polypropylene

*Requires Ball Valves

CHEMICAL	SHIPPING DATA		CHARACTERISTICS			
FORMULA COMMON NAME USE	GRADES OR AVAILABLE FORMS	CONTAINERS AND REQUIREMENTS	APPEARANCE AND PROPERTIES	WEIGHT lb/cu ft. (Bulk Density)	COMMERCIAL STRENGTH	SOLUBILITY IN WATER gm/100 ml
AMINES, NEUTRALIZING 1. Benzylamine (amino-toluene) C_7H_9N 2. Cyclohexylamine (Hagamin) $C_6H_{13}N$ 3. Morpholine (Morlex) C_4H_9NO (91% Soln.) Boiler water treatment Corrosion control	Liquids Solutions	1. 30 and 55 gal steel drums 2. 55 gal drums 3. 1 gal tin can; 5 and 55 gal iron drum	Colorless liquids, pH 10, caustic, toxic 1. Boiling Point is 184^0C 2. 135^0C Flash Point is 90^0F (Open cup) 3. 128^0C Flash Point is 100^0C (open cup) Hygroscopic liquid	Sp. G. (room temperature) 1. -0.97 2. -0.86. 3. -1.002	98 to 99 %	Completely Miscible
AMMONIA ANHYDROUS NH_3 (Ammonia) Chlorine - ammonia treatment Anaerobic digestion Nutrient	Colorless, liquified gas	Steel cylinders - 50, 100, 150 lb T/C - 50,000 lb Green Gas Label	Pungent , irritating odor; liquid causes burns Freezing Point is -107.9^0F Boiling Point is -28^0F. S.Gr. (gas) 0.59 at 70^0C and 1 atm. MCA Warning Label Visc. liquid = 0.27 cps at 33^0C	Sp. G. of liquid is 0.68 at -28^0F	99 to 100% NH_3	% by weight 47.3 at 0^0C 40.6 at 10^0C 34.6 at 20^0C 29.1 at 30^0C
AMMONIA, AQUA NH_4OH (Ammonium hydroxide, Ammonia water, Ammonium hydrate) Chlorine - ammonia treatment pH control Nutrient	Technial Certified Pure Solution 16^0Be 20^0Be 26^0BE	Carboys - 5, 10 gal Drums - 375, 750 lb T/C - 8000 gal	Water white solution, strongly alkaline, causes burns; irratating vapor Unstable, store in cool place and tight container MCA Warning Label Vent feeding systems	At 60^0F 26^0Be S. Gr. 0.8974	16^0Be 10.28% NH_3 20^0Be 17.76% NH_3 26^0Be 29.4% NH_3	Completely miscible
AMMONIUM CHLORIDE NH_4Cl (Sal ammoniac, Ammonium muriate) Source of NH_3 Chloramines Nutrient	Fine crystals	Barrels Multiwall paper sack	White crystals Saline taste Hygroscopic Sublimes at 350^0C Endothermic pH 5.5 (1% Soln.)	45 to 57	99.5%	29.4 at 0^0C 33.3 at 10^0C 37.2 at 20^0C 41.4 at 30^0C
AMMONIUM SILICO-FLUORIDE $(NH_4)SiF_6$ (Ammonium Fluosilicate) Fluoridation for dose 1 ppm 13.3 lb/million gal H_2O	Free flowing fine crystals Commercial grade fine crystals	Pkgs. - 4 to 26.5 lb Kegs - 100 lb Bbl. or Drums - 400 lb.	Odorless, colorless crystals Poison pH 3.5 (1% Soln.)	70 to 80	98% (62.7% Fluoride)	12 at 0^0C 19 at 20^0C 50 at 100^0C
AMMONIUM SULFATE $(NH_4)_2SO_4$ (Sulfate of Ammonia) Chlorine - ammonia treatment Anaerobic digestion Activation of Silica	Small crystals Superdry crystals Damp crystals	Boxes - 25 lb Kegs - 100 lb Bags - 100 lb Bbl. - 300, 400 lb Bulk	White to brown sugar size crystals Tends to cake (Add $CaSO_4$ to prevent caking) Decomposes above 280^0C PH0.1M = 5.5	Dry 49 - 64 41 - 54 Damp 60	99% (21% NH_3-N)	70.6 at 0^0C 75.4 at 20^0C 78.0 at 30^0C
BARIUM CARBONATE $BaCO_3$ (Witherite) Boiler water treatment to precipitate out sulfates	Powder	Boxes - 25 lb Kegs - 100 lb Bags - 200 lb	White, heavy co-hesive powder Poisonous	52 to 78	98 to 99%	0.0022 at 18^0C 0.0065 at 100^0C
BENTONITE $H_2O(Al_2O_3.Fe_2O_3.3MgO)$. $4SiO_2.nH_2O$ (Colloidal clay, Wilkinite, Volclay) Coagulation aid as a weighting agent	Powder Fine granules	Bags - 50, 100 lb Bulk - C/L	Tannish powder Can form a viscous colloidal solution 3% colloidal solution Viscosity = 4 to 10 cps at 72^0F	Powder 45 to 60 Granules 65 to 75	—	Forms a colloidal solution

		FEEDING RECOMMENDATIONS		
BEST FEEDING FORM	CHEMICAL TO WATER RATIO FOR CONTINUOUS DISSOLVING	TYPES OF FEEDERS	ACCESSORY EQUIPMENT REQUIRED	SUITABLE HANDLING MATERIALS FOR SOLUTIONS
As received	Amines are used in small amounts mixed with other additives, etc.	Solution Diaphragm Metering Pump Series 1200 Series 1700 Series 5700	—	Steel, stainless steel, TFE
Dry gas or as aqueous solution - see ammonia, aqua	—	Gas feeder	Scales	Steel, Ni-Resist, Monel, 316 ss, Neoprene
Full strength	—	Solution Diaphragm Metering Pump Series 1200 & 1700 & 5700	Scales Drum handling equipment or storage tanks Transfer pumps	Iron, steel, rubber, Hypalon, 316 ss, Tyril (room temp. to 28%)
Crystals (No lumps or massed crystals) Add TCP or other anti-caking agent	—	Volumetric Helix Solution Diaphragm Metering Pump - Series 1200 & 1700 & 5700	Special agitators if material becomes massed	316 (pits) rubber, PVC-1, Saran, Vinyl, FRP, Halar, Glass, Carp. 20 SS Hypalon, Ti, Noryl (140°F), TFE, Polypropylene
Free flowing fine crystals or solution	Dry feed 0.5 lb/gal Solution feed up to 1.5 lb/gal	Volumetric Helix Solution Diaphragm Metering Pump - Series 1200 & 1700 & 5700 Rotodip	Dustless bag loading hopper or dust filter Non-flood feeder, if hopper holds more than 2 bags, with tight seal	316 ss, rubber, Vinyl, PVC-1, Saran, Uscolite, Hypalon, Monel, Halar
Superdried, or CaSO₄ coated If high moisture content, use solution feed Solutions "climb"	Dry feed 0.5 lb/gal maximum Solution feed 1.0 lb/gal	Gravimetric L-I-W Volumetric Helix Solution Diaphragm Metering Pump - Series 1200 & 1700 & 5700 Rotodip	Hopper agitators (spec.)- if dampish	Damp crystals or solution: 316 ss, Hastelloy C, D, rubber, FRP, PVC-1, Vinyl, Hypalon, Tyril, Monel, Uscolite, Ti, Halar, Polypropylene
Powder plus H-Sil(4) or a slurry	Slurry feed, 15 to 20% to a 25% maximum slurry.	Gravimetric Belt Volumetric Helix Rotodip (slurry)	Hopper agitators (heavy) and /or vibrator	Slurry Rubber, Glass, Steel, Ceramic, 316 ss, Hypalon Uscolite, PVC-1, Tyril
Granules	0.258 lb/gal (a 3% colloidal solution)	Gravimetric Belt Volumetric Helix Solution (Colloidal) Diaphragm Metering Pump - Series 1200 & 1700 & 5700	Hopper agitators for powder Non-flood rotor mechanism for powder Polypak	Iron, Steel, Hypalon, Tyril, TFE

(4)Hi-Sil is a hydrated silica additive – for improved flow

CHEMICAL	SHIPPING DATA		CHARACTERISTICS			
FORMULA COMMON NAME USE	GRADES OR AVAILABLE FORMS	CONTAINERS AND REQUIREMENTE	APPEARANCE AND PROPERTIES	WEIGHT lb/cu ft (Bulk Density)	COMMERCIAL STRENGHT	SOLUBILITY IN WATER gm/100 ml
BROMINE Br$_2$ Swimming pool disinfection Algae control	Liquid	Glass bottles 6.5 lb Earthenware bottles White Label	Dark red liquid Boiling point is 58°C Burns skin Irritating fumes Fire hazard Corrosive Vaporizes at room temperature	Sp. G. 3.119 at 20/15°C 26 lb/gal	99.5%	4.17 at 0°C 3.13 at 30°C
CALCIUM CARBONATE CaCO$_3$ (Calcite, Limestone, Whiting, Chalk) Stabilization of water Neutralizing agent Corrosion prevention Waste treatment	Powder Granules	Bags - 50 lb Drums Bulk - C/L	White amorphous powder	Powder 35 to 50 Granules 100 to 115	96 to 99%	0.0013 at 28°C 0.022 at 100°C (In CO$_2$ free H$_2$O)
CALCIUM FLUORIDE CaF$_2$ (Fluorspar, Fluorite) Fluoridation–For 1 ppm Fluoride, use 18 lb/ million gal H$_2$O	Granules and powder Granular powder Powder Powder (micronized)	Bags - 125 lb Bbl - 500 lb Bulk - C/L	Tan or white granules and powder M.P. 1350°C Toxic Do not breathe dust	Bulk Density varies: 95 to 123 85 to 117 62 to 83 51 to 69	85 to 98%	0.0016 at 18°C 0.0017 at 26°C (Hard to wet the fine powder)
CALCIUM HYDROXIDE Ca(OH)$_2$ (Hydrated Lime, slaked Lime) Coagulation, softening, pH adjustment Waste neutralization Sludge conditioning Precipitate PO$_4$	Light powder Powder	Bags - 50 lb Bbl - 100 lb Bulk - C/L (Store in dry place)	White, 200 - 400 mesh powder, free from lumps Caustic, irritant, dusty Sat. Soln. - pH 12.4 Absorbs H$_2$O and CO$_2$ from air to re- vert back to CaCO$_3$ 10% Slurry — 50 to 150 cp S. Gr. = 1.06	20 to 30 and 30 to 50 To calcu- late hopper capacity, use 25 to 35	Ca(OH)$_2$ 82 to 95% CaO 62 to 72%	0.18 at 0°C 0.16 at 20°C 0.15 at 30°C
CALCIUM HYPOCHLO- RITE Ca(OCl)$_2$.4H$_2$O (H.T.H., Perchloron, Pittchlor) Disinfection, slime control, deodorization	Granules Powder Pellets	Bbl - 415 lb Cans - 5, 15, 100, 300 lb Drums - 800 lb (Store dry and cool; avoid cotact with organic matter)	White or yellowish white, hygroscopic corrosive Strong chlorine odor (Alkaine pH) Yellow Label - oxidiz- ing agent	Granules 68 to 80 Powder 32 to 50	70% Available Cl$_2$	21.88 at 0°C 22.7 at 20°C 23.4 at 40°C
CALCIUM OXIDE CaO (Quicklime, Burnt Lime, Chemical lime, Unslaked Lime) Coagulation, softening, pH adjustment Waste neutralization Sludge conditioning Precipitate PO$_4$	Pebble Lump Ground Pulverized Pellet Granules Crushed	Moisture-proof bags - 100 lb Wood barrel Bulk - C/L (Storedry - Max. 60 days; keep container closed)	White (light gray, tan) lumps to powder Unstable, caustic, irritant Slakes to hydroxide slurry evolving heat Air slakes to form CaCO$_3$ Sat. Soln. pH is 12.4	55 to 70 To calculate hopper ca- pacity, use 60 Pulverized is 43 to 65	70 to 96% CaO (Below 85% can be poor quality)	Reacts to form Ca(OH)$_2$. See Ca(OH)$_2$ above (See footnote 7 below)
CARBON DIOXIDE CO$_2$ (Carbonic Acid Gas) (Dry Ice) Recarbonation in water softening Activation of Silica	Generated by burning fuel oil or gas, or from stack gases on site Dry ice Liquefied gas under pressure	Steel cylinders for compressed gas - 150 lb Dry ice delivered as required and evaporated on site in large steel cylinders Green Label	Colorless, Odorless gas (liquefied under pressure) Snowlike solid below - 78°C Sat. Soln. pH 4.0 at 20°C	Sp. G. with respect to air is 1.5 (gas)	99.5% in Cyl- inders 10 to 14% in combustion gases 100% from dry	0.348 at 0°C 0.145 at 25°C 0.097 at 40°C

(7)Each pound of CaO will slake to form 1.16 to 1.32 lb of Ca(OH)$_2$(depending on purity) and from 2 to 12% grit.

FEEDING RECOMMENDATIONS				
BEST FEEDING FORM	CHEMICAL TO WATER RATIO FOR CONTINUOUS DISSOLVING	TYPES OF FEEDERS	ACCESSORY EQUIPMENT REQUIRED	SUITABLE HANDLING MATERIALS FOR SOLUTION
Liquid, Anhydrous	———	Liquid Special feeder req'd.	———	Dry ($<$ 30 ppm H_2O): Monel, Carp. 20 ss, Lead (Chemical), Nickel Wet and Dry Teflon, Glass, Ceramic Tantalum
According to use, powder may flood	Slurry - 10% (Light to 20% conc. max.) (heavy to 25% conc. max.)	Gravimetric Belt Volumetric Helix Slurry • Diaphram Metering Pump - Series 1200 & 1700	Hopper agitators Non - flood rotor - for fluid-like or floodable grades	Slurry: Iron, Steel, Rubber
Granules and powder	To a 10% max. slurry Very difficult to disperse the fine powder For use with alum, see Footnote 5 (is soluble in alum solution)	Volumetric Helix Slurry • Diaphram Metering Pump - Series 1200 & 1700	Washdown type wetting tank or Vortex mixer Dust collector Dust machinery Agitators	316 ss, Rubber, PVC-1 Saran, FRP
Finer particle sizes more reactive but more difficult to handle and feed, maintain slurry velocity greater than 2 FPS to avoid settling	Dry feed 0.5 lb/gal - max. Slurry - 0.93 lb/gal; i.e. a 10% slurry (Light to a 20% conc. max.) (Heavy to a 25% conc. max.)	Gravimetric Belt Volumetric Helix Slurry Rotodip • Diaphram Metering Pump - Series 1200 & 1700 (See footnote 6)	Hopper agitators Non-flood rotor under large hoppers Dust collectors	Rubber hose, Iron, Steel Cocrete, Hypalon, PVC-1, No lead, Noryl. Derakane
Up to 3% Soln. max. (practical)	0.125 lb/gal makes 1% Soln. of avaivable Cl_2	Liquid Diaphragm Metering Pump - 1200 & 1700 & 5700 Rotodip	Dissolving tanks in pairs with drains to draw off sediment Injection nozzle Foot Valve	Ceramic, Glass, Rubber-lined tanks, PVC-1 Tyril (rm. temp.), Hypalon, Vinyl, Uscolite (rm. temp.) Saran Hastelloy C (good). No tin. Ti, (RM), Halar, Polypropylene, TFE
1/4 to 3/4 in. pebble lime. Pellets Ground lime arches and is floodable Pulverized will arch and is floodable Soft burned, porous best for slaking	2.1 lb/gal (range from 1.4 to 3.3 lb/gal (according to slaker etc.) Dilute after slaking to 0.93 lb/gal which is (10%) average slurry (15% is max. Dilution unless residue carry-over is not critical)	Gravimetric Belt Volumetric Helix	Hopper agitator and non-flood rotor for ground and pulverized lime Lime slaker	Rubber ,Iron, Steel, Concrete, Hypalon PCV-1
Stack gases and generators are most economical Dry ice or compressed gas involves least equiptment	Gas diffused in water under treatment	Generators and gas meter Underwater burners Gas feeders	Compressors and scrubbers for stack gases Diffusers	Dry Gas Iron, Steel Wet: 316 ss, FRP, Monel, Hypalon, PVC - 1 Saran, Viton, Vinyl, Tyril (rm. temp.)

(5) Advances in the use of Fluorspar - F. J. Maier, JAWWA', January 1960
(6) When feeding rates exceed 100 lb/hr, economic factors may dictate use of calcium oxide (quicklime)
    • Requires Ball Valves

CHEMICAL	SHIPPING DATA		CHARACTERISTICS			
FORMULAR COMMON NAME	GRADES OR AVAILABLE FORMS	CONTAINERS AND REQUIREMENTS	APPEARANCE AND PROPERTIES	WEIGHT lb/cu ft. (Bulk Density)	COMMERCIAL STRENGTH	SOLUBILITY IN WATER gm/100 ml
CHLORINATED COPPERAS $Fe_2(SO_4)_3$-$FeCl_3$ (Chlorinated Ferrous Sulfate) Coagulation Sludge Conditioning	Produced at site by reaction of chlorine and $FeSO_4$ in solution	See chlorine and ferrous sulfate as reagents needed	Yellow solution Corrosive	---	---	---
CHLORINATED LIME CaO-$2CaOCl_2$.$3H_2O$ -Variable formula- (Chloride of Lime, Bleaching Powder) Disinfection Slime Control	Powder White or Yellow Label	Drums - 100, 300, 800 lb Store cool, dry	White or yellowish-white, deteriorates; chlorine odor Tends to precipitate incrustants in hard water in feeders Do not mix dry mat with oil, grease, etc.	45 to 50	25 to 37% Available Cl_2	Forms hypochlorous acid (HOCl) with H_2O
CHLORINE Cl_2 (Chlorine Gas, Liquid Chlorine) Disinfection Slime Control Taste and odor control Waste treatment Activation of Silica (See Footnote[8])	Liquified gas under pressure	Steel Cylinders 100, 150 lb Ton containers T/C - 15 ton containers T/C - 16, 30, 55 tons Green Label	Greenish yellow gas liquified under pressure Pungent, noxious, corrosive gas heavier than air; health hazard	Sp. G. with respect to air 2.49	99.8% Cl_2	0.98 at 10°C 0.716 at 20°C 0.57 at 30°C
CHLORINE DIOXIDE ClO_2 Disinfection Taste and odor control (especially phenol) Waste treatment 0.5 to 5 lb $NaClO_2$ per million gal H_2O dosage	Generated as used from Cl_2 and $NaClO_2$ or from $NaClO_2$ and $NaOCl$ plus acid Dissolved as generated	26.3% Avail. Cl_2	Yellow solution when generated in water Yellow-red gas, unstable, irritating, poisonous, explosive; keep cool - keep from light	---	Use 2 lb of $NaClO_2$ to 1 lb of Cl_2, or equal conc. of $NaClO_2$ and $NaOCl$ plus acid (max. 2% each plus Diln. water)	0.29 at 21°C
COAGULANT AIDS (See Footnote[9]) Dry - potable H_2O High M.W. Synthetic Polymers Separan NP10 potable grade, Magnifloc 990, Purifloc N17 Ave. dosage - 0.1 to 1ppm	Powdered, flakey granules	Multiwall paper bags	White flake powder pH varies	27 to 35	---	Colloidal solution
COAGULANT AIDS (See Footnote[9]) Dry-potable H_2O High M.W. synthetic Polymers Separan NP10 & 20 AP30, Magnifloc 972, Superfloc 16 Average Dosage 0.1 to 5 ppm	Powdered, flakey granules	Multiwall paper bags	White flake powder ph varies	NP10 27 to 35 30 to 37 NP20 12 to 15 AP30 40 to 50 Superfloc 16 26 to 33 Magnifloc 972 28 to 34	---	Colloidal solution

(8) For small dosages of chlorine use calcium hypochlorite or sodium hypochlorite

(9) Information available on many other coagulant aids (or flocculant aids) from, BIF "Polypak" BIF "Polypak" Polymer selection Guide Ref. 28.21-1, Nalco, Calgon, Drew, Betz, North American Mogul, American Cyanamid, Dow, etc.

FEEDING RECOMMENDATIONS				
BEST FEEDING FORM	CHEMICAL TO WATER RATIO FOR CONTINUOUS DISSOLVING	TYPES OF FEEDERS	ACCESSORY EQUIPMENT REQUIRED	SUITABLE HANDLING MATERIALS FOR SOLUTION
1 part Cl_2 to 7.8 parts $FeSO_4$ by weight Mix solution discharge of dry feed machine ($FeSO_4$) with solution discharge from Chlorinizer in reaction chamber	Ferrous sulfate 0.5 lb/gal Chlorine 1 lb to 40-50 gal or more	See ferrous sulfate See chlorine	Rubber lined reaction tank	Rubber, Ceramic, Hypalon, PVC-1, Vinyl, TFE, Polypropylene
2% solution of available Cl_2 - Maximum Settle out solids before pumping	About 0.25 lb/gal makes 1% solution of available Cl_2	Liquid Diaphragm metering pump - Series 1200 & 1700 Rotodip	Disolving tanks in pairs with drains to remove sludge	Rubber, Stoneware, Glass, PVC - 1 Vinyl. No tin.
Gas - Vaporized from liquid	1 lb to 45-50 gal or more	Gas chlorinator	Vaporizers for high capacities Scales Residue Analyzer	Anhydrous Liquid or Gas Steel, Copper, Black Iron Wet Gas: Viton, Hastelloy C, PVC - 1, (good), silver, Tantalum, Ti Chlorinated H_2O: Saran, Stoneware, Carp. 20 ss Hastelloy C, FRP [3] PVC-1, Viton, Uscolite Ti, Derakane
Solution from generator - Mix discharge from Chlorinizer and $NaClO_2$ solution or add acid to mixture of $NaClO_2$ and NaOCl. Use equal concentrations - 2% max.	Chlorine water must contain 500 ppm or over of Cl_2 and have a pH or 3.5 or less. Water use depends on method of preparation.	Solution Diaphragm metering pump. Series 1200 & 1700 & 5700	Dissolving tanks or crocks	For solutions with 3% ClO_2: Ceramic, Glass, Hypalon, PVC-1, Saran, Vinyl, Teflon, Ti (fair), TFE, Polypropylene
Powdered, flattish granules	Max. conc. - 1% Feed even stream to vigorous vortex - (too fast mixing will break up colloidal growth) 1/2 hour detention	Volumetric Helix Solution (Colloidal) Diaphragm metering pump. Series 1200 & 1700 & 5700	Special dispersing procedure	Steel, Rubber, TFE, Hypalon, Tyril, Derakane, Non-corrosive, but no zinc Same as for H_2O of similar pH or according to its pH, Polypropylene
Powdered, flattish granules	Same as potable water grades	Same as potable water grades	Same as potable water grades	Same as potable water grades

CHEMICAL	CHARACTERISTICS		SHIPPING DATA			
FORMULA COMMON NAME USE	GRADES OR AVAILABLE FORMS	CONTAINERS AND REQUIREMENTS	APPEARANCE AND PROPERTIES	WEIGHT lb/cu ft (Bulk Density)	COMMERCIAL STRENGTH	SOLUBILITY IN WATER gm/100 ml
COPPER SULFATE $CuSO_4.5H_2O$ (Blue Vitriol, Blue Stone, Cupric Sulfate) Algae control in reservoirs Root control in sewers 0.1 to 10 ppm dosage	Lump Crystal Powder	Bags - 100 lb Bbl - 450 lb Drums	Clear blue crystals or pale blue powder Slowly efflorescing Poisonous pH about 3	Crystal 75 to 90 Powder 60 to 68	99%	19.1 at $0^{\circ}C$ 25.2 at $15^{\circ}C$ 26.3 at $20^{\circ}C$ 31.1 at $30^{\circ}C$
DIATOMACEOUS EARTH SiO_2 Plus Oxides Infusorial earth, Diatoms, Celite, Dicalite, Celatom, Supercel, Speedex, Speed Flow Filter Aid	Natural Calcined Flux Calcined	Bags - 50 lb Bulk	Natural Gray-white bulky powder Calcined Pinkish bulky powder Flux Calcine White bulky powder Do not breathe excessive amount of dust	Natural 5 to 12 8 to 18 Calcined 6 to 13 Flux Calcined 10 to 18 15 to 25	----	Insoluble- forms a slurry
o- DICHLOROBENZENE $C_6H_4Cl_2$ (1, 2 - Dichlorobenzene) (Emulsified grade availabe) Ozene, tech. 80% Cifon 88.5% Chloroben, Benichlor Odor control in sewage and waste treatment	Liquid Emulsified liquid	55 gal steel drum Tank car	Clear liquid Sp. G. 1.3 at 15^0 C/-15^0 C Keep 32°F Flash point (pure liquid) (Open Cup) is 167°F IgnitionPoint is 1199°F Freezing Point is -18 to 22^0 C Toxic	About 10 lb/gal	Regular ortho - 80% para - 20% Pure ortho - 99.5% para - 0.5%	Insoluble (emulsified grade is miscible with H_2O)
DI-SODIUM PHOSPHATE $Na_2HPO_4.12H_2O$ (DSP, Secondary Sodium Phosphate, Dibasic Sod. Phos.) Also Na_2HPO_4—Anhydrous Sodium Phosphate Scale and corrosion control Boiler water conditioning, precipitates Ca and Mg	Crystals Anhydrous	Bags - 100 lb Kegs - 125 lb Drums - 25, 100, 125 lb Bbl - 325, 350 lb	White, hygroscopic M.P. is 34.6°C for (94°F) hydrate Store dry in tightly closed container 1% Soln. -pH 8.4 to 9.0 (approx.) Anhydrous is hygroscopic Hydrate is efflorescent	Crystal Hydrate 80 to 90 Anhydrous 53 to 62	Crystal 19 to 19.5% P_2O_5 Anhydrous 48% P_2O_5	Hydrate 4.2 at $0^{\circ}C$ 19.0 at $20^{\circ}C$ 52.0 at $30^{\circ}C$ Anhydrous 1.7 at $0^{\circ}C$ 3.6 at $10^{\circ}C$ 7.7 at $20^{\circ}C$ 20.8 at $30^{\circ}C$
DOLOMITIC HYDRATED LIME $Ca(OH)_2.Mg(OH)_2$ (dihydrated) $Ca(OH)_2.MgO$ (monohydrated) Silica removal Acid waste neutralization	Powder (Light or heavy)	Bags - 50 lb Bbl Bulk - C/L	White powder, free from lumps Caustic, dusty Store dry Sat. Soln. - pH 12.4	30 to 50 To calculate hopper capacity use 40	Ave Ca $(OH)_2$ 60% MgO 35% Variable Comp.	See Ca$(OH)_2$
DOLOMITIC LIME CaO.MgO (MgO content varies) Dolomitic Quicklime Silica removal Acid neutralization, including waste treatment	Pebble Crushed Lump Ground Pulverized	Bags - 50, 60 lb Bbl Bulk - C/L	White (light gray, tan) lumps to powder Unstable, caustic irritant Slakes to hydroxide slurry evolving heat Sat. Soln. - pH 12.4	Pebble 60 to 65 Ground 50 to 75 Lump 50 to 65 Powder 37 to 65 Average 60	CaO 55 to 57.5% MgO 37.6 to 40.5%	Slakes to form hydrated lime slurry plus MgO. (MgO slakes at>212°F) See Ca$(OH)_2$ solubility
DOLOMITIC LIMESTONE $CaCO_3.MgCO_3$ (Dolomite) Neutralization	Lump or crushed Granule Ground Powder	Bags - 50 lb Drums Bulk- C/L	White, gray, tan Sat. Soln. - pH9 to 9.5	Granule 93to 124 or 85 to 101 Powder 43 to 75	Varies	Approx. same as $CaCO_2$

		FEEDING RECOMMENDATIONS		
BEST FEEDING FORM	CHEMICAL TO WATER RATIO FOR CONTINUOUS DISSOLVING	TYPES OF FEEDERS	ACCESSORY EQUIPMENT REQUIRED	SUITABLE HANDLING MATERIALS FOR SOLUTIONS
Granules or Ground (sugar) or power for Boat application	0.25 lb/gal Retention time: 3/8" crystals - 10.0 minutes Fine crystals - 5.0 minutes Powder - 2.3 minutes Temp. 50°F and efficient mixing	Volumetric Helix Solution Diaphragm metering pump - series 1200 & 1700 & 5700 Rotodip	Dissolving tanks For direct application from boats; Dissolving tank, pump, spray or crystal spreader, or bags	Rubber, ceramic, 316 ss PVC-1, Vinyl, Hypalon FRP, Viton, Uscolite, Saran, Epoxy, Tyril, Polypropylene, TFE
Flux calcined, with higher bulk density	1 lb/gal, depending on bulkiness, a 15 - 20% slurry is the maximum possible	Volumetric Helix Gravimetric Belt Slurry •Diaphragm pump - series 1200 & 1700	Slurry tank Positive agitation	Iron, Steel, Rubber, FRP, Tyril, Hypalon
Liquid	——	Liquid Diaphragm metering pump - series 1200 & 1700 & 5700		Steel, 440 ss, 316 ss, Viton (slight swelling with emulsified grade. Could be less resistance with pure liquid). TFE (RM) (very slight swelling also)
Solution (crystal grade) 0.3 lb/gal Anhydrous grade 0.12 lb/gal	Good, efficient mixing needed for anhydrous (tends to ball)	Solution Diaphragm metering pump - series 1200 & 1700 &5700	Solution tank	Iron, Steel, Rubber, 304 ss, 316 ss, PVC-1 Uscolite, Ni-Resist, Hypalon, Tyril, TFE, Polypropylene
Finer particle sizes more efficient but more difficult to handle and feed	Dry feed (continuous) 0.5 lb/gal max. Slurry (stock) 0.93 lb/gal - 10% Slurry (stock) Light to a 20% max. Heavy to a 25% max.	Gravimetric Volumetric Belt Helix Slurry Rotodip •Diaphragm metering pump - series 1200 & 1700	Hopper agitators Non-flood rotor under large hopper Dust collectors	Iron, Steel, Rubber hose, Concrete, Hypalon, PVC-1, No lead. Requires carbide valves.
Pebble (small) or crushed to pass 3/4 in. mesh; fines included	2lb/gal (range 1.6 to 4.2 lb/gal Dilute after slaking to 0.93 lb/gal (10% slurry) max. Detention, temp., amount of water are critical for efficient slaking	Gravimetric Belt Volumetric Helix	Hopper agitator and non-flood rotor for ground and pulverized Lime slaker	See dolomitic hydrated lime
Granule	——	Gravimetric Belt Volumetric Helix	Agitators (for finer grades) Dust collector (powder) Possibly a rotor	For slurry (powder); Steel, Iron, Rubber

•Requires Ball Valve

1049

CHEMICAL	SHIPPING DATA		CHARACTERISTICS			
FORMULA COMMON NAME USE	GRADES OR AVAILABLE FORMS	CONTAINES AND REQUIREMENTS	APPEARANCE AND PROPERTIES	WEIGHT lb/cu ft (Bulk Density)	COMMERCIAL STRENGTH	SOLUBILITY IN WATER gm/100 ml
FERRIC CHLORIDE $FeCl_3$ - Anhydrous $FeCl_3$ - $6H_2O$ = Crystal $FeCl_3$ Solution (Ferrichlor, Chloride of Iron) Coagulation pH 4 to 11 Dosage - 0.3 to 3 gpg (Sludge cond. 1.5 to 4.5% $FeCl_3$ by weight of solids Precipitate PO_4	Solution Lumps - sticks (crystals) Granules	Solution Carboys - 5, 13 gal Truck, T/C Crystal Keg - 100, 400, 450 lb Drums - 150, 350, 630 lbs	Solution - Dark brown syrup. Crystals - Yellow brown lumps Anhydrous - Green black. Very hygroscopic, *staining corrosive in liquid form. 1% Soln. -pH 2.0 40% Soln. = 7 cp visc. & Sp. G. = 1.42	Solution 11.2 to 12.4 lb Crystal 60 to 64 Anhy. 45 to 60	Solution 35 to 45% $FeCl_3$ Crystal 60% $FeCl_3$ Anhy. 96 to 97% $FeCl_3$	Solution Completely Miscible Crystals 64.42 81.9 at 10°C 91.1 at 20°C Anhy. 74.4 at 0°C
FERRIC SULFATE $Fe_2(SO_4)_3.3H_2O$ (Ferrifloc) $Fe_2(SO_4)_3.2H_2O$ (Ferriclear) (Iron Sulfate) Coag. pH 4-6 & 8.8 - 9.2 Dosage - 0.3 to 3 gpg Precipitate PO_4	Granules	Bags - 100 lb Drums - 400 425 lb Bulk - C/L	$2H_2O$; Red brown $3H_2O$; Red Gray Cakes at high R. H. Corrosive in Soln; Store dry in tight containers Stains	70 - 72	$3H_2O$ 68% $Fe_2(SO_4)_3$ 18.5% Fe $2H_2O$ 76% $Fe_2(SO_4)_3$ 21% Fe	Very Soluble
EERROUS SULFATE $FeSO_4.7H_2O$ (Copperas, Iron Sulfate, Sugar Sulfate, Green Vitriol) Coagulation at pH 8.8 to 9.2 Chrome reduction in waste treatment. Sewage odor control. Precipitate PO_4	Granules Crystals Powder Lumps	Bags - 100 lb Bbl - 400 lb Bulk	Fine greenish crystals M. P. is 64°C. Oxidizes in moist air. Efflorescent in dry air. Masses in storage at higher temp. Soln. is acid.	63 to 66	55% $FeSO_4$ 20% Fe	32.8 at 0°C 37.5 at 10°C 48.5 at 20°C 60.2 at 30°C
FLUOSILICIC ACID H_2SiF_6. (Hydrofluosilicic Acid) Fluoridation: for 1 ppm use 35.1 lb. per 1.0 million gal H_2O (of 30% H_2SiF_6) or 3.3 gal./mil. - gal.	Liquid (Tech & C.P.)	Kegs - 5 gal. pitch-lined Drums- 50 gal. rubber-lined Bbl-420 lb T/C rubber-lined White Label	Clear, colorless, slightly fuming, corrosive, toxic liquid. 1% Soln. - pH 1.2 Attacks glass Vent storage tanks	30% is 10.4 to 10.5 lb/ gal and Sp. G. = 1.27	15 to 30% (12 to 24% F)	Miscible
HYDRAZINE H_2NNH_2 + Water. (Diamine Hydrazine Base) Corrosion inhibitor Scavenger for O_2	Solutions of Hydrazine Hydrazine 35% Tech, 44.8% Purified 54.4% Purified 64% (All hydrazine hydrates)	Glass carboys Steel drums	Colorless fuming hygroscopic liquid M. P. 2°C. Flash Pt. (OC) 126°F Reducing agent. Vapor flammable Toxic. Keep from eyes. Corrosive liquid. White Label. Fire hazard with organic mat. Keep from oxidizers and acids.	Anhy density 1.008 at 20°C. 8.38 lb/gal	Anhy. to 99% pure Also a 35% water solution (Deoxy-sol.)	Miscible with water in all proportions
HYDROCHLORIC ACID HC1 (Muriatic Acid) Neutralization of alkaline waste. Cleaner Regenerate deionizer	Concentrated - 37% 18, 20, and 22° 22°Be Tech. U.S.P. XVI N.F. C.P.	Bottles Carboys Drums Tank cars	Clear, colorless or slightly yellow, fuming, pungent liquid. Highly corrosive to metals. Avoid skin contact. Toxic, pH of 1 normal is 0.1 White Label Releases H_2 gas from metals.	Conc. - 37% 9.95 lb/gal 18°Be 9.5 lb/gal 20°Be 9.65 lb/gal 22°Be 9.8 lb/gal	Concentrated 37% - 38% HC1 18°Be 27.9% HC1 20°Be 31.5% HC1 22°Be 35.2% HC1	Miscible with water in all proportions
HYDROFLUORIC ACID HF Fluoridation - not usually recommended. for use	Liquid	Drums - steel 20, 30, 100 gal. Bulk - T/C White Label	Clear, colorless, toxic, dangerous liquid. Corrosive to steel below 60% Conc., pH approx. 2. 0 Attacks glass. Anhy. B.P. = 68°F.	-----	30% 48% 52% 60% Anhy. 99%	Miscible

1050

BEST FEEDING FORM	CHEMICAL TO WATER RATIO FOR CONTINUOUS DISSOLVING	TYPES OF FEEDERS	ACCESSORY EQUIPMENT REQUIRED	SUITABLE HANDLING MATERIALS FOR SOLUTIONS
Solution or any dilution up to 45% FeCl$_3$ content (Anhy. form has a high heat of soln.)	Anhydrous to form 45% - 5.59 lb/gal 40% - 4.75 lb/gal 35% - 3.96 lb/gal 30% - 3.24 lb/gal 20% - 1.98 lb/gal 10% - .91 lb/gal (Multiply FeCl$_3$ by 1.666 to obtain FeCl$_3$.6H$_2$O) at 20°C	Solution Diaphragm Metering Pump - Series 1200 & 1700 & 5700 Rotodip	Storage tanks for Liquid Dissolving tanks for lumps or granules	Rubber, Glass, Ceramics Hypalon, Saran, PVC-1, (3) FRP, Vinyl Hast C (good to fair- below 40%) Uscolite, Epoxy, Tyril (Rm) Ti, Noryl, Kynar, Derakane, Halar, Polypropylene, TFE
Granules	2 lb/gal (Range) 1.4 to 2.4 lb/gal for 20 min. detention (Warm water permits shorter detention) Water insolubles can be high	Volumetric Helix Solution Diaphragm Meterng Pump - Series 1200 & 1700 & 5700 Rotodip	Dissolver w/motor - driven mixer and water control. Vapor remover solution tank.	316 SS, Rubber, Glass, Ceramics, Hypalon Saran, PVC-1, Vinyl, Carp 20 SS (3) FRP, Epoxy, Tyril, Ti, Halar Derakane, Polyethylene, Polypropylene, TFE
Granules	0.5 lb/gal (Dissolve detention time 5 min. Minimum)	Gravimetric L-I-W Volumetric Helix Solution Diaphragm Metering Pump - Series 1200 & 1700 & 5700	Dissolvers Scales	Rubber, (3)FRP, PVC-1 Vinyl, Epoxy, Hypalon, Uscolite, Ceramic, Carp. 20 SS Tyril, Polypropylene, TFE
Liquid direct from containers do not dilute (unless investigated - depends on free silica in acid, etc.) (10)	———	Liquid Diaphragm Metering Pump - Series 1200 &1700 & 5700 Rotodip	Scales to check weight fed	Rubber, PVCI Saran, Vinyl, TFE, Hypalon, Carp. Hypalon, Carp. 20 Hast. C, Viton, Halar, Polypropylene
Dilute Solution	Can dilute to 0.2 lb/gal stock or feed solution	Solution (Special) Diaphragm Metering Pump - Series 1200 & 1700	———	No molybdenum or Cu (will catalyze and decomp.) Glass, 304 SS, Ceramic, 347 SS, Poly- ethylene, Teflon, (no welding slag fillings either). 316 SS Suitable for low concentration less than 3%, TFE (RM) Hypalon (to 35%) (to 100°F)
As received (less fuming with diluted acid)	———	Liquid Diaphragm Metering Pump - Series 1200 & 1700 & 5700		Metals not generally recommended. - Hast. C (caution) Rubber, Poylethylene (3) FRP, Vinyl, Hypalon, (to 120°F), Viton, Halar, Tyril (RM), Ti (RM) Noryl (20% Derakane, TFE Polypropylene (RM)
Solution (60%) as received	———	Diaphragm Metering Pump-Series 1200 & 1700 Special precautions for equipment and materials, chemical not generally recommended because of handling problems		Monel (50°C, air free), Rubber (to 52%) Derakane (10%), Epoxy, Carp. 20 (50%fair) Hast C (10 & 60%) Viton, Tantalum (Steel> 60%) Hypalon (to 48%) Saran, Polyethylene

(10) Bellack, E & Maier, F. "Dilution or Fluosilic Acid," Feb. 1956, JAWWA, PG. 199

CHEMICAL	SHIPPING DATA		CHARACTERISTICS			
FORMULAR COMMON NAME USE	GRADES OR AVAILABLE FORMS	CONTAINERS AND REQUIREMENTS	APPEARANCE AND PROPERTIES	WEIGHT lb/cu ft. (Bulk Density)	COMMERCIAL STRENGTH	SOLUBILITY IN WATER gm/100 ml
IODINE I_2 Disinfection-Swimming Pool 1 - 3 ppm	Lumps Granules Plates	Bottles 100 -200 lk kegs Drums	Blue, black, lustrous volatile flattish granules. M.P. is .113ºC Toxic Sp. G. - 4.93 Violet vapor Readily sublimes	164 - 184 (granules)	----	0.03 at 25ºC 0.09 at 60ºC (soluble in KI soln).
MAGNESIUM HYDROXIDE $Mg(OH)_2$ Defluoridation	Heavy powder Light powder	Wooden barrel or drums Glass bottles Carboys	White powder Sp. G. = 2.36 Absorbs CO_2 In presence of H_2O. Suspension is called milk of magnesia	Heavy Powder 30 - 45 24 - 34 Light Powder 14 - 23	----	0.0009 at 18ºC
MAGNESIUM OXIDE MgO (Magnesia, Magox, cacined Magnesite) Defluoridation Removal Fe (pH 7.1) 2 ppm MgO each ppm Fe plus diatomite Boiler water treatment Water softening	Heavy powder Light powder Granule	Fiber drums Multiwall paper sacks Tonnage lots Bulk	White powder M.P. 2800ºC Absorbs CO_2 and moisture. pH is alkaline	Heavy Powder 57 - 91 44 - 70 40 - 58 Light Powder 17 - 29 16 - 25 7.6 - 11.6	Reagent is 98% min.	0.00062 (cold H_2O) Hydrates very slowly to $Mg(OH)_2$
OCTADECYLAMINE $C_{18}H_{37}NH_2$ Filming amine Hagafilm, Nalco 353 & 354 Octafilm 80 &120, Permacal 90 & 110 Calgon FL-20 Boiler Water - corrosion control, condensate and feed-water systems	Solution Liquid Flake	Liquids 5 & 50 gal. drums Dry 55 lb drum	Liquids or solutions (mainly) Sp. G. $<$ 1.0 Visc. 600 to 6000 cp pH 8.0	$<$ 8.34 lb/ gal.	----	Most are dispersible in Hot H_2O
OZONE O_3 Taste and Oder Control Disinfection, Waste treatment 1 to 5 ppm-odor 0.5 to 1 ppm-disenfection Color Removal CN & Phenol Oxidation	Gas Liquid	Generated at site by action of electric discharge through dry air- 0.5 to 1% produced.	Colorless-bluish gas or blue liquid Toxic - do not breathe-Explosive fire hazard-keep from oil or readily combustable materials.	Density of Gas is 2.1 Liquid Sp. G is 1.71 G. is 1.71 at-183ºC	1 - 2%	49.4 cc at 0ºC
PHOSPHORIC ACID, ORTHO H_3PO_4 Boiler water softening Reduce alkalinity Cleaning boilers. Nutrient Feeding	50%, 75%, 85% 90% Anhy. Commercial Tech. Food N.F.	Bottles, 1-5 lb. 5, 6-1/2, 13 gal. carboys 55 gal drums & barrels. Tank cars and trucks	Clear, colorless liquid. F.P. (50%) is -40% B.P. (50%) is 108ºC pH (0.1N) is 1.5 15 to 30 cp. viscosity accord to % Avoid skin contact MCA warning label Can form H_2 with some metals.	50% 11.2 lb/ gal 75% 13.3 lb gal 85% 14.1 lb/ gal	50, 75 and 85% conc.	Liquid miscible with water in all proportions
POTASSIUM IODIDE KI With I_2 Disinfection	Crystal Powder	Fiber drums	White crystals M.P. = 680ºC Toxic	Crystal 84 - 100 Powder 62 - 85	Reagent is 99.5% min.	127.5 at 0ºC 136.0 at 10ºC 144.0 at 20ºC 152.0 at 30ºC

	FEEDING RECOMMENDATIONS			
BEST FEEDING FORM	CHEMICAL TO WATER RATIO FOR CONTINUOUS DISSOLVING	TYPES OF FEEDERS	ACCESSORY EQUIPMENT REQUIRED	SUITABLE HANDLING MATERIALS FOR SOLUTIONS
	I_2+KI-1 to 1.33 ratio-dilute (batch soln.) Investigate (Dissolve in K1 soln; tablets in alcohol, etc.)	Solution Diaphragm Metering Pump - Series 1200 & 1700	Vibrator (if needed)	10% Soln. 316 St. (pits) Carp. 20 SS, Alum. (3)FRP, Glass, Stoneware, Vinyl lining. Viton.
Same as MgO	Slurry Feed: 5 - 10% conc. (Light to a 20% maximum) (Heavy to a 30% maximum)	Gravimetric Belt Volumetric Helix Slurry *Diaphragm Metering Pump Series 1200 & 1700	Agitators Rotor Dust Collector Vortex Mixer Eductor or Dissolver-mixer.	Steel, Iron, Rubber, Hypalon,
Heavy powder (feeding) (Light powder is more reactive) (Fresh MgO important for Fe removal)	Slurry Feed: 10% conc. (Light - 10 to 15% maximum) (Heavy to a 25% maximum)	Gravimetric Belt Volumetric Helix *Diaphragm Metering Pump Series 1200 & 1700	Agitators Rotor Dust Collector. Vortex Mixer Eductor or equivalent or Dissolver-mixer.	Steel, Iron, Rubber Requires Carbide Valves.
Diluted Solutions	---------	Solution Diaphragm Metering Pump Series 1200 & 1700 & 5700	---------	Steel; St. Steel, TFE
As generated. Approx. 1% Ozone in air	Gas diffused in water under treatment.	Ozonator	Air drying equipment Diffusers	Glass, 316 SS, Ceramics, Aluminum Teflon
50 to 75% conc. (85% is syrupy) (100% is crystaline)	---------	Liquid Diaphragm Metering Pump Series 1200 & 1700 & 5700		316 St. (no F), Rubber, (3)FRP, Vinyl, PVC-1, Hypalon, Viton, Carp 20 SS Hast. C. Noryl (to 75%), TFE, Polypropylene
Crystal	See I_2	---------	---------	Vinyl lining, (Pits) 316 St., Hast C., Saran, Hypalon, PVC-1, (3)FRP, Carp. 20

*Requires Ball Valves

CHEMICAL	SHIPPING DATA		CHARACTERISTICS			
FORMULA COMMON NAME USE	GRADES OR AVAILABLE FORMS	CONTAINERS AND REQUIREMENTS	APPEARANCE AND PROPERTIES	WEIGHT lb/cu ft. (Bulk Density)	COMMERCIAL STRENGTH	SOLUBILITY IN WATER gm/100 ml
POTASSIUM PERMANGANATE $KMnO_4$ Cairox Taste odor control 0.5 to 4.0 ppm Removes Fe & Mn at 1-1 and up to 2-1 ratio, Respectively	Crystal	U.S.P. 25, 110, 125 lb steel keg Tech. 25, 110, 600 lb steel drum	Purple crystals Sp. G. - 2.7 Decomposes 240°C Can cake up at high relative humidity Strong oxidant. Toxic Keep from organics. Yellow Label.	86 - 102	Tech. is 97% min. Reagent is 99% min.	2.8 at 0°C 3.3 at 10°C 5.0 at 20°C 7.5 at 30°C
SODIUM ALUMINATE $Na_2Al_2O_4$, Anhy. (soda alum) Ratio Na_2O/Al_2O_3 1/1 or 1.15/1 (high purity) Also $Na_2Al_2O_4 \cdot 3H_2O$ hydrated form Coagulation Boiler H_2O treatment.	Ground (pulverized) Crystals Liquid, 27°Be Hydrated Anhydrous	Ground Bags - 50, 100 lb. Drums Liquid Drums	High Purity White Standard Gray. Hygroscopic Aq. Soln. is alkaline Exothermic heat of solution pH = 11 32% is 110°P at 70°F and Sp.G =1.47	High Purity 50 Standard 60	High Purity Al_2O_3 45% $Na_2Al_2O_4$ 72% Standard Al_2O_3 55% $Na_2Al_2O_4$ 88-90%	Hydrated 80 at 75°F Standard 6 - 8% Insolubles Anhydrous 3 lb/gal at 60°F 3 lb/gal at 60°F
SODIUM BICARBONATE $NaHCO_3$ (Baking Soda) Activation of Silica pH adjustment Treat acidic wastes Digester pH	U.S.P. C.P. Commercial Pure Powder Granules	Bags - 100 lbs Bbl- 112, 400 lb Drums- 25 lb Kegs	White powder Slightly alkaline 1% soln. -pH 8.2 Unstable in soln. (Decomposes into CO_2, and Na_2CO_3) Decomposes 100°F.	59 - 62	99% $NaHCO_3$	6.9 at 0°C 8.2 at 10°C 9.6 at 20°C 10.0 at 30°C (3.5% is practical)
SODIUM BISULFITE, ANHYDROUS $Na_2S_2O_5$ ($NaHSO_3$) Sodium Pyrosulfite, Sodium metabisulfite) Dechlorination-About 1.4 ppm for each ppm Cl_2 Reducing agent in waste treatment (as Cr)	Crystals Crystals plus powder Solution (3.25 to 44.9%)	Bags - 100 lb. Drums - 100 400 lb.	White to slight yellow Sulfurous odor Slightly hygroscopic Store dry in tight container Forms $NaHSO_3$ in soln. 1% soln. pH 4.6 Vent Soln. tanks	74 - 85 and 55 - 70	97.5% to 99% $Na_2S_2O_5$ SO_2 65.8%	54 at 20°C 81 at 100°C
SODIUM CARBONATE Na_2CO_3 (Soda Ash -58% Na_2O) Water Softening, pH Adjustment Neutralization	Dense granules Med Gran. & Pwd. Light Powder	Bags - 100 lb Bbl - 100 lb Drums - 25, 100 lb Bulk - C/L	White, alkaline Hygroscopic- can cake up. 1% soln. -pH 11.2	Dense 65 Medium 40 Light 30	99.2% Na_2CO_3 58% Na_2O	7.0 at 0°C 12.5 at 10°C 21.5 at 20°C 38.8 at 30°C
SODIUM CHLORIDE $NaCl$ (Salt, Common Salt) Regeneration of Ion exchange materials (zeolites) in water softening	Rock Powder Crystal Granules	Bags - 100 lb. Bbl Drums 25 lb. Bulk - C/L	Colorless crystals or white powder, slightly hygroscopic, corrosive when moist- Impure form very hygroscopic - will cake up.	Rock 50 to 60 Crystal 58 to 70 Powder 48 - 66	98%	35.7 at 0°C 35.8 at 10°C 36.0 at 20°C 36.3 at 30°C
SODIUM CHLORITE $NaClO_2$ (Technical Sodium Chlorite) Disinfection, Taste and Odor Control, Ind. Waste Treatment (with Cl_2 produces ClO_2)	Powder Flakes Crystals (Tech. and Analytical) Solution (about 40%) Crystalizes about 95°F.	Drums 100 lb. (Do not let $NaClO_2$ dry out on combustible materials)	Tan or white crystals or powder, Hygroscopic, poisonous Powerful oxidizing agent. Explosive on contact with organic matter. Store in metal containers only, Oxidizer Liq. - White Label Solid - Yellow Label	65 - 75	Technical 81% 78% (min.) 124% Available Cl_2 Analytical 98.5% 153% Available Cl_2	34 at 5°C 39 at 17°C 46 at 30°C 55 at 60°C

BEST FEEDING FORM	CHEMICAL TO WATER RATIO FOR CONTINU-OUS DISSOLVING	TYPES OF FEEDERS	ACCESSORY EQUIPMENT REQUIRED	SUITABLE HANDLING MATERIALS FOR SOLUTIONS
Crystals plus anticaking additive	1.0% conc. (2.0% max.) 5-min. Retention with Mech. Mixer & Water > 55°F. If water < 55°F, 10 min.	Volumetric Helix Solution Diaphragm Metering Pump - Series 1200 & 1700 & 5700	Dissolving Tank, Mixer, mechanical	Steel, Iron (neutral & alkaline) 316 St. PVC-1, (3)FRP, Hypalon Lucite, Rubber (alkaline) Noryl, Kynar, Halar Derakane, TFE, Polypropylene
Granular or soln. as received Std. grade produces sludge on dissolving	Dry 0.5 lb/gal Soln. dilute as desired	Gravimetric Belt Volumetric Helix Solution Rotodip Diaphragm - Metering Pump - Series 1200 & 1700 & 5700	Hopper agitators for dry form	Iron, Steel, Rubber, 316 St. s., Concrete, Hypalon, TFE, Polypropylene
Granules or Powder plus TCP (0.4%)	0.3 lb/gal.	Gravimetric Belt Volumetric Helix Solution Diaphragm Metering Pump - Series 1200 & 1700 & 5700	Hopper agitators and Non-flood rotor for powder, if large storage hopper.	Iron & Steel (dilute solns), (caution), Rubber, St. Steel, Saran, Hypalon, Tyril, TFE, Polypropylene
Crystals (Do not let set.) Storage difficult	0.5 lb/gal.	Volumetric Helix Solution Rotodip Diaphragm Metering Pump - Series 1200 & 1700 & 5700	Hopper Agitators for powdered grades Vent dissolver to outside	Glass Carp. 20 SS, PVC-1 Uscolite, 316 st., (3)FRP, Tyril, Hypalon, TFE, Polypropylene
Dense	Dry Feed 0.25 lb/gal for 10 min. detention time & 0.5 lb/gal for 20 min. Solution Feed 1.0 lb/gal Warm H₂0 and/or efficient mixing can reduce detention time if mat. has not set around too long and formed lumps - to 5 min.	Gravimetric Belt Volumetric Helix Solution Diaphragm Metering Pump - Series 1200 & 1700 & 5700 Rotodip	Rotor for light forms to prevent flooding. Large dissolvers Bin agitators for medium or light grades & very light grades.	Iron, Steel, Rubber, Hypalon, Tyril, T1 Halar, TFE, Polypropylene
— — —	Saturated brine usually made up for regeneration of ion-exchange material.	Gravimetric Volumetric Belt Helix Solution Diaphragm Metering Pump - Series 1200 & 1700 & 5700	Scales Solution Tanks	(3)FRP, Glass, PVC-1, Saran, Vinyl, Hast. C, Hypalon, Halar 316 St. (Fair), Derafane Carp. 20 SS, Tyril, Ti, Noryl, Kynar, TFE, Polypropylene
Solution as rec'd.	Batch Solutions 0.12 to 2 lb/gal.	Solution Diaphragm Metering Pump - Series 1200 & 1700 & 5700 Rotodip	Chlorine feeder and chlorine dioxide generator	Glass, Saran, PVC-1, Vinyl Tygon (3) FRP, Has. (C (fair), Hypalon, Tyril, Ti, Noryl, Derakane, Halar (100°F), TFE, Polypropylene

CHEMICAL	SHIPPING DATA		CHARACTERISTICS			
FORMULA COMMON NAME USE	GRADES OR AVAILABLE FORMS	CONTAINERS AND REQUIREMENTS	APPEARANCE AND PROPERTIES	WEIGHT lb/cu ft. (Bulk Density)	COMMERCIAL STRENGTH	SOLUBILITY IN WATER gm/100 ml
SODIUM CHROMATE Na_2CrO_4 $Na_2CrO_4.4H_2O$ Sodium Dichromate $Na_2Cr_2O_7$ $Na_2Cr_2O_72H_2O$ (Sodium Bichromate) Corrosion control	Granules Powder Solution Tech. Reagent	Chromate 100 lb bags 100, 200, 400, lb steel drums Dichromate 50, 100 lb bags 100, 400 lb drums	Chromate, Anhy. Yellow crystals Deliquesces M.P. is 792°C pH 8.4 to 9.2 Dichromate Anhy. Orange Xis M.P. 357°C pH 3.1 to 4.4 Dichromate 2H₂O Orange Xis Deliquescent pH 3.1 to 4.4 All are toxic	Chromate Anhy. 93-113 Hydrate 55-63 Dichromate Anhy. 8-100 Hydrate 63-99	Chromate Anhy. 39-42% soln. 98.8% tech. Dichromate (2H₂O) 69% soln. 99.8% tech. 99.9% reagent	Na_2CrO_4 57 at 60°F $Na_2CrO_4.4H_2O$ 88 at 60°F $Na_2Cr_2O_7$ 111 at 60°F $Na_2Cr_2O_7.2H_2O$ 126 at 60°F
SODIUM FLUORIDE NaF (Fluoride) Flouridation - For 1 ppm F-use 18.8 lb. per 1 Million gallons H₂O (98% NaF)	Granules Crystal Powder	Bags - 100 lb Drums - 25, 125, 375 lb (Bags shipped only by car load or truck, LCL by drums only) White Label	White or tinted Nile Blue for identification Store dry and separately from other chemicals. Poisonous 1% Soln. is pH 6.5 4% Soln. is pH 7.6	Powder 65 to 100 Granules Crystals 90 to 106	NaF 95 to 98% F 43 to 44%	4.0 at 0°C 4.03 at 15.5°C 4.05 at 20°C 4.1 at 25°C 5.0 at 100°C
SODIUM HEXA-META PHOSPHATE $(NaPO_3)_6$ (Calgon, Glassy, Phosphate Sodium Polyphos, Micromet Metafos, Nalco 519, Quadrafos) Corrosion and Red Water control , Stabilization, Water Well Yield Adjustment. Sequester Ions	"Glass" (Powder or flake more expensive) (Unadjusted form less expensive)	Bag - 100 lb Drums - 100, 300, 320 lb	Powderes & Granular grades. Also broken glass grade. slightly hygroscopic 0.25% soln. -pH 6.8 - 7.0 (unadjusted) Adjusted with soda ash to > pH 8	Glass 64 to 100 Powder & Granular 44-60	67% P_2O_5 Min.	Infinitely Soluble.
SODIUM HYDROXIDE NaOH (Caustic Soda, Soda Lye) pH Adjustment, Neutralization	Flakes Lumps Powder Solution	Drums - 25, 50, 350, 400, 700 lb Bulk-Solution in T/C Liquid White label	White flakes, Granules or pellets, Deliquescent, caustic poison. Dangerous to handle. 1% Soln. -pH 12.9 50% Soln. will crystalize at 54°F and Viscosity = 80 cp (Rm) & Sp. G. = 1.53	Pellets 60 to 70 Flakes 46 to 62	Solid 98.9% NaOH 74.76% Na₂0 Solution 12 to 50% NaOH	42 at 0°C 51.5 at 10°C 109 at 20°C 119 at 30°C
SODIUM HYPOCHLORITE NaOCl (Javelle Water, Bleach Liquor, Chlorine Bleach) Disinfection, Slime control, Bleaching	Solution White or Yellow Label	Carboys - 5, 13, gal. Drums - 30 gal. Bulk - 1300, 1800, 2000 gal % T	Yellow liquid strongly alkaline, store in cool place, protect from light and vent containers at intervels. Can be stored about 60 days under proper conditions. Sp.G. =1.21 (15%)	15% 10.2 lb/gal 12.5% 10 lb/gal	15% NaOCl= 1.25 lb. Cl₂/gl 12.5% NaOCl= 1.04 lb. Cl₂/gallon	Completely Miscible
SODIUM NITRATE $NaNO_3$ (Soda, Niter, Chile Saltpeter) Waste Treatment Source of Oxygen Boiler H₂O Treatment Odor Control	Crude Purified C.P. U.S.P.	Multiwall Paper Sacks Bags up to 100 lb	Colorless crystals or white pellets Sp.G. - 2.26 M.P. - 308°C Deliquescent Fire Hazard-oxidizing agent Yellow Label	Crystals 82 - 97 Pellets 75 - 83	Crude 92% C.P. 99%+	73 at 0°C 80 at 20°C 96 at 30°C (when dissolved lowers temp. of water.)
SODIUM PENTACHLO-ROPHENATE $C_6Cl_5ONa.C_6Cl_5ONa.H_2O$ Pentachlorophenate "Santobrite" ; "Dowicide G" (monohydrate) Slime & Algae Control Biocide	Anhydrous Flakes Pellets Briquettes Monohydrates beads Flakes Ground	Flake 50 lb Burlap 60 lb Fiber drums Pellets 200 lb Fiber drum Ground 100 lb drum	Tan Flakes, etc. pH 9.0 Stable when stored dry. Converted to insoluble phenol at pH 6.8 TOXIC - Keep from skin, eyes. Do not breathe dust.	Powder 20 - 25 Pellets 48 - 55 Bisquettes 50 - 58 Monohydrate 20- 25	88% Min.	Anhy. 15.0 at 4°C 20.0 at 9.5°C 40.0 at 18°C Monohydrate 33.0 at 25°C

1056

FEEDING RECOMMENDATIONS				
BEST FEEDING FORM	CHEMICAL TO WATER RATIO FOR CONTINUOUS DISSOLVING	TYPES OF FEEDERS	ACCESSORY EQUIPMENT REQUIRED	SUITABLE HANDLING MATERIALS FOR SOLUTIONS
Solution	———	Solution Diaphragm metering Pump - Series 1200 & 1700 & 5700	———	316 S. St., Glass, Rubber, PVC-1, Uscolite, Hypalon, Tyril, Steel (Caution), TFE Polypropylene
Granular (Powder arches, floods, is dustier)	Dry Feed 1 lb to 12 gal (1%) 5-10 min. detention Solution Feed 1 lb to 3 gal or more	Volumetric Helix Solution Diaphragm metering Pump Series 1200 & 1700 & 5700	Vibration or Agitator for powder grades. Dust Control Equipment Large Dissolvers. Softeners for water used in making solutions for pumping, if hardness $>$ 50 ppm (Hard H_2O will form CaF_2)	Rubber, 316 st., PVC-1, Saran, (3)FRP, Hypalon, Tyril, TFE, Polypropylene
Solution	1 lb/gal in solution feeder (Above 2 lb/gal becomes syrupy)	Solution Diaphragm metering Pump - Series 1200 & 1700 & 5700	Tray or basket below surface of dissolving tank. Agitation after solution.	316 st., Rubber Ceramics, Vinyl, PVC-1, (3)FRP, Saran, Hypalon, Tyril, TFE, Polypropylene
Solution feed	NaOH has a high heat of solution	Solution *Diaphragm metering Pump - Series 1200 & 1700 & 5700 Rotodip		Cast Iron, Steel, For no contam, use, Rubber, PVC-1, 316 St., Hypalon, TFE, Polypropylene (RM)
Solution up to 16% Available Cl_2 Conc.	1.0 gal of 12.5% (Avail. Cl_2) Soln. to 12.5 gal. of water gives a 1% Available Cl_2 soln.	Solution Diaphragm metering Pump - Series 1200 & 1700 & 5700 Rotodip	Solution tanks Foot Valves Water Meters Injection Nozzles	Rubber, Glass, Tyril, Saran, PVC-1, Vinyl, PVC-1, Vinyl, T1, Hast. C., Hypalon. Derakane (170°F) Halar (100°F), TFE, Polypropylene (RM)
Crystals or pellets (coated) (keep dry)	10% conc. 5 min. detention	Gravimetric Belt Volumetric Helix Solution Diaphragm metering Pump - Series 1200 & 1700 & 5700	Dissolving Tank, Mixer	Steel (dry), St. steel Hypalon, Tyril, TFE, Polypropylene
Solution	Anhydrous Do not exceed a 20% Solution. Generally used in a 1/4 to 2% soln.	Solution Diaphragm metering Pump - Series 1200 & 1700 & 5700	———	Iron, Steel, Brass, Hypalon, Tyril, TFE, Polypropylene

*Requires Ball Valves

CHEMICAL	SHIPPING DATA		CHARACTERISTICS			
FORMULA COMMON USE NAME	GRADES OR AVAILABLE FORMS	CONTAINERS AND REQUIREMENTS	APPEARANCE AND PROPERTIES	WEIGHT lb/cu ft. (Bulk Density)	COMMERCIAL STRENGTH	SOLUBILITY IN WATER gm/100 ml
SODIUM PHOSPHATE, MONO Anhydrous - NaH_2PO_4 Boiler H_2O Treatment pH Control	Powder	100 lb. paper bag 125 lb. drum 350 lb. barrel	White powder Deliquescent pH 4.5	51 - 71	98% (58% P_2O_5)	57.9 at 0^oC 69.9 at 10^oC 85.2 at 20^oC 107.0 at 30^oC
SODIUM PHOSPHATE, MONO Hydrated - $NaH_2PO_4H_2O$ Boiler H_2O Treatment pH control	Crystal	Same as mono. phosphate, anhy.	White crystals Loses H_2O at 100^oC Cakes up at room R.H. pH 4.5	49 - 73	97% (50% P_2O_5)	66.6 at 0^oC 80.4 at 10^oC 98.0 at 20^oC 123.0 at 30^oC
SODIUM SILICATE Na_2SiO_3 $41^oBe1Na_2O:3.22S_1O_2$ (Water Glass) Coagulation aid as activated silica. Corrosion control	Liquid (Various ratios of Na_2O. SiO_2 are available)	Drums - 1, 5, 55 gal Bulk - T/T, T/C	Opaque, viscous, Conc. solution may freeze about 25^oF. 1% Soln. -pH 12.7 Viscosity = 180 cp Sp. G=1.39 (68^oF)	41^oBe 11.6 lb/gal 42.2^oBe 11.73 lb/gal	41^oBe 8.9% Na_2O 28.7% SiO_2 42.2^oBe 9.16% Na_2O 29.5% SiO_2	Completely miscible
SODIUM SILICOFLUORIDE Na_2SiF_6 (Sodium Fluosilicate) Fluoridation - for 1 ppm F use 13.0 lbs per 1 million gallons H_2O	Granular Powdered Granular Granular, Powd.	Bags - 100 lb Drums - 25, 125 375 lb. Bags shipped only in carloads or trucks LCL in drums, only White Label	White or tinted Nile blue for identification Poison; store separately from other chemicals. Sat. Soln. -pH 3.4	Granular 85 - 105 Pwd., Gran. 60 - 96 Gran. & Pwd. 70 - 100	98.5 (60%F)	0.44 at 0^oC 0.762 at 25^oC 0.98 at 37.8^oC 1.52 at 65.6^oC 2.45 at 100°
SODIUM SULFITE Na_2SO_3 (Sulfite) Dechlorination in disinfection, reducing agent in waste treatment. Oxygen removal in boiler water treatment. About 1.8 ppm for each ppm Cl_2 in dechlorination.	Granules (crystals) Powder	Bags Bbl Drums Kegs	White crystals or powder Saline Sulfurous taste 1% Soln. -pH is 9.8 Soln. gives off SO_2	Powder 54 - 81 Granules 84 - 107	93 to 99%	14 at 0^oC 20 at 10^oC 27 at 20^oC 36 at 30^oC
SODIUM THIOSULFATE $Na_2S_2O_3.5H_2O$ (Hypo, Sodium Hyposulfite) Dechlorination in disinfection, reducing agent for Cl. 1 to 1.5 ppm for each ppm Cl_2 (dechlorination)	Standard Crystals Rice Granules	Bags Bbl Drums Kegs	White translucent crystals; cooling bitter taste, 1% Soln. pH 6.5 8.0 Efflorescent - dry air & slightly deliq. in damp air. Loses all H_2O at 700^oC M.P. about 110^oF.	53 - 60	98 to 99%	74.7 at 0^oC 301 at 60^oC
SULFAMIC ACID HSO_3NH_2 pH control Cleaning scale	Granule Crystallin (Damp)	100 lb Fiber drum (Polyethylene liner)	White granules, M.P. is 176^oC pH is 5.2	71 - 81	Crystallin 99.% plus Granular is less pure.	16.0 at 0^oC 18.0 at 10^oC 21.0 at 20^oC 25.0 at 30^oC
SULFUR DIOXIDE SO_2 Dechlorination in disinfection. Filter bed cleaning. About 1 ppm SO_2 for each ppm Cl_2, in declorizing. Waste treatment - Cr +6 reduction.	Liquified Gas under pressure	Steel cylinders 100, 150, 200 lb. Green Label	Heavier than air Colorless gas; suffocating odor, corrosive, poison Acid in solution dissolves to form H_2SO_3 Density of liquid is 89.60 lb/cu ft. at 32^oF.	———	100% SO_2	760mm 22.8 at 0^oC 16.2 at 10^oC 11.3 at 20^oC 7.8 at 30^oC

1058

		FEEDING RECOMMENDATIONS		
BEST FEEDING FORM	**CHEMICAL TO WATER RATIO FOR CONTINUOUS DISSOLVING**	**TYPES OF FEEDERS**	**ACCESSORY EQUIPMENT REQUIRED**	**SUITABLE HANDLING MATERIALS FOR SOLUTIONS**
Powder dry	Good, efficient mixing required. ("Tend to Ball")	Gravimetric Volumetric Helix Belt Solution Diaphragm Metering Pump - Series 1200 & 1700 & 5700	Agitators Rotor-possibly Dust Collector	316 SS, Carp. 20 SS Hypalon Rubber, FRP PVC-1, Vinyl, Tyril, TFE, Polyproplene
Crystals, dry	———	Volumetric Helix Solution Diaphragm Metering Pump - Series 1200 & 1700 & 5700	Vibrator-possible	Same as for the anhydrous Form.
Solution Feed as received to dilution tanks See Activated Silica	See Activated Silica	Solution Diaphragm Metering Pump - Series 1200 & 1700 & 5700 Rotodip	Scales to check loss in weight of drums as silicate is fed	Cast Iron, Steel, Rubber, Hypalon, TFE, Polypropylene
Optimum Mesh Mesh %On 60 - 1 (max.) 100 - 9 (Max.) 200 - 35 325 - 15 - 30 pan - 15 - 30	1 lb to 120 gal of water (0.1%) 10 min. detention with $< 60°F\ H_2O$ mechanical mixer (5 min. with warm H_2O) with right cond. 1 lb to 60 gal H_2O (0.2%)	Volumetric Helix Solution Diaphragm Metering Pump - Series 1200 & 1700 & 5700	Hopper Agitators for fine powder Vibrator for heavier, powdered grade. Large dissolving chamber, or special dissolving arrangement	316 St. Rubber, PVC1 Vinyl, Saran. Hypalon, Viton Tyril, Halar, Derakane TFE, Polypropylene
Solution 2 to 12.5%	3% Solution 5 - 10 min. setention	Solution Diaphragm Metering Pump - Series 1200 & 1700 & 5700 Rotodip	Tanks Vent. Soln. Tanks	Cast Iron, steel For no contam: 316 st., PVC-1, Saran, FRP, Vinyl, Hypalon, Tyril TFE, Polypropylene
Solution	———	Solution Diaphragm Metering Pump - Series 1200 & 1700 & 5700	———	Cast Iron, Steel No Contam: 316 st., PVC-1 Saran, Tyril, [3]FRP, Vinyl, Hypalon, TFE, Polypropylene
Granule	1 - 3% made up, can be made stronger	Volumetric - Helix Solution Diaphragm Metering Pump - Series 1200 & 1700 & 5700	Damp-spec. agitation or possibly vibrate. No aid needed for dry granules.	316 st., wood, Polyethylene, glass, Teflon, Hypalon, Carp.20 S.S. (Pumps), Tyril, TFE, Polypropylene
Gas	1 lb to 1 gal. ———	Gas Rotameter SO_2 Feeder		Wet gas: Glass, Carp. 20 S.S. PVC-1 Ceramics, 316 (g), Viton, Hypalon

CHEMICALS	SHIPPING DATA		CHARACTERISTICS			
FORMULA COMMON NAME USE	GRADES OR AVAILABLE FORMS	CONTAINERS AND REQUIREMENTS	APPEARANCE AND PROPERTIES	WEIGHT lb/cu ft, (Bulk Density)	COMMERCIAL STRENGTH	SOLUBILITY IN WATER gm/100 ml
SULFURIC ACID H_2SO_4 (Oil of Vitriol, Vitriole) pH Adjustment, Activation of Silica Neutralization of Alkaline wastes.	Liquid 60°Be 60°Be 50°Be	Bottles Carboys 5, 13 gal Drums 55, 100 gal Bulk T/T, T/C White Label	Syrupy liquid, very corrosive, hygroscopic Sore dry and cool in tight container, pH = 1.2 for o.IN Soln. + 0.3 for IN Soln. Visc. = 22 cp Sp. G. = 1.8	66°Be 15.1 lb/gal 60°Be 14.2 lb/gal 66°Be	66°Be 93.2% H_2SO_4 60°Be 77.7% H_2SO_4 50°Be 62.2% H_2SO_4	Completely miscible
TETRA-SODIUM PYROPHOSPHATE $Na_4P_2O_7$ $10H_2O$ (TSPP, Alkaline Sodium Pyrosophosphate) $Na_4P_2O_7$ - Anhydrous Boiler Water Treatment	Crystal Powder C.P. Pure	Bags - 100, 200 lb Bbl - 350 lb Drums - 25, 100, 300 & 350 lb Kegs - 125 lb	Crystal hydrate is efflorescent in dry air. Anhy. is slightly hygroscopic. Crystal ioses water at 100° C.1% Soln. pH 10.8 (Approx.)	Crystal 50-70 Powder 46-66	Crystal Hydrate 34% P_2O_5 Anhydrous 53-57% P_2O_5	Hydrate 5.4 at 0°C 7.0 at 26.7°C Anhy. 2.6 at 25°C
TRICALCIUM PHOSPHATE $Ca_3(PO_4)_2$ $[Ca_5OH(PO_4)_3]$ Calcium Phosphate, tribasic Defluoridation	Granular Powder Tech. C.P. N.F. Pure	60 lb bag 150 lb drum	White powder Sp. G. = 3.18 M.P. = 1670°C pH 6-7	21 - 33	96%	0.0025 (Cold H_2O) (d ecomposes in hot H_2O)
TRI-SODIUM PHOSPHATE Anhy. Na_3PO_4 (TSP) Na_3PO_4 $12H_2O$ Na_3PO_4 H_2O Boiler water treatment Cleaner	Crystals, coarse medium, std. Powder C.P. Commercial Highest Purity	Bags - 100, 200 lb Bbl - 325, 400 lb Kegs - 125 lb	Colorless crystals, Crystal, efflorescent May cake under high R.H. Monohydrate, Anhy. are hygroscopic. Loses 11 H_2O at 100°C 1% Soln. - pH 11.9 Crystal M.P. = 73°C	Crystal 55 to 60 Mono- Hydrate 65 Anhy. 50-70	Crystal 19% P_2O_5 Anhy. 42% P_2O_5 Hydrate, Mono 38.5% P_2O_5	Hydrated (12H₂0) ($12H_2O$) 3.5 at 0°C 9.5 at 10°C 25.5 at 20°C 46.4 at 30°C Anhy. 1.5 at 0°C 4.1 at 10°C 11.0 at 20°C 20.0 at 30°C

	FEEDING RECOMMENDATIONS			
BEST FEEDING FORM	**CHEMICAL TO WATER RATIO FOR CONTINUOUS DISSOLVING**	**TYPE OF FEEDERS**	**ACESSORY EQUIPMENT REQUIRED**	**SUITABLE HANDLING MATERIALS FOR SOLUTIONS**
Solution at desired dilution. H_2SO_4 has a high heat of solution	Dilute to any desired concentration, but always add acid to water. NEVER water to acid	Liquid Diaphragm Metering Pump - Series 1200 & 1700 Rotodip & 5700	Dilution tanks	Conc. >85% Steel, Iron, PVC-1 (Good), TFE (RM) Viton 40 to 85%; Carp. 20, PVC-1 Viton, TFE 2 to 40%; Carp. 20, PVC-1 glass, PVC1, Viton, Noryl, Derakane (70%), TFE
Crystal, powder or solution (0.3 lb/gal)	0.3 lb/gal. 20 min. detention. Good, efficient mixing for Anhy. product. ("Tends to ball")	Gravimetric - Belt Volumetric - Helix Solution Diaphragm Metering Pump - Series 1200 & 1700 & 5700 Rotodip	Hopper agitators and non-flood rotor for Anhy. powder. Solution tank for solution feed. Cont. soln. feed not to practical unless warm H_2O used	Cast Iron, steel rubber, Hypalon, Tyril, TFE, Polypropylene
Powder	20% Slurry	Gravimetric - Belt Volumetric - Helix Slurry Diaphragm Metering Pump Series 1200 & 1700	Agitation Dust collector Rotor, possibly	Steel, Iron, rubber, TFE
Crystal, powder or solution Do not store crystal grade over one day.	0.3 lb/gal. Good, efficient mixing needed for Anhy. & mono. ("Tends to ball") Warm water will help.	Volumetric - Helix Solution Diaphragm Metering Pump - Series 1200 & 1700 & 5700 Rotodip	Hopper agitators for powder. Solution tank for solution feed.	Cast Iron, steel, rubber, Hypalon, Tyril, TFE

CHEMICALS USED IN WATER, SEWAGE AND WASTE TREATMENT PROCESSES

*Checks indicate normal uses; Checks in parenthesis (x) indicate potential applications, usually dependent on the economics involved.

**Followed by sedimentation.

1 - Chlorine - ammonia treatment for disinfection and tastes and odor control.

2 - Activation of silica.

3 - Sludge digestion control.

CHEMICAL	WATER TREATMENT													SEWAGE TREATMENT						INDUSTRIAL WASTE TREATMENT										
	Algae Control	Boiler Water Treatment	Coagulation**	Color Removal	Corrosion and Scale Control	Dechlorination	Disinfection	Fluoridation	Iron & Manganese Removal	pH Control	Softening	Taste and Odor Control	Miscellaneous	B.O.D. Removal	Coagulation**	Condition-Dewater Sludge	Disinfection	Odor Control	Miscellaneous	B.O.D. Removal	Coagulation**	Chlorination	Condition-Dewater Sludge	Dechlorination	Flotation	Neutralization-Acid	Neutralization-Alkali	Oxidation	pH Control	Reduction
Activated Carbon	x					x						x																		
Activated Silica			x	x									2		x						x									
Alum. Ammonium Sulfate			x									1																		
Aluminum Chloride Soln.			(x)				1								(x)	x					(x)		x							
Alum. Potassium Sulfate			x																											
Aluminum Sulfate			x	x					x						x						x									
Alum. Liquid			x												x						x									
Amines Neutr.		x								x			2																	
Ammonia, Anhydrous							1					1							3											
Ammonia, Aqua							1					1							3											
Ammon. Chloride							1												17											
Ammonium Silicofluoride							1	x																						
Ammonium Sulfate							1					1	2						3											
Barium Carbonate		x	x																											
Bentonite			x																		x									
Bromine	(x)					4																								

Table of water treatment chemicals and their applications.

Legend:

4 - Swimming pool disinfection.

5 - Slurry up-flow coagulation and sedimentation for solids and B.O.D. removal.

6 - Slime control.

7 - Recarbonation in water softening.

8 - Ponding control in activated sludge.

9 - Bulking control in activated sludge.

10 - Silica removal.

11 - Fluoride removal. Also accomplished by magnesium oxide (or hydroxide) and tri-calcium phosphate.

12 - Regeneration of ion exchange materials (zeolites).

13 - To produce chlorine dioxide.

14 - For activated silica coagulation aid.

Chemicals listed:

- Calcium Fluoride
- Calcium Hydroxide
- Calcium Hypochlorite
- Calcium Oxide
- Carbon Dioxide
- Chlorinated Copperas
- Chlorinated Lime
- Chlorine
- Chlorine Dioxide
- Copper Sulfate
- Diatomaceous Earth
- Disodium Phosphate
- Dolomitic Hydrated Lime
- Dolomitic Lime
- Dolomitic Limestone
- Ferric Chloride
- Ferrous Sulfate
- Fluosilicic Acid
- Hydrazine
- Hydrochloric Acid
- Hydrofluoric Acid
- Iodine
- Mag. Hydroxide
- Mag. Oxide
- o-Dichlorobenzene
- Octadecylamine

CHEMICALS USED IN WATER SEWAGE AND WASTE TREATMENT PROCESSES

CHEMICAL	WATER TREATMENT													SEWAGE TREATMENT						INDUSTRIAL WASTE TREATMENT										
	Algae Control	Boiler Water Treatment	Coagulation**	Color Removal	Corrosion and Scale Control	Dechlorination	Disinfection	Fluoridation	Iron & Manganese Removal	pH Control	Softening	Taste and Odor Control	Miscellaneous	B.O.D. Removal	Coagulation**	Condition-Dewater Sludge	Disinfection	Odor Control	Miscellaneous	B.O.D. Removal	Coagulation**	Chlorination	Condition-Dewater Sludge	Dechlorination	Flotation	Neutralization-Acid	Neutralization-Alkali	Oxidation	pH Control	Reduction
Ozone				x			x					x		x						x										
Phosphoric Acid		(x)									16								17								(x)			
Potassium Iodide							4																							
Pot. Permanganate									x			x																		
Sodium Aluminate		x	x	x									(10)		x						x									
Sodium Bicarbonate										x			2												(x)				(x)	
Sodium Bisulfite						x							19											x						x
Sodium Carbonate					x					x	x															x			x	
Sodium Chloride											12																			
Sodium Chlorite (13)				x			x		x			x	13	(x)			(x)			(x)								(x)		
Sodium Chromate					x																									
Sodium Dichromate					x																									
Sodium Fluoride								x																						
Sod. Hexametaphosphate					x							x	18																	
Sodium Hydroxide		x	(x)		x					x	x				x	x					(x)		(x)		(x)	(x)			x	

15 - Source of O_2
16 - Boiler H_2O Soft
17 - Nutrient
a - See Notes 1 and 11
b - See Notes 2,6 & 13
c - See Notes, 3,6,8 & 9
18 - Sequester ions
19 - O_2 Scavenger

Compound																	
Sodium Hypochlorite	(x)		x					x	x	(x)	x	(x)	x				
Sodium Nitrate							x	1 x		1							
Sodium Pentachlorophenate x																	
Sod. Phos. Mono Anhy.	x			x				6	6							x	
Sod. Phos. Mono Hyd.	x			x											14	x	
Sodium Silicate	14	x	x				2										
Sodium Silicofluoride			x											x			x
Sodium Sulfite	x	x x	x				19							x			(x
Sodium Thiosulfate		x															
Sulfamic Acid		x		x			(2)							x			x
Sulfur Dioxide							2									x	
Sulfuric Acid				x			18					x					
Tetra-Sod. Pyrophosphate	x		11				11										
Trical. Phosphate				x x			16										
Tri-Sodium Phosphate	x	x															

CHEMICALS	SHIPPING DATA		CHARACTERISTICS			
FORMULA COMMON NAME	GRADES OR AVAILABLE FORMS	CONTAINERS AND EQUIPMENT	APPEARANCE AND PROPERTIES	WEIGHT lb/cu ft (Bulk Density)	COMMERCIAL STRENGTH	SOLUBILITY IN WATER gm/100 ml
Hydrogen Peroxide H_2O_2 Odor Control. 2 part H_2O_2 = 1 part H_2S, Lessens Sludge Bulking (60 to 400ppm) Treat Phenols SO_2, CN, Mercaptans and Refractory Organics Hypochlorite destruction, Lowers COD. Prevents denitrification. BOD removal control H_2S odor wastewater treatment	Liquid USP 3% Tech. 3, 6, 27.5, 30, 35.50 and 90%	30 - gallon Aluminum Drums Polyethylene (for up to 37% with fiber and steel over pack Amber glass bottles Carboys Tank Trucks Tank Cars	Colorless heavy liquid Unstable. Decomposition Catalyzed by metallic impurities, usually has inhibitor. pH 9 to 10. F.P. 35% = -27.4°F. F.P. 40% = -62°F. Visc. 50% 1.2 cp @20°C 40°C	35% Sp.G.= 1.132 9.4 lb/gal 50% Sp.G= 1.196 10 lb/gal (20°C)	% 35 or 50	Miscible
Oxygen O_2 Oxidation of municiple and industrial wastes. Activated Sludge Treatment	Gas Low Purity High Purity U.S.P. Research (99.99%) Extra Dry (99.6%)	Steel Cylinders Tank Cars	Colorless Odorless Tasteless Gas. Liquid at -183°C to slightly bluish. Sp.G. gas = 1.105 Sp.G. liquid=1.14 Nonliquified gas at 2200 psig at 70°F	Sp. Vol. 121 cu. ft/lb	High Purity 99.95% Research Grade 99.99%	1 Vol./32 Vol. Water at 32°F, or 9 mg/l at 20°C and 7.5 mg/l at 30°C and 11.2 mg/l at 100°C

FEEDING RECOMMENDATIONS				
BEST FEEDING FORM	CHEMICAL TO WATER RATIO FOR CONTINUOUS DISSOLVING	TYPES OF FEEDERS	ACCESSORY EQUIPMENT REQUIRED	SUITABLE HANDLING MATERIALS FOR SOLUTIONS
35 and 50% Liquid	———	Solution Diaphragm Pump 1700 Pump Series (glycol as hydraulic fluid)	Goggles Neoprene Gloves Protective Aprons of Dacron or Orlon	Al Alloy 1160, 1100, Tygon, Koroseal, Polyethylene, Derakane (to 30%) Viton (to 50%) 316 st. steel, to 50%, under certain conditions (Rm), Hast. C (Good)
On Site Generation, especially if using 50 T/D Smaller Amounts Delivered if source close by	Gas Diffused in Water under treatment	Unox Process, Air Products Systems: Airco, F30 System.	———	Glass, 316, Ceramics, Al Teflon. (no grease or oil or easily combustible substance

Appendix C
Miscellaneous Tables

TABLE C-1. PHYSICAL PROPERTIES OF WATER (U.S. UNITS)

Temperature °F	Specific Weight γ lb/ft^3	Density* ρ slug/ft^3	Vapor Pressure p_v lb$_f$/in^2	Kinematic Viscosity $\nu \times 10^5$ ft^2/s	Dynamic Viscosity $\mu \times 10^5$ lb·s/ft^2	Surface Tension** σ lb/ft	Modulus of Elasticity* $E/10^3$ lb$_f$/in^2
32	62.42	1.940	0.09	1.931	3.746	0.00518	287
40	62.43	1.940	0.12	1.664	3.229	0.00614	296
50	62.41	1.940	0.18	1.410	2.735	0.00509	305
60	62.37	1.938	0.26	1.217	2.359	0.00504	313
70	62.30	1.936	0.36	1.059	2.050	0.00498	319
80	62.22	1.934	0.51	0.930	1.799	0.00492	324
90	62.11	1.931	0.70	0.826	1.595	0.00486	328
100	62.00	1.927	0.95	0.739	1.424	0.00480	331
110	61.86	1.923	1.27	0.667	1.284	0.00473	332
120	61.71	1.918	1.69	0.609	1.168	0.00467	332
130	61.55	1.913	2.22	0.558	1.069	0.00460	331
140	61.38	1.908	2.89	0.514	0.981	0.00454	330
150	61.20	1.902	3.72	0.476	0.905	0.00447	328
160	61.00	1.896	4.74	0.442	0.838	0.00441	326
170	60.80	1.890	5.99	0.413	0.780	0.00434	322
180	60.58	1.883	7.51	0.385	0.726	0.00427	318
190	60.36	1.876	9.34	0.362	0.678	0.00420	313
200	60.12	1.868	11.52	0.341	0.637	0.00413	308
212	59.83	1.860	14.70	0.319	0.593	0.00404	300

* At atmospheric pressure.
**In contact with air.

TABLE C-2. PHYSICAL PROPERTIES OF WATER (SI UNITS)

Temperature °C	Specific Weight γ kN/m^3	Density ρ kg/m^3	Vapor Pressure p_v kN/m^2	Kinematic Viscosity $\nu \times 10^6$ m^2/s	Dynamic Viscosity $\mu \times 10^3$ N·s/m^2	Surface Tension** σ N/m	Modulus of Elasticity* $E/10^6$ kN/m^2
0	9.805	999.8	0.61	1.785	1.781	0.0765	1.98
5	9.807	1000.0	0.87	1.519	1.518	0.0749	2.05
10	9.804	999.7	1.23	1.306	1.307	0.0742	2.10
15	9.798	999.1	1.70	1.139	1.139	0.0735	2.15
20	9.789	998.2	2.34	1.003	1.002	0.0728	2.17
25	9.777	997.0	3.17	0.893	0.890	0.0720	2.22
30	9.764	995.7	4.24	0.800	0.798	0.0712	2.25
40	9.730	992.2	7.38	0.658	0.653	0.0696	2.28
50	9.689	988.0	12.33	0.553	0.547	0.0679	2.29
60	9.642	983.2	19.92	0.474	0.466	0.0662	2.28
70	9.589	977.8	31.16	0.413	0.404	0.0644	2.25
80	9.530	971.8	47.34	0.364	0.354	0.0626	2.20
90	9.466	965.3	70.10	0.326	0.315	0.0608	2.14
100	9.399	958.4	101.33	0.294	0.282	0.0589	2.07

* At atmospheric pressure.
**In contact with air.

TABLE C-3

VAPOR PRESSURE OF WATER BELOW 100°C

Pressure of aqueous vapor over water in mm of Hg for temperatures from
−15.8 to 100°C. Values for fractional degrees between 50 and 89 were
obtained by interpolation.

Temp. °C	0.0	0.2	0.4	0.6	0.8	Temp. °C	0.0	0.2	0.4	0.6	0.8
−15	1.436	1.414	1.390	1.368	1.345	42	61.50	62.14	62.80	63.46	64.12
−14	1.560	1.534	1.511	1.485	1.460	43	64.80	65.48	66.16	66.86	67.56
−13	1.691	1.665	1.637	1.611	1.585	44	68.26	68.97	69.69	70.41	71.14
−12	1.834	1.804	1.776	1.748	1.720						
−11	1.987	1.955	1.924	1.893	1.863	45	71.88	72.62	73.36	74.12	74.88
						46	75.65	76.43	77.21	78.00	78.80
−10	2.149	2.116	2.084	2.050	2.018	47	79.60	80.41	81.23	82.05	82.87
− 9	2.326	2.289	2.254	2.219	2.184	48	83.71	84.56	85.42	86.28	87.14
− 8	2.514	2.475	2.437	2.399	2.362	49	88.02	88.90	89.79	90.69	91.59
− 7	2.715	2.674	2.633	2.593	2.553						
− 6	2.931	2.887	2.843	2.800	2.757	50	92.51	93.5	94.4	95.3	96.3
						51	97.20	98.2	99.1	100.1	101.1
− 5	3.163	3.115	3.069	3.022	2.976	52	102.09	103.1	104.1	105.1	106.2
− 4	3.410	3.359	3.309	3.259	3.211	53	107.20	108.2	109.3	110.4	111.4
− 3	3.673	3.620	3.567	3.514	3.461	54	112.51	113.6	114.7	115.8	116.9
− 2	3.956	3.898	3.841	3.785	3.730						
− 1	4.258	4.196	4.135	4.075	4.016	55	118.04	119.1	120.3	121.5	122.6
						56	123.80	125.0	126.2	127.4	128.6
− 0	4.579	4.513	4.448	4.385	4.320	57	129.82	131.0	132.3	133.5	134.7
						58	136.08	137.3	138.5	139.9	141.2
0	4.579	4.647	4.715	4.785	4.855	59	142.60	143.9	145.2	146.6	148.0
1	4.926	4.998	5.070	5.144	5.219						
2	5.294	5.370	5.447	5.525	5.605	60	149.38	150.7	152.1	153.5	155.0
3	5.685	5.766	5.848	5.931	6.015	61	156.43	157.8	159.3	160.8	162.3
4	6.101	6.187	6.274	6.363	6.453	62	163.77	165.2	166.8	168.3	169.8
						63	171.38	172.9	174.5	176.1	177.7
5	6.543	6.635	6.728	6.822	6.917	64	179.31	180.9	182.5	184.2	185.8
6	7.013	7.111	7.209	7.309	7.411						
7	7.513	7.617	7.722	7.828	7.936	65	187.54	189.2	190.9	192.6	194.3
8	8.045	8.155	8.267	8.380	8.494	66	196.09	197.8	199.5	201.3	203.1
9	8.609	8.727	8.845	8.965	9.086	67	204.96	206.8	208.6	210.5	212.3
						68	214.17	216.0	218.0	219.9	221.8
10	9.209	9.333	9.458	9.585	9.714	69	223.73	225.7	227.7	229.7	231.7
11	9.844	9.976	10.109	10.244	10.380						
12	10.518	10.658	10.799	10.941	11.085	70	233.7	235.7	237.7	239.7	241.8
13	11.231	11.379	11.528	11.680	11.833	71	243.9	246.0	248.2	250.3	252.4
14	11.987	12.144	12.302	12.462	12.624	72	254.6	256.8	259.0	261.2	263.4
						73	265.7	268.0	270.2	272.6	274.8
15	12.788	12.953	13.121	13.290	13.461	74	277.2	279.4	281.8	284.2	286.6
16	13.634	13.809	13.987	14.166	14.347						
17	14.530	14.715	14.903	15.092	15.284	75	289.1	291.5	294.0	296.4	298.8
18	15.477	15.673	15.871	16.071	16.272	76	301.4	303.8	306.4	308.9	311.4
19	16.477	16.685	16.894	17.105	17.319	77	314.1	316.6	319.2	322.0	324.6
						78	327.3	330.0	332.8	335.6	338.2
20	17.535	17.753	17.974	18.197	18.422	79	341.0	343.8	346.6	349.4	352.2
21	18.650	18.880	19.113	19.349	19.587						
22	19.827	20.070	20.316	20.565	20.815	80	355.1	358.0	361.0	363.8	366.8
23	21.068	21.324	21.583	21.845	22.110	81	369.7	372.6	375.6	378.8	381.8
24	22.377	22.648	22.922	23.198	23.476	82	384.9	388.0	391.2	394.4	397.4
						83	400.6	403.8	407.0	410.2	413.6
25	23.756	24.039	24.326	24.617	24.912	84	416.8	420.2	423.6	426.8	430.2
26	25.209	25.509	25.812	26.117	26.426						
27	26.739	27.055	27.374	27.696	28.021	85	433.6	437.0	440.4	444.0	447.5
28	28.349	28.680	29.015	29.354	29.697	86	450.9	454.4	458.0	461.6	465.2
29	30.043	30.392	30.745	31.102	31.461	87	468.7	472.4	476.0	479.8	483.4
						88	487.1	491.0	494.7	498.5	502.2
30	31.824	32.191	32.561	32.934	33.312	89	506.1	510.0	513.9	517.8	521.8
31	33.695	34.082	34.471	34.864	35.261						
32	35.663	36.068	36.477	36.891	37.308	90	525.76	529.77	533.80	537.86	541.95
33	37.729	38.155	38.584	39.018	39.457	91	546.05	550.18	554.35	558.53	562.75
34	39.898	40.344	40.796	41.251	41.710	92	566.99	571.26	575.55	579.87	584.22
						93	588.60	593.00	597.43	601.89	606.38
35	42.175	42.644	43.117	43.595	44.078	94	610.90	615.44	620.01	624.61	629.24
36	44.563	45.054	45.549	46.050	46.556						
37	47.067	47.582	48.102	48.627	49.157	95	633.90	638.59	643.30	648.05	652.82
38	49.692	50.231	50.774	51.323	51.879	96	657.62	662.45	667.31	672.20	677.12
39	52.442	53.009	53.580	54.156	54.737	97	682.07	687.04	692.05	697.10	702.17
						98	707.27	712.40	717.56	722.75	727.98
40	55.324	55.91	56.51	57.11	57.72	99	733.24	738.53	743.85	749.20	754.58
41	58.34	58.96	59.58	60.22	60.86						
						100	760.00	765.45	770.93	776.44	782.00
						101	787.57	793.18	798.82	804.50	810.21

Source: Handbook of Chemistry and Physics, 52nd Edition
CRC Press, Inc., Boca Raton, Florida
1971

TABLE C-4
PROPERTIES OF WATER AT VARIOUS TEMPERATURES FROM 32° TO 705.4° F

Temp. F	Temp. C	Specific Volume Cu Ft/Lb	SPECIFIC GRAVITY			Wt in Lb/Cu Ft	Vapor Pressure Psi Abs
			39.2 F Reference	60 F Reference	68 F Reference		
32	0	.01602	1.000	1.001	1.002	62.42	0.088
35	1.7	.01602	1.000	1.001	1.002	62.42	0.100
40	4.4	.01602	1.000	1.001	1.002	62.42	0.1217
50	10.0	.01603	.999	1.001	1.002	62.38	0.1781
60	15.6	.01604	.999	1.000	1.001	62.34	0.2563
70	21.1	.01606	.998	.999	1.000	62.27	0.3631
80	26.7	.01608	.996	.998	.999	62.19	0.5069
90	32.2	.01610	.995	.996	.997	62.11	0.6982
100	37.8	.01613	.993	.994	.995	62.00	0.9492
120	48.9	.01620	.989	.990	.991	61.73	1.692
140	60.0	.01629	.983	.985	.986	61.39	2.889
160	71.1	.01639	.977	.979	.979	61.01	4.741
180	82.2	.01651	.970	.972	.973	60.57	7.510
200	93.3	.01663	.963	.964	.966	60.13	11.526
212	100.0	.01672	.958	.959	.960	59.81	14.696
220	104.4	.01677	.955	.956	.957	59.63	17.186
240	115.6	.01692	.947	.948	.949	59.10	24.97
260	126.7	.01709	.938	.939	.940	58.51	35.43
280	137.8	.01726	.928	.929	.930	58.00	49.20
300	148.9	.01745	.918	.919	.920	57.31	67.01
320	160.0	.01765	.908	.909	.910	56.66	89.66
340	171.1	.01787	.896	.898	.899	55.96	118.01
360	182.2	.01811	.885	.886	.887	55.22	153.04
380	193.3	.01836	.873	.874	.875	54.47	195.77
400	204.4	.01864	.859	.860	.862	53.65	247.31
420	215.6	.01894	.846	.847	.848	52.80	308.83
440	226.7	.01926	.832	.833	.834	51.92	381.59
460	237.8	.0196	.817	.818	.819	51.02	466.9
480	248.9	.0200	.801	.802	.803	50.00	566.1
500	260.0	.0204	.785	.786	.787	49.02	680.8
520	271.1	.0209	.765	.766	.767	47.85	812.4
540	282.2	.0215	.746	.747	.748	46.51	962.5
560	293.3	.0221	.726	.727	.728	45.3	1133.1
580	304.4	.0228	.703	.704	.704	43.9	1325.8
600	315.6	.0236	.678	.679	.680	42.3	1542.9
620	326.7	.0247	.649	.650	.650	40.5	1786.6
640	337.8	.0260	.617	.618	.618	38.5	2059.7
660	348.9	.0278	.577	.577	.578	36.0	2365.4
680	360.0	.0305	.525	.526	.527	32.8	2708.1
700	371.1	.0369	.434	.435	.435	27.1	3093.7
705.4	374.1	.0503	.319	.319	.320	19.9	3206.2

Computed from Keenan & Keyes' Steam Table.

Source: Hydraulic Institute Engineering Data Book
Hydraulic Institute, Cleveland, Ohio
1979

TABLE C-5. ATMOSPHERIC PRESSURE BAROMETER READING AND BOILING POINT OF WATER AT VARIOUS ALTITUDES

Altitude		Barometer Reading		Atmos. Pressure		Boiling Point of Water
Feet	Meters	In.Hg.	Mm.Hg.	Ft. Water	psia	°F
− 1000	− 304.8	31.0	788	35.2	15.2	213.8
− 500	− 152.4	30.5	775	34.6	15.0	212.9
0	0.0	29.9	760	33.9	14.7	212.0
+ 500	+ 152.4	29.4	747	33.3	14.4	211.1
+ 1000	304.8	28.9	734	32.8	14.2	210.2
1500	457.2	28.3	719	32.1	13.9	209.3
2000	609.6	27.8	706	31.5	13.7	208.4
2500	762.0	27.3	694	31.0	13.4	207.4
3000	914.4	26.8	681	30.4	13.2	206.5
3500	1066.8	26.3	668	29.8	12.9	205.6
4000	1219.2	25.8	655	29.2	12.7	204.7
4500	1371.6	25.4	645	28.8	12.4	203.8
5000	1524.0	24.9	633	28.2	12.2	202.9
5500	1676.4	24.4	620	27.6	12.0	201.9
6000	1828.8	24.0	610	27.2	11.8	201.0
6500	1981.2	23.5	597	26.7	11.5	200.1
7000	2133.6	23.1	587	26.2	11.3	199.2
7500	2286.0	22.7	577	25.7	11.1	198.3
8000	2438.4	22.2	564	25.2	10.9	197.4
8500	2590.8	21.8	554	24.7	10.7	196.5
9000	2743.2	21.4	544	24.3	10.5	195.5
9500	2895.6	21.0	533	23.8	10.3	194.6
10000	3048.0	20.6	523	23.4	10.1	193.7
15000	4572.0	16.9	429	19.2	8.3	184.0

TABLE C-6
DENSITY OF VARIOUS SOLIDS

The approximate density of various solids at ordinary atmospheric temperature.
In the case of substances with voids such as paper or leather the bulk density is indicated rather than the density of the solid portion.

(Selected principally from the Smithsonian Tables.)

Substance	Grams per cu. cm	Pounds per cu. ft.	Substance	Grams per cu. cm	Pound per cu. ft.	Substance	Grams per cu. cm	Pounds per cu. ft.
Agate	2.5–2.7	156–168	Glass, common	2.4–2.8	150–175	Tallow, beef	0.94	59
Alabaster, carbon-			flint	2.9–5.9	180–370	mutton	0.94	59
ate	2.69–2.78	168–173	Glue	1.27	79	Tar	1.02	66
sulfate	2.26–2.32	141–145	Granite	2.64–2.76	165–172	Topaz	3.5–3.6	219–223
Albite	2.62–2.65	163–165	Graphite*	2.30–2.72	144–170	Tourmaline	3.0–3.2	190–200
Amber	1.06–1.11	66–69	Gum arabic	1.3–1.4	81–87	Wax, sealing	1.8	112
Amphiboles	2.9–3.2	180–200	Gypsum	2.31–2.33	144–145	Wood (seasoned)		
Anorthite	2.74–2.76	171–172	Hematite	4.9–5.3	306–330	alder	0.42–0.68	26–42
Asbestos	2.0–2.8	125–175	Hornblende	3.0	187	apple	0.66–0.84	41–52
Ash		27–33	Ice	0.917	57.2	ash	0.65–0.85	40–53
Asphalt	1.1–1.5	69–94	Ivory	1.83–1.92	114–120	balsa	0.11–0.14	7–9
Basalt	2.4–3.1	150–190	Leather, dry	0.86	54	bamboo	0.31–0.40	19–25
Beeswax	0.96–0.97	60–61	Lime, slaked	1.3–1.4	81–87	basswood	0.32–0.59	20–37
Beryl	2.69–2.7	168–169	Limestone	2.68–2.76	167–171	beech	0.70–0.90	43–56
Biotite	2.7–3.1	170–190	Linoleum	1.18	74	birch	0.51–0.77	32–48
Bone	1.7–2.0	106–125	Magnetite	4.9–5.2	306–324	blue gum	1.00	62
Brick	1.4–2.2	87–137	Malachite	3.7–4.1	231–256	box	0.95–1.16	59–72
Butter	0.86–0.87	53–54	Marble	2.6–2.84	160–177	butternut	0.38	24
Calamine	4.1–4.5	255–280	Meerschaum	0.99–1.28	62–80	cedar	0.49–0.57	30–35
Calcspar	2.6–2.8	162–175	Mica	2.6–3.2	165–200	cherry	0.70–0.90	43–56
Camphor	0.99	62	Muscovite	2.76–3.00	172–187	dogwood	0.76	47
Caoutchouc	0.92–0.99	57–62	Ochre	3.5	218	ebony	1.11–1.33	69–83
Cardboard	0.69	43	Opal	2.2	137	elm	0.54–0.60	34–37
Celluloid	1.4	87	Paper	0.7–1.15	44–72	hickory	0.60–0.93	37–58
Cement, set	2.7–3.0	170.190	Paraffin	0.87–0.91	54–57	holly	0.76	47
Chalk	1.9–2.8	118–175	Peat blocks	0.84	52	juniper	0.56	35
Charcoal, oak	0.57	35	Pitch	1.07	67	larch	0.50–0.56	31–35
pine	0.28–0.44	18–28	Porcelain	2.3–2.5	143–156	lignum vitae	1.17–1.33	73–83
Cinnabar	8.12	507	Porphyry	2.6–2.9	162–181	locust	0.67–0.71	42–44
Clay	1.8–2.6	112–162	Pressed wood			logwood	0.91	57
Coal, anthracite	1.4–1.8	87–112	pulp board	0.19	12	mahogany		
bituminous	1.2–1.5	75–94	Pyrite	4.95–5.1	309–318	Honduras	0.66	41
Cocoa butter	0.89–0.91	56–57	Quartz	2.65	165	Spanish	0.85	53
Coke	1.0–1.7	62–105	Resin	1.07	67	maple	0.62–0.75	39–47
Copal	1.04–1.14	65–71	Rock salt	2.18	136	oak	0.60–0.90	37–56
Cork	0.22–0.26	14–16	Rubber, hard	1.19	74	pear	0.61–0.73	38–45
Cork linoleum	0.54	34	Rubber, soft			pine, pitch	0.83–0.85	52–53
Corundum	3.9–4.0	245–250	commercial	1.1	69	white	0.35–0.50	22–31
Diamond	3.01–3.52	188–220	pure gum	0.91–0.93	57–58	yellow	0.37–0.60	23–37
Dolomite	2.84	177	Sandstone	2.14–2.36	134–147	plum	0.66–0.78	41–49
Ebonite	1.15	72	Serpentine	2.50–2.65	156–165	poplar	0.35–0.5	22–31
Emery	4.0	250	Silica, fused trans-			satinwood	0.95	59
Epidote	3.25–3.50	203–218	parent	2.21	138	spruce	0.48–0.70	30–44
Feldspar	2.55–2.75	159–172	translucent	2.07	129	sycamore	0.40–0.60	24–37
Flint	2.63	164	Slag	2.0–3.9	125–240	teak, Indian	0.66–0.88	41–55
Fluorite	3.18	198	Slate	2.6–3.3	162–205	African	0.98	61
Galena	7.3–7.6	460–470	Soapstone	2.6–2.8	162–175	walnut	0.64–0.70	40–43
Gamboge	1.2	75	Spermaceti	0.95	59	water gum	1.00	62
Garnet	3.15–4.3	197–268	Starch	1.53	95	willow	0.40–0.60	24–37
Gas carbon	1.88	117	Sugar	1.59	99			
Gelatin	1.27	79	Talc	2.7–2.8	168–174			

*Some values reported as low as 1.6.

Source: Handbook of Chemistry and Physics, 52nd Edition
CRC Press, Inc., Boca Raton, Florida
1971

TABLE C-7 ENERGY AND WORK CONVERSION FACTORS

Unit Measure	Equivalent Units		
	Horsepower-Hours	Foot-Pounds	Btu
1 International steam table calorie (IT cal)	1.5596×10^{-6}	3.088	3.968×10^{-3}
1 kilocalorie (kcal)	1.5596×10^{-3}	3.088×10^3	3.968
1 gigacalorie (Gcal)	1559.6	3.088×10^9	3.968×10^6
1 British thermal unit (Btu)	0.3930×10^{-3}	778.17	1.0
1 therm (thm)	39.3	77.817×10^6	100,000
1 joule (J)	0.3725×10^{-6}	0.7376	$.9478 \times 10^{-3}$
1 kilojoule (kJ)	0.3725×10^{-3}	737.6	0.9478
1 megajoule (MJ)	0.3725	737.6×10^3	$.9478 \times 10^3$
1 gigajoule (GJ)	372.5	737.6×10^6	$.9478 \times 10^6$
1 kilowatthour (kWh)	1.341	2.655×10^6	3412
1 megawatthour (MWh)	1341	2.655×10^9	3412×10^3
1 gigawatthour (GWh)	1341×10^3	2655×10^9	3412×10^6
1 horsepower-hour (hp-h)	1.0	1.98×10^6	2544.43
1 foot-pound (ft-lb)	0.50505×10^{-6}	1.0	1.2851×10^{-3}

TABLE C-8 POWER CONVERSION FACTORS

Unit Measure	Equivalent Units	
	kW	hp
1 foot-pound per second (ft-lb/s) =	1.355×10^{-3}	1.818×10^{-3}
1 kilogram force meter per second		
(kgf-m/s) =	9.803×10^{-3}	0.01315
1 kilowatt (kW) =	1.0	1.341
1 horsepower (hp) =	0.7457	1.0
1 metric horsepower (cv) =	0.7353	0.9862

TABLE C-9 ENERGY EQUIVALENTS OF CRUDE OIL

Unit Measure	Energy Equivalents
One 42-gallon barrel of oil	5.8×10^6 Btu
	5.6×10^3 cubic feet of natural gas
	1.70 thermal megawatthours
	0.58 electrical megawatthour*
	0.232 ton bituminous coal
	0.42 ton lignite
	6.119×10^9 joules

*At a heat rate of 10,000 Btu/kWh of electricity.

TABLE C-10. CHLORINE CONTAINER SIZING

| TYPE OF CONTAINER | NET WEIGHT | APPROXIMATE | | | |
		TARE WEIGHT	GROSS WEIGHT	OUTSIDE DIAMETER	LENGTH
CYLINDERS	100 lb	63-115 lb	163-215 lb	8 1/4"-8 1/2"	4'5"-4'11"
		63-105 lb	163-205 lb	10 1/2"-10 3/4"	3'3 1/2"-3'7"
	150 lb	85-140 lb	235-290 lb	10 1/4"-10 3/4"	4'5"-4'8"
TON CONTAINER	2000 lb	1300-1650 lb	3300-3650 lb	2'6"	6'7 3/4"-6'10 1/2'
TMU TON CONTAINER CAR	15 tons	-	-	HEIGHT ABOVE TRACK 6'8"-7'6"	42'4"-47'0"
SINGLE UNIT TANK CARS	16 tons	-	-	10'5"-12'0"	32'2"-33'3"
	30 tons	-	-	12'4 1/2"-13'7"	33'10"-35'11 1/2"
	55 tons	-	-	14'3"-15'1"	29'9"-43'0"
	85 tons	-	-	14'11"-15'1"	43'7"-50'0"
	90 tons	-	-	14'11"-15'1"	45'8"-47'2"

Dimensional Data from Chlorine Manual, Fourth Edition

1078

TABLE C-11
ALTITUDE-DENSITY TABLE FOR AIR

Altitudes in Feet—Standard Air at 0 Alt. (29.92 In. Bar.) = 1

Alt.	Den.	Bar.	Alt.	Den.	Bar.
0	1.00	29.92	3000	.891	26.68
100	.966	29.81	3200	.885	26.48
200	.992	29.70	3400	.878	26.28
300	.989	29.58	3600	.872	26.08
400	.985	29.47	3800	.865	25.88
500	.981	29.36	4000	.858	25.68
600	.977	29.25	4200	.852	25.49
700	.974	29.14	4400	.846	25.30
800	.970	29.02	4600	.839	25.10
900	.966	28.91	4800	.833	24.91
1000	.962	28.80	5000	.826	24.72
1100	.959	28.69	5200	.820	24.53
1200	.955	28.58	5400	.814	24.35
1300	.952	28.47	5600	.808	24.16
1400	.948	28.36	5800	.802	23.98
1500	.944	28.26	6000	.795	23.79
1600	.941	28.15	6200	.789	23.61
1700	.937	28.04	6400	.784	23.43
1800	.933	27.93	6600	.778	23.26
1900	.930	27.83	6800	.772	23.08
2000	.926	27.72	7000	.766	22.90
2100	.923	27.62	7200	.760	22.73
2200	.919	27.51	7400	.754	22.56
2300	.916	27.41	7600	.748	22.38
2400	.912	27.30	7800	.743	22.21
2500	.909	27.20	8000	.737	22.04
2600	.905	27.09	8200	.731	21.87
1700	.902	26.99	8400	.726	21.70
2800	.898	26.89	8600	.720	21.54
2900	.895	26.78	8800	.714	21.37

Source: Handbook of Chemistry and Physics, 14th Edition
The Chemical Rubber Co., Cleveland, Ohio
1929

TABLE C-12

PROPERTIES OF DRY AIR

BAROMETRIC PRESSURE 29.921 INCHES

Temperature, degrees Fahr.	Weight per cu. ft., pounds	Per cent of volume at 70°	B.t.u. absorbed by one cu. ft. dry air per degree F.	Cu. ft. dry air warmed one degree per B.t.u.	Temperature, degrees Fahr.	Weight per cu. ft., pounds	Per cent of volume at 70°	B.t.u. absorbed by one cu. ft. dry air per degree F.	Cu. ft. dry air warmed one degree per B.t.u.
0	.08636	.8680	.02080	48.08	130	.06732	1.1133	.01631	61.32
5	.08544	.8772	.02060	48.55	135	.06675	1.1230	.01618	61.81
10	.08453	.8867	.02039	49.05	140	.06620	1.1320	.01605	62.31
15	.08363	.8962	.02018	49.56	145	.06565	1.1417	.01592	62.82
20	.08276	.9057	.01998	50.05	150	.06510	1.1512	.01578	63.37
25	.08190	.9152	.01977	50.58	160	.06406	1.1700	.01554	64.35
30	.08107	.9246	.01957	51.10	170	.06304	1.1890	.01530	64.36
35	.08025	.9340	.01938	51.60	180	.06205	1.2080	.01506	66.40
40	.07945	.9434	.01919	52.11	190	.06110	1.2270	.01484	67.40
45	.07866	.9530	.01900	52.64	200	.06018	1.2455	.01462	68.41
50	.07788	.9624	.01881	53.17	220	.05840	1.2833	.01419	70.48
55	.07713	.9718	.01863	53.68	240	.05673	1.3212	.01380	72.46
60	.07640	.9811	.01846	54.18	260	.05516	1.3590	.01343	74.46
65	.07567	.9905	.01829	54.68	280	.05367	1.3967	.01308	76.46
70	.07495	1.0000	.01812	55.19	300	.05225	1.4345	.01274	78.50
75	.07424	1.0095	.01795	55.72	350	.04903	1.5288	.01197	83.55
80	.07356	1.0190	.01779	56.21	400	.04618	1.6230	.01130	88.50
85	.07289	1.0283	.01763	56.72	450	.04364	1.7177	.01070	93.46
90	.07222	1.0380	.01747	57.25	500	.04138	1.8113	.01018	98.24
95	.07157	1.0472	.01732	57.74	550	.03932	1.9060	.00967	103.42
100	.07093	1.0570	.01716	58.28	600	.03746	2.0010	.00923	108.35
105	.07030	1.0660	.01702	58.76	700	.03423	2.1900	.00847	118.07
110	.06968	1.0756	.01687	59.28	800	.03151	2.3785	.00782	127.88
115	.06908	1.0850	.01673	59.78	900	.02920	2.5670	.00728	137.37
120	.06848	1.0945	.01659	60.28	1000	.02720	2.7560	.00680	147.07
125	.06790	1.1040	.01645	60.79	1200	.02392	3.1335	.00603	165.83

Source: Handbook of Chemistry and Physics, 14th Edition
The Chemical Rubber Co., Cleveland, Ohio
1929

TABLE C-13

PROPERTIES OF SATURATED AIR

WEIGHTS OF AIR, VAPOR OF WATER, AND SATURATED MIXTURE
OF AIR AND VAPOR AT DIFFERENT TEMPERATURES, UNDER
STANDARD ATMOSPHERIC PRESSURE
OF 29.921 INCHES OF MERCURY

		Weight in a cubic foot of mixture				
Temperature, degrees Fahr.	Vapor pressure, inches of mercury	Weight of the dry air, pounds	Weight of the vapor, pounds	Total weight of the mixture, pounds	B.t.u. absorbed by one cubic foot sat. air per degree F.	Cubic feet sat. air warmed one degree per B.t.u.
1	2	3	4	5	6	7
0	.0383	.08625	.000069	.08632	.02082	48.04
10	.0631	.08433	.000111	.08444	.02039	49.05
20	.1030	.08247	.000177	.08265	.01998	50.05
30	.1640	.08063	.000276	.08091	.01955	51.15
40	.2477	.07880	.000409	.07921	.01921	52.06
50	.3625	.07694	.000587	.07753	.01883	53.11
60	.5220	.07506	.000829	.07589	.01852	54.00
70	.7390	.07310	.001152	.07425	.01811	55.22
80	1.0290	.07095	.001576	.07253	.01788	55.93
90	1.4170	.06881	.002132	.07094	.01763	56.72
100	1.9260	.06637	.002848	.06922	.01737	57.57
110	2.5890	.06367	.003763	.06743	.01716	58.27
120	3.4380	.06062	.004914	.06553	.01696	58.96
130	4.5200	.05716	.006357	.06352	.01681	59.50
140	5.8800	.05319	.008140	.06133	.01669	59.92
150	7.5700	.04864	.010310	.05894	.01663	60.14
160	9.6500	.04341	.012956	.05673	.01664	60.10
170	12.2000	.03735	.016140	.05349	.01671	59.85
180	15.2900	.03035	.019940	.05029	.01682	59.45
190	19.0000	.02227	.024465	.04674	.01706	58.80
200	23.4700	.01297	.029780	.04275	.01750	57.15

Source: Handbook of Chemistry and Physics, 14th Edition
The Chemical Rubber Co., Cleveland, Ohio
1929

TABLE C–14 PIPE DATA
Carbon and Alloy Steel — Stainless Steel

Nominal Pipe Size Inches	Outside Diam. (D) Inches	Steel Iron Pipe Size	Steel Sched. No.	Stainless Steel Sched. No.	Wall Thickness (t) Inches	Inside Diameter (d) Inches	Area of Metal (a) Square Inches	Transverse Internal Area Square Inches	Transverse Internal Area Square Feet	Moment of Inertia (I) $Inches^4$	Weight Pipe Pounds per foot	Weight Water Pounds per foot of pipe	External Surface Sq. Ft. per foot of pipe	Section Modulus $\left(\frac{2I}{D}\right)$
1/8	0.405	...	...	10S	.049	.307	.0548	.0740	.00051	.00088	.19	.032	.106	.00437
		STD	40	40S	.068	.269	.0720	.0568	.00040	.00106	.24	.025	.106	.00523
		XS	80	80S	.095	.215	.0925	.0364	.00025	.00122	.31	.016	.106	.00602
1/4	0.540	...	...	10S	.065	.410	.0970	.1320	.00091	.00279	.33	.057	.141	.01032
		STD	40	40S	.088	.364	.1250	.1041	.00072	.00331	.42	.045	.141	.01227
		XS	80	80S	.119	.302	.1574	.0716	.00050	.00377	.54	.031	.141	.01395
3/8	0.675	...	...	10S	.065	.545	.1246	.2333	.00162	.00586	.42	.101	.178	.01736
		STD	40	40S	.091	.493	.1670	.1910	.00133	.00729	.57	.083	.178	.02160
		XS	80	80S	.126	.423	.2173	.1405	.00098	.00862	.74	.061	.178	.02554
1/2	0.840	...	...	5S	.065	.710	.1583	.3959	.00275	.01197	.54	.172	.220	.02849
		...	...	10S	.083	.674	.1974	.3568	.00248	.01431	.67	.155	.220	.03407
		STD	40	40S	.109	.622	.2503	.3040	.00211	.01709	.85	.132	.220	.04069
		XS	80	80S	.147	.546	.3200	.2340	.00163	.02008	1.09	.102	.220	.04780
		XXS	160	...	.187	.466	.3836	.1706	.00118	.02212	1.31	.074	.220	.05267
			...	...	.294	.252	.5043	.050	.00035	.02424	1.71	.022	.220	.05772
3/4	1.050	...	...	5S	.065	.920	.2011	.6648	.00462	.02450	.69	.288	.275	.04667
		...	...	10S	.083	.884	.2521	.6138	.00426	.02969	.86	.266	.275	.05655
		STD	40	40S	.113	.824	.3326	.5330	.00371	.03704	1.13	.231	.275	.07055
		XS	80	80S	.154	.742	.4335	.4330	.00300	.04479	1.47	.188	.275	.08531
		XXS	160	...	.219	.612	.5698	.2961	.00206	.05269	1.94	.128	.275	.10036
			...	...	.308	.434	.7180	.148	.00103	.05792	2.44	.064	.275	.11032

Nominal size	Outside diameter	Ident.	Sched. No.	Sched.	Wall thickness	Inside diameter	Metal area	Inside area	Transverse internal area (sq ft)	Moment of inertia	Weight per ft, pipe	Weight per ft, water	External surface (sq ft/ft)	Section modulus
1	1.315			5S	.065	1.185	.2553	1.1029	.00766	.04999	.8	.478	.344	.0663
				10S	.109	1.097	.4130	.9452	.00656	.07569	1.40	.409	.344	.11512
		STD	40	40S	.133	1.049	.4939	.8640	.00600	.08734	1.68	.375	.344	.1328
		XS	80	80S	.179	.957	.6388	.7190	.00499	.1056	2.17	.312	.344	.1606
			160		.250	.815	.8365	.5217	.00362	.1251	2.84	.230	.344	.1903
		XXS			.358	.599	1.0760	.282	.00196	.1405	3.66	.122	.344	.2136
1¼	1.660			5S	.065	1.530	.3257	1.839	.01277	.1038	1.11	.797	.435	.1250
				10S	.109	1.442	.4717	1.633	.01134	.1605	1.81	.708	.435	.1934
		STD	40	40S	.140	1.380	.6685	1.495	.01040	.1947	2.27	.649	.435	.2346
		XS	80	80S	.191	1.278	.8815	1.283	.00891	.2418	3.00	.555	.435	.2913
			160		.250	1.160	1.1070	1.057	.00734	.2839	3.76	.458	.435	.3421
		XXS			.382	.896	1.534	.630	.00438	.3411	5.21	.273	.435	.4110
1½	1.900			5S	.065	1.770	.3747	2.461	.01709	.1579	1.28	1.066	.497	.1662
				10S	.109	1.682	.6133	2.222	.01543	.2468	2.09	.963	.497	.2598
		STD	40	40S	.145	1.610	.7995	2.036	.01414	.3099	2.72	.882	.497	.3262
		XS	80	80S	.200	1.500	1.068	1.767	.01225	.3912	3.63	.765	.497	.4118
			160		.281	1.338	1.429	1.406	.00976	.4824	4.86	.608	.497	.5078
		XXS			.400	1.100	1.885	.950	.00660	.5678	6.41	.42	.497	.5977
2	2.375			5S	.065	2.245	.4717	3.958	.02749	.3149	1.61	1.72	.622	.2652
				10S	.109	2.157	.7760	3.654	.02538	.4992	2.64	1.58	.622	.4204
		STD	40	40S	.154	2.067	1.075	3.355	.02330	.6657	3.65	1.45	.622	.5606
		XS	80	80S	.218	1.939	1.477	2.953	.02050	.8679	5.02	1.28	.622	.7309
			160		.344	1.687	2.190	2.241	.01556	1.162	7.46	.97	.622	.979
		XXS			.436	1.503	2.656	1.774	.01232	1.311	9.03	.77	.622	1.104
2½	2.875			5S	.083	2.709	.7280	5.764	.04002	.7100	2.48	2.50	.753	.4939
				10S	.120	2.635	1.039	5.453	.03787	.9873	3.53	2.36	.753	.6868
		STD	40	40S	.203	2.469	1.704	4.788	.03322	1.530	5.79	2.07	.753	1.064
		XS	80	80S	.276	2.323	2.254	4.238	.02942	1.924	7.66	1.87	.753	1.339
			160		.375	2.125	2.945	3.546	.02463	2.353	10.01	1.54	.753	1.638
		XXS			.552	1.771	4.028	2.464	.01710	2.871	13.69	1.07	.753	1.997
3	3.500			5S	.083	3.334	.8910	8.730	.06063	1.301	3.03	3.78	.916	.7435
				10S	.120	3.260	1.274	8.347	.05796	1.822	4.33	3.62	.916	1.041
		STD	40	40S	.216	3.068	2.228	7.393	.05130	3.017	7.58	3.20	.916	1.724
		XS	80	80S	.300	2.900	3.016	6.605	.04587	3.894	10.25	2.86	.916	2.225
			160		.438	2.624	4.205	5.408	.03755	5.032	14.32	2.35	.916	2.876
		XXS			.600	2.300	5.466	4.155	.02885	5.993	18.58	1.80	.916	3.424

Transverse internal area values listed in "square feet" also represent volume in cubic feet per foot of pipe length.

Identification, wall thickness and weights are extracted from ANSI B36.10 and B36.19. The notations STD, XS, and XXS indicate Standard, Extra Strong, and Double Extra Strong pipe respectively.

Source: Crane, Engineering Data
Crane Co., New York, New York
1976

TABLE C-14 PIPE DATA — cont.

Nominal Pipe Size Inches	Outside Diam. (D) Inches	Identification — Steel — Iron Pipe Size	Identification — Steel — Sched. No.	Identification — Stainless Steel Sched. No.	Wall Thickness (t) Inches	Inside Diameter (d) Inches	Area of Metal (a) Square Inches	Transverse Internal Area — Square Inches	Transverse Internal Area — Square Feet	Moment of Inertia (I) Inches⁴	Weight Pipe Pounds per foot	Weight Water Pounds per foot of pipe	External Surface Sq. Ft. per foot of pipe	Section Modulus $\left(2\frac{I}{D}\right)$
3½	4.000	...	...	5S	.083	3.834	1.021	11.545	.08017	1.960	3.48	5.00	1.047	.9799
		...	...	10S	.120	3.760	1.463	11.104	.07711	2.755	4.97	4.81	1.047	1.378
		STD	40	40S	.226	3.548	2.680	9.886	.06870	4.788	9.11	4.29	1.047	2.394
		XS	80	80S	.318	3.364	3.678	8.888	.06170	6.280	12.50	3.84	1.047	3.140
4	4.500	...	...	5S	.083	4.334	1.152	14.75	.10245	2.810	3.92	6.39	1.178	1.249
		...	...	10S	.120	4.260	1.651	14.25	.09898	3.963	5.61	6.18	1.178	1.761
		STD	40	40S	.237	4.026	3.174	12.73	.08840	7.233	10.79	5.50	1.178	3.214
		XS	80	80S	.337	3.826	4.407	11.50	.07986	9.610	14.98	4.98	1.178	4.271
		...	120	...	.438	3.624	5.595	10.31	.0716	11.65	19.00	4.47	1.178	5.178
		...	160	...	.531	3.438	6.621	9.28	.0645	13.27	22.51	4.02	1.178	5.898
		XXS	...	...	.674	3.152	8.101	7.80	.0542	15.28	27.54	3.38	1.178	6.791
5	5.563	...	...	5S	.109	5.345	1.868	22.44	.1558	6.947	6.36	9.72	1.456	2.498
		...	...	10S	.134	5.295	2.285	22.02	.1529	8.425	7.77	9.54	1.456	3.029
		STD	40	40S	.258	5.047	4.300	20.01	.1390	15.16	14.62	8.67	1.456	5.451
		XS	80	80S	.375	4.813	6.112	18.19	.1263	20.67	20.78	7.88	1.456	7.431
		...	120	...	.500	4.563	7.953	16.35	.1136	25.73	27.04	7.09	1.456	9.250
		...	160	...	.625	4.313	9.696	14.61	.1015	30.03	32.96	6.33	1.456	10.796
		XXS	...	...	.750	4.063	11.340	12.97	.0901	33.63	38.55	5.61	1.456	12.090
6	6.625	...	...	5S	.109	6.407	2.231	32.24	.2239	11.85	7.60	13.97	1.734	3.576
		...	...	10S	.134	6.357	2.733	31.74	.2204	14.40	9.29	13.75	1.734	4.346
		STD	40	40S	.280	6.065	5.581	28.89	.2006	28.14	18.97	12.51	1.734	8.496
		XS	80	80S	.432	5.761	8.405	26.07	.1810	40.49	28.57	11.29	1.734	12.22
		...	120	...	.562	5.501	10.70	23.77	.1650	49.61	36.39	10.30	1.734	14.98
		...	160	...	.719	5.187	13.32	21.15	.1469	58.97	45.35	9.16	1.734	17.81
		XXS	...	...	.864	4.897	15.64	18.84	.1308	66.33	53.16	8.16	1.734	20.02

Nominal	OD	Identification	Schedule No.	S-series	Wall thickness	Inside diam.								
8	8.625			5S	.109	8.407	2.916	55.51	.3855	26.44	9.93	24.06	2.258	6.131
				10S	.148	8.329	3.941	54.48	.3784	35.41	13.40	23.61	2.258	8.212
			20		.250	8.125	6.57	51.85	.3601	57.72	22.36	22.47	2.258	13.39
			30		.277	8.071	7.26	51.16	.3553	63.35	24.70	22.17	2.258	14.69
		STD	40	40S	.322 •	7.981	8.40	50.03	.3474	72.49	28.55	21.70	2.258	16.81
			60		.406	7.813	10.48	47.94	.3329	88.73	35.64	20.77	2.258	20.58
		XS	80	80S	.500	7.625	12.76	45.66	.3171	105.7	43.39	19.78	2.258	24.51
			100		.594	7.437	14.96	43.46	.3018	121.3	50.95	18.83	2.258	28.14
			120		.719	7.187	17.84	40.59	.2819	140.5	60.71	17.59	2.258	32.58
			140		.812	7.001	19.93	38.50	.2673	153.7	67.76	16.68	2.258	35.65
		XXS			.875	6.875	21.30	37.12	.2578	162.0	72.42	16.10	2.258	37.56
			160		.906	6.813	21.97	36.46	.2532	165.9	74.69	15.80	2.258	38.48
10	10.750			5S	.134	10.482	4.36	86.29	.5992	63.0	15.19	37.39	2.814	11.71
				10S	.165	10.420	5.49	85.28	.5922	76.9	18.65	36.95	2.814	14.30
			20		.250	10.250	8.24	82.52	.5731	113.7	28.04	35.76	2.814	21.15
			30		.307	10.136	10.07	80.69	.5603	137.4	34.24	34.96	2.814	25.57
		STD	40	40S	.365	10.020	11.90	78.86	.5475	160.7	40.48	34.20	2.814	29.90
		XS	60	80S	.500	9.750	16.10	74.66	.5185	212.0	54.74	32.35	2.814	39.43
			80		.594	9.562	18.92	71.84	.4989	244.8	64.43	31.13	2.814	45.54
			100		.719	9.312	22.63	68.13	.4732	286.1	77.03	29.53	2.814	53.22
			120		.844	9.062	26.24	64.53	.4481	324.2	89.29	27.96	2.814	60.32
			140		1.000	8.750	30.63	60.13	.4176	367.8	104.13	26.06	2.814	68.43
			160		1.125	8.500	34.02	56.75	.3941	399.3	115.64	24.59	2.814	74.29
12	12.75			5S	.156	12.438	6.17	121.50	.8438	122.4	20.98	52.65	3.338	19.2
				10S	.180	12.390	7.11	120.57	.8373	140.4	24.17	52.25	3.338	22.0
			20		.250	12.250	9.82	117.86	.8185	191.8	33.38	51.07	3.338	30.2
			30		.330	12.090	12.87	114.80	.7972	248.4	43.77	49.74	3.338	39.0
		STD			.375	12.000	14.58	113.10	.7854	279.3	49.56	49.00	3.338	43.8
			40	40S	.406	11.938	15.77	111.93	.7773	300.3	53.52	48.50	3.338	47.1
		XS		80S	.500	11.750	19.24	108.43	.7528	361.5	65.42	46.92	3.338	56.7
			60		.562	11.626	21.52	106.16	.7372	400.4	73.15	46.00	3.338	62.8
			80		.688	11.374	26.03	101.64	.7058	475.1	88.63	44.04	3.338	74.6
			100		.844	11.062	1.53	96.14	.6677	561.6	107.32	41.66	3.338	88.1
			120		1.000	10.750	36.91	90.76	.6303	641.6	125.49	39.33	3.338	100.7
			140		1.125	10.500	41.08	86.59	.6013	700.5	139.67	37.52	3.338	109.9
			160		1.312	10.126	47.14	80.53	.5592	781.1	160.27	34.89	3.338	122.6

Identification, wall thickness and weights are extracted from ANSI B36.10 and B36.19. The notations STD, XS, and XXS indicate Standard, Extra Strong, and Double Extra Strong pipe respectively.

Transverse internal area values listed in "square feet" also represent volume in cubic feet per foot of pipe length.

TABLE C-14 PIPE DATA—cont.

Nominal Pipe Size Inches	Outside Diam. (D) Inches	Identification Steel Iron Pipe Size	Identification Steel Sched. No.	Identification Stainless Steel Sched. No.	Wall Thickness (t) Inches	Inside Diameter (d) Inches	Area of Metal (a) Square Inches	Transverse Internal Area Square Inches	Transverse Internal Area Square Feet	Moment of Inertia (I) Inches⁴	Weight Pipe Pounds per foot	Weight Water Pounds per foot of pipe	External Surface Sq. Ft. per foot of pipe	Section Modulus $\left(2\frac{I}{D}\right)$
14	14.00			5S	.156	13.688	6.78	147.15	1.0219	162.6	23.07	63.77	3.665	23.2
				10S	.188	13.624	8.16	145.78	1.0124	194.6	27.73	63.17	3.665	27.8
			10		.210	13.580	10.80	143.14	.9940	255.3	30.93	62.03	3.665	36.6
			20		.312	13.376	13.42	140.52	.9758	314.4	45.61	60.89	3.665	45.0
		STD	30		.375	13.250	16.05	137.88	.9575	372.8	54.57	59.75	3.665	53.2
			40		.438	13.124	18.66	135.28	.9394	429.1	63.44	58.64	3.665	61.3
		XS			.500	13.000	21.21	132.73	.9217	483.8	72.09	57.46	3.665	69.1
			60		.594	12.812	24.98	128.96	.8956	562.3	85.05	55.86	3.665	80.3
			80		.750	12.500	31.22	122.72	.8522	678.3	106.13	53.18	3.665	98.2
			100		.938	12.124	38.45	115.49	.8020	824.4	130.85	50.04	3.665	117.8
			120		1.094	11.812	44.32	109.62	.7612	929.6	150.79	47.45	3.665	132.8
			140		1.250	11.500	50.07	103.87	.7213	1027.0	170.28	45.01	3.665	146.8
			160		1.406	11.188	55.63	98.31	.6827	1117.0	189.11	42.60	3.665	159.6
16	16.00			5S	.165	15.670	8.21	192.85	1.3393	257.3	27.90	83.57	4.189	32.2
				10S	.188	15.624	9.34	191.72	1.3314	291.9	31.75	83.08	4.189	36.5
			10		.250	15.500	12.37	188.69	1.3103	383.7	42.05	81.74	4.189	48.0
			20		.312	15.376	15.38	185.69	1.2895	473.2	52.27	80.50	4.189	59.2
		STD	30		.375	15.250	18.41	182.65	1.2684	562.1	62.58	79.12	4.189	70.3
		XS	40		.500	15.000	24.35	176.72	1.2272	731.9	82.77	76.58	4.189	91.5
			60		.656	14.688	31.62	169.44	1.1766	932.4	107.50	73.42	4.189	116.6
			80		.844	14.312	40.14	160.92	1.1175	1155.8	136.61	69.73	4.189	144.5
			100		1.031	13.938	48.48	152.58	1.0596	1364.5	164.82	66.12	4.189	170.5
			120		1.219	13.562	56.56	144.50	1.0035	1555.8	192.43	62.62	4.189	194.5
			140		1.438	13.124	65.78	135.28	.9394	1760.3	223.64	58.64	4.189	220.0
			160		1.594	12.812	72.10	128.96	.8956	1893.5	245.25	55.83	4.189	236.7

Nom. pipe size	OD, in.	Identification		Wall thickness, in.	ID, in.	Metal area, in.²	Internal area, in.²	Transverse internal area, ft²	Moment of inertia, in.⁴	Weight of pipe, lb/ft	Weight of water, lb/ft	External surface, ft²/ft	Section modulus, in.³
18	18.00	5S		.165	17.670	9.25	245.22	1.7029	366.6	31.43	106.26	4.712	40.8
		10S		.188	17.624	10.52	243.95	1.6941	417.3	35.76	105.71	4.712	46.4
			10	.250	17.500	13.94	240.53	1.6703	549.1	47.39	104.21	4.712	61.1
			20	.312	17.376	17.34	237.13	1.6467	678.2	58.94	102.77	4.712	75.5
		STD		.375	17.250	20.76	233.71	1.6230	806.7	70.59	101.18	4.712	89.6
			30	.438	17.124	24.17	230.30	1.5990	930.3	82.15	99.84	4.712	103.4
		XS	40	.500	17.000	27.49	226.98	1.5763	1053.2	93.45	98.27	4.712	117.0
				.562	16.876	30.79	223.68	1.5533	1171.5	104.67	96.93	4.712	130.1
			60	.750	16.500	40.64	213.83	1.4849	1514.7	138.17	92.57	4.712	168.3
			80	.938	16.124	50.23	204.24	1.4183	1833.0	170.92	88.50	4.712	203.8
			100	1.156	15.688	61.17	193.30	1.3423	2180.0	207.96	83.76	4.712	242.3
			120	1.375	15.250	71.81	182.66	1.2684	2498.1	244.14	79.07	4.712	277.6
			140	1.562	14.876	80.66	173.80	1.2070	2749.0	274.22	75.32	4.712	305.5
			160	1.781	14.438	90.75	163.72	1.1369	3020.0	308.50	70.88	4.712	335.6
20	20.00	5S		.188	19.624	11.70	302.46	2.1004	574.2	39.78	131.06	5.236	57.4
		10S		.218	19.564	13.55	300.61	2.0876	662.8	46.06	130.27	5.236	66.3
			10	.250	19.500	15.51	298.65	2.0740	765.4	52.73	129.42	5.236	75.6
		STD	20	.375	19.250	23.12	290.04	2.0142	1113.0	78.60	125.67	5.236	111.3
		XS	30	.500	19.000	30.63	283.53	1.9690	1457.0	104.13	122.87	5.236	145.7
			40	.594	18.812	36.15	278.00	1.9305	1703.0	123.11	120.46	5.236	170.4
			60	.812	18.376	48.95	265.21	1.8417	2257.0	166.40	114.92	5.236	225.7
			80	1.031	17.938	61.44	252.72	1.7550	2772.0	208.87	109.51	5.236	277.1
			100	1.281	17.438	75.33	238.83	1.6585	3315.2	256.10	103.39	5.236	331.5
			120	1.500	17.000	87.18	226.98	1.5762	3754.0	296.37	98.35	5.236	375.5
			140	1.750	16.500	100.33	213.82	1.4849	4216.0	341.09	92.66	5.236	421.7
			160	1.969	16.062	111.49	202.67	1.4074	4585.5	379.17	87.74	5.236	458.5
22	22.00	5S		.188	21.624	12.88	367.25	2.5503	766.2	43.80	159.14	5.760	69.7
		10S		.218	21.564	14.92	365.21	2.5362	884.8	50.71	158.26	5.760	80.4
			10	.250	21.500	17.08	363.05	2.5212	1010.3	58.07	157.32	5.760	91.8
		STD	20	.375	21.250	25.48	354.66	2.4629	1489.7	86.61	153.68	5.760	135.4
		XS	30	.500	21.000	33.77	346.36	2.4053	1952.5	114.81	150.09	5.760	177.5
			60	.875	20.250	58.07	322.06	2.2365	3244.9	197.41	139.56	5.760	295.0
			80	1.125	19.75	73.78	306.35	2.1275	4030.4	250.81	132.76	5.760	366.4
			100	1.375	19.25	89.09	291.04	2.0211	4758.5	302.88	126.12	5.760	432.6
			120	1.625	18.75	104.02	276.12	1.9175	5432.0	353.61	119.65	5.760	493.8
			140	1.875	18.25	118.55	261.59	1.8166	6053.7	403.00	113.36	5.760	550.3
			160	2.125	17.75	132.68	247.45	1.7184	6626.4	451.06	107.23	5.760	602.4

Identification, wall thickness and weights are extracted from ANSI B36.10 and B36.19. The notations STD, XS, and XXS indicate Standard, Extra Strong, and Double Extra Strong pipe respectively.

Transverse internal area values listed in "square feet"; also represent volume in cubic feet per foot of pipe length.

TABLE C-14 PIPE DATA—cont.

Nominal Pipe Size Inches	Outside Diam. (D) Inches	Identification Steel — Iron Pipe Size	Identification Steel — Sched. No.	Identification Stainless Steel Sched. No.	Wall Thickness (t) Inches	Inside Diameter (d) Inches	Area of Metal (a) Square Inches	Transverse Internal Area Square Inches	Transverse Internal Area Square Feet	Moment of Inertia (I) Inches^4	Weight Pipe Pounds per foot	Weight Water Pounds per foot of pipe	External Surface Sq. Ft. per foot of pipe	Section Modulus $\left(\dfrac{I}{2 \ D}\right)$
24	24.00	...	...	5S	.218	23.564	16.29	436.10	3.0285	1151.6	55.37	188.98	6.283	96.0
		...	10	10S	.250	23.500	18.65	433.74	3.0121	1315.4	63.41	187.95	6.283	109.6
		STD	20	...	.375	23.250	27.83	424.56	2.9483	1942.0	94.62	183.95	6.283	161.9
		XS	...	...	.500	23.000	36.91	415.48	2.8853	2549.5	125.49	179.87	6.283	212.5
		...	30	...	.562	22.876	41.39	411.00	2.8542	2843.0	140.68	178.09	6.283	237.0
		...	40	...	.688	22.624	50.31	402.07	2.7921	3421.3	171.29	174.23	6.283	285.1
		...	60	...	.969	22.062	70.04	382.35	2.6552	4652.8	238.35	165.52	6.283	387.7
		...	80	...	1.219	21.562	87.17	365.22	2.5362	5672.0	296.58	158.26	6.283	472.8
		...	100	...	1.531	20.938	108.07	344.32	2.3911	6849.9	367.39	149.06	6.283	570.8
		...	120	...	1.812	20.376	126.31	326.08	2.2645	7825.0	429.39	141.17	6.283	652.1
		...	140	...	2.062	19.876	142.11	310.28	2.1547	8625.0	483.12	134.45	6.283	718.9
		...	160	...	2.344	19.312	159.41	292.98	2.0346	9455.9	542.13	126.84	6.283	787.9
26	26.00	...	10	...	.312	25.376	25.18	505.75	3.5122	2077.2	85.60	219.16	6.806	159.8
		STD	...	...	.375	25.250	30.19	500.74	3.4774	2478.4	102.63	216.99	6.806	190.6
		XS	20	...	.500	25.000	40.06	490.87	3.4088	3257.0	136.17	212.71	6.806	250.5

Nom. pipe size (in.)	Outside diameter (in.)	Identification	Schedule No.	Wall thickness (in.)	Inside diameter (in.)	Area of metal (sq in.)	Inside area (sq in.)	Transverse internal area (sq ft)	Moment of inertia (in.^{4})	Weight per ft, pipe (lb)	Weight per ft, water (lb)	External surface (sq ft per ft)	Section modulus (in.^{3})
28	28.00	…	10	.312	27.376	27.14	588.61	4.0876	2601.0	92.26	255.07	7.330	185.8
		STD	…	.375	27.250	32.54	583.21	4.0501	3105.1	110.64	252.73	7.330	221.8
		XS	20	.500	27.000	43.20	572.56	3.9761	4084.8	146.85	248.11	7.330	291.8
		…	30	.625	26.750	53.75	562.00	3.9028	5037.7	182.73	243.53	7.330	359.8
30	30.00	5S	…	.250	29.500	23.37	683.49	4.7465	2585.2	79.43	296.18	7.854	172.3
		10S	10	.312	29.376	29.10	677.76	4.7067	3206.3	98.93	293.70	7.854	213.8
		STD	…	.375	29.250	34.90	671.96	4.6664	3829.4	118.65	291.18	7.854	255.3
		XS	20	.500	29.000	46.34	660.52	4.5869	5042.2	157.53	286.22	7.854	336.1
		…	30	.625	28.750	57.68	649.18	4.5082	6224.0	196.08	281.31	7.854	414.9
32	32.00	…	10	.312	31.376	31.06	773.19	5.3694	3898.9	105.59	335.05	8.378	243.7
		STD	…	.375	31.250	37.26	766.99	5.3263	4658.5	126.66	332.36	8.378	291.2
		XS	20	.500	31.000	49.48	754.77	5.2414	6138.6	168.21	327.06	8.378	383.7
		…	30	.625	30.750	61.60	742.64	5.1572	7583.4	209.43	321.81	8.378	474.0
		…	40	.688	30.624	67.68	736.57	5.1151	8298.3	230.08	319.18	8.378	518.6
34	34.00	…	10	.344	33.312	36.37	871.55	6.0524	5150.5	123.65	377.67	8.901	303.0
		STD	…	.375	33.250	39.61	868.31	6.0299	5599.3	134.67	376.27	8.901	329.4
		XS	20	.500	33.000	52.62	855.30	5.9396	7383.5	178.89	370.63	8.901	434.3
		…	30	.625	32.750	65.53	842.39	5.8499	9127.6	222.78	365.03	8.901	536.9
		…	40	.688	32.624	72.00	835.92	5.8050	9991.6	244.77	362.23	8.901	587.7
36	36.00	…	10	.312	35.376	34.98	982.90	6.8257	5569.5	118.92	425.92	9.425	309.4
		STD	…	.375	35.250	41.97	975.91	6.7771	6658.9	142.68	422.89	9.425	369.9
		XS	20	.500	35.000	55.76	962.11	6.6813	8786.2	189.57	416.91	9.425	488.1
		…	30	.625	34.750	69.46	948.42	6.5862	10868.4	236.13	417.22	9.425	603.8
		…	40	.750	34.500	83.06	934.82	6.4918	12906.1	282.35	405.09	9.425	717.0

Identification, wall thickness and weights are extracted from ANSI B36.10 and B36.19. The notations STD, XS, and XXS indicate Standard, Extra Strong, and Double Extra Strong pipe respectively.

Transverse internal area values listed in "square feet" also represent volume in cubic feet per foot of pipe length.

TABLE C-15 HEAT CONTENT OF VARIOUS ENERGY SOURCES

Energy Source	Btu*	Per Unit
Coal		
Anthracite (Pa.)	25,400,000	ton
Bituminous	26,200,000	ton
Blast furnace gas	100	ft^3
Briquettes and package fuels	28,000,000	ton
Coke	24,800,000	ton
Coke-breeze	20,000,000	ton
Coke-oven gas	550	ft^3
Coal tar	150,000	gal
Electricity†	3,412	kWh
Natural gas (dry)	1,035	ft^3
Natural gas liquids (average)	4,011,000	bbl
Butane	4,284,000	bbl
Propane	3,843,000	bbl
Petroleum		
Asphalt	6,640,000	bbl
Coke	6,024,000	bbl
Crude oil	5,800,000	bbl
Diesel	5,806,000	bbl
Gasoline, aviation	5,048,000	bbl
Gasoline, motor fuel	5,253,000	bbl
Jet fuel		
Commercial	5,670,000	bbl
Military	5,355,000	bbl
Kerosene	5,670,000	bbl
Lubricants	6,060,000	bbl
Miscellaneous oils	5,588,000	bbl
Heavy fuel oil	6,287,000	bbl
Road oils	6,640,000	bbl
Shale oil	5,800,000	bbl

*Btu is the amount of heat required to raise the temperature of 1 lb of water 1°F.
†Because of conversion losses in generation of electric power from heat, about 10,000 to 11,000 Btu is required to produce 1 kilowatthour (kWh).

Index

Geyer, J. C., 599
Giardia lamblia, 381, 384
Giardiasis, 108
Gildard, F. A., 599
Gottlieb, S., 645
Gouy, G., 176
Gouy-Chapman layer, 176, 177
Grading, water storage structures, 879
Gram weight
  atomic, 141
  equivalent, 144
  molecular, 141
Granular activated alumina, 654
Granular activated carbon (GAC), 7, 59, 133,
    228–229, 628
  adsorption, 133
  VOC removal, 543, 544, 548
  See also Activated carbon, granular
Gravel, 345
  mounding, 369
  placement, 349
Gravity
  acceleration, 283
  dewatering. *See* Sludge handling
  filters, 337, 339, 700
  isothermal center, 168
  thickening, 736
Great Plains (mottled enamel), 645
Green algae, 601
Ground-level storage, 876
Groundwater
  aquifer type, 83
  artesian conditions, 85
  contaminants, 25, 26
  quality, 99, 133
  recharge, 135
  reservoirs, 81
  storage, 81, 83
  supply source, 64–67, 81
  THM monitoring, 496
  VOC removal, 154, 519
  withdrawal points, 100
  zones
    aeration, 83
    mixing, 165
    saturation, 81–85
G-value, 189
Gt, 190, 202

Half-cell equation, 154
Halogens, 408
Ham, R. K., 191

Hardness, 4, 40–45, 659
  ratings, 660
Hardy-Cross method, 897
Hazardous material, 115
Hazen, A., 2, 299
Hazen-Williams formula, 845, 852
Hazen-Williams friction factor, 846
Headloss, 360, 834–869
  coefficient, 852, 855, 858
  control devices, 858
  filters, 866–869
  manifolds, 854
  nozzle, 853
  orfices, 853
  piping system, 836, 844–857
  venturi tubes, 853
  see also Hydraulics
Heavy metal
  discharges, 6
  removal, 226
Helical screw conveyor, 746
Hem, J. D., 57
Hematopoietic system, 24
Henry's constant, 529
Henry's law, 147, 148, 527
Hepatitis, 25, 108, 383
Heptachlor epoxide, 28
Hexachlorocyclopentadiene, 28
Hexametaphosphate. *See* Sodium hexameta-
    phosphate
High pressure sodium lamp, 989
Homopolymer, 185
Hopkins, E. S., 620, 625, 635
Horizontal basins, 290–298, 315, 317
Horizontal collection wells, 94
Howell, D. H., 643
Hudson, H. E., Jr., 194, 294, 353
Humic acid, 112, 174, 489
Humic substances, 16, 105, 106
HVAC, 977–992
Hydrated lime, 661
Hydraulics, 834–869
  analysis, 814, 897–899
  available head, 835
  backdrive, 749
  capacity, 834
  carbon handling, 436
  characteristics, 66
  clutches, 970
  computer programs, 898
  design, 834, 835, 894
  filters, 866–869

ATOMIC WEIGHTS

Based on the assigned relative atomic mass of $^{12}C = 12$

The following values apply to elements as they exist in materials of terrestrial origin and to certain artificial elements. When used with the footnotes, they are reliable to ±1 in the last digit, or ±3 if that digit is in small type.

Name	Sym-bol	At. No.	At. wt.	M.P.°C	B.P.°C
Actinium	Ac	89	(227)	1050	3200 ± 300
Aluminum	Al	13	26.9815ᵃ	660.37	2467
Americium	Am	95	(243)	994 ± 4	2607
Antimony, stibium	Sb	51	121.7₅	630.74	1750
Argon	Ar	18	39.948ᵇ,ᶜ,ᵈ,ᵉ	−189.2	−185.7
Arsenic	As	33	74.9216ᵃ	817	613
Astatine	At	85	~210	302	337
Barium	Ba	56	137.3₄	725	1640
Berkelium	Bk	97	(247)	—	—
Beryllium	Be	4	9.01218ᵃ	1278 ± 5	2970
Bismuth	Bi	83	208.9806ᵃ	271.3	1560 ± 5⁷⁶⁰
Boron	B	5	10.81ᶜ,ᵈ,ᵉ	2300	2550
Bromine	Br	35	79.904ᶜ	−7.2	58.78
Cadmium	Cd	48	112.40	320.9	765
Calcium	Ca	20	40.08	839 ± 2	1484
Californium	Cf	98	(251)	—	—
Carbon	C	6	12.011ᵇ,ᵈ	3550	4827
Cerium	Ce	58	140.12	798 ± 3	3257
Cesium	Cs	55	132.9055ᵃ	28.40 ± 0.01	678.4
Chlorine	Cl	17	35.453ᶜ	−100.98	−34.6
Chromium	Cr	24	51.996ᶜ	1857 ± 20	2672
Cobalt	Co	27	58.9332ᵃ	1495	2870
Copper	Cu	29	63.546ᶜ,ᵈ	1083.4 ± 0.2	2567
Curium	Cm	96	(247)	1340 ± 40	—
Dysprosium	Dy	66	162.50	14 09	2335
Einsteinium	Es	99	(254)	—	—
Erbium	Er	68	167.26	1522	2510
Europium	Eu	63	151.96	822 ± 5	1597
Fermium	Fm	100	(257)	—	—
Fluorine	F	9	18.9984ᵃ	−219.62	−188.14
Francium	Fr	87	(223)	(27)	(677)
Gadolinium	Gd	64	157.2₅	1311 ± 1	3233
Gallium	Ga	31	69.72	29.78	2403
Germanium	Ge	32	72.5₉	937.4	2830
Gold	Au	79	196.9665ᵃ	1064.43	2807
Hafnium	Hf	72	178.4₉	2227 ± 20	4602
Helium	He	2	4.00260ᵇ,ᶜ	−272.2²⁶ ᵃᵗᵐ	−268.934
Holmium	Ho	67	164.9303ᵃ	1470	2720
Hydrogen	H	1	1.0080ᵇ,ᵈ	−259.14	−252.87
Indium	In	49	114.82	156.61	2080
Iodine	I	53	126.9045ᵃ	113.5	184.35
Iridium	Ir	77	192.2₂	2410	4130
Iron	Fe	26	55.84₇	1535	2750
Krypton	Kr	36	83.80	−156.6	−152.30 ± 0.10
Lanthanum	La	57	138.9055ᵇ	920 ± 5	3454
Lawrencium	Lr	103	(257)	—	—
Lead	Pb	82	207.2ᵈ,ᵉ	327.502	1740
Lithium	Li	3	6.941ᶜ,ᵈ,ᵉ	180.54	1347
Lutetium	Lu	71	174.97	1656 ± 5	3315
Magnesium	Mg	12	24.305ᶜ	648.8 ± 0.5	1090
Manganese	Mn	25	54.9380ᵃ	1244 ± 3	1962
Mendelevium	Md	101	(256)	—	—
Mercury	Hg	80	200.5₉	−38.87	356.58

ᵃ Mononuclidic element.
ᵇ Element with one predominant isotope (about 99 to 100 % abundance).
ᶜ Element for which the atomic weight is based on calibrated measurements.
ᵈ Element for which variation in isotopic abundance in terrestrial samples limits the precision of the atomic weight given.
ᵉ Element for which users are cautioned against the possibility of large